Classical and Modern Fourier Analysis
(013-035399-X)
Loukas Grafakos

A First Course in Fourier Analysis
(013-578782-3)
David Kammler

A First Course in Wavelets with Fourier Analysis
(013-022809-5)
Albert Boggess and Fran Narcowich

Mathematics of Medical Imaging
(013-067548-2)
Charles Epstein

Introduction to Mathematical Statistics 6/E
(013-008507-3)
Robert Hogg, Joe McKean, and Allen Craig

John E. Freund's Mathematical Statistics with Applications 7/E
(013-142706-7)
Irwin Miller and Marylees Miller

Probability and Statistical Inference 6/E
(013-027294-9)
Robert Hogg and Elliot Tanis

An Introduction to Mathematical Statistics and Its Applications 3/E
(013-922303-7)
Richard Larsen and Morris Marx

Statistics and Data Analysis
(013-744426-5)
Ajit Tamhane and Dorothy Dunlop

Probability and Statistics for Engineers and Scientists 7/E
(013-041529-4)
Ron Walpole, Ray Myers, Sharon Myers, and Keying Ye

Miller & Freund's Probability and Statistics for Engineers 7/E
(013-143745-3)
Richard Johnson

Applied Statistics for Engineers and Scientists: Using Microsoft Excel and Minitab
(013-488801-4)
David Levine, Patricia Ramsey, and Robert Smidt

THIRD EDITION

FUNDAMENTALS OF PROBABILITY
WITH STOCHASTIC PROCESSES

SAEED GHAHRAMANI

Western New England College

Upper Saddle River, New Jersey 07458

Library of Congress Cataloging-in-Publication Data

Ghahramani, Saeed.
   Fundamentals of probability with stochastic processes/ Saeed Ghahramani.—3rd edition.
     p.   cm.
     Includes Index.
     ISBN: 0-13-145340-8
     1. Probabilities.   I. Title.

     QA273.G464 2005
     519.2—dc22   2004048541

Executive Editor: **George Lobell**
Editor-in-Chief: **Sally Yagan**
Production Editor: **Jeanne Audino**
Assistant Managing Editor: **Bayani Mendoza DeLeon**
Senior Managing Editor: **Linda Mihatov Behrens**
Executive Managing Editor: **Kathleen Schiaparelli**
Vice-President/Director of Production and Manufacturing: **David W. Riccardi**
Assistant Manufacturing Manager/Buyer: **Michael Bell**
Manufacturing Manager: **Trudy Pisciotti**
Marketing Manager: **Halee Dinsey**
Marketing Assistant: **Rachel Beckman**
Art Director: **Jayne Conte**
Cover Designer: **Bruce Kenselaar**
Cover Image Specialist: **Rita Wenning**
Cover Photo: **PhotoLibrary.com**
Back Cover Photo: **Benjamin Shear / Taxi / Getty Images, Inc.**
Compositor: **Saeed Ghahramani**
Composition: $\mathcal{AMS}$-LaTeX

Printed in the United States of America

10

ISBN 0-13-145340-8

Pearson Education LTD., *London*
Pearson Education of Australia PTY, Limited, *Sydney*
Pearson Education Singapore, Pte. Ltd.
Pearson Education North Asia Ltd, *Hong Kong*
Pearson Education Canada, Ltd., *Toronto*
Pearson Educación de Mexico, S.A. de C.V.
Pearson Education – Japan, *Tokyo*
Pearson Education Malaysia, Pte. Ltd

To Lili, Adam, and Andrew

Contents

► **11   Sums of Independent Random Variables and Limit Theorems**   **457**

► **12   Stochastic Processes**   **511**

Preface

This one- or two-term basic probability text is written for majors in mathematics, physical sciences, engineering, statistics, actuarial science, business and finance, operations research, and computer science. It can also be used by students who have completed a basic calculus course. Our aim is to present probability in a natural way: through interesting and instructive examples and exercises that motivate the theory, definitions, theorems, and methodology. Examples and exercises have been carefully designed to arouse curiosity and hence encourage the students to delve into the theory with enthusiasm.

Authors are usually faced with two opposing impulses. One is a tendency to put too much into the book, because *everything* is important and *everything* has to be said the author's way! On the other hand, authors must also keep in mind a clear definition of the focus, the level, and the audience for the book, thereby choosing carefully what should be "in" and what "out." Hopefully, this book is an acceptable resolution of the tension generated by these opposing forces.

Instructors should enjoy the versatility of this text. They can choose their favorite problems and exercises from a collection of 1558 and, if necessary, omit some sections and/or theorems to teach at an appropriate level.

Exercises for most sections are divided into two categories: A and B. Those in category A are routine, and those in category B are challenging. However, not all exercises in category B are uniformly challenging. Some of those exercises are included because students find them somewhat difficult.

I have tried to maintain an approach that is mathematically rigorous and, at the same time, closely matches the historical development of probability. Whenever appropriate, I include historical remarks, and also include discussions of a number of probability problems published in recent years in journals such as *Mathematics Magazine* and *American Mathematical Monthly*. These are interesting and instructive problems that deserve discussion in classrooms.

Chapter 13 concerns computer simulation. That chapter is divided into several sections, presenting algorithms that are used to find approximate solutions to complicated probabilistic problems. These sections can be discussed independently when relevant materials from earlier chapters are being taught, or they can be discussed concurrently, toward the end of the semester. Although I believe that the emphasis should remain on concepts, methodology, and the mathematics of the subject, I also think that students should be asked to read the material on simulation and perhaps do some projects. Computer simulation is an excellent means to acquire insight into the nature of a problem, its functions, its magnitude, and the characteristics of the solution.

Other Continuing Features

- The historical roots and applications of many of the theorems and definitions are presented in detail, accompanied by suitable examples or counterexamples.

- As much as possible, examples and exercises for each section do not refer to exercises in other chapters or sections—a style that often frustrates students and instructors.

- Whenever a new concept is introduced, its relationship to preceding concepts and theorems is explained.

- Although the usual analytic proofs are given, simple probabilistic arguments are presented to promote deeper understanding of the subject.

- The book begins with discussions on probability and its definition, rather than with combinatorics. I believe that combinatorics should be taught after students have learned the preliminary concepts of probability. The advantage of this approach is that the need for methods of counting will occur naturally to students, and the connection between the two areas becomes clear from the beginning. Moreover, combinatorics becomes more interesting and enjoyable.

- Students beginning their study of probability have a tendency to think that sample spaces always have a finite number of sample points. To minimize this proclivity, the concept of *random selection of a point from an interval* is introduced in Chapter 1 and applied where appropriate throughout the book. Moreover, since the basis of simulating indeterministic problems is selection of random points from $(0, 1)$, in order to understand simulations, students need to be thoroughly familiar with that concept.

- Often, when we think of a collection of events, we have a tendency to think about them in either temporal or logical sequence. So, if, for example, a sequence of events $A_1, A_2, \ldots, A_n$ occur in time or in some logical order, we can usually immediately write down the probabilities $P(A_1)$, $P(A_2 \mid A_1)$, $\ldots$, $P(A_n \mid A_1 A_2 \cdots A_{n-1})$ without much computation. However, we may be interested in probabilities of the intersection of events, or probabilities of events unconditional on the rest, or probabilities of earlier events, given later events. These three questions motivated the need for the law of multiplication, the law of total probability, and Bayes' theorem. I have given the law of multiplication a section of its own so that each of these fundamental uses of conditional probability would have its full share of attention and coverage.

- The concepts of expectation and variance are introduced early, because important concepts should be defined and used as soon as possible. One benefit of this practice is that, when random variables such as Poisson and normal are studied, the associated parameters will be understood immediately rather than remaining ambiguous until expectation and variance are introduced. Therefore, from the beginning, students will develop a natural feeling about such parameters.

- Special attention is paid to the Poisson distribution; it is made clear that this distribution is frequently applicable, for two reasons: first, because it approximates the binomial distribution and, second, it is the mathematical model for an enormous class of phenomena. The comprehensive presentation of the Poisson process and its applications can be understood by junior- and senior-level students.

- Students often have difficulties understanding functions or quantities such as the density function of a continuous random variable and the formula for mathematical expectation. For example, they may wonder why $\int x f(x)\,dx$ is the appropriate definition for $E(X)$ and why correction for continuity is necessary. I have explained the reason behind such definitions, theorems, and concepts, and have demonstrated why they are the natural extensions of discrete cases.

- The first six chapters include many examples and exercises concerning selection of random points from intervals. Consequently, in Chapter 7, when discussing uniform random variables, I have been able to calculate the distribution and (by differentiation) the density function of X, a random point from an interval (a, b). In this way the concept of a uniform random variable and the definition of its density function are readily motivated.

- In Chapters 7 and 8 the usefulness of uniform densities is shown by using many examples. In particular, applications of uniform density in *geometric probability theory* are emphasized.

- Normal density, arguably the most important density function, is readily motivated by De Moivre's theorem. In Section 7.2, I introduce the standard normal density, the elementary version of the central limit theorem, and the normal density just as they were developed historically. Experience shows this to be a good pedagogical approach. When teaching this approach, the normal density becomes natural and does not look like a strange function appearing out of the blue.

- Exponential random variables naturally occur as *times between consecutive events of Poisson processes*. The time of occurrence of the nth event of a Poisson process has a gamma distribution. For these reasons I have motivated exponential and gamma distributions by Poisson processes. In this way we can obtain many examples of exponential and gamma random variables from the abundant examples of Poisson processes already known. Another advantage is that it helps us visualize memoryless random variables by looking at the interevent times of Poisson processes.

- Joint distributions and conditioning are often trouble areas for students. A detailed explanation and many applications concerning these concepts and techniques make these materials somewhat easier for students to understand.

- The concepts of covariance and correlation are motivated thoroughly.

- A subsection on pattern appearance is presented in Section 10.1. Even though the method discussed in this subsection is intuitive and probabilistic, it should help the students understand such paradoxical-looking results as the following. On the average, it takes almost twice as many flips of a fair coin to obtain a sequence of five successive heads as it does to obtain a tail followed by four heads.

- The answers to the odd-numbered exercises are included at the end of the book.

New To This Edition

Since 2000, when the second edition of this book was published, I have received much additional correspondence and feedback from faculty and students in this country and abroad. The comments, discussions, recommendations, and reviews helped me to improve the book in many ways. All detected errors were corrected, and the text has been fine-tuned for accuracy. More explanations and clarifying comments have been added to almost every section. In this edition, 278 new exercises and examples, mostly of an applied nature, have been added. More insightful and better solutions are given for a number of problems and exercises. For example, I have discussed Borel's normal number theorem, and I have presented a version of a famous set which is not an event. If a fair coin is tossed a very large number of times, the general perception is that heads occurs as often as tails. In a new subsection, in Section 11.4, I have explained what is meant by "heads occurs as often as tails."

Some of the other features of the present revision are the following:

- An introductory chapter on stochastic processes is added. That chapter covers more in-depth material on Poisson processes. It also presents the basics of Markov chains, continuous-time Markov chains, and Brownian motion. The topics are covered in some depth. Therefore, the current edition has enough material for a second course in probability as well. The level of difficulty of the chapter on stochastic processes is consistent with the rest of the book. I believe the explanations in the new edition of the book make some challenging material more easily accessible to undergraduate and beginning graduate students. We assume only calculus as a prerequisite. Throughout the chapter, as examples, certain important results from such areas as queuing theory, random walks, branching processes, superposition of Poisson processes, and compound Poisson processes are discussed. I have also explained what the famous theorem, PASTA, *Poisson Arrivals See Time Average*, states. In short, the chapter on stochastic processes is laying the foundation on which students' further pure and applied probability studies and work can build.

- Some practical, meaningful, nontrivial, and relevant applications of probability and stochastic processes in finance, economics, and actuarial sciences are presented.

- Ever since 1853, when Gregor Johann Mendel (1822–1884) began his breeding experiments with the garden pea *Pisum sativum*, probability has played an important role in the understanding of the principles of heredity. In this edition, I have included more genetics examples to demonstrate the extent of that role.

- To study the risk or rate of "failure," per unit of time of "lifetimes" that have already survived a certain length of time, I have added a new section, Survival Analysis and Hazard Functions, to Chapter 7.

- For random sums of random variables, I have discussed Wald's equation and its analogous case for variance. Certain applications of Wald's equation have been discussed in the exercises, as well as in Chapter 12, Stochastic Processes.

- To make the order of topics more natural, the previous editions' Chapter 8 is broken into two separate chapters, Bivariate Distributions and Multivariate Distributions. As a result, the section Transformations of Two Random Variables has been covered earlier along with the material on bivariate distributions, and the convolution theorem has found a better home as an example of transformation methods. That theorem is now presented as a motivation for introducing moment-generating functions, since it cannot be extended so easily to many random variables.

Sample Syllabi

For a one-term course on probability, instructors have been able to omit many sections without difficulty. The book is designed for students with different levels of ability, and a variety of probability courses, applied and/or pure, can be taught using this book. A typical one-semester course on probability would cover Chapters 1 and 2; Sections 3.1–3.5; Chapters 4, 5, 6; Sections 7.1–7.4; Sections 8.1–8.3; Section 9.1; Sections 10.1–10.3; and Chapter 11.

A follow-up course on introductory stochastic processes, or on a more advanced probability would cover the remaining material in the book with an emphasis on Sections 8.4, 9.2–9.3, 10.4 and, especially, the entire Chapter 12.

A course on **discrete probability** would cover Sections 1.1–1.5; Chapters 2, 3, 4, and 5; The subsections Joint Probability Mass Functions, Independence of Discrete Random Variables, and Conditional Distributions: Discrete Case, from Chapter 8; the subsection Joint Probability Mass Functions, from Chapter 9; Section 9.3; selected discrete topics from Chapters 10 and 11; and Section 12.3.

Web Site

For the issues concerning this book, such as reviews and errata, the Web site

http://mars.wnec.edu/~sghahram/probabilitybooks.html

is established. In this Web site, I may also post new examples, exercises, and topics that I will write for future editions.

Solutions Manual

I have written an *Instructor's Solutions Manual* that gives detailed solutions to virtually all of the 1224 exercises of the book. This manual is available, directly from Prentice Hall, only for those instructors who teach their courses from this book.

Acknowledgments

While writing the manuscript, many people helped me either directly or indirectly. Lili, my beloved wife, deserves an accolade for her patience and encouragement; as do my wonderful children.

According to Ecclesiastes 12:12, "of the making of books, there is no end." Improvements and advancement to different levels of excellence cannot possibly be achieved without the help, criticism, suggestions, and recommendations of others. I have been blessed with so many colleagues, friends, and students who have contributed to the improvement of this textbook. One reason I like writing books is the pleasure of receiving so many suggestions and so much help, support, and encouragement from colleagues and students all over the world. My experience from writing the three editions of this book indicates that collaboration and camaraderie in the scientific community is truly overwhelming.

For the third edition of this book and its solutions manual, my brother, Dr. Soroush Ghahramani, a professor of architecture from Sinclair College in Ohio, using AutoCad, with utmost patience and meticulosity, resketched each and every one of the figures. As a result, the illustrations are more accurate and clearer than they were in the previous editions. I am most indebted to my brother for his hard work.

For the third edition, I wrote many new $\mathcal{A}_{\mathcal{M}}S$-LATEX files. My assistants, Ann Guyotte and Avril Couture, with utmost patience, keen eyes, positive attitude, and eagerness put these hand-written files onto the computer. My colleague, Professor Ann Kizanis, who is known for being a perfectionist, read, very carefully, these new files and made many good suggestions. While writing about the application of genetics to probability, I had several discussions with Western New England's distinguished geneticist, Dr. Lorraine Sartori. I learned a lot from Lorraine, who also read my material on genetics carefully and made valuable suggestions. Dr. Michael Meeropol, the Chair of our Economics Department, read parts of my manuscripts on financial applications and mentioned some new ideas. Dr. David Mazur was teaching from my book even before we were colleagues. Over the past four years, I have enjoyed hearing his comments and suggestions about my book. It gives me a distinct pleasure to thank Ann Guyotte, Avril, Ann Kizanis, Lorraine, Michael, and Dave for their help.

Professor Jay Devore from California Polytechnic Institute—San Luis Obispo, made excellent comments that improved the manuscript substantially for the first edition. From

Boston University, Professor Mark E. Glickman's careful review and insightful suggestions and ideas helped me in writing the second edition. I was very lucky to receive thorough reviews of the third edition from Professor James Kuelbs of University of Wisconsin, Madison, Professor Robert Smits of New Mexico State University, and Ms. Ellen Gundlach from Purdue University. The thoughtful suggestions and ideas of these colleagues improved the current edition of this book in several ways. I am most grateful to Drs. Devore, Glickman, Kuelbs, Smits, and Ms. Gundlach.

For the first two editions of the book, my colleagues and friends at Towson University read or taught from various revisions of the text and offered useful advice. In particular, I am grateful to Professors Mostafa Aminzadeh, Raouf Boules, Jerome Cohen, James P. Coughlin, Geoffrey Goodson, Sharon Jones, Ohoe Kim, Bill Rose, Martha Siegel, Houshang Sohrab, Eric Tissue, and my late dear friend Sayeed Kayvan. I want to thank my colleagues Professors Coughlin and Sohrab, especially, for their kindness and the generosity with which they spent their time carefully reading the entire text every time it was revised.

I am also grateful to the following professors for their valuable suggestions and constructive criticisms: Todd Arbogast, The University of Texas at Austin; Robert B. Cooper, Florida Atlantic University; Richard DeVault, Northwestern State University of Louisiana; Bob Dillon, Aurora University; Dan Fitzgerald, Kansas Newman University; Sergey Fomin, Massachusetts Institute of Technology; D. H. Frank, Indiana University of Pennsylvania; James Frykman, Kent State University; M. Lawrence Glasser, Clarkson University; Moe Habib, George Mason University; Paul T. Holmes, Clemson University; Edward Kao, University of Houston; Joe Kearney, Davenport College; Eric D. Kolaczyk, Boston University; Philippe Loustaunau, George Mason University; John Morrison, University of Delaware; Elizabeth Papousek, Fisk University; Richard J. Rossi, California Polytechnic Institute—San Luis Obispo; James R. Schott, University of Central Florida; Siavash Shahshahani, Sharif University of Technology, Tehran, Iran; Yang Shangjun, Anhui University, Hefei, China; Kyle Siegrist, University of Alabama—Huntsville; Loren Spice, my former advisee, a prodigy who became a Ph.D. student at age 16 and a faculty member at the University of Michigan at age 21; Olaf Stackelberg, Kent State University; and Don D. Warren, Texas Legislative Council.

Special thanks are due to Prentice Hall's visionary editor, George Lobell, for his encouragement and assistance in seeing this effort through. I also appreciate the excellent job Jeanne Audino has done as production editor for this edition.

Last, but not least, I want to express my gratitude for all the technical help I received, for 17 years, from my good friend and colleague Professor Howard Kaplon of Towson University, and all technical help I regularly receive from Kevin Gorman and John Willemain, my friends and colleagues at Western New England College. I am also grateful to Professor Nakhlé Asmar, from the University of Missouri, who generously shared with me his experiences in the professional typesetting of his own beautiful book.

Saeed Ghahramani
sghahram@wnec.edu

Chapter 1

Axioms of Probability

1.1 INTRODUCTION

In search of natural laws that govern a phenomenon, science often faces "events" that may or may not occur. The event of *disintegration of a given atom of radium* is one such example because, in any given time interval, such an atom may or may not disintegrate. The event of *finding no defect during inspection of a microwave oven* is another example, since an inspector may or may not find defects in the microwave oven. The event that an *orbital satellite in space is at a certain position* is a third example. In any experiment, an event that may or may not occur is called **random**. If the occurrence of an event is inevitable, it is called **certain**, and if it can never occur, it is called **impossible**. For example, the event that an object travels faster than light is impossible, and the event that in a thunderstorm flashes of lightning precede any thunder echoes is certain.

Knowing that an event is random determines only that the existing conditions under which the experiment is being performed do not guarantee its occurrence. Therefore, the knowledge obtained from randomness itself is hardly decisive. It is highly desirable to determine quantitatively the exact value, or an estimate, of the chance of the occurrence of a random event . The theory of probability has emerged from attempts to deal with this problem. In many different fields of science and technology, it has been observed that, under a long series of experiments, the proportion of the time that an event occurs may appear to approach a constant. It is these constants that probability theory (and statistics) aims at predicting and describing as quantitative measures of the chance of occurrence of events. For example, if a fair coin is tossed repeatedly, the proportion of the heads approaches 1/2. Hence probability theory postulates that the number 1/2 be assigned to the event of *getting heads in a toss of a fair coin*.

Historically, from the dawn of civilization, humans have been interested in games of chance and gambling. However, the advent of probability as a mathematical discipline is relatively recent. Ancient Egyptians, about 3500 B.C., were using astragali, a four-sided die-shaped bone found in the heels of some animals, to play a game now called *hounds and jackals*. The ordinary six-sided die was made about 1600 B.C. and since then has been used in all kinds of games. The ordinary deck of playing cards, probably the most popular tool in games and gambling, is much more recent than dice.

Although it is not known where and when dice originated, there are reasons to believe that they were invented in China sometime between the seventh and tenth centuries. Clearly, through gambling and games of chance people have gained intuitive ideas about the frequency of occurrence of certain events and, hence, about probabilities. But surprisingly, studies of the chances of events were not begun until the fifteenth century. The Italian scholars Luca Paccioli (1445–1514), Niccolò Tartaglia (1499–1557), Girolamo Cardano (1501–1576), and especially Galileo Galilei (1564–1642) were among the first prominent mathematicians who calculated probabilities concerning many different games of chance. They also tried to construct a mathematical foundation for probability. Cardano even published a handbook on gambling, with sections discussing methods of cheating. Nevertheless, real progress started in France in 1654, when Blaise Pascal (1623–1662) and Pierre de Fermat (1601–1665) exchanged several letters in which they discussed general methods for the calculation of probabilities. In 1655, the Dutch scholar Christian Huygens (1629–1695) joined them. In 1657 Huygens published the first book on probability, *De Ratiocinates in Aleae Ludo (On Calculations in Games of Chance)*. This book marked the birth of probability. Scholars who read it realized that they had encountered an important theory. Discussions of solved and unsolved problems and these new ideas generated readers interested in this challenging new field.

After the work of Pascal, Fermat, and Huygens, the book written by James Bernoulli (1654–1705) and published in 1713 and that by Abraham de Moivre (1667–1754) in 1730 were major breakthroughs. In the eighteenth century, studies by Pierre-Simon Laplace (1749–1827), Siméon Denis Poisson (1781–1840), and Karl Friedrich Gauss (1777–1855) expanded the growth of probability and its applications very rapidly and in many different directions. In the nineteenth century, prominent Russian mathematicians Pafnuty Chebyshev (1821–1894), Andrei Markov (1856–1922), and Aleksandr Lyapunov (1857–1918) advanced the works of Laplace, De Moivre, and Bernoulli considerably. By the early twentieth century, probability was already a developed theory, but its foundation was not firm. A major goal was to put it on firm mathematical grounds. Until then, among other interpretations perhaps the **relative frequency interpretation** of probability was the most satisfactory. According to this interpretation, to define p, the probability of the occurrence of an event A of an experiment, we study a series of sequential or simultaneous performances of the experiment and observe that the proportion of times that A occurs approaches a constant. Then we count $n(A)$, the number of times that A occurs during n performances of the experiment, and we define $p = \lim_{n \to \infty} n(A)/n$. This definition is mathematically problematic and cannot be the basis of a rigorous probability theory. Some of the difficulties that this definition creates are as follows:

1. In practice, $\lim_{n \to \infty} n(A)/n$ cannot be computed since it is impossible to repeat an experiment infinitely many times. Moreover, if for a large n, $n(A)/n$ is taken as an approximation for the probability of A, there is no way to analyze the error.

2. There is no reason to believe that the limit of $n(A)/n$, as $n \to \infty$, exists. Also, if the existence of this limit is accepted as an axiom, many dilemmas arise that cannot

be solved. For example, there is no reason to believe that, in a different series of experiments and for the same event A, this ratio approaches the same limit. Hence the uniqueness of the probability of the event A is not guaranteed.

3.   By this definition, probabilities that are based on our personal belief and knowledge are not justifiable. Thus statements such as the following would be meaningless.

- The probability that the price of oil will be raised in the next six months is 60%.

- The probability that the 50,000th decimal figure of the number π is 7 exceeds 10%.

- The probability that it will snow next Christmas is 30%.

- The probability that Mozart was poisoned by Salieri is 18%.

In 1900, at the International Congress of Mathematicians in Paris, David Hilbert (1862–1943) proposed 23 problems whose solutions were, in his opinion, crucial to the advancement of mathematics. One of these problems was the axiomatic treatment of the theory of probability. In his lecture, Hilbert quoted Weierstrass, who had said, "The final object, always to be kept in mind, is to arrive at a correct understanding of the foundations of the science." Hilbert added that a thorough understanding of special theories of a science is necessary for successful treatment of its foundation. Probability had reached that point and was studied enough to warrant the creation of a firm mathematical foundation. Some work toward this goal had been done by Émile Borel (1871–1956), Serge Bernstein (1880–1968), and Richard von Mises (1883–1953), but it was not until 1933 that Andrei Kolmogorov (1903–1987), a prominent Russian mathematician, successfully axiomatized the theory of probability. In Kolmogorov's work, which is now universally accepted, three self-evident and indisputable properties of probability (discussed later) are taken as **axioms,** and the entire theory of probability is developed and rigorously based on these axioms. In particular, the existence of a constant p, as the limit of the proportion of the number of times that the event A occurs when the number of experiments increases to ∞, in some sense, is shown. Subjective probabilities based on our personal knowledge, feelings, and beliefs may also be modeled and studied by this axiomatic approach.

In this book we study the mathematics of probability based on the axiomatic approach. Since in this approach the concepts of **sample space** and **event** play a central role, we now explain these concepts in detail.

1.2   SAMPLE SPACE AND EVENTS

If the outcome of an experiment is not certain but all of its possible outcomes are predictable in advance, then the set of all these possible outcomes is called the **sample space** of the experiment and is usually denoted by S. Therefore, the sample space of an experiment consists of all possible outcomes of the experiment. These outcomes are sometimes called **sample points,** or simply **points,** of the sample space. In the language of probability, certain subsets of S are referred to as **events**. So events are sets of points of the sample space. Some examples follow.

Example 1.1   For the experiment of *tossing a coin once*, the sample space S consists of two points (outcomes), "heads" (H) and "tails" (T). Thus $S = \{H, T\}$.   ♦

Example 1.2   Suppose that an experiment consists of two steps. First a coin is flipped. If the outcome is tails, a die is tossed. If the outcome is heads, the coin is flipped again. The sample space of this experiment is $S = \{T1, T2, T3, T4, T5, T6, HT, HH\}$. For this experiment, the event of *heads in the first flip of the coin* is $E = \{HT, HH\}$, and the event of *an odd outcome* when the die is tossed is $F = \{T1, T3, T5\}$.   ♦

Example 1.3   Consider measuring the lifetime of a light bulb. Since any nonnegative real number can be considered as the lifetime of the light bulb (in hours), the sample space is $S = \{x : x \geq 0\}$. In this experiment, $E = \{x : x \geq 100\}$ is the event that *the light bulb lasts at least 100 hours*, $F = \{x : x \leq 1000\}$ is the event that *it lasts at most 1000 hours*, and $G = \{505.5\}$ is the event that *it lasts exactly 505.5 hours*.   ♦

Example 1.4   Suppose that a study is being done on all families with one, two, or three children. Let the outcomes of the study be the genders of the children in descending order of their ages. Then

$$S = \{b, g, bg, gb, bb, gg, bbb, bgb, bbg, bgg, ggg, gbg, ggb, gbb\}.$$

Here the outcome b means that the child is a boy, and g means that it is a girl. The events $F = \{b, bg, bb, bbb, bgb, bbg, bgg\}$ and $G = \{gg, bgg, gbg, ggb\}$ represent families where the eldest child is a boy and families with exactly two girls, respectively.   ♦

Example 1.5   A bus with a capacity of 34 passengers stops at a station some time between 11:00 A.M. and 11:40 A.M. every day. The sample space of the experiment, consisting of counting the number of passengers on the bus and measuring the arrival time of the bus, is

$$S = \left\{(i, t) : 0 \leq i \leq 34, \quad 11 \leq t \leq 11\frac{2}{3}\right\}, \tag{1.1}$$

where i represents the number of passengers and t the arrival time of the bus in hours and fractions of hours. The subset of S defined by $F = \left\{(27, t) : 11\frac{1}{3} < t < 11\frac{2}{3}\right\}$ is the event that the bus arrives between 11:20 A.M. and 11:40 A.M. with 27 passengers.   ♦

Remark 1.1   Different manifestations of outcomes of an experiment might lead to different representations for the sample space of the same experiment. For instance, in Example 1.5, the outcome that the *bus arrives at t with i passengers* is represented by (i, t), where t is expressed in hours and fractions of hours. By this representation, (1.1) is the sample space of the experiment. Now if the same outcome is denoted by (i, t), where t is the number of minutes after 11 A.M. that the bus arrives, then the sample space takes the form

$$S_1 = \{(i, t) : 0 \le i \le 34, \ 0 \le t \le 40\}.$$

To the outcome that the *bus arrives at* 11:20 A.M. *with* 31 *passengers*, in S the corresponding point is $\left(31, 11\frac{1}{3}\right)$, while in S_1 it is $(31, 20)$.   ◆

Example 1.6   (Round-Off Error)   Suppose that each time Jay charges an item to his credit card, he will round the amount to the nearest dollar in his records. Therefore, the round-off error, which is the true value charged minus the amount recorded, is random, with the sample space

$$S = \{0, 0.01, 0.02, \dots, 0.49, -0.50, -0.49, \dots, -0.01\},$$

where we have assumed that for any integer dollar amount a, Jay rounds $a.50$ to $a + 1$. The event of rounding off at most 3 cents in a random charge is given by

$$\{0, 0.01, 0.02, 0.03, -0.01, -0.02, -0.03\}.   ◆$$

If the outcome of an experiment belongs to an event E, we say that the event E has **occurred**. For example, if we draw two cards from an ordinary deck of 52 cards and observe that one is a spade and the other a heart, all of the events $\{sh\}$, $\{sh, dd\}$, $\{cc, dh, sh\}$, $\{hc, sh, ss, hh\}$, and $\{cc, hh, sh, dd\}$ have occurred because sh, the outcome of the experiment, belongs to all of them. However, none of the events $\{dh, sc\}$, $\{dd\}$, $\{ss, hh, cc\}$, and $\{hd, hc, dc, sc, sd\}$ have occurred because sh does not belong to any of them.

In the study of probability theory the relations between different events of an experiment play a central role. In the remainder of this section we study these relations. In all of the following definitions the events belong to a fixed sample space S.

Subset   An event E is said to be a **subset** of the event F if, whenever E occurs, F also occurs. This means that all of the sample points of E are contained in F. Hence considering E and F solely as two sets, E is a subset of F in the usual set-theoretic sense: that is, $E \subseteq F$.

Equality   Events E and F are said to be **equal** if the occurrence of E implies the occurrence of F, and vice versa; that is, if $E \subseteq F$ and $F \subseteq E$, hence $E = F$.

Intersection   An event is called the **intersection** of two events E and F if it occurs only whenever E and F occur simultaneously. In the language of sets this event is denoted by EF or $E \cap F$ because it is the set containing exactly the common points of E and F.

Union   An event is called the **union** of two events E and F if it occurs whenever at least one of them occurs. This event is $E \cup F$ since all of its points are in E or F or both.

Complement   An event is called the **complement** of the event E if it only occurs whenever E does not occur. The complement of E is denoted by E^c.

Difference   An event is called the **difference** of two events E and F if it occurs whenever E occurs but F does not. The difference of the events E and F is denoted by $E - F$. It is clear that $E^c = S - E$ and $E - F = E \cap F^c$.

Certainty   An event is called **certain** if its occurrence is inevitable. Thus the sample space is a certain event.

Impossibility   An event is called **impossible** if there is certainty in its nonoccurrence. Therefore, the empty set $\emptyset$, which is S^c, is an impossible event.

Mutually Exclusiveness   If the joint occurrence of two events E and F is impossible, we say that E and F are **mutually exclusive**. Thus E and F are mutually exclusive if the occurrence of E precludes the occurrence of F, and vice versa. Since the event representing the joint occurrence of E and F is EF, their intersection, E and F, are mutually exclusive if $EF = \emptyset$. A set of events $\{E_1, E_2, \ldots\}$ is called **mutually exclusive** if the joint occurrence of any two of them is impossible, that is, if $\forall i \neq j$, $E_i E_j = \emptyset$. Thus $\{E_1, E_2, \ldots\}$ is mutually exclusive if and only if every pair of them is mutually exclusive.

The events $\bigcup_{i=1}^{n} E_i$, $\bigcap_{i=1}^{n} E_i$, $\bigcup_{i=1}^{\infty} E_i$, and $\bigcap_{i=1}^{\infty} E_i$ are defined in a way similar to $E_1 \cup E_2$ and $E_1 \cap E_2$. For example, if $\{E_1, E_2, \ldots, E_n\}$ is a set of events, by $\bigcup_{i=1}^{n} E_i$ we mean the event in which at least one of the events E_i, $1 \leq i \leq n$, occurs. By $\bigcap_{i=1}^{n} E_i$ we mean an event that occurs only when all of the events E_i, $1 \leq i \leq n$, occur.

Sometimes Venn diagrams are used to represent the relations among events of a sample space. The sample space S of the experiment is usually shown as a large rectangle and, inside S, circles or other geometrical objects are drawn to indicate the events of interest. Figure 1.1 presents Venn diagrams for EF, $E \cup F$, E^c, and $(E^c G) \cup F$. The shaded regions are the indicated events.

Example 1.7   At a busy international airport, arriving planes land on a first-come, first-served basis. Let

$$E = \text{there are at least five planes waiting to land,}$$

$$F = \text{there are at most three planes waiting to land,}$$

$$H = \text{there are exactly two planes waiting to land.}$$

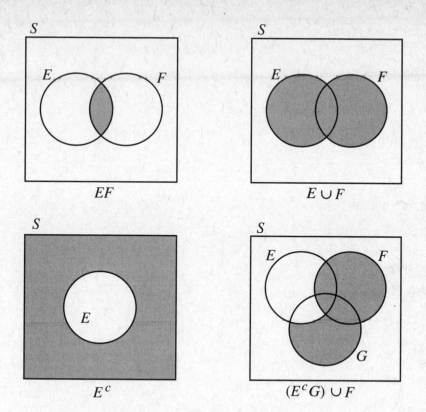

Figure 1.1   Venn diagrams of the events specified.

Then

1. E^c is the event that at most four planes are waiting to land.

2. F^c is the event that at least four planes are waiting to land.

3. E is a subset of F^c; that is, if E occurs, then F^c occurs. Therefore, $EF^c = E$.

4. H is a subset of F; that is, if H occurs, then F occurs. Therefore, $FH = H$.

5. E and F are mutually exclusive; that is, $EF = \emptyset$. E and H are also mutually exclusive since $EH = \emptyset$.

6. FH^c is the event that the number of planes waiting to land is zero, one, or three.

◆

Unions, intersections, and complementations satisfy many useful relations between events. A few of these relations are as follows:

$$(E^c)^c = E, \quad E \cup E^c = S, \quad \text{and} \, EE^c = \emptyset.$$

Commutative laws:

$$E \cup F = F \cup E, \quad EF = FE.$$

Associative laws:

$$E \cup (F \cup G) = (E \cup F) \cup G, \quad E(FG) = (EF)G.$$

Distributive laws:

$$(EF) \cup H = (E \cup H)(F \cup H), \quad (E \cup F)H = (EH) \cup (FH).$$

De Morgan's first law:

$$(E \cup F)^c = E^c F^c, \quad \left(\bigcup_{i=1}^{n} E_i\right)^c = \bigcap_{i=1}^{n} E_i^c, \quad \left(\bigcup_{i=1}^{\infty} E_i\right)^c = \bigcap_{i=1}^{\infty} E_i^c.$$

De Morgan's second law:

$$(EF)^c = E^c \cup F^c, \quad \left(\bigcap_{i=1}^{n} E_i\right)^c = \bigcup_{l=1}^{n} E_i^c, \quad \left(\bigcap_{i=1}^{\infty} E_i\right)^c = \bigcup_{i=1}^{\infty} E_i^c.$$

Another useful relation between E and F, two arbitrary events of a sample space S, is

$$E = EF \cup EF^c.$$

This equality readily follows from $E = ES$ and distributivity:

$$E = ES = E(F \cup F^c) = EF \cup EF^c.$$

These and similar identities are usually proved by the **elementwise method**. The idea is to show that the events on both sides of the equation are formed of the same sample points. To use this method, we prove set inclusion in both directions. That is, sample points belonging to the event on the left also belong to the event on the right, and vice versa. An example follows.

Example 1.8   Prove De Morgan's first law: For E and F, two events of a sample space S, $(E \cup F)^c = E^c F^c$.

Proof: First we show that $(E \cup F)^c \subseteq E^c F^c$; then we prove the reverse inclusion $E^c F^c \subseteq (E \cup F)^c$. To show that $(E \cup F)^c \subseteq E^c F^c$, let x be an outcome that belongs to $(E \cup F)^c$. Then x does not belong to $E \cup F$, meaning that x is neither in E nor in F. So x belongs to both E^c and F^c and hence to $E^c F^c$. To prove the reverse inclusion, let $x \in E^c F^c$. Then $x \in E^c$ and $x \in F^c$, implying that $x \notin E$ and $x \notin F$. Therefore, $x \notin E \cup F$ and thus $x \in (E \cup F)^c$.   ◆

Note that Venn diagrams are an excellent way to give intuitive justification for the validity of relations or to create counterexamples and show invalidity of relations. However, they are not appropriate to prove relations. This is because of the large number of cases that must be considered (particularly if more than two events are involved). For example, suppose that by means of Venn diagrams, we want to prove the identity $(EF)^c = E^c \cup F^c$. First we must draw appropriate representations for all possible ways that E and F can be related: cases such as $EF = \emptyset$, $EF \neq \emptyset$, $E = F$, $E = \emptyset$, $F = S$, and so on. Then in each particular case we should find the regions that represent $(EF)^c$ and $E^c \cup F^c$ and observe that they are the same. Even if these two sets have different representations in only one case, the identity would be false.

EXERCISES

A

1. A deck of six cards consists of three black cards numbered 1, 2, 3, and three red cards numbered 1, 2, 3. First, Vann draws a card at random and without replacement. Then Paul draws a card at random and without replacement from the remaining cards. Let A be the event that Paul's card has a larger number than Vann's card. Let B be the event that Vann's card has a larger number than Paul's card.

    (a) Are A and B mutually exclusive?

    (b) Are A and B complements of one another?

2. A box contains three red and five blue balls. Define a sample space for the experiment of recording the colors of three balls that are drawn from the box, one by one, with replacement.

3. Define a sample space for the experiment of choosing a number from the interval $(0, 20)$. Describe the event that such a number is an integer.

4. Define a sample space for the experiment of putting three different books on a shelf in random order. If two of these three books are a two-volume dictionary, describe the event that these volumes stand in increasing order side-by-side (i.e., volume I precedes volume II).

5. Two dice are rolled. Let E be the event that the sum of the outcomes is odd and F be the event of at least one 1. Interpret the events EF, $E^c F$, and $E^c F^c$.

6. Define a sample space for the experiment of drawing two coins from a purse that contains two quarters, three nickels, one dime, and four pennies. For the same experiment describe the following events:

    (a)   drawing 26 cents;

    (b)   drawing more than 9 but less than 25 cents;

    (c)   drawing 29 cents.

7.   A telephone call from a certain person is received some time between 7:00 A.M. and 9:10 A.M. every day. Define a sample space for this phenomenon, and describe the event that the call arrives within 15 minutes of the hour.

8.   Let E, F, and G be three events; explain the meaning of the relations $E \cup F \cup G = G$ and $EFG = G$.

9.   A limousine that carries passengers from an airport to three different hotels just left the airport with two passengers. Describe the sample space of the stops and the event that both of the passengers get off at the same hotel.

10. Find the simplest possible expression for the following events.

    (a) $(E \cup F)(F \cup G)$.

    (b) $(E \cup F)(E^c \cup F)(E \cup F^c)$.

11. At a certain university, every year eight to 12 professors are granted University Merit Awards. This year among the nominated faculty are Drs. Jones, Smith, and Brown. Let A, B, and C denote the events, respectively, that these professors will be given awards. In terms of A, B, and C, find an expression for the event that the award goes to (a) only Dr. Jones; (b) at least one of the three; (c) none of the three; (d) exactly two of them; (e) exactly one of them; (f) Drs. Jones or Smith but not both.

12. Prove that the event B is impossible if and only if for every event A,

$$A = (B \cap A^c) \cup (B^c \cap A).$$

13. Let E, F, and G be three events. Determine which of the following statements are correct and which are incorrect. Justify your answers.

    (a) $(E - EF) \cup F = E \cup F$.

    (b) $F^c G \cup E^c G = G(F \cup E)^c$.

    (c) $(E \cup F)^c G = E^c F^c G$.

    (d) $EF \cup EG \cup FG \subset E \cup F \cup G$.

14. In an experiment, cards are drawn, one by one, at random and successively from an ordinary deck of 52 cards. Let A_n be the event that no face card or ace appears on the first $n - 1$ drawings, and the nth draw is an ace. In terms of A_n's, find an expression for the event that an ace appears before a face card, (a) if the cards are drawn with replacement; (b) if they are drawn without replacement.

B

15. Prove De Morgan's second law, $(AB)^c = A^c \cup B^c$, (a) by elementwise proof; (b) by applying De Morgan's first law to A^c and B^c.

16. Let A and B be two events. Prove the following relations by the elementwise method.

    (a) $(A - AB) \cup B = A \cup B$.

    (b) $(A \cup B) - AB = AB^c \cup A^c B$.

17. Let $\{A_n\}_{n=1}^{\infty}$ be a sequence of events. Prove that for every event B,

    (a) $B\left(\bigcup_{i=1}^{\infty} A_i\right) = \bigcup_{i=1}^{\infty} BA_i$.

    (b) $B\bigcup\left(\bigcap_{i=1}^{\infty} A_i\right) = \bigcap_{i=1}^{\infty}(B \cup A_i)$.

18. Define a sample space for the experiment of putting in a random order seven different books on a shelf. If three of these seven books are a three-volume dictionary, describe the event that these volumes stand in increasing order side by side (i.e., volume I precedes volume II and volume II precedes volume III).

19. Let $\{A_1, A_2, A_3, \dots\}$ be a sequence of events. Find an expression for the event that infinitely many of the A_i's occur.

20. Let $\{A_1, A_2, A_3, \dots\}$ be a sequence of events of a sample space S. Find a sequence $\{B_1, B_2, B_3, \dots\}$ of mutually exclusive events such that for all $n \geq 1$, $\bigcup_{i=1}^{n} A_i = \bigcup_{i=1}^{n} B_i$.

1.3   AXIOMS OF PROBABILITY

In mathematics, the goals of researchers are to obtain new results and prove their correctness, create simple proofs for already established results, discover or create connections between different fields of mathematics, construct and solve mathematical models for real-world problems, and so on. To discover new results, mathematicians use trial and error, instinct and inspired guessing, inductive analysis, studies of special cases, and other methods. But when a new result is discovered, its validity remains subject to skepticism until it is rigorously proven. Sometimes attempts to prove a result fail and contradictory examples are found. Such examples that invalidate a result are called **counterexamples**. No mathematical proposition is settled unless it is either proven or refuted by a counterexample. If a result is false, a counterexample exists to refute it. Similarly, if a result is valid, a proof must be found for its validity, although in some cases it might take years, decades, or even centuries to find it.

Proofs in probability theory (and virtually any other theory) are done in the framework of the **axiomatic method**. By this method, if we want to convince any rational person, say Sonya, that a statement L_1 is correct, we will show her how L_1 can be deduced logically from another statement L_2 that might be acceptable to her. However, if Sonya does not accept L_2, we should demonstrate how L_2 can be deduced logically from a simpler statement L_3. If she disputes L_3, we must continue this process until, somewhere along the way we reach a statement that, without further justification, is acceptable to her. This statement will then become the basis of our argument. Its existence is necessary since otherwise the process continues ad infinitum without any conclusions. Therefore, in the axiomatic method, first we adopt certain simple, indisputable, and consistent statements without justifications. These are **axioms** or **postulates**. Then we agree on how and when one statement is a logical consequence of another one and, finally, using the terms that are already clearly understood, axioms and definitions, we obtain new results. New results found in this manner are called **theorems**. Theorems are statements that can be proved. Upon establishment, they are used for discovery of new theorems, and the process continues and a theory evolves.

In this book, our approach is based on the axiomatic method. There are three axioms upon which probability theory is based and, except for them, everything else needs to be proved. We will now explain these axioms.

Definition   (Probability Axioms)   *Let S be the sample space of a random phe-nomenon. Suppose that to each event A of S, a number denoted by $P(A)$ is associated with A. If P satisfies the following axioms, then it is called a **probability** and the number $P(A)$ is said to be the **probability of A**.*

Axiom 1 $P(A) \geq 0$.

Axiom 2 $P(S) = 1$.

Axiom 3   *If $\{A_1, A_2, A_3, \dots\}$ is a sequence of mutually exclusive events (i.e., the joint occurrence of every pair of them is impossible: $A_i A_j = \emptyset$ when $i \neq j$), then*

$$P\left(\bigcup_{i=1}^{\infty} A_i\right) = \sum_{i=1}^{\infty} P(A_i).$$

Note that the axioms of probability are a set of rules that must be satisfied before S and P can be considered a probability model.

Axiom 1 states that the probability of the occurrence of an event is always nonnegative. Axiom 2 guarantees that the probability of the occurrence of the event S that is certain is 1. Axiom 3 states that for a sequence of mutually exclusive events the probability of the occurrence of at least one of them is equal to the sum of their probabilities.

Axiom 2 is merely a convenience to make things definite. It would be equally reasonable to have $P(S) = 100$ and interpret probabilities as percentages (which we frequently do).

Let S be the sample space of an experiment. Let A and B be events of S. We say that A and B are **equally likely** if $P(A) = P(B)$. Let ω_1 and ω_2 be sample points of S. We say that ω_1 and ω_2 are **equally likely** if the events $\{\omega_1\}$ and $\{\omega_2\}$ are equally likely, that is, if $P(\{\omega_1\}) = P(\{\omega_2\})$.

We will now prove some immediate implications of the axioms of probability.

Theorem 1.1   *The probability of the empty set $\emptyset$ is 0. That is, $P(\emptyset) = 0$.*

Proof: Let $A_1 = S$ and $A_i = \emptyset$ for $i \geq 2$; then $A_1, A_2, A_3, \ldots$ is a sequence of mutually exclusive events. Thus, by Axiom 3,

$$P(S) = P\left(\bigcup_{i=1}^{\infty} A_i\right) = \sum_{i=1}^{\infty} P(A_i) = P(S) + \sum_{i=2}^{\infty} P(\emptyset),$$

implying that $\sum_{i=2}^{\infty} P(\emptyset) = 0$. This is possible only if $P(\emptyset) = 0$.   ◆

Axiom 3 is stated for a *countably infinite* collection of mutually exclusive events. For this reason, it is also called the *axiom of countable additivity*. We will now show that the same property holds for a *finite* collection of mutually exclusive events as well. That is, P also satisfies *finite additivity*.

Theorem 1.2   *Let $\{A_1, A_2, \ldots, A_n\}$ be a mutually exclusive set of events. Then*

$$P\left(\bigcup_{i=1}^{n} A_i\right) = \sum_{i=1}^{n} P(A_i).$$

Proof: For $i > n$, let $A_i = \emptyset$. Then $A_1, A_2, A_3, \ldots$ is a sequence of mutually exclusive events. From Axiom 3 and Theorem 1.1 we get

$$P\left(\bigcup_{i=1}^{n} A_i\right) = P\left(\bigcup_{i=1}^{\infty} A_i\right) = \sum_{i=1}^{\infty} P(A_i)$$

$$= \sum_{i=1}^{n} P(A_i) + \sum_{i=n+1}^{\infty} P(A_i) = \sum_{i=1}^{n} P(A_i) + \sum_{i=n+1}^{\infty} P(\emptyset)$$

$$= \sum_{i=1}^{n} P(A_i).$$ ◆

Letting $n = 2$, Theorem 1.2 implies that if A_1 and A_2 are mutually exclusive, then

$$P(A_1 \cup A_2) = P(A_1) + P(A_2). \tag{1.2}$$

The intuitive meaning of (1.2) is that if an experiment can be repeated indefinitely, then for two mutually exclusive events A_1 and A_2, the proportion of times that $A_1 \cup A_2$ occurs is equal to the sum of the proportion of times that A_1 occurs and the proportion of times that A_2 occurs. For example, for the experiment of tossing a fair die, $S = \{1, 2, 3, 4, 5, 6\}$ is the sample space. Let A_1 be the event that the outcome is 6, and A_2 be the event that the outcome is odd. Then $A_1 = \{6\}$ and $A_2 = \{1, 3, 5\}$. Since all sample points are equally likely to occur (by the definition of a fair die) and the number of sample points of A_1 is 1/6 of the number of sample points of S, we expect that $P(A_1) = 1/6$. Similarly, we expect that $P(A_2) = 3/6$. Now $A_1 A_2 = \emptyset$ implies that the number of sample points of $A_1 \cup A_2$ is $(1/6 + 3/6)$th of the number of sample points of S. Hence we should expect that $P(A_1 \cup A_2) = 1/6 + 3/6$, which is the same as $P(A_1) + P(A_2)$. This and many other examples suggest that (1.2) is a reasonable relation to be taken as Axiom 3. However, if we do this, difficulties arise when a sample space contains infinitely many sample points, that is, when the number of possible outcomes of an experiment is not finite. For example, in successive throws of a die let A_n be the event that the first 6 occurs on the nth throw. Then we would be unable to find the probability of $\bigcup_{n=1}^{\infty} A_n$, which represents the event that eventually a 6 occurs. For this reason, Axiom 3, which is the infinite analog of (1.2), is required. It by no means contradicts our intuitive ideas of probability, and one of its great advantages is that Theorems 1.1 and 1.2 are its immediate consequences.

A significant implication of (1.2) is that for any event A, $P(A) \leq 1$. To see this, note that

$$P(A \cup A^c) = P(A) + P(A^c).$$

Now, by Axiom 2,

$$P(A \cup A^c) = P(S) = 1;$$

therefore, $P(A) + P(A^c) = 1$. This and Axiom 1 imply that $P(A) \leq 1$. Hence

> The probability of the occurrence of an event is always some number between 0 and 1. That is,

$$0 \leq P(A) \leq 1.$$

★ **Remark 1.2**[†]   Let S be the sample space of an experiment. The set of all subsets of S is denoted by $\mathcal{P}(S)$ and is called the **power set** of S. Since the aim of probability theory is to associate a number between 0 and 1 to every subset of the sample space, probability is a function P from $\mathcal{P}(S)$ to $[0, 1]$. However, in theory, there is one exception to this: If the sample space S is not countable, not all of the elements of $\mathcal{P}(S)$ are events. There are elements of $\mathcal{P}(S)$ that are not (in a sense defined in more advanced probability texts) **measurable**. These elements are not events (see Example 1.21). In other words, it is a curious mathematical fact that the Kolmogorov axioms are inconsistent with the notion that every subset of every sample space has a probability. Since in real-world problems we are only dealing with those elements of $\mathcal{P}(S)$ that are measurable, we are

---

[†]Throughout the book, items that are optional and can be skipped are identified by ★'s.

not concerned with these exceptions. We must also add that, in general, if the domain of a function is a collection of sets, it is called a **set function**. Hence probability is a real-valued, nonnegative, countably additive set function.   ◆

Example 1.9   A coin is called **unbiased** or **fair** if, whenever it is flipped, the probability of obtaining heads equals that of obtaining tails. Suppose that in an experiment an unbiased coin is flipped. The sample space of such an experiment is $S = \{T, H\}$. Since the events $\{H\}$ and $\{T\}$ are equally likely to occur, $P(\{T\}) = P(\{H\})$, and since they are mutually exclusive,

$$P(\{T, H\}) = P(\{T\}) + P(\{H\}).$$

Hence Axioms 2 and 3 imply that

$$1 = P(S) = P(\{H, T\}) = P(\{H\}) + P(\{T\}) = P(\{H\}) + P(\{H\}) = 2P(\{H\}).$$

This gives that $P(\{H\}) = 1/2$ and $P(\{T\}) = 1/2$. Now suppose that an experiment consists of flipping a biased coin where the outcome of tails is twice as likely as heads; then $P(\{T\}) = 2P(\{H\})$. Hence

$$1 = P(S) = P(\{H, T\}) = P(\{H\}) + P(\{T\}) = P(\{H\}) + 2P(\{H\}) = 3P(\{H\}).$$

This shows that $P(\{H\}) = 1/3$; thus $P(\{T\}) = 2/3$.   ◆

Example 1.10   Sharon has baked five loaves of bread that are identical except that one of them is underweight. Sharon's husband chooses one of these loaves at random. Let B_i, $1 \le i \le 5$, be the event that he chooses the ith loaf. Since all five loaves are equally likely to be drawn, we have

$$P(\{B_1\}) = P(\{B_2\}) = P(\{B_3\}) = P(\{B_4\}) = P(\{B_5\}).$$

But the events $\{B_1\}$, $\{B_2\}$, $\{B_3\}$, $\{B_4\}$, and $\{B_5\}$ are mutually exclusive, and the sample space is $S = \{B_1, B_2, B_3, B_4, B_5\}$. Therefore, by Axioms 2 and 3,

$$1 = P(S) = P(\{B_1\}) + P(\{B_2\}) + P(\{B_3\}) + P(\{B_4\}) + P(\{B_5\}) = 5 \cdot P(\{B_1\}).$$

This gives $P(\{B_1\}) = 1/5$ and hence $P(\{B_i\}) = 1/5$, $1 \le i \le 5$. Therefore, the probability that Sharon's husband chooses the underweight loaf is $1/5$.   ◆

From Examples 1.9 and 1.10 it should be clear that if a sample space contains N points that are equally likely to occur, then the probability of each outcome (sample point) is $1/N$. In general, this can be shown as follows. Let $S = \{s_1, s_2, \dots, s_N\}$ be the sample space of an experiment; then, if all of the sample points are equally likely to occur, we have

$$P(\{s_1\}) = P(\{s_2\}) = \cdots = P(\{s_N\}).$$

But $P(S) = 1$, and the events $\{s_1\}, \{s_2\}, \ldots, \{s_N\}$ are mutually exclusive. Therefore,

$$
\begin{aligned}
1 = P(S) &= P\big(\{s_1, s_2, \ldots, s_N\}\big) \\
&= P\big(\{s_1\}\big) + P\big(\{s_2\}\big) + \cdots + P\big(\{s_N\}\big) = N P\big(\{s_1\}\big).
\end{aligned}
$$

This shows that $P\big(\{s_1\}\big) = 1/N$. Thus $P\big(\{s_i\}\big) = 1/N$ for $1 \le i \le N$.

One simple consequence of the axioms of probability is that, if the sample space of an experiment contains N points that are all equally likely to occur, then the probability of the occurrence of any event A is equal to the number of points of A, say $N(A)$, divided by N. Historically, until the introduction of the axiomatic method by A. N. Kolmogorov in 1933, this fact was taken as the definition of the probability of A. It is now called the **classical definition of probability**. The following theorem, which shows that the classical definition is a simple result of the axiomatic approach, is also an important tool for the computation of probabilities of events for experiments with finite sample spaces.

Theorem 1.3   *Let S be the sample space of an experiment. If S has N points that are all equally likely to occur, then for any event A of S,*

$$
P(A) = \frac{N(A)}{N},
$$

where $N(A)$ is the number of points of A.

Proof: Let $S = \{s_1, s_2, \ldots, s_N\}$, where each s_i is an outcome (a sample point) of the experiment. Since the outcomes are equiprobable, $P\big(\{s_i\}\big) = 1/N$ for all i, $1 \le i \le N$. Now let $A = \{s_{i_1}, s_{i_2}, \ldots, s_{i_{N(A)}}\}$, where $s_{i_j} \in S$ for all i_j. Since $\{s_{i_1}\}, \{s_{i_2}\}, \ldots, \{s_{i_{N(A)}}\}$ are mutually exclusive, Axiom 3 implies that

$$
\begin{aligned}
P(A) &= P\big(\{s_{i_1}, s_{i_2}, \ldots, s_{i_{N(A)}}\}\big) \\
&= P\big(\{s_{i_1}\}\big) + P\big(\{s_{i_2}\}\big) + \cdots + P\big(\{s_{i_{N(A)}}\}\big) \\
&= \underbrace{\frac{1}{N} + \frac{1}{N} + \cdots + \frac{1}{N}}_{N(A)\ \text{terms}} = \frac{N(A)}{N}. \quad \blacklozenge
\end{aligned}
$$

Example 1.11   Let S be the sample space of flipping a fair coin three times and A be the event of at least two heads; then

$$
S = \big\{\text{HHH, HTH, HHT, HTT, THH, THT, TTH, TTT}\big\}
$$

and $A = \{\text{HHH, HTH, HHT, THH}\}$. So $N = 8$ and $N(A) = 4$. Therefore, the probability of at least two heads in flipping a fair coin three times is $N(A)/N = 4/8 = 1/2$. $\quad \blacklozenge$

Example 1.12   An elevator with two passengers stops at the second, third, and fourth floors. If it is equally likely that a passenger gets off at any of the three floors, what is the probability that the passengers get off at different floors?

Solution: Let a and b denote the two passengers and a_2b_4 mean that a gets off at the second floor and b gets off at the fourth floor, with similar representations for other cases. Let A be the event that the passengers get off at different floors. Then

$$S = \{a_2b_2, a_2b_3, a_2b_4, a_3b_2, a_3b_3, a_3b_4, a_4b_2, a_4b_3, a_4b_4\}$$

and $A = \{a_2b_3, a_2b_4, a_3b_2, a_3b_4, a_4b_2, a_4b_3\}$. So $N = 9$ and $N(A) = 6$. Therefore, the desired probability is $N(A)/N = 6/9 = 2/3$.   ♦

Example 1.13   A number is selected at random from the set of integers $\{1, 2, \ldots, 1000\}$. What is the probability that the number is divisible by 3?

Solution: Here the sample space contains 1000 points, so $N = 1000$. Let A be the set of all numbers between 1 and 1000 that are divisible by 3. Then $A = \{3m : 1 \leq m \leq 333\}$. So $N(A) = 333$. Therefore, the probability that a random natural number between 1 and 1000 is divisible by 3 is equal to $333/1000$.   ♦

Example 1.14   A number is selected at random from the set $\{1, 2, \ldots, N\}$. What is the probability that the number is divisible by k, $1 \leq k \leq N$?

Solution: Here the sample space contains N points. Let A be the event that the outcome is divisible by k. Then $A = \{km : 1 \leq m \leq [N/k]\}$, where $[N/k]$ is the greatest integer less than or equal to N/k (to compute $[N/k]$, just divide N by k and round down). So $N(A) = [N/k]$ and $P(A) = [N/k]/N$.   ♦

Remark 1.3   As explained in Remark 1.1, different manifestations of outcomes of an experiment may lead to different representations for the sample space of the same experiment. Because of this, different sample points of a representation might not have the same probability of occurrence. For example, suppose that a study is being done on families with three children. Let the outcomes of the study be the number of girls and the number of boys in a randomly selected family. Then

$$S = \{bbb, bgg, bgb, bbg, ggb, gbg, gbb, ggg\}$$

and

$$\Omega = \{bbb, bbg, bgg, ggg\}$$

are both reasonable sample spaces for the genders of the children of the family. In S, for example, bgg means that the first child of the family is a boy, the second child is a girl, and the third child is also a girl. In Ω, bgg means that the family has one boy and two girls. Therefore, in S all sample points occur with the same probability, namely, 1/8. In Ω, however, probabilities associated to the sample points are not equal:

$$P(\{bbb\}) = P(\{ggg\}) = 1/8,$$

but

$$P(\{bbg\}) = P(\{bgg\}) = 3/8. \quad \blacklozenge$$

Finally, we should note that for finite sample spaces, if nonnegative probabilities are assigned to sample points so that they sum to 1, then the probability axioms hold. Let $S = \{w_1, w_2, \ldots, w_n\}$ be a sample space. Let $p_1, p_2, \ldots, p_n$ be n nonnegative real numbers with $\sum_{i=1}^{n} p_i = 1$. Let P be defined on subsets of S by $P(\{w_i\}) = p_i$, and

$$P(\{w_{i_1}, w_{i_2}, \ldots, w_{i_\ell}\}) = p_{i_1} + p_{i_2} + \cdots + p_{i_\ell}.$$

It is straightforward to verify that P satisfies the probability axioms. Hence P defines a probability on the sample space S.

1.4   BASIC THEOREMS

Theorem 1.4   *For any event A, $P(A^c) = 1 - P(A)$.*

Proof: Since $AA^c = \emptyset$, A and A^c are mutually exclusive. Thus

$$P(A \cup A^c) = P(A) + P(A^c).$$

But $A \cup A^c = S$ and $P(S) = 1$, so

$$1 = P(S) = P(A \cup A^c) = P(A) + P(A^c).$$

Therefore, $P(A^c) = 1 - P(A)$. $\blacklozenge$

This theorem states that the probability of nonoccurrence of the event A is 1 minus the probability of its occurrence. For example, consider $S = \{(i, j): 1 \le i \le 6, \ 1 \le j \le 6\}$, the sample space of tossing two fair dice. If A is the event of getting a sum of 4, then $A = \{(1, 3), (2, 2), (3, 1)\}$ and $P(A) = 3/36$. Theorem 1.4 states that the probability of A^c, the event of not getting a sum of 4, which is harder to count, is $1 - 3/36 = 33/36$. As another example, consider the experiment of selecting a random number from the set $\{1, 2, 3, \ldots, 1000\}$. By Example 1.13, the probability that the number selected is divisible by 3 is $333/1000$. Thus by Theorem 1.4, the probability that it is not divisible by 3, a quantity harder to find directly, is $1 - 333/1000 = 667/1000$.

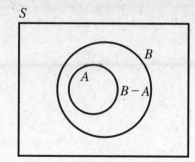

Figure 1.2 $A \subseteq B$ implies that $B = (B - A) \cup A$.

Theorem 1.5   *If $A \subseteq B$, then*

$$P(B - A) = P(BA^c) = P(B) - P(A).$$

Proof: $A \subseteq B$ implies that $B = (B - A) \cup A$ (see Figure 1.2). But $(B - A)A = \emptyset$. So the events $B - A$ and A are mutually exclusive, and $P(B) = P\big((B - A) \cup A\big) = P(B - A) + P(A)$. This gives $P(B - A) = P(B) - P(A)$. ♦

Corollary   *If $A \subseteq B$, then $P(A) \le P(B)$.*

Proof: By Theorem 1.5, $P(B - A) = P(B) - P(A)$. Since $P(B - A) \ge 0$, we have that $P(B) - P(A) \ge 0$. Hence $P(B) \ge P(A)$. ♦

This corollary says that, for instance, it is less likely that a computer has one defect than it has at least one defect. Note that in Theorem 1.5, the condition of $A \subseteq B$ is necessary. The relation $P(B - A) = P(B) - P(A)$ is not true in general. For example, in rolling a fair die, let $B = \{1, 2\}$ and $A = \{3, 4, 5\}$, then $B - A = \{1, 2\}$. Therefore, $P(B - A) = 1/3$, $P(B) = 1/3$, and $P(A) = 1/2$. Hence $P(B - A) \ne P(B) - P(A)$.

Theorem 1.6 $P(A \cup B) = P(A) + P(B) - P(AB)$.

Proof: $A \cup B = A \cup (B - AB)$ (see Figure 1.3) and $A(B - AB) = \emptyset$, so A and $B - AB$ are mutually exclusive events and

$$P(A \cup B) = P\big(A \cup (B - AB)\big) = P(A) + P(B - AB). \qquad (1.3)$$

Now since $AB \subseteq B$, Theorem 1.5 implies that

$$P(B - AB) = P(B) - P(AB).$$

Therefore, (1.3) gives

$$P(A \cup B) = P(A) + P(B) - P(AB). \quad ♦$$

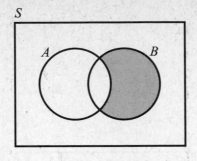

Figure 1.3   The shaded region is $B - AB$.
Thus $A \cup B = A \cup (B - AB)$.

Example 1.15   Suppose that in a community of 400 adults, 300 bike or swim or do both, 160 swim, and 120 swim and bike. What is the probability that an adult, selected at random from this community, bikes?

Solution: Let A be the event that the person swims and B be the event that he or she bikes; then $P(A \cup B) = 300/400$, $P(A) = 160/400$, and $P(AB) = 120/400$. Hence the relation $P(A \cup B) = P(A) + P(B) - P(AB)$ implies that

$$P(B) = P(A \cup B) + P(AB) - P(A)$$
$$= \frac{300}{400} + \frac{120}{400} - \frac{160}{400} = \frac{260}{400} = 0.65. \quad \blacklozenge$$

Example 1.16   A number is chosen at random from the set of integers $\{1, 2, \ldots, 1000\}$. What is the probability that it is divisible by 3 or 5 (i.e., either 3 or 5 or both)?

Solution: The number of integers between 1 and N that are divisible by k is computed by dividing N by k and then rounding down (see Examples 1.13 and 1.14). Therefore, if A is the event that the outcome is divisible by 3 and B is the event that it is divisible by 5, then $P(A) = 333/1000$ and $P(B) = 200/1000$. Now AB is the event that the outcome is divisible by both 3 and 5. Since a number is divisible by 3 and 5 if and only if it is divisible by 15 (3 and 5 are prime numbers), $P(AB) = 66/1000$ (divide 1000 by 15 and round down to get 66). Thus the desired probability is computed as follows:

$$P(A \cup B) = P(A) + P(B) - P(AB)$$
$$= \frac{333}{1000} + \frac{200}{1000} - \frac{66}{1000} = \frac{467}{1000} = 0.467. \quad \blacklozenge$$

Theorem 1.6 gives a formula to calculate the probability that at least one of A and B occurs. We may also calculate the probability that at least one of the events A_1, A_2, A_3, $\ldots$, and A_n occurs. For three events A_1, A_2, and A_3,

$$P(A_1 \cup A_2 \cup A_3) = P(A_1) + P(A_2) + P(A_3) - P(A_1 A_2)$$
$$- P(A_1 A_3) - P(A_2 A_3) + P(A_1 A_2 A_3).$$

For four events,

$$P(A_1 \cup A_2 \cup A_3 \cup A_4) = P(A_1) + P(A_2) + P(A_3) + P(A_4) - P(A_1 A_2)$$
$$- P(A_1 A_3) - P(A_1 A_4) - P(A_2 A_3) - P(A_2 A_4)$$
$$- P(A_3 A_4) + P(A_1 A_2 A_3) + P(A_1 A_2 A_4)$$
$$+ P(A_1 A_3 A_4) + P(A_2 A_3 A_4) - P(A_1 A_2 A_3 A_4).$$

We now explain a procedure, which will enable us to find $P(A_1 \cup A_2 \cup \cdots \cup A_n)$, the probability that at least one of the events $A_1, A_2, \cdots, A_n$ occurs.

Inclusion-Exclusion Principle   *To calculate $P(A_1 \cup A_2 \cup \cdots \cup A_n)$, first find all of the possible intersections of events from $A_1, A_2, \ldots, A_n$ and calculate their probabilities. Then add the probabilities of those intersections that are formed of an odd number of events, and subtract the probabilities of those formed of an even number of events.*

The following formula is an expression for the principle of inclusion-exclusion. It follows by induction. (For an intuitive proof, see Example 2.29.)

$$P\left(\bigcup_{i=1}^{n} A_i\right) = \sum_{i=1}^{n} P(A_i) - \sum_{i=1}^{n-1} \sum_{j=i+1}^{n} P(A_i A_j) + \sum_{i=1}^{n-2} \sum_{j=i+1}^{n-1} \sum_{k=j+1}^{n} P(A_i A_j A_k)$$
$$- \cdots + (-1)^{n-1} P(A_1 A_2 \cdots A_n).$$

Example 1.17   Suppose that 25% of the population of a city read newspaper A, 20% read newspaper B, 13% read C, 10% read both A and B, 8% read both A and C, 5% read B and C, and 4% read all three. If a person from this city is selected at random, what is the probability that he or she does not read any of these newspapers?

Solution: Let E, F, and G be the events that the person reads A, B, and C, respectively. The event that the person reads at least one of the newspapers A, B, or C is $E \cup F \cup G$. Therefore, $1 - P(E \cup F \cup G)$ is the probability that he or she reads none of them. Since

$$P(E \cup F \cup G) = P(E) + P(F) + P(G) - P(EF) - P(EG)$$
$$- P(FG) + P(EFG)$$
$$= 0.25 + 0.20 + 0.13 - 0.10 - 0.08 - 0.05 + 0.04 = 0.39,$$

the desired probability equals $1 - 0.39 = 0.61$.   ◆

Example 1.18   Dr. Grossman, an internist, has 520 patients, of which (1) 230 are hypertensive, (2) 185 are diabetic, (3) 35 are hypochondriac and diabetic, (4) 25 are all three, (5) 150 are none, (6) 140 are only hypertensive, and finally, (7) 15 are hypertensive and hypochondriac but not diabetic. Find the probability that Dr. Grossman's next appointment is hypochondriac but neither diabetic nor hypertensive. Assume that appointments are all random. This implies that even hypochondriacs do not make more visits than others.

Solution: Let T, C, and D denote the events that the next appointment of Dr. Grossman is hypertensive, hypochondriac, and diabetic, respectively. The Venn diagram of Figure 1.4 shows that the number of patients with only hypochondria is 30. Therefore, the desired probability is $30/520 \approx 0.06$.   ♦

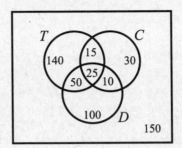

Figure 1.4   Venn diagram of Example 1.18.

Theorem 1.7 $P(A) = P(AB) + P(AB^c)$.

Proof: Clearly, $A = AS = A(B \cup B^c) = AB \cup AB^c$. Since AB and AB^c are mutually exclusive,

$$P(A) = P(AB \cup AB^c) = P(AB) + P(AB^c).$$ ♦

Example 1.19   In a community, 32% of the population are male smokers; 27% are female smokers. What percentage of the population of this community smoke?

Solution: Let A be the event that a randomly selected person from this community smokes. Let B be the event that the person is male. By Theorem 1.7,

$$P(A) = P(AB) + P(AB^c) = 0.32 + 0.27 = 0.59.$$

Therefore, 59% of the population of this community smoke.   ♦

Remark 1.4   In the real world, especially for games of chance, it is common to express probability in terms of **odds**. We say that the *odds in favor of an event A are r to s if* $P(A) = r/(r+s)$. Similarly, the *odds against an event A are r to s if* $P(A) = s/(r+s)$. Therefore, if the odds in favor of an event A are r to s, then the odds against A are s to r. For example, in drawing a card at random from an ordinary deck of 52 cards, the odds against drawing an ace are 48 to 4 or, equivalently, 12 to 1. The odds in favor of an ace are 4 to 48 or, equivalently, 1 to 12. If for an event A, $P(A) = p$, then the odds in favor of A are p to $1 - p$. Therefore, for example, in flipping a fair coin three times, by Example 1.11, the odds in favor of HHH are 1/8 to 7/8 or, equivalently, 1 to 7.   ♦

EXERCISES

A

1. Gottfried Wilhelm Leibniz (1646–1716), the German mathematician, philosopher, statesman, and one of the supreme intellects of the seventeenth century, believed that in a throw of a pair of fair dice, the probability of obtaining the sum 11 is equal to that of obtaining the sum 12. Do you agree with Leibniz? Explain.

2. Suppose that 33% of the people have O^+ blood and 7% have O^-. What is the probability that the next president of the United States has type O blood?

3. The probability that an earthquake will damage a certain structure during a year is 0.015. The probability that a hurricane will damage the same structure during a year is 0.025. If the probability that *both* an earthquake and a hurricane will damage the structure during a year is 0.0073, what is the probability that next year the structure will not be damaged by a hurricane or an earthquake?

4. Suppose that the probability that a driver is a male, and has at least one motor vehicles accident during a one-year period, is 0.12. Suppose that the corresponding probability for a female is 0.06. What is the probability of a randomly selected driver having at least one accident during the next 12 months?

5. Suppose that 75% of all investors invest in traditional annuities and 45% of them invest in the stock market. If 85% invest in the stock market and/or traditional annuities, what percentage invest in both?

6. In a horse race, the odds in favor of the first horse winning in an 8-horse race are 2 to 5. The odds against the second horse winning are 7 to 3. What is the probability that one of these horses will win?

7. Excerpt from the TV show *The Rockford Files:*

   > Rockford: There are only two doctors in town. The chances of both autopsies being performed by the same doctor are 50–50.

   > Reporter: No, that is only for one autopsy. For two autopsies, the chances are 25–75.

   > Rockford: You're right.

   Was Rockford right to agree with the reporter? Explain why or why not?

8. A company has only one position with three highly qualified applicants: John, Barbara, and Marty. However, because the company has only a few women employees, Barbara's chance to be hired is 20% higher than John's and 20% higher than Marty's. Find the probability that Barbara will be hired.

9. In a psychiatric hospital, the number of patients with schizophrenia is three times the number with psychoneurotic reactions, twice the number with alcohol addictions, and 10 times the number with involutional psychotic reaction. If a patient is selected randomly from the list of all patients with one of these four diseases, what is the probability that he or she suffers from schizophrenia? Assume that none of these patients has more than one of these four diseases.

10. Let A and B be two events. Prove that

$$P(AB) \geq P(A) + P(B) - 1.$$

11. A card is drawn at random from an ordinary deck of 52 cards. What is the probability that it is (a) a black ace or a red queen; (b) a face or a black card; (c) neither a heart nor a queen?

12. Which of the following statements is true? If a statement is true, prove it. If it is false, give a counterexample.

(a) If $P(A) + P(B) + P(C) = 1$, then the events A, B, and C are mutually exclusive.

(b) If $P(A \cup B \cup C) = 1$, then A, B, and C are mutually exclusive events.

13. Suppose that in the Baltimore metropolitan area 25% of the crimes occur during the day and 80% of the crimes occur in the city. If only 10% of the crimes occur outside the city during the day, what percent occur inside the city during the night? What percent occur outside the city during the night?

14. Let A, B, and C be three events. Prove that

$$P(A \cup B \cup C)$$
$$= P(A) + P(B) + P(C) - P(AB) - P(AC) - P(BC) + P(ABC).$$

15. Let A, B, and C be three events. Show that exactly two of these events will occur with probability

$$P(AB) + P(AC) + P(BC) - 3P(ABC).$$

16. Eleven chairs are numbered 1 through 11. Four girls and seven boys sit on these chairs at random. What is the probability that chair 5 is occupied by a boy?

17. A ball is thrown at a square that is divided into n^2 identical squares. The probability that the ball hits the square of the ith column and jth row is p_{ij}, where $\sum_{i=1}^{n} \sum_{j=1}^{n} p_{ij} = 1$. In terms of p_{ij}'s, find the probability that the ball hits the jth horizontal strip.

18. Among 33 students in a class, 17 of them earned A's on the midterm exam, 14 earned A's on the final exam, and 11 did not earn A's on either examination. What is the probability that a randomly selected student from this class earned an A on both exams?

19. From a small town 120 persons were selected at random and asked the following question: Which of the three shampoos, A, B, or C, do you use? The following results were obtained: 20 use A and C, 10 use A and B but not C, 15 use all three, 30 use only C, 35 use B but not C, 25 use B and C, and 10 use none of the three. If a person is selected at random from this group, what is the probability that he or she uses (a) only A; (b) only B; (c) A and B? (Draw a Venn diagram.)

20. The coefficients of the quadratic equation $x^2 + bx + c = 0$ are determined by tossing a fair die twice (the first outcome is b, the second one is c). Find the probability that the equation has real roots.

21. Two integers m and n are called *relatively prime* if 1 is their only common positive divisor. Thus 8 and 5 are relatively prime, whereas 8 and 6 are not. A number is selected at random from the set $\{1, 2, 3, \ldots, 63\}$. Find the probability that it is relatively prime to 63.

22. A number is selected randomly from the set $\{1, 2, \ldots, 1000\}$. What is the probability that (a) it is divisible by 3 but not by 5; (b) it is divisible neither by 3 nor by 5?

23. The secretary of a college has calculated that from the students who took calculus, physics, and chemistry last semester, 78% passed calculus, 80% physics, 84% chemistry, 60% calculus and physics, 65% physics and chemistry, 70% calculus and chemistry, and 55% all three. Show that these numbers are not consistent, and therefore the secretary has made a mistake.

B

24. From an ordinary deck of 52 cards, we draw cards at random and without replacement until only cards of one suit are left. Find the probability that the cards left are all spades.

25. A number is selected at random from the set of natural numbers $\{1, 2, \ldots, 1000\}$. What is the probability that it is divisible by 4 but neither by 5 nor by 7 ?

26. For a Democratic candidate to win an election, she must win districts I, II, and III. Polls have shown that the probability of winning I and III is 0.55, losing II but not I is 0.34, and losing II and III but not I is 0.15. Find the probability that this candidate will win all three districts. (Draw a Venn diagram.)

27. Two numbers are successively selected at random and with replacement from the set $\{1, 2, \ldots, 100\}$. What is the probability that the first one is greater than the second?

28. Let $A_1, A_2, A_3, \ldots$ be a sequence of events of a sample space. Prove that

$$P\left(\bigcup_{n=1}^{\infty} A_n\right) \leq \sum_{n=1}^{\infty} P(A_n).$$

This is called **Boole's inequality**.

29. Let $A_1, A_2, A_3, \ldots$ be a sequence of events of an experiment. Prove that

$$P\left(\bigcap_{n=1}^{\infty} A_n\right) \geq 1 - \sum_{n=1}^{\infty} P(A_n^c).$$

Hint: Use Boole's inequality, discussed in Exercise 28.

30. In a certain country, the probability is 49/50 that a randomly selected fighter plane returns from a mission without mishap. Mia argues that this means there is one mission with a mishap in every 50 consecutive flights. She concludes that if a fighter pilot returns safely from 49 consecutive missions, he should return home before his fiftieth mission. Is Mia right? Explain why or why not.

31. Let P be a probability defined on a sample space S. For events A of S define $Q(A) = \left[P(A)\right]^2$ and $R(A) = P(A)/2$. Is Q a probability on S? Is R a probability on S? Why or why not?

In its general case, the following exercise has important applications in coding theory, telecommunications, and computer science.

32. (The Hat Problem)   A game begins with a team of three players entering a room one at a time. For each player, a fair coin is tossed. If the outcome is heads, a red hat is placed on the player's head, and if it is tails, a blue hat is placed on the player's head. The players are allowed to communicate before the game begins to decide on a strategy. However, no communication is permitted after the game begins. Players cannot see their own hats. But each player can see the other two players' hats. Each player is given the option to guess the color of his or her hat or to pass. The game ends when the three players simultaneously make their choices. The team wins if no player's guess is incorrect and at least one player's guess is correct. Obviously, the team's goal is to develop a strategy that maximizes the probability of winning. A trivial strategy for the team would be for two of its players to pass and the third player to guess red or blue as he or she wishes. This gives the team a 50% chance to win. Can you think of a strategy that improves the chances of the team winning?

1.5   CONTINUITY OF PROBABILITY FUNCTIONS

Let $\mathbf{R}$ denote(here and everywhere else throughout the book) the set of all real numbers. We know from calculus that a function $f : \mathbf{R} \to \mathbf{R}$ is called continuous at a point $c \in \mathbf{R}$ if $\lim_{x \to c} f(x) = f(c)$. It is called continuous on $\mathbf{R}$ if it is continuous at all points $c \in \mathbf{R}$. We also know that this definition is equivalent to the sequential criterion $f : \mathbf{R} \to \mathbf{R}$ *is continuous on* $\mathbf{R}$ *if and only if, for every convergent sequence* $\{x_n\}_{n=1}^{\infty}$ *in* $\mathbf{R}$,

$$\lim_{n \to \infty} f(x_n) = f(\lim_{n \to \infty} x_n). \tag{1.4}$$

This property, in some sense, is shared by the probability function. To explain this, we need to introduce some definitions. But first recall that probability is a set function from $\mathcal{P}(S)$, the set of all possible events of the sample space S, to $[0, 1]$.

A sequence $\{E_n, n \geq 1\}$ of events of a sample space is called **increasing** if

$$E_1 \subseteq E_2 \subseteq E_3 \subseteq \cdots \subseteq E_n \subseteq E_{n+1} \cdots ;$$

it is called **decreasing** if

$$E_1 \supseteq E_2 \supseteq E_3 \supseteq \cdots \supseteq E_n \supseteq E_{n+1} \supseteq \cdots .$$

For an increasing sequence of events $\{E_n, n \geq 1\}$, by $\lim_{n \to \infty} E_n$ we mean the event that at least one $E_i, 1 \leq i < \infty$ occurs. Therefore,

$$\lim_{n \to \infty} E_n = \bigcup_{n=1}^{\infty} E_n.$$

Similarly, for a decreasing sequence of events $\{E_n, n \geq 1\}$, by $\lim_{n \to \infty} E_n$ we mean the event that every E_i occurs. Thus in this case

$$\lim_{n \to \infty} E_n = \bigcap_{n=1}^{\infty} E_n.$$

The following theorem expresses the property of probability function that is analogous to (1.4).

Theorem 1.8   (Continuity of Probability Function)   *For any increasing or decreasing sequence of events,* $\{E_n, n \geq 1\}$,

$$\lim_{n \to \infty} P(E_n) = P(\lim_{n \to \infty} E_n).$$

Figure 1.5   The circular disks are the E_i's and the shaded circular annuli are the F_i's, except for F_1, which equals E_1.

Proof: For the case where $\{E_n, \ n \geq 1\}$ is increasing, let $F_1 = E_1$, $F_2 = E_2 - E_1$, $F_3 = E_3 - E_2, \ldots, F_n = E_n - E_{n-1}, \ldots$. Clearly, $\{F_i, \ i \geq 1\}$ is a mutually exclusive set of events that satisfies the following relations:

$$\bigcup_{i=1}^{n} F_i = \bigcup_{i=1}^{n} E_i = E_n, \qquad n = 1, 2, 3, \ldots,$$

$$\bigcup_{i=1}^{\infty} F_i = \bigcup_{i=1}^{\infty} E_i$$

(see Figure 1.5). Hence

$$P(\lim_{n\to\infty} E_n) = P\left(\bigcup_{i=1}^{\infty} E_i\right) = P\left(\bigcup_{i=1}^{\infty} F_i\right) = \sum_{i=1}^{\infty} P(F_i) = \lim_{n\to\infty} \sum_{i=1}^{n} P(F_i)$$

$$= \lim_{n\to\infty} P\left(\bigcup_{i=1}^{n} F_i\right) = \lim_{n\to\infty} P\left(\bigcup_{i=1}^{n} E_i\right) = \lim_{n\to\infty} P(E_n),$$

where the last equality follows since $\{E_n, \ n \geq 1\}$ is increasing, and hence $\bigcup_{i=1}^{n} E_i = E_n$. This establishes the theorem for increasing sequences.

If $\{E_n, \ n \geq 1\}$ is decreasing, then $E_n \supseteq E_{n+1}, \forall n$, implies that $E_n^c \subseteq E_{n+1}^c, \forall n$. Therefore, the sequence $\{E_n^c, \ n \geq 1\}$ is increasing and

$$P(\lim_{n\to\infty} E_n) = P\left(\bigcap_{i=1}^{\infty} E_i\right) = 1 - P\left[\left(\bigcap_{i=1}^{\infty} E_i\right)^c\right] = 1 - P\left(\bigcup_{i=1}^{\infty} E_i^c\right)$$

$$= 1 - P(\lim_{n\to\infty} E_n^c) = 1 - \lim_{n\to\infty} P(E_n^c) = 1 - \lim_{n\to\infty} \left[1 - P(E_n)\right]$$

$$= 1 - 1 + \lim_{n\to\infty} P(E_n) = \lim_{n\to\infty} P(E_n). \quad \blacklozenge$$

Example 1.20   Suppose that some individuals in a population produce offspring of the same kind. The offspring of the initial population are called second generation, the offspring of the second generation are called third generation, and so on. If with probability $\exp\left[-(2n^2+7)/(6n^2)\right]$ the entire population completely dies out by the nth generation before producing any offspring, what is the probability that such a population survives forever?

Solution: Let E_n denote the event of extinction of the entire population by the nth generation; then

$$E_1 \subseteq E_2 \subseteq E_3 \subseteq \cdots \subseteq E_n \subseteq E_{n+1} \subseteq \cdots$$

because if E_n occurs, then E_{n+1} also occurs. Hence, by Theorem 1.8,

$$P\{\text{population survives forever}\} = 1 - P\{\text{population eventually dies out}\}$$

$$= 1 - P\left(\bigcup_{i=1}^{\infty} E_i\right) = 1 - \lim_{n\to\infty} P(E_n)$$

$$= 1 - \lim_{n\to\infty} \exp\left(-\frac{2n^2+7}{6n^2}\right) = 1 - e^{-1/3}. \quad \blacklozenge$$

1.6   PROBABILITIES 0 AND 1

Events with probabilities 1 and 0 should not be misinterpreted. If E and F are events with probabilities 1 and 0, respectively, it is not correct to say that E is the sample space S and F is the empty set $\emptyset$. In fact, there are experiments in which there exist infinitely many events each with probability 1, and infinitely many events each with probability 0. An example follows.

   Suppose that an experiment consists of selecting a random point from the interval $(0, 1)$. Since every point in $(0, 1)$ has a decimal representation such as

$$0.529387043219721\cdots,$$

the experiment is equivalent to picking an endless decimal from $(0, 1)$ at random (note that if a decimal terminates, all of its digits from some point on are 0). In such an experiment we want to compute the probability of selecting the point $1/3$. In other words, we want to compute the probability of choosing $0.333333\cdots$ in a random selection of an endless decimal. Let A_n be the event that the selected decimal has 3 as its first n digits; then

$$A_1 \supset A_2 \supset A_3 \supset A_4 \supset \cdots \supset A_n \supset A_{n+1} \supset \cdots,$$

since the occurrence of A_{n+1} guarantees the occurrence of A_n. Now $P(A_1) = 1/10$ because there are 10 choices $0, 1, 2, \ldots, 9$ for the first digit, and we want only one of them, namely 3, to occur. $P(A_2) = 1/100$ since there are 100 choices $00, 01, \ldots, 09,$ $10, 11, \ldots, 19, 20, \ldots, 99$ for the first two digits, and we want only one of them, 33, to occur. $P(A_3) = 1/1000$ because there are 1000 choices $000, 001, \ldots, 999$ for the first

three digits, and we want only one of them, 333, to occur. Continuing this argument, we
have $P(A_n) = (1/10)^n$. Since $\bigcap_{n=1}^{\infty} A_n = \{1/3\}$, by Theorem 1.8,

$$P\left(\frac{1}{3} \text{ is selected}\right) = P\left(\bigcap_{n=1}^{\infty} A_n\right) = \lim_{n\to\infty} P(A_n) = \lim_{n\to\infty} \left(\frac{1}{10}\right)^n = 0.$$

Note that there is nothing special about the point $1/3$. For any other point $0.\alpha_1\alpha_2\alpha_3\alpha_4 \cdots$
from $(0, 1)$, the same argument could be used to show that the probability of its occur-
rence is 0 (define A_n to be the event that the first n digits of the selected decimal are
$\alpha_1, \alpha_2, \dots, \alpha_n$, respectively, and repeat the same argument). We have shown that in random selection of points from $(0, 1)$, the probability of the occurrence of any particular
point is 0. Now for $t \in (0, 1)$, let $B_t = (0, 1) - \{t\}$. Then $P(\{t\}) = 0$ implies that

$$P(B_t) = P(\{t\}^c) = 1 - P(\{t\}) = 1.$$

Therefore, there are infinitely many events, B_t's, each with probability 1 and none equal
to the sample space $(0, 1)$.

1.7   RANDOM SELECTION OF POINTS FROM INTERVALS

In Section 1.6, we showed that the probability of the occurrence of any particular point in
a random selection of points from an interval (a, b) is 0. This implies immediately that if
$[\alpha, \beta] \subseteq (a, b)$, then the events that the point falls in $[\alpha, \beta]$, (α, β), $[\alpha, \beta)$, and $(\alpha, \beta]$ are

all equiprobable. Now consider the intervals $\left(a, \dfrac{a+b}{2}\right)$ and $\left(\dfrac{a+b}{2}, b\right)$; since $\dfrac{a+b}{2}$

is the midpoint of (a, b), it is reasonable to assume that

$$p_1 = p_2, \tag{1.5}$$

where p_1 is the probability that the point belongs to $\left(a, \dfrac{a+b}{2}\right)$ and p_2 is the probability

that it belongs to $\left(\dfrac{a+b}{2}, b\right)$. The events that the random point belongs to $\left(a, \dfrac{a+b}{2}\right)$

and $\left(\dfrac{a+b}{2}, b\right)$ are mutually exclusive and

$$\left(a, \frac{a+b}{2}\right) \cup \left[\frac{a+b}{2}, b\right) = (a, b);$$

therefore,

$$p_1 + p_2 = 1.$$

This relation and (1.5) imply that

$$p_1 = p_2 = 1/2.$$

Hence *the probability that a random point selected from (a, b) falls into the interval*

$\left(a, \dfrac{a+b}{2}\right)$ *is* $1/2$. *The probability that it falls into* $\left[\dfrac{a+b}{2}, b\right)$ *is also* $1/2$. Note that the

length of each of these intervals is $1/2$ of the length of (a, b). Now consider the intervals

$\left(a, \dfrac{2a+b}{3}\right]$, $\left(\dfrac{2a+b}{3}, \dfrac{a+2b}{3}\right]$, and $\left(\dfrac{a+2b}{3}, b\right)$. Since $\dfrac{2a+b}{3}$ and $\dfrac{a+2b}{3}$ are the

points that divide the interval (a, b) into three subintervals with equal lengths, we can
assume that

$$p_1 = p_2 = p_3, \tag{1.6}$$

where p_1, p_2, and p_3 are the probabilities that the point falls into $\left(a, \dfrac{2a+b}{3}\right]$,
$\left(\dfrac{2a+b}{3}, \dfrac{a+2b}{3}\right]$, and $\left(\dfrac{a+2b}{3}, b\right)$, respectively. On the other hand, these three

intervals are mutually disjoint and

$$\left(a, \dfrac{2a+b}{3}\right] \cup \left(\dfrac{2a+b}{3}, \dfrac{a+2b}{3}\right] \cup \left(\dfrac{a+2b}{3}, b\right) = (a, b).$$

Hence

$$p_1 + p_2 + p_3 = 1.$$

This relation and (1.6) imply that

$$p_1 = p_2 = p_3 = 1/3.$$

Therefore, *the probability that a random point selected from (a, b) falls into the interval*

$\left(a, \dfrac{2a+b}{3}\right]$ *is* $1/3$. *The probability that it falls into* $\left(\dfrac{2a+b}{3}, \dfrac{a+2b}{3}\right]$ *is* $1/3$, *and the*

probability that it falls into $\left(\dfrac{a+2b}{3}, b\right)$ *is* $1/3$. Note that the length of each of these

intervals is $1/3$ of the length of (a, b). These and other similar observations indicate that
the probability of the event that a random point from (a, b) falls into a subinterval (α, β)
is equal to $(\beta - \alpha)/(b - a)$.

Note that in this discussion we have assumed that subintervals of equal lengths are
equiprobable. Even though two subintervals of equal lengths may differ by a finite or a
countably infinite set (or even a set of measure zero), this assumption is still consistent
with our intuitive understanding of choosing random points from intervals. This is
because in such an experiment, the probability of the occurrence of a finite or countably
infinite set (or a set of measure zero) is 0.

Thus far, we have based our discussion of selecting random points from intervals on our intuitive understanding of this experiment and not on a mathematical definition. Such discussions are often necessary for the creation of appropriate mathematical meanings for unclear concepts. The following definition, which is based on our intuitive analysis, gives an exact mathematical meaning to the experiment of random selection of points from intervals.

Definition   *A point is said to be **randomly selected from an interval** (a, b) if any two subintervals of (a, b) that have the same length are equally likely to include the point. The probability associated with the event that the subinterval (α, β) contains the point is defined to be $(\beta - \alpha)/(b - a)$.*

As explained before, choosing a random number from $(0, 1)$ is equivalent to choosing randomly all the decimal digits of the number successively. Since in practice this is impossible, choosing exact random points or numbers from $(0, 1)$ or any other interval is only a theoretical matter. Approximate random numbers, however, can be generated by computers. Most of the computer languages, some scientific computer software, and some calculators are equipped with subroutines that generate approximate random numbers from intervals. However, since it is difficult to construct good random number generators, there are computer languages, software programs, and calculators that are equipped with poor random number generator algorithms. An excellent reference for construction of good random number generators is *The Art of Computer Programming*, Volume 2, *Seminumerical Algorithms,* third edition, by Donald E. Knuth (Addison Wesley, 1998). Simple mechanical tools can also be used to find such approximations. For example, consider a spinner mounted on a wheel of *unit circumference (radius $1/2\pi$)*. Let A be a point on the perimeter of the wheel. Each time that we flick the spinner, it stops, pointing toward some point B on the wheel's circumference. The length of the arc AB (directed, say, counterclockwise) is an approximate random number between 0 and 1 if the spinner does not have any "sticky" spots (see Figure 1.6).

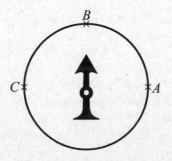

Figure 1.6   Spinner, a model to generate random numbers.

Finally, when selecting a random point from an interval (a, b), we may think of an extremely large hypothetical box that contains infinitely many indistinguishable balls.

Imagine that each ball is marked by a number from (a, b), each number of (a, b) is marked on exactly one ball, and the balls are completely mixed up, so that in a random selection of balls, any two of them have the same chance of being drawn. With this transcendental model in mind, choosing a random number from (a, b) is then equivalent to drawing a random ball from such a box and looking at its number.

★ **Example 1.21 (A Set that Is Not an Event)**   Let an experiment consist of selecting a point at random from the interval $[-1, 2]$. We will construct a set that is not an event (i.e., it is impossible to associate a probability with the set). We begin by defining an equivalence relation on $[0, 1]$: $x \sim y$ if $x - y$ is a rational number. Let $\mathbf{Q} = \{r_1, r_2, \ldots\}$ be the set of rational numbers in $[-1, 1]$. Clearly, x is equivalent to y if $x - y \in \mathbf{Q}$. The fact that this relation is reflexive, symmetric, and transitive is trivial. Therefore, being an equivalence relation, it partitions the interval $[0, 1]$ into disjoint equivalence classes (Λ_α). These classes are such that if x and $y \in \Lambda_\alpha$ for some α, then $x - y$ is rational. However, if $x \in \Lambda_\alpha$ and $y \in \Lambda_\beta$, and $\alpha \neq \beta$, then $x - y$ is irrational. These observations imply that for each α, the equivalence class Λ_α is countable. Since $\bigcup_\alpha \Lambda_\alpha = [0, 1]$ is uncountable, the number of equivalence classes is uncountable. Let E be a set consisting of *exactly* one point from each equivalence class Λ_α. The existence of such a set is guaranteed by the Axiom of Choice. We will show, by contradiction, that E is not an event. Suppose that E is an event, and let p be the probability associated with E. For each positive integer n, let $E_n = \{r_n + x : x \in E\} \subseteq [-1, 2]$. For each $r_n \in \mathbf{Q}$, E_n is simply a translation of E. Thus, for all n, the set E_n is also an event, and $P(E_n) = P(E) = p$.

We now make two more observations: (1) For $n \neq m$, $E_n \cap E_m = \emptyset$, (2) $[0, 1] \subset \bigcup_{n=1}^{\infty} E_n$. To prove (1), let $t \in E_n \cap E_m$. We will show that $E_n = E_m$. If $t \in E_n \cap E_m$, then for some $r_n, r_m \in \mathbf{Q}$, and $x, y \in E$, we have that $t = r_n + x = r_m + y$. That is, $x - y = r_m - r_n$ is rational, and $x - y$ belongs to the same equivalence class. Since E has exactly one point from each equivalence class, we must have $x = y$, hence $r_n = r_m$, hence $E_n = E_m$. To prove (2), let $x \in [0, 1]$. Then $x \sim y$ for some $y \in E$. This implies that $x - y$ is a rational number in $\mathbf{Q}$. That is, for some n, $x - y = r_n$, or $x = y + r_n$, or $x \in E_n$. Thus $x \in \bigcup_{n=1}^{\infty} E_n$.

Putting (1) and (2) together, we obtain

$$1/3 = P([0, 1]) \leq P\left(\bigcup_{n=1}^{\infty} E_n\right) \leq 1,$$

or

$$1/3 \leq \sum_{n=1}^{\infty} P(E_n) = \sum_{n=1}^{\infty} p \leq 1.$$

This is a contradiction because $\sum_{n=1}^{\infty} p$ is either 0 or ∞. Hence E is not an event, and we cannot associate a probability with this set. ◆

EXERCISES

A

1. A bus arrives at a station every day at a random time between 1:00 P.M. and 1:30 P.M. What is the probability that a person arriving at this station at 1:00 P.M. will have to wait at least 10 minutes?

2. Past experience shows that every new book by a certain publisher captures randomly between 4 and 12% of the market. What is the probability that the next book by this publisher captures at most 6.35% of the market?

3. Which of the following statements are true? If a statement is true, prove it. If it is false, give a counterexample.

   (a) If A is an event with probability 1, then A is the sample space.

   (b) If B is an event with probability 0, then $B = \emptyset$.

4. Let A and B be two events. Show that if $P(A) = 1$ and $P(B) = 1$, then $P(AB) = 1$.

5. A point is selected at random from the interval $(0, 2000)$. What is the probability that it is an integer?

6. Suppose that a point is randomly selected from the interval $(0, 1)$. Using the definition in Section 1.7, show that all numerals are equally likely to appear as the first digit of the decimal representation of the selected point.

B

7. Is it possible to define a probability on a countably infinite sample space so that the outcomes are equally probable?

8. Let $A_1, A_2, \ldots, A_n$ be n events. Show that if

$$P(A_1) = P(A_2) = \cdots = P(A_n) = 1,$$

   then $P(A_1 A_2 \cdots A_n) = 1$.

9. (a) Prove that $\bigcap_{n=1}^{\infty}(1/2 - 1/2n, 1/2 + 1/2n) = \{1/2\}$.

   (b) Using part (a), show that the probability of selecting $1/2$ in a random selection of a point from $(0, 1)$ is 0.

10. A point is selected at random from the interval $(0, 1)$. What is the probability that it is rational? What is the probability that it is irrational?

11. Suppose that a point is randomly selected from the interval $(0, 1)$. Using the definition in Section 1.7, show that all numerals are equally likely to appear as the nth digit of the decimal representation of the selected point.

12. Let $\{A_1, A_2, A_3, \ldots\}$ be a sequence of events. Prove that if the series $\sum_{n=1}^{\infty} P(A_n)$ converges, then $P\left(\bigcap_{m=1}^{\infty} \bigcup_{n=m}^{\infty} A_n\right) = 0$. This is called the **Borel-Cantelli lemma.** It says that if $\sum_{n=1}^{\infty} P(A_n) < \infty$, the probability that infinitely many of the A_n's occur is 0.

    Hint: Let $B_m = \bigcup_{n=m}^{\infty} A_n$ and apply Theorem 1.8 to $\{B_m, m \geq 1\}$.

13. Show that the result of Exercise 8 is not true for an infinite number of events. That is, show that if $\{E_t : 0 < t < 1\}$ is a collection of events for which $P(E_t) = 1$, it is not necessarily true that $P\left(\bigcap_{t \in (0,1)} E_t\right) = 1$.

14. Let A be the set of rational numbers in $(0, 1)$. Since A is countable, it can be written as a sequence $\left(\text{i.e.}, A = \{r_n : n = 1, 2, 3, \ldots\}\right)$. Prove that for any $\varepsilon > 0$, A can be covered by a sequence of open balls whose total length is less than ε. That is, $\forall \varepsilon > 0$, there exists a sequence of open intervals (α_n, β_n) such that $r_n \in (\alpha_n, \beta_n)$ and $\sum_{n=1}^{\infty} (\beta_n - \alpha_n) < \varepsilon$. This important result explains why in a random selection of points from $(0, 1)$ the probability of choosing a rational is zero.

    Hint: Let $\alpha_n = r_n - \varepsilon/2^{n+2}, \quad \beta_n = r_n + \varepsilon/2^{n+2}$.

■

REVIEW PROBLEMS

---

1. The number of minutes it takes for a certain animal to react to a certain stimulus is a random number between 2 and 4.3. Find the probability that the reaction time of such an animal to this stimulus is no longer than 3.25 minutes.

2. Let $\mathcal{P}$ be the set of all subsets of $A = \{1, 2\}$. We choose two distinct sets randomly from $\mathcal{P}$. Define a sample space for this experiment, and describe the following events:

    (a)   The intersection of the sets chosen at random is empty.

    (b)   The sets are complements of each other.

    (c)   One of the sets contains more elements than the other.

3. In a certain experiment, whenever the event A occurs, the event B also occurs. Which of the following statements is true and why?

    (a)   If we know that A has not occurred, we can be sure that B has not occurred as well.

(b)   If we know that B has not occurred, we can be sure that A has not occurred as well.

4.   The following relations *are not* always true. In each case give an example to refute them.

(a) $P(A \cup B) = P(A) + P(B)$.

(b) $P(AB) = P(A)P(B)$.

5.   A coin is tossed until, for the first time, the same result appears twice in succession. Define a sample space for this experiment.

6.   The number of the patients now in a hospital is 63. Of these 37 are male and 20 are for surgery. If among those who are for surgery 12 are male, how many of the 63 patients are neither male nor for surgery?

7.   Let A, B, and C be three events. Prove that

$$P(A \cup B \cup C) \le P(A) + P(B) + P(C).$$

8.   Let A, B, and C be three events. Show that

$$P(A \cup B \cup C) = P(A) + P(B) + P(C)$$

if and only if $P(AB) = P(AC) = P(BC) = 0$.

9.   Suppose that 40% of the people in a community drink or serve white wine, 50% drink or serve red wine, and 70% drink or serve red or white wine. What percentage of the people in this community drink or serve both red and white wine?

10.   Answer the following question, asked of Marilyn Vos Savant in the "Ask Marilyn" column of *Parade Magazine,* March 3, 1996.

> My dad heard this story on the radio. At Duke University, two students had received A's in chemistry all semester. But on the night before the final exam, they were partying in another state and didn't get back to Duke until it was over. Their excuse to the professor was that they had a flat tire, and they asked if they could take a make-up test. The professor agreed, wrote out a test and sent the two to separate rooms to take it. The first question (on one side of the paper) was worth 5 points, and they answered it easily. Then they flipped the paper over and found the second question, worth 95 points: 'Which tire was it?' What was the probability that both students would say the same thing? My dad and I think it's 1 in 16. Is that right?

11.   Let A and B be two events. Suppose that $P(A)$, $P(B)$, and $P(AB)$ are given. What is the probability that neither A nor B will occur?

12. Let A and B be two events. The event $(A - B) \cup (B - A)$ is called the **symmetric difference** of A and B and is denoted by $A \triangle B$. Clearly, $A \triangle B$ is the event that exactly one of the two events A and B occurs. Show that

$$P(A \triangle B) = P(A) + P(B) - 2P(AB).$$

13. A bookstore receives six boxes of books per month on six random days of each month. Suppose that two of those boxes are from one publisher, two from another publisher, and the remaining two from a third publisher. Define a sample space for the possible orders in which the boxes are received in a given month by the bookstore. Describe the event that the last two boxes of books received last month are from the same publisher.

14. Suppose that in a certain town the number of people with blood type O and blood type A are approximately the same. The number of people with blood type B is 1/10 of those with blood type A and twice the number of those with blood type AB. Find the probability that the next baby born in the town has blood type AB.

15. A number is selected at random from the set of natural numbers $\{1, 2, 3, \ldots, 1000\}$. What is the probability that it is not divisible by 4, 7, or 9?

16. A number is selected at random from the set $\{1, 2, 3, \ldots, 150\}$. What is the probability that it is relatively prime to 150? See Exercise 21, Section 1.4, for the definition of relatively prime numbers.

17. Suppose that each day the price of a stock moves up 1/8 of a point, moves down 1/8 of a point, or remains unchanged. For $i \geq 1$, let U_i and D_i be the events that the price of the stock moves up and down on the ith trading day, respectively. In terms of U_i's and D_i's, find an expression for the event that the price of the stock

  (a) remains unchanged on the ith trading day;

  (b) moves up every day of the next n trading days;

  (c) remains unchanged on at least one of the next n trading days;

  (d) is the same as today after three trading days;

  (e) does not move down on any of the next n trading days.

18. A bus traveling from Baltimore to New York has breaks down at a random location. What is the probability that the breakdown occurred after passing through Philadelphia? The distances from New York and Philadelphia to Baltimore are, respectively, 199 and 96 miles.

19. The coefficient of the quadratic equation $ax^2 + bx + c = 0$ are determined by tossing a fair die three times (the first outcome is a, the second one b, and the third one c). Find the probability that the equation has no real roots.

Chapter 2

Combinatorial Methods

2.1 INTRODUCTION

The study of probability includes many applications, such as games of chance, occupancy and order problems, and sampling procedures. In some of such applications, we deal with finite sample spaces in which all sample points are equally likely to occur. Theorem 1.3 shows that, in such cases, the probability of an event A is evaluated simply by dividing the number of points of A by the total number of sample points. Therefore, some probability problems can be solved simply by counting the total number of sample points and the number of ways that an event can occur.

In this chapter we study a few rules that enable us to count systematically. **Combinatorial analysis** deals with methods of counting: a very broad field with applications in virtually every branch of applied and pure mathematics. Besides probability and statistics, it is used in information theory, coding and decoding, linear programming, transportation problems, industrial planning, scheduling production, group theory, foundations of geometry, and other fields. Combinatorial analysis, as a formal branch of mathematics, began with Tartaglia in the sixteenth century. After Tartaglia, Pascal, Fermat, Chevalier Antoine de Méré (1607–1684), James Bernoulli, Gottfried Leibniz, and Leonhard Euler (1707–1783) made contributions to this field. The mathematical development of the twentieth century accelerated development by combinatorial analysis.

2.2 COUNTING PRINCIPLES

Suppose that there are n routes from town A to town B, and m routes from B to a third town, C. If we decide to go from A to C via B, then for each route that we choose from A to B, we have m choices from B to C. Therefore, altogether we have nm choices to go from A to C via B. This simple example motivates the following principle, which is the basis of this chapter.

Theorem 2.1 (Counting Principle) *If the set E contains n elements and the set F contains m elements, there are nm ways in which we can choose, first, an element of E and then an element of F.*

Proof: Let $E = \{a_1, a_2, \ldots, a_n\}$ and $F = \{b_1, b_2, \ldots, b_m\}$; then the following rectangular array, which consists of nm elements, contains all possible ways that we can choose, first, an element of E and then an element of F.

$$(a_1, b_1), \quad (a_1, b_2), \quad \ldots, \quad (a_1, b_m)$$
$$(a_2, b_1), \quad (a_2, b_2), \quad \ldots, \quad (a_2, b_m)$$
$$\vdots$$
$$(a_n, b_1), \quad (a_n, b_2), \quad \ldots, \quad (a_n, b_m) \quad \blacklozenge$$

Now suppose that a fourth town, D, is connected to C by ℓ routes. If we decide to go from A to D, passing through C after B, then for each pair of routes that we choose from A to C, there are ℓ possibilities from C to D. Therefore, by the counting principle, the total number of ways we can go from A to D via B and C is the number of ways we can go from A to C through B times ℓ, that is, $nm\ell$. This concept motivates a generalization of the counting principle.

Theorem 2.2   (Generalized Counting Principle)   *Let $E_1, E_2, \ldots, E_k$ be sets with $n_1, n_2, \ldots, n_k$ elements, respectively. Then there are $n_1 \times n_2 \times n_3 \times \cdots \times n_k$ ways in which we can, first, choose an element of E_1, then an element of E_2, then an element of $E_3, \ldots$, and finally an element of E_k.*

In probability, this theorem is used whenever we want to compute the total number of possible outcomes when k experiments are performed. Suppose that the first experiment has n_1 possible outcomes, the second experiment has n_2 possible outcomes, $\ldots$, and the kth experiment has n_k possible outcomes. If we define E_i to be the set of all possible outcomes of the ith experiment, then the total number of possible outcomes coincides with the number of ways that we can, first, choose an element of E_1, then an element of E_2, then an element of $E_3, \ldots$, and finally an element of E_k; that is, $n_1 \times n_2 \times \cdots \times n_k$.

Example 2.1   How many outcomes are there if we throw five dice?

Solution: Let E_i, $1 \leq i \leq 5$, be the set of all possible outcomes of the ith die. Then $E_i = \{1, 2, 3, 4, 5, 6\}$. The number of the outcomes of throwing five dice equals the number of ways we can, first, choose an element of E_1, then an element of $E_2, \ldots$, and finally an element of E_5. Thus we get $6 \times 6 \times 6 \times 6 \times 6 = 6^5$. $\blacklozenge$

Remark 2.1   Consider experiments such as flipping a fair coin several times, tossing a number of fair dice, drawing a number of cards from an ordinary deck of 52 cards at random and with replacement, and drawing a number of balls from an urn at random and with replacement. In Section 3.5, discussing the concept of *independence*, we will show that all the possible outcomes in such experiments are equiprobable. Until then, however, in all the problems dealing with these kinds of experiments, we *assume*, without explicitly so stating in each case, that the sample points of the sample space of the experiment under consideration are all equally likely. $\blacklozenge$

Example 2.2   In tossing four fair dice, what is the probability of at least one 3?

Solution: Let A be the event of at least one 3. Then A^c is the event of no 3 in tossing the four dice. $N(A^c)$ and N, the number of sample points of A^c and the total number of sample points, respectively, are given by $5 \times 5 \times 5 \times 5 = 5^4$ and $6 \times 6 \times 6 \times 6 = 6^4$. Therefore, $P(A^c) = N(A^c)/N = 5^4/6^4$. Hence $P(A) = 1 - P(A^c) = 1 - 625/1296 = 671/1296 \approx 0.52$.   ◆

Example 2.3   Virginia wants to give her son, Brian, 14 different baseball cards within a 7-day period. If Virginia gives Brian cards no more than once a day, in how many ways can this be done?

Solution: Each of the baseball cards can be given on 7 different days. Therefore, in $7 \times 7 \times \cdots \times 7 = 7^{14} \approx 6.78 \times 10^{11}$ ways Virginia can give the cards to Brian.   ◆

Example 2.4   Rose has invited n friends to her birthday party. If they all attend, and each one shakes hands with everyone else at the party exactly once, what is the number of handshakes?

Solution 1:   There are $n + 1$ people at the party and each of them shakes hands with the other n people. This is a total of $(n + 1)n$ handshakes, but that is an overcount since it counts "A shakes hands with B" as one handshake and "B shakes hand with A" as a second. Since each handshake is counted exactly twice, the actual number of handshakes is $(n + 1)n/2$.

Solution 2: Suppose that guests arrive one at a time. Rose will shake hands with all the n guests. The first guest to appear will shake hands with Rose and all the remaining guests. Since we have already counted his or her handshake with Rose, there will be $n - 1$ additional handshakes. The second guest will also shake hands with $n - 1$ fellow guests and Rose. However, we have already counted his or her handshakes with Rose and the first guest. So this will add $n - 2$ additional handshakes. Similarly, the third guest will add $n - 3$ additional handshakes, and so on. Therefore, the total number of handshakes will be $n + (n - 1) + (n - 2) + \cdots + 3 + 2 + 1$. Comparing solutions 1 and 2, we have the well-known relation

$$1 + 2 + 3 \cdots + (n - 2) + (n - 1) + n = \frac{n(n + 1)}{2}. \quad ◆$$

Example 2.5   At a state university in Maryland, there is hardly enough space for students to park their cars in their own lots. Jack, a student who parks in the faculty parking lot every day, noticed that none of the last 10 tickets he got was issued on a Monday or on a Friday. Is it wise for Jack to conclude that the campus police do not patrol the faculty parking lot on Mondays and on Fridays? Assume that police give no tickets on weekends.

Solution: Suppose that the answer is negative and the campus police patrol the parking lot randomly; that is, the parking lot is patrolled every day with the same probability. Let A be the event that out of 10 tickets given on random days, none is issued on a Monday or on a Friday. If $P(A)$ is very small, we can conclude that the campus police do not patrol the parking lot on these two days. Otherwise, we conclude that what happened is accidental and police patrol the parking lot randomly. To find $P(A)$, note that since each ticket has five possible days of being issued, there are 5^{10} possible ways for all tickets to have been issued. Of these, in only 3^{10} ways no ticket is issued on a Monday or on a Friday. Thus $P(A) = 3^{10}/5^{10} \approx 0.006$, a rather small probability. Therefore, it is reasonable to assume that the campus police do not patrol the parking lot on these two days.   ◆

Example 2.6   (Standard Birthday Problem)   What is the probability that at least two students of a class of size n have the same birthday? Compute the numerical values of such probabilities for $n = 23, 30, 50,$ and 60. Assume that the birth rates are constant throughout the year and that each year has 365 days.

Solution: There are 365 possibilities for the birthdays of each of the n students. Therefore, the sample space has 365^n points. In $365 \times 364 \times 363 \times \cdots \times [365 - (n-1)]$ ways the birthdays of no two of the n students coincide. Hence $P(n)$, the probability that no two students have the same birthday, is

$$P(n) = \frac{365 \times 364 \times 363 \times \cdots \times [365 - (n-1)]}{365^n},$$

and therefore the desired probability is $1 - P(n)$. For $n = 23, 30, 50,$ and 60 the answers are 0.507, 0.706, 0.970, and 0.995, respectively.   ◆

Remark 2.2   In probability and statistics studies, birthday problems similar to Example 2.6 have been very popular since 1939, when introduced by von Mises. This is probably because when solving such problems, numerical values obtained are often surprising. Persi Diaconis and Frederick Mosteller, two Harvard professors, have mentioned that they "find the utility of birthday problems impressive as a tool for thinking about coincidences." Diaconis and Mosteller have illustrated basic statistical techniques for studying the fascinating, curious, and complicated "Theory of Coincidences" in the December 1989 issue of the *Journal of the American Statistical Association*. In their study, they have used birthday problems "as examples which make the point that in *many problems our intuitive grasp of the odds is far off.*" Throughout this book, where appropriate, we will bring up some interesting versions of these problems, but now that we have cited "coincidence," let us read a few sentences from the abstract of the aforementioned paper to get a better feeling for its meaning.

Once we set aside coincidences having apparent causes, four principles account for large numbers of remaining coincidences: hidden cause; psychology, including memory and perception; multiplicity of endpoints, including the counting of "close" or nearly alike events as if they were identical; and the law of truly large numbers, which says that when enormous numbers of events and people and their interactions cumulate over time, almost any outrageous event is bound to occur. These sources account for much of the force of synchronicity. ◆

Number of Subsets of a Set

Let A be a set. The set of all subsets of A is called the **power set** of A. As an important application of the generalized counting principle, we now prove that the power set of a set with n elements has 2^n elements. This important fact has lots of good applications.

Theorem 2.3   *A set with n elements has 2^n subsets.*

Proof: Let $A = \{a_1, a_2, a_3, \ldots, a_n\}$ be a set with n elements. Then there is a one-to-one correspondence between the subsets of A and the sequences of 0's and 1's of length n: To a subset B of A we associate a sequence $b_1 b_2 b_3 \cdots b_n$, where $b_i = 0$ if $a_i \notin B$, and $b_i = 1$ if $a_i \in B$. For example, if $n = 3$, we associate to the empty subset of A the sequence 000, to $\{a_2, a_3\}$ the sequence 011, and to $\{a_1\}$ the sequence 100. Now, by the generalized counting principle, the number of sequences of 0's and 1's of length n is $2 \times 2 \times 2 \times \cdots \times 2 = 2^n$. Thus the number of subsets of A is also 2^n. ◆

Example 2.7   A restaurant advertises that it offers over 1000 varieties of pizza. If, at the restaurant, it is possible to have on a pizza any combination of pepperoni, mushrooms, sausage, green peppers, onions, anchovies, salami, bacon, olives, and ground beef, is the restaurant's advertisement true?

Solution: Any combination of the 10 ingredients that the restaurant offers can be put on a pizza. Thus the number of different types of pizza that it is possible to make is equal to the number of subsets of the set {pepperoni, mushrooms, sausage, green peppers, onions, anchovies, salami, bacon, olives, ground beef}, which is $2^{10} = 1024$. Therefore, the restaurant's advertisement is true. Note that the empty subset of the set of ingredients corresponds to a plain cheese pizza. ◆

Tree Diagrams

Tree diagrams are useful pictorial representations that break down a complex counting problem into smaller, more tractable ones. They are used in situations where the number of possible ways an experiment can be performed is finite. The following examples illustrate how tree diagrams are constructed and why they are useful. A great advantage of tree diagrams is that they systematically identify all possible cases.

Example 2.8   Bill and John keep playing chess until one of them wins two games in a row or three games altogether. In what percent of all possible cases does the game end because Bill wins three games without winning two in a row?

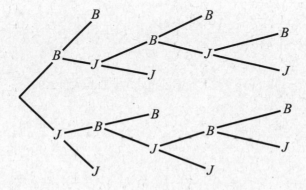

Figure 2.1   Tree diagram of Example 2.8.

Solution: The tree diagram of Figure 2.1 illustrates all possible cases. The total number of possible cases is equal to the number of the endpoints of the branches, which is 10. The number of cases in which Bill wins three games without winning two in a row, as seen from the figure, is one. So the answer is 10%. Note that the probability of this event is not 0.10 because not all of the branches of the tree are equiprobable.   ◆

Example 2.9   Mark has $4. He decides to bet $1 on the flip of a fair coin four times. What is the probability that (a) he breaks even; (b) he wins money?

Solution: The tree diagram of the Figure 2.2 illustrates various possible outcomes for Mark. The diagram has 16 endpoints, showing that the sample space has 16 elements. In six of these 16 cases, Mark breaks even and in five cases he wins money, so the desired probabilities are 6/16 and 5/16, respectively.   ◆

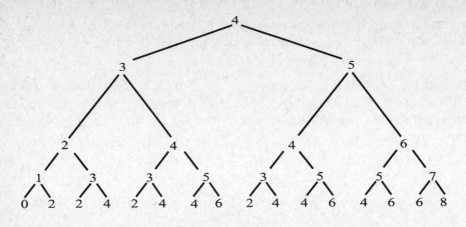

Figure 2.2   Tree diagram of Example 2.9.

EXERCISES

A

1. How many six-digit numbers are there? How many of them contain the digit 5? Note that the first digit of an n-digit number is nonzero.

2. How many different five-letter codes can be made using a, b, c, d, and e? How many of them start with ab?

3. The population of a town is 20,000. If each resident has three initials, is it true that at least two people have the same initials?

4. In how many different ways can 15 offices be painted with four different colors?

5. In flipping a fair coin 23 times, what is the probability of all heads or all tails?

6. In how many ways can we draw five cards from an ordinary deck of 52 cards (a) with replacement; (b) without replacement?

7. Two fair dice are thrown. What is the probability that the outcome is a 6 and an odd number?

8. Mr. Smith has 12 shirts, eight pairs of slacks, eight ties, and four jackets. Suppose that four shirts, three pairs of slacks, two ties, and two jackets are blue. (a) What is the probability that an all-blue outfit is the result of a random selection? (b) What is the probability that he wears at least one blue item tomorrow?

9. A multiple-choice test has 15 questions, each having four possible answers, of which only one is correct. If the questions are answered at random, what is the probability of getting all of them right?

10. Suppose that in a state, license plates have three letters followed by three numbers, in a way that no letter or number is repeated in a single plate. Determine the number of possible license plates for this state.

11. A library has 800,000 books, and the librarian wants to encode each by using a code word consisting of three letters followed by two numbers. Are there enough code words to encode all of these books with different code words?

12. How many $n \times m$ arrays (matrices) with entries 0 or 1 are there?

13. How many divisors does 55,125 have?
    Hint: $55,125 = 3^2 5^3 7^2$.

14. A delicatessen has advertised that it offers over 500 varieties of sandwiches. If at this deli it is possible to have any combination of salami, turkey, bologna, corned beef, ham, and cheese on French bread with the possible additions of lettuce, tomato, and mayonnaise, is the deli's advertisement true? Assume that a sandwich necessarily has bread and at least one type of meat or cheese.

15. How many four-digit numbers can be formed by using only the digits 2, 4, 6, 8, and 9? How many of these have some digit repeated?

16. In a mental health clinic there are 12 patients. A therapist invites all these patients to join her for group therapy. How many possible groups could she get?

17. Suppose that four cards are drawn successively from an ordinary deck of 52 cards, with replacement and at random. What is the probability of drawing at least one king?

18. A campus telephone extension has four digits. How many different extensions with no repeated digits exist? Of these, (a) how many do not start with a 0; (b) how many do not have 01 as the first two digits?

19. There are N types of drugs sold to reduce acid indigestion. A random sample of n drugs is taken with replacement. What is the probability that brand A is included?

20. Jenny, a probability student, having seen Example 2.6 and its solution, becomes convinced that it is a nearly even bet that someone among the next 22 people she meets randomly will have the same birthday as she does. What is the fallacy in Jenny's thinking? What is the minimum number of people that Jenny must meet before the chances are better than even that someone shares her birthday?

21. A salesperson covers islands $A, B, \ldots, I$. These islands are connected by the bridges shown in the Figure 2.3. While on an island, the salesperson takes one of the possible bridges at random and goes to another one. She does her business

on this new island and then takes a bridge at random to go to the next one. She continues this until she reaches an island for the second time on the same day. She stays there overnight and then continues her trips the next day. If she starts her trip from island I tomorrow, in what percent of all possible trips will she end up staying overnight again at island I?

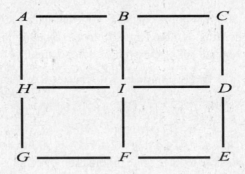

Figure 2.3   Islands and connecting bridges of Exercise 21.

B

22. In a large town, Kennedy Avenue is a long north-south avenue with many intersections. A drunken man is wandering along the avenue and does not really know which way he is going. He is currently at an intersection O somewhere in the middle of the avenue. Suppose that, at the end of each block, he either goes north with probability 1/2, or he goes south with probability 1/2. Draw a tree diagram to find the probability that, after walking four blocks, (a) he is back at intersection O; (b) he is only one block away from intersection O.

23. An integer is selected at random from the set $\{1, 2, \dots, 1,000,000\}$. What is the probability that it contains the digit 5?

24. How many divisors does a natural number N have?
    Hint: A natural number N can be written as $p_1^{n_1} p_2^{n_2} \cdots p_k^{n_k}$, where $p_1, p_2, \dots, p_k$ are distinct primes.

25. In tossing four fair dice, what is the probability of tossing, at most, one 3?

26. A delicatessen advertises that it offers over 3000 varieties of sandwiches. If at this deli it is possible to have any combination of salami, turkey, bologna, corned beef, and ham with or without Swiss and/or American cheese on French, white, or whole wheat bread, and possible additions of lettuce, tomato, and mayonnaise, is the deli's advertisement true? Assume that a sandwich necessarily has bread and at least one type of meat or cheese.

27. One of the five elevators in a building leaves the basement with eight passengers and stops at all of the remaining 11 floors. If it is equally likely that a passenger gets off at any of these 11 floors, what is the probability that no two of these eight passengers will get off at the same floor?

28. The elevator of a four-floor building leaves the first floor with six passengers and stops at all of the remaining three floors. If it is equally likely that a passenger gets off at any of these three floors, what is the probability that, at each stop of the elevator, at least one passenger departs?

29. A number is selected randomly from the set {0000, 0001, 0002, ... , 9999}. What is the probability that the sum of the first two digits of the number selected is equal to the sum of its last two digits?

30. What is the probability that a random r-digit number ($r \geq 3$) contains at least one 0, at least one 1, and at least one 2?

■

2.3   PERMUTATIONS

To count the number of outcomes of an experiment or the number of possible ways an event can occur, it is often useful to look for special patterns. Sometimes patterns help us develop techniques for counting. Two simple cases in which patterns enable us to count easily are **permutations** and **combinations**. We study these two patterns in this section and the next.

Definition   *An ordered arrangement of r objects from a set A containing n objects* ($0 < r \leq n$) *is called an **r-element permutation** of A, or a permutation of the elements of A taken r at a time. The number of r-element permutations of a set containing n objects is denoted by $_nP_r$.*

By this definition, if three people, Brown, Smith, and Jones, are to be scheduled for job interviews, any possible order for the interviews is a three-element permutation of the set {Brown, Smith, Jones}.

If, for example, $A = \{a, b, c, d\}$, then ab is a two-element permutation of A, acd is a three-element permutation of A, and $adcb$ is a four-element permutation of A. The order in which objects are arranged is important. For example, ab and ba are considered different two-element permutations, abc and cba are distinct three-element permutations, and $abcd$ and $cbad$ are different four-element permutations.

To compute $_nP_r$, the number of permutations of a set A containing n elements taken r at a time ($1 \leq r \leq n$), we use the generalized counting principle: Since A has n elements, the number of choices for the first object in the r-element permutation is n. For the second object, the number of choices is the remaining $n - 1$ elements of A. For the third one, the number of choices is the remaining $n - 2, \ldots$, and, finally, for the rth object the number of choices is $n - (r - 1) = n - r + 1$. Hence

$$_nP_r = n(n-1)(n-2) \cdots (n-r+1). \tag{2.1}$$

An n-element permutation of a set with n objects is simply called a **permutation**. The number of permutations of a set containing n elements, $_nP_n$, is evaluated from (2.1) by putting $r = n$.

$$_nP_n = n(n-1)(n-2) \cdots (n-n+1) = n!. \tag{2.2}$$

The formula $n!$ (the number of permutations of a set of n objects) has been well known for a long time. Although it first appeared in the works of Persian and Arab mathematicians in the twelfth century, there are indications that the mathematicians of India were aware of this rule a few hundred years before Christ. However, the surprise notation ! used for "factorial" was introduced by Christian Kramp in 1808. He chose this symbol perhaps because $n!$ gets *surprisingly* large even for small numbers. For example, we have that $18! \approx 6.402373 \times 10^{15}$, a number that, according to Karl Smith,[†] is greater than six times "the number of all words ever printed."

There is a popular alternative for relation (2.1). It is obtained by multiplying both sides of (2.1) by $(n - r)! = (n - r)(n - r - 1) \cdots 3 \cdot 2 \cdot 1$. We get

$$_nP_r \cdot (n-r)!$$
$$= \big[n(n-1)(n-2) \cdots (n-r+1)\big] \cdot \big[(n-r)(n-r-1) \cdots 3 \cdot 2 \cdot 1\big].$$

This gives $_nP_r \cdot (n - r)! = n!$. Therefore,

> The number of r-element permutations of a set containing n objects is given by
>
> $$_nP_r = \frac{n!}{(n-r)!}. \tag{2.3}$$

Note that for $r = n$, this relation implies that $_nP_n = n!/0!$. But by (2.2), $_nP_n = n!$. Therefore, for $r = n$, to make (2.3) consistent with (2.2), we define $0! = 1$.

Example 2.10   Three people, Brown, Smith, and Jones, must be scheduled for job interviews. In how many different orders can this be done?

Solution: The number of different orders is equal to the number of permutations of the set {Brown, Smith, Jones}. So there are $3! = 6$ possible orders for the interviews.   ◆

Example 2.11   Suppose that two anthropology, four computer science, three statistics, three biology, and five music books are put on a bookshelf with a random arrangement. What is the probability that the books of the same subject are together?

[†]Karl J. Smith, *The Nature of Mathematics,* 9th ed., Brooks/Cole, Pacific Grove, Calif., 2001, p. 39.

Solution: Let A be the event that all the books of the same subject are together. Then $P(A) = N(A)/N$, where $N(A)$ is the number of arrangements in which the books dealing with the same subject are together and N is the total number of possible arrangements. Since there are 17 books and each of their arrangements is a permutation of the set of these books, $N = 17!$. To calculate $N(A)$, note that there are $2! \times 4! \times 3! \times 3! \times 5!$ arrangements in which anthropology books are first, computer science books are next, then statistics books, after that biology, and finally, music. Also, there are the same number of arrangements for each possible ordering of the subjects. Since the subjects can be ordered in $5!$ ways, $N(A) = 5! \times 2! \times 4! \times 3! \times 3! \times 5!$. Hence

$$P(A) = \frac{5! \times 2! \times 4! \times 3! \times 3! \times 5!}{17!} \approx 6.996 \times 10^{-8}. \quad \blacklozenge$$

Example 2.12   If five boys and five girls sit in a row in a random order, what is the probability that no two children of the same sex sit together?

Solution: There are $10!$ ways for 10 persons to sit in a row. In order that no two of the same sex sit together, boys must occupy positions 1, 3, 5, 7, 9, and girls positions 2, 4, 6, 8, 10, or vice versa. In each case there are $5! \times 5!$ possibilities. Therefore, the desired probability is equal to

$$\frac{2 \times 5! \times 5!}{10!} \approx 0.008. \quad \blacklozenge$$

We showed that the number of permutations of a set of n objects is $n!$. This formula is valid only if all of the objects of the set are distinguishable from each other. Otherwise, the number of permutations is different. For example, there are $8!$ permutations of the eight letters $STANFORD$ because all of these letters are distinguishable from each other. But the number of permutations of the 8 letters $BERKELEY$ is less than $8!$ since the second, the fifth, and the seventh letters in $BERKELEY$ are indistinguishable. If in any permutation of these letters we change the positions of these three indistinguishable E's with each other, no new permutations would be generated. Let us calculate the number of permutations of the letters $BERKELEY$. Suppose that there are x such permutations and we consider any particular one of them, say $BYERELEK$. If we label the E's: $BYE_1RE_2LE_3K$ so that all of the letters are distinguishable, then by arranging E's among themselves we get $3!$ new permutations, namely,

$$
\begin{array}{ll}
BYE_1RE_2LE_3K & BYE_2RE_3LE_1K \\
BYE_1RE_3LE_2K & BYE_3RE_1LE_2K \\
BYE_2RE_1LE_3K & BYE_3RE_2LE_1K
\end{array}
$$

which are otherwise all the same. Therefore, if all the letters were different, then for each one of the x permutations we would have $3!$ times as many. That is, the total number of permutations would have been $x \times 3!$. But since eight different letters generate exactly

8! permutations, we must have $x \times 3! = 8!$. This gives $x = 8!/3!$. We have shown that *the number of distinguishable permutations of the letters BERKELEY is $8!/3!$.* This sort of reasoning leads us to the following general theorem.

Theorem 2.4   *The number of distinguishable permutations of n objects of k different types, where n_1 are alike, n_2 are alike, ... , n_k are alike and $n = n_1 + n_2 + \cdots + n_k$, is*

$$\frac{n!}{n_1! \times n_2! \times \cdots \times n_k!}.$$

Example 2.13   How many different 10-letter codes can be made using three a's, four b's, and three c's?

Solution: By Theorem 2.4, the number of such codes is $10!/(3! \times 4! \times 3!) = 4200.$ ♦

Example 2.14   In how many ways can we paint 11 offices so that four of them will be painted green, three yellow, two white, and the remaining two pink?

Solution: Let "*ggypgwpygwy*" represent the situation in which the first office is painted green, the second office is painted green, the third one yellow, and so on, with similar representations for other cases. Then the answer is equal to the number of distinguishable permutations of "*ggggyyywwpp*," which by Theorem 2.4 is $11!/(4! \times 3! \times 2! \times 2!) = 69,300.$ ♦

Example 2.15   A fair coin is flipped 10 times. What is the probability of obtaining exactly three heads?

Solution: The set of all sequences of H (heads) and T (tails) of length 10 forms the sample space and contains 2^{10} elements. Of all these, those with three H, and seven T, are desirable. But the number of distinguishable sequences with three H's and seven T's is equal to $10!/(3! \times 7!)$. Therefore, the probability of exactly three heads is $\left(\dfrac{10!}{3! \times 7!}\right) \Big/ 2^{10} \approx 0.12.$ ♦

EXERCISES

A

1. In the popular TV show *Who Wants to Be a Millionaire*, contestants are asked to sort four items in accordance with some norm: for example, landmarks in geographical order, movies in the order of date of release, singers in the order of date of birth. What is the probability that a contestant can get the correct answer solely by guessing?

2. How many permutations of the set $\{a, b, c, d, e\}$ begin with a and end with c?

3. How many different messages can be sent by five dashes and three dots?

4. Robert has eight guests, two of whom are Jim and John. If the guests will arrive in a random order, what is the probability that John will not arrive right after Jim?

5. Let A be the set of all sequences of 0's, 1's, and 2's of length 12.

   (a) How many elements are there in A?

   (b) How many elements of A have exactly six 0's and six 1's?

   (c) How many elements of A have exactly three 0's, four 1's, and five 2's?

6. Professor Haste is somewhat familiar with six languages. To translate texts from one language into another directly, how many one-way dictionaries does he need?

7. In an exhibition, 20 cars of the same style that are distinguishable only by their colors, are to be parked in a row, all facing a certain window. If four of the cars are blue, three are black, five are yellow, and eight are white, how many choices are there?

8. At various yard sales, a woman has acquired five forks, of which no two are alike. The same applies to her four knives and seven spoons. In how many different ways can three place settings be chosen if each place setting consists of exactly one fork, one knife, and one spoon? Assume that the arrangement of the place settings on the table is unimportant.

9. In a conference, Dr. Richman's lecture is related to Dr. Chollet's and should not precede it. If there are six more speakers, how many schedules could be arranged? *Warning:* Dr. Richman's lecture is not necessarily scheduled right after Dr. Chollet's lecture.

10. A dancing contest has 11 competitors, of whom three are Americans, two are Mexicans, three are Russians, and three are Italians. If the contest result lists only the nationality of the dancers, how many outcomes are possible?

11. Six fair dice are tossed. What is the probability that at least two of them show the same face?

12. (a) Find the number of distinguishable permutations of the letters $MISSISSIPPI$. (b) In how many of these permutations P's are together? (c) In how many I's are together? (d) In how many P's are together, and I's are together? (e) In a random order of the letters $MISSISSIPPI$, what is the probability that all S's are together?

13. A fair die is tossed eight times. What is the probability of exactly two 3's, three 1's, and three 6's?

14. In drawing nine cards with replacement from an ordinary deck of 52 cards, what is the probability of three aces of spades, three queens of hearts, and three kings of clubs?

15. At a party, n men and m women put their drinks on a table and go out on the floor to dance. When they return, none of them recognizes his or her drink, so everyone takes a drink at random. What is the probability that each man selects his own drink?

16. There are 20 chairs in a room numbered 1 through 20. If eight girls and 12 boys sit on these chairs at random, what is the probability that the thirteenth chair is occupied by a boy?

17. There are 12 students in a class. What is the probability that their birthdays fall in 12 different months? Assume that all months have the same probability of including the birthday of a randomly selected person.

18. If we put five math, six biology, eight history, and three literature books on a bookshelf at random, what is the probability that all the math books are together?

19. One of the five elevators in a building starts with seven passengers and stops at nine floors. Assuming that it is equally likely that a passenger gets off at any of these nine floors, find the probability that at least two of these passengers will get off at the same floor.

20. Five boys and five girls sit at random in a row. What is the probability that the boys are together and the girls are together?

21. If n balls are randomly placed into n cells, what is the probability that each cell will be occupied?

22. A town has six parks. On a Saturday, six classmates, who are unaware of each other's decision, choose a park at random and go there at the same time. What is the probability that at least two of them go to the same park? Convince yourself that this exercise is the same as Exercise 11, only expressed in a different context.

23. A club of 136 members is in the process of choosing a president, a vice president, a secretary, and a treasurer. If two of the members are not on speaking terms and do not serve together, in how many ways can these four people be chosen?

B

24. Let S and T be finite sets with n and m elements, respectively.

   (a) How many functions $f: S \to T$ can be defined?

   (b) If $m \geq n$, how many injective (one-to-one) functions $f: S \to T$ can be defined?

   (c) If $m = n$, how many surjective (onto) functions $f: S \to T$ can be defined?

25. A fair die is tossed eight times. What is the probability of exactly two 3's, exactly three 1's, and exactly two 6's?

26. Suppose that 20 sticks are broken, each into one long and one short part. By pairing them randomly, the 40 parts are then used to make 20 new sticks. (a) What is the probability that long parts are all paired with short ones? (b) What is the probability that the new sticks are exactly the same as the old ones?

27. At a party, 15 married couples are seated at random at a round table. What is the probability that all men are sitting next to their wives? Suppose that of these married couples, five husbands and their wives are older than 50 and the remaining husbands and wives are all younger than 50. What is the probability that all men over 50 are sitting next to their wives? Note that when people are sitting around a round table, only their seats relative to each other matters. The exact position of a person is not important.

28. A box contains five blue and eight red balls. Jim and Jack start drawing balls from the box, respectively, one at a time, at random, and without replacement until a blue ball is drawn. What is the probability that Jack draws the blue ball?

■

2.4 COMBINATIONS

In many combinatorial problems, unlike permutations, the order in which elements are arranged is immaterial. For example, suppose that in a contest there are 10 semifinalists and we want to count the number of possible ways that three contestants enter the finals. If we argue that there are $10 \times 9 \times 8$ such possibilities, we are wrong since the contestants cannot be ordered. If A, B, and C are three of the semifinalists, then ABC, BCA, ACB, BAC, CAB, and CBA are all the same event and have the same meaning: "A, B, and C are the finalists." The technique known as **combinations** is used to deal with such problems.

Definition *An unordered arrangement of r objects from a set A containing n objects $(r \leq n)$ is called an r-element combination of A, or a combination of the elements of A taken r at a time.*

Therefore, two combinations are different only if they differ in composition. Let x be the number of r-element combinations of a set A of n objects. If all the permutations of each r-element combination are found, then all the r-element permutations of A are found. Since for each r-element combination of A there are $r!$ permutations and the total number of r-element permutations is $_nP_r$, we have

$$x \cdot r! = {}_nP_r.$$

Hence $x \cdot r! = n!/(n - r)!$, so $x = n!/[(n - r)! \, r!]$. Therefore, we have shown that

The number of r-element combinations of n objects is given by

$$_nC_r = \frac{n!}{(n - r)! \, r!}.$$

Historically, a formula equivalent to $n!/[(n - r)! \, r!]$ turned up in the works of the Indian mathematician Bhaskara II (1114–1185) in the middle of the twelfth century. Bhaskara II used his formula to calculate the number of possible medicinal preparations using six ingredients. Therefore, the rule for calculation of the number of r-element combinations of n objects has been known for a long time.

It is worthwhile to observe that $_nC_r$ is the number of subsets of size r that can be constructed from a set of size n. By Theorem 2.3, a set with n elements has 2^n subsets. Therefore, of these 2^n subsets, the number of those that have exactly r elements is $_nC_r$.

Notation: By the symbol $\binom{n}{r}$ (read: n choose r) we mean the number of all r-element combinations of n objects. Therefore, for $r \leq n$,

$$\binom{n}{r} = \frac{n!}{r! \, (n - r)!}.$$

Observe that $\binom{n}{0} = \binom{n}{n} = 1$ and $\binom{n}{1} = \binom{n}{n-1} = n$. Also, for any $0 \leq r \leq n$,

$$\binom{n}{r} = \binom{n}{n - r}$$

and

$$\binom{n + 1}{r} = \binom{n}{r} + \binom{n}{r - 1}. \tag{2.4}$$

These relations can be proved algebraically or verified combinatorially. Let us prove (2.4) by a combinatorial argument. Consider a set of $n+1$ objects, $\{a_1, a_2, \ldots, a_n, a_{n+1}\}$. There are $\binom{n + 1}{r}$ r-element combinations of this set. Now we separate these r-element combinations into two disjoint classes: one class consisting of all r-element combinations of $\{a_1, a_2, \ldots, a_n\}$ and another consisting of all $(r - 1)$-element combinations of $\{a_1, a_2, \ldots, a_n\}$ attached to a_{n+1}. The latter class contains $\binom{n}{r - 1}$ elements and the former contains $\binom{n}{r}$ elements, showing that (2.4) is valid.

Example 2.16   In how many ways can two mathematics and three biology books be selected from eight mathematics and six biology books?

Solution: There are $\binom{8}{2}$ possible ways to select two mathematics books and $\binom{6}{3}$ possible ways to select three biology books. Therefore, by the counting principle,

$$\binom{8}{2} \times \binom{6}{3} = \frac{8!}{6!\,2!} \times \frac{6!}{3!\,3!} = 560$$

is the total number of ways in which two mathematics and three biology books can be selected. ◆

Example 2.17   A random sample of 45 instructors from different state universities were selected randomly and asked whether they are happy with their teaching loads. The responses of 32 were negative. If Drs. Smith, Brown, and Jones were among those questioned, what is the probability that all three of them gave negative responses?

Solution: There are $\binom{45}{32}$ different possible groups with negative responses. If three of them are Drs. Smith, Brown, and Jones, the other 29 are from the remaining 42 faculty members questioned. Hence the desired probability is

$$\frac{\binom{42}{29}}{\binom{45}{32}} \approx 0.35. \quad ◆$$

Example 2.18   In a small town, 11 of the 25 schoolteachers are against abortion, eight are for abortion, and the rest are indifferent. A random sample of five schoolteachers is selected for an interview. What is the probability that (a) all of them are for abortion; (b) all of them have the same opinion?

Solution:

(a)   There are $\binom{25}{5}$ different ways to select random samples of size 5 out of 25 teachers. Of these, only $\binom{8}{5}$ are all for abortion. Hence the desired probability is

$$\frac{\binom{8}{5}}{\binom{25}{5}} \approx 0.0011.$$

(b)   By an argument similar to part (a), the desired probability equals

$$\frac{\binom{11}{5} + \binom{8}{5} + \binom{6}{5}}{\binom{25}{5}} \approx 0.0099. \quad ◆$$

Example 2.19   In Maryland's lottery, players pick six different integers between 1 and 49, order of selection being irrelevant. The lottery commission then randomly selects six of these as the *winning numbers*. A player wins the grand prize if all six numbers that he or she has selected match the winning numbers. He or she wins the second prize if exactly five, and the third prize if exactly four of the six numbers chosen match with the winning ones. Find the probability that a certain choice of a bettor wins the grand, the second, and the third prizes, respectively.

Solution: The probability of winning the grand prize is

$$\frac{1}{\binom{49}{6}} = \frac{1}{13,983,816}.$$

The probability of winning the second prize is

$$\frac{\binom{6}{5}\binom{43}{1}}{\binom{49}{6}} = \frac{258}{13,983,816} \approx \frac{1}{54,200},$$

and the probability of winning the third prize is

$$\frac{\binom{6}{4}\binom{43}{2}}{\binom{49}{6}} = \frac{13,545}{13,983,816} \approx \frac{1}{1032}. \quad \blacklozenge$$

Example 2.20   From an ordinary deck of 52 cards, seven cards are drawn at random and without replacement. What is the probability that at least one of the cards is a king?

Solution: In $\binom{52}{7}$ ways seven cards can be selected from an ordinary deck of 52 cards. In $\binom{48}{7}$ of these, none of the cards selected is a king. Therefore, the desired probability is

$$P(\text{at least one king}) = 1 - P(\text{no kings}) = 1 - \frac{\binom{48}{7}}{\binom{52}{7}} = 0.4496.$$

Warning: A common mistake is to calculate this and similar probabilities as follows: To make sure that there is at least one king among the seven cards drawn, we will first choose a king; there are $\binom{4}{1}$ possibilities. Then we choose the remaining six cards from the remaining 51 cards; there are $\binom{51}{6}$ possibilities for this. Thus the answer is

$$\frac{\binom{4}{1}\binom{51}{6}}{\binom{52}{7}} = 0.5385.$$

This solution is wrong because it counts some of the possible outcomes several times. For example, the hand K_H, 5_C, 6_D, 7_H, K_D, J_C, and 9_S is counted twice: once when K_H is selected as the first card from the kings and 5_C, 6_D, 7_H, K_D, J_C, and 9_S from the remaining 51, and once when K_D is selected as the first card from the kings and K_H, 5_C, 6_D, 7_H, J_C, and 9_S from the remaining 51 cards.   ◆

Example 2.21   What is the probability that a poker hand is a full house? A poker hand consists of five randomly selected cards from an ordinary deck of 52 cards. It is a full house if three cards are of one denomination and two cards are of another denomination: for example, three queens and two 4's.

Solution: The number of different poker hands is $\binom{52}{5}$. To count the number of full houses, let us call a hand of type (Q,4) if it has three queens and two 4's, with similar representations for other types of full houses. Observe that (Q,4) and (4,Q) are different full houses, and types such as (Q,Q) and (K,K) do not exist. Hence there are 13×12 different types of full houses. Since for every particular type, say (4,Q), there are $\binom{4}{3}$ ways to select three 4's and $\binom{4}{2}$ ways to select two Q's, the desired probability equals

$$\frac{13 \times 12 \times \binom{4}{3} \times \binom{4}{2}}{\binom{52}{5}} \approx 0.0014.   ◆$$

Example 2.22   Show that the number of different ways n indistinguishable objects can be placed into k distinguishable cells is

$$\binom{n+k-1}{n} = \binom{n+k-1}{k-1}.$$

Solution: Let the n indistinguishable objects be represented by n identical oranges, and the k distinguishable cells be represented by k people. We want to count the number of different ways that n identical oranges can be divided among k people. To do this, add $k - 1$ identical apples to the oranges. Then take the $n + k - 1$ apples and oranges and line them up in some random order. Give all of the oranges preceding the first apple to the first person, all of the oranges between the first and second apples to the second person, all of the oranges between the second and third apples to the third person, and so on. Note that if, for example, an apple appears in the beginning of the line, then the first person does not receive any oranges. Similarly, if two apples appear next to each other, say, at the ith and $(i + 1)$st positions, then the $(i + 1)$st person does not receive any oranges. This process establishes a one-to-one correspondence between the ways n identical oranges can be divided among k people, and the number of distinguishable permutations of $n + k - 1$ apples and oranges, of which the n oranges are identical and the $k - 1$ apples are identical. By Theorem 2.4, the answer to this problem is

$$\frac{(n + k - 1)!}{n!\,(k - 1)!} = \binom{n + k - 1}{n} = \binom{n + k - 1}{k - 1}. \quad \blacklozenge$$

Example 2.23   Let n be a positive integer, and let $x_1 + x_2 + \cdots + x_k = n$ be a given equation. A vector $(x_1, x_2, \ldots, x_k)$ satisfying $x_1 + x_2 + \cdots + x_k = n$ is said to be a *nonnegative integer solution* of the equation if for each i, $1 \le i \le k$, x_i is a nonnegative integer. It is said to be a *positive integer solution* of the equation if for each i, $1 \le i \le k$, x_i is a positive integer.

(a)   How many distinct nonnegative integer solutions does the equation $x_1 + x_2 + \cdots + x_k = n$ have?

(b)   How many distinct positive integer solutions does the equation $x_1 + x_2 + \cdots + x_k = n$ have?

Solution:

(a)   If we think of $x_1, x_2, \ldots, x_k$ as k cells, then the problem reduces to that of dividing n identical objects (namely, n 1's) into k cells. Hence, by Example 2.22, the answer is

$$\binom{n + k - 1}{n} = \binom{n + k - 1}{k - 1}.$$

(b)   For $1 \le i \le k$, let $y_i = x_i - 1$. Then, for each positive integer solution $(x_1, x_2, \ldots, x_k)$ of $x_1 + x_2 + \cdots + x_k = n$, there is exactly one nonnegative integer solution $(y_1, y_2, \ldots, y_k)$ of $y_1 + y_2 + \cdots + y_k = n - k$, and conversely. Therefore, the number of positive integer solutions of $x_1 + x_2 + \cdots + x_k = n$ is equal to the number of nonnegative integer solutions of $y_1 + y_2 + \cdots + y_k = n - k$, which, by part (a), is

$$\binom{(n - k) + k - 1}{n - k} = \binom{n - 1}{n - k} = \binom{n - 1}{k - 1}. \quad \blacklozenge$$

Example 2.24 An absentminded professor wrote n letters and sealed them in envelopes before writing the addresses on the envelopes. Then he wrote the n addresses on the envelopes at random. What is the probability that at least one letter was addressed correctly?

Solution: The total number of ways that one can write n addresses on n envelopes is $n!$; thus the sample space contains $n!$ points. Now we calculate the number of outcomes in which at least one envelope is addressed correctly. To do this, let E_i be the event that the ith letter is addressed correctly; then $E_1 \cup E_2 \cup \cdots \cup E_n$ is the event that at least one letter is addressed correctly. To calculate $P(E_1 \cup E_2 \cup \cdots \cup E_n)$, we use the inclusion-exclusion principle. To do so we must calculate the probabilities of all possible intersections of the events from $E_1, \ldots, E_n$, add the probabilities that are obtained by intersecting an odd number of the events, and subtract all the probabilities that are obtained by intersecting an even number of the events. Therefore, we need to know the number of elements of E_i's, $E_i \cap E_j$'s, $E_i \cap E_j \cap E_k$'s, and so on. Now E_i contains $(n-1)!$ points, because when the ith letter is addressed correctly, there are $(n-1)!$ ways to address the remaining $n-1$ envelopes. So $P(E_i) = (n-1)!/n!$. Similarly, $E_i \cap E_j$ contains $(n-2)!$ points, because if the ith and jth envelopes are addressed correctly, the remaining $n-2$ envelopes can be addressed in $(n-2)!$ ways. Thus $P(E_i \cap E_j) = (n-2)!/n!$. Similarly, $P(E_i \cap E_j \cap E_k) = (n-3)!/n!$, and so on. Now in computing $P(E_1 \cup E_2 \cup \cdots \cup E_n)$, there are n terms of the form $P(E_i)$, $\binom{n}{2}$ terms of the form $P(E_i \cap E_j)$, $\binom{n}{3}$ terms of the form $P(E_i \cap E_j \cap E_k)$, and so on. Hence

$$P(E_1 \cup E_2 \cup \cdots \cup E_n) = n\frac{(n-1)!}{n!} - \binom{n}{2}\frac{(n-2)!}{n!} + \cdots +$$

$$(-1)^{n-2}\binom{n}{n-1}\frac{[n-(n-1)]!}{n!} + (-1)^{n-1}\binom{n}{n}\frac{1}{n!}.$$

This expression simplifies to

$$P(E_1 \cup E_2 \cup \cdots \cup E_n) = 1 - \frac{1}{2!} + \frac{1}{3!} - \frac{1}{4!} + \cdots + \frac{(-1)^{n-1}}{n!}.$$

Remark: Note that since $e^x = \sum_{n=0}^{\infty}(x^n/n!)$, if $n \to \infty$, then

$$P\left(\bigcup_{i=1}^{\infty} E_i\right) = 1 - \frac{1}{2!} + \frac{1}{3!} - \frac{1}{4!} + \cdots + \frac{(-1)^{n-1}}{n!} + \cdots$$

$$= 1 - \left(1 - 1 + \frac{1}{2!} - \frac{1}{3!} + \frac{1}{4!} - \cdots + \frac{(-1)^n}{n!} + \cdots\right)$$

$$= 1 - \sum_{n=0}^{\infty} \frac{(-1)^n}{n!} = 1 - \frac{1}{e} \approx 0.632.$$

Hence even if the number of the envelopes is very large, there is still a very good chance for at least one envelope to be addressed correctly.   ◆

One of the most important applications of combinatorics is that the formula for the r-element combinations of n objects enables us to find an algebraic expansion for $(x+y)^n$.

Theorem 2.5   (Binomial Expansion)   *For any integer $n \geq 0$,*

$$(x + y)^n = \sum_{i=0}^{n} \binom{n}{i} x^{n-i} y^i.$$

Proof: By looking at some special cases, such as

$$(x + y)^2 = (x + y)(x + y) = x^2 + xy + yx + y^2$$

and

$$(x + y)^3 = (x + y)(x + y)(x + y)$$
$$= x^3 + x^2 y + yx^2 + xy^2 + yx^2 + xy^2 + y^2 x + y^3,$$

it should become clear that when we carry out the multiplication

$$(x + y)^n = \underbrace{(x + y)(x + y) \cdots (x + y)}_{n \text{ times}},$$

we obtain only terms of the form $x^{n-i} y^i$, $0 \leq i \leq n$. Therefore, all we have to do is to find out how many times the term $x^{n-i} y^i$ appears, $0 \leq i \leq n$. This is seen to be $\binom{n}{n-i} = \binom{n}{i}$ because $x^{n-i} y^i$ emerges only whenever the x's of $n - i$ of the n factors of $(x + y)$ are multiplied by the y's of the remaining i factors of $(x + y)$. Hence

$$(x + y)^n = \binom{n}{0} x^n + \binom{n}{1} x^{n-1} y + \binom{n}{2} x^{n-2} y^2 + \cdots$$
$$+ \binom{n}{n-1} xy^{n-1} + \binom{n}{n} y^n. \quad ◆$$

The binomial coefficients $_nC_r$ have a long history. Chinese mathematicians were using them as early as the late eleventh century. The famous mathematician and poet of Persia, Omar Khayyam, also rediscovered them (twelfth century), as did Nasir ad-Din Tusi in the following century; they were rediscovered in Europe in the sixteenth century by the German mathematician Stifel as well as by the Italian mathematicians Tartaglia and Cardano. But perhaps the most exhaustive study of these numbers was made by Pascal in the seventeenth century, and for that reason they are usually associated with him.

Example 2.25   What is the coefficient of $x^2 y^3$ in the expansion of $(2x + 3y)^5$?

Solution: Let $u = 2x$ and $v = 3y$; then $(2x + 3y)^5 = (u + v)^5$. The coefficient of $u^2 v^3$ in the expansion of $(u + v)^5$ is $\binom{5}{3}$ and $u^2 v^3 = (2^2 \cdot 3^3) x^2 y^3$; therefore, the coefficient of $x^2 y^3$ in the expansion of $(2x + 3y)^5$ is $\binom{5}{3}(2^2 \cdot 3^3) = 1080$.   ◆

Example 2.26   Evaluate the sum $\binom{n}{0} + \binom{n}{1} + \binom{n}{2} + \binom{n}{3} + \cdots + \binom{n}{n}$.

Solution: A set containing n elements has $\binom{n}{i}$, $0 \le i \le n$, subsets with i elements. So the given expression is the total number of the subsets of a set of n elements, and therefore it equals 2^n. A second way to see this is to note that, by the binomial expansion, the given expression equals

$$\sum_{i=0}^{n} \binom{n}{i} 1^{n-i} 1^i = (1 + 1)^n = 2^n. \quad ◆$$

Example 2.27   Evaluate the sum $\binom{n}{1} + 2\binom{n}{2} + 3\binom{n}{3} + \cdots + n\binom{n}{n}$.

Solution:

$$i\binom{n}{i} = i \cdot \frac{n!}{i!\,(n-i)!} = \frac{n \cdot (n-1)!}{(i-1)!\,(n-i)!} = n\binom{n-1}{i-1}.$$

So

$$\binom{n}{1} + 2\binom{n}{2} + 3\binom{n}{3} + \cdots + n\binom{n}{n}$$
$$= n\left[\binom{n-1}{0} + \binom{n-1}{1} + \binom{n-1}{2} + \cdots + \binom{n-1}{n-1}\right] = n \cdot 2^{n-1},$$

by Example 2.26.   ◆

Example 2.28   Prove that $\binom{2n}{n} = \sum_{i=0}^{n} \binom{n}{i}^2$.

Solution: We show this by using a combinatorial argument. For an analytic proof, see Exercise 49. Let $A = \{a_1, a_2, \ldots, a_n\}$ and $B = \{b_1, b_2, \ldots, b_n\}$ be two disjoint sets. The number of subsets of $A \cup B$ with n elements is $\binom{2n}{n}$. On the other hand, any subset of $A \cup B$ with n elements is the union of a subset of A with i elements and a subset of

B with $n - i$ elements for some $0 \le i \le n$. Since for each i there are $\binom{n}{i}\binom{n}{n-i}$ such subsets, we have that the total number of subsets of $A \cup B$ with n elements is $\sum_{i=0}^{n} \binom{n}{i}\binom{n}{n-i}$. But since $\binom{n}{n-i} = \binom{n}{i}$, we have the identity. ◆

Example 2.29 In this example, we present an *intuitive* proof for the inclusion-exclusion principle explained in Section 1.4:

$$P\left(\bigcup_{i=1}^{n} A_i\right) = \sum_{i=1}^{n} P(A_i) - \sum_{i=1}^{n-1}\sum_{j=i+1}^{n} P(A_i A_j) + \sum_{i=1}^{n-2}\sum_{j=i+1}^{n-1}\sum_{k=j+1}^{n} P(A_i A_j A_k)$$
$$- \cdots + (-1)^{n-1} P(A_1 A_2 \cdots A_n).$$

Let S be the sample space. If an outcome $\omega \notin \bigcup_{i=1}^{n} A_i$, then $\omega \notin A_i$, $1 \le i \le n$. Therefore, the probability of ω is not added to either side of the preceding equation. Suppose that $\omega \in \bigcup_{i=1}^{n} A_i$; then ω belongs to k of the A_i's for some k, $1 \le k \le n$. Now, to the left side of the equation the probability of ω is added exactly once. We will show that the same is true on the right. Clearly, the first term on the right side contains the probability of ω exactly $k = \binom{k}{1}$ times. The second term subtracts this probability $\binom{k}{2}$ times, the third term adds it back, this time $\binom{k}{3}$ times, and so on until the kth term. Depending on whether k is even or odd, the kth term either subtracts or adds the probability of ω exactly $\binom{k}{k}$ times. The remaining terms do not contain the probability of ω. Therefore, the total number of times the probability of ω is added to the right side is

$$\binom{k}{1} - \binom{k}{2} + \binom{k}{3} - \cdots (-1)^{k-1}\binom{k}{k}$$
$$= -\sum_{i=1}^{k} \binom{k}{i} 1^{k-i}(-1)^i = \binom{k}{0} - \sum_{i=0}^{k} \binom{k}{i} 1^{k-i}(-1)^i$$
$$= 1 - (1-1)^k = 1,$$

where the next-to-the-last equation follows from the binomial expansion. This establishes the inclusion-exclusion principle. ◆

Example 2.30 Suppose that we want to distribute n distinguishable balls into k distinguishable cells so that n_1 balls are distributed into the first cell, n_2 balls into the second cell, ... , n_k balls into the kth cell, where $n_1 + n_2 + n_3 + \cdots + n_k = n$. To count the number of ways that this is possible, note that we have $\binom{n}{n_1}$ choices to distribute n_1

balls into the first cell; for each choice of n_1 balls in the first cell, we then have $\binom{n - n_1}{n_2}$ choices to distribute n_2 balls into the second cell; for each choice of n_1 balls in the first cell and n_2 balls in the second cell, we have $\binom{n - n_1 - n_2}{n_3}$ choices to distribute n_3 balls into the third cell; and so on. Hence by the generalized counting principle the total number of such distributions is

$$\binom{n}{n_1}\binom{n - n_1}{n_2}\binom{n - n_1 - n_2}{n_3} \cdots \binom{n - n_1 - n_2 - \cdots - n_{k-1}}{n_k}$$

$$= \frac{n!}{(n - n_1)!\,n_1!} \times \frac{(n - n_1)!}{(n - n_1 - n_2)!\,n_2!} \times \frac{(n - n_1 - n_2)!}{(n - n_1 - n_2 - n_3)!\,n_3!} \times \cdots$$

$$\times \frac{(n - n_1 - n_2 - n_3 - \cdots - n_{k-1})!}{(n - n_1 - n_2 - n_3 - \cdots - n_{k-1} - n_k)!\,n_k!} = \frac{n!}{n_1!\,n_2!\,n_3! \cdots n_k!}$$

because $n_1 + n_2 + n_3 + \cdots + n_k = n$ and $(n - n_1 - n_2 - \cdots - n_k)! = 0! = 1$.   ◆

As an application of Example 2.30, we state a generalization of the binomial expansion (Theorem 2.5) and leave its proof as an exercise (see Exercise 28).

Theorem 2.6   (Multinomial Expansion)   *In the expansion of*

$$(x_1 + x_2 + \cdots + x_k)^n,$$

the coefficient of the term $x_1^{n_1} x_2^{n_2} x_3^{n_3} \cdots x_k^{n_k}, n_1 + n_2 + \cdots + n_k = n$, *is*

$$\frac{n!}{n_1!\,n_2! \cdots n_k!}.$$

Therefore,

$$(x_1 + x_2 + \cdots + x_k)^n = \sum_{n_1 + n_2 + \cdots + n_k = n} \frac{n!}{n_1!\,n_2! \cdots n_k!}\, x_1^{n_1} x_2^{n_2} x_3^{n_3} \cdots x_k^{n_k}.$$

Note that the sum is taken over all nonnegative integers $n_1, n_2, \ldots, n_k$ *such that* $n_1 + n_2 + \cdots + n_k = n$.

EXERCISES

A

1.   Jim has 20 friends. If he decides to invite six of them to his birthday party, how many choices does he have?

2. Each state of the 50 in the United States has two senators. In how many ways may a majority be achieved in the U.S. Senate? Ignore the possibility of absence or abstention. Assume that all senators are present and voting.

3. A panel consists of 20 men and 25 women. How many choices do we have to obtain a jury of six men and six women from this panel?

4. From an ordinary deck of 52 cards, five are drawn randomly. What is the probability of drawing exactly three face cards?

5. A random sample of n elements is taken from a population of size N without replacement. What is the probability that a fixed element of the population is included? Simplify your answer.

6. Judy puts one piece of fruit in her child's lunch bag every day. If she has three oranges and two apples for the next five days, in how many ways can she do this?

7. Ann puts at most one piece of fruit in her child's lunch bag every day. If she has only three oranges and two apples for the next eight lunches of her child, in how many ways can she do this?

8. Lili has 20 friends. Among them are Kevin and Gerry, who are husband and wife. Lili wants to invite six of her friends to her birthday party. If neither Kevin nor Gerry will go to a party without the other, how many choices does Lili have?

9. In front of Jeff's office there is a parking lot with 13 parking spots in a row. When cars arrive at this lot, they park randomly at one of the empty spots. Jeff parks his car in the only empty spot that is left. Then he goes to his office. On his return he finds that there are seven empty spots. If he has not parked his car at either end of the parking area, what is the probability that both of the parking spaces that are next to Jeff's car are empty?

10. Find the coefficient of x^9 in the expansion of $(2 + x)^{12}$.

11. Find the coefficient of $x^3 y^4$ in the expansion of $(2x - 4y)^7$.

12. A team consisting of three boys and four girls must be formed from a group of nine boys and eight girls. If two of the girls are feuding and refuse to play on the same team, how many possibilities do we have?

13. A fair coin is tossed 10 times. What is the probability of (a) five heads; (b) at least five heads?

14. If five numbers are selected at random from the set $\{1, 2, 3, \ldots, 20\}$, what is the probability that their minimum is larger than 5?

15. From a faculty of six professors, six associate professors, ten assistant professors, and twelve instructors, a committee of size six is formed randomly. What is the probability that (a) there are exactly two professors on the committee; (b) all committee members are of the same rank?

16. A lake contains 200 trout; 50 of them are caught randomly, tagged, and returned. If, again, we catch 50 trout at random, what is the probability of getting exactly five tagged trout?

17. Find the values of $\sum_{i=0}^{n} 2^i \binom{n}{i}$ and $\sum_{i=0}^{n} x^i \binom{n}{i}$.

18. A fair die is tossed six times. What is the probability of getting exactly two 6's?

19. Suppose that 12 married couples take part in a contest. If 12 persons each win a prize, what is the probability that from every couple one of them is a winner? Assume that all of the $\binom{24}{12}$ possible sets of winners are equally probable.

20. Poker hands are classified into the following 10 nonoverlapping categories in increasing order of likelihood. Calculate the probability of the occurrence of each class separately. Recall that a poker hand consists of five cards selected randomly from an ordinary deck of 52 cards.

> **Royal flush:** The 10, jack, queen, king, and ace of the same suit.
>
> **Straight flush:** All cards in the same suit, with consecutive denominations except for the royal flush.
>
> **Four of a kind:** Four cards of one denomination and one card of a second denomination: for example, four 8's and a jack.
>
> **Full house:** Three cards of one denomination and two cards of a second denomination: for example, three 4's and two queens.
>
> **Flush:** Five cards all in one suit but not a straight or royal flush.
>
> **Straight:** Cards of distinct consecutive denominations, not all in one suit: for example, 3 of hearts, 4 of hearts, 5 of spades, 6 of hearts, and 7 of clubs.
>
> **Three of a kind:** Three cards of one denomination, a fourth card of a second denomination, and a fifth card of a third denomination.
>
> **Two pairs:** Two cards from one denomination, another two from a second denomination, and the fifth card from a third denomination.
>
> **One pair:** Two cards from one denomination, with the third, fourth, and fifth cards from a second, third, and fourth denomination, respectively: for example, two 8's, a king, a 5, and a 4.
>
> **None of the above.**

21. There are 12 nuts and 12 bolts in a box. If the contents of the box are divided between two handymen, what is the probability that each handyman will get six nuts and six bolts?

22. A history professor who teaches three sections of the same course every semester decides to make several tests and use them for the next 10 years (20 semesters) as final exams. The professor has two policies: (1) not to give the same test to more than one class in a semester, and (2) not to repeat the same combination of three tests for any two semesters. Determine the minimum number of different tests that the professor should prepare.

23. A four-digit number is selected at random. What is the probability that its ones place is less than its tens place, its tens place is less than its hundreds place, and its hundreds place is less than its thousands place? Note that the first digit of an n-digit number is nonzero.

24. Using Theorem 2.6, expand $(x + y + z)^2$.

25. What is the coefficient of $x^2 y^3 z^2$ in the expansion of $(2x - y + 3z)^7$?

26. What is the coefficient of $x^3 y^7$ in the expansion of $(2x - y + 3)^{13}$?

27. An ordinary deck of 52 cards is dealt, 13 each, to four players at random. What is the probability that each player receives 13 cards of the same suit?

28. Using induction, binomial expansion, and the identity

$$\binom{n}{n_1} \frac{(n - n_1)!}{n_2!\, n_3! \, \cdots \, n_k!} = \frac{n!}{n_1!\, n_2! \, \cdots \, n_k!},$$

prove the formula of multinomial expansion.

29. A staircase is to be constructed between M and N (see Figure 2.4). The distances from M to L, and from L to N, are 5 and 2 meters, respectively. If the height of a step is 25 centimeters and its width can be any integer multiple of 50 centimeters, how many different choices do we have?

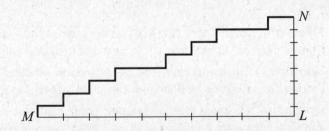

Figure 2.4   Staircase of Exercise 29.

30. Each state of the 50 in the United States has two senators. What is the probability that in a random committee of 50 senators (a) Maryland is represented; (b) all states are represented?

31. According to the 1998 edition of *Encyclopedia Britannica*, "there are at least 15,000 to as many as 35,000 species of orchids." These species have been found naturally and are distinct from each other. Suppose that hybrids can be created by crossing any two existing species. Furthermore, suppose that hybrids themselves can be continued to be hybridized with each other or with an original species. Orchid lovers develop thousands and thousands of hybrids for use as garden or greenhouse ornamental and for the commercial flower trade. Suppose that all species are crossed, two at a time, to create the first generation of hybrids. Then the first generation of hybrids are crossed with each other and with the original species, two at a time, to develop the second generation of hybrids. The second generation of hybrids are crossed with each other, with the first generation hybrids, and with the original species, two at a time, to generate the third generation of hybrids, and so on. Let n be the total number of original species of orchids. Let n_k be the number of hybrids in the kth generation.

   (a) Find n_k in terms of $n, n_1, \ldots, n_{k-1}$.

   (b) For $n = 25,000$, find the largest possible total number of all hybrids in the first four generations.

B

32. Prove the binomial expansion formula by induction.
Hint: Use the identity $\binom{n}{k-1} + \binom{n}{k} = \binom{n+1}{k}$.

33. A class contains 30 students. What is the probability that there are six months each containing the birthdays of two students, and six months each containing the birthdays of three students? Assume that all months have the same probability of including the birthday of a randomly selected person.

34. In a closet there are 10 pairs of shoes. If six shoes are selected at random, what is the probability of (a) no complete pairs; (b) exactly one complete pair; (c) exactly two complete pairs; (d) exactly three complete pairs?

35. An ordinary deck of 52 cards is divided into two equal sets randomly. What is the probability that each set contains exactly 13 red cards?

36. In Maryland's lottery, players pick six different integers between 1 and 49, the order of selection being irrelevant. The lottery commission then randomly selects six of these as the *winning numbers*. What is the probability that at least two consecutive integers are selected among the winning numbers?
Hint: Let $\mathcal{A}$ be the set of all 6-element combinations of $\{1, 2, \ldots, 49\}$ with no consecutive integers. Let $\mathcal{B}$ be the set of all 6-element combinations of $\{1, 2, \ldots, 44\}$. Begin by showing that there is a one-to-one correspondence between $\mathcal{A}$ and $\mathcal{B}$.

37. A train consists of n cars. Each of m passengers ($m > n$) will choose a car at random to ride in. What is the probability that (a) there will be at least one passenger in each car; (b) exactly r $(r < n)$ cars remain unoccupied?

38. Suppose that n indistinguishable balls are placed at random into n distinguishable cells. What is the probability that exactly one cell remains empty?

39. Prove that

$$\binom{n}{0} - \binom{n}{1} + \binom{n}{2} - \cdots + (-1)^k \binom{n}{k} + \cdots + (-1)^n \binom{n}{n} = 0.$$

40. Show that

$$\binom{n}{0} + \binom{n+1}{1} + \cdots + \binom{n+r}{r} = \binom{n+r+1}{r}.$$

Hint: $\binom{n}{r} = \binom{n+1}{r} - \binom{n}{r-1}.$

41. By a combinatorial argument, prove that for $r \le n$ and $r \le m$,

$$\binom{n+m}{r} = \binom{m}{0}\binom{n}{r} + \binom{m}{1}\binom{n}{r-1} + \cdots + \binom{m}{r}\binom{n}{0}.$$

42. Evaluate the following sum:

$$\binom{n}{0} + \frac{1}{2}\binom{n}{1} + \frac{1}{3}\binom{n}{2} + \cdots + \frac{1}{n+1}\binom{n}{n}.$$

43. Suppose that five points are selected at random from the interval $(0, 1)$. What is the probability that exactly two of them are between 0 and $1/4$?
Hint: For any point there are four equally likely possibilities: to fall into $(0, 1/4)$, $[1/4, 1/2)$, $[1/2, 3/4)$, and $[3/4, 1)$.

44. A lake has N trout, and t of them are caught at random, tagged, and returned. We catch n trout at a later time randomly and observe that m of them are tagged.

    (a) Find P_N, the probability of what we observed actually happen.

    (b) To estimate the number of trout in the lake, statisticians find the value of N that maximizes P_N. Such a value is called the **maximum likelihood estimator** of N. Show that the maximum of P_N is $[nt/m]$, where by $[nt/m]$ we mean the greatest integer less than or equal to nt/m. That is, prove that the maximum likelihood estimator of the number of trout in the lake is $[nt/m]$.

    Hint: Investigate for what values of N the probability P_N is increasing and for what values it is decreasing.

45. Let n be a positive integer. A random sample of four elements is taken from the set $\{0, 1, 2, \ldots, n\}$, one at a time and *with replacement*. What is the probability that the sum of the first two elements is equal to the sum of the last two elements?

46. For a given position with n applicants, m applicants are equally qualified and $n - m$ applicants are not qualified at all. Assume that a recruitment process is considered to be fair if the probability that a qualified applicant is hired is $1/m$, and the probability that an unqualified applicant is hired is 0. One fair recruitment process is to interview all applicants, identify the qualified ones, and then hire one of the m qualified applicants randomly. A second fair process, which is more efficient, is to interview applicants in a random order and employ the first qualified applicant encountered. For the fairness of the second recruitment process,

   (a)   present an *intuitive* argument;

   (b)   give a rigorous combinatorial proof.

47. In how many ways can 10 different photographs be placed in six different envelopes, no envelope remaining empty?
Hint: An easy way to do this problem is to use the following version of the inclusion-exclusion principle: Let $A_1, A_2, \ldots, A_n$ be n subsets of a finite set Ω with N elements. Let $N(A_i)$ be the number of elements of A_i, $1 \le i \le n$, and $N(A_i^c) = N - N(A_i)$. Let S_k be the sum of the elements of all those intersections of $A_1, A_2, \ldots, A_n$ that are formed of exactly k sets. That is,

$$S_1 = N(A_1) + N(A_2) + \cdots + N(A_n),$$

$$S_2 = N(A_1 A_2) + N(A_1 A_3) + \cdots + N(A_{n-1} A_n),$$

and so on. Then

$$N(A_1^c A_2^c \cdots A_n^c) = N - S_1 + S_2 - S_3 + \cdots + (-1)^n S_n.$$

To solve the problem, let N be the number of ways that 10 different photographs can be placed in six different envelopes, allowing for the possibility of empty envelopes. Let A_i be the set of all situations in which envelope i is empty. Then the desired quantity is $N(A_1^c A_2^c \cdots A_6^c)$.

48. We are given n ($n > 5$) points in space, no three of which lie on the same straight line. Let Ω be the family of planes defined by any three of these points. Suppose that the points are situated in a way that no four of them are coplanar, and no two planes of Ω are parallel. From the set of the lines of the intersections of the planes of Ω, a line is selected at random. What is the probability that it passes through none of the n points?
Hint: For $i = 0, 1, 2$, let A_i be the set of all lines of the intersections that are determined by planes having i of the given n points in common. If $|A_i|$ denotes the number of elements of A_i, the answer is $|A_0|/(|A_0| + |A_1| + |A_2|)$.

49. Using the binomial theorem, calculate the coefficient of x^n in the expansion of $(1 + x)^{2n} = (1 + x)^n (1 + x)^n$ to prove that

$$\binom{2n}{n} = \sum_{i=0}^{n} \binom{n}{i}^2.$$

For a combinatorial proof of this relation, see Example 2.28.

50. An absentminded professor wrote n letters and sealed them in envelopes without writing the addresses on them. Then he wrote the n addresses on the envelopes at random. What is the probability that exactly k of the envelopes were addressed correctly?

Hint: Consider a particular set of k letters. Let M be the total number of ways that only these k letters can be addressed correctly. The desired probability is the quantity $\binom{n}{k} M/n!$; using Example 2.24, argue that M satisfies $\sum_{i=2}^{n-k} (-1)^i / i! = M/(n-k)!$.

51. A fair coin is tossed n times. Calculate the probability of getting no successive heads.

Hint: Let x_i be the number of sequences of H's and T's of length i with no successive H's. Show that x_i satisfies $x_i = x_{i-1} + x_{i-2}$, $i \geq 2$, where $x_0 = 1$ and $x_1 = 2$. The answer is $x_n/2^n$. Note that $\{x_i\}_{i=1}^{\infty}$ is a Fibonacci-type sequence.

52. What is the probability that the birthdays of at least two students of a class of size n are at most k days apart? Assume that the birthrates are constants throughout the year and that each year has 365 days.

◼

2.5   STIRLING's FORMULA

To estimate $n!$ for large values of n, the following formula is used. Discovered in 1730 by James Stirling (1692–1770), it appeared in his book *Methodus Differentialis* (Bowyer, London, 1730). Also, independently, a slightly less general version of the formula was discovered by De Moivre, who used it to prove the central limit theorem.

Theorem 2.7   (Stirling's Formula)

$$n! \sim \sqrt{2\pi n}\, n^n e^{-n},$$

where the sign $\sim$ means

$$\lim_{n \to \infty} \frac{n!}{\sqrt{2\pi n}\, n^n e^{-n}} = 1.$$

Stirling's formula, which usually gives excellent approximations in numerical computations, is also often used to prove theoretical problems. Note that although the ratio $R(n) = n!/(\sqrt{2\pi n}\, n^n e^{-n})$ becomes 1 at ∞, it is still close to 1, even for very small values of n. The following table shows this fact.

n	$n!$	$\sqrt{2\pi n}\, n^n e^{-n}$	$R(n)$
1	1	0.922	1.084
2	2	1.919	1.042
5	120	118.019	1.017
8	40,320	39,902.396	1.010
10	3,628,800	3,598,695.618	1.008
12	479,001,600	475,687,486.474	1.007

Be aware that, even though $\lim_{n\to\infty} R(n)$ is 1, the difference between $n!$ and $\sqrt{2\pi n}\, n^n e^{-n}$ increases as n gets larger. In fact, it goes to ∞ as $n \to \infty$.

Example 2.31   Approximate the value of $\left[2^n (n!)^2\right]/(2n)!$ for large n.

Solution: By Stirling's formula, we have

$$\frac{2^n (n!)^2}{(2n)!} \sim \frac{2^n 2\pi n (n^n)^2 e^{-2n}}{\sqrt{4\pi n}\,(2n)^{2n} e^{-2n}} = \frac{\sqrt{\pi n}}{2^n}. \quad \blacklozenge$$

EXERCISE

---

1.   Use Stirling's formula to approximate $\dbinom{2n}{n} \dfrac{1}{2^{2n}}$ and $\dfrac{\left[(2n)!\right]^3}{\left[(4n)!\,(n!)^2\right]}$ for large n.   ■

REVIEW PROBLEMS

---

1.   Albert goes to the grocery store to buy fruit. There are seven different varieties of fruit, and Albert is determined to buy no more than one of any variety. How many different orders can he place?

2. Virginia has 1 one-dollar bill, 1 two-dollar bill, 1 five-dollar bill, 1 ten-dollar bill, and 1 twenty-dollar bill. She decides to give some money to her son Brian without asking for change. How many choices does she have?

3. If four fair dice are tossed, what is the probability that they will show four different faces?

4. From the 10 points that are placed on a circumference, two are selected randomly. What is the probability that they are adjacent?

5. A father buys nine different toys for his four children. In how many ways can he give one child three toys and the remaining three children two toys each?

6. A window dresser has decided to display five different dresses in a circular arrangement. How many choices does she have?

7. Judy has three sets of classics in literature, each set having four volumes. In how many ways can she put them on a bookshelf so that books of each set are not separated?

8. Suppose that 30 lawn mowers, of which seven have defects, are sold to a hardware store. If the store manager inspects six of the lawn mowers randomly, what is the probability that he finds at least one defective lawn mower?

9. In how many ways can 23 identical refrigerators be allocated among four stores so that one store gets eight refrigerators, another four, a third store five, and the last one six refrigerators?

10. In how many arrangements of the letters $BERKELEY$ are all three E's adjacent?

11. Bill and John play in a backgammon tournament. A player is the winner if he wins three games in a row or four games altogether. In what percent of all possible cases does the tournament end because John wins four games without winning three in a row?

12. In a small town, both of the accidents that occurred during the week of June 8, 1988, were on Friday the 13th. Is this a good excuse for a superstitious person to argue that Friday the 13th's are inauspicious?

13. How many eight-digit numbers without two identical successive digits are there?

14. A **palindrome** is a sequence of characters that reads the same forward or backward. For example, *rotator, madam, Hannah*, the German name *Otto*, and an Indian language, *Malayalam*, are palindromes. So are the following expressions: "Put up," "Madam I'm Adam," "Was it a cat I saw?" and these two sentences in Latin concerning St. Martin, Bishop of Tours: "Signa te, signa; temere me tangis et angis. Roma tibi subito motibus ibit amor." (Respectively: Cross, cross yourself; you annoy and vex me needlessly. Through my exertions, Rome, your desire, will soon be near.) Determine the number of palindromes containing 11 characters that

can be made with (a) no letters repeating more than twice; (b) one letter repeating three times and no other letter more than twice (the chemical term *detartrated* is one such palindrome); (c) one letter repeating three times, two each repeating two times, and one repeating four times.

Historical Remark: It is said that the palindrome was invented by a Greek poet named Sotades in the third Century B.C. It is also said that Sotades was drowned by order of a king of the Macedonian dynasty, the reigning Ptolemy, who found him a real bore.

15. By mistake, a student who is running a computer program enters with negative signs two of the six positive numbers and with positive signs two of the four negative numbers. If at some stage the program chooses three distinct numbers from these 10 at random and multiplies them, what is the probability that at that stage no mistake will occur?

16. Suppose that four women and two men enter a restaurant and sit at random around a table that has four chairs on one side and another four on the other side. What is the probability that the men are not all sitting on one side?

17. Cyrus and 27 other students are taking a course in probability this semester. If their professor chooses eight students at random and with replacement to ask them eight different questions, what is the probability that one of them is Cyrus?

18. If five Americans, five Italians, and five Mexicans sit randomly at a round table, what is the probability that the persons of the same nationality sit together?

19. The chairperson of the industry-academic partnership of a town invites all 12 members of the board and their spouses to his house for a Christmas party. If a board member may attend without his spouse, but not vice versa, how many different groups can the chairperson get?

20. In a bridge game, each of the four players gets 13 random cards. What is the probability that every player has an ace?

21. From a faculty of six professors, six associate professors, ten assistant professors, and twelve instructors, a committee of size six is formed randomly. What is the probability that there is at least one person from each rank on the committee?
Hint: Be careful, the answer is not

$$\frac{\binom{6}{1}\binom{6}{1}\binom{10}{1}\binom{12}{1}\binom{30}{2}}{\binom{34}{6}} = 1.397.$$

To find the correct answer, use the inclusion-exclusion principle explained in Section 1.4.

22. An urn contains 15 white and 15 black balls. Suppose that 15 persons each draw

two balls blindfolded from the urn without replacement. What is the probability that each of them draws one white ball and one black ball?

23. In a lottery the tickets are numbered 1 through N. A person purchases n ($1 \leq n \leq N$) tickets at random. What is the probability that the ticket numbers are consecutive? (This is a special case of a problem posed by Euler in 1763.)

24. An ordinary deck of 52 cards is dealt, 13 each, at random among A, B, C, and D. What is the probability that (a) A and B together get two aces; (b) A gets all the face cards; (c) A gets five hearts and B gets the remaining eight hearts?

25. To test if a computer program works properly, we run it with 12 different data sets, using four computers, each running three data sets. If the data sets are distributed randomly among different computers, how many possibilities are there?

26. A fair die is tossed eight times. What is the probability that the eighth outcome is not a repetition?

27. A four-digit number is selected at random. What is the probability that its ones place is greater than its tens place, its tens place is greater than its hundreds place, and its hundreds place is greater than its thousands place? Note that the first digit of an n-digit number is nonzero.

28. From the set of integers $\{1, 2, 3, \dots, 100000\}$ a number is selected at random. What is the probability that the sum of its digits is 8?

   Hint: Establish a one-to-one correspondence between the set of integers from $\{1, 2, \dots, 100000\}$ the sum of whose digits is 8, and the set of possible ways 8 identical objects can be placed into 5 distinguishable cells. Then use Example 2.22. ∎

Chapter 3

Conditional Probability and Independence

3.1 CONDITIONAL PROBABILITY

To introduce the notion of conditional probability, let us first examine the following question: Suppose that all of the freshmen of an engineering college took calculus and discrete math last semester. Suppose that 70% of the students passed calculus, 55% passed discrete math, and 45% passed both. If a randomly selected freshman is found to have passed calculus last semester, what is the probability that he or she also passed discrete math last semester? To answer this question, let A and B be the events that the randomly selected freshman passed discrete math and calculus last semester, respectively. Note that the quantity we are asked to find is not $P(A)$, which is 0.55; it would have been if we were not aware that B has occurred. Knowing that B has occurred changes the chances of the occurrence of A. To find the desired probability, denoted by the symbol $P(A \mid B)$ [read: probability of A given B] and called the **conditional probability** of A given B, let n be the number of all the freshmen in the engineering college. Then $(0.7)n$ is the number of freshmen who passed calculus, and $(0.45)n$ is the number of those who passed both calculus and discrete math. Therefore, of the $(0.7)n$ freshmen who passed calculus, $(0.45)n$ of them passed discrete math as well. It is given that the randomly selected student is one of the $(0.7)n$ who passed calculus; we also want to find the probability that he or she is one of the $(0.45)n$ who passed discrete math. This is obviously equal to $(0.45)n/(0.7)n = 0.45/0.7$. Hence

$$P(A \mid B) = \frac{0.45}{0.7}.$$

But 0.45 is $P(AB)$ and 0.7 is $P(B)$. Therefore, this example suggests that

$$P(A \mid B) = \frac{P(AB)}{P(B)}. \tag{3.1}$$

As another example, suppose that a bus arrives at a station every day at a random

75

time between 1:00 P.M. and 1:30 P.M. Suppose that at 1:10 the bus has not arrived and we want to calculate the probability that it will arrive in not more than 10 minutes from now. Let A be the event that the bus will arrive between 1:10 and 1:20 and B be the event that it will arrive between 1:10 and 1:30. Then the desired probability is

$$P(A \mid B) = \frac{20 - 10}{30 - 10} = \frac{\dfrac{20 - 10}{30 - 0}}{\dfrac{30 - 10}{30 - 0}} = \frac{P(AB)}{P(B)}. \tag{3.2}$$

The relation $P(A \mid B) = P(AB)/P(A)$ discovered in (3.1) and (3.2) can also be verified for other types of conditional probability problems. It is compatible with our intuition and is a logical generalization of the existing relation between the conditional relative frequency of the event A with respect to the condition B and the ratio of the relative frequencies of AB and the relative frequency of B. For these reasons it is taken as a definition.

Definition   If $P(B) > 0$, *the **conditional probability** of A given B, denoted by* $P(A \mid B)$, *is*

$$P(A \mid B) = \frac{P(AB)}{P(B)}. \tag{3.3}$$

If $P(B) = 0$, formula (3.3) makes no sense, so the conditional probability is defined only for $P(B) > 0$. This formula may be expressed in words by saying that the conditional probability of A given that B has occurred is the ratio of the probability of joint occurrence of A and B and the probability of B. It is important to note that (3.3) is neither an axiom nor a theorem. It is a definition. As explained previously, this definition is fully justified and is not made arbitrarily. Indeed, historically, it was used implicitly even before it was formally introduced by De Moivre in the book *The Doctrine of Chance*.

Example 3.1   In a certain region of Russia, the probability that a person lives at least 80 years is 0.75, and the probability that he or she lives at least 90 years is 0.63. What is the probability that a randomly selected 80-year-old person from this region will survive to become 90?

Solution: Let A and B be the events that the person selected survives to become 90 and 80 years old, respectively. We are interested in $P(A \mid B)$. By definition,

$$P(A \mid B) = \frac{P(AB)}{P(B)}.$$

Now, $P(AB) = P(A)$ because "at least 80" and "at least 90" overlap with "at least 90," which is A. Hence

$$P(A \mid B) = \frac{P(AB)}{P(B)} = \frac{P(A)}{P(B)} = \frac{0.63}{0.75} = 0.84. \quad \blacklozenge$$

Example 3.2   From the set of all families with two children, *a family* is selected at random and is found to have a girl. What is the probability that the other child of the family is a girl? Assume that in a two-child family all sex distributions are equally probable.

Solution: Let B and A be the events that the family has a girl and the family has two girls, respectively. We are interested in $P(A \mid B)$. Now, in a family with two children there are four equally likely possibilities: (boy, boy), (girl, girl), (boy, girl), (girl, boy), where by, say, (girl, boy), we mean that the older child is a girl and the younger is a boy. Thus $P(B) = 3/4$, $P(AB) = 1/4$. Hence

$$P(A \mid B) = \frac{P(AB)}{P(B)} = \frac{1/4}{3/4} = \frac{1}{3}. \quad \blacklozenge$$

Example 3.3   From the set of all families with two children, *a child* is selected at random and is found to be a girl. What is the probability that the second child of this girl's family is also a girl? Assume that in a two-child family all sex distributions are equally probable.

Solution: Let B be the event that the child selected at random is a girl, and let A be the event that the second child of her family is also a girl. We want to calculate $P(A \mid B)$. Now the set of all possibilities is as follows: The child is a girl with a sister, the child is a girl with a brother, the child is a boy with a sister, and the child is a boy with a brother. Thus $P(B) = 2/4$ and $P(AB) = 1/4$. Hence

$$P(A \mid B) = \frac{P(AB)}{P(B)} = \frac{1/4}{2/4} = \frac{1}{2}. \quad \blacklozenge$$

Remark 3.1   There is a major difference between Examples 3.2 and 3.3. In Example 3.2, a family is selected and found to have a girl. Therefore, we have three equally likely possibilities: (girl, girl), (boy, girl), and (girl, boy), of which only one is desirable: (girl, girl). So the required probability is 1/3. In Example 3.3, since a child rather than a family is selected, families with (girl, girl), (boy, girl), and (girl, boy) are not equally likely to be chosen. In fact, the probability that a family with two girls is selected equals twice the probability that a family with one girl is selected. That is,

$$P(\text{girl, girl}) = 2P(\text{girl, boy}) = 2P(\text{boy, girl}).$$

This and the fact that the sum of the probabilities of (girl, girl), (girl, boy), (boy, girl) is 1 give

$$P(\text{girl, girl}) + \frac{1}{2}P(\text{girl, girl}) + \frac{1}{2}P(\text{girl, girl}) = 1,$$

which implies that

$$P(\text{girl, girl}) = \frac{1}{2}. \quad \blacklozenge$$

Example 3.4 An English class consists of 10 Koreans, 5 Italians, and 15 Hispanics. A paper is found belonging to one of the students of this class. If the name on the paper is not Korean, what is the probability that it is Italian? Assume that names completely identify ethnic groups.

Solution: Let A be the event that the name on the paper is Italian, and let B be the event that it is not Korean. To calculate $P(A \mid B)$, note that $AB = A$ because "the name being Italian," and "the name not being Korean" overlap with "the name being Italian," which is A. Hence $P(AB) = P(A) = 5/30$. Therefore,

$$P(A \mid B) = \frac{P(AB)}{P(B)} = \frac{5/30}{20/30} = \frac{1}{4}. \quad \blacklozenge$$

Example 3.5 We draw eight cards at random from an ordinary deck of 52 cards. Given that three of them are spades, what is the probability that the remaining five are also spades?

Solution: Let B be the event that at least three of them are spades, and let A denote the event that all of the eight cards selected are spades. Then $P(A \mid B)$ is the desired quantity and is calculated from $P(A \mid B) = P(AB)/P(B)$ as follows:

$$P(AB) = \frac{\binom{13}{8}}{\binom{52}{8}}$$

since AB is the event that all eight cards selected are spades.

$$P(B) = \sum_{x=3}^{8} \frac{\binom{13}{x}\binom{39}{8-x}}{\binom{52}{8}}$$

because at least three of the cards selected are spades if exactly three of them are spades, or exactly four of them are spades, . . . , or all eight of them are spades. Hence

$$P(A \mid B) = \frac{P(AB)}{P(B)} = \frac{\dfrac{\dbinom{13}{8}}{\dbinom{52}{8}}}{\displaystyle\sum_{x=3}^{8} \frac{\dbinom{13}{x}\dbinom{39}{8-x}}{\dbinom{52}{8}}} \approx 5.44 \times 10^{-6}. \quad \blacklozenge$$

An important feature of conditional probabilities (stated in the following theorem) is that they satisfy the same axioms that ordinary probabilities satisfy. This enables us to use the theorems that are true for probabilities for conditional probabilities as well.

Theorem 3.1   *Let S be the sample space of an experiment, and let B be an event of S with $P(B) > 0$. Then*

(a)   *$P(A \mid B) \geq 0$ for any event A of S.*

(b)   *$P(S \mid B) = 1$.*

(c)   *If $A_1, A_2, \ldots$ is a sequence of mutually exclusive events, then*

$$P\left(\bigcup_{i=1}^{\infty} A_i \mid B\right) = \sum_{i=1}^{\infty} P(A_i \mid B).$$

The proof of this theorem is left as an exercise.

Reduction of Sample Space

Let B be an event of a sample space S with $P(B) > 0$. For a subset A of B, define $Q(A) = P(A \mid B)$. Then Q is a function from the set of subsets of B to $[0, 1]$. Clearly, $Q(A) \geq 0$, $Q(B) = P(B \mid B) = 1$ and, by Theorem 3.1, if $A_1, A_2, \ldots$ is a sequence of mutually exclusive subsets of B, then

$$Q\left(\bigcup_{i=1}^{\infty} A_i\right) = P\left(\bigcup_{i=1}^{\infty} A_i \mid B\right) = \sum_{i=1}^{\infty} P(A_i \mid B) = \sum_{i=1}^{\infty} Q(A_i).$$

Thus Q satisfies the axioms that probabilities do, and hence it is a probability function. Note that while P is defined for all subsets of S, the probability function Q is defined only for subsets of B. Therefore, for Q, the sample space is reduced from S to B. This reduction of sample space is sometimes very helpful in calculating conditional probabilities. Suppose that we are interested in $P(E \mid B)$, where $E \subseteq S$. One way to calculate this quantity is to reduce S to B and find $Q(EB)$. It is usually easier to compute the unconditional probability $Q(EB)$ rather than the conditional probability $P(E \mid B)$. Examples follow.

Example 3.6   A child mixes 10 good and three dead batteries. To find the dead batteries, his father tests them one-by-one and without replacement. If the first four batteries tested are all good, what is the probability that the fifth one is dead?

Solution: Using the information that the first four batteries tested are all good, we rephrase the problem in the reduced sample space: Six good and three dead batteries are mixed. A battery is selected at random: What is the probability that it is dead? The solution to this trivial question is 3/9 = 1/3. Note that without reducing the sample space, the solution of this problem is not so easy.   ♦

Example 3.7   A farmer decides to test four fertilizers for his soybean fields. He buys 32 bags of fertilizers, eight bags from each kind, and tries them randomly on 32 plots, eight plots from each of fields A, B, C, and D, one bag per plot. If from type I fertilizer one bag is tried on field A and three on field B, what is the probability that two bags of type I fertilizer are tried on field D?

Solution: Using the information that one and three bags of type I fertilizer are tried on fields A and B, respectively, we rephrase the problem in the reduced sample space: 16 bags of fertilizers (of which four are type I) are tried randomly, eight on field C and eight on field D. What is the probability that two bags of type I fertilizer are tried on field D? The solution to this easier unconditional version of the problem is

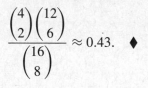

$$\frac{\binom{4}{2}\binom{12}{6}}{\binom{16}{8}} \approx 0.43. \quad ♦$$

Example 3.8   On a TV game show, there are three curtains. Behind two of the curtains there is nothing, but behind the third there is a prize that the player might win. The probability that the prize is behind a given curtain is 1/3. The game begins with the contestant randomly guessing a curtain. The host of the show (master of ceremonies), who *knows* behind which curtain the prize is, will then pull back a curtain *other than* the one chosen by the player and reveal that the prize is not behind that curtain. The host will not pull back the curtain selected by the player, nor will he pull back the one with the prize, if different from the player's choice. At this point, the host gives the player the opportunity to change his choice of curtain and select the other one. The question is whether the player should change his choice. That is, has the probability of the prize being behind the curtain chosen changed from 1/3 to 1/2 or it is still 1/3? If it is still 1/3, the contestant should definitely change his choice. Otherwise, there is no point in doing so.

Solution: We will show that the conditional probability that the contestant wins, given that he always changes his original choice is 2/3. Therefore, the contestant should definitely change his choice. To show this, suppose that the prize is behind curtain 1, and the contestant always changes his choice. These two assumptions reduce the sample space. The elements of the reduced sample space can be described by 3-tuples (x, y, z), where x is the curtain the contestant guesses first, y is the curtain the master of ceremonies pulls back, and z is the curtain the contestant switches to. For example, $(3, 2, 1)$ represents a game in which the contestant guesses curtain 3, the master of ceremonies pulls back curtain 2, and the contestant switches to curtain 1. Therefore, given that the prize is behind curtain 1 and the contestant changes his choice, the reduced sample space S is

$$S = \{(1, 2, 3), (1, 3, 2), (2, 3, 1), (3, 2, 1)\}.$$

Note that the event that the contestant guesses curtain 2 is $\{(2, 3, 1)\}$, the event that he guesses curtain 3 is $\{(3, 2, 1)\}$, whereas the event that he guesses curtain 1 is $\{(1, 2, 3), (1, 3, 2)\}$. Given that the prize is behind curtain 1 and the contestant always switches his original choice, the event that the contestant wins is $\{(2, 3, 1), (3, 2, 1)\}$. Since the contestant guesses a curtain with probability 1/3,

$$P(\{(2, 3, 1)\}) = P(\{(3, 2, 1)\}) = P(\{(1, 2, 3), (1, 3, 2)\}) = \frac{1}{3}.$$

This shows that no matter what values are assigned to $P(\{(1, 2, 3)\})$ and $P(\{(1, 3, 2)\})$, as long as their sum is 1/3, the conditional probability that the contestant wins, given that he always changes his original choice, is

$$P(\{(2, 3, 1)\}) + P(\{(3, 2, 1)\}) = \frac{2}{3}. \quad \blacklozenge$$

By Theorem 3.1, the function Q defined above by $Q(A) = P(A \mid B)$ is a probability function. Therefore, $Q(A) = P(A \mid B)$ satisfies the theorems stated for P. In particular, for all choices of B, $P(B) > 0$,

1. $P(\emptyset \mid B) = 0$.

2. $P(A^c \mid B) = 1 - P(A \mid B)$.

3. If $C \subseteq A$, then $P(AC^c \mid B) = P(A - C \mid B) = P(A \mid B) - P(C \mid B)$.

4. If $C \subseteq A$, then $P(C \mid B) \le P(A \mid B)$.

5. $P(A \cup C \mid B) = P(A \mid B) + P(C \mid B) - P(AC \mid B)$.

6. $P(A \mid B) = P(AC \mid B) + P(AC^c \mid B)$.

7. To calculate $P(A_1 \cup A_2 \cup A_3 \cup \cdots \cup A_n \mid B)$, we calculate conditional probabilities of all possible intersections of events from $\{A_1, A_2, \ldots, A_n\}$, given B, add the conditional probabilities obtained by intersecting an odd number of the events, and subtract the conditional probabilities obtained by intersecting an even number of events.

8. For any increasing or decreasing sequences of events $\{A_n, n \geq 1\}$,

$$\lim_{n \to \infty} P(A_n \mid B) = P(\lim_{n \to \infty} A_n \mid B).$$

EXERCISES

A

1. Suppose that 15% of the population of a country are unemployed women, and a total of 25% are unemployed. What percent of the unemployed are women?

2. Suppose that 41% of Americans have blood type A, and 4% have blood type AB. If in the blood of a randomly selected American soldier the A antigen is found, what is the probability that his blood type is A? The A antigen is found only in blood types A and AB.

3. In a technical college all students are required to take calculus and physics. Statistics show that 32% of the students of this college get A's in calculus, and 20% of them get A's in both calculus and physics. Gino, a randomly selected student of this college, has passed calculus with an A. What is the probability that he got an A in physics?

4. Suppose that two fair dice have been tossed and the total of their top faces is found to be divisible by 5. What is the probability that both of them have landed 5?

5. A bus arrives at a station every day at a random time between 1:00 P.M. and 1:30 P.M. A person arrives at this station at 1:00 and waits for the bus. If at 1:15 the bus has not yet arrived, what is the probability that the person will have to wait at least an additional 5 minutes?

6. Prove that $P(A \mid B) > P(A)$ if and only if $P(B \mid A) > P(B)$. In probability, if for two events A and B, $P(A \mid B) > P(A)$, we say that A and B are **positively correlated.** If $P(A \mid B) < P(A)$, A and B are said to be **negatively correlated.**

7. A spinner is mounted on a wheel of unit circumference (radius $1/2\pi$). Arcs A, B, and C of lengths $1/3$, $1/2$, and $1/6$, respectively, are marked off on the wheel's perimeter (see Figure 3.1). The spinner is flicked and we know that it is not pointing toward C. What is the probability that it points toward A?

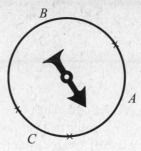

Figure 3.1   Spinner of Exercise 7.

8. In throwing two fair dice, what is the probability of a sum of 5 if they land on different numbers?

9. In a small lake, it is estimated that there are approximately 105 fish, of which 40 are trout and 65 are carp. A fisherman caught eight fish; what is the probability that exactly two of them are trout if we know that at least three of them are not?

10. From 100 cards numbered $00, 01, \ldots, 99$, one card is drawn. Suppose that α and β are the sum and the product, respectively, of the digits of the card selected. Calculate $P(\{\alpha = i \mid \beta = 0\}), i = 0, 1, 2, 3, \ldots, 18$.

11. From families with three children, a family is selected at random and found to have a boy. What is the probability that the boy has (a) an older brother and a younger sister; (b) an older brother; (c) a brother and a sister? Assume that in a three-child family all gender distributions have equal probabilities.

12. Show that if $P(A) = 1$, then $P(B \mid A) = P(B)$.

13. Prove that if $P(A) = a$ and $P(B) = b$, then $P(A \mid B) \geq (a + b - 1)/b$.

14. Prove Theorem 3.1.

15. Prove that if $P(E \mid F) \geq P(G \mid F)$ and $P(E \mid F^c) \geq P(G \mid F^c)$, then $P(E) \geq P(G)$.

16. The theaters of a town are showing seven comedies and nine dramas. Marlon has seen five of the movies. If the first three movies he has seen are dramas, what is the probability that the last two are comedies? Assume that Marlon chooses the shows at random and sees each movie at most once.

17. Suppose that 28 crayons, of which four are red, are divided randomly among Jack, Marty, Sharon, and Martha (seven each). If Sharon has exactly one red crayon, what is the probability that Marty has the remaining three?

B

18. A number is selected at random from the set $\{1, 2, \ldots, 10{,}000\}$ and is observed to be odd. What is the probability that it is (a) divisible by 3; (b) divisible by neither 3 nor 5?

19. A retired person chooses randomly one of the six parks of his town everyday and goes there for hiking. We are told that he was seen in one of these parks, Oregon Ridge, once during the last 10 days. What is the probability that during this period he has hiked in this park two or more times?

20. A big urn contains 1000 red chips, numbered 1 through 1000, and 1750 blue chips, numbered 1 through 1750. A chip is removed at random, and its number is found to be divisible by 3. What is the probability that its number is also divisible by 5?

21. There are three types of animals in a laboratory: 15 type I, 13 type II, and 12 type III. Animals of type I react to a particular stimulus in 5 seconds, animals of types II and III react to the same stimulus in 4.5 and 6.2 seconds, respectively. A psychologist selects 10 of these animals at random and finds that exactly four of them react to this stimulus in 6.2 seconds. What is the probability that at least two of them react to the same stimulus in 4.5 seconds?

22. In an international school, 60 students, of whom 15 are Korean, 20 are French, eight are Greek, and the rest are Chinese, are divided randomly into four classes of 15 each. If there are a total of eight French and six Korean students in classes A and B, what is the probability that class C has four of the remaining 12 French and three of the remaining nine Korean students?

23. From the set of all families with two children, *a family* is selected at random and is found to have a girl called Mary. We want to know the probability that both children of the family are girls. By Example 3.2, the probability should apparently be 1/3 because presumably knowing the name of the girl should not make a difference. However, if we ask a friend of that family whether Mary is the older or the younger child of the family, upon receiving the answer, we can conclude that the probability of the other child being a girl is 1/2. Therefore, there is no need to ask the friend, and we conclude already that the probability is 1/2! What is the flaw in this argument? Explain.

■

Often, when we think of a collection of events, we have a tendency to think about them in either temporal or logical sequence. So, if, for example, a sequence of events $A_1, A_2, \ldots, A_n$ occur in time or in some logical order, we can usually immediately write down the probabilities $P(A_1)$, $P(A_2 \mid A_1)$, $\ldots$, $P(A_n \mid A_1 A_2 \cdots A_{n-1})$ without much computation. However, we may be interested in probabilities of intersection of events, or probabilities of events unconditional on the rest, or probabilities of earlier events, given later events. In the next three sections, we will develop techniques for calculating such probabilities.

Suppose that, in a random phenomenon, events occur in either temporal or logical sequence. In Section 3.2 (Law of Multiplication), we will discuss the probabilities of the intersection of events in terms of conditional probabilities of later events given earlier events. In Section 3.3 (Law of Total Probability), we will find the probabilities of unconditional events in terms of conditional probabilities of later events given the earlier ones. In Section 3.4 (Bayes' Formula), we will discuss a formula relating probabilities of earlier events, given later events, to the conditional probabilities of later events given the earlier ones.

3.2   LAW OF MULTIPLICATION

The relation

$$P(A \mid B) = \frac{P(AB)}{P(B)}$$

is also useful for calculating $P(AB)$. If we multiply both sides of this relation by $P(B)$ $\left[\text{note that } P(B) > 0\right]$, we get

$$P(AB) = P(B)P(A \mid B), \tag{3.4}$$

which means that the probability of the joint occurrence of A and B is the product of the probability of B and the conditional probability of A given that B has occurred. If $P(A) > 0$, then by letting $A = B$ and $B = A$ in (3.4), we obtain

$$P(BA) = P(A)P(B \mid A).$$

Since $P(BA) = P(AB)$, this relation gives

$$P(AB) = P(A)P(B \mid A). \tag{3.5}$$

Thus, to calculate $P(AB)$, depending on which of the quantities $P(A \mid B)$ and $P(B \mid A)$ is known, we may use (3.4) or (3.5), respectively. The following example clarifies the usefulness of these relations.

Example 3.9   Suppose that five good fuses and two defective ones have been mixed up. To find the defective fuses, we test them one-by-one, at random and without replacement. What is the probability that we are lucky and find both of the defective fuses in the first two tests?

Solution: Let D_1 and D_2 be the events of finding a defective fuse in the first and second tests, respectively. We are interested in $P(D_1 D_2)$. Using (3.5), we get

$$P(D_1 D_2) = P(D_1)P(D_2 \mid D_1) = \frac{2}{7} \times \frac{1}{6} = \frac{1}{21}. \quad \blacklozenge$$

Relation (3.5) can be generalized for calculating the probability of the joint occurrence of several events. For example, if $P(AB) > 0$, then

$$P(ABC) = P(A)P(B \mid A)P(C \mid AB). \tag{3.6}$$

To see this, note that $P(AB) > 0$ implies that $P(A) > 0$; therefore,

$$P(A)P(B \mid A)P(C \mid AB) = P(A)\frac{P(AB)}{P(A)}\frac{P(ABC)}{P(AB)} = P(ABC).$$

The following theorem will generalize (3.5) and (3.6) to n events and can be shown in the same way as (3.6) is shown.

Theorem 3.2   *If $P(A_1 A_2 A_3 \cdots A_{n-1}) > 0$, then*

$$P(A_1 A_2 A_3 \cdots A_{n-1} A_n)$$
$$= P(A_1)P(A_2 \mid A_1)P(A_3 \mid A_1 A_2) \cdots P(A_n \mid A_1 A_2 A_3 \cdots A_{n-1}).$$

Example 3.10   A consulting firm is awarded 43% of the contracts it bids on. Suppose that Nordulf works for a division of the firm that gets to do 15% of the projects contracted for. If Nordulf directs 35% of the projects submitted to his division, what percentage of all bids submitted by the firm will result in contracts for projects directed by Nordulf?

Solution: Let A_1 be the event that the firm is awarded a contract for some randomly selected bid. Let A_2 be the event that the contract will be sent to Nordulf's division. Let A_3 be the event that Nordulf will direct the project. The desired probability is calculated as follows:

$$P(A_1 A_2 A_3) = P(A_1)P(A_2 \mid A_1)P(A_3 \mid A_1 A_2) = (0.43)(0.15)(0.35) = 0.0226.$$

Hence 2.26% of all bids submitted by the firm will be directed by Nordulf.   ♦

Example 3.11   Suppose that five good and two defective fuses have been mixed up. To find the defective ones, we test them one by one, at random and without replacement. What is the probability that we find both of the defective fuses in exactly three tests?

Solution: Let D_1, D_2, and D_3 be the events that the first, second, and third fuses tested are defective, respectively. Let G_1, G_2, and G_3 be the events that the first, second, and third fuses tested are good, respectively. We are interested in the probability of the event $G_1 D_2 D_3 \cup D_1 G_2 D_3$. We have

$$P(G_1 D_2 D_3 \cup D_1 G_2 D_3)$$
$$= P(G_1 D_2 D_3) + P(D_1 G_2 D_3)$$
$$= P(G_1)P(D_2 \mid G_1)P(D_3 \mid G_1 D_2) + P(D_1)P(G_2 \mid D_1)P(D_3 \mid D_1 G_2)$$
$$= \frac{5}{7} \times \frac{2}{6} \times \frac{1}{5} + \frac{2}{7} \times \frac{5}{6} \times \frac{1}{5} \approx 0.095.   ♦$$

EXERCISES

A

1. In a trial, the judge is 65% sure that Susan has committed a crime. Robert is a witness who knows whether Susan is innocent or guilty. However, Robert is Susan's friend and will lie with probability 0.25 if Susan is guilty. He will tell the truth if she is innocent. What is the probability that Robert will commit perjury?

2. There are 14 marketing firms hiring new graduates. Kate randomly found the recruitment ads of six of these firms and sent them her resume. If three of these marketing firms are in Maryland, what is the probability that Kate did not apply to a marketing firm in Maryland?

3. In a game of cards, two cards of the same color and denomination form a pair. For example, 8 of hearts and 8 of diamonds is one pair, king of spades and king of clubs is another. If six cards are selected at random and without replacement, what is the probability that there will be no pairs?

4. If eight defective and 12 nondefective items are inspected one by one, at random and without replacement, what is the probability that (a) the first four items inspected are defective; (b) from the first three items at least two are defective?

5. There are five boys and six girls in a class. For an oral exam, their teacher calls them one by one and randomly. (a) What is the probability that the boys and the girls alternate? (b) What is the probability that the boys are called first? Compare the answers to parts (a) and (b).

6. An urn contains five white and three red chips. Each time we draw a chip, we look at its color. If it is red, we replace it along with two new red chips, and if it is white, we replace it along with three new white chips. What is the probability that, in successive drawing of chips, the colors of the first four chips alternate?

7. Solve the following problem, asked of Marilyn Vos Savant in the "Ask Marilyn" column of *Parade Magazine,* January 3, 1999.

    You're at a party with 199 other guests when robbers break in and announce that they are going to rob one of you. They put 199 blank pieces of paper in a hat, plus one marked "you lose." Each guest must draw, and the person who draws "you lose" will get robbed. The robbers offer you the option of drawing first, last, or at any time in between. When would you take your turn?

    Assume that those who draw earlier will not announce whether they drew the "you lose" paper.

8. Suppose that 75% of all people with credit records improve their credit ratings within three years. Suppose that 18% of the population at large have poor credit records, and of those only 30% will improve their credit ratings within three years. What percentage of the people who will improve their credit records within the next three years are the ones who currently have good credit ratings?

B

9. Cards are drawn at random from an ordinary deck of 52, one by one and without replacement. What is the probability that no heart is drawn before the ace of spades is drawn?

10. From an ordinary deck of 52 cards, cards are drawn one by one, at random and without replacement. What is the probability that the fourth heart is drawn on the tenth draw?
   Hint: Let F denote the event that in the first nine draws there are exactly three hearts, and E be the event that the tenth draw is a heart. Use $P(FE) = P(F)P(E \mid F)$.

11. In a series of games, the winning number of the nth game, $n = 1, 2, 3, \ldots$, is a number selected at random from the set of integers $\{1, 2, \ldots, n + 2\}$. Don bets on 1 in each game and says that he will quit as soon as he wins. What is the probability that he has to play indefinitely?
   Hint: Let A_n be the event that Don loses the first n games. To calculate the desired probability, $P\left(\bigcap_{i=1}^{\infty} A_i\right)$, use Theorem 1.8.   ∎

3.3   LAW OF TOTAL PROBABILITY

Sometimes it is not possible to calculate directly the probability of the occurrence of an event A, but it is possible to find $P(A \mid B)$ and $P(A \mid B^c)$ for some event B. In such cases, the following theorem, which is conceptually rich and has widespread applications, is used. It states that $P(A)$ is the weighted average of the probability of A given that B has occurred and probability of A given that it has not occurred.

Theorem 3.3   (Law of Total Probability)   *Let B be an event with $P(B) > 0$ and $P(B^c) > 0$. Then for any event A,*

$$P(A) = P(A \mid B)P(B) + P(A \mid B^c)P(B^c).$$

Proof: By Theorem 1.7,

$$P(A) = P(AB) + P(AB^c). \tag{3.7}$$

Now $P(B) > 0$ and $P(B^c) > 0$. These imply that $P(AB) = P(A \mid B)P(B)$ and $P(AB^c) = P(A \mid B^c)P(B^c)$. Putting these in (3.7), we have proved the theorem.   ♦

Example 3.12   An insurance company rents 35% of the cars for its customers from agency I and 65% from agency II. If 8% of the cars of agency I and 5% of the cars of agency II break down during the rental periods, what is the probability that a car rented by this insurance company breaks down?

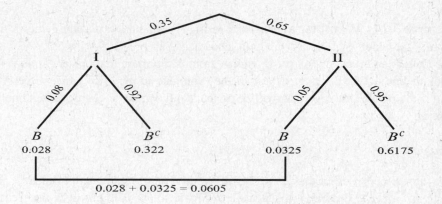

Figure 3.2   Tree diagram of Example 3.12.

Solution: Let A be the event that a car rented by this insurance company breaks down. Let I and II be the events that it is rented from agencies I and II, respectively. Then by the law of total probability,

$$P(A) = P(A \mid I)P(I) + P(A \mid II)P(II)$$
$$= (0.08)(0.35) + (0.05)(0.65) = 0.0605.$$

Tree diagrams facilitate solutions to this kind of problem. Let B and B^c stand for breakdown and not breakdown during the rental period, respectively. Then, as Figure 3.2 shows, to find the probability that a car breaks down, all we need to do is to compute, by multiplication, the probability of each path that leads to a point B and then add them up. So, as seen from the tree, the probability that the car breaks down is $(0.35)(0.08) + (0.65)(0.05) = 0.06$, and the probability that it does not break down is $(0.35)(0.92) + (0.65)(0.95) = 0.94$.   ♦

Example 3.13   In a trial, the judge is 65% sure that Susan has committed a crime. Julie and Robert are two witnesses who know whether Susan is innocent or guilty. However,

Robert is Susan's friend and will lie with probability 0.25 if Susan is guilty. He will tell the truth if she is innocent. Julie is Susan's enemy and will lie with probability 0.30 if Susan is innocent. She will tell the truth if Susan is guilty. What is the probability that, in the course of the trial, Robert and Julie will give conflicting testimony?

Solution: Let I be the event that Susan is innocent. Let C be the event that Robert and Julie will give conflicting testimony. By the law of total probability,

$$P(C) = P(C \mid I)P(I) + P(C \mid I^c)P(I^c)$$
$$= (0.30)(.35) + (0.25)(.65) = 0.2675. \quad \blacklozenge$$

Example 3.14   (Gambler's Ruin Problem)   Two gamblers play the game of "heads or tails," in which each time a fair coin lands heads up player A wins \$1 from B, and each time it lands tails up, player B wins \$1 from A. Suppose that player A initially has a dollars and player B has b dollars. If they continue to play this game successively, what is the probability that (a) A will be ruined; (b) the game goes forever with nobody winning?

Solution:

(a)   Let E be the event that A will be ruined if he or she starts with i dollars, and let $p_i = P(E)$. Our aim is to calculate p_a. To do so, we define F to be the event that A wins the first game. Then

$$P(E) = P(E \mid F)P(F) + P(E \mid F^c)P(F^c).$$

In this formula, $P(E \mid F)$ is the probability that A will be ruined, given that he wins the first game; so $P(E \mid F)$ is the probability that A will be ruined if his capital is $i + 1$; that is, $P(E \mid F) = p_{i+1}$. Similarly, $P(E \mid F^c) = p_{i-1}$. Hence

$$p_i = p_{i+1} \cdot \frac{1}{2} + p_{i-1} \cdot \frac{1}{2}. \tag{3.8}$$

Now $p_0 = 1$ because if A starts with 0 dollars, he or she is already ruined. Also, if the capital of A reaches $a + b$, then B is ruined; thus $p_{a+b} = 0$. Therefore, we have to solve the system of recursive equations (3.8), subject to the boundary conditions $p_0 = 1$ and $p_{a+b} = 0$. To do so, note that (3.8) implies that

$$p_{i+1} - p_i = p_i - p_{i-1}.$$

Hence, letting $p_1 - p_0 = \alpha$, we get

$$p_i - p_{i-1} = p_{i-1} - p_{i-2} = p_{i-2} - p_{i-3} = \cdots = p_2 - p_1 = p_1 - p_0 = \alpha.$$

Thus

$$p_1 = p_0 + \alpha$$
$$p_2 = p_1 + \alpha = p_0 + \alpha + \alpha = p_0 + 2\alpha$$
$$p_3 = p_2 + \alpha = p_0 + 2\alpha + \alpha = p_0 + 3\alpha$$
$$\vdots$$
$$p_i = p_0 + i\alpha$$
$$\vdots$$

Now $p_0 = 1$ gives $p_i = 1 + i\alpha$. But $p_{a+b} = 0$; thus $0 = 1 + (a+b)\alpha$. This gives $\alpha = -1/(a+b)$; therefore,

$$p_i = 1 - \frac{i}{a+b} = \frac{a+b-i}{a+b}.$$

In particular, $p_a = b/(a+b)$. That is, the probability that A will be ruined is $b/(a+b)$.

(b) The same method can be used with obvious modifications to calculate q_i, the probability that B is ruined if he or she starts with i dollars. The result is

$$q_i = \frac{a+b-i}{a+b}.$$

Since B starts with b dollars, he or she will be ruined with probability $q_b = a/(a+b)$. Thus the probability that the game goes on forever with nobody winning is $1 - (q_b + p_a)$. But $1 - (q_b + p_a) = 1 - a/(a+b) - b/(a+b) = 0$. Therefore, if this game is played successively, eventually either A is ruined or B is ruined.   ◆

Remark 3.2   If the game is not fair and on each play gambler A wins \$1 from B with probability p, $0 < p < 1$, $p \neq 1/2$, and loses \$1 to B with probability $q = 1 - p$, relation (3.8) becomes

$$p_i = pp_{i+1} + qp_{i-1},$$

but $p_i = (p+q)p_i$, so $(p+q)p_i = pp_{i+1} + qp_{i-1}$. This gives

$$q(p_i - p_{i-1}) = p(p_{i+1} - p_i),$$

which reduces to

$$p_{i+1} - p_i = \frac{q}{p}(p_i - p_{i-1}).$$

Using this relation and following the same line of argument lead us to

$$p_i = \left[\left(\frac{q}{p}\right)^{i-1} + \left(\frac{q}{p}\right)^{i-2} + \cdots + \left(\frac{q}{p}\right) + 1 \right]\alpha + p_0,$$

where $\alpha = p_1 - p_0$. Using $p_0 = 1$, we obtain

$$p_i = \frac{(q/p)^i - 1}{(q/p) - 1}\,\alpha + 1.$$

After finding α from $p_{a+b} = 0$ and substituting in this equation, we get

$$p_i = \frac{(q/p)^i - (q/p)^{a+b}}{1 - (q/p)^{a+b}} = \frac{1 - (p/q)^{a+b-i}}{1 - (p/q)^{a+b}},$$

where the last equality is obtained by multiplying the numerator and denominator of the function by $(p/q)^{a+b}$. In particular,

$$p_a = \frac{1 - (p/q)^b}{1 - (p/q)^{a+b}},$$

and, similarly,

$$q_b = \frac{1 - (q/p)^a}{1 - (q/p)^{a+b}}.$$

In this case also, $p_a + q_b = 1$, meaning that eventually either A or B will be ruined and the game does not go on forever. For comparison, suppose that A and B both start with \$10. If they play a fair game, the probability that A will be ruined is 1/2, and the probability that B will be ruined is also 1/2. If they play an unfair game with $p = 3/4$, $q = 1/4$, then p_{10}, the probability that A will be ruined, is almost 0.00002.   ◆

To generalize Theorem 3.3, we will now state a definition.

Definition   *Let $\{B_1, B_2, \ldots, B_n\}$ be a set of nonempty subsets of the sample space S of an experiment. If the events $B_1, B_2, \ldots, B_n$ are mutually exclusive and $\bigcup_{i=1}^{n} B_i = S$, the set $\{B_1, B_2, \ldots, B_n\}$ is called a **partition** of S.*

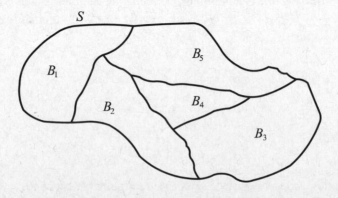

Figure 3.3   Partition of the given sample space S.

For example, if S of Figure 3.3 is the sample space of an experiment, then B_1, B_2, B_3, B_4, and B_5 of the same figure form a partition of S. As another example, consider the experiment of drawing a card from an ordinary deck of 52 cards. Let B_1, B_2, B_3, and B_4 denote the events that the card is a spade, a club, a diamond, and a heart, respectively. Then $\{B_1, B_2, B_3, B_4\}$ is a partition of the sample space of this experiment. If A_i, $1 \leq i \leq 10$, denotes the event that the value of the card drawn is i, and A_{11}, A_{12}, and A_{13} are the events of jack, queen, and king, respectively, then $\{A_1, A_2, \ldots, A_{13}\}$ is another partition of the same sample space. Note that

> For an experiment with sample space S, for any event A, A and A^c both nonempty, the set $\{A, A^c\}$ is a partition.

As observed, Theorem 3.3 is used whenever it is not possible to calculate $P(A)$ directly, but it is possible to find $P(A \mid B)$ and $P(A \mid B^c)$ for some event B. In many situations, it is neither possible to find $P(A)$ directly, nor possible to find a single event B that enables us to use

$$P(A) = P(A \mid B)P(B) + P(A \mid B^c)P(B^c).$$

In such situations Theorem 3.4, which is a generalization of Theorem 3.3 and is also called the **law of total probability**, might be applicable.

Theorem 3.4   (Law of Total Probability)   *If $\{B_1, B_2, \ldots, B_n\}$ is a partition of the sample space of an experiment and $P(B_i) > 0$ for $i = 1, 2, \ldots, n$, then for any event A of S,*

$$P(A) = P(A \mid B_1)P(B_1) + P(A \mid B_2)P(B_2) + \cdots + P(A \mid B_n)P(B_n)$$

$$= \sum_{i=1}^{n} P(A \mid B_i)P(B_i).$$

More generally, let $\{B_1, B_2, \ldots\}$ be a sequence of mutually exclusive events of S such that $\bigcup_{i=1}^{\infty} B_i = S$. Suppose that, for all $i \geq 1$, $P(B_i) > 0$. Then for any event A of S,

$$P(A) = \sum_{i=1}^{\infty} P(A \mid B_i)P(B_i).$$

Proof: Since $B_1, B_2, \ldots, B_n$ are mutually exclusive, $B_i B_j = \emptyset$ for $i \neq j$. Thus $(AB_i)(AB_j) = \emptyset$ for $i \neq j$. Hence $\{AB_1, AB_2, \ldots, AB_n\}$ is a set of mutually exclusive events. Now

$$S = B_1 \cup B_2 \cup \cdots \cup B_n$$

gives

$$A = AS = AB_1 \cup AB_2 \cup \cdots \cup AB_n;$$

therefore,

$$P(A) = P(AB_1) + P(AB_2) + \cdots + P(AB_n).$$

But $P(AB_i) = P(A \mid B_i)P(B_i)$ for $i = 1, 2, \ldots, n$, so

$$P(A) = P(A \mid B_1)P(B_1) + P(A \mid B_2)P(B_2) + \cdots + P(A \mid B_n)P(B_n).$$

The proof of the more general case is similar.   ♦

When using this theorem, one should be very careful to choose B_1, B_2, B_3, ... , so that they form a partition of the sample space.

Example 3.15   Suppose that 80% of the seniors, 70% of the juniors, 50% of the sophomores, and 30% of the freshmen of a college use the library of their campus frequently. If 30% of all students are freshmen, 25% are sophomores, 25% are juniors, and 20% are seniors, what percent of all students use the library frequently?

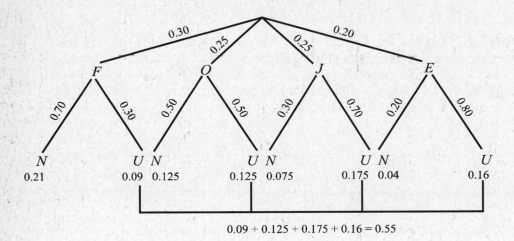

Figure 3.4   Tree diagram of Example 3.15.

Solution: Let U be the event that a randomly selected student is using the library frequently. Let F, O, J, and E be the events that he or she is a freshman, sophomore, junior, or senior, respectively. Then $\{F, O, J, E\}$ is a partition of the sample space. Thus

$$P(U) = P(U \mid F)P(F) + P(U \mid O)P(O) + P(U \mid J)P(J) + P(U \mid E)P(E)$$
$$= (0.30)(0.30) + (0.50)(0.25) + (0.70)(0.25) + (0.80)(0.20) = 0.55.$$

Therefore, 55% of these students use the campus library frequently. The same calculation can be carried out readily from the tree diagram of Figure 3.4, where U means they use the library frequently and N means that they do not.   ♦

Example 3.16   Suppose that 16% of an insurance company's automobile policyholders are male and under the age of 25, while 12% are female and under the age of 25. The following table lists the percentages of various groups of policyholders who were involved in a car accident last a year.

Group	Male Under 25	Female Under 25	Between 25 and 65	Over 65
Percentage of Accidents	20%	8%	5%	10%

Find the range of the percentages of this company's policyholders who got involved in a car accident during the previous year.

Solution: Let M, F, B, and O be, respectively, the events that a randomly selected automobile policyholder of this company is a male and under the age of 25, a female and under the age of 25, between the ages 25 and 65, and over the age of 65. Let A be the event that he or she got involved in an accident last year. By the law of total probability,

$$P(A) = P(A \mid M)P(M) + P(A \mid F)P(F) + P(A \mid B)P(B) + P(A \mid O)P(O)$$
$$= (0.20)(0.16) + (0.08)(0.12) + (0.05)P(B) + (0.10)P(O)$$
$$= 0.0416 + (0.05)P(B) + (0.10)P(O).$$

The lower bound of the range of percentages of all policyholders of this company who got involved in a car accident last year is obtained, from this relation, by putting $P(B) = 1$ and $P(O) = 0$. It gives $P(A) = 0.0916$. The upper bound of the range is obtained by putting $P(B) = 0$ and $P(O) = 1$, which gives $P(A) = 0.1416$. Therefore, the range of the percentages we are interested in is between 9.16% and 14.16%.   ◆

Example 3.17   An urn contains 10 white and 12 red chips. Two chips are drawn at random and, without looking at their colors, are discarded. What is the probability that a third chip drawn is red?

Solution: For $i \geq 1$, let R_i be the event that the ith chip drawn is red and W_i be the event that it is white. Intuitively, it should be clear that the two discarded chips provide no information, so $P(R_3) = 12/22$, the same as if it were the first chip drawn from the urn. To prove this mathematically, note that $\{R_2W_1, W_2R_1, R_2R_1, W_2W_1\}$ is a partition of the sample space; therefore,

$$P(R_3) = P(R_3 \mid R_2W_1)P(R_2W_1) + P(R_3 \mid W_2R_1)P(W_2R_1)$$
$$+ P(R_3 \mid R_2R_1)P(R_2R_1) + P(R_3 \mid W_2W_1)P(W_2W_1). \qquad (3.9)$$

Now

$$P(R_2 W_1) = P(R_2 \mid W_1)P(W_1) = \frac{12}{21} \times \frac{10}{22} = \frac{20}{77},$$

$$P(W_2 R_1) = P(W_2 \mid R_1)P(R_1) = \frac{10}{21} \times \frac{12}{22} = \frac{20}{77},$$

$$P(R_2 R_1) = P(R_2 \mid R_1)P(R_1) = \frac{11}{21} \times \frac{12}{22} = \frac{22}{77},$$

and

$$P(W_2 W_1) = P(W_2 \mid W_1)P(W_1) = \frac{9}{21} \times \frac{10}{22} = \frac{15}{77}.$$

Substituting these values in (3.9), we get

$$P(R_3) = \frac{11}{20} \times \frac{20}{77} + \frac{11}{20} \times \frac{20}{77} + \frac{10}{20} \times \frac{22}{77} + \frac{12}{20} \times \frac{15}{77} = \frac{12}{22}. \quad \blacklozenge$$

EXERCISES

A

1. If 5% of men and 0.25% of women are color blind, what is the probability that a randomly selected person is color blind?

2. Suppose that 40% of the students of a campus are women. If 20% of the women and 16% of the men of this campus are A students, what percent of all of them are A students?

3. Jim has three cars of different models: A, B, and C. The probabilities that models A, B, and C use over 3 gallons of gasoline from Jim's house to his work are 0.25, 0.32, and 0.53, respectively. On a certain day, all three of Jim's cars have 3 gallons of gasoline each. Jim chooses one of his cars at random, and without paying attention to the amount of gasoline in the car drives it toward his office. What is the probability that he makes it to the office?

4. One of the cards of an ordinary deck of 52 cards is lost. What is the probability that a random card drawn from this deck is a spade?

5. Two cards from an ordinary deck of 52 cards are missing. What is the probability that a random card drawn from this deck is a spade?

6. Of the patients in a hospital, 20% of those with, and 35% of those without myocardial infarction have had strokes. If 40% of the patients have had myocardial infarction, what percent of the patients have had strokes?

7. Suppose that 37% of a community are at least 45 years old. If 80% of the time a person who is 45 or older tells the truth, and 65% of the time a person below 45 tells the truth, what is the probability that a randomly selected person answers a question truthfully?

8. A person has six guns. The probability of hitting a target when these guns are properly aimed and fired is 0.6, 0.5, 0.7, 0.9, 0.7, and 0.8, respectively. What is the probability of hitting a target if a gun is selected at random, properly aimed, and fired?

9. A factory produces its entire output with three machines. Machines I, II, and III produce 50%, 30%, and 20% of the output, but 4%, 2%, and 4% of their outputs are defective, respectively. What fraction of the total output is defective?

10. Solve the following problem, from the "Ask Marilyn" column of *Parade Magazine*, October 29, 2000.

> I recently returned from a trip to China, where the government is so concerned about population growth that it has instituted strict laws about family size. In the cities, a couple is permitted to have only one child. In the countryside, where sons traditionally have been valued, if the first child is a son, the couple may have no more children. But if the first child is a daughter, the couple may have another child. Regardless of the sex of the second child, no more are permitted. How will this policy affect the mix of males and females?

    To pose the question mathematically, what is the probability that a randomly selected child from the countryside is a boy?

11. Suppose that five coins, of which exactly three are gold, are distributed among five persons, one each, at random, and one by one. Are the chances of getting a gold coin equal for all participants? Why or why not?

12. In a town, 7/9th of the men and 3/5th of the women are married. In that town, what fraction of the adults are married? Assume that all married adults are the residents of the town.

13. A child gets lost in the Disneyland at the Epcot Center in Florida. The father of the child believes that the probability of his being lost in the east wing of the center is 0.75 and in the west wing is 0.25. The security department sends an officer to the east and an officer to the west to look for the child. If the probability that a security officer who is looking in the correct wing finds the child is 0.4, find the probability that the child is found.

14. Suppose that there exist N families on the earth and that the maximum number of children a family has is c. Let α_j $\left(0 \leq j \leq c, \ \sum_{j=0}^{c} \alpha_j = 1\right)$ be the fraction of families with j children. Find the fraction of all children in the world who are the kth born of their families $(k = 1, 2, \ldots, c)$.

15. Let B be an event of a sample space S with $P(B) > 0$. For a subset A of S, define $Q(A) = P(A \mid B)$. By Theorem 3.1 we know that Q is a probability function. For E and F, events of S $\left[P(FB) > 0\right]$, show that $Q(E \mid F) = P(E \mid FB)$.

B

16. Suppose that 40% of the students on a campus, who are married to students on the same campus, are female. Moreover, suppose that 30% of those who are married, but not to students at this campus, are also female. If one-third of the married students on this campus are married to other students on this campus, what is the probability that a randomly selected married student from this campus is a woman? *Hint:* Let M, C, and F denote the events that the random student is married, is married to a student on the same campus, and is female. For any event A, let $Q(A) = P(A \mid M)$. Then, by Theorem 3.1, Q satisfies the same axioms that probabilities satisfy. Applying Theorem 3.3 to Q and, using the result of Exercise 15, we obtain

$$P(F \mid M) = P(F \mid MC)P(C \mid M) + P(F \mid MC^c)P(C^c \mid M).$$

17. Suppose that the probability that a new seed planted in a specific farm germinates is equal to the proportion of all planted seeds that germinated in that farm previously. Suppose that the first seed planted in the farm germinated, but the second seed planted did not germinate. For positive integers n and k $(k < n)$, what is the probability that of the first n seeds planted in the farm exactly k germinated?

18. Suppose that 10 good and three dead batteries are mixed up. Jack tests them one by one, at random and without replacement. But before testing the fifth battery he realizes that he does not remember whether the first one tested is good or is dead. All he remembers is that the last three that were tested were all good. What is the probability that the first one is also good?

19. A box contains 18 tennis balls, of which eight are new. Suppose that three balls are selected randomly, played with, and after play are returned to the box. If another three balls are selected for play a second time, what is the probability that they are all new?

20. From families with three children, *a child* is selected at random and found to be a girl. What is the probability that she has an older sister? Assume that in a three-child family all sex distributions are equally probable. *Hint:* Let G be the event that the randomly selected child is a girl, A be the event

that she has an older sister, and O, M, and Y be the events that she is the oldest, the middle, and the youngest child of the family, respectively. For any subset B of the sample space let $Q(B) = P(B \mid G)$; then apply Theorem 3.3 to Q. (See also Exercises 15 and 16.)

21. Suppose that three numbers are selected one by one, at random and without replacement from the set of numbers $\{1, 2, 3, \ldots, n\}$. What is the probability that the third number falls between the first two if the first number is smaller than the second?

22. Avril has certain standards for selecting her future husband. She has n suitors and knows how to compare any two and rank them. She decides to date one suitor at a time randomly. When she knows a suitor well enough, she can marry or reject him. If she marries the suitor, she can never know the ones not dated. If she rejects the suitor, she will not be able to reconsider him. In this process, Avril will not choose a suitor she ranks lower than at least one of the previous ones dated. Avril's goal is to maximize the probability of selecting the best suitor. To achieve this, she adopts the following strategy: For some m, $0 \le m < n$, she dumps the first m suitors she dates after knowing each of them well no matter how good they are. Then she marries the first suitor she dates who is better than all those preceding him. In terms of n, find the value of m which maximizes the probability of selecting the best suitor.

Remark: This exercise is a model for many real-world problems in which we have to reject "choices" or "offers" until the one we think is the "best." Note that it is quite possible that we reject the best "offer" or "choice" in the process. If this happens, we will not be able to make a selection.

23. (Shrewd Prisoner's Dilemma)   Because of a prisoner's constant supplication, the king grants him this favor: He is given $2N$ balls, which differ from each other only in that half of them are green and half are red. The king instructs the prisoner to divide the balls between two identical urns. One of the urns will then be selected at random, and the prisoner will be asked to choose a ball at random from the urn chosen. If the ball turns out to be green, the prisoner will be freed. How should he distribute the balls in the urn to maximize his chances of freedom?

Hint: Let g be the number of green balls and r be the number of red balls in the first urn. The corresponding numbers in the second urn are $N - g$ and $N - r$. The probability that a green ball is drawn is $f(g, r)$:

$$f(g, r) = \frac{1}{2}\left(\frac{g}{g + r} + \frac{N - g}{2N - g - r}\right).$$

Find the maximum of this function of two variables (r and g). Note that the maximum need not occur at an interior point of the domain.

3.4   BAYES' FORMULA

To introduce Bayes' formula, let us first examine the following problem. In a bolt factory, 30, 50, and 20% of production is manufactured by machines I, II, and III, respectively. If 4, 5, and 3% of the output of these respective machines is defective, what is the probability that a randomly selected bolt that is found to be defective is manufactured by machine III? To solve this problem, let A be the event that a random bolt is defective and B_3 be the event that it is manufactured by machine III. We are asked to find $P(B_3 \mid A)$. Now

$$P(B_3 \mid A) = \frac{P(B_3 A)}{P(A)}, \tag{3.10}$$

so we need to know the quantities $P(B_3 A)$ and $P(A)$. But neither of these is given. To find $P(B_3 A)$, note that since $P(A \mid B_3)$ and $P(B_3)$ are known we can use the relation

$$P(B_3 A) = P(A \mid B_3) P(B_3). \tag{3.11}$$

To calculate $P(A)$, we must use the law of total probability. Let B_1 and B_2 be the events that the bolt is manufactured by machines I and II, respectively. Then $\{B_1, B_2, B_3\}$ is a partition of the sample space; hence

$$P(A) = P(A \mid B_1) P(B_1) + P(A \mid B_2) P(B_2) + P(A \mid B_3) P(B_3). \tag{3.12}$$

Substituting (3.11) and (3.12) in (3.10), we arrive at **Bayes' formula**:

$$P(B_3 \mid A) = \frac{P(B_3 A)}{P(A)}$$

$$= \frac{P(A \mid B_3) P(B_3)}{P(A \mid B_1) P(B_1) + P(A \mid B_2) P(B_2) + P(A \mid B_3) P(B_3)} \tag{3.13}$$

$$= \frac{(0.03)(0.20)}{(0.04)(0.30) + (0.05)(0.50) + (0.03)(0.20)} \approx 0.14. \tag{3.14}$$

Relation (3.13) is a particular case of Bayes' formula (Theorem 3.5). We will now explain how a tree diagram is used to write relation (3.14). Figure 3.5, in which D stands for "defective" and N for "not defective," is a tree diagram for this problem. To find the desired probability all we need do is find (by multiplication) the probability of the required path, the path from III to D, and divide it by the sum of the probabilities of the paths that lead to D's.

In general, modifying the argument from which (3.13) was deduced, we arrive at the following theorem.

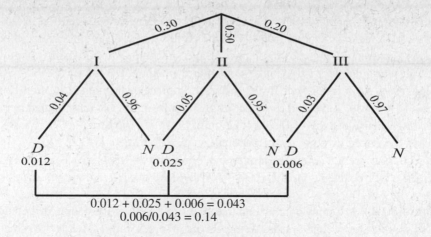

Figure 3.5   Tree diagram for relation (3.14).

Theorem 3.5   (Bayes' Theorem)   *Let $\{B_1, B_2, \ldots, B_n\}$ be a partition of the sample space S of an experiment. If for $i = 1, 2, \ldots, n$, $P(B_i) > 0$, then for any event A of S with $P(A) > 0$,*

$$P(B_k \mid A) = \frac{P(A \mid B_k)P(B_k)}{P(A \mid B_1)P(B_1) + P(A \mid B_2)P(B_2) + \cdots + P(A \mid B_n)P(B_n)}.$$

In statistical applications of Bayes' theorem, $B_1, B_2, \ldots, B_n$ are called **hypotheses**, $P(B_i)$ is called the **prior** probability of B_i, and the conditional probability $P(B_i \mid A)$ is called the **posterior** probability of B_i after the occurrence of A.

Note that, for any event B of S, B and B^c both nonempty, the set $\{B, B^c\}$ is a partition of S. Thus, by Theorem 3.5,

If $P(B) > 0$ and $P(B^c) > 0$, then for any event A of S with $P(A) > 0$,

$$P(B \mid A) = \frac{P(A \mid B)P(B)}{P(A \mid B)P(B) + P(A \mid B^c)P(B^c)}.$$

Similarly,

$$P(B^c \mid A) = \frac{P(A \mid B^c)P(B^c)}{P(A \mid B)P(B) + P(A \mid B^c)P(B^c)}.$$

These are the simplest forms of Bayes' formula. They are used whenever the quantities $P(A \mid B)$, $P(A \mid B^c)$, and $P(B)$ are given or can be calculated. The typical situation is that A happens logically or temporally after B, so that the probabilities $P(B)$ and $P(A \mid B)$ can be readily computed. Bayes' formula is applicable when we know the probability of the more recent event, given that the earlier event has occurred, $P(A \mid B)$, and we wish to calculate the probability of the earlier event, given that the more recent event has occurred, $P(B \mid A)$. In practice, Bayes' formula is used when we know the effect of a cause and we wish to make some inference about the cause.

Theorem 3.5, in its present form, is due to Laplace, who named it after Thomas Bayes (1701–1761). Bayes, a prominent English philosopher and an ordained minister, did a comprehensive study of the calculation of $P(B \mid A)$ in terms of $P(A \mid B)$. His work was continued by other mathematicians such as Laplace and Gauss.

We will now present several examples concerning Bayes' theorem. To emphasize the convenience of tree diagrams, in Example 3.21, we will use a tree diagram as well.

Example 3.18   In a study conducted three years ago, 82% of the people in a randomly selected sample were found to have "good" financial credit ratings, while the remaining 18% were found to have "bad" financial credit ratings. Current records of the people from that sample show that 30% of those with bad credit ratings have since improved their ratings to good, while 15% of those with good credit ratings have since changed to having a bad credit rating. What percentage of people with good credit ratings now had bad ratings three years ago?

Solution: Let G be the event that a randomly selected person from the sample has a good credit rating now. Let B be the event that he or she had a bad credit rating three years ago. The desired quantity is $P(B \mid G)$. By Bayes' formula,

$$P(B \mid G) = \frac{P(G \mid B)P(B)}{P(G \mid B)P(B) + P(G \mid B^c)P(B^c)}$$

$$= \frac{(0.30)(.18)}{(0.30)(.18) + (.85)(.82)} = 0.072,$$

where $P(G \mid B^c) = 0.85$, because the probability is $1 - 0.15 = 0.85$ that a person with good credit rating three years ago has a good credit rating now. Therefore, 7.2% of people with good credit ratings now had bad ratings three years ago.   ◆

Example 3.19   During a double homicide murder trial, based on circumstantial evidence alone, the jury becomes 15% certain that a suspect is guilty. DNA samples recovered from the murder scene are then compared with DNA samples extracted from the suspect. Given the size and conditions of the recovered samples, a forensic scientist estimates that the probability of the sample having come from someone other than the suspect is 10^{-9}. With this new information, how certain should the jury be of the suspect's guilt?

Solution: Let G and I be the events that the suspect is guilty and innocent, respectively. Let D be the event that the recovered DNA samples from the murder scene match with the DNA samples extracted from the suspect. Since $\{G, I\}$ is a partition of the sample space, we can use Bayes' formula to calculate $P(G \mid D)$, the probability that the suspect is the murderer in view of the new evidence.

$$P(G \mid D) = \frac{P(D \mid G)P(G)}{P(D \mid G)P(G) + P(D \mid I)P(I)}$$

$$= \frac{1(.15)}{1(.15) + 10^{-9}(.85)} = 0.9999999943.$$

This shows that $P(I \mid D) = 1 - P(G \mid D)$ is approximately 5.67×10^{-9}, leaving no reasonable doubt for the innocence of the suspect.

In some trials, prosecutors have argued that if $P(D \mid I)$ is "small enough," then there is no reasonable doubt for the guilt of the defendant. Such an argument, called the *prosecutor's fallacy*, probably stems from confusing $P(D \mid I)$ for $P(I \mid D)$. One should pay attention to the fact that $P(D \mid I)$ is infinitesimal regardless of the suspect's guilt or innocence. To elaborate this further, note that in this example, $P(I \mid D)$ is approximately 5.67 times larger than $P(D \mid I)$, which is 10^{-9}. This is because even without the DNA evidence, there is a 15% chance that the suspect is guilty.

We will now present a situation, which clearly demonstrates that $P(D \mid I)$ should not be viewed as the probability of guilt in evaluating reasonable doubt. Suppose that the double homicide in this example occurred in California, and there is no suspect identified. Suppose that a DNA data bank identifies a person in South Dakota whose DNA matches what was recovered at the California crime scene. Furthermore, suppose that the forensic scientist estimates that the probability of the sample recovered at the crime scene having come from someone other than the person in South Dakota is still 10^{-9}. If there is no evidence that this person ever traveled to California, or had any motive for committing the double homicide, it is doubtful that he or she would be indicted. In such a case $P(D \mid I)$ is still 10^{-9}, but this quantity hardly constitutes the probability of guilt for the person in South Dakota. The argument that since $P(D \mid I)$ is infinitesimal there is no reasonable doubt for the guilt of the person from South Dakota does not seem convincing at all. In such a case, the quantity $P(I \mid D)$, which can be viewed as the probability of guilt, cannot even be estimated if nothing beyond DNA evidence exists.   ♦

Example 3.20   On the basis of reconnaissance reports, Colonel Smith decides that the probability of an enemy attack against the left is 0.20, against the center is 0.50, and against the right is 0.30. A flurry of enemy radio traffic occurs in preparation for the attack. Since deception is normal as a prelude to battle, Colonel Brown, having intercepted the radio traffic, tells General Quick that if the enemy wanted to attack on the left, the probability is 0.20 that he would have sent this particular radio traffic. He tells the general that the corresponding probabilities for an attack on the center or the right are 0.70 and 0.10, respectively. How should General Quick use these two equally reliable staff members' views to get the best probability profile for the forthcoming attack?

Solution: Let A be the event that the attack would be against the left, B be the event that the attack would be against the center, and C be the event that the attack would be against the right. Let Δ be the event that this particular flurry of radio traffic occurs. Colonel Brown has provided information of *conditional* probabilities of a particular flurry of radio traffic given that the enemy is preparing to attack against the left, the center, and the right. However, Colonel Smith has presented *unconditional* probabilities, on the basis of reconnaissance reports, for the enemy attacking against the left, the center, and the right. Because of these, the general should take the opinion of Colonel Smith as prior probabilities for A, B, and C. That is, $P(A) = 0.20$, $P(B) = 0.50$, and $P(C) = 0.30$. Then he should calculate $P(A \mid \Delta)$, $P(B \mid \Delta)$, and $P(C \mid \Delta)$ based on Colonel Brown's view, using Bayes' theorem, as follows.

$$P(A \mid \Delta) = \frac{P(\Delta \mid A)P(A)}{P(\Delta \mid A)P(A) + P(\Delta \mid B)P(B) + P(\Delta \mid C)P(C)}$$

$$= \frac{(0.2)(0.2)}{(0.2)(0.2) + (0.7)(0.5) + (0.1)(0.3)} = \frac{(0.2)(0.2)}{0.42} \approx 0.095.$$

Similarly,

$$P(B \mid \Delta) = \frac{(0.7)(0.5)}{0.42} \approx 0.83 \text{ and } P(C \mid \Delta) = \frac{(0.1)(0.3)}{0.42} \approx 0.071. \quad \blacklozenge$$

Example 3.21   A box contains seven red and 13 blue balls. Two balls are selected at random and are discarded without their colors being seen. If a third ball is drawn randomly and observed to be red, what is the probability that both of the discarded balls were blue?

Solution: Let BB, BR, and RR be the events that the discarded balls are blue and blue, blue and red, red and red, respectively. Also, let R be the event that the third ball drawn is red. Since $\{BB, BR, RR\}$ is a partition of sample space, Bayes' formula can be used to calculate $P(BB \mid R)$.

$$P(BB \mid R) = \frac{P(R \mid BB)P(BB)}{P(R \mid BB)P(BB) + P(R \mid BR)P(BR) + P(R \mid RR)P(RR)}.$$

Now

$$P(BB) = \frac{13}{20} \times \frac{12}{19} = \frac{39}{95}, \qquad P(RR) = \frac{7}{20} \times \frac{6}{19} = \frac{21}{190},$$

and

$$P(BR) = \frac{13}{20} \times \frac{7}{19} + \frac{7}{20} \times \frac{13}{19} = \frac{91}{190},$$

where the last equation follows since BR is the union of two disjoint events: namely, the first ball discarded was blue, the second was red, and vice versa. Thus

$$P(BB \mid R) = \frac{\dfrac{7}{18} \times \dfrac{39}{95}}{\dfrac{7}{18} \times \dfrac{39}{95} + \dfrac{6}{18} \times \dfrac{91}{190} + \dfrac{5}{18} \times \dfrac{21}{190}} \approx 0.46.$$

This result can be found from the tree diagram of Figure 3.6 as well. Alternatively, reducing sample space, given that the third ball was red, there are 13 blue and six red balls which could have been discarded. Thus

$$P(BB \mid R) = \frac{13}{19} \times \frac{12}{18} \approx 0.46. \quad \blacklozenge$$

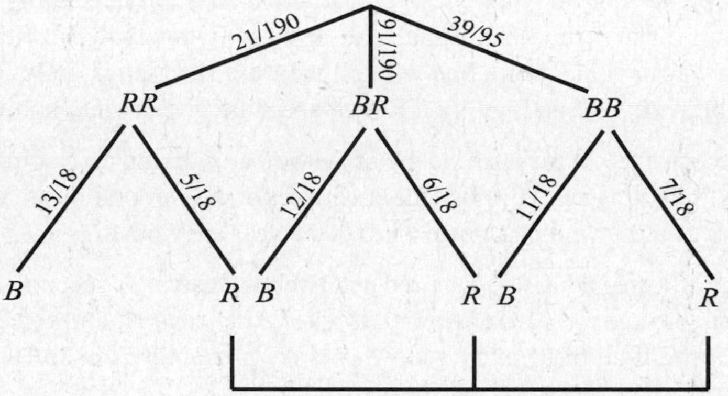

$$\frac{5}{18} \cdot \frac{21}{190} + \frac{6}{18} \cdot \frac{91}{190} + \frac{7}{18} \cdot \frac{39}{95} \approx 0.35,$$

$$\frac{7}{18} \cdot \frac{39}{95} \approx 0.16, \quad \frac{0.16}{0.35} \approx 0.46.$$

Figure 3.6   Tree diagram for Example 3.21.

EXERCISES

A

1. In transmitting dot and dash signals, a communication system changes 1/4 of the dots to dashes and 1/3 of the dashes to dots. If 40% of the signals transmitted are dots and 60% are dashes, what is the probability that a dot received was actually a transmitted dot?

2. On a multiple-choice exam with four choices for each question, a student either knows the answer to a question or marks it at random. If the probability that he or she knows the answers is 2/3, what is the probability that an answer that was marked correctly was not marked randomly?

3. A judge is 65% sure that a suspect has committed a crime. During the course of the trial, a witness convinces the judge that there is an 85% chance that the criminal is left-handed. If 23% of the population is left-handed and the suspect is also left-handed, with this new information, how certain should the judge be of the guilt of the suspect?

4. In a trial, the judge is 65% sure that Susan has committed a crime. Julie and Robert are two witnesses who know whether Susan is innocent or guilty. However, Robert is Susan's friend and will lie with probability 0.25 if Susan is guilty. He will tell the truth if she is innocent. Julie is Susan's enemy and will lie with probability 0.30 if Susan is innocent. She will tell the truth if Susan is guilty. What is the probability that Susan is guilty if Robert and Julie give conflicting testimony?

5. Suppose that 5% of the men and 2% of the women working for a corporation make over \$120,000 a year. If 30% of the employees of the corporation are women, what percent of those who make over \$120,000 a year are women?

6. A stack of cards consists of six red and five blue cards. A second stack of cards consists of nine red cards. A stack is selected at random and three of its cards are drawn. If all of them are red, what is the probability that the first stack was selected?

7. A certain cancer is found in one person in 5000. If a person does have the disease, in 92% of the cases the diagnostic procedure will show that he or she actually has it. If a person does not have the disease, the diagnostic procedure in one out of 500 cases gives a false positive result. Determine the probability that a person with a positive test result has the cancer.

8. Urns I, II, and III contain three pennies and four dimes, two pennies and five dimes, three pennies and one dime, respectively. One coin is selected at random from each urn. If two of the three coins are dimes, what is the probability that the coin selected from urn I is a dime?

9. In a study it was discovered that 25% of the paintings of a certain gallery are not original. A collector in 15% of the cases makes a mistake in judging if a painting is authentic or a copy. If she buys a piece thinking that it is original, what is the probability that it is not?

10. There are three identical cards that differ only in color. Both sides of one are black, both sides of the second one are red, and one side of the third card is black and its other side is red. These cards are mixed up and one of them is selected at random. If the upper side of this card is red, what is the probability that its other side is black?

11. With probability of 1/6 there are i defective fuses among 1000 fuses ($i = 0, 1, 2, 3, 4, 5$). If among 100 fuses selected at random, none was defective, what is the probability of no defective fuses at all?

12. Solve the following problem, asked of Marilyn Vos Savant in the "Ask Marilyn" column of *Parade Magazine,* February 18, 1996.

> Say I have a wallet that contains either a $2 bill or a $20 bill (with equal likelihood), but I don't know which one. I add a $2 bill. Later, I reach into my wallet (without looking) and remove a bill. It's a $2 bill. There's one bill remaining in the wallet. What are the chances that it's a $2 bill?

B

13. There are two stables on a farm, one that houses 20 horses and 13 mules, the other with 25 horses and eight mules. Without any pattern, animals occasionally leave their stables and then return to their stables. Suppose that during a period when all the animals are in their stables, a horse comes out of a stable and then returns. What is the probability that the next animal coming out of the same stable will also be a horse?

14. An urn contains five red and three blue chips. Suppose that four of these chips are selected at random and transferred to a second urn, which was originally empty. If a random chip from this second urn is blue, what is the probability that two red and two blue chips were transferred from the first urn to the second urn?

15. The advantage of a certain blood test is that 90% of the time it is positive for patients having a certain disease. Its disadvantage is that 25% of the time it is also positive in healthy people. In a certain location 30% of the people have the disease, and anybody with a positive blood test is given a drug that cures the disease. If 20% of the time the drug produces a characteristic rash, what is the probability that a person from this location who has the rash had the disease in the first place?

3.5   INDEPENDENCE

Let A and B be two events of a sample space S, and assume that $P(A) > 0$ and $P(B) > 0$. We have seen that, in general, the conditional probability of A given B is not equal to the probability of A. However, if it is, that is, if $P(A \mid B) = P(A)$, we say that A is **independent** of B. This means that if A is independent of B, knowledge regarding the occurrence of B does not change the chance of the occurrence of A. The relation $P(A \mid B) = P(A)$ is equivalent to the relations $P(AB)/P(B) = P(A)$, $P(AB) = P(A)P(B)$, $P(BA)/P(A) = P(B)$, and $P(B \mid A) = P(B)$. The equivalence of the first and last of these relations implies that if A is independent of B, then B is independent of A. In other words, if knowledge regarding the occurrence of B does not change the chance of occurrence of A, then knowledge regarding the occurrence of A does not change the chance of occurrence of B. Hence independence is a symmetric relation on the set of all events of a sample space. As a result of this property, instead of making the definitions "A is independent of B" and "B is independent of A," we simply define the concept of the "independence of A and B." To do so, we take $P(AB) = P(A)P(B)$ as the definition. We do this because a symmetrical definition relating A and B does not readily follow from either of the other relations given [i.e., $P(A \mid B) = P(A)$ or $P(B \mid A) = P(B)$]. Moreover, these relations require either that $P(B) > 0$ or $P(A) > 0$, whereas our definition does not.

Definition   *Two events A and B are called **independent** if*

$$P(AB) = P(A)P(B).$$

*If two events are not independent, they are called **dependent**. If A and B are independent, we say that $\{A, B\}$ is an independent set of events.*

Note that in this definition we did not require $P(A)$ or $P(B)$ to be strictly positive. Hence by this definition *any* event A with $P(A) = 0$ or 1 is independent of *every* event B (see Exercise 13).

Example 3.22   In the experiment of tossing a fair coin twice, let A and B be the events of getting heads on the first and second tosses, respectively. Intuitively, it is clear that A and B are independent. To prove this mathematically, note that $P(A) = 1/2$ and $P(B) = 1/2$. But since the sample space of this experiment consists of the four equally probable events: HH, HT, TH, and TT, we have $P(AB) = P(\text{HH}) = 1/4$. Hence $P(AB) = P(A)P(B)$ is valid, implying the independence of A and B. It is interesting to know that Jean Le Rond d'Alembert (1717–1783), a French mathematician, had argued that, since in the experiment of tossing a fair coin twice the possible number of heads is 0, 1, and 2, the probability of no heads, one heads, and two heads, each is 1/3.   ◆

Example 3.23   In the experiment of drawing a card from an ordinary deck of 52 cards, let A and B be the events of getting a heart and an ace, respectively. Whether A and B are independent cannot be answered easily on the basis of intuition alone. However, using the defining formula, $P(AB) = P(A)P(B)$, this can be answered at once since $P(AB) = 1/52$, $P(A) = 1/4$, $P(B) = 1/13$, and $1/52 = 1/4 \times 1/13$. Hence A and B are independent events.   ◆

Example 3.24   An urn contains five red and seven blue balls. Suppose that two balls are selected at random and with replacement. Let A and B be the events that the first and the second balls are red, respectively. Then, using the counting principle, we get $P(AB) = (5 \times 5)/(12 \times 12)$. Now $P(AB) = P(A)P(B)$ since $P(A) = 5/12$ and $P(B) = 5/12$. Thus A and B are independent. If we do the same experiment without replacement, then $P(B \mid A) = 4/11$ while

$$P(B) = P(B \mid A)P(A) + P(B \mid A^c)P(A^c)$$
$$= \frac{4}{11} \times \frac{5}{12} + \frac{5}{11} \times \frac{7}{12} = \frac{5}{12},$$

which might be quite surprising to some. But it is true. If no information is given on the outcome of the first draw, there is no reason for the probability of second ball being red to differ from 5/12. Thus $P(B \mid A) \neq P(B)$, implying that A and B are dependent.   ◆

Example 3.25   In the experiment of selecting a random number from the set of natural numbers $\{1, 2, 3, \ldots, 100\}$, let A, B, and C denote the events that they are divisible by 2, 3, and 5, respectively. Clearly, $P(A) = 1/2$, $P(B) = 33/100$, $P(C) = 1/5$, $P(AB) = 16/100$, and $P(AC) = 1/10$. Hence A and B are dependent while A and C are independent. Note that if the random number is selected from $\{1, 2, 3, \ldots, 300\}$, then each of $\{A, B\}$, $\{A, C\}$, and $\{B, C\}$ is an independent set of events. This is because 300 is divisible by 2, 3, and 5, but 100 is not divisible by 3.   ◆

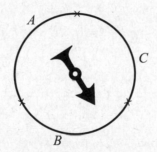

Figure 3.7   Spinner of Example 3.26.

Example 3.26   A spinner is mounted on a wheel. Arcs A, B, and C, of equal length, are marked off on the wheel's perimeter (see Figure 3.7). In a game of chance, the spinner is flicked, and depending on whether it stops on A, B, or C, the player wins 1, 2, or 3 points, respectively. Suppose that a player plays this game twice. Let E denote the event that he wins 1 point in the first game and any number of points in the second. Let F be the event that he wins a total of 3 points in both games, and G be the event that he wins a total of 4 points in both games. The sample space of this experiment has $3 \times 3 = 9$ elements, $E = \{(1, 1), (1, 2), (1, 3)\}$, $F = \{(1, 2), (2, 1)\}$, and $G = \{(1, 3), (2, 2), (3, 1)\}$. Therefore, $P(E) = 1/3$, $P(G) = 1/3$, $P(F) = 2/9$, $P(FE) = 1/9$, and $P(GE) = 1/9$. These show that E and G are independent, whereas E and F are not. To justify these intuitively, note that if we are interested in a total of 3 points, then getting 1 point in the first game is good luck, since obtaining 3 points in the first game makes it impossible to win a sum of 3. However, if we are interested in a sum of 4 points, it does not matter what we win in the first game. We have the same chance of obtaining a sum of 4 if the first game results in any of the numbers 1, 2, or 3. ♦

We now prove that if A and B are independent events, so are A and B^c.

Theorem 3.6   *If A and B are independent, then A and B^c are independent as well.*

Proof: By Theorem 1.7,

$$P(A) = P(AB) + P(AB^c).$$

Therefore,

$$P(AB^c) = P(A) - P(AB) = P(A) - P(A)P(B)$$
$$= P(A)\big[1 - P(B)\big] = P(A)P(B^c). \quad ♦$$

Corollary   *If A and B are independent, then A^c and B^c are independent as well.*

Proof: A and B are independent, so by Theorem 3.6 the events A and B^c are independent. Now, using the same theorem again, we have that A^c and B^c are independent.
♦

Thus, if A and B are independent, knowledge about the occurrence or nonoccurrence of A does not change the chances of the occurrence or nonoccurrence of B, and vice versa.

Remark 3.3   *If A and B are mutually exclusive events and $P(A) > 0$, $P(B) > 0$, then they are dependent. This is because, if we are given that one has occurred, the chance of the occurrence of the other one is zero.* That is, the occurrence of one of them precludes the occurrence of the other. For example, let A be the event that the next president of the United States is a Democrat and B be the event that he or she is a Republican. Then A and B are mutually exclusive; hence they are dependent. If A occurs, that is, if the next president is a Democrat, the probability that B occurs, that is, he or she is a Republican is zero, and vice versa. ♦

The following example shows that if A is independent of B and if A is independent of C, then A is not necessarily independent of BC or of $B \cup C$.

Example 3.27   Dennis arrives at his office every day at a random time between 8:00 A.M. and 9:00 A.M. Let A be the event that Dennis arrives at his office tomorrow between 8:15 A.M. and 8:45 A.M. Let B be the event that he arrives between 8:30 A.M. and 9:00 A.M., and let C be the event that he arrives either between 8:15 A.M. and 8:30 A.M. or between 8:45 A.M. and 9:00 A.M. Then AB, AC, BC, and $B \cup C$ are the events that Dennis arrives at his office between 8:30 and 8:45, 8:15 and 8:30, 8:45 and 9:00, and 8:15 and 9:00, respectively. Thus $P(A) = P(B) = P(C) = 1/2$ and $P(AB) = P(AC) = 1/4$. So $P(AB) = P(A)P(B)$ and $P(AC) = P(A)P(C)$; that is, A is independent of B and it is independent of C. However, since BC and A are mutually exclusive, they are dependent. Also, $P(A \mid B \cup C) = 2/3 \neq P(A)$. Thus $B \cup C$ and A are dependent as well.   ◆

Example 3.28   (Jailer's Paradox)   The jailer of a prison in which Alex, Ben, and Tim are held is the only person, other than the judge, who knows which of these three prisoners is condemned to death, and which *two* will be freed. The prisoners know that exactly two of them will go free; they do not know which two. Alex has written a letter to his fiancée. Just in case he is not one of the two who will be freed, he wants to give the letter to a prisoner who goes free to deliver. So Alex asks the jailer to tell him the name of one of the two prisoners who will go free. The jailer refuses to give that information to Alex. To begin with, he is not allowed to tell Alex whether Alex goes free or not. Putting Alex aside, he argues that, if he reveals the name of a prisoner who will go free, then the probability of Alex dying increases from 1/3 to 1/2. He does not want to do that.

As Zweifel notes in the June 1986 issue of *Mathematics Magazine*, page 156, "this seems intuitively suspect, since the jailer is providing no new information to Alex, so why should his probability of dying change?" Zweifel is correct. Just revealing to Alex that Ben goes free, or just revealing to him that Tim goes free is not the type of information that changes the probability of Alex dying. What changes the probability of Alex dying is telling him whether both Ben and Tim are going free or exactly one of them is going free.

To explain this paradox, we will show that, under suitable conditions, if the jailer tells Alex that Tim goes free, still the probability is 1/3 that Alex dies. Telling Alex that Tim is going free reveals to Alex that the probability of Ben dying is 2/3. Similarly, telling Alex that Ben is going free reveals to Alex that the probability of Tim dying is 2/3. To show these facts, let A, B, and T be the events that "Alex dies," "Ben dies," and "Tim dies." Let

$$\omega_1 = \text{Tim dies, and the jailer tells Alex that Ben goes free}$$
$$\omega_2 = \text{Ben dies, and the jailer tells Alex that Tim goes free}$$
$$\omega_3 = \text{Alex dies, and the jailer tells Alex that Ben goes free}$$
$$\omega_4 = \text{Alex dies, and the jailer tells Alex that Tim goes free.}$$

The sample space of all possible episodes is $S = \{\omega_1, \omega_2, \omega_3, \omega_4\}$. Now, if Tim dies, with probability 1, the jailer will tell Alex that Ben goes free. Therefore, ω_1 occurs if and only if Tim dies. This implies that $P(\omega_1) = 1/3$. Similarly, if Ben dies, with probability 1, the jailer will tell Alex that Tim goes free. Therefore, ω_2 occurs if and only if Ben dies. This shows that $P(\omega_2) = 1/3$. To assign probabilities to ω_3 and ω_4, we will make two assumptions: (1) If Alex is the one who is scheduled to die, then the event that the jailer will tell Alex that Ben goes free is *independent* of the event that the jailer will tell Alex that Tim goes free, and (2) if Alex is the one who is scheduled to die, then the probability that the jailer will tell Alex that Ben goes free is 1/2; hence the probability that he will tell

Alex that Tim goes free is 1/2 as well. Under these conditions, $P(\omega_3) = P(\omega_4) = 1/6$. Let J be the event that "the jailer tells Alex that Tim goes free;" then

$$P(A \mid J) = \frac{P(AJ)}{P(J)} = \frac{P(\omega_4)}{P(\omega_2) + P(\omega_4)} = \frac{\dfrac{1}{6}}{\dfrac{1}{3} + \dfrac{1}{6}} = \frac{1}{3},$$

This shows that if the jailer reveals no information about the fate of Alex, telling Alex the name of one prisoner who goes free does not change the probability of Alex dying; it remains to be 1/3.

Note that the decision as which of Ben, Tim, or Alex is condemned to death, and which two will be freed, has been made by the judge. The jailer has no option to control Alex's fate. With probability 1, the jailer knows which one of the three dies and which two go free. It is Alex who doesn't know any of these probabilities. Alex can only analyze these probabilities based on the information he receives from the jailer. If Alex is dying, and the jailer disobeys the conditions (1) and (2), then he could decide to tell Alex, with arbitrary probabilities, which of Ben or Tim goes free. In such a case, $P(A \mid J)$ will no longer equal 1/3. It will vary depending on the probabilities of J and J^c. If the jailer insists on not giving extra information to Alex, he should obey conditions (1) and (2). Furthermore, the jailer should tell Alex what his rules are when he reveals information. Alex can only analyze the probability of dying based on the full information he receives from the jailer. If the jailer does not reveal his rules to Alex, there is no way for Alex to know whether the probability that he is the unlucky prisoner has changed or not.

Zweifel analyzes this paradox by using Bayes' formula:

$$P(A \mid J) = \frac{P(J \mid A)P(A)}{P(J \mid A)P(A) + P(J \mid B)P(B) + P(J \mid T)P(T)}$$

$$= \frac{\dfrac{1}{2} \times \dfrac{1}{3}}{\dfrac{1}{2} \times \dfrac{1}{3} + 1 \times \dfrac{1}{3} + 0 \times \dfrac{1}{3}} = \frac{1}{3}.$$

Similarly, if D is the event that "the jailer tells Alex that Ben goes free," then $P(A \mid D) = 1/3$. In his explanation of this paradox Zweifel writes

> Aside from the formal application of Bayes' theorem, one would like to understand this "paradox" from an intuitive point of view. The crucial point which can perhaps make the situation clear is the discrepancy between $P(J \mid A)$ and $P(J \mid B)$ noted above. Bridge players call this the "Principle of Restricted Choice." The probability of a restricted choice is obviously greater than that of a free choice, and a common error made by those who attempt to solve such problems intuitively is to overlook this point. In the case of the jailer's paradox, if the jailer says "Tim will go free" this is twice as likely to

occur when Ben is scheduled to die (restricted choice; jailer *must* say "Tim") as when Alex is scheduled to die (free choice; jailer could say either "Tim" or "Ben").

The jailer's paradox and its solution have been around for a long time. Despite this, when the same problem in a different context came up in the "Ask Marilyn" column of *Parade Magazine* on September 9, 1990 (see Example 3.8) and Marilyn Vos Savant[†] gave the correct answer to the problem in the December 2, 1990 issue, she was taken to task by three mathematicians. By the time the February 17, 1991 issue of *Parade Magazine* was published, Vos Savant had received 2000 letters on the problem, of which 92% of general respondents and 65% of university respondents opposed her answer. This story reminds us of the statement of De Moivre in his dedication of *The Doctrine of Chance*:

> Some of the Problems about Chance having a great appearance of Simplicity, the Mind is easily drawn into a belief, that their Solution may be attained by the mere Strength of natural good Sense; which generally proving otherwise and the Mistakes occasioned thereby being not unfrequent, 'tis presumed that a Book of this Kind, which teaches to distinguish Truth from what seems so nearly to resemble it, will be looked upon as a help to good Reasoning.   ♦

We now extend the concept of independence to three events: A, B, and C are called **independent** if knowledge about the occurrence of any of them, or the joint occurrence of any two of them, does not change the chances of the occurrence of the remaining events. That is, A, B, and C are independent if $\{A, B\}$, $\{A, C\}$, $\{B, C\}$, $\{A, BC\}$, $\{B, AC\}$, and $\{C, AB\}$ are all independent sets of events. Hence A, B, and C are independent if

$$P(AB) = P(A)P(B),$$
$$P(AC) = P(A)P(C),$$
$$P(BC) = P(B)P(C),$$
$$P(A(BC)) = P(A)P(BC),$$
$$P(B(AC)) = P(B)P(AC),$$
$$P(C(AB)) = P(C)P(AB).$$

Now note that these relations can be reduced since the first three and the relation $P(ABC) = P(A)P(B)P(C)$ imply the last three relations. Hence the definition of the independence of three events can be shortened as follows.

[†] Ms. Vos Savant is the writer of a reader-correspondence column and is listed in the *Guinness Book of Records* Hall of Fame for "highest IQ."

Definition   *The events A, B, and C are called **independent** if*

$$P(AB) = P(A)P(B),$$
$$P(AC) = P(A)P(C),$$
$$P(BC) = P(B)P(C),$$
$$P(ABC) = P(A)P(B)P(C).$$

If A, B, and C are independent events, we say that {A, B, C} is an independent set of events.

The following example demonstrates that $P(ABC) = P(A)P(B)P(C)$, in general, does not imply that $\{A, B, C\}$ is a set of independent events.

Example 3.29   Let an experiment consist of throwing a die twice. Let A be the event that in the second throw the die lands 1, 2, or 5; B the event that in the second throw it lands 4, 5 or 6; and C the event that the sum of the two outcomes is 9. Then $P(A) = P(B) = 1/2$, $P(C) = 1/9$, and

$$P(AB) = \frac{1}{6} \neq \frac{1}{4} = P(A)P(B),$$
$$P(AC) = \frac{1}{36} \neq \frac{1}{18} = P(A)P(C),$$
$$P(BC) = \frac{1}{12} \neq \frac{1}{18} = P(B)P(C),$$

while

$$P(ABC) = \frac{1}{36} = P(A)P(B)P(C).$$

Thus the validity of $P(ABC) = P(A)P(B)P(C)$ is not sufficient for the independence of A, B, and C.   ◆

If A, B, and C are three events and the occurrence of any of them does not change the chances of the occurrence of the remaining two, we say that A, B, and C are **pairwise independent.** Thus $\{A, B, C\}$ forms a set of pairwise independent events if $P(AB) = P(A)P(B)$, $P(AC) = P(A)P(C)$, and $P(BC) = P(B)P(C)$. The difference between pairwise independent events and independent events is that, in the former, knowledge about the joint occurrence of any two of them may change the chances of the occurrence of the remaining one, but in the latter it would not. The following example illuminates the difference between pairwise independence and independence.

Example 3.30   A regular tetrahedron is a body that has four faces and, if it is tossed, the probability that it lands on any face is 1/4. Suppose that one face of a regular tetrahedron has three colors: red, green, and blue. The other three faces each have only one color:

red, blue, and green, respectively. We throw the tetrahedron once and let R, G, and B be the events that the face on which it lands contains red, green, and blue, respectively. Then $P(R \mid G) = 1/2 = P(R)$, $P(R \mid B) = 1/2 = P(R)$, and $P(B \mid G) = 1/2 = P(B)$. Thus the events R, B, and G are pairwise independent. However, R, B, and G are not independent events since $P(R \mid GB) = 1 \neq P(R)$.   ◆

The independence of more than three events may be defined in a similar manner. A set of n events $A_1, A_2, \ldots, A_n$ is said to be **independent** if knowledge about the occurrence of any of them or the joint occurrence of any number of them does not change the chances of the occurrence of the remaining events. If we write this definition in terms of formulas, we get many equations. Similar to the case of three events, if we reduce the number of these equations to the minimum number that can be used to have all of the formulas satisfied, we reach the following definition.

Definition   *The set of events $\{A_1, A_2, \ldots, A_n\}$ is called **independent** if for every subset $\{A_{i_1}, A_{i_2}, \ldots, A_{i_k}\}$, $k \geq 2$, of $\{A_1, A_2, \ldots, A_n\}$,*

$$P(A_{i_1} A_{i_2} \cdots A_{i_k}) = P(A_{i_1}) P(A_{i_2}) \cdots P(A_{i_k}). \tag{3.15}$$

This definition is not in fact limited to finite sets and is extended to infinite sets of events (countable or uncountable) in the obvious way. For example, the sequence of events $\{A_i\}_{i=1}^{\infty}$ is called independent if for any of its subsets $\{A_{i_1}, A_{i_2}, \ldots, A_{i_k}\}$, $k \geq 2$, (3.15) is valid.

By definition, events $A_1, A_2, \ldots, A_n$ are independent if, for all combinations $1 \leq i < j < k < \cdots \leq n$, the relations

$$P(A_i A_j) = P(A_i) P(A_j),$$

$$P(A_i A_j A_k) = P(A_i) P(A_j) P(A_k),$$

$$\vdots$$

$$P(A_1 A_2 \cdots A_n) = P(A_1) P(A_2) \cdots P(A_n)$$

are valid. Now we see that the first line stands for $\binom{n}{2}$ equations, the second line stands for $\binom{n}{3}$ equations, $\ldots$, and the last line stands for $\binom{n}{n}$ equations. Therefore, $A_1, A_2, \ldots, A_n$ are independent if all of the above $\binom{n}{2} + \binom{n}{3} + \cdots + \binom{n}{n}$ relations are satisfied. Note that, by the binomial expansion (see Theorem 2.5 and Example 2.26),

$$\binom{n}{2} + \binom{n}{3} + \cdots + \binom{n}{n} = (1+1)^n - \binom{n}{1} - \binom{n}{0} = 2^n - n - 1.$$

Thus the number of these equations is $2^n - n - 1$. Although these equations seem to be cumbersome to check, it usually turns out that they are obvious and checking is not necessary.

Example 3.31 We draw cards, one at a time, at random and successively from an ordinary deck of 52 cards with replacement. What is the probability that an ace appears before a face card?

Solution: We will explain two different techniques that may be used to solve this type of problems. For a third technique, see Exercise 32, Section 12.3.

   Technique 1:   Let E be the event of an ace appearing before a face card. Let A, F, and B be the events of ace, face card, and neither in the first experiment, respectively. Then, by the law of total probability,

$$P(E) = P(E \mid A)P(A) + P(E \mid F)P(F) + P(E \mid B)P(B).$$

Thus

$$P(E) = 1 \times \frac{4}{52} + 0 \times \frac{12}{52} + P(E \mid B) \times \frac{36}{52}. \tag{3.16}$$

Now note that since the outcomes of successive experiments are all independent of each other, when the second experiment begins, the whole probability process starts all over again. Therefore, if in the first experiment neither a face card nor an ace are drawn, the probability of E before doing the first experiment and after it would be the same; that is, $P(E \mid B) = P(E)$. Thus Equation (3.16) gives

$$P(E) = \frac{4}{52} + P(E) \times \frac{36}{52}.$$

Solving this equation for $P(E)$, we obtain $P(E) = 1/4$, a quantity expected because the number of face cards is three times the number of aces (see also Exercise 36).

   Technique 2:   Let A_n be the event that no face card or ace appears on the first $(n-1)$ drawings, and the nth draw is an ace. Then the event of "an ace before a face card" is $\bigcup_{n=1}^{\infty} A_n$. Now $\{A_n, n \geq 1\}$ forms a sequence of mutually exclusive events because, if $n \neq m$, simultaneous occurrence of A_n and A_m is the impossible event that an ace appears for the first time in the nth and mth draws. Hence

$$P\left(\bigcup_{n=1}^{\infty} A_n\right) = \sum_{n=1}^{\infty} P(A_n).$$

To compute $P(A_n)$, note that P(an ace on any draw) $= 1/13$ and P(no face card and no ace in any trial) $= 9/13$. By the independence of trials we obtain

$$P(A_n) = \left(\frac{9}{13}\right)^{n-1}\left(\frac{1}{13}\right).$$

Therefore,

$$P\left(\bigcup_{n=1}^{\infty} A_n\right) = \sum_{n=1}^{\infty} \left(\frac{9}{13}\right)^{n-1}\left(\frac{1}{13}\right) = \frac{1}{13}\sum_{n=1}^{\infty} \left(\frac{9}{13}\right)^{n-1}$$

$$= \frac{1}{13} \cdot \frac{1}{1 - 9/13} = \frac{1}{4}.$$

Here $\sum$ is calculated from the geometric series theorem: For $a \neq 0$, $|r| < 1$, the geometric series $\sum_{n=m}^{\infty} ar^n$ converges to $ar^m/(1-r)$.

It is interesting to observe that, in this problem, if cards are drawn *without replacement*, even though the trials are no longer independent, the answer would still be the same. For $1 \leq n \leq 37$, let E_n be the event that no face card or ace appears on the first $n-1$ drawings; let F_n be the event that the nth draw is an ace. Then the event of "an ace appearing before a face card" is $\bigcup_{n=1}^{37} E_n F_n$. Clearly, $\{E_n F_n, \ 1 \leq n \leq 37\}$ forms a sequence of mutually exclusive events. Hence

$$P\left(\bigcup_{n=1}^{37} E_n F_n\right) = \sum_{n=1}^{37} P(E_n F_n) = \sum_{n=1}^{37} P(E_n) P(F_n \mid E_n)$$

$$= \sum_{n=1}^{37} \frac{\dbinom{36}{n-1}}{\dbinom{52}{n-1}} \times \frac{4}{52 - (n-1)} = \frac{1}{4}. \quad \blacklozenge$$

Example 3.32   Figure 3.8 shows an electric circuit in which each of the switches located at 1, 2, 3, and 4 is independently closed or open with probabilities p and $1-p$, respectively. If a signal is fed to the input, what is the probability that it is transmitted to the output?

Solution: Let E_i be the event that the switch at location i is closed, $1 \leq i \leq 4$. A signal fed to the input will be transmitted to the output if at least one of the events $E_1 E_2$ and $E_3 E_4$ occurs. Hence the desired probability is

$$P(E_1 E_2 \cup E_3 E_4) = P(E_1 E_2) + P(E_3 E_4) - P(E_1 E_2 E_3 E_4)$$
$$= p^2 + p^2 - p^4 = p^2(2 - p^2). \quad \blacklozenge$$

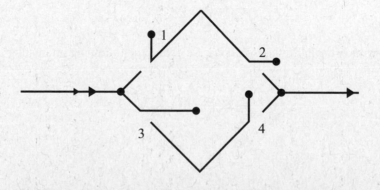

Figure 3.8   Electric circuit of Example 3.32.

Example 3.33   Adam tosses a fair coin $n + 1$ times, Andrew tosses the same coin n times. What is the probability that Adam gets more heads than Andrew?

Solution: Let H_1 and H_2 be the number of heads obtained by Adam and Andrew, respectively. Also, let T_1 and T_2 be the number of tails obtained by Adam and Andrew, respectively. Since the coin is fair,

$$P(H_1 > H_2) = P(T_1 > T_2).$$

But

$$P(T_1 > T_2) = P(n + 1 - H_1 > n - H_2) = P(H_1 \le H_2).$$

Therefore, $P(H_1 > H_2) = P(H_1 \le H_2)$. So

$$P(H_1 > H_2) + P(H_1 \le H_2) = 1$$

implies that

$$P(H_1 > H_2) = P(H_1 \le H_2) = \frac{1}{2}.$$

Note that a combinatorial solution to this problem is neither elegant nor easy to handle:

$$P(H_1 > H_2) = \sum_{i=0}^{n} P(H_1 > H_2 \mid H_2 = i) P(H_2 = i)$$

$$= \sum_{i=0}^{n} \sum_{j=i+1}^{n+1} P(H_1 = j) P(H_2 = i)$$

$$= \sum_{i=0}^{n} \sum_{j=i+1}^{n+1} \frac{\dfrac{(n + 1)!}{j!\,(n + 1 - j)!}}{2^{n+1}} \, \frac{\dfrac{n!}{i!\,(n - i)!}}{2^{n}}$$

$$= \frac{1}{2^{2n+1}} \sum_{i=0}^{n} \sum_{j=i+1}^{n+1} \binom{n + 1}{j} \binom{n}{i}.$$

However, comparing these two solutions, we obtain the following interesting identity.

$$\sum_{i=0}^{n} \sum_{j=i+1}^{n+1} \binom{n + 1}{j} \binom{n}{i} = 2^{2n}. \quad \blacklozenge$$

EXERCISES

A

1. Jean le Rond d'Alembert, a French mathematician, believed that in successive flips of a fair coin, after a long run of heads, a tail is more likely. Do you agree with d'Alembert on this? Explain.

2. Clark and Anthony are two old friends. Let A be the event that Clark will attend Anthony's funeral. Let B be the event that Anthony will attend Clark's funeral. Are A and B independent? Why or why not?

3. In a certain country, the probability that a fighter plane returns from a mission without mishap is 49/50, *independent* of other missions. In a conversation, Mia concluded that any pilot who flew 49 consecutive missions without mishap should be returned home before the fiftieth mission. But, on considering the matter, Jim concluded that the probability of a randomly selected pilot being able to fly 49 consecutive missions safely is $(49/50)^{49} = 0.3716017$. In other words, the odds are almost two to one against an ordinary pilot performing the feat that this pilot has already performed. Hence the pilot would seem to be more skillful than most and thus has a better chance of surviving the 50th mission. Who is right, Mia, Jim, or neither of them? Explain.

4. A fair die is rolled twice. Let A denote the event that the sum of the outcomes is odd, and B denote the event that it lands 2 on the first toss. Are A and B independent? Why or why not?

5. An urn has three red and five blue balls. Suppose that eight balls are selected at random and with replacement. What is the probability that the first three are red and the rest are blue balls?

6. Suppose that two points are selected at random and independently from the interval $(0, 1)$. What is the probability that the first one is less than 3/4, and the second one is greater than 1/4?

7. According to a recent mortality table, the probability that a 35-year-old U.S. citizen will live to age 65 is 0.725. (a) What is the probability that John and Jim, two 35-year-old Americans who are not relatives, both live to age 65? (b) What is the probability that neither John nor Jim lives to that age?

8. The Italian mathematician Giorlamo Cardano once wrote that if the odds in favor of an event are 3 to 1, then the odds in favor of the occurrence of that event in two consecutive independent experiments are 9 to 1. (He squared 3 and 1 to obtain 9 to 1.) Was Cardano correct?

9.   Consider the four "unfolded" dice in Figure 3.9 designed by Stanford professor Bradley Effron. Clearly, none of these dice is an ordinary die with sides numbered 1 through 6. A game consists of two players each choosing one of these four dice and rolling it. The player rolling a larger number is the winner. Show that if all four dice are fair, it is twice as likely for die A to beat die B, twice as likely for die B to beat die C, twice as likely for die C to beat D, and *surprisingly*, it is twice as likely for die D to beat die A.

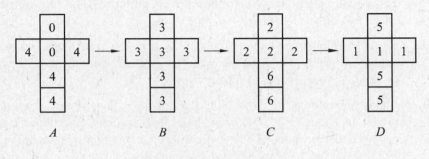

Figure 3.9   Effron dice.

10. **(Chevalier de Méré's Paradox[†])**   In the seventeenth century in France there were two popular games, one to obtain at least one 6 in four throws of a fair die and the other to bet on at least one double 6 in 24 throws of two fair dice. French nobleman and mathematician Chevalier de Méré argued that the probabilities of a 6 in one throw of a fair die and a double 6 in one throw of two fair dice are 1/6 and 1/36, respectively. Therefore, the probability of at least one 6 in four throws of a fair die and at least one double 6 in 24 throws of two fair dice are $4 \times 1/6 = 2/3$ and $24 \times 1/36 = 2/3$, respectively. However, even then experience had convinced gamblers that the probability of winning the first game was higher than winning the second game, a contradiction. Explain why de Méré's argument is false, and calculate the correct probabilities of winning in these two games.

11. In data communications, a message transmitted from one end is subject to various sources of distortion and may be received erroneously at the other end. Suppose that a message of 64 bits (a bit is the smallest unit of information and is either 1 or 0) is transmitted through a medium. If each bit is received incorrectly with probability 0.0001 independently of the other bits, what is the probability that the message received is free of error?

12. Find an example in which $P(AB) < P(A)P(B)$.

13. **(a)**   Show that if $P(A) = 1$, then $P(AB) = P(B)$.

     (b)   Prove that *any* event A with $P(A) = 0$ or $P(A) = 1$ is independent of *every* event B.

[†]Some scholars consider this problem to be the inception of modern probability theory.

14. Show that if an event A is independent of itself, then $P(A) = 0$ or 1.

15. Show that if A and B are independent and $A \subseteq B$, then either $P(A) = 0$ or $P(B) = 1$.

16. Suppose that 55% of the customers of a shoestore buy black shoes. Find the probability that at least one of the next six customers who purchase a pair of shoes from this store will buy black shoes. Assume that these customers decide independently.

17. Three missiles are fired at a target and hit it independently, with probabilities 0.7, 0.8, and 0.9, respectively. What is the probability that the target is hit?

18. In his book, *Probability* 1, published by *Harcourt Brace and Company*, 1998, Amir Aczel estimates that the probability of life for any one given star in the known universe is 0.00000000000005 independently of life for any other star. Assuming that there are 100 billion galaxies in the universe and each galaxy has 300 billion stars, what is the probability of life on at least one other star in the known universe? How does this probability change if there were only a billion galaxies, each having 10 billion stars?

19. In a tire factory, the quality control inspector examines a randomly chosen sample of 15 tires. When more than one defective tire is found, production is halted, the existing tires are recycled, and production is then resumed. The purpose of this process is to ensure that the defect rate is no higher than 6%. Occasionally, by a stroke of bad luck, even if the defect rate is no higher than 6%, more than one defective tire occurs among the 15 chosen. Determine the fraction of the time that this happens.

20. In a community of M men and W women, m men and w women smoke ($m \leq M$, $w \leq W$). If a person is selected at random and A and B are the events that the person is a man and smokes, respectively, under what conditions are A and B independent?

21. Prove that if A, B, and C are independent, then A and $B \cup C$ are independent. Also show that $A - B$ and C are independent.

22. A fair die is rolled six times. If on the ith roll, $1 \leq i \leq 6$, the outcome is i, we say that a *match* has occurred. What is the probability that at least one match occurs?

23. There are n cards in a box numbered 1 through n. We draw cards successively and at random with replacement. If the ith draw is the card numbered i, we say that a match has occurred. (a) What is the probability of at least one match in n trials? (b) What happens if n increases without bound?

24. In a certain county, 15% of patients suffering heart attacks are younger than 40, 20% are between 40 and 50, 30% are between 50 and 60, and 35% are above 60. On a certain day, 10 unrelated patients suffering heart attacks are transferred to a county hospital. If among them there is at least one patient younger than 40, what is the probability that there are two or more patients younger than 40?

25. If the events A and B are independent and the events B and C are independent, is it true that the events A and C are also independent? Why or why not?

26. From the set of all families with three children *a family* is selected at random. Let A be the event that "the family has children of both sexes" and B be the event that "there is at most one girl in the family." Are A and B independent? Answer the same question for families with two children and families with four children. Assume that for any family size all sex distributions have equal probabilities.

27. An event occurs at least once in four independent trials with probability 0.59. What is the probability of its occurrence in one trial?

28. Let $\{A_1, A_2, \dots , A_n\}$ be an independent set of events and $P(A_i) = p_i$, $1 \le i \le n$.

   (a) What is the probability that at least one of the events $A_1, A_2, \dots , A_n$ occurs?

   (b) What is the probability that none of the events $A_1, A_2, \dots , A_n$ occurs?

29. Figure 3.10 shows an electric circuit in which each of the switches located at 1, 2, 3, 4, 5, and 6 is independently closed or open with probabilities p and $1 - p$, respectively. If a signal is fed to the input, what is the probability that it is transmitted to the output?

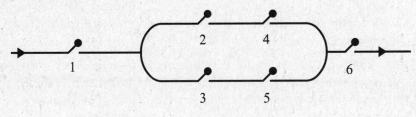

Figure 3.10   Electric circuit of Exercise 29.

B

30. An urn contains two red and four white balls. Balls are drawn from the urn successively, at random and with replacement. What is the probability that exactly three whites occur in the first five trials?

31. A fair coin is tossed n times. Show that the events "at least two heads" and "one or two tails" are independent if $n = 3$ but dependent if $n = 4$.

32. A fair coin is flipped indefinitely. What is the probability of (a) at least one head in the first n flips; (b) exactly k heads in the first n flips; (c) getting heads in all of the flips indefinitely?

33. If two fair dice are tossed six times, what is the probability that the sixth sum obtained is not a repetition?

34. Suppose that an airplane passenger whose itinerary requires a change of airplanes in Ankara, Turkey, has a 4% chance of losing each piece of his or her luggage independently. Suppose that the probability of losing each piece of luggage in this way is 5% at Da Vinci airport in Rome, 5% at Kennedy airport in New York, and 4% at O'Hare airport in Chicago. Dr. May travels from Bombay to San Francisco with one piece of luggage in the baggage compartment. He changes airplanes in Ankara, Rome, New York, and Chicago.

    (a) What is the probability that his luggage does not reach his destination with him?

    (b) If the luggage does not reach his destination with him, what is the probability that it was lost at Da Vinci airport in Rome?

35. An experiment consists of first tossing a fair coin and then drawing a card randomly from an ordinary deck of 52 cards with replacement. If we perform this experiment successively, what is the probability of obtaining heads on the coin before an ace from the cards?
Hint: See Example 3.31.

36. Let S be the sample space of a repeatable experiment. Let A and B be mutually exclusive events of S. Prove that, in independent trials of this experiment, the event A occurs before the event B with probability $P(A)/[P(A) + P(B)]$.
Hint: See Example 3.31; this exercise can be done the same way.

37. In the experiment of rolling two fair dice successively, what is the probability that a sum of 5 appears before a sum of 7?
Hint: See Exercise 36.

38. An urn contains nine red and one blue balls. A second urn contains one red and five blue balls. One ball is removed from each urn at random and without replacement, and all of the remaining balls are put into a third urn. If we draw two balls randomly from the third urn, what is the probability that one of them is red and the other one is blue?

39. Suppose that $n \geq 2$ missiles are fired at a target and hit it independently. If the probability that the ith missile hits it is p_i, $i = 1, 2, \ldots, n$, find the probability that at least two missiles will hit the target.

40. From a population of people with unrelated birthdays, 30 people are selected at random. What is the probability that exactly four people of this group have the same birthday and that all the others have different birthdays (exactly 27 birthdays altogether)? Assume that the birthrates are constant throughout the first 365 days of a year but that on the 366th day it is one-fourth that of the other days.

41. Figure 3.11 shows an electric circuit in which each of the switches located at 1, 2, 3, 4, and 5 is independently closed or open with probabilities p and $1 - p$, respectively. If a signal is fed to the input, what is the probability that it is transmitted to the output?

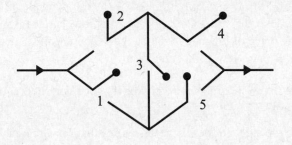

Figure 3.11   Electric circuit of Exercise 41.

42. In a contest, contestants A, B, and C are each asked, in turn, a general scientific question. If a contestant gives a wrong answer to a question, he drops out of the game. The remaining two will continue to compete until one of them drops out. The last person remaining is the winner. Suppose that a contestant knows the answer to a question independently of the other contestants, with probability p. Let CAB represent the event that C drops out first, A next, and B wins, with similar representations for other cases. Calculate and compare the probabilities of ABC, BCA, and CAB.

43. Hemophilia is a hereditary disease. If a mother has it, then with probability 1/2, any of her sons independently will inherit it. Otherwise, none of the sons becomes hemophilic. Julie is the mother of two sons, and from her family's medical history it is known that, with the probability 1/4, she is hemophilic. What is the probability that (a) her first son is hemophilic; (b) her second son is hemophilic; (c) none of her sons are hemophilic?
Hint: Let H, H_1, and H_2 denote the events that the mother, the first son, and the second son are hemophilic, respectively. Note that H_1 and H_2 are **conditionally independent** given H. That is, if we are given that the mother is hemophilic, knowledge about one son being hemophilic does not change the chance of the other son being hemophilic. However, H_1 and H_2 are not independent. This is because if we know that one son is hemophilic, the mother is hemophilic and therefore with probability 1/2 the other son is also hemophilic.

44. **(Laplace's Law of Succession)**   Suppose that $n + 1$ urns are numbered 0 through n, and the ith urn contains i red and $n - i$ white balls, $0 \leq i \leq n$. An urn is selected at random, and then the balls it contains are removed one by one, at random, and with replacement. If the first m balls are all red, what is the probability that the $(m + 1)$st ball is also red?

Hint: Let U_i be the event that the ith urn is selected, R_m the event that the first m balls drawn are all red, and R the event that the $(m + 1)$st ball drawn is red. Note that R and R_m are conditionally independent given U_i; that is, given that the ith urn is selected, R and R_m are independent. Hence

$$P(R \mid R_m U_i) = P(R \mid U_i) = \frac{i}{n}.$$

To find $P(R \mid R_m)$, use

$$P(R \mid R_m) = \sum_{i=0}^{n} P(R \mid R_m U_i) P(U_i \mid R_m),$$

where $P(U_i \mid R_m)$ is obtained from Bayes' theorem. The final answer is

$$P(R \mid R_m) = \frac{\sum_{i=0}^{n} \left(\frac{i}{n}\right)^{m+1}}{\sum_{k=0}^{n} \left(\frac{k}{n}\right)^{m}}.$$

Note that if we consider the function $f(x) = x^{m+1}$ on the interval $[0, 1]$, the right Riemann sum of this function on the partition $0 = x_0 < x_1 < x_2 < \cdots < x_n = 1$, $x_i = i/n$ is

$$\sum_{i=1}^{n} f(x_i)(x_i - x_{i-1}) = \sum_{i=1}^{n} x_i^{m+1} \frac{1}{n} = \frac{1}{n} \sum_{i=0}^{n} \left(\frac{i}{n}\right)^{m+1},$$

Therefore, for large n, $(1/n) \sum_{i=0}^{n} (i/n)^{m+1}$ is approximately equal to $\int_0^1 x^{m+1} dx = 1/(m+2)$. Similarly, $(1/n) \sum_{k=0}^{n} (k/n)^m$ is approximately $\int_0^1 x^m dx = 1/(m+1)$. Hence

$$P(R \mid R_m) = \frac{\frac{1}{n} \sum_{i=0}^{n} \left(\frac{i}{n}\right)^{m+1}}{\frac{1}{n} \sum_{k=0}^{n} \left(\frac{k}{n}\right)^{m}} \approx \frac{1/(m+2)}{1/(m+1)} = \frac{m+1}{m+2}.$$

Laplace designed and solved this problem for philosophical reasons. He used it to argue that the sun will rise tomorrow with probability $(m+1)/(m+2)$ if we know that it has risen in the m preceding days. Therefore, according to this, as time passes by, the probability that the sun rises again becomes higher, and the event of the sun rising becomes more and more certain day after day. It is clear that, to argue this way, we should accept the problematic assumption that the phenomenon of the sun rising is "random," and the model of this problem is identical to its random behavior.

■

In the following section, no prior knowledge of genetics is assumed. If it is skipped, then all exercises and examples in this and future chapters marked "(Genetics)" should be skipped as well.

★ 3.6   APPLICATIONS OF PROBABILITY TO GENETICS

In 1854, the Austrian monk Greger Johann Mendel (1822–1884) proposed a mechanism for the inheritance of physical traits. Since then probability has played a central role in the understanding of certain aspects of genetics. In this section, we will examine some elementary applications of probability in the foundations of modern genetics.

Biologists estimate that the human body contains approximately 60 trillion cells. The width of a cell is approximately one-hundredth of a millimeter. Within each cell there is a large oval structure called the **nucleus**. In the nucleus, the control center of the cell, are threadlike bodies called **chromosomes**. Chromosomes contain genes. In the human, there are about 100,000 genes. A **gene** is the fundamental physical unit of heredity. Genes code for characteristics of the individual and are passed on from one generation to another. For example, a person with green eyes has genes that code for green eyes, and a person with a bent little finger has genes that code for that.

Human beings have 46 chromosomes (23 pairs) in each cell. This number is constant for a particular species. For example, chickens have 36 and peas have 14 chromosomes in each of their cells, respectively. In humans, the reproductive cells are the egg and sperm cells. Except for the reproductive cells, in each human cell, 23 chromosomes come from the father and 23 from the mother. Each member of a pair, is the same length and has the same appearance. It is important to know that the genes that control a certain characteristic occupy the same location on the paired chromosomes. Of all the genetic material on the chromosomes, there are between 20,000 and 50,000 genes that code for all characteristics of the individual. For nonreproductive cells, every cell carries every single gene. Therefore, all of the genetic material of individuals can be found in each and every cell. In each nonreproductive cell, there is one pair of chromosomes called the *sex chromosome,* which determines a person's gender. Unlike all other pairs of chromosomes, in males, the sex chromosomes do not look alike. One is called the X chromosome and one is called the Y. They differ in shape and size. Females have two X chromosomes. Furthermore, in males, the X chromosome has no corresponding genetic information on the Y chromosome. The reproductive cells contain *only* 23 unpaired chromosomes. This is so that during fertilization each parent can contribute 23 chromosomes to the zygote or fertilized egg. We will discuss sex chromosomes and reproductive cells later. For now, we concentrate on nonreproductive cells and nonsex chromosomes.

Genes appear in alternate forms. For example, for seed color of the garden pea, the gene appears in one form for yellow and another form for green. Similarly, the gene for the stem height of the garden pea has one form for "tall" and another form for "short." Alternate forms of genes are called **alleles**. Therefore, the gene for the seed color of the garden pea has two alleles, the allele for yellow and the allele for green. The gene for the stem height of a garden pea has an allele for tall and an allele for short.

During fertilization, for each trait, each parent contributes only one of a pair of alleles. That is how the paired conditions of chromosomes and genes, discussed previously, are transferred from one generation to another. The genes contributed by the parents may or may not be the same allele. For the seed color of the garden pea, during reproduction, both parents might contribute their yellow producing alleles, both might contribute their green producing alleles, or one might contribute its green and the other its yellow producing allele. In the former case, after reproduction, it does not matter which parent contributed the yellow producing allele and which the green producing allele. The result is the same: The offspring has one yellow and one green producing allele. It often happens that when the genes contributed by the parents are different alleles, then, in the offspring, only one allele is fully expressed and the other one is masked. That is, one allele is **dominant** over the other. In such cases, one can observe the full effect of the dominant allele, but no effect of other alleles can be detected. The allele that is masked is called **recessive**.

As Mendel did, suppose that a biologist takes a number of pea plants that have inherited one dominant allele for purple flower color from father and one dominant allele for purple flower color from mother. Suppose that the biologist crosses each of these plants with a pea plant that has inherited two recessive alleles for white flower color, one from each parent. Then she will observe that the resulting first generation offspring all will have purple flowers and, as a result of fertilization, the colors will not change to lighter purple. This is an observation that proves that inheritance is not *necessarily* a blending phenomenon. At this stage, one might think that the absence of white flowers among the first generation offspring indicates that its heritable factor is annihilated. However, if the biologist continues her experiment and crosses the new generation of the plants between each other, she will observe that approximately 1/4th of their offspring will have white flowers. This phenomenon demonstrates the presence of the recessive white flower allele in the first generation of the plants.

An observable characteristic such as physical appearance of an organism is called its **phenotype**. The organism's genetic constitution (makeup) is its **genotype**. To discuss applications of probability to genetics, we use uppercase letters to represent dominant alleles and lowercase to represent recessive alleles. For the pea plants, let A represent the dominant allele for purple flowers and a represent the recessive allele for white flowers. Then each of the pea plants in the experiment above belongs to one of the three genotypes AA, Aa, aa. Recall that Aa and aA are identical. They both mean that the pea plant has inherited a dominant purple flower allele from one parent and a recessive white flower allele from another parent. AA and aa are defined similarly.

Inheritance patterns usually involve exactly two alleles per gene. However, there are genes that have multiple alleles. It is important to know that even when, in a population, multiple alleles of a gene exist, still, for that gene, each individual carries at most two different alleles. As an example, for blood groups in humans, there are three alleles: A, B, and O. Since each person inherits one of these alleles from each parent, there are six genotypes. Blood groups A and B are dominant to the O. Therefore, people who inherit type A from one parent and type O from another will have type A blood. Similarly, people who inherit type B blood from one parent and type O from another will have type B blood. Obviously, inheriting type A blood from both parents results in type A blood, inheriting type B blood from both parents results in type B blood, and inheriting type O blood from both parents results in type O blood. Interestingly enough, neither blood group A is dominant to blood group B nor vice versa. Individuals who inherit type A blood from one parent and type B blood from another have a blood type in which both A and B alleles are expressed. That blood type is called type AB blood. A and B are said to be **codominant** to each other.

In general, during fertilization, for each characteristic, each parent contributes *randomly* one member of its alleles *independently* of other characteristics and *independently* of the reproduction of other offspring. Therefore, in the pea plant example discussed previously, where A represents the dominant allele for purple flowers and a represents the recessive allele for white flowers, when the biologist crosses a pea plant of genotype AA with a pea plant of genotype aa, the resulting plant is Aa with probability 1. However, when crossing a plant of genotype Aa with another plant of genotype Aa, the resulting plant is AA with probability 1/4, Aa with probability 1/2, and aa with probability 1/4. That is why, in the second generation, approximately 1/4th of the pea plants will have white flowers.

Example 3.34   In guinea pigs, having long hair is a recessive trait. Suppose that an offspring of two short-haired guinea pigs has long hair. What is the probability that at least one of the next three offspring of these two guinea pigs also has long hair?

Solution: Let A represent the dominant allele for short hair and a represent the recessive allele for long hair. Since the offspring with long hair is aa, both parents must be Aa. Hence any other offspring is Aa with probability 1/2, is AA with probability 1/4, and is aa with probability 1/4. Therefore, the probability is 3/4 that any offspring is short-haired. The probability that the next three offspring are all short-haired is $(3/4)^3$. The probability that at least one of them is long-haired is $1 - (3/4)^3 = 37/64 = 0.578$.   ♦

Example 3.35   Consider the three alleles A, B, and O for blood types. Suppose that a man of genotype AB marries a woman of genotype AO. If they have six children, what is the probability that none will have type AB blood?

Solution: The genotype of a child of these parents will be one of the following with equal probabilities: AA, AB, AO, BO. Since the alleles A and B are codominant to each other and dominant to O, a child of these parents will have blood type A with probability 1/2, blood type AB with probability 1/4, and blood type B with probability 1/4. Therefore, by independence, the probability that none of the six children will have blood type AB is $(3/4)^6 \approx 0.178$.   ◆

Example 3.36   Consider the three alleles A, B, and O for the human blood types. Suppose that, in a certain population, the frequencies of these alleles are 0.45, 0.20, and 0.35, respectively. Kim and John, two members of the population, are married and have a son named Dan. Kim and Dan both have blood types AB. John's blood type is B. What is the probability that John's genotype is BB?

Solution: Clearly, John's genotype is either BB or BO. Let E be the event that it is BB. Then E^c is the event that John's genotype is BO. Let F be the event that Dan's genotype is AB. By Bayes' formula, the probability we are interested in is

$$P(E \mid F) = \frac{P(F \mid E)P(E)}{P(F \mid E)P(E) + P(F \mid E^c)P(E^c)}$$

$$= \frac{(1/2)(0.20)^2}{(1/2)(0.20)^2 + (1/4)(0.20)(0.35)} = 0.53.$$

Note that $P(E) = (0.20)^2$ and $P(E^c) = (0.20)(0.35)$ since for the two B alleles in John's blood type, each is coming from one parent *independently* of the other parent.
◆

Certain characteristics such as human height or human skin color result from **polygenic inheritance**. Unlike the color of the pea plant flowers, such characteristics are not controlled by a single gene. For example, it is the additive effect of several genes that determines the level of darkness of human skin. Apart from the environmental factors such as nutrition and light, in humans, variation in darkness level of the skin is continuous mainly because, as a heritable character, skin color is controlled by several genes. In humans, the same is true for the height and most other quantitative inheritance. Suppose that human skin color is controlled by four genes independently inherited. For those genes, let A, B, C, and D represent the dominant alleles for dark skin. For the respective genes, let a, b, c, and d represent the recessive alleles for light skin. Then individuals who are $AABBCCDD$ are the darkest members of the population, and individuals who are $aabbccdd$ are the lightest. Moreover, all the dominant alleles contribute to the skin color. For example, an individual who is $AAbbccdd$ is darker than an individual who is $Aabbccdd$, and those who are $Aabbccdd$ are darker than those with genotypes $aabbccdd$. Depending on the degree to which each of the four genes contributes to the darkness of the skin, there is a very wide range of skin colors among the members of the population. As an example, assuming that each gene contributes the same amount of darkness to the skin, people with the following genotypes are ordered from lightest to darkest:

$$Aabbccdd, \quad aaBbCcdd, \quad AABbccDd, \quad AaBbCcDD,$$

whereas people with the following genotypes are indistinguishable in skin color phenotype:

$$AAbbCcDd, \quad AaBbCcDd, \quad aaBbCcDD, \quad AABBccdd.$$

As another example, suppose that on an island, the zeroth generation is the entire population, which consists of a male population all members having the lightest skin and a female population all members having the darkest skin. Suppose that a member belongs to the second generation if all of his or her grandparents belong to the zeroth generation. To find the probability that a member of the second generation has the lightest possible skin, of the four genes that control skin color, consider the gene whose dominant allele for dark skin is denoted by A. In the zeroth generation, all females are AA and all males are aa. Therefore, in the first generation, all members, regardless of gender, are Aa. In the second generation, a member is AA with probability 1/4, Aa with probability 1/2, and aa with probability 1/4. Similarly, the probability is 1/4 that in the second generation a member is bb, the probability is 1/4 that he or she is cc, and the probability is 1/4 that he or she is dd. Since genes are independently inherited, the probability that, in the second generation, a person has the lightest color (i.e., has genotype $aabbccdd$) is $(1/4)^4 = 1/256$.

Hardy-Weinberg Law

In organisms having two sets of chromosomes, called diploid organisms, which we all are, each hereditary character of each individual is carried by a pair of genes. A gene has alternate alleles that usually are dominant A or recessive a, so that the possible pairs of genes are AA, Aa (same as aA), and aa. Under certain conditions, by the Hardy-Weinberg law, to be discussed later, the probabilities of AA, Aa, and Aa in all future generations remain the same as those in the second generation. To show this important fact, we will first discuss what is meant by *natural selection, mutation, random mating*, and *migration*.

Roughly speaking, **natural selection** is the process by which forms of life best adapted to the environment are the most fit. They live longer and are most likely to reproduce and hence perpetuate the survival of their type. Lack of natural selection implies that, in a form of life, all types are equally adapted to the environment. There is no type that survives longer and produces in greater number.

A *heritable* change to the genetic material is called **mutation**. It can happen by radiation, chemicals, or any carcinogen. A **mutagen** is a substance or preparation that can cause mutation.

In certain populations, individuals choose mates on the basis of their own genotypes. For example, among humans, short and tall individuals may marry more often within their own groups. In other situations, individuals might often choose to mate with individuals of different genotypes. In such cases, mating is not random. We say that, in a population, **random mating** occurs if mating between genotypes occur totally at random. Under random mating, between genotypes, mating will occur in proportion to the frequencies of the genotypes. Therefore, the probability of a mating between any two genotypes is the product of the frequencies of those genotypes in the population.

To prevent gene flow between populations of the same species, individuals should be prevented from **migration**. Clearly, immigrants and emigrants can change the balance of the frequencies of the genotypes. Existing alleles may be removed from the population by emigration. New alleles might be added by immigration. In both cases, the frequencies of the remaining alleles might increase or decrease.

Let r be the probability that a randomly selected person in the population is AA. Let s be the probability that he or she is Aa, and let t be the probability that he or she is aa. Then $r + s + t = 1$. Let E be the event that a randomly selected father transmits A to a child. Then

$$
\begin{aligned}
P(E) = {} & P(E \mid \text{the father is } AA)P(\text{the father is } AA) \\
& + P(E \mid \text{the father is } Aa)P(\text{the father is } Aa) \\
& + P(E \mid \text{the father is } aa)P(\text{the father is } aa) \\
= {} & 1 \cdot r + \frac{1}{2} \cdot s + 0 \cdot t = r + \frac{s}{2}.
\end{aligned}
$$

Similarly, the probability that the father transmits a is $t + s/2$, the probability that the mother transmits A is $r + s/2$, and the probability that the mother transmits a is $t + s/2$. Now suppose that we have a very large population in which natural selection does not occur and is free from mutation and migration. Then, under random mating, the contributions of the parents to the genetic heritage of their offspring are independent of each other. Therefore, the probability that an offspring selected at random is AA is $r_1 = (r + s/2)^2$, the probability that he or she is Aa is $s_1 = 2(r + s/2)(t + s/2)$, and the probability that he or she is aa is $t_1 = (t + s/2)^2$. These formulas imply that, r_2, the probability that a randomly selected offspring from the third generation is AA is given by $(r_1 + s_1/2)^2$, the probability that he or she is Aa is $s_2 = 2(r_1 + s_1/2)(t_1 + s_1/2)$, and the probability that he or she is aa is $t_2 = (t_1 + s_1/2)^2$. Some algebra shows that $r_2 = r_1$, $t_2 = t_1$ and $s_2 = s_1$. For example,

$$
\begin{aligned}
r_2 = \left(r_1 + \frac{s_1}{2}\right)^2 &= \left[\left(r + \frac{s}{2}\right)^2 + \left(r + \frac{s}{2}\right)\left(t + \frac{s}{2}\right)\right]^2 \\
&= \left[\left(r + \frac{s}{2}\right)\left(r + \frac{s}{2} + t + \frac{s}{2}\right)\right]^2 \\
&= \left[\left(r + \frac{s}{2}\right)(r + t + s)\right]^2 = \left(r + \frac{s}{2}\right)^2 = r_1.
\end{aligned}
$$

Therefore, the proportions of AA, Aa, and aa in the third generation are the same as in the second generation. Let $p = r + \dfrac{s}{2}$ and $q = t + \dfrac{s}{2}$. Then p is the frequency of allele A and q is the frequency of allele a in the population. We have shown that, under the conditions explained previously, when the population is in equilibrium, the genotype frequencies of AA, Aa, and aa are p^2, $2pq$, and q^2, respectively. Note that $p + q = 1$

implies that

$$p^2 + 2pq + q^2 = (p + q)^2 = 1,$$

as expected. We have established the following important law:

The Hardy-Weinberg Law: *For a diploid organism with alternate dominant allele A and recessive allele a, the possible pairs of genes are AA, Aa, and aa. Under random mating conditions, in a very large population of a diploid organism in which natural selection does not occur and is free from mutation and migration, the proportions of AA, Aa, and aa in all future generations remain the same as those in the second generation. Moreover, if the gene frequency of A is p and the gene frequency of a is q, then, in all future generations, the genotype frequencies of AA, Aa, and aa remain in the proportions p^2, $2pq$, and q^2.*

Example 3.37   Suppose that 64% of the sheep in a certain region have straight horn. If, for the sheep in that region, the allele for curly horn is dominant to the allele for straight horn, under the conditions of Hardy-Weinberg law, find the percentage of sheep who have two different alleles of the gene for horn shape.

Solution: Let the dominant allele for the curly horn be denoted by H and the recessive allele for straight horn be h. Since the genotype of a sheep with straight horn is hh, applying the Hardy-Weinberg's law, we have that $q^2 = 0.64$, or $q = 0.8$. Therefore, $p = 0.2$, and the percentage of sheep who have Hh is $2pq = 2(0.2)(0.8) = 0.32$.   ◆

Sex-Linked Genes

In this subsection, we will discuss sex chromosomes and reproductive cells. Recall that the reproductive cells are the egg and sperm cells, collectively called **gametes**. We know that except for gametes, in each of the human cells, 23 chromosomes come from the father and 23 from the mother. Among the 23 pairs of chromosomes in each nongamete cell, there is one pair of chromosomes, called the **sex chromosomes**, which determines a person's gender. The pair of sex chromosomes in human females are called X **chromosomes**. Similar to other pairs of chromosomes, the pair of female sex chromosomes come one from each parent and look alike. The genes also appear in pairs, one from each parent, and those controlling the same characteristic occupy the same position on the paired X chromosomes inherited from father and mother. Concerning dominance and recessivity, the rules for the paired X chromosomes are the same as those followed by the 22 pairs of nonsex chromosomes of nongametes. The pair of sex chromosomes in human males do not look alike. They differ in shape and size. One of the chromosomes, inherited from the mother, is an X chromosome. The other one, inherited from the father, is said to be a Y **chromosome**. For most genes lying on the X chromosome, there is no corresponding gene on the Y chromosome. The only genes that lie on the Y chromosomes are those that are important in determining the male structures and organs. All genes on the X chromosome are called X-**linked genes**. Similarly, all genes on the Y chromosome are called **Y-linked genes**. For X-linked genes in males, we do not have recessivity and dominance. This is because males have exactly one X chromosome. Therefore, males carry only one allele of each X-linked gene. As a result, for example, if a male is color-blind, since the gene for color blindness is on the X chromosome, he must have inherited it from the mother. Some people bleed excessively when injured. They might have a genetic disease called hemophilia. If a male is a hemophiliac, he must have inherited it from his mother, since the genes for hemophilia lie on the X chromosome. There are many other characteristics such as X-linked baldness where a male inherits the characteristic from his mother.

Like humans, in all mammals, the sex chromosomes determine a mammal's gender. Mammalian males have one X chromosome and one Y chromosome. Mammalian females have two X chromosomes. In such cases, males are represented as XY, females as XX. It is important to know that the XY and XX patterns are not true for all males and females of all species. For example, female birds carry two different chromosomes, whereas male birds carry two identical chromosomes.

We will now discuss gametes. Unlike nonproductive cells, which have 23 pairs of chromosomes, gametes carry 23 single chromosomes. Therefore, no gamete contains any paired chromosomes. A gamete carries every single gene of the individual, but in only one copy. When gametes are formed, with equal probabilities and independent of other genes, only one of the alleles of each gene will end up in the gamete. Clearly, a gamete has only one sex chromosome. In human females, the unique chromosome is necessarily an X chromosome. However, in males, it is either an X chromosome or a Y chromosome with equal probabilities. This follows because, in females, both of the sex chromosomes in nonreproductive cells are X and, in males, one of them is X and the other one is Y. Therefore, an egg contains only one X chromosome, while a sperm cell contains either an X chromosome or a Y chromosome with equal probability. If the sperm cell that fertilizes the egg carries an X chromosome, then the offspring is female. If it carries a Y chromosome, then the offspring is male. That explains why in humans it is equally likely that an offspring is a boy or a girl.

Example 3.38   What is the probability that 12 of the chromosomes in a randomly selected human gamete are paternal and 11 are maternal?

Solution: There is a one-to-one correspondence between the set of all possible combinations of 23 chromosomes of which 12 are paternal (P) and 11 are maternal (M) and the set of all distinguishable sequences of 12 P's and 11 M's. By the generalized counting principle, the number of sequences of P and M of length 23 is 2^{23}. By Theorem 2.4, the number of distinguishable permutations of 23 letters of which 12 are P and 11 are M is $\dfrac{23!}{12!\,11!}$. Therefore, the desired probability is $\left(\dfrac{23!}{12!\,11!}\right)/2^{23} = 0.16.$ ◆

Example 3.39   Hemophilia is a sex-linked disease with normal allele H dominant to the mutant allele h. Kim and John are married and their son, Dan, has hemophilia. If the frequencies of H and h are 0.98 and 0.02, respectively, what is the probability that Kim has hemophilia?

Solution: Dan has inherited all of his sex-linked genes from his mother. Since Dan is hemophiliac, Kim is either Hh or hh. Let E be the event that she is hh. Then E^c is the event that she is Hh. Let F be the event that Dan has hemophilia. By Bayes' formula, the desired probability is

$$P(E \mid F) = \frac{P(F \mid E)P(E)}{P(F \mid E)P(E) + P(F \mid E^c)P(E^c)}$$

$$= \frac{1 \cdot (0.02)(0.02)}{1 \cdot (0.02)(0.02) + (1/2) \cdot \left[2(0.98)(0.02)\right]} = 0.02.$$

Hence the conditional probability that Kim has hemophilia given that her son, Dan, has hemophilia is 0.02. Note that the unconditional probability that Kim has hemophilia is $(0.02)(0.02) = 0.0004$.   ♦

EXERCISES

A

1. Kim has blood type O and John has blood type A. The blood type of their son, Dan, is O. What is the probability that John's genotype is AO?

2. Suppose that a gene has k $(k > 2)$ alleles. For that gene, how many genotypes can an individual have?

3. For the shape of pea seed, the allele for round shape, denoted by R, is dominant to the allele for wrinkled shape, denoted by r. A pea plant with round seed shape is crossed with a pea plant with wrinkled seed shape. If half of the offspring have round seed shape and half have wrinkled seed shape, what were the genotypes of the parents?

4. In humans, the presence of freckles and having free earlobes are independently inherited dominant traits. Kim and John both have free earlobes and both have freckles, but their son, Dan, has attached earlobes and no freckles. What is the probability that Kim and John's next child has free earlobes and no freckles?

5. For *Drosophila* (a kind of fruit fly), B, the gray body, is dominant over b, the black body, and V, the wild-type wing is dominant over v, *vestigial* (a very small wing). A geneticist, T. H. Morgan, when mating *Drosophila* of genotype $BbVv$ with *Drosophila* of genotype $bbvv$, observed that 42% of the offspring were $BbVv$, 41% were $bbvv$, 9% were $Bbvv$, and 8% were $bbVv$. Based on these results,

should we expect that the body color and wild-type genes of *Drosophila* are on different chromosomes (that is, are unlinked and hence independent)? Why or why not?

6. In a population, 1% of the people suffer from hereditary genetic deafness. The hearing allele is dominant and is denoted by D. The recessive allele for deafness is denoted by d. Kim and John, from the population, are married and have a son named Dan. If Kim and Dan are both deaf, what is the probability that John is also deaf?

7. In the United States, among the Caucasian population, cystic fibrosis is a serious genetic disease. For cystic fibrosis, the normal allele is dominant to the mutant allele. Suppose that, in a certain region of the country, 5.29% of the people have cystic fibrosis. Under the Hardy-Weinberg conditions, determine the percentage of people in that region who carry at least one mutant allele of the disease.

8. Hemophilia is a sex-linked disease with normal allele H dominant to the mutant allele h. Kim and John are both phenotypically normal. If the frequencies of H and h are 0.98 and 0.02, respectively, what is the probability that Dan, their son, has hemophilia?

9. Color blindness is a sex-linked hereditary disease with normal allele C dominant to the mutant allele c. Kim and John are married and their son, Dan, is color-blind. If, in the population, the frequencies of C and c are 0.83 and 0.17, respectively, what is the probability that Kim is color-blind?

10. Hemophilia is a sex-linked hereditary disease with normal allele H dominant to the mutant allele h. Kim and John are married and their daughter, Ann, has hemophilia. If John has hemophilia, and the frequencies of H and h, in the population, are 0.98 and 0.02, respectively, what is the probability that Kim does not have hemophilia but she is a carrier for that disease?

B

11. In a certain country, all children with cystic fibrosis die before reaching adulthood. For this lethal disease, the normal allele, denoted by C, is dominant to the mutant allele, denoted by c. In that country, the probability is p that a person is only a carrier; that is, he or she is Cc. If Mr. J, a resident of the country had a brother who died of cystic fibrosis, what is the probability that Mr. J's next child will have cystic fibrosis?

12. Let p and q be positive numbers with $p + q = 1$. For a gene with dominant allele A and recessive allele a, let p^2, $2pq$, and q^2 be the probabilities that a randomly selected person has genotype AA, Aa, and aa, respectively. If a man is of genotype AA, what is the probability that his brother is also of genotype AA?

REVIEW PROBLEMS

1. Two persons arrive at a train station, independently of each other, at random times between 1:00 P.M. and 1:30 P.M. What is the probability that one will arrive between 1:00 P.M. and 1:12 P.M., and the other between 1:17 P.M. and 1:30 P.M.?

2. A polygraph operator detects innocent suspects as being guilty 3% of the time. If during a crime investigation six innocent suspects are examined by the operator, what is the probability that at least one of them is detected as guilty?

3. In statistical surveys where individuals are selected randomly and are asked questions, experience has shown that only 48% of those under 25 years of age, 67% between 25 and 50, and 89% above 50 will respond. A social scientist is about to send a questionnaire to a group of randomly selected people. If 30% of the population are younger than 25 and 17% are older than 50, what percent will answer her questionnaire?

4. Diseases D_1, D_2, and D_3 cause symptom A with probabilities 0.5, 0.7, and 0.8, respectively. If 5% of a population have disease D_1, 2% have disease D_2, and 3.5% have disease D_3, what percent of the population have symptom A? Assume that the only possible causes of symptom A are D_1, D_2, and D_3 and that no one carries more than one of these three diseases.

5. Professor Stern has three cars. The probability that on a given day car 1 is operative is 0.95, that car 2 is operative is 0.97, and that car 3 is operative is 0.85. If Professor Stern's cars operate independently, find the probability that on next Thanksgiving day (a) all three of his cars are operative; (b) at least one of his cars is operative; (c) at most two of his cars are operative; (d) none of his cars is operative.

6. A bus traveling from Baltimore to New York breaks down at a random location. If the bus was seen running at Wilmington, what is the probability that the breakdown occurred after passing through Philadelphia? The distances from New York, Philadelphia, and Wilmington to Baltimore are, respectively, 199, 96, and 67 miles.

7. Roads A, B, and C are the only escape routes from a state prison. Prison records show that, of the prisoners who tried to escape, 30% used road A, 50% used road B, and 20% used road C. These records also show that 80% of those who tried to escape via A, 75% of those who tried to escape via B, and 92% of those who tried to escape via C were captured. What is the probability that a prisoner who succeeded in escaping used road C?

8. From an ordinary deck of 52 cards, 10 cards are drawn at random. If exactly four of them are hearts, what is the probability of at least one spade being among them?

9. A fair die is thrown twice. If the second outcome is 6, what is the probability that the first one is 6 as well?

10. Suppose that 10 dice are thrown and we are told that among them at least one has landed 6. What is the probability that there are two or more sixes?

11. Urns I and II contain three pennies and four dimes, and two pennies and five dimes, respectively. One coin is selected at random from each urn. If exactly one of them is a dime, what is the probability that the coin selected from urn I is the dime?

12. An experiment consists of first tossing an unbiased coin and then rolling a fair die. If we perform this experiment successively, what is the probability of obtaining a heads on the coin before a 1 or 2 on the die?

13. Six fair dice are tossed independently. Find the probability that the number of 1's minus the number of 2's will be 3.

14. Urn I contains 25 white and 20 black balls. Urn II contains 15 white and 10 black balls. An urn is selected at random and one of its balls is drawn randomly and observed to be black and then returned to the same urn. If a second ball is drawn at random from this urn, what is the probability that it is black?

15. An urn contains nine red and one blue balls. A second urn contains one red and five blue balls. One ball is removed from each urn at random and without replacement, and all of the remaining balls are put into a third urn. What is the probability that a ball drawn randomly from the third urn is blue?

16. A fair coin is tossed. If the outcome is heads, a red hat is placed on Lorna's head. If it is tails, a blue hat is placed on her head. Lorna cannot see the hat. She is asked to guess the color of her hat. Is there a strategy that maximizes Lorna's chances of guessing correctly?
    Hint: Suppose that Lorna chooses the color red with probability α and blue with probability $1 - \alpha$. Find the probability that she guesses correctly.

17. A child is lost at Epcot Center in Florida. The father of the child believes that the probability of his being lost in the east wing of the center is 0.75, and in the west wing 0.25. The security department sends three officers to the east wing and two to the west to look for the child. Suppose that an officer who is looking in the correct wing (east or west) finds the child, independently of the other officers, with probability 0.4. Find the probability that the child is found.

18. Solve the following problem, asked of Marilyn Vos Savant in the "Ask Marilyn" column of *Parade Magazine*, August 9, 1992.

    > Three of us couples are going to Lava Hot Springs next weekend. We're staying two nights, and we've rented two studios, because each holds a maximum of only four people. One couple will get their own studio on Friday, a different couple on Saturday, and one couple will be out of

luck. We'll draw straws to see which are the two lucky couples. I told my wife we should just draw once, and the loser would be the couple out of luck both nights. I figure we'll have a two-out-of-three ($66\frac{2}{3}$%) chance of winning one of the nights to ourselves. But she contends that we should draw straws *twice*—first on Friday and then, for the remaining two couples only, on Saturday—reasoning that a one-in-three ($33\frac{1}{3}$%) chance for Friday and a one-in-two (50%) chance for Saturday will give us better odds. Which way should we go?

19.   A student at a certain university will pass the oral Ph.D. qualifying examination if at least two of the three examiners pass her or him. Past experience shows that (a) 15% of the students who take the qualifying exam are not prepared, and (b) each examiner will independently pass 85% of the prepared and 20% of the unprepared students. Kevin took his Ph.D. qualifying exam with Professors Smith, Brown, and Rose. What is the probability that Professor Rose has passed Kevin if we know that neither Professor Brown nor Professor Smith has passed him? Let S, B, and R be the respective events that Professors Smith, Brown, and Rose have passed Kevin. Are these three events independent? Are they conditionally independent given that Kevin is prepared? (See Exercises 43 and 44, Section 3.5.)

20.   Adam and three of his friends are playing bridge. (a) If, holding a certain hand, Adam announces that he has a king, what is the probability that he has at least one more king? (b) If, for some other hand, Adam announces that he has the king of diamonds, what is the probability that he has at least one more king? Compare parts (a) and (b) and explain why the answers are not the same.   ■

Chapter 4

Distribution Functions and Discrete Random Variables

4.1 RANDOM VARIABLES

In real-world problems we are often faced with one or more quantities that do not have fixed values. The values of such quantities depend on random actions, and they usually change from one experiment to another. For example, the number of babies born in a certain hospital each day is not a fixed quantity. It is a complicated function of many random factors that vary from one day to another. So are the following quantities: the arrival time of a bus at a station, the sum of the outcomes of two dice when thrown, the amount of rainfall in Seattle during a given year, the number of earthquakes that occur in California per month, and the weight of grains of wheat grown on a certain plot of land (it varies from one grain to another). In probability, quantities introduced in these diverse examples are called **random variables**. The numerical values of random variables are unknown. They depend on random elements occurring at the time of the experiment and over which we have no control. For example, if in rolling two fair dice, X is the sum, then X can only assume the values 2, 3, 4, ... , 12 with the following probabilities:

$$P(X = 2) = P(\{(1, 1)\}) = 1/36,$$
$$P(X = 3) = P(\{(1, 2), (2, 1)\}) = 2/36,$$
$$P(X = 4) = P(\{(1, 3), (2, 2), (3, 1)\}) = 3/36,$$

and, similarly,

Sum, i	5	6	7	8	9	10	11	12
$P(X = i)$	4/36	5/36	6/36	5/36	4/36	3/36	2/36	1/36

Clearly, $\{2, 3, 4, \ldots, 12\}$ is the set of possible values of X. Since $X \in \{2, 3, 4, \ldots, 12\}$, we should have $\sum_{i=2}^{12} P(X = i) = 1$, which is readily verified. The numerical value of a random variable depends on the outcome of the experiment. In this example, for instance, if the outcome is $(2, 3)$, then X is 5, and if it is $(5, 6)$, then X is 11. X is not defined for points that do not belong to S, the sample space of the experiment. Thus X is a real-valued function on S. However, not all real-valued functions on S are considered to be random variables. For theoretical reasons, it is necessary that the inverse image of an interval in $\mathbf{R}$ be an event of S, which motivates the following definition.

Definition · *Let S be the sample space of an experiment. A real-valued function $X: S \to \mathbf{R}$ is called a **random variable** of the experiment if, for each interval $I \subseteq \mathbf{R}$, $\{s: X(s) \in I\}$ is an event.*

In probability, the set $\{s: X(s) \in I\}$ is often abbreviated as $\{X \in I\}$, or simply as $X \in I$.

Example 4.1   Suppose that three cards are drawn from an ordinary deck of 52 cards, one by one, at random and with replacement. Let X be the number of spades drawn; then X is a random variable. If an outcome of spades is denoted by s, and other outcomes are represented by t, then X is a real-valued function defined on the sample space

$$S = \{(s, s, s), (t, s, s), (s, t, s), (s, s, t), (s, t, t), (t, s, t), (t, t, s), (t, t, t)\},$$

by $X(s, s, s) = 3$, $X(s, t, s) = 2$, $X(s, s, t) = 2$, $X(s, t, t) = 1$, and so on. Now we must determine the values that X assumes and the probabilities that are associated with them. Clearly, X can take the values 0, 1, 2, and 3. The probabilities associated with these values are calculated as follows:

$$P(X = 0) = P\big(\{(t, t, t)\}\big) = \frac{3}{4} \times \frac{3}{4} \times \frac{3}{4} = \frac{27}{64},$$

$$P(X = 1) = P\big(\{(s, t, t), (t, s, t), (t, t, s)\}\big)$$

$$= \left(\frac{1}{4} \times \frac{3}{4} \times \frac{3}{4}\right) + \left(\frac{3}{4} \times \frac{1}{4} \times \frac{3}{4}\right) + \left(\frac{3}{4} \times \frac{3}{4} \times \frac{1}{4}\right) = \frac{27}{64},$$

$$P(X = 2) = P\big(\{(s, s, t), (s, t, s), (t, s, s)\}\big)$$

$$= \left(\frac{1}{4} \times \frac{1}{4} \times \frac{3}{4}\right) + \left(\frac{1}{4} \times \frac{3}{4} \times \frac{1}{4}\right) + \left(\frac{3}{4} \times \frac{1}{4} \times \frac{1}{4}\right) = \frac{9}{64},$$

$$P(X = 3) = P\big(\{(s, s, s)\}\big) = \frac{1}{64}.$$

If the cards are drawn without replacement, the probabilities associated with the values 0, 1, 2, and 3 are

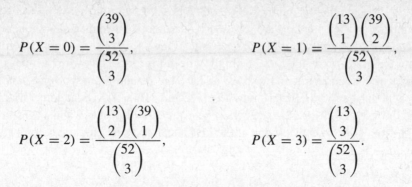

$$P(X = 0) = \frac{\binom{39}{3}}{\binom{52}{3}}, \qquad P(X = 1) = \frac{\binom{13}{1}\binom{39}{2}}{\binom{52}{3}},$$

$$P(X = 2) = \frac{\binom{13}{2}\binom{39}{1}}{\binom{52}{3}}, \qquad P(X = 3) = \frac{\binom{13}{3}}{\binom{52}{3}}.$$

Therefore,

$$P(X = i) = \frac{\binom{13}{i}\binom{39}{3-i}}{\binom{52}{3}}, \qquad i = 0, 1, 2, 3. \quad \blacklozenge$$

Example 4.2   A bus stops at a station every day at some random time between 11:00 A.M. and 11:30 A.M. If X is the actual arrival time of the bus, X is a random variable. It is a function defined on the sample space $S = \{t : 11 < t < 11\frac{1}{2}\}$ by $X(t) = t$. As we know from Section 1.7, $P(X = t) = 0$ for any $t \in S$, and $P\big(X \in (\alpha, \beta)\big) = \dfrac{\beta - \alpha}{11\frac{1}{2} - 11} = 2(\beta - \alpha)$ for any subinterval (α, β) of $(11, 11\frac{1}{2})$. $\blacklozenge$

Example 4.3   In the United States, the number of twin births is approximately 1 in 90. Let X be the number of births in a certain hospital until the first twins are born. X is a random variable. Denote twin births by T and single births by N. Then X is a real-valued function defined on the sample space

$$S = \{T, NT, NNT, NNNT, \dots\} \quad \text{by} \quad X(\underbrace{NNN \cdots N}_{i-1} T) = i.$$

The set of all possible values of X is $\{1, 2, 3, \dots\}$ and

$$P(X = i) = P(\underbrace{NNN \cdots N}_{i-1} T) = \Big(\frac{89}{90}\Big)^{i-1}\Big(\frac{1}{90}\Big). \quad \blacklozenge$$

Example 4.4   In a certain country, the draft-status priorities of eligible men are determined according to their birthdays. Suppose that numbers 1 to 31 are assigned to men with birthdays on January 1 to January 31, numbers 32 to 60 to men with birthdays on February 1 to February 29, numbers 61 to 91 to men with birthdays on March 1 to

March 31, ... , and finally, numbers 336 to 366 to those with birthdays on December 1 to December 31. Then numbers are selected at random, one by one, and without replacement, from 1 to 366 until all of them are chosen. Those with birthdays corresponding to the first number drawn would have the highest draft priority, those with birthdays corresponding to the second number drawn would have the second-highest priority, and so on. Let X be the largest of the first 10 numbers selected. Then X is a random variable that assumes the values 10, 11, 12, ... , 366. The event $X = i$ occurs if the largest number among the first 10 is i, that is, if one of the first 10 numbers is i and the other 9 are from 1 through $i - 1$. Thus

$$P(X = i) = \frac{\binom{i-1}{9}}{\binom{366}{10}}, \qquad i = 10, 11, 12, \ldots, 366.$$

As an application, let us calculate $P(X \geq 336)$, the probability that, among individuals having one of the 10 highest draft priorities, there are some with birthdays in December. We have that

$$P(X \geq 336) = \sum_{i=336}^{366} \frac{\binom{i-1}{9}}{\binom{366}{10}} \approx 0.592. \quad \blacklozenge$$

Remark: On December 1, 1969 the Selective Service headquarters in Washington, D.C. determined the draft priorities of 19-year-old males by using a method very similar to that of Example 4.4.   ♦

We will now explain how, from given random variables, new ones can be formed. Let X and Y be two random variables over the same sample space S; then $X : S \to \mathbf{R}$ and $Y : S \to \mathbf{R}$ are real-valued functions having the same domain. Therefore, we can form the functions $X + Y$; $X - Y$; $aX + bY$, where a and b are constants; XY; and X/Y, where $Y \neq 0$. Since the domain of these real-valued functions is S, they are also random variables defined on S. Hence the sum, the difference, linear combinations, the product, and the quotients (if they exist) of random variables are themselves random variables. Similarly, if $f : \mathbf{R} \to \mathbf{R}$ is an ordinary real-valued function, the composition of f and X, $f \circ X : S \to \mathbf{R}$, is also a random variable. For example, let $f : \mathbf{R} \to \mathbf{R}$ be defined by $f(x) = x^2$; then $f \circ X : S \to \mathbf{R}$ is X^2. Hence X^2 is a random variable as well. Similarly, functions such as $\sin X$, $\cos X^2$, e^X, and $X^3 - 2X$ are random variables. So are the following functions: $X^2 + Y^2$; $(X^2 + Y^2)/(2Y + 1)$, $Y \neq -1/2$; $\sin X + \cos Y$; $\sqrt{X^2 + Y^2}$; and so on. Functions of random variables appear naturally in probability problems. As an example, suppose that we choose a point at random from the unit disk in the plane. If X and Y are the x and the y coordinates of the point, X and Y are random variables. The distance of (X, Y) from the origin is $\sqrt{X^2 + Y^2}$, a random variable that is a function of both X and Y.

Example 4.5   The diameter of a flat metal disk manufactured by a factory is a random number between 4 and 4.5. What is the probability that the area of such a flat disk chosen at random is at least 4.41π?

Solution: Let D be the diameter of the metal disk selected at random. D is a random variable, and the area of the metal disk, $\pi (D/2)^2$, which is a function of D, is also a random variable. We are interested in the probability of the event $\pi D^2/4 > 4.41\pi$, which is

$$P\left(\frac{\pi D^2}{4} > 4.41\pi\right) = P(D^2 > 17.64) = P(D > 4.2).$$

Now to calculate $P(D > 4.2)$, note that, since the length of D is a random number in the interval $(4, 4.5)$, the probability that it falls into the subinterval $(4.2, 4.5)$ is $(4.5 - 4.2)/(4.5 - 4) = 3/5$. Hence $P(\pi D^2/4 > 4.41\pi) = 3/5$. ♦

Example 4.6   A random number is selected from the interval $(0, \pi/2)$. What is the probability that its sine is greater than its cosine?

Solution: Let the number selected be X; then X is a random variable and therefore $\sin X$ and $\cos X$, which are functions of X, are also random variables. We are interested in the probability of the event $\sin X > \cos X$:

$$P(\sin X > \cos X) = P(\tan X > 1) = P\left(X > \frac{\pi}{4}\right) = \frac{\dfrac{\pi}{2} - \dfrac{\pi}{4}}{\dfrac{\pi}{2} - 0} = \frac{1}{2},$$

where the first equality holds since in the interval $(0, \pi/2)$, $\cos X > 0$, and the second equality holds since, in this interval, $\tan X$ is strictly increasing. ♦

4.2   DISTRIBUTION FUNCTIONS

Random variables are often used for the calculation of the probabilities of events. For example, in the experiment of throwing two dice, if we are interested in a sum of at least 8, we define X to be the sum and calculate $P(X > 8)$. Other examples are the following:

1.   If a bus arrives at a random time between 10:00 A.M. and 10:30 A.M. at a station, and X is the arrival time, then $X < 10\frac{1}{6}$ is the event that the bus arrives before 10:10 A.M.

2.   If X is the price of gold per troy ounce on a random day, then $X \leq 400$ is the event that the price of gold remains at or below \$400 per troy ounce.

3. If X is the number of votes that the next Democratic presidential candidate will get, then $X \geq 5 \times 10^7$ is the event that he or she will get at least 50 million votes.

4. If X is the number of heads in 100 tosses of a coin, then $40 < X \leq 60$ is the event that the number of heads is at least 41 and at most 60.

Usually, when dealing with a random variable X, for constants a and b ($b < a$), computation of one or several of the probabilities $P(X = a)$, $P(X < a)$, $P(X \leq a)$, $P(X > b)$, $P(X \geq b)$, $P(b \leq X \leq a)$, $P(b < X \leq a)$, $P(b \leq X < a)$, and $P(b < X < a)$ is our ultimate goal. For this reason we calculate $P(X \leq t)$ for all $t \in (-\infty, +\infty)$. As we will show shortly, if $P(X \leq t)$ is known for all $t \in \mathbf{R}$, then for any a and b, all of the probabilities that are mentioned above can be calculated. In fact, since the real-valued function $P(X \leq t)$ characterizes X, it tells us almost everything about X. This function is called the **distribution function** of X.

Definition   *If X is a random variable, then the function F defined on $(-\infty, +\infty)$ by $F(t) = P(X \leq t)$ is called the **distribution function** of X.*

Since F "accumulates" all of the probabilities of the values of X up to and including t, sometimes it is called the **cumulative distribution function of X**. The most important properties of the distribution functions are as follows:

1. *F is nondecreasing*; *that is, if $t < u$, then $F(t) \leq F(u)$.* To see this, note that the occurrence of the event $\{X \leq t\}$ implies the occurrence of the event $\{X \leq u\}$. Thus $\{X \leq t\} \subseteq \{X \leq u\}$ and hence $P(X \leq t) \leq P(X \leq u)$. That is, $F(t) \leq F(u)$.

2. $\lim_{t \to \infty} F(t) = 1$. To prove this, it suffices to show that for any increasing sequence $\{t_n\}$ of real numbers that converges to ∞, $\lim_{n \to \infty} F(t_n) = 1$. This follows from the continuity property of the probability function (see Theorem 1.8). The events $\{X \leq t_n\}$ form an increasing sequence that converges to the event $\bigcup_{n=1}^{\infty}\{X \leq t_n\} = \{X < \infty\}$; that is, $\lim_{n \to \infty}\{X \leq t_n\} = \{X < \infty\}$. Hence

$$\lim_{n \to \infty} P(X \leq t_n) = P\left(\bigcup_{n=1}^{\infty}\{X \leq t_n\}\right) = P(X < \infty) = 1,$$

which means that

$$\lim_{n \to \infty} F(t) = 1.$$

3. $\lim_{t \to -\infty} F(t) = 0$. The proof of this is similar to the proof that $\lim_{t \to \infty} F(t) = 1$.

4. *F is right continuous.* *That is, for every $t \in \mathbf{R}$, $F(t+) = F(t)$. This means that if t_n is a decreasing sequence of real numbers converging to t, then*

$$\lim_{n \to \infty} F(t_n) = F(t).$$

To prove this, note that since t_n decreases to t, the events $\{X \leq t_n\}$ form a decreasing sequence that converges to the event $\bigcap_{n=1}^{\infty}\{X \leq t_n\} = \{X \leq t\}$. Thus, by the continuity property of the probability function,

$$\lim_{n \to \infty} P(X \leq t_n) = P\left(\bigcap_{n=1}^{\infty}\{X \leq t_n\}\right) = P(X \leq t),$$

which means that

$$\lim_{n \to \infty} F(t_n) = F(t).$$

As mentioned previously, by means of F, the distribution function of a random variable X, a wide range of probabilistic questions concerning X can be answered. Here are some examples.

1. To calculate $P(X > a)$, note that $P(X > a) = 1 - P(X \leq a)$, thus

$$P(X > a) = 1 - F(a).$$

2. To calculate $P(a < X \leq b)$, $b > a$, note that $\{a < X \leq b\} = \{X \leq b\} - \{X \leq a\}$ and $\{X \leq a\} \subseteq \{X \leq b\}$. Hence, by Theorem 1.5,

$$P(a < X \leq b) = P(X \leq b) - P(X \leq a) = F(b) - F(a).$$

3. To calculate $P(X < a)$, note that the sequence of the events $\{X \leq a - 1/n\}$ is an increasing sequence that converges to $\bigcup_{n=1}^{\infty}\{X \leq a - 1/n\} = \{X < a\}$. Therefore, by the continuity property of the probability function (Theorem 1.8),

$$\lim_{n \to \infty} P\left(X \leq a - \frac{1}{n}\right) = P\left(\bigcup_{n=1}^{\infty}\left\{X \leq a - \frac{1}{n}\right\}\right) = P(X < a),$$

which means that

$$P(X < a) = \lim_{n \to \infty} F\left(a - \frac{1}{n}\right).$$

Hence $P(X < a)$ is the left-hand limit of the function F as $x \to a$; that is,

$$P(X < a) = F(a-).$$

4. To calculate $P(X \geq a)$, note that $P(X \geq a) = 1 - P(X < a)$. Thus

$$P(X \geq a) = 1 - F(a-).$$

5. Since $\{X = a\} = \{X \leq a\} - \{X < a\}$ and $\{X < a\} \subseteq \{X \leq a\}$, we can write

$$P(X = a) = P(X \leq a) - P(X < a) = F(a) - F(a-).$$

Note that since F is right continuous, $F(a)$ is the right-hand limit of F. This implies the following important fact:

> Let F be the distribution function of a random variable X; $P(X = a)$ is the difference between the right- and left-hand limits of F at a. If the function F is continuous at a, these limits are the same and equal to $F(a)$. Hence $P(X = a) = 0$. Otherwise, F has a *jump* at a, and the magnitude of the jump, $F(a) - F(a-)$, is the probability that $X = a$.

As in cases 1 to 5, we can establish similar cases to obtain the following table.

Event concerning X	Probability of the event in terms of F	Event concerning X	Probability of the event in terms of F
$X \leq a$	$F(a)$	$a < X \leq b$	$F(b) - F(a)$
$X > a$	$1 - F(a)$	$a < X < b$	$F(b-) - F(a)$
$X < a$	$F(a-)$	$a \leq X \leq b$	$F(b) - F(a-)$
$X \geq a$	$1 - F(a-)$	$a \leq X < b$	$F(b-) - F(a-)$
$X = a$	$F(a) - F(a-)$		

Example 4.7   The distribution function of a random variable X is given by

$$F(x) = \begin{cases} 0 & x < 0 \\ x/4 & 0 \leq x < 1 \\ 1/2 & 1 \leq x < 2 \\ \frac{1}{12}x + \frac{1}{2} & 2 \leq x < 3 \\ 1 & x \geq 3, \end{cases}$$

where the graph of F is shown in Figure 4.1. Compute the following quantities:

(a) $P(X < 2)$;   (b) $P(X = 2)$;   (c) $P(1 \leq X < 3)$;
(d) $P(X > 3/2)$;   (e) $P(X = 5/2)$;   (f) $P(2 < X \leq 7)$.

Solution:

(a) $P(X < 2) = F(2-) = 1/2$.

(b) $P(X = 2) = F(2) - F(2-) = (2/12 + 1/2) - 1/2 = 1/6$.

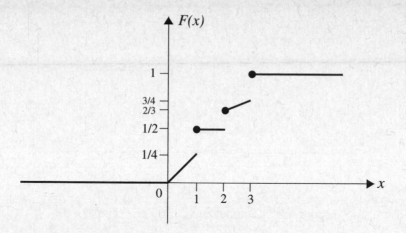

Figure 4.1   Distribution function of Example 4.7.

(c) $P(1 \leq X < 3) = P(X < 3) - P(X < 1) = F(3-) - F(1-) =$
$(3/12 + 1/2) - 1/4 = 1/2.$

(d) $P(X > 3/2) = 1 - F(3/2) = 1 - 1/2 = 1/2.$

(e) $P(X = 5/2) = 0$ since F is continuous at 5/2 and has no jumps.

(f) $P(2 < X \leq 7) = F(7) - F(2) = 1 - (2/12 + 1/2) = 1/3.$ ◆

Example 4.8   For the experiment of flipping a fair coin twice, let X be the number of
tails and calculate $F(t)$, the distribution function of X, and then sketch its graph.

Solution: Since X assumes only the values 0, 1, and 2, we have $F(t) = P(X \leq t) = 0$,
if $t < 0$. If $0 \leq t < 1$, then

$$F(t) = P(X \leq t) = P(X = 0) = P(\{HH\}) = 1/4.$$

If $1 \leq t < 2$, then

$$F(t) = P(X \leq t) = P(X = 0 \text{ or } X = 1) = P(\{HH, HT, TH\}) = 3/4,$$

and if $t \geq 2$, then $P(X \leq t) = 1$. Hence

$$F(t) = \begin{cases} 0 & t < 0 \\ 1/4 & 0 \leq t < 1 \\ 3/4 & 1 \leq t < 2 \\ 1 & t \geq 2. \end{cases}$$

Figure 4.2 shows the graph of F. ◆

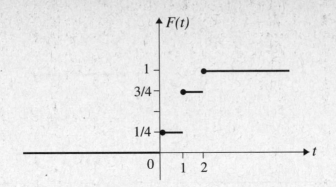

Figure 4.2   Distribution function of Example 4.8.

Example 4.9   Suppose that a bus arrives at a station every day between 10:00 A.M. and 10:30 A.M., at random. Let X be the arrival time; find the distribution function of X and sketch its graph.

Solution: The bus arrives at the station at random, between 10 and $10\frac{1}{2}$, so if $t \leq 10$, $F(t) = P(X \leq t) = 0$. Now if $t \in \left(10, 10\frac{1}{2}\right)$, then

$$F(t) = P(X \leq t) = \frac{t - 10}{10\frac{1}{2} - 10} = 2(t - 10),$$

and if $t \geq 10\frac{1}{2}$, then $F(t) = P(X \leq t) = 1$. Thus

$$F(t) = \begin{cases} 0 & t < 10 \\ 2(t - 10) & 10 \leq t < 10\frac{1}{2} \\ 1 & t \geq 10\frac{1}{2}. \end{cases}$$

The graph of F is shown in Figure 4.3.   ◆

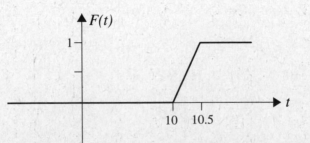

Figure 4.3   Distribution function of Example 4.9.

Example 4.10   The sales of a convenience store on a randomly selected day are X thousand dollars, where X is a random variable with a distribution function of the following form:

$$F(t) = \begin{cases} 0 & t < 0 \\ (1/2)\,t^2 & 0 \leq t < 1 \\ k(4t - t^2) & 1 \leq t < 2 \\ 1 & t \geq 2. \end{cases}$$

Suppose that this convenience store's total sales on any given day are less than $2000.

(a)   Find the value of k.

(b)   Let A and B be the events that tomorrow the store's total sales are between 500 and 1500 dollars, and over 1000 dollars, respectively. Find $P(A)$ and $P(B)$.

(c)   Are A and B independent events?

Solution:

(a)   Since $X < 2$, we have that $P(X < 2) = 1$, so $F(2-) = 1$. This gives $k(8-4) = 1$, so $k = 1/4$.

(b) $$P(A) = P\left(\frac{1}{2} \leq X \leq \frac{3}{2}\right) = F\left(\frac{3}{2}\right) - F\left(\frac{1}{2} - \right)$$

$$= F\left(\frac{3}{2}\right) - F\left(\frac{1}{2}\right) = \frac{15}{16} - \frac{1}{8} = \frac{13}{16},$$

$$P(B) = P(X > 1) = 1 - F(1) = 1 - \frac{3}{4} = \frac{1}{4}.$$

(c) $P(AB) = P\left(1 < X \leq \frac{3}{2}\right) = F\left(\frac{3}{2}\right) - F(1) = \frac{15}{16} - \frac{3}{4} = \frac{3}{16}$. Since $P(AB) \neq P(A)P(B)$, A and B are not independent.   ◆

Remark 4.1   Suppose that F is a right-continuous, nondecreasing function on $(-\infty, \infty)$ that satisfies $\lim_{t \to \infty} F(t) = 1$ and $\lim_{t \to -\infty} F(t) = 0$. It can be shown that there exists a sample space S with a probability function and a random variable X over S such that the distribution function of X is F. Therefore, a function is a distribution function if it satisfies the conditions specified in this remark.   ◆

EXERCISES

A

1. Two fair dice are rolled and the absolute value of the difference of the outcomes is denoted by X. What are the possible values of X, and the probabilities associated with them?

2. From an urn that contains five red, five white, and five blue chips, we draw two chips at random. For each blue chip we win \$1, for each white chip we win \$2, but for each red chip we lose \$3. If X represents the amount that we either win or we lose, what are the possible values of X and probabilities associated with them?

3. In a society of population N, the probability is p that a person has a certain rare disease independently of others. Let X be the number of people who should be tested until a person with the disease is found, $X = 0$ if no one with the disease is found. What are the possible values of X? Determine the probabilities associated with these values.

4. The side measurement of a plastic die, manufactured by factory A, is a random number between 1 and $1\frac{1}{4}$ centimeters. What is the probability that the volume of a randomly selected die manufactured by this company is greater than 1.424? Assume that the die will always be a cube.

5. F, the distribution function of a random variable X, is given by

$$F(t) = \begin{cases} 0 & t < -1 \\ (1/4)t + 1/4 & -1 \le t < 0 \\ 1/2 & 0 \le t < 1 \\ (1/12)t + 7/12 & 1 \le t < 2 \\ 1 & t \ge 2. \end{cases}$$

   (a) Sketch the graph of F.

   (b) Calculate the following quantities: $P(X < 1)$, $P(X = 1)$, $P(1 \le X < 2)$, $P(X > 1/2)$, $P(X = 3/2)$, and $P(1 < X \le 6)$.

6. From families with three children *a family* is chosen at random. Let X be the number of girls in the family. Calculate and sketch the distribution function of X. Assume that in a three-child family all gender distributions are equally probable.

7. A grocery store sells X hundred kilograms of rice every day, where the distribution of the random variable X is of the following form:

$$F(x) = \begin{cases} 0 & x < 0 \\ kx^2 & 0 \le x < 3 \\ k(-x^2 + 12x - 3) & 3 \le x < 6 \\ 1 & x \ge 6. \end{cases}$$

Suppose that this grocery store's total sales of rice do not reach 600 kilograms on any given day.

(a) Find the value of k.

(b) What is the probability that the store sells between 200 and 400 kilograms of rice next Thursday?

(c) What is the probability that the store sells over 300 kilograms of rice next Thursday?

(d) We are given that the store sold at least 300 kilograms of rice last Friday. What is the probability that it did not sell more than 400 kilograms on that day?

8. Let X be a random variable with distribution function F. For p $(0 < p < 1)$, Q_p is said to be a **quantile** of order p if

$$F(Q_p-) \le p \le F(Q_p).$$

In a certain country, the rate at which the price of oil per gallon changes from one year to another has the following distribution function:

$$F(x) = \frac{1}{1 + e^{-x}}, \qquad -\infty < x < \infty.$$

Find $Q_{0.50}$, called the **median** of F; $Q_{0.25}$, called the **first quartile** of F; and $Q_{0.75}$, called the **third quartile** of F. Interpret these quantities.

9. A random variable X is called **symmetric about 0** if for all $x \in \mathbf{R}$,

$$P(X \ge x) = P(X \le -x).$$

Prove that if X is symmetric about 0, then for all $t > 0$ its distribution function F satisfies the following relations:

(a) $P(|X| \le t) = 2F(t) - 1$.

(b) $P(|X| > t) = 2[1 - F(t)]$.

(c) $P(X = t) = F(t) + F(-t) - 1$.

10. Determine if the following is a distribution function.

$$F(t) = \begin{cases} 1 - \dfrac{1}{\pi}e^{-t} & \text{if } t \ge 0 \\ 0 & \text{if } t < 0. \end{cases}$$

11. Determine if the following is a distribution function.

$$F(t) = \begin{cases} \dfrac{t}{1+t} & \text{if } t \ge 0 \\ 0 & \text{if } t < 0. \end{cases}$$

12. Determine if the following is a distribution function.

$$F(t) = \begin{cases} (1/2)e^t & t < 0 \\ 1 - (3/4)e^{-t} & t \ge 0. \end{cases}$$

13. Airline A has commuter flights every 45 minutes from San Francisco airport to Fresno. A passenger who wants to take one of these flights arrives at the airport at a random time. Suppose that X is the waiting time for this passenger; find the distribution function of X. Assume that seats are always available for these flights.

B

14. A scientific calculator can generate two-digit random numbers. That is, it can choose a number at random from the set $\{00, 01, 02, \dots, 99\}$. To obtain a random number from the set $\{4, 5, \dots, 18\}$, show that we have to keep generating two-digits random numbers until we obtain one between 4 and 18.

15. In a small town there are 40 taxis, numbered 1 to 40. Three taxis arrive at random at a station to pick up passengers. What is the probability that the number of at least one of the taxis is less than 5?

16. Let X be a randomly selected point from the interval $(0, 3)$. What is the probability that $X^2 - 5X + 6 > 0$?

17. Let X be a random point selected from the interval $(0, 1)$. Calculate F, the distribution function of $Y = X/(1 + X)$, and sketch its graph.

18. In the United States, the number of twin births is approximately 1 in 90. At a certain hospital let X be the number of births until the first twins are born. Find the first quartile, the median, and the third quartile of X. See Exercise 8 for the definitions of these quantities.

19. Let the time until a new car breaks down be denoted by X, and let

$$Y = \begin{cases} X & \text{if } X \leq 5 \\ 5 & \text{if } X > 5. \end{cases}$$

Then Y is the life of the car, if it lasts less than 5 years, and is 5 if it lasts longer than 5 years. Calculate the distribution function of Y in terms of F, the distribution function of X. ∎

4.3 DISCRETE RANDOM VARIABLES

In Section 4.1 we observed that the set of possible values of a random variable might be finite, infinite but countable, or uncountable. For example, let X, Y, and Z be three random variables representing the respective number of tails in flipping a coin twice, the number of flips of until the first heads, and the amount of next year's rainfall. Then the sets of possible values for X, Y, and Z are the finite set $\{0, 1, 2\}$, the countable set $\{1, 2, 3, 4, \ldots\}$, and the uncountable set $\{x : x \geq 0\}$, respectively. Whenever the set of possible values that a random variable X can assume is at most countable, X is called **discrete**. Therefore, X is discrete if either the set of its possible values is finite or it is countably infinite. To each discrete random variable, a real-valued function $p : \mathbf{R} \to \mathbf{R}$, defined by $p(x) = P(X = x)$, is assigned and is called the **probability mass function** of X. (It is also called the **probability function** of X or the **discrete probability function** of X.) Since the set of values of X is countable, $p(x)$ is positive at most for a countable set. It is zero elsewhere; that is, if possible values of X are $x_1, x_2, x_3, \ldots$, then $p(x_i) \geq 0$ $(i = 1, 2, 3, \ldots)$ and $p(x) = 0$ if $x \notin \{x_1, x_2, x_3, \ldots\}$. Now, clearly, the occurrence of one of the events $X = x_1$, $X = x_2$, $X = x_3$, ..., is certain. Therefore, $\sum_{i=1}^{\infty} P(X = x_i) = 1$ or, equivalently, $\sum_{i=1}^{\infty} p(x_i) = 1$.

Definition   *The **probability mass function** p of a random variable X whose set of possible values is $\{x_1, x_2, x_3, \ldots\}$ is a function from $\mathbf{R}$ to $\mathbf{R}$ that satisfies the following properties.*

(a) *$p(x) = 0$ if $x \notin \{x_1, x_2, x_3, \ldots\}$.*

(b) *$p(x_i) = P(X = x_i)$ and hence $p(x_i) \geq 0$ $(i = 1, 2, 3, \ldots)$.*

(c) *$\sum_{i=1}^{\infty} p(x_i) = 1$.*

Because of this definition, if, for a set $\{x_1, x_2, x_3, \ldots\}$, there exists a function $p : \mathbf{R} \to \mathbf{R}$ such that $p(x_i) \geq 0$ $(i = 1, 2, 3, \ldots)$, $p(x) = 0$, $x \notin \{x_1, x_2, x_3, \ldots\}$, and $\sum_{i=1}^{\infty} p(x_i) = 1$, then p is called a **probability mass function**.

The probability mass function of a random variable is often demonstrated geometrically by a set of vertical lines connecting the points $(x_i, 0)$ and $(x_i, p(x_i))$. For example, if X is the number of heads in two flips of a fair coin, then $X = 0, 1, 2$ with $p(0) = 1/4$, $p(1) = 1/2$, and $p(2) = 1/4$. Hence the graphical representation of p is as shown in Figure 4.4.

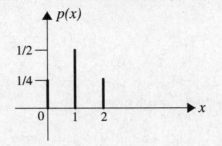

Figure 4.4   Graph of the number of heads in two flips of a fair coin.

The distribution function F of a discrete random variable X, with the set of possible values $\{x_1, x_2, x_3, \dots\}$, is a step function. Assuming that $x_1 < x_2 < x_3 < \cdots$, we have that if $t < x_1$, then

$$F(t) = 0;$$

if $x_1 \leq t < x_2$, then

$$F(t) = P(X \leq t) = P(X = x_1) = p(x_1);$$

if $x_2 \leq t < x_3$, then

$$F(t) = P(X \leq t) = P(X = x_1 \text{ or } X = x_2) = p(x_1) + p(x_2);$$

and in general, if $x_{n-1} \leq t < x_n$, then

$$F(t) = \sum_{i=1}^{n-1} p(x_i).$$

Thus F is constant in the intervals $[x_{n-1}, x_n)$ with jumps at $x_1, x_2, x_3, \dots$. The magnitude of the jump at x_i is $p(x_i)$.

Example 4.11   In the experiment of rolling a balanced die twice, let X be the maximum of the two numbers obtained. Determine and sketch the probability mass function and the distribution function of X.

Solution: The possible values of X are 1, 2, 3, 4, 5, and 6. The sample space of this experiment consists of 36 equally likely outcomes. Hence the probability of any of them is 1/36. Thus

$$p(1) = P(X = 1) = P(\{(1, 1)\}) = 1/36,$$
$$p(2) = P(X = 2) = P(\{(1, 2), (2, 2), (2, 1)\}) = 3/36,$$
$$p(3) = P(X = 3) = P(\{(1, 3), (2, 3), (3, 3), (3, 2), (3, 1)\}) = 5/36.$$

Similarly, $p(4) = 7/36$, $p(5) = 9/36$, and $p(6) = 11/36$; $p(x) = 0$ for $x \notin \{1, 2, 3, 4, 5, 6\}$. The graphical representation of p is shown in Figure 4.5. The distribution function of X, F, is as follows (its graph is shown in Figure 4.6):

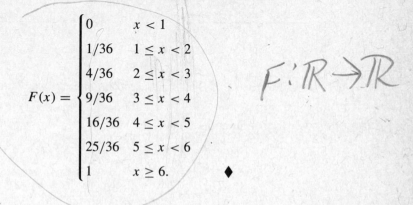

$$F(x) = \begin{cases} 0 & x < 1 \\ 1/36 & 1 \le x < 2 \\ 4/36 & 2 \le x < 3 \\ 9/36 & 3 \le x < 4 \\ 16/36 & 4 \le x < 5 \\ 25/36 & 5 \le x < 6 \\ 1 & x \ge 6. \end{cases}$$

$F : \mathbb{R} \to \mathbb{R}$

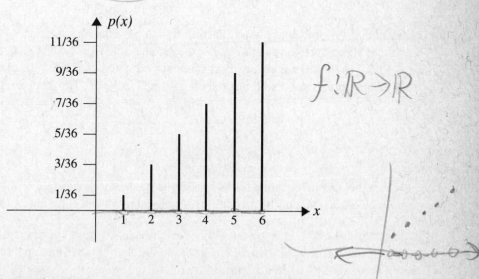

$f : \mathbb{R} \to \mathbb{R}$

Figure 4.5   Probability mass function of Example 4.11.

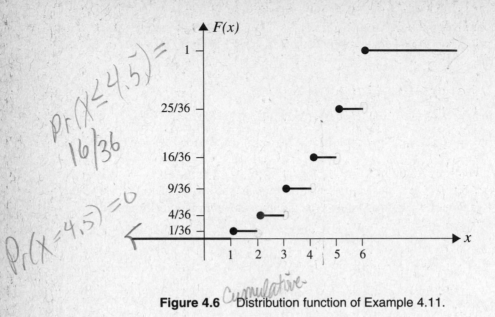

$\Pr(X \leq 4.5) =$
$16/36$

$\Pr(X = 4.5) = 0$

Figure 4.6 ~~Cumulative~~ Distribution function of Example 4.11.

Example 4.12 Can a function of the form

$$p(x) = \begin{cases} c\left(\dfrac{2}{3}\right)^x & x = 1, 2, 3, \ldots \\ 0 & \text{elsewhere} \end{cases}$$

be a probability mass function?

Solution: A probability mass function should have three properties: (1) $p(x)$ must be zero at all points except on a finite or a countable set. Clearly, this property is satisfied. (2) $p(x)$ should be nonnegative. This is satisfied if and only if $c \geq 0$. (3) $\sum p(x_i) = 1$. This condition is satisfied if and only if $\sum_{i=1}^{\infty} c(2/3)^i = 1$. This happens precisely when

$$c = \frac{1}{\displaystyle\sum_{i=1}^{\infty} (2/3)^i} = \frac{1}{\dfrac{2/3}{1 - 2/3}} = \frac{1}{2},$$

where the second equality follows from the geometric series theorem. Thus only for $c = 1/2$, a function of the given form is a probability mass function. ◆

Example 4.13 Let X be the number of births in a hospital until the first girl is born. Determine the probability mass function and the distribution function of X. Assume that the probability is 1/2 that a baby born is a girl.

Solution: X is a random variable that can assume any positive integer i. $p(i) = P(X = i)$, and $X = i$ occurs if the first $i - 1$ births are all boys and the ith birth is a girl. Thus $p(i) = (1/2)^{i-1}(1/2) = (1/2)^i$ for $i = 1, 2, 3, \ldots$, and $p(x) = 0$ if $x \neq 1, 2, 3, \ldots$. To determine $F(t)$, note that for $t < 1$, $F(t) = 0$; for $1 \leq t < 2$, $F(t) = 1/2$; for $2 \leq t < 3$, $F(t) = 1/2 + 1/4 = 3/4$; for $3 \leq t < 4$, $F(t) = 1/2 + 1/4 + 1/8 = 7/8$; and in general for $n - 1 \leq t < n$,

$$F(t) = \frac{1}{2} + \frac{1}{2^2} + \frac{1}{2^3} + \cdots + \frac{1}{2^{n-1}} = \sum_{i=1}^{n-1} \left(\frac{1}{2}\right)^i$$

$$= \frac{1 - (1/2)^n}{1 - 1/2} - 1 = 1 - \left(\frac{1}{2}\right)^{n-1},$$

by the partial sum formula for geometric series. Thus

$$F(t) = \begin{cases} 0 & t < 1 \\ 1 - (1/2)^{n-1} & n - 1 \leq t < n, \quad n = 2, 3, 4, \ldots. \end{cases} \quad \blacklozenge$$

EXERCISES

A

1. Let $p(x) = x/15$, $x = 1, 2, 3, 4, 5$ be probability mass function of a random variable X. Determine F, the distribution function of X, and sketch its graph.

2. In the experiment of rolling a balanced die twice, let X be the minimum of the two numbers obtained. Determine the probability mass function and the distribution function of X and sketch their graphs.

3. In the experiment of rolling a balanced die twice, let X be the sum of the two numbers obtained. Determine the probability mass function of X.

4. The distribution function of a random variable X is given by

$$F(x) = \begin{cases} 0 & \text{if } x < -2 \\ 1/2 & \text{if } -2 \leq x < 2 \\ 3/5 & \text{if } 2 \leq x < 4 \\ 8/9 & \text{if } 4 \leq x < 6 \\ 1 & \text{if } x \geq 6. \end{cases}$$

Determine the probability mass function of X and sketch its graph.

5. Let X be the number of random numbers selected from $\{0, 1, 2, \ldots, 9\}$ independently until 0 is chosen. Find the probability mass functions of X and $Y = 2X + 1$.

6. A value i is said to be the **mode** of a discrete random variable X if it maximizes $p(x)$, the probability mass function of X. Find the modes of random variables X and Y with probability mass functions

$$p(x) = \left(\frac{1}{2}\right)^x, \qquad x = 1, 2, 3, \ldots ,$$

and

$$q(y) = \frac{4!}{y!\,(4-y)!}\left(\frac{1}{4}\right)^y\left(\frac{3}{4}\right)^{4-y}, \qquad y = 0, 1, 2, 3, 4,$$

respectively.

7. For each of the following, determine the value(s) of k for which p is a probability mass function. Note that in parts (d) and (e), n is a positive integer.

   (a) $p(x) = kx, x = 1, 2, 3, 4, 5.$

   (b) $p(x) = k(1+x)^2, x = -2, 0, 1, 2.$

   (c) $p(x) = k(1/9)^x, x = 1, 2, 3, \ldots .$

   (d) $p(x) = kx, x = 1, 2, 3, \ldots , n.$

   (e) $p(x) = kx^2, x = 1, 2, 3, \ldots , n.$

   Hint: Recall that

   $$\sum_{i=1}^{n} i = \frac{n(n+1)}{2}, \qquad \sum_{i=1}^{n} i^2 = \frac{n(n+1)(2n+1)}{6}.$$

8. From 18 potential women jurors and 28 potential men jurors, a jury of 12 is chosen at random. Let X be the number of women selected. Find the probability mass function of X.

9. Let $p(x) = 3/4(1/4)^x, x = 0, 1, 2, 3, \ldots ,$ be probability mass function of a random variable X. Find F, the distribution function of X, and sketch its graph.

10. In successive rolls of a fair die, let X be the number of rolls until the first 6 appears. Determine the probability mass function and the distribution function of X.

11. A *binary digit* or *bit* is a zero or one. A computer assembly language can generate independent random bits. Let X be the number of independent random bits to be generated until both 0 and 1 are obtained. Find the probability mass function of X.

B

12. Every Sunday, Bob calls Liz to see if she will play tennis with him on that day. If Liz has not played tennis with Bob since i Sundays ago, the probability that she

will say yes to him is i/k, $k \geq 2$, $i = 1, 2, \ldots, k$. Therefore, if, for example, Liz does not play tennis with Bob for $k - 1$ consecutive Sundays, then she will play with him next Sunday with probability 1. Let Z be the number of weeks it takes Liz to play again with Bob since they last played. Find the probability mass function of Z.

13. Let X be the number of vowels (not necessarily distinct) among the first five letters of a random arrangement of the following expression.

ELIZABETHTAYLOR

Find the probability mass function of X. Count the letter Y as a consonant.

14. From a drawer that contains 10 pair of gloves, six gloves are selected randomly. Let X be the number of pairs of gloves obtained. Find the probability mass function of X.

15. A fair die is tossed successively. Let X denote the number of tosses until each of the six possible outcomes occurs at least once. Find the probability mass function of X.

Hint: For $1 \leq i \leq 6$, let E_i be the event that the outcome i does not occur during the first n tosses of the die. First calculate $P(X > n)$ by writing the event $X > n$ in terms of $E_1, E_2, \ldots, E_6$.

16. To an engineering class containing 23 male and three female students, there are 13 work stations available. To assign each work station to two students, the professor forms 13 teams one at a time, each consisting of two randomly selected students. In this process, let X be the total number of students selected when the first team consisting of a male and a female appears. Find the probability mass function of X.

■

4.4 EXPECTATIONS OF DISCRETE RANDOM VARIABLES

To clarify the concept of expectation, consider a casino game in which the probability of losing \$1 per game is 0.6, and the probabilities of winning \$1, \$2, and \$3 per game are 0.3, 0.08, and 0.02, respectively. The gain or loss of a gambler who plays this game only a few times depends on his luck more than anything else. For example, in one play of the game, a lucky gambler might win \$3, but he has a 60% chance of losing \$1. However, if a gambler decides to play the game a large number of times, his loss or gain depends more on the number of plays than on his luck. A calculating player argues that if he plays the game n times, for a large n, then in approximately $(0.6)n$ games he will lose \$1 per game, and in approximately $(0.3)n$, $(0.08)n$, and $(0.02)n$ games he will win \$1, \$2, and

$3, respectively. Therefore, his total gain is

$$(0.6)n \cdot (-1) + (0.3)n \cdot 1 + (0.08)n \cdot 2 + (0.02)n \cdot 3 = (-0.08)n. \qquad (4.1)$$

This gives an average of $ − 0.08, or about 8 cents of loss per game. The more the gambler plays, the less luck interferes and the closer his loss comes to $0.08 per game. If X is the random variable denoting the gain in one play, then the number -0.08 is called the **expected value** of X. We write $E(X) = -0.08$. $E(X)$ is the average value of X. That is, if we play the game n times and find the average of the values of X, then as $n \to \infty$, $E(X)$ is obtained. Since, for this game, $E(X) < 0$, we have that, on the average, the more we play, the more we lose. If for some game $E(X) = 0$, then in long run the player neither loses nor wins. Such games are called **fair**. In this example, X is a discrete random variable with the set of possible values $\{-1, 1, 2, 3\}$. The probability mass function of X, $p(x)$, is given by

i	-1	1	2	3
$p(i) = P(X = i)$	0.6	0.3	0.08	0.02

and $p(x) = 0$ if $x \notin \{-1, 1, 2, 3\}$. Dividing both sides of (4.1) by n, we obtain

$$(0.6) \cdot (-1) + (0.3) \cdot 1 + (0.08) \cdot 2 + (0.02) \cdot 3 = -0.08.$$

Hence

$$-1 \cdot p(-1) + 1 \cdot p(1) + 2 \cdot p(2) + 3 \cdot p(3) = -0.08,$$

a relation showing that the expected value of X can be calculated directly by summing up the product of possible values of X by their probabilities. This and similar examples motivate the following general definition, which was first used casually by Pascal but introduced formally by Huygens in the late seventeenth century.

Definition   *The* ***expected value*** *of a discrete random variable X with the set of possible values A and probability mass function p(x) is defined by*

$$E(X) = \sum_{x \in A} x p(x).$$

We say that $E(X)$ exists if this sum converges absolutely.

The expected value of a random variable X is also called the **mean**, or the **mathematical expectation**, or simply the **expectation** of X. It is also occasionally denoted by $E[X]$, EX, μ_X, or μ.

Note that if each value x of X is weighted by $p(x) = P(X = x)$, then $\sum_{x \in A} x p(x)$ is nothing but the weighted average of X. Similarly, if we think of a unit mass distributed along the real line at the points of A so that the mass at $x \in A$ is $P(X = x)$, then $E(X)$ is the center of gravity.

Here are some examples to illuminate the notion of expected value, a fundamental concept in probability and statistics.

Example 4.14   We flip a fair coin twice and let X be the number of heads obtained. What is the expected value of X?

Solution: The possible values of X are 0, 1, and 2, and the probability mass function of X is given by $p(0) = P(X = 0) = 1/4$, $p(1) = P(X = 1) = 1/2$, $p(2) = P(X = 2) = 1/4$, and $p(x) = 0$ if $x \notin \{0, 1, 2\}$. Thus

$$E(X) = 0 \cdot p(0) + 1 \cdot p(1) + 2 \cdot p(2) = 0 \cdot \frac{1}{4} + 1 \cdot \frac{1}{2} + 2 \cdot \frac{1}{4} = 1.$$

Therefore, we can expect an average of one head in every two flips.   ♦

Example 4.15   We write the numbers $a_1, a_2, \ldots, a_n$ on n identical balls and mix them in a box. What is the expected value of a ball selected at random?

Solution: The set of possible values of X, the numbers written on the balls selected, is $\{a_1, a_2, \ldots, a_n\}$. The probability mass function of X is given by

$$p(a_1) = p(a_2) = \cdots = p(a_n) = 1/n,$$

and $p(x) = 0$ if $x \notin \{a_1, a_2, \ldots, a_n\}$. Thus

$$E(X) = \sum_{x \in \{a_1, a_2, \ldots, a_n\}} x p(x) = a_1 \cdot \frac{1}{n} + a_2 \cdot \frac{1}{n} + \cdots + a_n \cdot \frac{1}{n}$$

$$= \frac{a_1 + a_2 + \cdots + a_n}{n}.$$

Therefore, as expected, $E(X)$ coincides with the average of the values $a_1, a_2, \ldots, a_n$. That is, if for a large number of times we draw balls at random and with replacement, record their values, and then find their average, the result obtained is the arithmetic mean of the numbers $a_1, a_2, \ldots, a_n$.   ♦

Example 4.16   A college mathematics department sends 8 to 12 professors to the annual meeting of the American Mathematical Society, which lasts five days. The hotel at which the conference is held offers a bargain rate of a dollars per day per person if reservations are made 45 or more days in advance, but charges a cancellation fee of $2a$ dollars per person. The department is not certain how many professors will go. However, from past experience it is known that the probability of the attendance of i professors is $1/5$ for $i = 8, 9, 10, 11$, and 12. If the regular rate of the hotel is $2a$ dollars per day per person, should the department make any reservations? If so, how many?

Solution: For $i = 8, 9, 10, 11$, and 12, let X_i be the total cost in dollars if the department makes reservations for i professors. We should compute $E(X_i)$ for $i = 8, 9, 10, 11$, and 12. If $E(X_j)$ is the smallest of all, the department should make reservations for j professors. To calculate $E(X_8)$, note that X_8 only assumes the values $40a$, $50a$, $60a$, $70a$, and $80a$, which correspond to the cases that 8, 9, 10, 11, and 12 professors attend, respectively, while reservations are made only for eight professors. Since the probability of any of these is 1/5, we have

$$E(X_8) = (40a)\frac{1}{5} + (50a)\frac{1}{5} + (60a)\frac{1}{5} + (70a)\frac{1}{5} + (80a)\frac{1}{5} = 60a.$$

Similarly,

$$E(X_9) = (42a)\frac{1}{5} + (45a)\frac{1}{5} + (55a)\frac{1}{5} + (65a)\frac{1}{5} + (75a)\frac{1}{5} = 56.4a,$$

$$E(X_{10}) = (44a)\frac{1}{5} + (47a)\frac{1}{5} + (50a)\frac{1}{5} + (60a)\frac{1}{5} + (70a)\frac{1}{5} = 54.2a,$$

$$E(X_{11}) = (46a)\frac{1}{5} + (49a)\frac{1}{5} + (52a)\frac{1}{5} + (55a)\frac{1}{5} + (65a)\frac{1}{5} = 53.4a,$$

$$E(X_{12}) = (48a)\frac{1}{5} + (51a)\frac{1}{5} + (54a)\frac{1}{5} + (57a)\frac{1}{5} + (60a)\frac{1}{5} = 54a.$$

We see that X_{11} has the smallest expected value. Thus making 11 reservations is the most reasonable policy.   ◆

Example 4.17   In the lottery of a certain state, players pick six different integers between 1 and 49, the order of selection being irrelevant. The lottery commission then selects six of these numbers at random as the winning numbers. A player wins the grand prize of \$1,200,000 if all six numbers that he has selected match the winning numbers. He wins the second and third prizes of \$800 and \$35, respectively, if exactly five and four of his six selected numbers match the winning numbers. What is the expected value of the amount a player wins in one game?

Solution: Let X be the amount that a player wins in one game. Then the possible values of X are 1,200,000; 800; 35; and 0. The probabilities associated with these values are

$$P(X = 1,200,000) = \frac{1}{\binom{49}{6}} \approx 0.000,000,072,$$

$$P(X = 800) = \frac{\binom{6}{5}\binom{43}{1}}{\binom{49}{6}} \approx 0.000,018, \quad P(X = 35) = \frac{\binom{6}{4}\binom{43}{2}}{\binom{49}{6}} \approx 0.000,97,$$

and

$$P(X = 0) = 1 - 0.000,000,072 - 0.000,018 - 0.000,97 = 0.999,011,928.$$

Therefore,

$$E(X) \approx 1,200,000(0.000,000,072) + 800(0.000,018) + 35(0.000,97)$$
$$+ 0(0.999,011,928) \approx 0.13.$$

This shows that on the average players will win 13 cents per game. If the cost per game is 50 cents, then, on the average, a player will lose 37 cents per game. Therefore, a player who plays 10,000 games over several years will lose approximately $3700. ♦

Let X be a discrete random variable with a set of possible values A and probability mass function p. We say that $E(X)$ exists if the sum $\sum_{x \in A} x p(x)$ converges, that is, if $\sum_{x \in A} x p(x) < \infty$. We now present two examples of random variables whose mathematical expectations do not exist.

Example 4.18   (St. Petersburg Paradox)   In a game, the player flips a fair coin successively until he gets a heads. If this occurs on the kth flip, the player wins 2^k dollars. Therefore, if the outcome of the first flip is heads, the player wins $2. If the outcome of the first flip is tails but that of the second flip is heads, he wins $4. If the outcomes of the first two are tails but the third one heads, he will win $8, and so on. The question is, to play this game, how much should a person, who is willing to play a fair game, pay? To answer this question, let X be the amount of money the player wins. Then X is a random variable with the set of possible values $\{2, 4, 8, \ldots, 2^k, \ldots\}$ and

$$P(X = 2^k) = \left(\frac{1}{2}\right)^k, \qquad k = 1, 2, 3, \ldots .$$

Therefore,

$$E(X) = \sum_{k=1}^{\infty} 2^k \left(\frac{1}{2}\right)^k = \sum_{k=1}^{\infty} 1 = 1 + 1 + 1 + \cdots = \infty.$$

This result shows that the game remains unfair even if a person pays the largest possible amount to play it. In other words, this is a game in which one always wins no matter how expensive it is to play. To see what the flaw is, note that theoretically this game is not feasible to play because it requires an enormous amount of money. In practice, however, the probability that a gambler wins 2^k for a large k is close to zero. Even for small values of k, winning is highly unlikely. For example, to win $2^{30} = 1,073,741,824$, you should get 29 tails in a row followed by a head. The chance of this happening is 1 in 1,073,741,824, much less than 1 in a billion.   ♦

The following example serves to illuminate further the concept of expected value. At the same time, it shows the inadequacy of this concept as a central measure.

Example 4.19   Let X_0 be the amount of rain that will fall in the United States on the next Christmas day. For $n > 0$, let X_n be the amount of rain that will fall in the United States on Christmas n years later. Let N be the smallest number of years that elapse before we get a Christmas rainfall greater than X_0. Suppose that $P(X_i = X_j) = 0$ if $i \neq j$, the events concerning the amount of rain on Christmas days of different years are all independent, and the X_n's are identically distributed. Find the expected value of N.

Solution: Since N is the first value for n for which $X_n > X_0$,

$$P(N > n) = P(X_0 > X_1, X_0 > X_2, \ldots, X_0 > X_n)$$

$$= P\big(\max(X_0, X_1, X_2, \ldots, X_n) = X_0\big) = \frac{1}{n+1};$$

where the last equality follows from the symmetry, there is no more reason for the maximum to be at X_0 than there is for it to be at X_i, $0 \leq i \leq n$. Therefore,

$$P(N = n) = P(N > n - 1) - P(N > n) = \frac{1}{n} - \frac{1}{n+1} = \frac{1}{n(n+1)}.$$

From this it follows that

$$E(N) = \sum_{n=1}^{\infty} n P(N = n) = \sum_{n=1}^{\infty} \frac{n}{n(n+1)} = \sum_{n=1}^{\infty} \frac{1}{n+1} = \infty.$$

Note that $P(N > n - 1) = 1/n$ gives the probability that, in the United States, we will have to wait more than, say, three years for a Christmas rainfall that is greater than X_0 is only 1/4, and the probability that we must wait more than nine years is only 1/10. Even with such low probabilities, on average, it will still take infinitely many years before we will have more rain on a Christmas day than we will have on next Christmas day.   ♦

Example 4.20   The tanks of a country's army are numbered 1 to N. In a war this country loses n random tanks to the enemy, who discovers that the captured tanks are numbered. If $X_1, X_2, \ldots, X_n$ are the numbers of the captured tanks, what is $E(\max X_i)$? How can the enemy use $E(\max X_i)$ to find an estimate of N, the total number of this country's tanks?

Solution: Let $Y = \max X_i$; then

$$P(Y = k) = \frac{\binom{k-1}{n-1}}{\binom{N}{n}} \qquad \text{for } k = n, n+1, n+2, \ldots, N,$$

because if the maximum of X_i's is k, the numbers of the remaining $n - 1$ tanks are from 1 to $k - 1$. Now

$$E(Y) = \sum_{k=n}^{N} k P(Y = k) = \sum_{k=n}^{N} \frac{k\binom{k-1}{n-1}}{\binom{N}{n}}$$

$$= \frac{1}{\binom{N}{n}} \sum_{k=n}^{N} \frac{k\,(k-1)!}{(n-1)!\,(k-n)!} = \frac{n}{\binom{N}{n}} \sum_{k=n}^{N} \frac{k!}{n!\,(k-n)!}$$

$$= \frac{n}{\binom{N}{n}} \sum_{k=n}^{N} \binom{k}{n}. \tag{4.2}$$

To calculate $\displaystyle\sum_{k=n}^{N} \binom{k}{n}$, note that $\binom{k}{n}$ is the coefficient of x^n in the polynomial $(1 + x)^k$.

Therefore, $\displaystyle\sum_{k=n}^{N} \binom{k}{n}$ is the coefficient of x^n in the polynomial $\displaystyle\sum_{k=n}^{N}(1 + x)^k$. Since

$$\sum_{k=n}^{N}(1 + x)^k = (1 + x)^n \sum_{k=0}^{N-n}(1 + x)^k = (1 + x)^n \frac{(1 + x)^{N-n+1} - 1}{(1 + x) - 1}$$

$$= \frac{1}{x}\big[(1 + x)^{N+1} - (1 + x)^n\big],$$

and the coefficient of x^n in the polynomial $\dfrac{1}{x}\big[(1 + x)^{N+1} - (1 + x)^n\big]$ is $\dbinom{N+1}{n+1}$, we have

$$\sum_{k=n}^{N} \binom{k}{n} = \binom{N+1}{n+1}.$$

Substituting this result in (4.2), we obtain

$$E(Y) = \frac{n\dbinom{N+1}{n+1}}{\dbinom{N}{n}} = \frac{n\dfrac{(N+1)!}{(n+1)!\,(N-n)!}}{\dfrac{N!}{n!\,(N-n)!}} = \frac{n(N+1)}{n+1}.$$

To estimate N, the total number of this country's tanks, we solve $E(Y) = \dfrac{n(N+1)}{n+1}$ for

N. We obtain $N = \dfrac{n+1}{n} E(Y) - 1$. Therefore, if, for example, the enemy captures 12 tanks and the maximum of the numbers of the tanks captured is 117, then assuming that $E(Y)$ is approximately equal to the value of Y observed, we get $N \approx (13/12) \times 117 - 1 \approx$ 126. ◆

The solutions to Examples 4.21 and 4.22 were given by Professor James Frykman from Kent State University.

Example 4.21   An urn contains w white and b blue chips. A chip is drawn at random and then is returned to the urn along with $c > 0$ chips of the same color. This experiment is then repeated successively. Let X_n be the number of white chips drawn during the first n draws. Show that $E(X_n) = nw/(w + b)$.

Solution: For $n \geq 1$, let p_n be the probability mass function of X_n. We will prove, by induction, that $E(X_n) = nw/(w + b)$. For $n = 1$,

$$E(X_1) = 0 \cdot \frac{b}{w + b} + 1 \cdot \frac{w}{w + b} = \frac{1 \cdot w}{w + b}.$$

Suppose that $E(X_n) = nw/(w + b)$ for any integer $n \geq 1$. To demonstrate that $E(X_{n+1}) = (n + 1)w/(w + b)$, note that

$$E(X_{n+1}) = \sum_{k=0}^{n+1} k p_{n+1}(k) = (n + 1) p_{n+1}(n + 1) + \sum_{k=1}^{n} k p_{n+1}(k). \qquad (4.3)$$

Now

$$\begin{aligned}
p_{n+1}(n + 1) &= P(X_{n+1} = n + 1) \\
&= P(X_{n+1} = n + 1 \mid X_n = n) P(X_n = n) \\
&= \frac{w + nc}{w + b + nc} \, p_n(n),
\end{aligned} \qquad (4.4)$$

and for $1 \leq k \leq n$,

$$\begin{aligned}
p_{n+1}(k) &= P(X_{n+1} = k) = P(X_{n+1} = k \mid X_n = k) P(X_n = k) \\
&\quad + P(X_{n+1} = k \mid X_n = k - 1) P(X_n = k - 1) \\
&= \frac{b + (n - k)c}{w + b + nc} \, p_n(k) + \frac{w + (k - 1)c}{w + b + nc} \, p_n(k - 1).
\end{aligned} \qquad (4.5)$$

In relation (4.3), substituting (4.4) for $p_{n+1}(n + 1)$ and (4.5) for $p_{n+1}(k)$, we obtain

$$\begin{aligned}
E(X_{n+1}) &= \frac{(n + 1)(w + nc)}{w + b + nc} \, p_n(n) + \sum_{k=1}^{n} \frac{k[b + (n - k)c]}{w + b + nc} \, p_n(k) \\
&\quad + \sum_{k=1}^{n} \frac{k[w + (k - 1)c]}{w + b + nc} \, p_n(k - 1).
\end{aligned} \qquad (4.6)$$

A shift in index of the last sum in (4.6) gives

$$\sum_{k=1}^{n} \frac{k[w+(k-1)c]}{w+b+nc} p_n(k-1)$$

$$= \frac{1}{w+b+nc} \sum_{k=0}^{n-1} (k+1)(w+kc)p_n(k)$$

$$= \frac{1}{w+b+nc} \left[\sum_{k=0}^{n-1} k(w+kc)p_n(k) + \sum_{k=0}^{n-1} (w+kc)p_n(k) \right]$$

$$= \frac{1}{w+b+nc} \left[\sum_{k=1}^{n-1} k(w+kc)p_n(k) + \sum_{k=0}^{n-1} wp_n(k) + \sum_{k=1}^{n-1} kcp_n(k) \right]$$

$$= \frac{1}{w+b+nc} \left[\sum_{k=1}^{n-1} k(w+kc+c)p_n(k) + w \sum_{k=0}^{n-1} p_n(k) \right]$$

$$= \frac{1}{w+b+nc} \left[\sum_{k=1}^{n} k(w+kc+c)p_n(k) + w \sum_{k=1}^{n} p_n(k) \right.$$

$$\left. - n(w+nc+c)p_n(n) - wp_n(n) \right]. \tag{4.7}$$

Relations (4.6) and (4.7) combined yield what is needed to complete the proof:

$$E(X_{n+1}) = \frac{(n+1)(w+nc) - n(w+nc+c) - w}{w+b+nc} p_n(n)$$

$$+ \sum_{k=1}^{n} \frac{k[b+(n-k)c] + k(w+kc+c)}{w+b+nc} p_n(k) + \frac{w}{w+b+nc}$$

$$= \frac{w+b+nc+c}{w+b+nc} \sum_{k=1}^{n} kp_n(k) + \frac{w}{w+b+nc}$$

$$= \frac{w+b+nc+c}{w+b+nc} E(X_n) + \frac{w}{w+b+nc}$$

$$= \frac{w+b+nc+c}{w+b+nc} \cdot \frac{nw}{w+b} + \frac{w}{w+b+nc} = \frac{(n+1)w}{w+b}. \quad \blacklozenge$$

Example 4.22   (Pólya's Urn Model)   An urn contains w white and b blue chips. A chip is drawn at random and then is returned to the urn along with $c > 0$ chips of the same color. Prove that if $n = 2, 3, 4, \ldots$, such experiments are made, then at each draw the probability of a white chip is still $w/(w+b)$ and the probability of a blue chip is $b/(w+b)$.

This model was first introduced in preliminary studies of "contagious diseases" and the spread of epidemics as well as "accident proneness" in actuarial mathematics.

Solution: For all $n \geq 1$, let W_n be the event that the nth draw is white and B_n be the event that it is blue. We will show that $P(W_n) = w/(w + b)$. This implies that $P(B_n) = 1 - P(W_n) = b/(w + b)$. For all $n \geq 1$, let X_n be the number of white chips drawn during the first n draws. Let p_n be the probability mass function of X_n. Clearly, $P(W_1) = w/(w+b)$. To show that for $n \geq 2$, $P(W_n) = w/(w+b)$, note that the events $\{X_{n-1} = 0\}, \{X_{n-1} = 1\}, \ldots, \{X_{n-1} = n - 1\}$ form a partition of the sample space. Therefore, by the law of total probability (Theorem 3.4),

$$
\begin{aligned}
P(W_n) &= \sum_{k=0}^{n-1} P(W_n \mid X_{n-1} = k) P(X_{n-1} = k) \\
&= \sum_{k=0}^{n-1} P(X_n = k + 1 \mid X_{n-1} = k) P(X_{n-1} = k) \\
&= \sum_{k=0}^{n-1} \frac{w + kc}{w + b + (n - 1)c} p_{n-1}(k) \\
&= \frac{w}{w + b + (n - 1)c} \sum_{k=0}^{n-1} p_{n-1}(k) + \frac{c}{w + b + (n - 1)c} \sum_{k=0}^{n-1} k p_{n-1}(k) \\
&= \frac{w}{w + b + (n - 1)c} + \frac{c}{w + b + (n - 1)c} E(X_{n-1}).
\end{aligned}
$$

Now by Example 4.21, $E(X_{n-1}) = (n - 1)w/(w + b)$. Hence, for $n \geq 2$,

$$
P(W_n) = \frac{w}{w + b + (n - 1)c} + \frac{c}{w + b + (n - 1)c} \cdot \frac{(n - 1)w}{w + b} = \frac{w}{w + b}. \quad \blacklozenge
$$

We now discuss some elementary properties of the expectation of a discrete random variable. Further properties of expectations are discussed in subsequent chapters.

Theorem 4.1   *If X is a constant random variable, that is, if $P(X = c) = 1$ for a constant c, then $E(X) = c$.*

Proof: There is only one possible value for X and that is c; hence $E(X) = c \cdot P(X = c) = c \cdot 1 = c$. $\blacklozenge$

Let $g: \mathbf{R} \to \mathbf{R}$ be a real-valued function and X be a discrete random variable with set of possible values A and probability mass function $p(x)$. Similar to $E(X) = \sum_{x \in A} x p(x)$, there is the important relation $E[g(X)] = \sum_{x \in A} g(x) p(x)$, known as the **law of the unconscious statistician**, which we now prove. This relation enables us to calculate the expected value of the random variable $g(X)$ without deriving its probability mass function. It implies that, for example,

$$E(X^2) = \sum_{x \in A} x^2 p(x),$$

$$E(X^2 - 2X + 4) = \sum_{x \in A} (x^2 - 2x + 4) p(x),$$

$$E(X \cos X) = \sum_{x \in A} (x \cos x) p(x),$$

$$E(e^X) = \sum_{x \in A} e^x p(x).$$

Theorem 4.2   *Let X be a discrete random variable with set of possible values A and probability mass function $p(x)$, and let g be a real-valued function. Then $g(X)$ is a random variable with*

$$E\big[g(X)\big] = \sum_{x \in A} g(x) p(x).$$

Proof: Let S be the sample space. We are given that $g \colon \mathbf{R} \to \mathbf{R}$ is a real-valued function and $X, \colon S \to A \subseteq \mathbf{R}$ is a random variable with the set of possible values A. As we know, $g(X)$, the composition of g and X, is a function from S to the set $g(A) = \{g(x) \colon x \in A\}$. Hence $g(X)$ is a random variable with the possible set of values $g(A)$. Now, by the definition of expectation,

$$E\big[g(X)\big] = \sum_{z \in g(A)} z P\{g(X) = z\}.$$

Let $g^{-1}(\{z\}) = \{x \colon g(x) = z\}$, and notice that we are not claiming that g has an inverse function. We are simply considering the set $\{x \colon g(x) = z\}$, which is called *the inverse image* of z and is denoted by $g^{-1}(\{z\})$. Now

$$P\big(g(X) = z\big) = P\Big(X \in g^{-1}(\{z\})\Big) = \sum_{\{x \colon x \in g^{-1}(\{z\})\}} P(X = x) = \sum_{\{x \colon g(x) = z\}} p(x).$$

Thus

$$E\big[g(X)\big] = \sum_{z \in g(A)} z P\big(g(X) = z\big) = \sum_{z \in g(A)} z \sum_{\{x \colon g(x) = z\}} p(x)$$

$$= \sum_{z \in g(A)} \sum_{\{x \colon g(x) = z\}} z p(x) = \sum_{z \in g(A)} \sum_{\{x \colon g(x) = z\}} g(x) p(x)$$

$$= \sum_{x \in A} g(x) p(x),$$

where the last equality follows from the fact that the sum over A can be performed in two stages: We can first sum over all x with $g(x) = z$, and then over all z.   ◆

Corollary   *Let X be a discrete random variable; $g_1, g_2, \ldots, g_n$ be real-valued functions, and let $\alpha_1, \alpha_2, \ldots, \alpha_n$ be real numbers. Then*

$$E\big[\alpha_1 g_1(X) + \alpha_2 g_2(X) + \cdots + \alpha_n g_n(X)\big]$$
$$= \alpha_1 E\big[g_1(X)\big] + \alpha_2 E\big[g_2(X)\big] + \cdots + \alpha_n E\big[g_n(X)\big].$$

Proof: Let the set of possible values of X be A, and its probability mass function be $p(x)$. Then, by Theorem 4.2,

$$E\big[\alpha_1 g_1(X) + \alpha_2 g_2(X) + \cdots + \alpha_n g_n(X)\big]$$
$$= \sum_{x \in A} \big[\alpha_1 g_1(x) + \alpha_2 g_2(x) + \cdots + \alpha_n g_n(x)\big] p(x)$$
$$= \alpha_1 \sum_{x \in A} g_1(x) p(x) + \alpha_2 \sum_{x \in A} g_2(x) p(x) + \cdots + \alpha_n \sum_{x \in A} g_n(x) p(x)$$
$$= \alpha_1 E\big[g_1(X)\big] + \alpha_2 E\big[g_2(X)\big] + \cdots + \alpha_n E\big[g_n(X)\big]. ◆$$

By this corollary, for example, we have relations such as the following:

$$E(2X^3 + 5X^2 + 7X + 4) = 2E(X^3) + 5E(X^2) + 7E(X) + 4$$
$$E(e^X + 2\sin X + \log X) = E(e^X) + 2E(\sin X) + E(\log X).$$

Moreover, this corollary implies that $E(X)$ is linear. That is, if $\alpha, \beta \in \mathbf{R}$, then

$$E(\alpha X + \beta) = \alpha E(X) + \beta.$$

Example 4.23   The probability mass function of a discrete random variable X is given by

$$p(x) = \begin{cases} x/15 & x = 1, 2, 3, 4, 5 \\ 0 & \text{otherwise.} \end{cases}$$

What is the expected value of $X(6 - X)$?

Solution: By Theorem 4.2,

$$E\big[X(6 - X)\big] = 5 \cdot \frac{1}{15} + 8 \cdot \frac{2}{15} + 9 \cdot \frac{3}{15} + 8 \cdot \frac{4}{15} + 5 \cdot \frac{5}{15} = 7. ◆$$

Example 4.24   A box contains 10 disks of radii 1, 2, ..., and 10, respectively. What is the expected value of the area of a disk selected at random from this box?

Solution: Let the radius of the disk be R; then R is a random variable with the probability mass function $p(x) = 1/10$ if $x = 1, 2, \ldots, 10$, and $p(x) = 0$ otherwise. $E(\pi R^2)$, the desired quantity is calculated as follows:

$$E(\pi R^2) = \pi E(R^2) = \pi\left(\sum_{i=1}^{10} i^2 \frac{1}{10}\right) = 38.5\pi. \quad \blacklozenge$$

★ **Example 4.25   (Investment)** [†]   In business, **financial assets** are instruments of trade or commerce valued only in terms of the national medium of exchange (i.e., money) having no other intrinsic value. These instruments can be freely bought and sold. They are, in general, divided into three categories. Those guaranteed with a fixed income or income based on a specific formula are **fixed-income securities.** An example would be a bond. Those that depend on a firm's success through the purchase of shares are called **equity.** An example would be common stock. If the financial value of an asset is derived from other assets or depends on the values of other assets, then it is called a **derivative security.** For example, the interest rate of an adjustable-rate mortgage depends on a combination of factors concerning various other interest rates that determine an interest rate index, which in turn determines the periodical adjustments to the mortgage rate.

For the purposes of simplicity we are going to assume that the fixed-income securities are in the form of zero-coupon bonds, meaning they pay no annual interest but provide a return to the investor based on the capital gain over the purchase price. Similarly, we will assume that the instruments based on the firm's success represent businesses that pay no annual dividends but instead promise a return to the investor based on the growth of the company and thus the value of the stock.

Let X be the amount paid to purchase an asset, and let Y be the amount received from the sale of the same asset. Putting fixed-income securities aside, the ratio Y/X is a random variable called the **total return** and is denoted by R. Obviously, $Y = RX$. The ratio $r = (Y - X)/X$ is a random variable called the **rate of return**. If we purchase an asset for $100 and sell it for $120, the total return is 1.2, whereas the rate of return is 0.2. The latter quantity shows that the value of the asset was increased 20%, whereas the former shows that its value reached 120% of its original price. Clearly, $r = (Y/X) - 1 = R - 1$, or $R = 1 + r$.

Every investor has a collection of financial assets, which can be kept, added to, or sold. This collection of financial assets forms the investor's **portfolio.** Let X be the total investment. Suppose that the portfolio of the investor consists of a total of n financial assets. Furthermore, let w_i be the *fraction* of investment in the ith financial asset. Then

---

[†]If this example is skipped, then all exercises and examples in this and future chapters marked "(investment)" should be skipped as well.

$X_i = w_i X$ is the amount invested in the ith financial asset, and w_i is called the **weight** of asset i. Clearly,

$$\sum_{i=1}^{n} X_i = \sum_{i=1}^{n} w_i X = X$$

implies that $\sum_{i}^{n} w_i = 1$. If R_i is the total return of financial asset i, then asset i sells for $R_i w_i X$, and R, the total return is obtained from

$$R = \frac{\displaystyle\sum_{i=1}^{n} R_i w_i X}{X} = \sum_{i=1}^{n} w_i R_i. \tag{4.8}$$

Similarly, r, the rate of return of the portfolio is

$$r = \frac{\displaystyle\sum_{i=1}^{n} R_i w_i X - X}{X} = \sum_{i=1}^{n} R_i w_i X - 1 = \sum_{i=1}^{n} R_i w_i - \sum_{i=1}^{n} w_i$$

$$= \sum_{i=1}^{n} w_i (R_i - 1) = \sum_{i=1}^{n} w_i r_i. \tag{4.9}$$

We have made the following important observations:

> The total return of the portfolio is the weighted sum of the total returns from each financial asst of the portfolio.

> The rate of return of the portfolio is the weighted sum of the rates from the assets of the portfolio.

Putting fixed-income securities aside, for each i, R_i and r_i are random variables, and we have the following formulas for the expected values of R and r:

$$E(R) = \sum_{i=1}^{n} w_i E(R_i), \tag{4.10}$$

$$E(r) = \sum_{i=1}^{n} w_i E(r_i). \tag{4.11}$$

These formulas are simple results of Theorem 10.1 proved in Chapter 10. That theorem is a generalization of the corollary following Theorem 4.2.   ♦

EXERCISES

A

1. There is a story about Charles Dickens (1812–1870), the English novelist and one of the most popular writers in the history of literature. It is known that Dickens was interested in practical applications of mathematics. On the final day in March during a year in the second half of the nineteenth century, he was scheduled to leave London by train and travel about an hour to visit a very good friend. However, Mr. Dickens was aware of the fact that in England there were, on the average, two serious train accidents each month. Knowing that there had been only one serious accident so far during the month of March, Dickens thought that the probability of a serious train accident on the last day of March would be very high. Thus he called his friend and postponed his visit until the next day. He boarded the train on April 1, feeling much safer and believing that he had used his knowledge of mathematics correctly by leaving the next day. He *did* arrive safely! Is there a fallacy in Dickens, argument? Explain.

2. In a certain part of downtown Baltimore parking lots charge $7 per day. A car that is illegally parked on the street will be fined $25 if caught, and the chance of being caught is 60%. If money is the only concern of a commuter who must park in this location every day, should he park at a lot or park illegally?

3. In a lottery every week, 2,000,000 tickets are sold for $1 apiece. If 4000 of these tickets pay off $30 each, 500 pay off $800 each, one ticket pays off $1,200,000, and no ticket pays off more than one prize, what is the expected value of the winning amount for a player with a single ticket?

4. In a lottery, a player pays $1 and selects four distinct numbers from 0 to 9. Then, from an urn containing 10 identical balls numbered from 0 to 9, four balls are drawn at random and without replacement. If the numbers of three or all four of these balls matches the player's numbers, he wins $5 and $10, respectively. Otherwise, he loses. On the average, how much money does the player *gain* per game? (Gain = win − loss.)

5. An urn contains five balls, two of which are marked $1, two $5, and one $15. A game is played by paying $10 for winning the sum of the amounts marked on two balls selected randomly from the urn. Is this a fair game?

6. A box contains 20 fuses, of which five are defective. What is the expected number of defective items among three fuses selected randomly?

7. The demand for a certain weekly magazine at a newsstand is a random variable with probability mass function $p(i) = (10 - i)/18$, $i = 4, 5, 6, 7$. If the magazine

sells for $\$a$ and costs $\$2a/3$ to the owner, and the unsold magazines cannot be returned, how many magazines should be ordered every week to maximize the profit in the long run?

8. It is well known that $\sum_{x=1}^{\infty} 1/x^2 = \pi^2/6$.

   (a) Show that $p(x) = 6/(\pi x)^2$, $x = 1, 2, 3, \ldots$ is the probability mass function of a random variable X.

   (b) Prove that $E(X)$ does not exist.

9. (a) Show that $p(x) = (|x| + 1)^2/27$, $x = -2, -1, 0, 1, 2$, is the probability mass function of a random variable X.

   (b) Calculate $E(X)$, $E(|X|)$, and $E(2X^2 - 5X + 7)$.

10. A box contains 10 disks of radii 1, 2, ... , 10, respectively. What is the expected value of the circumference of a disk selected at random from this box?

11. The distribution function of a random variable X is given by

$$F(x) = \begin{cases} 0 & \text{if } x < -3 \\ 3/8 & \text{if } -3 \leq x < 0 \\ 1/2 & \text{if } 0 \leq x < 3 \\ 3/4 & \text{if } 3 \leq x < 4 \\ 1 & \text{if } x \geq 4. \end{cases}$$

   Calculate $E(X)$, $E(X^2 - 2|X|)$, and $E(X|X|)$.

12. If X is a random number selected from the first 10 positive integers, what is $E[X(11 - X)]$?

13. Let X be the number of different birthdays among four persons selected randomly. Find $E(X)$.

14. A newly married couple decides to continue having children until they have one of each sex. If the events of having a boy and a girl are independent and equiprobable, how many children should this couple expect?
    Hint: Note that $\sum_{i=1}^{\infty} ir^i = r/(1 - r)^2$, $|r| < 1$.

B

15. Suppose that there exist N families on the earth and that the maximum number of children a family has is c. For $j = 0, 1, 2, \ldots, c$, let α_j be the fraction of families with j children $\left(\sum_{j=0}^{c} \alpha_j = 1 \right)$. A *child* is selected at random from the set of all children in the world. Let this child be the Kth born of his or her family; then K is a random variable. Find $E(K)$.

16. An ordinary deck of 52 cards is well-shuffled, and then the cards are turned face up one by one until an ace appears. Find the expected number of cards that are face up.

17. Suppose that n random integers are selected from $\{1, 2, \ldots, N\}$ with replacement. What is the expected value of the largest number selected? Show that for large N the answer is approximately $nN/(n+1)$.

18. **(a)** Show that

$$p(n) = \frac{1}{n(n+1)}, \qquad n \geq 1,$$

is a probability mass function.

(b) Let X be a random variable with probability mass function p given in part (a); find $E(X)$.

19. To an engineering class containing $2n - 3$ male and three female students, there are n work stations available. To assign each workstation to two students, the professor forms n teams one at a time, each consisting of two randomly selected students. In this process, let X be the number of students selected until a team of a male and a female is formed. Find the expected value of X.

∎

4.5 VARIANCES AND MOMENTS OF DISCRETE RANDOM VARIABLES

Thus far, through many examples, we have explained the importance of mathematical expectation in detail. For instance, in Example 4.16, we have shown how expectation is applied in decision making. Also, in Example 4.17, concerning lottery, we showed that the expectation of the winning amount per game gives an excellent estimation for the total amount a player will win if he or she plays a large number of times. In these and many other situations, mathematical expectation is the only quantity one needs to calculate. However, very frequently we face situations in which the expectation by itself does not say much. In such cases more information should be extracted from the probability mass function. As an example, suppose that we are interested in measuring a certain quantity. Let X be the true value[†] of the quantity minus the value obtained by measurement. Then X is the error of measurement. It is a random variable with expected value zero, the reason being that in measuring a quantity a very large number of times, positive and negative errors of the same magnitudes occur with equal probabilities. Now consider an experiment in which a quantity is measured several times, and the average of the errors is obtained to be a number close to zero. Can we conclude that the measurements

---

[†]True value is a nebulous concept. Here we shall use it to mean the average of a *large* number of measurements.

are very close to the true value and thus are accurate? The answer is no because they might differ from the true value by relatively large quantities but be scattered both in positive and negative directions, resulting in zero expectation. Thus in this and similar cases, expectation by itself does not give adequate information, so additional measures for decision making are needed. One such quantity is the **variance** of a random variable.

Variance measures the average magnitude of the fluctuations of a random variable from its expectation. This is particularly important because random variables fluctuate from their expectations. To mathematically define the variance of a random variable X, the first temptation is to consider the expectation of the difference of X from its expectation, that is, $E[X - E(X)]$. But the difficulty with this quantity is that the positive and negative deviations of X from $E(X)$ cancel each other, and we always get 0. This can be seen mathematically from the corollary of Theorem 4.2: Let $E(X) = \mu$; then

$$E[X - E(X)] = E(X - \mu) = E(X) - \mu = E(X) - E(X) = 0.$$

Hence $E[X - E(X)]$ is not an appropriate measure for the variance. However, if we consider $E(|X - E(X)|)$ instead, the problem of negative and positive deviations canceling each other disappears. Since this quantity is the true average magnitude of the fluctuations of X from $E(X)$, it seems that it is the best candidate for an expression for the variance of X. But mathematically, $E(|X - E(X)|)$ is difficult to handle; for this reason the quantity $E[(X - E(X))^2]$, analogous to Euclidean distance in geometry, is used instead and is called the **variance** of X. The square root of $E[(X - E(X))^2]$ is called the **standard deviation** of X.

Definition   *Let X be a discrete random variable with a set of possible values A, probability mass function $p(x)$, and $E(X) = \mu$. Then σ_X and $Var(X)$, called the **standard deviation** and the **variance** of X, respectively, are defined by*

$$\sigma_X = \sqrt{E[(X - \mu)^2]} \quad \text{and} \quad Var(X) = E[(X - \mu)^2].$$

Note that by this definition and Theorem 4.2,

$$\mathbf{Var}(X) = E[(X - \mu)^2] = \sum_{x \in A}(x - \mu)^2 p(x).$$

Let X be a discrete random variable with the set of possible values A and probability mass function $p(x)$. Suppose that the prediction of the value of X is in order, and if the value t is predicted for X, then based on the error $X - t$, a penalty is charged. To minimize the penalty, it seems reasonable to minimize $E[(X - t)^2]$. But

$$E[(X - t)^2] = \sum_{x \in A}(x - t)^2 p(x).$$

Assuming that this series converges $\left[\text{i.e., } E(X^2) < \infty\right]$, we differentiate it to find the minimum value of $E\left[(X - t)^2\right]$:

$$\frac{d}{dt}E\left[(X - t)^2\right] = \frac{d}{dt}\sum_{x \in A}(x - t)^2 p(x) = \sum_{x \in A}-2(x - t)p(x) = 0.$$

This gives

$$\sum_{x \in A} xp(x) = t \sum_{x \in A} p(x) = t.$$

Therefore, $E\left[(X - t)^2\right]$ is a minimum for $t = \sum_{x \in A} xp(x) = E(X)$ and the minimum value is $E\left[\left(X - E(X)\right)^2\right] = \text{Var}(X)$. So the smaller that $\text{Var}(X)$ is, the better $E(X)$ predicts X. We have that

$$\text{Var}(X) = \min_{t} E\left[(X - t)^2\right].$$

Earlier we mentioned that, if we think of a unit mass distributed along the real line at the points of A so that the mass at $x \in A$ is $p(x) = P(X = x)$, then $E(X)$ is the center of gravity. As we know, since center of gravity does not provide any information about how the mass is distributed around this center, the concept of moment of inertia is introduced. Moment of inertia is a measure of dispersion (spread) of the mass distribution about the center of gravity. $E(X)$ is analogous to the center of gravity, and it too does not provide any information about the distribution of X about this center of location. However, variance, the analog of the moment of inertia, measures the dispersion, or spread, of a distribution about its expectation.

Related Historical Remark: In 1900, the Wright brothers were looking for a private location with consistent wind to test their gliders. The data they got from the U.S. Weather Bureau indicated that Kill Devil Hill, near Kitty Hawk, North Carolina, had, on average, suitable winds, so they chose that location for their tests. However, the wind was not consistent. There were many calm days and many days with strong winds that were not suitable for their tests. The summary of the data was obtained by averaging undesirable extreme wind conditions. The Weather Bureau statistics were misleading because they failed to utilize the standard deviation. If the Wright brothers had been provided with the standard deviation of the wind speed at Kitty Hawk, they would not have chosen that location for their tests.   ◆

Example 4.26   Karen is interested in two games, Keno and Bolita. To play Bolita, she buys a ticket for \$1, draws a ball at random from a box of 100 balls numbered 1 to 100. If the ball drawn matches the number on her ticket, she wins \$75; otherwise, she loses. To play Keno, Karen bets \$1 on a single number that has a 25% chance to win. If she wins, they will return her dollar plus two dollars more; otherwise, they keep the dollar. Let B and K be the amounts that Karen gains in one play of Bolita and Keno, respectively. Then

$$E(B) = (74)(0.01) + (-1)(0.99) = -0.25,$$
$$E(K) = (2)(0.25) + (-1)(0.75) = -0.25.$$

Therefore, in the long run, it does not matter which of the two games Karen plays. Her gain would be about the same. However, by virtue of

$$\text{Var}(B) = E\big[(B - \mu)^2\big] = (74 + 0.25)^2(0.01) + (-1 + 0.25)^2(0.99) = 55.69$$

and

$$\text{Var}(K) = E\big[(K - \mu)^2\big] = (2 + 0.25)^2(0.25) + (-1 + 0.25)^2(0.75) = 1.6875,$$

we can say that in Bolita, on average, the deviation of the gain from the expectation is much higher than in Keno. In other words, the risk with Keno is far less than the risk with Bolita. In Bolita, the probability of winning is very small, but the amount one might win is high. In Keno, players win more often but in smaller amounts. ◆

The following theorem states another useful formula for $\text{Var}(X)$.

Theorem 4.3 $\text{Var}(X) = E(X^2) - \big[E(X)\big]^2.$

Proof: By the definition of variance,

$$\begin{aligned}
\text{Var}(X) = E\big[(X - \mu)^2\big] &= E(X^2 - 2\mu X + \mu^2) \\
&= E(X^2) - 2\mu E(X) + \mu^2 \\
&= E(X^2) - 2\mu^2 + \mu^2 = E(X^2) - \mu^2 \\
&= E(X^2) - \big[E(X)\big]^2. \quad ◆
\end{aligned}$$

One immediate application of this formula is that, since $\text{Var}(X) \geq 0$, for any discrete random variable X,

$$\big[E(X)\big]^2 \leq E(X^2).$$

The formula $\text{Var}(X) = E(X^2) - \big[E(X)\big]^2$ is usually a better alternative for computing the variance of X. Here is an example.

Example 4.27   What is the variance of the random variable X, the outcome of rolling a fair die?

Solution: The probability mass function of X is given by $p(x) = 1/6$; $x = 1, 2, 3, 4,$ 5, 6, and $p(x) = 0$, otherwise. Hence

$$E(X) = \sum_{x=1}^{6} xp(x) = \frac{1}{6} \sum_{x=1}^{6} x = \frac{1}{6}(1 + 2 + 3 + 4 + 5 + 6) = \frac{7}{2},$$

$$E(X^2) = \sum_{x=1}^{6} x^2 p(x) = \frac{1}{6} \sum_{x=1}^{6} x^2 = \frac{1}{6}(1 + 4 + 9 + 16 + 25 + 36) = \frac{91}{6}.$$

Thus

$$\text{Var}(X) = E(X^2) - \left[E(X)\right]^2 = \frac{91}{6} - \frac{49}{4} = \frac{35}{12}. \quad \blacklozenge$$

Suppose that a random variable X is constant; then $E(X) = X$ and the deviations of X from $E(X)$ are 0. Therefore, the average deviation of X from $E(X)$ is also 0. We have the following theorem.

Theorem 4.4   *Let X be a discrete random variable with the set of possible values A, and mean μ. Then $\text{Var}(X) = 0$ if and only if X is a constant with probability 1.*

Proof: We will show that $\text{Var}(X) = 0$ implies that $X = \mu$ with probability 1. Suppose not; then there exists some $k \neq \mu$ such that $p(k) = P(X = k) > 0$. But then

$$\text{Var}(X) = (k - \mu)^2 p(k) + \sum_{x \in A - \{k\}} (x - \mu)^2 p(x) > 0,$$

which is a contradiction to $\text{Var}(X) = 0$. Conversely, if X is a constant c with probability 1, then $X = E(X) = c = \mu$ with probability 1. This implies that $\text{Var}(X) = E\left[(X - \mu)^2\right] = 0.$ $\blacklozenge$

For constants a and b, a linear relation similar to $E(aX + b) = aE(X) + b$ does not exist for variance and for standard deviation. However, other important relations exist and are given by the following theorem.

Theorem 4.5   *Let X be a discrete random variable; then for constants a and b we have that*

$$\mathbf{Var}(aX + b) = a^2 \mathbf{Var}(X),$$

$$\sigma_{aX+b} = |a|\sigma_X.$$

Proof: To see this, note that

$$\begin{aligned}
\text{Var}(aX + b) &= E\big[(aX + b) - E(aX + b)\big]^2 \\
&= E\big[(aX + b) - \big(aE(X) + b\big)\big]^2 \\
&= E\big[a\big(X - E(X)\big)\big]^2 \\
&= E\big[a^2\big(X - E(X)\big)^2\big] \\
&= a^2 E\big[\big(X - E(X)\big)^2\big] \\
&= a^2 \text{Var}(X).
\end{aligned}$$

Taking the square roots of both sides of this relation, we find that $\sigma_{aX+b} = |a|\sigma_X$.   ◆

Example 4.28   Suppose that, for a discrete random variable X, $E(X) = 2$ and $E\big[X(X - 4)\big] = 5$. Find the variance and the standard deviation of $-4X + 12$.

Solution: By the Corollary of Theorem 4.2, $E(X^2 - 4X) = 5$ implies that

$$E(X^2) - 4E(X) = 5.$$

Substituting $E(X)$ in this relation gives $E(X^2) = 13$. Hence, by Theorem 4.3,

$$\text{Var}(X) = E(X^2) - \big[E(X)\big]^2 = 13 - 4 = 9,$$
$$\sigma_X = \sqrt{9} = 3.$$

By Theorem 4.5,

$$\text{Var}(-4X + 12) = 16\,\text{Var}(X) = 16 \times 9 = 144,$$
$$\sigma_{-4X+12} = |-4|\sigma_X = 4 \times 3 = 12.   ◆$$

Optional

As we know, variance measures the dispersion, or spread, of a distribution about its expectation. One way to find out which one of the two given random variables X and Y is more dispersed, or spread, about an arbitrary point ω is to see which one is more *concentrated* about ω. The following is a mathematical definition to this concept.

Definition   *Let X and Y be two random variables and ω be a given point. If for all $t > 0$,*

$$P\big(|Y - \omega| \le t\big) \le P\big(|X - \omega| \le t\big),$$

*then we say that X is more **concentrated** about ω than is Y.*

A useful consequence of this definition is the following theorem, the proof of which we leave as an exercise. This theorem should be intuitively clear.

Theorem 4.6   *Suppose that X and Y are two random variables with $E(X) = E(Y) = \mu$. If X is more concentrated about μ than is Y, then $\mathrm{Var}(X) \leq \mathrm{Var}(Y)$.*

MOMENTS

Let X be a random variable with expected value μ. Let c be a constant, $n \geq 0$ be an integer, and $r > 0$ be any real number, integral or not. The expected value of X, $E(X)$, is also called the **first moment** of X. In practice, expected values of some important functions of X have also numerical and theoretical significance. Some of these functions are $g(X) = X^n, |X|^n, X - c, (X - c)^n$, and $(X - \mu)^n$. Provided that $E\big(|g(X)|\big) < \infty$, $E\big[g(X)\big]$ in each of these cases is defined as follows.

$E\big[g(X)\big]$	Definition
$E(X^n)$	The nth moment of X
$E\big(|X|^r\big)$	The rth absolute moment of X
$E(X - c)$	The first moment of X about c
$E\big[(X - c)^n\big]$	The nth moment of X about c
$E\big[(X - \mu)^n\big]$	The nth central moment of X

★ **Remark 4.2**   Let X be a discrete random variable with probability mass function $p(x)$ and set of possible values A. Let n be a positive integer. It is important to know that if $E(X^{n+1})$ exists, then $E(X^n)$ also exists. That is, *the existence of higher moments implies the existence of lower moments.* In particular, this implies that if $E(X^2)$ exists, then $E(X)$ and, hence, $\mathrm{Var}(X)$ exist. To prove this fact, note that, by definition, $E(X^{n+1})$ exists if $\sum_{x \in A} |x|^{n+1} p(x) < \infty$. Let $B = \{x \in A : |x| < 1\}$; then $B^c = \{x \in A : |x| \geq 1\}$. We have

$$\sum_{x \in B} |x|^n p(x) \leq \sum_{x \in B} p(x) \leq \sum_{x \in A} p(x) = 1;$$

$$\sum_{x \in B^c} |x|^n p(x) \leq \sum_{x \in B^c} |x|^{n+1} p(x) \leq \sum_{x \in A} |x|^{n+1} p(x) < \infty.$$

By these inequalities,

$$\sum_{x \in A} |x|^n p(x) = \sum_{x \in B} |x|^n p(x) + \sum_{x \in B^c} |x|^n p(x) \leq 1 + \sum_{x \in A} |x|^{n+1} p(x) < \infty,$$

showing that $E(X^n)$ also exists.   ◆

EXERCISES

A

1. Mr. Jones is about to purchase a business. There are two businesses available. The first has a daily expected profit of $150 with standard deviation $30, and the second has a daily expected profit of $150 with standard deviation $55. If Mr. Jones is interested in a business with a steady income, which should he choose?

2. The temperature of a material is measured by two devices. Using the first device, the expected temperature is t with standard deviation 0.8; using the second device, the expected temperature is t with standard deviation 0.3. Which device measures the temperature more precisely?

3. Find the variance of X, the random variable with probability mass function

$$p(x) = \begin{cases} (|x-3|+1)/28 & x = -3, -2, -1, 0, 1, 2, 3 \\ 0 & \text{otherwise.} \end{cases}$$

4. Find the variance and the standard deviation of a random variable X with distribution function

$$F(x) = \begin{cases} 0 & x < -3 \\ 3/8 & -3 \le x < 0 \\ 3/4 & 0 \le x < 6 \\ 1 & x \ge 6. \end{cases}$$

5. Let X be a random integer from the set $\{1, 2, \ldots, N\}$. Find $E(X)$, $\text{Var}(X)$, and σ_X.

6. What are the expected number, the variance, and the standard deviation of the number of spades in a poker hand? (A poker hand is a set of five cards that are randomly selected from an ordinary deck of 52 cards.)

7. Suppose that X is a discrete random variable with $E(X) = 1$ and $E[X(X-2)] = 3$. Find $\text{Var}(-3X + 5)$.

8. In a game, Emily gives Harry three well-balanced quarters to flip. Harry will get to keep all the ones that will land heads. He will return those landing tails. However, if all three coins land tails, Harry must pay Emily two dollars. Find the expected value and the variance of Harry's net gain.

9. Let X be a random variable defined by

$$P(X = -1) = P(X = 1) = 1/2.$$

Let Y be a random variable defined by

$$P(Y = -10) = P(Y = 10) = 1/2.$$

Which one of X and Y is more concentrated about 0 and why?

B

10. A drunken man has n keys, one of which opens the door to his office. He tries the keys at random, one by one, and independently. Compute the mean and the variance of the number of trials required to open the door if the wrong keys (a) are not eliminated; (b) are eliminated.

11. For $n = 1, 2, 3, \ldots$, let $x_n = (-1)^n \sqrt{n}$. Let X be a discrete random variable with the set of possible values $A = \{x_n : n = 1, 2, 3, \ldots\}$ and probability mass function

$$p(x_n) = P(X = x_n) = \frac{6}{(\pi n)^2}.$$

Show that even though $\sum_{n=1}^{\infty} x_n^3 \, p(x_n) < \infty$, $E(X^3)$ does not exist.

12. Let X be a discrete random variable; let $0 < s < r$. Show that if the rth absolute moment of X exists, then the absolute moment of order s of X also exists.

13. Let X and Y be two discrete random variables with the identical set of possible values $A = \{a, b, c\}$, where a, b, and c are three *different* real numbers. Show that if $E(X) = E(Y)$ and Var$(X) =$Var(Y), then X and Y are identically distributed. That is,

$$P(X = t) = P(Y = t) \quad \text{for } t = a, b, c.$$

14. Let X and Y be two discrete random variables with the identical set of possible values $A = \{a_1, a_2, \ldots, a_n\}$, where $a_1, a_2, \ldots, a_n$ are n *different* real numbers. Show that if

$$E(X^r) = E(Y^r), \quad r = 1, 2, \ldots, n - 1,$$

then X and Y are identically distributed. That is,

$$P(X = t) = P(Y = t) \quad \text{for } t = a_1, a_2, \ldots, a_n.$$

4.6   STANDARDIZED RANDOM VARIABLES

Let X be a random variable with mean μ and standard deviation σ. The random variable $X^* = (X - \mu)/\sigma$ is called the **standardized** X. We have that

$$E(X^*) = E\left(\frac{1}{\sigma}X - \frac{\mu}{\sigma}\right) = \frac{1}{\sigma}E(X) - \frac{\mu}{\sigma} = \frac{\mu}{\sigma} - \frac{\mu}{\sigma} = 0,$$

$$\text{Var}(X^*) = \text{Var}\left(\frac{1}{\sigma}X - \frac{\mu}{\sigma}\right) = \frac{1}{\sigma^2}\text{Var}(X) = \frac{\sigma^2}{\sigma^2} = 1.$$

When standardizing a random variable X, we change the origin to μ and the scale to the units of standard deviation. The value that is obtained for X^* is independent of the units in which X is measured. It is the number of standard deviation units by which X differs from $E(X)$. For example, let X be a random variable with mean 10 feet and standard deviation 2 feet. Suppose that in a random observation we obtain $X = 16$; then $X^* = (16 - 10)/2 = 3$. This shows that the distance of X from its mean is 3 standard deviation units regardless of the scale of measurement. That is, if the same quantities are measured, say, in inches (12 inches = 1 foot), then we will get the same standardized value:

$$X^* = \frac{16 \times 12 - 10 \times 12}{2 \times 12} = 3.$$

Standardization is particularly useful if two or more random variables with different distributions must be compared. Suppose that, for example, a student's grade in a probability test is 72 and that her grade in a history test is 85. At first glance these grades suggest that the student is doing much better in the history course than in the probability course. However, this might not be true—the relative grade of the student in probability might be better than that in history. To illustrate, suppose that the mean and standard deviation of all grades in the history test are 82 and 7, respectively, while these quantities in the probability test are 68 and 4. If we convert the student's grades to their standard deviation units, we find that her standard scores on the probability and history tests are given by $(72 - 68)/4 = 1$ and $(85 - 82)/7 = 0.43$, respectively. These show that her grade in probability is 1 and in history is 0.43 standard deviation unit higher than their respective averages. Therefore, she is doing relatively better in the probability course than in the history course. This comparison is most useful when only the means and standard deviations of the random variables being studied are known. If the distribution functions of these random variables are given, better comparisons might be possible.

We now prove that, for a random variable X, the standardized X, denoted by X^*, is independent of the units in which X is measured. To do so, let X_1 be the observed value of X when a different scale of measurement is used. Then for some $\alpha > 0$, we have that

$X_1 = \alpha X + \beta$, and

$$X_1^* = \frac{X_1 - E(X_1)}{\sigma_{X_1}} = \frac{(\alpha X + \beta) - \left[\alpha E(X) + \beta\right]}{\sigma_{\alpha X + \beta}}$$

$$= \frac{\alpha\left[X - E(X)\right]}{\alpha \sigma_X} = \frac{X - E(X)}{\sigma_X} = X^*.$$

EXERCISES

1. Mr. Norton owns two appliance stores. In store 1 the number of TV sets sold by a salesperson is, on average, 13 per week with a standard deviation of five. In store 2 the number of TV sets sold by a salesperson is, on average, seven with a standard deviation of four. Mr. Norton has a position open for a person to sell TV sets. There are two applicants. Mr. Norton asked one of them to work in store 1 and the other in store 2, each for one week. The salesperson in store 1 sold 10 sets, and the salesperson in store 2 sold six sets. Based on this information, which person should Mr. Norton hire?

2. The mean and standard deviation in midterm tests of a probability course are 72 and 12, respectively. These quantities for final tests are 68 and 15. What final grade is comparable to Velma's 82 in the midterm.   ■

REVIEW PROBLEMS

1. An urn contains 10 chips numbered from 0 to 9. Two chips are drawn at random and without replacement. What is the probability mass function of their total?

2. A word is selected at random from the following poem of Persian poet and mathematician Omar Khayyām (1048–1131), translated by English poet Edward Fitzgerald (1808–1883). Find the expected value of the length of the word.

> The moving finger writes and, having writ,
> Moves on; nor all your Piety nor Wit
> Shall lure it back to cancel half a line,
> Nor all your tears wash out a word of it.

3. A statistical survey shows that only 2% of secretaries know how to use the highly sophisticated word processor language T$_E$X. If a certain mathematics department prefers to hire a secretary who knows T$_E$X, what is the least number of applicants that should be interviewed so as to have at least a 50% chance of finding one such secretary?

4. An electronic system fails if both of its components fail. Let X be the time (in hours) until the system fails. Experience has shown that

$$P(X > t) = \left(1 + \frac{t}{200}\right)e^{-t/200}, \qquad t \geq 0.$$

What is the probability that the system lasts at least 200 but not more than 300 hours?

5. A professor has prepared 30 exams of which 8 are difficult, 12 are reasonable, and 10 are easy. The exams are mixed up, and the professor selects four of them at random to give to four sections of the course he is teaching. How many sections would be expected to get a difficult test?

6. The annual amount of rainfall (in centimeters) in a certain area is a random variable with the distribution function

$$F(x) = \begin{cases} 0 & x < 5 \\ 1 - (5/x^2) & x \geq 5. \end{cases}$$

What is the probability that next year it will rain (a) at least 6 centimeters; (b) at most 9 centimeters; (c) at least 2 and at most 7 centimeters?

7. Let X be the amount (in fluid ounces) of soft drink in a randomly chosen bottle from company A, and Y be the amount of soft drink in a randomly chosen bottle from company B. A study has shown that the probability distributions of X and Y are as follows:

x	15.85	15.9	16	16.1	16.2
$P(X = x)$	0.15	0.21	0.35	0.15	0.14
$P(Y = x)$	0.14	0.05	0.64	0.08	0.09

Find $E(X)$, $E(Y)$, Var(X), and Var(Y) and interpret them.

8. The fasting blood-glucose levels of 30 children are as follows.

```
58  62  80  58  64  76  80  80  80  58
62  64  76  76  58  64  62  80  58  58
80  64  58  62  76  62  64  80  62  76
```

Let X be the fasting blood-glucose level of a child chosen randomly from this group. Find the distribution function of X.

9. Experience shows that X, the number of customers entering a post office, during any period of length t, is a random variable the probability mass function of which is of the form

$$p(i) = k\frac{(2t)^i}{i!}, \qquad i = 0, 1, 2, \ldots .$$

   (a) Determine the value of k.

   (b) Compute $P(X < 4)$ and $P(X > 1)$.

10. From the set of families with three children *a family* is selected at random, and the number of its boys is denoted by the random variable X. Find the probability mass function and the probability distribution functions of X. Assume that in a three-child family all gender distributions are equally probable.

   The following exercise, a *truly challenging* one, is an example of a game in which despite a low probability of winning, the expected length of the play is high.

11. **(The Clock Solitaire)**   An ordinary deck of 52 cards is well shuffled and dealt face down into 13 equal piles. The first 12 piles are arranged in a circle like the numbers on the face of a clock. The 13th pile is placed at the center of the circle. Play begins by turning over the bottom card in the center pile. If this card is a king, it is placed face up on the top of the center pile, and a new card is drawn from the bottom of this pile. If the card drawn is not a king, then (counting the jack as 11 and the queen as 12) it is placed face up on the pile located in the hour position corresponding to the number of the card. Whichever pile the card drawn is placed on, a new card is drawn from the bottom of that pile. This card is placed face up on the pile indicated (either the hour position or the center depending on whether the card is or is not a king) and the play is repeated. The game ends when the 4th king is placed on the center pile. If that occurs on the last remaining card, the player wins. The number of cards turned over until the 4th king appears determines the length of the game. Therefore, the player wins if the length of the game is 52.

   (a) Find $p(j)$, the probability that the length of the game is j. That is, the 4th king will appear on the jth card.

   (b) Find the probability that the player wins.

   (c) Find the expected length of the game.

Chapter 5

Special Discrete Distributions

In this chapter we study some examples of discrete random variables. These random variables appear frequently in theory and applications of probability, statistics, and branches of science and engineering.

5.1 BERNOULLI AND BINOMIAL RANDOM VARIABLES

Bernoulli trials, named after the Swiss mathematician James Bernoulli, are perhaps the simplest type of random variable. They have only two possible outcomes. One outcome is usually called a **success**, denoted by s. The other outcome is called a **failure,** denoted by f. The experiment of flipping a coin is a Bernoulli trial. Its only outcomes are "heads" and "tails." If we are interested in heads, we may call it a success; tails is then a failure. The experiment of tossing a die is a Bernoulli trial if, for example, we are interested in knowing whether the outcome is odd or even. An even outcome may be called a success, and hence an odd outcome a failure, or vice versa. If a fuse is inspected, it is either "defective" or it is "good." So the experiment of inspecting fuses is a Bernoulli trial. A good fuse may be called a success, a defective fuse a failure.

The sample space of a Bernoulli trial contains two points, s and f. The random variable defined by $X(s) = 1$ and $X(f) = 0$ is called a **Bernoulli random variable.** Therefore, a Bernoulli random variable takes on the value 1 when the outcome of the Bernoulli trial is a success and 0 when it is a failure. If p is the probability of a success, then $1 - p$ (sometimes denoted q) is the probability of a failure. Hence the probability mass function of X is

$$p(x) = \begin{cases} 1 - p \equiv q & \text{if } x = 0 \\ p & \text{if } x = 1 \\ 0 & \text{otherwise.} \end{cases} \tag{5.1}$$

188

Note that the same symbol p is used for the probability mass function and the Bernoulli parameter. This duplication should not be confusing since the p's used for the probability mass function often appear in the form $p(x)$.

An accurate mathematical definition for Bernoulli random variables is as follows.

Definition   *A random variable is called **Bernoulli** with parameter p if its probability mass function is given by equation (5.1).*

From (5.1) it follows that the expected value of a Bernoulli random variable X, with parameter p, is p, because

$$E(X) = 0 \cdot P(X = 0) + 1 \cdot P(X = 1) = P(X = 1) = p.$$

Also, since

$$E(X^2) = 0 \cdot P(X = 0) + 1 \cdot P(X = 1) = p,$$

we have

$$\text{Var}(X) = E(X^2) - \left[E(X)\right]^2 = p - p^2 = p(1 - p).$$

We will now summarize what we have shown.

For a Bernoulli random variable X with parameter p, $0 < p < 1$,

$$E(X) = p, \quad \text{Var}(X) = p(1 - p), \quad \sigma_X = \sqrt{p(1 - p)}.$$

Example 5.1   If in a throw of a fair die the event of obtaining 4 or 6 is called a success, and the event of obtaining 1, 2, 3, or 5 is called a failure, then

$$X = \begin{cases} 1 & \text{if 4 or 6 is obtained} \\ 0 & \text{otherwise} \end{cases}$$

is a Bernoulli random variable with the parameter $p = 1/3$. Therefore, its probability mass function is

$$p(x) = \begin{cases} 2/3 & \text{if } x = 0 \\ 1/3 & \text{if } x = 1 \\ 0 & \text{elsewhere.} \end{cases}$$

The expected value of X is given by $E(X) = p = 1/3$, and its variance by $\text{Var}(X) = 1/3(1 - 1/3) = 2/9$. $\blacklozenge$

Let $X_1, X_2, X_3, \ldots$ be a sequence of Bernoulli random variables. If, for all $j_i \in \{0, 1\}$, the sequence of events $\{X_1 = j_1\}, \{X_2 = j_2\}, \{X_3 = j_3\}, \ldots$ are independent, we say that $\{X_1, X_2, X_3, \ldots\}$ and the corresponding Bernoulli trials are independent.

Although Bernoulli trials are simple, if they are repeated independently, they may pose interesting and even sometimes complicated questions. Consider an experiment in which n Bernoulli trials are performed independently. The sample space of such an experiment, S, is the set of different sequences of length n with x ($x = 0, 1, \ldots, n$) successes (s's) and ($n - x$) failures (f's). For example, if, in an experiment, three Bernoulli trials are performed independently, then the sample space is

$$\{fff, sff, fsf, ffs, fss, sfs, ssf, sss\}.$$

If n Bernoulli trials all with probability of success p are performed independently, then X, the number of successes, is one of the most important random variables. It is called a **binomial with parameters n and p**. The set of possible values of X is $\{0, 1, 2, \ldots, n\}$, it is defined on the set S described previously, and its probability mass function is given by the following theorem.

Theorem 5.1   *Let X be a binomial random variable with parameters n and p. Then $p(x)$, the probability mass function of X, is*

$$p(x) = P(X = x) = \begin{cases} \binom{n}{x} p^x (1 - p)^{n-x} & \text{if } x = 0, 1, 2, \ldots, n \\ \\ 0 & \text{elsewhere.} \end{cases} \tag{5.2}$$

Proof: Observe that the number of ways that, in n Bernoulli trials, x ($x = 0, 1, 2, \ldots, n$) successes can occur is equal to the number of different sequences of length n with x successes (s's) and ($n - x$) failures (f's). But the number of such sequences is $\binom{n}{x}$ because the number of distinguishable permutations of n objects of two different types, where x are alike and $n - x$ are alike is $\dfrac{n!}{x!\,(n - x)!} = \binom{n}{x}$ (see Theorem 2.4). Since by the independence of the trials the probability of each of these sequences is $p^x (1 - p)^{n-x}$, we have $P(X = x) = \binom{n}{x} p^x (1 - p)^{n-x}$. Hence (5.2) follows.   ◆

Definition   *The function $p(x)$ given by equation (5.2) is called the **binomial** probability mass function with parameters (n, p).*

The reason for this name is that the binomial expansion theorem (Theorem 2.5) guarantees that p is a probability mass function:

$$\sum_{x=0}^{n} p(x) = \sum_{x=0}^{n} \binom{n}{x} p^x (1 - p)^{n-x} = \left[p + (1 - p) \right]^n = 1^n = 1.$$

Example 5.2   A restaurant serves 8 entrées of fish, 12 of beef, and 10 of poultry. If customers select from these entrées randomly, what is the probability that two of the next four customers order fish entrées?

Solution: Let X denote the number of fish entrées (successes) ordered by the next four customers. Then X is binomial with the parameters $(4, 8/30 = 4/15)$. Thus

$$P(X = 2) = \binom{4}{2}\left(\frac{4}{15}\right)^2\left(\frac{11}{15}\right)^2 = 0.23. \quad \blacklozenge$$

Example 5.3   In a county hospital 10 babies, of whom six were boys, were born last Thursday. What is the probability that the first six births were all boys? Assume that the events that a child born is a girl or is a boy are equiprobable.

Solution: Let A be the event that the first six births were all boys and the last four all girls. Let X be the number of boys; then X is binomial with parameters 10 and 1/2. The desired probability is

$$P(A \mid X = 6) = \frac{P(A \text{ and } X = 6)}{P(X = 6)} = \frac{P(A)}{P(X = 6)}$$

$$= \frac{\left(\frac{1}{2}\right)^{10}}{\binom{10}{6}\left(\frac{1}{2}\right)^6\left(\frac{1}{2}\right)^4} = \frac{1}{\binom{10}{6}} \approx 0.0048.$$

Remark: We can also argue as follows: *bbbbbbgggg*, the event that the first six were all boys, and the remainder girls, is one sequence out of the $\binom{10}{6}$ possible sequences of the sexes of the births. Since each of these sequences has the same probability of occurrence, the answer is $1 \Big/ \binom{10}{6}$. $\quad \blacklozenge$

Example 5.4   In a small town, out of 12 accidents that occurred in June 1986, four happened on Friday the 13th. Is this a good reason for a superstitious person to argue that Friday the 13th is inauspicious?

Solution: Suppose the probability that each accident occurs on Friday the 13th is 1/30, just as on any other day. Then the probability of at least four accidents on Friday the 13th is

$$1 - \sum_{i=0}^{3} \binom{12}{i}\left(\frac{1}{30}\right)^i\left(\frac{29}{30}\right)^{12-i} \approx 0.000,493.$$

Since the probability of four or more of these accidents occurring on Friday the 13th is very small, this is a good reason for superstitious persons to argue that Friday the 13th is inauspicious.   ◆

Example 5.5   A realtor claims that only 30% of the houses in a certain neighborhood are appraised at less than $200,000. A random sample of 20 houses from that neighborhood is selected and appraised. The results in (thousands of dollars) are as follows:

$$\begin{array}{ccccccc}
285 & 156 & 202 & 306 & 276 & 562 & 415 \\
245 & 185 & 143 & 186 & 377 & 225 & 192 \\
510 & 222 & 264 & 198 & 168 & 363 &
\end{array}$$

Based on these data, is the realtor's claim acceptable?

Solution: Suppose that the realtor's claim is acceptable, and $p = 0.30$ is the probability that a randomly selected house is appraised at less than $200,000. In a randomly selected sample of 20 houses, let X be the number that are less than $200,000. Since in the data given there are seven such houses, we will calculate $P(X \geq 7)$. If this probability is too small, we will reject the claim of the realtor; otherwise, we will accept it. To find $P(X \geq 7)$, note that X is binomial with parameters 20 and 0.30. Therefore,

$$P(X \geq 7) = 1 - P(X \leq 6) = 1 - \sum_{i=0}^{6} \binom{20}{i}(0.30)^i (0.70)^{20-i} \approx 0.392,$$

which shows that the event that seven or more houses are appraised at less than $200,000 is highly probable. So the claim of the realtor should be accepted.   ◆

Example 5.6   Suppose that jury members decide independently and that each with probability p $(0 < p < 1)$ makes the correct decision. If the decision of the majority is final, which is preferable: a three-person jury or a single juror?

Solution: Let X denote the number of persons who decide correctly among a three-person jury. Then X is a binomial random variable with parameters $(3, p)$. Hence the probability that a three-person jury decides correctly is

$$P(X \geq 2) = P(X = 2) + P(X = 3) = \binom{3}{2}p^2(1 - p) + \binom{3}{3}p^3(1 - p)^0$$
$$= 3p^2(1 - p) + p^3 = 3p^2 - 2p^3.$$

Since the probability is p that a single juror decides correctly, a three-person jury is preferable to a single juror if and only if

$$3p^2 - 2p^3 > p.$$

This is equivalent to $3p - 2p^2 > 1$, so $-2p^2 + 3p - 1 > 0$. But $-2p^2 + 3p - 1 = 2(1 - p)(p - 1/2)$. Since $1 - p > 0$, $2(1 - p)(p - 1/2) > 0$ if and only if $p > 1/2$. Hence a three-person jury is preferable if $p > 1/2$. If $p < 1/2$, the decision of a single juror is preferable. For $p = 1/2$ there is no difference.   ♦

Example 5.7   Let p be the probability that a randomly chosen person is against abortion, and let X be the number of persons against abortion in a random sample of size n. Suppose that, in a particular random sample of n persons, k are against abortion. Show that $P(X = k)$ is maximum for $\hat{p} = k/n$. That is, $\hat{p}$ is the value of p that makes the outcome $X = k$ *most probable*.

Solution: By definition of X,

$$P(X = k) = \binom{n}{k} p^k (1 - p)^{n-k}.$$

This gives

$$\frac{d}{dp} P(X = k) = \binom{n}{k} \left[k p^{k-1} (1 - p)^{n-k} - (n - k) p^k (1 - p)^{n-k-1} \right]$$

$$= \binom{n}{k} p^{k-1} (1 - p)^{n-k-1} \left[k(1 - p) - (n - k) p \right].$$

Letting $\dfrac{d}{dp} P(X = k) = 0$, we obtain $p = k/n$. Now since $\dfrac{d^2}{dp^2} P(X = k) < 0$, $\hat{p} = k/n$ is the maximum of $P(X = k)$, and hence it is an estimate of p that makes the outcome $x = k$ most probable.   ♦

Let X be a binomial random variable with parameters (n, p), $0 < p < 1$, and probability mass function $p(x)$. We will now find the value of X at which $p(x)$ is maximum. For any real number t, let $[t]$ denote the largest integer less than or equal to t. We will prove that $p(x)$ is maximum at $x = \left[(n + 1)p \right]$. To do so, we note that

$$\frac{p(x)}{p(x - 1)} = \frac{\dfrac{n!}{(n - x)! \, x!} \, p^x (1 - p)^{n-x}}{\dfrac{n!}{(n - x + 1)! \, (x - 1)!} \, p^{x-1} (1 - p)^{n-x+1}}$$

$$= \frac{(n - x + 1)p}{x(1 - p)} \tag{5.3}$$

$$= \frac{(n + 1)p - xp + x - x}{x(1 - p)} = \frac{\left[(n + 1)p - x \right] + x(1 - p)}{x(1 - p)}$$

$$= \frac{(n + 1)p - x}{x(1 - p)} + 1.$$

This equality shows that $p(x) > p(x-1)$ if and only if $(n+1)p-x > 0$, or, equivalently, if and only if $x < (n+1)p$. Hence *as x changes from 0 to $[(n+1)p]$, $p(x)$ increases. As x changes from $[(n+1)p]$ to n, $p(x)$ decreases. The maximum value of $p(x)$ [the peak of the graphical representation of $p(x)$] occurs at $[(n+1)p]$.*

What we have just shown is illustrated by the graphical representations of $p(x)$ in Figure 5.1 for binomial random variables with parameters $(5, 1/2)$, $(10, 1/2)$, and $(20, 1/2)$.

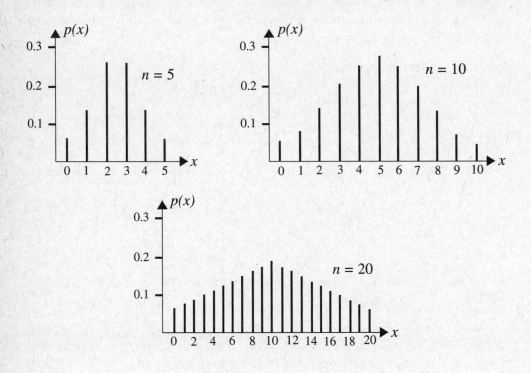

Figure 5.1   Examples of binomial probability mass functions.

Expectations and Variances of Binomial Random Variables

Let X be a binomial random variable with parameters (n, p). Intuitively, we expect that the expected value of X will be np. For example, if we toss a fair coin 100 times, we expect that the average number of heads will be 50, which is $100 \cdot 1/2 = np$. Also, if we choose 10 fuses from a set with 30% defective fuses, we expect the average number of defective fuses to be $np = 10(0.30) = 3$. The formula $E(X) = np$ can be verified directly from the definition of mathematical expectation as follows:

$$E(X) = \sum_{x=0}^{n} x \binom{n}{x} p^x (1-p)^{n-x} = \sum_{x=1}^{n} x \frac{n!}{x!\,(n-x)!} p^x (1-p)^{n-x}$$

$$= \sum_{x=1}^{n} \frac{n!}{(x-1)!\,(n-x)!} p^x (1-p)^{n-x}$$

$$= np \sum_{x=1}^{n} \frac{(n-1)!}{(x-1)!\,(n-x)!} p^{x-1}(1-p)^{n-x}$$

$$= np \sum_{x=1}^{n} \binom{n-1}{x-1} p^{x-1}(1-p)^{n-x}.$$

Letting $i = x - 1$ (by reindexing this sum), we obtain

$$E(X) = np \sum_{i=0}^{n-1} \binom{n-1}{i} p^i (1-p)^{(n-1)-i} = np\big[p + (1-p)\big]^{n-1} = np,$$

where the next-to-last equality follows from binomial expansion (Theorem 2.5). To calculate the variance of X, from a procedure similar to the one we used to compute $E(X)$, we obtain (see Exercise 25)

$$E(X^2) = \sum_{x=1}^{n} x^2 \binom{n}{x} p^x (1-p)^{n-x} = n^2 p^2 - np^2 + np.$$

Therefore,

$$\text{Var}(X) = E(X^2) - \big[E(X)\big]^2 = -np^2 + np = np(1-p).$$

We have established that

If X is a binomial random variable with parameters n and p, then

$$E(X) = np, \quad \text{Var}(X) = np(1-p), \quad \sigma_X = \sqrt{np(1-p)}.$$

Example 5.8   A town of 100,000 inhabitants is exposed to a contagious disease. If the probability that a person becomes infected is 0.04, what is the expected number of people who become infected?

Solution: Assuming that people become infected independently of each other, the number of those who will become infected is a binomial random variable with parameters 100,000 and 0.04. Thus the expected number of such people is $100,000 \times 0.04 = 4000$. Note that here "getting infected" is implicitly defined to be a success for mathematical simplicity! Therefore, in the context of Bernoulli trials, what is called a success may be a failure in real life.   ◆

Example 5.9   Two proofreaders, Ruby and Myra, read a book independently and found r and m misprints, respectively. Suppose that the probability that a misprint is noticed by

Ruby is p and the probability that it is noticed by Myra is q, where these two probabilities are independent. If the number of misprints noticed by *both* Ruby and Myra is b, estimate the number of unnoticed misprints. [This problem was posed and solved by George Pólya (1888–1985) in the January 1976 issue of the *American Mathematical Monthly*.]

Solution:　Let M be the total number of misprints in the book. The expected number of misprints that may be noticed by Ruby, Myra, and both of them are Mp, Mq, and Mpq, respectively. Assuming that the number of misprints found are approximately equal to the expected number, we have $Mp \approx r$, $Mq \approx m$, and $Mpq \approx b$. Therefore,

$$M = \frac{(Mp)(Mq)}{Mpq} \approx \frac{rm}{b}.$$

The number of unnoticed misprints is therefore estimated by

$$M - (r + m - b) \approx \frac{rm}{b} - (r + m - b) = \frac{(r - b)(m - b)}{b}. \quad \blacklozenge$$

EXERCISES

A

1. From an ordinary deck of 52 cards, cards are drawn at random and with replacement. What is the probability that, of the first eight cards drawn, four are spades?

2. The probability that a randomly selected person is female is 1/2. What is the expected number of the girls in the first-grade classes of an elementary school that has 64 first graders? What is the expected number of females in a family with six children?

3. A graduate class consists of six students. What is the probability that exactly three of them are born either in April or in October?

4. In a state where license plates contain six digits, what is the probability that the license number of a randomly selected car has two 9's? Assume that each digit of the license number is randomly selected from $\{0, 1, \ldots, 9\}$.

5. A box contains 30 balls numbered 1 through 30. Suppose that five balls are drawn at random, one at a time, with replacement. What is the probability that the numbers of two of them are prime?

6. A manufacturer of nails claims that only 3% of its nails are defective. A random sample of 24 nails is selected, and it is found that two of them are defective. Is it fair to reject the manufacturer's claim based on this observation?

7. Only 60% of certain kinds of seeds germinate when planted under normal conditions. Suppose that four such seeds are planted, and X denotes the number of those that will germinate. Find the probability mass functions of X and $Y = 2X + 1$.

8. Suppose that the Internal Revenue Service will audit 20% of income tax returns reporting an annual gross income of over $80,000. What is the probability that of 15 such returns, at most four will be audited?

9. If two fair dice are rolled 10 times, what is the probability of at least one 6 (on either die) in exactly five of these 10 rolls?

10. From the interval $(0, 1)$, five points are selected at random and independently. What is the probability that (a) at least two of them are less than 1/3; (b) the first decimal point of exactly two of them is 3?

11. Let X be a binomial random variable with parameters (n, p) and probability mass function $p(x)$. Prove that if $(n + 1)p$ is an integer, then $p(x)$ is maximum at two different points. Find both of these points.

12. On average, how many times should Ernie play poker in order to be dealt a straight flush (royal flush included)? (See Exercise 20 of Section 2.4 for definitions of a royal and a straight flush.)

13. Suppose that each day the price of a stock moves up 1/8th of a point with probability 1/3 and moves down 1/8th of a point with probability 2/3. If the price fluctuations from one day to another are independent, what is the probability that after six days the stock has its original price?

14. From the set $\{x : 0 \le x \le 1\}$, 100 independent numbers are selected at random and rounded to three decimal places. What is the probability that at least one of them is 0.345?

15. A certain basketball player makes a foul shot with probability 0.45. Determine for what value of k the probability of k baskets in 10 shots is maximum, and find this maximum probability.

16. What are the expected value and variance of the number of full house hands in n poker hands? (See Exercise 20 of Section 2.4 for a definition of a full house.)

17. What is the probability that at least two of the six members of a family are not born in the fall? Assume that all seasons have the same probability of containing the birthday of a person selected randomly.

18. A certain rare blood type can be found in only 0.05% of people. If the population of a randomly selected group is 3000, what is the probability that at least two persons in the group have this rare blood type?

19. Edward's experience shows that 7% of the parcels he mails will not reach their destination. He has bought two books for $20 each and wants to mail them to his

brother. If he sends them in one parcel, the postage is \$5.20, but if he sends them in separate parcels, the postage is \$3.30 for each book. To minimize the expected value of his expenses (loss + postage), which way is preferable to send the books, as two separate parcels or as a single parcel?

20. A woman and her husband want to have a 95% chance for at least one boy and at least one girl. What is the minimum number of children that they should plan to have? Assume that the events that a child is a girl and a boy are equiprobable and independent of the gender of other children born in the family.

21. A computer network consists of several stations connected by various media (usually cables). There are certain instances when no message is being transmitted. At such "suitable instances," each station will send a message with probability p independently of the other stations. However, if two or more stations send messages, a collision will corrupt the messages and they will be discarded. These messages will be retransmitted until they reach their destination. Suppose that the network consists of N stations.

   (a) What is the probability that at a "suitable instance" a message is initiated by one of the stations and will go through without a collision?

   (b) Show that, to maximize the probability of a message going through with no collisions, exactly one message, on average, should be initiated at each "suitable instance."

   (c) Find the limit of the maximum probability obtained in (b) as the number of stations of the network grows to ∞.

22. Consider the following problem posed by Michael Khoury, U.S. Math Olympiad Team Member, in "The Problem Solving Competition," Oklahoma Publishing Company and the American Society for the Communication of Mathematics, February 1999.

> Bob is teaching a class with n students. There are n desks in the classroom numbered from 1 to n. Bob has prepared a seating chart, but the students have already seated themselves randomly. Bob calls off the name of the person who belongs in seat 1. This person vacates the seat he or she is currently occupying and takes his or her rightful seat. If this displaces a person already in the seat, that person stands at the front of the room until he or she is assigned a seat. Bob does this for each seat in turn.

Let X be the number of students standing at the front of the room after k, $1 \leq k < n$, names have been called. Call a person "success" if he or she is standing. Is X a binomial random variable? Why or why not?

B

23. In a community, a persons are for abortion, b ($b < a$) are against it, and n ($n > a - b$) are undecided. Suppose that there will be a vote to determine the will of the majority with regard to legalizing abortion. If by then all of the undecided persons make up their minds, what is the probability that those against abortion will win? Assume that it is equally likely that an undecided person eventually votes against or for abortion.

24. A game often played in carnivals and gambling houses is called *chuck-a-luck*, where a player bets on any number 1 through 6. Then three fair dice are tossed. If one, two, or all three land the same number as the player's, then he or she receives one, two, or three times the original stake plus his or her original bet, respectively. Otherwise, the player loses his or her stake. Let X be the net gain of the player per unit of stake. First find the probability mass function of X; then determine the expected amount that the player will lose per unit of stake.

25. Let X be a binomial random variable with the parameters (n, p). Prove that

$$E(X^2) = \sum_{x=1}^{n} x^2 \binom{n}{x} p^x (1 - p)^{n-x} = n^2 p^2 - np^2 + np.$$

26. Suppose that an aircraft engine will fail in flight with probability $1 - p$ independently of the plane's other engines. Also suppose that a plane can complete the journey successfully if at least half of its engines do not fail.

    (a) Is it true that a four-engine plane is always preferable to a two-engine plane? Explain.

    (b) Is it true that a five-engine plane is always preferable to a three-engine plane? Explain.

27. The simplest error detection scheme used in data communication is **parity checking**. Usually messages sent consist of characters, each character consisting of a number of bits (a *bit* is the smallest unit of information and is either 1 or 0). In parity checking, a 1 or 0 is appended to the end of each character at the transmitter to make the total number of 1's even. The receiver checks the number of 1's in every character received, and if the result is odd it signals an error. Suppose that each bit is received correctly with probability 0.999, independently of other bits. What is the probability that a 7-bit character is received in error, but the error is not detected by the parity check?

28. In Exercise 27, suppose that a message consisting of six characters is transmitted. If each character consists of seven bits, what is the probability that the message is erroneously received, but none of the errors is detected by the parity check?

29. How many games of poker occur until a preassigned player is dealt at least one straight flush with probability of at least 3/4? (See Exercise 20 of Section 2.4 for a definition of a straight flush.)

30. (Genetics)   In a population of n diploid organisms with alternate dominant allele A and recessive allele a, those inheriting aa will not survive. Suppose that, in the population, the number of AA individuals is α, and the number of Aa individuals is $n - \alpha$. Suppose that, as a result of random mating, m ($m > 2$) offspring are produced.

   (a) What is the probability that at most two of the offspring are aa?

   (b) For $0 \le i \le m$, what is the probability that exactly i of the offspring are AA and the remaining are all Aa?

31. The postoffice of a certain small town has only one clerk to wait on customers. The probability that a customer will be served in any given minute is 0.6, regardless of the time that the customer has already taken. The probability of a new customer arriving is 0.45, regardless of the number of customers already in line. The chances of two customers arriving during the same minute are negligible. Similarly, the chances of two customers being served in the same minute are negligible. Suppose that we start with exactly two customers: one at the postal window and one waiting on line. After 4 minutes, what is the probability that there will be exactly four customers: one at the window and three waiting on line?

32. **(a)** What is the probability of an even number of successes in n independent Bernoulli trials?

   Hint: Let r_n be the probability of an even number of successes in n Bernoulli trials. By conditioning on the first trial and using the law of total probability (Theorem 3.3), show that for $n \ge 1$,

$$r_n = p(1 - r_{n-1}) + (1 - p)r_{n-1}.$$

   Then prove that $r_n = \dfrac{1}{2}\big[1 + (1 - 2p)^n\big]$.

   (b) Prove that

$$\sum_{k=0}^{[n/2]} \binom{n}{2k} p^{2k}(1 - p)^{n-2k} = \frac{1}{2}\big[1 + (1 - 2p)^n\big].$$

33. An urn contains n balls whose colors, red or blue, are equally probable. $\big[$For example, the probability that all of the balls are red is $(1/2)^n$.$\big]$ If in drawing k balls from the urn, successively with replacement and randomly, no red balls appear, what is the probability that the urn contains no red balls?
   Hint: Use Bayes' theorem.

34. While Rose always tells the truth, four of her friends, Albert, Brenda, Charles, and Donna, tell the truth randomly only in one out of three instances, independent of each other. Albert makes a statement. Brenda tells Charles that Albert's statement is the truth. Charles tells Donna that Brenda is right, and Donna says to Rose that Charles is telling the truth and Rose agrees. What is the probability that Albert's statement is the truth?

■

5.2   POISSON RANDOM VARIABLES

Poisson as an Approximation to Binomial

From Section 5.1, it should have become clear that in many natural phenomena and real-world problems, we are dealing with binomial random variables. Therefore, for possible values of parameters n, x, and p, we need to calculate $p(x) = \binom{n}{x} p^x (1 - p)^{n-x}$. However, in many cases, direct calculation of $p(x)$ from this formula is not possible, because even for moderate values of n, $n!$ exceeds the largest integer that a computer can store. To overcome this difficulty, indirect methods for calculation of $p(x)$ are developed. For example, we may use the recursive formula

$$p(x) = \frac{(n - x + 1)p}{x(1 - p)} \, p(x - 1),$$

obtained from (5.3), with the initial condition $p(0) = (1 - p)^n$, to evaluate $p(x)$. An important study concerning calculation of $p(x)$ is the one by the French mathematician Simeon-Denis Poisson in 1837 in his book concerning the applications of probability to law. Poisson introduced the following procedure to obtain the formula that approximates the binomial probability mass function when the number of trials is large ($n \to \infty$), the probability of success is small ($p \to 0$), and the average number of successes remains a fixed quantity of moderate value ($np = \lambda$ for some constant λ).

Let X be a binomial random variable with parameters (n, p); then

$$P(X = i) = \binom{n}{i} p^i (1 - p)^{n-i} = \frac{n!}{(n - i)! \, i!} \left(\frac{\lambda}{n}\right)^i \left(1 - \frac{\lambda}{n}\right)^{n-i}$$

$$= \frac{n(n - 1)(n - 2) \cdots (n - i + 1)}{n^i} \frac{\lambda^i}{i!} \frac{\left(1 - \frac{\lambda}{n}\right)^n}{\left(1 - \frac{\lambda}{n}\right)^i}.$$

Now, for large n and appreciable λ, $(1 - \lambda/n)^i$ is approximately 1, $(1 - \lambda/n)^n$ is approximately $e^{-\lambda}$ [from calculus we know that $\lim_{n\to\infty}(1 + x/n)^n = e^x$; thus

$\lim_{n \to \infty} (1 - \lambda/n)^n = e^{-\lambda}$], and $[n(n-1)(n-2) \cdots (n-i+1)]/n^i$ is approximately 1, because its numerator and denominator are both polynomials in n of degree i. Thus as $n \to \infty$,

$$P(X = i) \to \frac{e^{-\lambda}\lambda^i}{i!}.$$

Three years after his book was published, Poisson died. The importance of this approximation was left unknown until 1889, when the German-Russian mathematician L. V. Bortkiewicz demonstrated its significance for both in probability theory and its applications. Among other things, Bortkiewicz argued that since

$$\sum_{i=0}^{\infty} \frac{e^{-\lambda}\lambda^i}{i!} = e^{-\lambda} \sum_{i=0}^{\infty} \frac{\lambda^i}{i!} = e^{-\lambda}e^{\lambda} = 1,$$

Poisson's approximation by itself is a probability mass function. This and the introduction of the Poisson processes in the twentieth century made the Poisson probability mass function one of the three most important probability functions, the other two being the binomial and normal probability functions.

Definition   *A discrete random variable X with possible values* 0, 1, 2, 3, . . . *is called* **Poisson** *with parameter* λ, λ > 0, *if*

$$P(X = i) = \frac{e^{-\lambda}\lambda^i}{i!}, \quad i = 0, 1, 2, 3, \ldots .$$

Under the conditions specified in our discussion, binomial probabilities can be approximated by Poisson probabilities. Such approximations are generally good if $p < 0.1$ and $np \le 10$. If $np > 10$, it would be more appropriate to use normal approximation, discussed in Section 7.2.

Since a Poisson probability mass function is the limit of a binomial probability mass function, the expected value of a binomial random variable with parameters (n, p) is np, and $np = \lambda$, it is reasonable to expect that the mean of a Poisson random variable with parameter λ is λ. To prove that this is the case, observe that

$$E(X) = \sum_{i=0}^{\infty} i P(X = i) = \sum_{i=1}^{\infty} i \frac{e^{-\lambda}\lambda^i}{i!}$$

$$= \lambda e^{-\lambda} \sum_{i=1}^{\infty} \frac{\lambda^{i-1}}{(i-1)!} = \lambda e^{-\lambda} \sum_{i=0}^{\infty} \frac{\lambda^i}{i!} = \lambda e^{-\lambda}e^{\lambda} = \lambda.$$

The variance of a Poisson random variable X with parameter λ is also λ. To see this, note that

$$E(X^2) = \sum_{i=0}^{\infty} i^2 P(X = i) = \sum_{i=1}^{\infty} i^2 \frac{e^{-\lambda}\lambda^i}{i!}$$

$$= \lambda e^{-\lambda} \sum_{i=1}^{\infty} \frac{i\lambda^{i-1}}{(i-1)!} = \lambda e^{-\lambda} \sum_{i=1}^{\infty} \frac{1}{(i-1)!}\frac{d}{d\lambda}(\lambda^i)$$

$$= \lambda e^{-\lambda} \frac{d}{d\lambda}\Big[\sum_{i=1}^{\infty} \frac{\lambda^i}{(i-1)!}\Big] = \lambda e^{-\lambda} \frac{d}{d\lambda}\Big[\lambda \sum_{i=1}^{\infty} \frac{\lambda^{i-1}}{(i-1)!}\Big]$$

$$= \lambda e^{-\lambda} \frac{d}{d\lambda}(\lambda e^{\lambda}) = \lambda e^{-\lambda}(e^{\lambda} + \lambda e^{\lambda}) = \lambda + \lambda^2,$$

and hence

$$\mathrm{Var}(X) = (\lambda + \lambda^2) - \lambda^2 = \lambda.$$

We have shown that

If X is a Poisson random variable with parameter λ, then

$$E(X) = \mathbf{Var}(X) = \lambda, \qquad \sigma_X = \sqrt{\lambda}.$$

Some examples of binomial random variables that obey Poisson's approximation are as follows:

1. Let X be the number of babies in a community who grow up to at least 190 centimeters. If a baby is called a success, provided that he or she grows up to the height of 190 or more centimeters, then X is a binomial random variable. Since n, the total number of babies, is large, p, the probability that a baby grows to the height of 190 centimeters or more, is small, and np, the average number of such babies, is appreciable, X is approximately a Poisson random variable.

2. Let X be the number of winning tickets among the Maryland lottery tickets sold in Baltimore during one week. Then, calling winning tickets successes, we have that X is a binomial random variable. Since n, the total number of tickets sold in Baltimore, is large, p, the probability that a ticket wins, is small, and the average number of winning tickets is appreciable, X is approximately a Poisson random variable.

3. Let X be the number of misprints on a document page typed by a secretary. Then X is a binomial random variable if a word is called a success, provided that it is misprinted! Since misprints are rare events, the number of words is large, and np, the average number of misprints, is of moderate value, X is approximately a Poisson random variable.

In the same way we can argue that random variables such as "the number of defective fuses produced by a factory in a given year," "the number of inhabitants of a town who live at least 85 years," "the number of machine failures per day in a plant," and "the number of drivers arriving at a gas station per hour" are approximately Poisson random variables.

We will now give examples of problems that can be solved by Poisson's approximations. In the course of these examples, new Poisson random variables are also introduced.

Example 5.10   Every week the average number of wrong-number phone calls received by a certain mail-order house is seven. What is the probability that they will receive (a) two wrong calls tomorrow; (b) at least one wrong call tomorrow?

Solution: Assuming that the house receives a lot of calls, the number of wrong numbers received tomorrow, X, is approximately a Poisson random variable with $\lambda = E(X) = 1$. Thus

$$P(X = n) = \frac{e^{-1} \cdot (1)^n}{n!} = \frac{1}{e \cdot n!},$$

and hence the answers to (a) and (b) are, respectively,

$$P(X = 2) = \frac{1}{2e} \approx 0.18$$

and

$$P(X \geq 1) = 1 - P(X = 0) = 1 - \frac{1}{e} \approx 0.63. \quad \blacklozenge$$

Example 5.11   Suppose that, on average, in every three pages of a book there is one typographical error. If the number of typographical errors on a single page of the book is a Poisson random variable, what is the probability of at least one error on a specific page of the book?

Solution: Let X be the number of errors on the page we are interested in. Then X is a Poisson random variable with $E(X) = 1/3$. Hence $\lambda = E(X) = 1/3$, and thus

$$P(X = n) = \frac{(1/3)^n e^{-1/3}}{n!}.$$

Therefore,

$$P(X \geq 1) = 1 - P(X = 0) = 1 - e^{-1/3} \approx 0.28. \quad \blacklozenge$$

Example 5.12   The atoms of a radioactive element are randomly disintegrating. If every gram of this element, on average, emits 3.9 alpha particles per second, what is the probability that during the next second the number of alpha particles emitted from 1 gram is (a) at most 6; (b) at least 2; (c) at least 3 and at most 6?

Solution: Every gram of radioactive material consists of a large number n of atoms. If for these atoms the event of disintegrating and emitting an alpha particle during the next second is called a success, then X, the number of alpha particles emitted during the next second, is a binomial random variable. Now $E(X) = 3.9$, so $np = 3.9$ and $p = 3.9/n$. Since n is very large, p is very small, and hence, to a very close approximation, X has a Poisson distribution with the appreciable parameter $\lambda = 3.9$. Therefore,

$$P(X = n) = \frac{(3.9)^n e^{-3.9}}{n!},$$

and hence (a), (b), and (c) are calculated as follows.

(a) $P(X \leq 6) = \displaystyle\sum_{n=0}^{6} \frac{(3.9)^n e^{-3.9}}{n!} \approx 0.899.$

(b) $P(X \geq 2) = 1 - P(X = 0) - P(X = 1) = 0.901.$

(c) $P(3 \leq X \leq 6) = \displaystyle\sum_{n=3}^{6} \frac{(3.9)^n e^{-3.9}}{n!} \approx 0.646.$ ◆

Example 5.13   Suppose that n raisins are thoroughly mixed in dough. If we bake k raisin cookies of equal sizes from this mixture, what is the probability that a given cookie contains at least one raisin?

Solution: Since the raisins are thoroughly mixed in the dough, the probability that a given cookie contains any particular raisin is $p = 1/k$. If for the raisins the event of ending up in the given cookie is called a success, then X, the number of raisins in the given cookie, is a binomial random variable. For large values of k, $p = 1/k$ is small. If n is also large but n/k has a moderate value, it is reasonable to assume that X is approximately Poisson. Hence

$$P(X = i) = \frac{\lambda^i e^{-\lambda}}{i!},$$

where $\lambda = np = n/k$. Therefore, the probability that a given cookie contains at least one raisin is

$$P(X \neq 0) = 1 - P(X = 0) = 1 - e^{-n/k}.$$ ◆

For real-world problems, it is important to know that in most cases numerical examples show that, even for small values of n, the agreement between the binomial and the Poisson probabilities is surprisingly good (see Exercise 10).

Poisson Processes

Earlier, the Poisson distribution was introduced to approximate binomial distribution for very large n, very small p, and moderate np. On its own merit, the Poisson distribution appears in connection with the study of sequences of random events occurring over time. Examples of such sequences of random events are accidents occurring at an intersection, β-particles emitted from a radioactive substance, customers entering a post office, and California earthquakes. Suppose that, starting from a time point labeled $t = 0$, we begin counting the number of events. Then for each value of t, we obtain a number denoted by $N(t)$, which is the number of events that have occurred during $[0, t]$. For example, in studying the number of β-particles emitted from a radioactive substance, $N(t)$ is the number of β-particles emitted by time t. Similarly, in studying the number of California earthquakes, $N(t)$ is the number of earthquakes that have occurred in the time interval $[0, t]$. Clearly, for each value of t, $N(t)$ is a discrete random variable with the set of possible values $\{0, 1, 2, \ldots \}$. To study the distribution of $N(t)$, the number of events occurring in $[0, t]$, we make the following three simple and natural assumptions about the way in which events occur.

Stationarity: *For all $n \geq 0$, and for any two equal time intervals Δ_1 and Δ_2, the probability of n events in Δ_1 is equal to the probability of n events in Δ_2.*

Independent Increments: *For all $n \geq 0$, and for any time interval $(t, t + s)$, the probability of n events in $(t, t+s)$ is independent of how many events have occurred earlier or how they have occurred. In particular, suppose that the times $0 \leq t_1 < t_2 < \cdots < t_k$ are given. For $1 \leq i < k - 1$, let A_i be the event that n_i events of the process occur in $[t_i, t_{i+1})$. The independent increments mean that $\{A_1, A_2, \ldots, A_{k-1}\}$ is an independent set of events.*

Orderliness: *The occurrence of two or more events in a very small time interval is practically impossible. This condition is mathematically expressed by $\lim_{h \to 0} P\big(N(h) > 1\big)/h = 0$. This implies that as $h \to 0$, the probability of two or more events, $P\big(N(h) > 1\big)$, approaches 0 faster than h does. That is, if h is negligible, then $P\big(N(h) > 1\big)$ is even more negligible.*

Note that, by the stationarity property, the number of events in the interval $(t_1, t_2]$ has the same distribution as the number of events in $(t_1+s, t_2+s]$, $s \geq 0$. This means that the random variables $N(t_2) - N(t_1)$ and $N(t_2+s) - N(t_1+s)$ have the same probability mass function. In other words, the probability of occurrence of n events during the interval of time from t_1 to t_2 is a function of n and $t_2 - t_1$ and not of t_1 and t_2 independently. The number of events in $(t_i, t_{i+1}]$, $N(t_{i+1}) - N(t_i)$, is called the **increment** in $\{N(t)\}$ between t_i and t_{i+1}. That is why the second property in the preceding list is called independent-increments property. It is also worthwhile to mention that stationarity and orderliness together imply the following fact, proved in Section 12.2.

The simultaneous occurrence of two or more events is impossible. Therefore, under the aforementioned properties, events occur one at a time, not in pairs or groups.

It is for this reason that the third property is called the orderliness property.

Suppose that random events occur in time in a way that the preceding conditions—stationarity, independent increments, and orderliness—are always satisfied. Then, if for *some* interval of length $t > 0$, $P(N(t) = 0) = 0$, we have that in *any* interval of length t at least one event occurs. In such a case, it can be shown that in any interval of *arbitrary* length, with probability 1, at least one event occurs. Similarly, if for *some* interval of length $t > 0$, $P(N(t) = 0) = 1$, then in *any* interval of length t no event will occur. In such a case, it can be shown that in any interval of *arbitrary* length, with probability 1, no event occurs. To avoid these uninteresting, trivial cases, throughout the book, we assume that, for all $t > 0$,

$$0 < P(N(t) = 0) < 1.$$

We are now ready to state a celebrated theorem, for the validity of which we present a motivating argument. The theorem is presented again, with a rigorous proof, in Chapter 12 as Theorem 12.1.

Theorem 5.2   *If random events occur in time in a way that the preceding conditions—stationarity, independent increments, and orderliness—are always satisfied, $N(0) = 0$ and, for all $t > 0$, $0 < P(N(t) = 0) < 1$, then there exists a positive number λ such that*

$$P(N(t) = n) = \frac{(\lambda t)^n e^{-\lambda t}}{n!}.$$

That is, for all $t > 0$, $N(t)$ is a Poisson random variable with parameter λt. Hence $E[N(t)] = \lambda t$ and therefore $\lambda = E[N(1)]$.

A Motivating Argument: The property that under the stated conditions, $N(t)$ (the number of events that has occurred in $[0, t)$) is a Poisson random variable is not accidental. It is related to the fact that a Poisson random variable is approximately binomial for large n, small p, and moderate np. To see this, divide the interval $[0, t]$ into n subintervals of equal length. Then, as $n \to \infty$, the probability of two or more events in any of these subintervals is 0. Therefore, $N(t)$, the number of events in $[0, t]$, is equal to the number of subintervals in which an event occurs. If a subinterval in which an event occurs is called a success, then $N(t)$ is the number of successes. Moreover, the stationarity and independent-increments properties imply that $N(t)$ is a binomial random variable with parameters (n, p), where p is the probability of success (i.e., the probability that an event occurs in a subinterval). Now let λ be the expected number of events in an interval of unit length. Because of stationarity, events occur at a uniform rate over the entire time period. Therefore, the expected number of events in any period of length t is λt. Hence, in particular, the expected number of the events in $[0, t]$ is λt. But by the formula for the expectation of binomial random variables, the expected number of the events in $[0, t]$ is

np. Thus $np = \lambda t$ or $p = (\lambda t)/n$. Since n is extremely large ($n \to \infty$), we have that p is extremely small while $np = \lambda t$ is of moderate size. Therefore, $N(t)$ is a Poisson random variable with parameter λt.   ◆

In the study of sequences of random events occurring in time, suppose that $N(0) = 0$ and, for all $t > 0$, $0 < P\big(N(t) = 0\big) < 1$. Furthermore, suppose that the events occur in a way that the preceding conditions—stationarity, independent increments, and orderliness—are always satisfied. We argued that for each value of t, the discrete random variable $N(t)$, the number of events in $[0, t]$ and hence in any other time interval of length t, is a Poisson random variable with parameter λt. Any process with this property is called a **Poisson process with rate** λ and is often denoted by $\big\{N(t), \ t \geq 0\big\}$.

Theorem 5.2 is astonishing because it shows how three simple and natural physical conditions on $N(t)$ characterize the probability mass functions of random variables $N(t)$, $t > 0$. Moreover, λ, the only unknown parameter of the probability mass functions of $N(t)$'s, equals $E\big[N(1)\big]$. That is, it is the average number of events that occur in one unit of time. Hence in practice it can be measured readily. Historically, Theorem 5.2 is the most elegant and evolved form of several previous theorems. It was discovered in 1955 by the Russian mathematician Alexander Khinchin (1894–1959). The first major work in this direction was done by Thornton Fry (1892–1992) in 1929. But before Fry, Albert Einstein (1879–1955) and Roman Smoluchowski (1910–1996) had also discovered important results in connection with their work on the theory of Brownian motion.

Example 5.14   Suppose that children are born at a Poisson rate of five per day in a certain hospital. What is the probability that (a) at least two babies are born during the next six hours; (b) no babies are born during the next two days?

Solution: Let $N(t)$ denote the number of babies born at or prior to t. The assumption that $\big\{N(t), \ t \geq 0\big\}$ is a Poisson process is reasonable because it is stationary, it has independent increments, $N(0) = 0$, and simultaneous births are impossible. Thus $\big\{N(t), \ t \geq 0\big\}$ is a Poisson process. If we choose one day as time unit, then $\lambda = E\big[N(1)\big] = 5$. Therefore,

$$P\big(N(t) = n\big) = \frac{(5t)^n e^{-5t}}{n!}.$$

Hence the probability that at least two babies are born during the next six hours is

$$P\big(N(1/4) \geq 2\big) = 1 - P\big(N(1/4) = 0\big) - P\big(N(1/4) = 1\big)$$

$$= 1 - \frac{(5/4)^0 e^{-5/4}}{0!} - \frac{(5/4)^1 e^{-5/4}}{1!} \approx 0.36,$$

where $1/4$ is used since 6 hours is $1/4$ of a day. The probability that no babies are born during the next two days is

$$P\big(N(2) = 0\big) = \frac{(10)^0 e^{-10}}{0!} \approx 4.54 \times 10^{-5}.   ◆$$

Example 5.15   Suppose that earthquakes occur in a certain region of California, in accordance with a Poisson process, at a rate of seven per year.

(a)   What is the probability of no earthquakes in one year?

(b)   What is the probability that in exactly three of the next eight years no earthquakes will occur?

Solution:

(a)   Let $N(t)$ be the number of earthquakes in this region at or prior to t. We are given that $\{N(t), \ t \geq 0\}$ is a Poisson process. If we choose one year as the unit of time, then $\lambda = E[N(1)] = 7$. Thus

$$P(N(t) = n) = \frac{(7t)^n e^{-7t}}{n!}, \qquad n = 0, 1, 2, 3, \ldots .$$

Let p be the probability of no earthquakes in one year; then

$$p = P(N(1) = 0) = e^{-7} \approx 0.00091.$$

(b)   Suppose that a year is called a success if during its course no earthquakes occur. Of the next eight years, let X be the number of years in which no earthquakes will occur. Then X is a binomial random variable with parameters $(8, p)$. Thus

$$P(X = 3) \approx \binom{8}{3}(0.00091)^3(1 - 0.00091)^5 \approx 4.2 \times 10^{-8}. \quad \blacklozenge$$

Example 5.16   A fisherman catches fish at a Poisson rate of two per hour from a large lake with lots of fish. Yesterday, he went fishing at 10:00 A.M. and caught just one fish by 10:30 and a total of three by noon. What is the probability that he can duplicate this feat tomorrow?

Solution: Label the time the fisherman starts fishing tomorrow at $t = 0$. Let $N(t)$ denote the total number of fish caught at or prior to t. Clearly, $N(0) = 0$. It is reasonable to assume that catching two or more fish simultaneously is impossible. It is also reasonable to assume that $\{N(t), \ t \geq 0\}$ is stationary and has independent increments. Thus the assumption that $\{N(t), \ t \geq 0\}$ is a Poisson process is well grounded. Choosing 1 hour as the unit of time, we have $\lambda = E[N(1)] = 2$. Thus

$$P(N(t) = n) = \frac{(2t)^n e^{-2t}}{n!}, \qquad n = 0, 1, 2, \ldots .$$

We want to calculate the probability of the event

$$\{N(1/2) = 1 \text{ and } N(2) = 3\}.$$

But this event is the same as $\{N(1/2) = 1$ and $N(2) - N(1/2) = 2\}$. Thus by the independent-increments property,

$$P\big(N(1/2) = 1 \text{ and } N(2) - N(1/2) = 2\big) = P\big(N(1/2) = 1\big) \cdot P\big(N(2) - N(1/2) = 2\big).$$

Since stationarity implies that

$$P\big(N(2) - N(1/2) = 2\big) = P\big(N(3/2) = 2\big),$$

the desired probability equals

$$P\big(N(1/2) = 1\big) \cdot P\big(N(3/2) = 2\big) = \frac{1^1 e^{-1}}{1!} \cdot \frac{3^2 e^{-3}}{2!} \approx 0.082. \quad \blacklozenge$$

Example 5.17   Let $N(t)$ be the number of earthquakes that occur at or prior to time t worldwide. Suppose that $\{N(t) : t \geq 0\}$ is a Poisson process and the probability that the magnitude of an earthquake on the Richter scale is 5 or more is p. Find the probability of k earthquakes of such magnitudes at or prior to t worldwide.

Solution: Let $X(t)$ be the number of earthquakes of magnitude 5 or more on the Richter scale at or prior to t worldwide. Since the sequence of events $\{N(t) = n\}$, $n = 0, 1, 2, \ldots$ is mutually exclusive and $\bigcup_{n=0}^{\infty} \{N(t) = n\}$ is the sample space, by the law of total probability, Theorem 3.4,

$$P\big(X(t) = k\big) = \sum_{n=0}^{\infty} P\big(X(t) = k \mid N(t) = n\big) P\big(N(t) = n\big).$$

Now clearly, $P\big(X(t) = k \mid N(t) = n\big) = 0$ if $n < k$. If $n > k$, the conditional probability mass function of $X(t)$ given that $N(t) = n$ is binomial with parameters n and p. Thus

$$P\big(X(t) = k\big) = \sum_{n=k}^{\infty} \binom{n}{k} p^k (1 - p)^{n-k} \frac{e^{-\lambda t} (\lambda t)^n}{n!}$$

$$= \sum_{n=k}^{\infty} \frac{n!}{k! \, (n-k)!} p^k (1 - p)^{n-k} \frac{e^{-\lambda t} (\lambda t)^k (\lambda t)^{n-k}}{n!}$$

$$= \frac{e^{-\lambda t} (\lambda t p)^k}{k!} \sum_{n=k}^{\infty} \frac{1}{(n-k)!} \big[\lambda t (1 - p)\big]^{n-k}$$

$$= \frac{e^{-\lambda t} (\lambda t p)^k}{k!} \sum_{j=0}^{\infty} \frac{1}{(j)!} \big[\lambda t (1 - p)\big]^{j}$$

$$= \frac{e^{-\lambda t} (\lambda t p)^k}{k!} e^{\lambda t (1-p)}$$

$$= \frac{e^{-\lambda t p} (\lambda t p)^k}{k!}.$$

Therefore, $\{X(t) : t \geq 0\}$ is itself a Poisson process with mean λp. $\blacklozenge$

Remark 5.1   We only considered sequences of random events that occur in *time*. However, the restriction to time is not necessary. Random events that occur on the real line, the plane, or in space and satisfy the stationarity and orderliness conditions and possess independent increments also form Poisson processes. For example, suppose that a wire manufacturing company produces a wire that has various fracture sites where the wire will fail in tension. Let $N(t)$ be the number of fracture sites in the first t *meters* of wire. The process $\{N(t),\ t \geq 0\}$ may be modeled as a Poisson process. As another example, suppose that in a certain region S, the numbers of trees that grow in nonoverlapping subregions are independent of each other, the distributions of the number of trees in subregions of equal area are identical, and the probability of two or more trees in a very small subregion is negligible. Let λ be the expected number of trees in a region of area 1 and $A(R)$ be the area of a region R. Then $N(R)$, the number of trees in a subregion R is a Poisson random variable with parameter $\lambda A(R)$, and the set $\{N(R),\ R \subseteq S\}$ is a two-dimensional Poisson process.   ◆

EXERCISES

A

1. Jim buys 60 lottery tickets every week. If only 5% of the lottery tickets win, what is the probability that he wins next week?

2. Suppose that 3% of the families in a large city have an annual income of over \$60,000. What is the probability that, of 60 random families, at most three have an annual income of over \$60,000?

3. Suppose that 2.5% of the population of a border town are illegal immigrants. Find the probability that, in a theater of this town with 80 random viewers, there are at least two illegal immigrants.

4. By Example 2.21, the probability that a poker hand is a full house is 0.0014. What is the probability that in 500 random poker hands there are at least two full houses?

5. On a random day, the number of vacant rooms of a big hotel in New York City is 35, on average. What is the probability that next Saturday this hotel has at least 30 vacant rooms?

6. On average, there are three misprints in every 10 pages of a particular book. If every chapter of the book contains 35 pages, what is the probability that Chapters 1 and 5 have 10 misprints each?

7. Suppose that X is a Poisson random variable with $P(X = 1) = P(X = 3)$. Find $P(X = 5)$.

8. Suppose that n raisins have been carefully mixed with a batch of dough. If we bake k $(k > 4)$ raisin buns of equal size from this mixture, what is the probability that two out of four randomly selected buns contain no raisins?
    Hint: Note that, by Example 5.13, the number of raisins in a given bun is approximately Poisson with parameter n/k.

9. The children in a small town all own slingshots. In a recent contest, 4% of them were such poor shots that they did not hit the target even once in 100 shots. If the number of times a randomly selected child has hit the target is approximately a Poisson random variable, determine the percentage of children who have hit the target at least twice.

10. The department of mathematics of a state university has 26 faculty members. For $i = 0, 1, 2, 3$, find p_i, the probability that i of them were born on Independence Day (a) using the binomial distribution; (b) using the Poisson distribution. Assume that the birth rates are constant throughout the year and that each year has 365 days.

11. Suppose that on a summer evening, shooting stars are observed at a Poisson rate of one every 12 minutes. What is the probability that three shooting stars are observed in 30 minutes?

12. Suppose that in Japan earthquakes occur at a Poisson rate of three per week. What is the probability that the next earthquake occurs after two weeks?

13. Suppose that, for a telephone subscriber, the number of wrong numbers is Poisson, at a rate of $\lambda = 1$ per week. A certain subscriber has not received any wrong numbers from Sunday through Friday. What is the probability that he receives no wrong numbers on Saturday either?

14. In a certain town, crimes occur at a Poisson rate of five per month. What is the probability of having exactly two months (not necessarily consecutive) with no crimes during the next year?

15. Accidents occur at an intersection at a Poisson rate of three per day. What is the probability that during January there are exactly three days (not necessarily consecutive) without any accidents?

16. Customers arrive at a bookstore at a Poisson rate of six per hour. Given that the store opens at 9:30 A.M., what is the probability that exactly one customer arrives by 10:00 A.M. and 10 customers by noon?

17. A wire manufacturing company has inspectors to examine the wire for fractures as it comes out of a machine. The number of fractures is distributed in accordance with a Poisson process, having one fracture on the average for every 60 meters of wire. One day an inspector has to take an emergency phone call and is missing from his post for ten minutes. If the machine turns out 7 meters of wire per minute, what is the probability that the inspector will miss more than one fracture?

B

18. On a certain two-lane north-south highway, there is a T junction. Cars arrive at the junction according to a Poisson process, on the average four per minute. For cars to turn left onto the side street, the highway is widened by the addition of a left-turn lane that is long enough to accommodate three cars. If four or more cars are trying to turn left, the fourth car will effectively block north-bound traffic. At the junction for the left-turn lane there is a left-turn signal that allows cars to turn left for one minute and prohibits such turns for the next three minutes. The probability of a randomly selected car turning left at this T junction is 0.22. Suppose that during a green light for the left-turn lane all waiting cars were able to turn left. What is the probability that during the subsequent red light for the left-turn lane, the north-bound traffic will be blocked?

19. Suppose that, on the Richter scale, earthquakes of magnitude 5.5 or higher have probability 0.015 of damaging certain types of bridges. Suppose that such intense earthquakes occur at a Poisson rate of 1.5 per ten years. If a bridge of this type is constructed to last at least 60 years, what is the probability that it will be undamaged by earthquakes for that period of time?

20. According to the United States Postal Service, *http:www.usps.gov*, May 15, 1998,

> Dogs have caused problems for letter carriers for so long that the situation has become a cliché. In 1983, more than 7,000 letter carriers were bitten by dogs. ... However, the 2,795 letter carriers who were bitten by dogs last year represent less than one-half of 1 percent of *all reported* dog-bite victims.

Suppose that during a year 94% of the letter carriers are not bitten by dogs. Assuming that dogs bite letter carriers randomly, what percentage of those who sustained one bite will be bitten again?

21. Suppose that in Maryland, on a certain day, N lottery tickets are sold and M win. To have a probability of at least α of winning on that day, approximately how many tickets should be purchased?

22. Balls numbered $1, 2, \ldots,$ and n are randomly placed into cells numbered $1, 2, \ldots,$ and n. Therefore, for $1 \leq i \leq n$ and $1 \leq j \leq n$, the probability that ball i is in cell j is $1/n$. For each i, $1 \leq i \leq n$, if ball i is in cell i, we say that a *match* has occurred at cell i.

  (a) What is the probability of exactly k matches?

  (b) Let $n \to \infty$. Show that the probability mass function of the number of matches is Poisson with mean 1.

23. Let $\{N(t), \ t \geq 0\}$ be a Poisson process. What is the probability of (a) an even number of events in $(t, t + \alpha)$; (b) an odd number of events in $(t, t + \alpha)$?

24. Let $\{N(t),\ t \geq 0\}$ be a Poisson process with rate λ. Suppose that $N(t)$ is the total number of two types of events that have occurred in $[0, t]$. Let $N_1(t)$ and $N_2(t)$ be the total number of events of type 1 and events of type 2 that have occurred in $[0, t]$, respectively. If events of type 1 and type 2 occur independently with probabilities p and $1 - p$, respectively, prove that $\{N_1(t),\ t \geq 0\}$ and $\{N_2(t),\ t \geq 0\}$ are Poisson processes with respective rates λp and $\lambda(1 - p)$.

Hint: First calculate $P\big(N_1(t) = n$ and $N_2(t) = m\big)$ using the relation

$$P\big(N_1(t) = n \text{ and } N_2(t) = m\big)$$

$$= \sum_{i=0}^{\infty} P\big(N_1(t) = n \text{ and } N_2(t) = m \mid N(t) = i\big) P\big(N(t) = i\big).$$

(This is true because of Theorem 3.4.) Then use the relation

$$P\big(N_1(t) = n\big) = \sum_{m=0}^{\infty} P\big(N_1(t) = n \text{ and } N_2(t) = m\big).$$

25. Customers arrive at a grocery store at a Poisson rate of one per minute. If 2/3 of the customers are female and 1/3 are male, what is the probability that 15 females enter the store between 10:30 and 10:45?

Hint: Use the result of Exercise 24.

26. In a forest, the number of trees that grow in a region of area R has a Poisson distribution with mean λR, where λ is a given positive number.

   (a) Find the probability that the distance from a certain tree to the nearest tree is more than d.

   (b) Find the probability that the distance from a certain tree to the nth nearest tree is more than d.

27. Let X be a Poisson random variable with parameter λ. Show that the maximum of $P(X = i)$ occurs at $[\lambda]$, where $[\lambda]$ is the greatest integer less than or equal to λ.

Hint: Let p be the probability mass function of X. Prove that

$$p(i) = \frac{\lambda}{i}\, p(i - 1).$$

Use this to find the values of i at which p is increasing and the values of i at which it is decreasing.

5.3 OTHER DISCRETE RANDOM VARIABLES

Geometric Random Variables

Consider an experiment in which independent Bernoulli trials are performed until the first success occurs. The sample space for such an experiment is

$$S = \{s, fs, ffs, fffs, \ldots, ff \cdots fs, \ldots\}.$$

Now, suppose that a sequence of independent Bernoulli trials, each with probability of success p, $0 < p < 1$, are performed. Let X be the number of experiments until the first success occurs. Then X is a discrete random variable called **geometric**. It is defined on S, its set of possible values is $\{1, 2, \ldots\}$, and

$$P(X = n) = (1 - p)^{n-1}p, \qquad n = 1, 2, 3, \ldots.$$

This equation follows since (a) the first $(n - 1)$ trials are all failures, (b) the nth trial is a success, and (c) the successive Bernoulli trials are all independent.

Let $p(x) = (1 - p)^{x-1}p$ for $x = 1, 2, 3, \ldots$, and 0 elsewhere. Then, for all values of x in $\mathbf{R}$, $p(x) \geq 0$ and

$$\sum_{x=1}^{\infty} p(x) = \sum_{x=1}^{\infty} (1 - p)^{x-1}p = \frac{p}{1 - (1 - p)} = 1,$$

by the geometric series theorem. Hence $p(x)$ is a probability mass function.

Definition   *The probability mass function*

$$p(x) = \begin{cases} (1 - p)^{x-1}p & 0 < p < 1, \quad x = 1, 2, 3, \ldots, \\ 0 & \text{elsewhere} \end{cases}$$

is called **geometric**.

Let X be a geometric random variable with parameter p; then

$$E(X) = \sum_{x=1}^{\infty} xp(1 - p)^{x-1} = \frac{p}{1 - p} \sum_{x=1}^{\infty} x(1 - p)^x$$

$$= \frac{p}{1 - p} \frac{1 - p}{\left[1 - (1 - p)\right]^2} = \frac{1}{p},$$

where the third equality follows from the relation $\sum_{x=1}^{\infty} xr^x = r/(1 - r)^2$, $|r| < 1$. $E(X) = 1/p$ indicates that to get the first success, on average, $1/p$ independent Bernoulli trials are needed. The relation $\sum_{x=1}^{\infty} x^2 r^x = \left[r(r + 1)\right]/(1 - r)^3$, $|r| < 1$, implies that

$$E(X^2) = \sum_{x=1}^{\infty} x^2 p(1 - p)^{x-1} = \frac{p}{1 - p} \sum_{x=1}^{\infty} x^2(1 - p)^x = \frac{2 - p}{p^2}.$$

Hence

$$\text{Var}(X) = E(X^2) - \left[E(X)\right]^2 = \frac{2-p}{p^2} - \left(\frac{1}{p}\right)^2 = \frac{1-p}{p^2}.$$

We have established the following formulas:

Let X be a geometric random variable with parameter p, $0 < p < 1$.
Then

$$E(X) = \frac{1}{p}, \quad \text{Var}(X) = \frac{1-p}{p^2}, \quad \sigma_X = \frac{\sqrt{1-p}}{p}.$$

Example 5.18   From an ordinary deck of 52 cards we draw cards at random, *with replacement*, and successively until an ace is drawn. What is the probability that at least 10 draws are needed?

Solution: Let X be the number of draws until the first ace. The random variable X is geometric with the parameter $p = 1/13$. Thus

$$P(X = n) = \left(\frac{12}{13}\right)^{n-1}\left(\frac{1}{13}\right), \qquad n = 1, 2, 3, \ldots,$$

and so the probability that at least 10 draws are needed is

$$P(X \ge 10) = \sum_{n=10}^{\infty} \left(\frac{12}{13}\right)^{n-1}\left(\frac{1}{13}\right) = \frac{1}{13}\sum_{n=10}^{\infty}\left(\frac{12}{13}\right)^{n-1}$$

$$= \frac{1}{13} \cdot \frac{(12/13)^9}{1 - 12/13} = \left(\frac{12}{13}\right)^9 \approx 0.49.$$

Remark: There is a shortcut to the solution of this problem: The probability that at least 10 draws are needed to get an ace is the same as the probability that in the first nine draws there are no aces. This is equal to $(12/13)^9 \approx 0.49$.   ◆

Let X be a geometric random variable with parameter p, $0 < p < 1$. Then, for all positive integers n and m,

$$P(X > n + m \mid X > m) = \frac{P(X > n + m)}{P(X > m)} = \frac{(1-p)^{n+m}}{(1-p)^m} = (1-p)^n = P(X > n).$$

This is called the **memoryless property** of geometric random variables. It means that

In successive independent Bernoulli trials, the probability that the next n outcomes are all failures does not change if we are given that the previous m successive outcomes were all failures.

This is obvious by the independence of the trials. Interestingly enough, in the following sense, geometric random variable is the only memoryless discrete random variable.

> Let X be a discrete random variable with the set of possible values
> $\{1, 2, 3 \ldots\}$. If for all positive integers n and m,
>
> $$P(X > n + m \mid X > m) = P(X > n),$$
>
> then X is a geometric random variable. That is, there exists a number
> $p, 0 < p < 1$, such that
>
> $$P(X = n) = p(1 - p)^{n-1}, \qquad n \geq 1.$$

We leave the proof of this theorem as an exercise (see Exercise 22).

Example 5.19   A father asks his sons to cut their backyard lawn. Since he does not specify which of the three sons is to do the job, each boy tosses a coin to determine the odd person, who must then cut the lawn. In the case that all three get heads or tails, they continue tossing until they reach a decision. Let p be the probability of heads and $q = 1 - p$, the probability of tails.

(a)   Find the probability that they reach a decision in less than n tosses.

(b)   If $p = 1/2$, what is the minimum number of tosses required to reach a decision with probability 0.95?

Solution:

(a)   The probability that they reach a decision on a certain round of coin tossing is

$$\binom{3}{2} p^2 q + \binom{3}{2} q^2 p = 3pq(p + q) = 3pq.$$

The probability that they do not reach a decision on a certain round is $1 - 3pq$. Let X be the number of tosses until they reach a decision; then X is a geometric random variable with parameter $3pq$. Therefore,

$$P(X < n) = 1 - P(X \geq n) = 1 - (1 - 3pq)^{n-1},$$

where the second equality follows since $X \geq n$ if and only if none of the first $n - 1$ tosses results in a success.

(b)   We want to find the minimum n so that $P(X \leq n) \geq 0.95$. This gives $1 - P(X > n) \geq 0.95$ or $P(X > n) \leq 0.05$. But

$$P(X > n) = (1 - 3pq)^n = (1 - 3/4)^n = (1/4)^n.$$

Therefore, we must have $(1/4)^n \leq 0.05$, or $n \ln 1/4 \leq \ln 0.05$. This gives $n \geq 2.16$; hence the smallest n is 3.   ◆

Negative Binomial Random Variables

Negative binomial random variables are generalizations of geometric random variables. Suppose that a sequence of independent Bernoulli trials, each with probability of success p, $0 < p < 1$, is performed. Let X be the number of experiments until the rth success occurs. Then X is a discrete random variable called a **negative binomial.** Its set of possible values is $\{r, r + 1, r + 2, r + 3, \dots\}$ and

$$P(X = n) = \binom{n-1}{r-1} p^r (1-p)^{n-r}, \qquad n = r, r + 1, \dots . \tag{5.4}$$

This equation follows since if the outcome of the nth trial is the rth success, then in the first $(n - 1)$ trials exactly $(r - 1)$ successes have occurred and the nth trial is a success. The probability of the former event is

$$\binom{n-1}{r-1} p^{r-1}(1-p)^{(n-1)-(r-1)} = \binom{n-1}{r-1} p^{r-1}(1-p)^{n-r},$$

and the probability of the latter is p. Therefore, by the independence of the trials, (5.4) follows.

Definition *The probability mass function*

$$p(x) = \binom{x-1}{r-1} p^r (1-p)^{x-r}, \quad 0 < p < 1, \quad x = r, r + 1, r + 2, r + 3, \dots ,$$

*is called **negative binomial** with parameters* (r, p).

Note that a negative binomial probability mass function with parameters $(1, p)$ is geometric. In Chapter 10, Examples 10.7 and 10.16, we will show that

If X is a negative binomial random variable with parameters (r, p), then

$$E(X) = \frac{r}{p}, \quad \operatorname{Var}(X) = \frac{r(1-p)}{p^2}, \quad \sigma_X = \frac{\sqrt{r(1-p)}}{p}.$$

Example 5.20 Sharon and Ann play a series of backgammon games until one of them wins five games. Suppose that the games are independent and the probability that Sharon wins a game is 0.58.

(a) Find the probability that the series ends in seven games.

(b) If the series ends in seven games, what is the probability that Sharon wins?

Solution: (a) Let X be the number of games until Sharon wins five games. Let Y be the number of games until Ann wins five games. The random variables X and Y are negative binomial with parameters $(5, 0.58)$ and $(5, 0.42)$, respectively. The probability that the series ends in seven games is

$$P(X = 7) + P(Y = 7) = \binom{6}{4}(0.58)^5(0.42)^2 + \binom{6}{4}(0.42)^5(0.58)^2$$

$$\approx 0.17 + 0.066 \approx 0.24.$$

(b) Let A be the event that Sharon wins and B be the event that the series ends in seven games. Then the desired probability is

$$P(A \mid B) = \frac{P(AB)}{P(B)} = \frac{P(X = 7)}{P(X = 7) + P(Y = 7)} \approx \frac{0.17}{0.24} \approx 0.71. \quad \blacklozenge$$

The following example, given by Kaigh in January 1979 issue of *Mathematics Magazine*, is a modification of the gambler's ruin problem, Example 3.14.

Example 5.21   (Attrition Ruin Problem) Two gamblers play a game in which in each play gambler A beats B with probability p, $0 < p < 1$, and loses to B with probability $q = 1 - p$. Suppose that each play results in a forfeiture of \$1 for the loser and in no change for the winner. If player A initially has a dollars and player B has b dollars, what is the probability that B will be ruined?

Solution: Let E_i be the event that, in the first $b + i$ plays, B loses b times. Let A^* be the event that A wins. Then

$$P(A^*) = \sum_{i=0}^{a-1} P(E_i).$$

If every time that A wins is called a success, E_i is the event that the bth success occurs on the $(b + i)$th play. Using the negative binomial distribution, we have

$$P(E_i) = \binom{i + b - 1}{b - 1} p^b q^i.$$

Therefore,

$$P(A^*) = \sum_{i=0}^{a-1} \binom{i + b - 1}{b - 1} p^b q^i.$$

As a numerical illustration, let $a = b = 4$, $p = 0.6$, and $q = 0.4$. We get $P(A^*) = 0.710208$ and $P(B^*) = 0.289792$ exactly. $\quad \blacklozenge$

The following example is due to Hugo Steinhaus (1887–1972), who brought it up in a conference honoring Stefan Banach (1892–1945), a smoker and one of the greatest mathematicians of the twentieth century.

Example 5.22   (Banach Matchbox Problem)   A smoking mathematician carries two matchboxes, one in his right pocket and one in his left pocket. Whenever he wants to smoke, he selects a pocket at random and takes a match from the box in that pocket. If each matchbox initially contains N matches, what is the probability that when the mathematician for the first time *discovers* that one box is empty, there are exactly m matches in the other box, $m = 0, 1, 2, \ldots, N$?

Solution: Every time that the left pocket is selected we say that a success has occurred. When the mathematician discovers that the left box is empty, the right one contains m matches if and only if the $(N + 1)$st success occurs on the $(N - m) + (N + 1) = (2N - m + 1)$st trial. The probability of this event is

$$\binom{(2N - m + 1) - 1}{(N + 1) - 1}\left(\frac{1}{2}\right)^{N+1}\left(\frac{1}{2}\right)^{(2N-m+1)-(N+1)} = \binom{2N - m}{N}\left(\frac{1}{2}\right)^{2N-m+1}.$$

By symmetry, when the mathematician discovers that the right box is empty, with probability $\binom{2N - m}{N}\left(\frac{1}{2}\right)^{2N-m+1}$, the left box contains m matches. Therefore, the desired probability is

$$2\binom{2N - m}{N}\left(\frac{1}{2}\right)^{2N-m+1} = \binom{2N - m}{N}\left(\frac{1}{2}\right)^{2N-m}. \quad \blacklozenge$$

Hypergeometric Random Variables

Suppose that, from a box containing D defective and $N - D$ nondefective items, n are drawn at random and *without replacement*. Furthermore, suppose that the number of items drawn does not exceed the number of defective or the number of nondefective items. That is, suppose that $n \leq \min(D, N - D)$. Let X be the number of defective items drawn. Then X is a discrete random variable with the set of possible values $\{0, 1, \ldots n\}$, and a probability mass function

$$p(x) = P(X = x) = \frac{\binom{D}{x}\binom{N - D}{n - x}}{\binom{N}{n}}, \qquad x = 0, 1, 2, \ldots, n.$$

Any random variable X with such a probability mass function is called a **hypergeometric random variable.** The fact that $p(x)$ is a probability mass function is readily verified. Clearly, $p(x) \geq 0, \forall x$. To prove that $\sum_{x=0}^{n} p(x) = 1$, note that this is equivalent to

$$\sum_{x=0}^{n}\binom{D}{x}\binom{N - D}{n - x} = \binom{N}{n},$$

which can be shown by a simple combinatorial argument. (See Exercise 41 of Section 2.4; in that exercise let $m = D, n = N - D$, and $r = n$.)

Definition   *Let N, D, and n be positive integers with* $n \leq \min(D, N - D)$. *Then*

$$p(x) = P(X = x) = \begin{cases} \dfrac{\dbinom{D}{x}\dbinom{N - D}{n - x}}{\dbinom{N}{n}} & \text{if } x \in \{0, 1, 2, \ldots n\} \\ \\ 0 & \text{elsewhere} \end{cases}$$

is said to be a **hypergeometric** *probability mass function.*

For the following important results, see Example 10.8, and Exercise 27, Section 10.2.

For the hypergeometric random variable X, defined above,

$$E(X) = \frac{nD}{N}, \qquad \text{Var}(X) = \frac{nD(N - D)}{N^2}\left(1 - \frac{n - 1}{N - 1}\right).$$

Note that if the experiment of drawing n items from a box containing D defective and $N - D$ nondefective items is performed *with replacement*, then X is binomial with parameters n and D/N. Hence

$$E(X) = \frac{nD}{N}, \quad \text{Var}(X) = n\frac{D}{N}\left(1 - \frac{D}{N}\right) = \frac{nD(N - D)}{N^2}.$$

These show that if items are drawn *with replacement*, then the expected value of X will not change, but the variance will increase. However, if n is much smaller than N, then, as the variance formulae confirm, drawing with replacement is a good approximation for drawing without replacement.

Example 5.23   In 500 independent calculations a scientist has made 25 errors. If a second scientist checks seven of these calculations randomly, what is the probability that he detects two errors? Assume that the second scientist will definitely find the error of a false calculation.

Solution: Let X be the number of errors found by the second scientist. Then X is hypergeometric with $N = 500$, $D = 25$, and $n = 7$. We are interested in $P(X = 2)$, which is given by

$$p(2) = \frac{\dbinom{25}{2}\dbinom{500 - 25}{7 - 2}}{\dbinom{500}{7}} \approx 0.04. \quad \blacklozenge$$

Example 5.24   In a community of $a + b$ potential voters, a are for abortion and b $(b < a)$ are against it. Suppose that a vote is taken to determine the will of the majority with regard to legalizing abortion. If n $(n < b)$ random persons of these $a + b$ potential voters do not vote, what is the probability that those against abortion will win?

Solution: Let X be the number of those who do not vote and are for abortion. The persons against abortion will win if and only if

$$a - X < b - (n - X).$$

But this is true if and only if $X > (a - b + n)/2$. Since X is a hypergeometric random variable, we have that

$$P\left(X > \frac{a - b + n}{2}\right) = \sum_{i = [\frac{a-b+n}{2}]+1}^{n} P(X = i) = \sum_{i = [\frac{a-b+n}{2}]+1}^{n} \frac{\binom{a}{i}\binom{b}{n-i}}{\binom{a+b}{n}},$$

where $\left[\dfrac{a - b + n}{2}\right]$ is the greatest integer less than or equal to $\dfrac{a - b + n}{2}$.   ♦

Example 5.25   Professors Davidson and Johnson from the University of Victoria in Canada gave the following problem to their students in a finite math course:

> An urn contains N balls of which B are black and $N - B$ are white; n balls are chosen at random and without replacement from the urn. If X is the number of black balls chosen, find $P(X = i)$.

Using the hypergeometric formula, the solution to this problem is

$$P(X = i) = \frac{\binom{B}{i}\binom{N - B}{n - i}}{\binom{N}{n}}, \qquad i \leq \min(n, B).$$

However, by mistake, some students interchanged B with n and came up with the following answer:

$$P(X = i) = \frac{\binom{n}{i}\binom{N - n}{B - i}}{\binom{N}{B}},$$

which can be verified immediately as the correct answer for all values of i. Explain if this is accidental or if there is a probabilistic reason for its validity.

Solution: In their paper "Interchanging Parameters of Hypergeometric Distributions," *Mathematics Magazine*, December 1993, Volume 66, Davidson and Johnson indicate that what happened is not accidental and that there is a probabilistic reason for it. Their justification is elegant and is as follows:

> We begin with an urn that contains N white balls. When Ms. Painter arrives, she picks B balls at random without replacement from the urn, then paints each of the drawn balls black with instant dry paint, and finally returns these B balls to the urn. When Mr. Carver arrives, he chooses n balls at random without replacement from the urn, then engraves each of the chosen balls with the letter C and, finally, returns these n balls to the urn. Let the random variable X denote the number of black balls chosen (i.e., painted and engraved) when both Painter and Carver have finished their jobs. Since the tasks of Painter and Carver do not depend on which task is done first, the probability distribution of X is the same whether Painter does her job before or after Carver does his. If Painter goes first, Carver chooses n balls from an urn that contains B black balls and $N - B$ white balls, so
>
> $$P(X = i) = \frac{\binom{B}{i}\binom{N-B}{n-i}}{\binom{N}{n}} \qquad \text{for } i = 0, 1, 2, \ldots, \min(n, B).$$
>
> On the other hand, if Carver goes first, Painter draws B balls from an urn that contains n balls engraved with C and $N - n$ balls not engraved, so
>
> $$P(X = i) = \frac{\binom{n}{i}\binom{N-n}{B-i}}{\binom{N}{B}} \qquad \text{for } i = 0, 1, 2, \ldots, \min(n, B).$$
>
> Thus changing the order of who goes first provides an urn-drawing explanation of why the probability distribution of X remains the same when B is interchanged with n.   ♦

Let X be a hypergeometric random variable with parameters n, D, and N. Calculation of $E(X)$ and $\text{Var}(X)$ directly from $p(x)$, the probability mass function of X, is tedious. By the methods developed in Sections 10.1 and 10.2, calculation of these quantities is considerably easier. In Section 10.1 we will prove that $E(X) = nD/N$; it can be shown that

$$\text{Var}(X) = \frac{nD(N - D)}{N^2}\left(1 - \frac{n-1}{N-1}\right).$$

(See Exercise 27, Section 10.2.) Using these formulas, for example, we have that the average number of face cards in a bridge hand is $(13 \times 12)/52 = 3$ with variance

$$\frac{13 \times 12(52 - 12)}{52^2}\left(1 - \frac{13-1}{52-1}\right) = 1.765.$$

Remark 5.2   If the n items that are selected at random from the D defective and $N - D$ nondefective items are chosen *with replacement* rather than *without replacement*, then X, the number of defective items, is a binomial random variable with parameters n and D/N. Thus

$$P(X = x) = \binom{n}{x}\left(\frac{D}{N}\right)^x \left(1 - \frac{D}{N}\right)^{n-x}, \qquad x = 0, 1, 2, \dots, n.$$

Now, if N is very large, it is not that important whether a sample is taken with or without replacement. For example, if we have to choose 1000 persons from a population of 200,000,000, replacing persons that are selected back into the population will change the chance of other persons to be selected only insignificantly. Therefore, for large N, the binomial probability mass function is an excellent approximation for the hypergeometric probability mass function, which can be stated mathematically as follows.

$$\lim_{\substack{N \to \infty \\ D \to \infty \\ D/N \to p}} \frac{\binom{D}{x}\binom{N-D}{n-x}}{\binom{N}{n}} = \binom{n}{x} p^x (1-p)^{n-x}.$$

We leave its proof as an exercise.   ◆

EXERCISES

A

1. Define a sample space for the experiment that, from a box containing two defective and five nondefective items, three items are drawn at random and without replacement.

2. Define a sample space for the experiment of performing independent Bernoulli trials until the second success occurs.

3. An absentminded professor does not remember which of his 12 keys will open his office door. If he tries them at random and with replacement:

    (a)   On average, how many keys should he try before his door opens?

    (b)   What is the probability that he opens his office door after only three tries?

4. The probability is p that Marty hits target M when he fires at it. The probability is q that Alvie hits target A when he fires at it. Marty and Alvie fire one shot each at their targets. If both of them hit their targets, they stop; otherwise, they will continue.

    (a)   What is the probability that they stop after each has fired r times?

    (b)   What is the expected value of the number of the times that each of them has fired before stopping?

5.   Suppose that 20% of a group of people have hazel eyes. What is the probability that the eighth passenger boarding a plane is the third one having hazel eyes? Assume that passengers boarding the plane form a randomly chosen group.

6.   A certain basketball player makes a foul shot with probability 0.45. What is the probability that (a) his first basket occurs on the sixth shot; (b) his first and second baskets occur on his fourth and eighth shots, respectively?

7.   A store has 50 light bulbs available for sale. Of these, five are defective. A customer buys eight light bulbs randomly from this store. What is the probability that he finds exactly one defective light bulb among them?

8.   The probability is p that a randomly chosen light bulb is defective. We screw a bulb into a lamp and switch on the current. If the bulb works, we stop; otherwise, we try another and continue until a good bulb is found. What is the probability that at least n bulbs are required?

9.   Suppose that independent Bernoulli trials with parameter p are performed successively. Let N be the number of trials needed to get x successes, and X be the number of successes in the first n trials. Show that

$$P(N = n) = \frac{x}{n} P(X = x).$$

Remark: By this relation, in coin tossing, for example, we can state that the probability of getting a fifth head on the seventh toss is 5/7 of the probability of five heads in seven tosses.

10.   In rural Ireland, a century ago, the students had to form a line. The student at the front of the line would be asked to spell a word. If he spelled it correctly, he was allowed to sit down. If not, he received a whack on the hand with a switch and was sent to the end of the line. Suppose that a student could spell correctly 70% of the words in the lesson. What is the probability that the student would be able to sit down before receiving four whacks on the hand? Assume that the master chose the words to be spelled randomly and independently.

11.   The digits after the decimal point of a random number between 0 and 1 are numbers selected at random, with replacement, independently, and successively from the set $\{0, 1, \ldots, 9\}$. In a random number from $(0, 1)$, on the average, how many digits are there before the fifth 3?

12.   On average, how many games of bridge are necessary before a player is dealt three aces? A bridge hand is 13 randomly selected cards from an ordinary deck of 52 cards.

13. Solve the Banach matchbox problem (Example 5.22) for the case where the matchbox in the right pocket contains M matches, the matchbox in his left pocket contains N matches, and $m \leq \min(M, N)$.

14. Suppose that 15% of the population of a town are senior citizens. Let X be the number of nonsenior citizens who enter a mall before the tenth senior citizen arrives. Find the probability mass function of X. Assume that each customer who enters the mall is a random person from the entire population.

15. Florence is moving and wishes to sell her package of 100 computer diskettes. Unknown to her, 10 of those diskettes are defective. Sharon will purchase them if a random sample of 10 contains no more than one defective disk. What is the probability that she buys them?

16. In an annual charity drive, 35% of a population of 560 make contributions. If, in a statistical survey, 15 people are selected at random and without replacement, what is the probability that at least two persons have contributed?

17. The probability is p that a message sent over a communication channel is garbled. If the message is sent repeatedly until it is transmitted reliably, and if each time it takes 2 minutes to process it, what is the probability that the transmission of a message takes more than t minutes?

18. A vending machine contains cans of grapefruit juice that cost 75 cents each, but it is not working properly. The probability that it accepts a coin is 10%. Angela has a quarter and five dimes. Determine the probability that she should try the coins at least 50 times before she gets a can of grapefruit juice.

19. A computer network consists of several stations connected by various media (usually cables). There are certain instances when no message is being transmitted. At such "suitable instances," each station will send a message with probability p, independently of the other stations. However, if two or more stations send messages, a collision will corrupt the messages, and they will be discarded. These messages will be retransmitted until they reach their destination. If the network consists of N stations, on average, how many times should a certain station transmit and retransmit a message until it reaches its destination?

B

20. A fair coin is flipped repeatedly. What is the probability that the fifth tail occurs before the tenth head?

21. On average, how many independent games of poker are required until a preassigned player is dealt a straight? (See Exercise 20 of Section 2.4 for a definition of a straight. The cards have distinct consecutive values that are not of the same suit: for example, 3 of hearts, 4 of hearts, 5 of spades, 6 of hearts, and 7 of hearts.)

22. Let X be a geometric random variable with parameter p, and n and m be nonnegative integers.

(a)   For what values of n is $P(X = n)$ maximum?

(b)   What is the probability that X is even?

(c)   Show that the geometric is the only distribution on the positive integers with the memoryless property:

$$P(X > n + m \mid X > m) = P(X > n).$$

23.   In data communication, messages are usually combinations of characters, and each character consists of a number of bits. A bit is the smallest unit of information and is either 1 or 0. Suppose that the length of a character (in bits) is a geometric random variable with parameter p. Furthermore, suppose that the lengths of characters are independent random variables. What is the distribution of the total number of the bits forming a message of k random characters?

24.   Twelve hundred eggs, of which 200 are rotten, are distributed randomly in 100 cartons, each containing a dozen eggs. These cartons are then sold to a restaurant. How many cartons should we expect the chef of the restaurant to open before finding one without rotten eggs?

25.   In the Banach matchbox problem, Example 5.22, find the probability that when the first matchbox is emptied (not found empty) there are exactly m matches in the other box.

26.   In the Banach matchbox problem, Example 5.22, find the probability that the box which is emptied first is not the one that is first found empty.

27.   Adam rolls a well-balanced die until he gets a 6. Andrew rolls the same die until he rolls an odd number. What is the probability that Andrew rolls the die more than Adam does?

28.   To estimate the number of trout in a lake, we caught 50 trout, tagged and returned them. Later we caught 50 trout and found that four of them were tagged. From this experiment estimate n, the total number of trout in the lake.
      Hint: Let p_n be the probability of four tagged trout among the 50 trout caught. Find the value of n that maximizes p_n.

29.   Suppose that, from a box containing D defective and $N - D$ nondefective items, n ($\leq D$) are drawn one by one, at random and *without replacement*.

(a)   Find the probability that the kth item drawn is defective.

(b)   If the $(k - 1)$st item drawn is defective, what is the probability that the kth item is also defective?

REVIEW PROBLEMS

1. Of police academy applicants, only 25% will pass all the examinations. Suppose that 12 successful candidates are needed. What is the probability that, by examining 20 candidates, the academy finds all of the 12 persons needed?

2. The time between the arrival of two consecutive customers at a postoffice is 3 minutes, on average. Assuming that customers arrive in accordance with a Poisson process, find the probability that tomorrow during the lunch hour (between noon and 12:30 P.M.) fewer than seven customers arrive.

3. A restaurant serves 8 fish entrées, 12 beef, and 10 poultry. If customers select from these entrées randomly, what is the expected number of fish entrées ordered by the next four customers?

4. A university has n students, 70% of whom will finish an aptitude test in less than 25 minutes. If 12 students are selected at random, what is the probability that at most two of them will not finish in 25 minutes?

5. A doctor has five patients with migraine headaches. He prescribes for all five a drug that relieves the headaches of 82% of such patients. What is the probability that the medicine will not relieve the headaches of two of these patients?

6. In a community, the chance of a set of triplets is 1 in 1000 births. Determine the probability that the second set of triplets in this community occurs before the 2000th birth.

7. From a panel of prospective jurors, 12 are selected at random. If there are 200 men and 160 women on the panel, what is the probability that more than half of the jury selected are women?

8. Suppose that 10 trains arrive independently at a station every day, each at a random time between 10:00 A.M. and 11:00 A.M.. What is the expected number and the variance of those that arrive between 10:15 A.M. and 10:28 A.M.?

9. Suppose that a certain bank returns bad checks at a Poisson rate of three per day. What is the probability that this bank returns at most four bad checks during the next two days?

10. The policy of the quality control division of a certain corporation is to reject a shipment if more than 5% of its items are defective. A shipment of 500 items is received, 30 of them are randomly tested, and two have been found defective. Should that shipment be rejected?

11. A fair coin is tossed successively until a head occurs. If N is the number of tosses required, what are the expected value and the variance of N?

12. What is the probability that the sixth toss of a die is the first 6?

13. In data communication, one method for error control makes the receiver send a positive acknowledgment for every message received with no detected error and a negative acknowledgment for all the others. Suppose that (i) acknowledgments are transmitted error free, (ii) a negative acknowledgment makes the sender transmit the same message over, (iii) there is no upper limit on the number of retransmissions of a message, and (iv) errors in messages are detected with probability p, independently of other messages. On average, how many times is a message retransmitted?

14. Past experience shows that 30% of the customers entering Harry's Clothing Store will make a purchase. Of the customers who make a purchase, 85% use credit cards. Let X be the number of the next six customers who enter the store, make a purchase, and use a credit card. Find the probability mass function, the expected value, and the variance of X.

15. Of the 28 professors in a certain department, 18 drive foreign and 10 drive domestic cars. If five of these professors are selected at random, what is the probability that at least three of them drive foreign cars?

16. A bowl contains 10 red and six blue chips. What is the expected number of blue chips among five randomly selected chips?

17. A certain type of seed when planted fails to germinate with probability 0.06. If 40 of such seeds are planted, what is the probability that at least 36 of them germinate?

18. Suppose that 6% of the claims received by an insurance company are for damages from vandalism. What is the probability that at least three of the 20 claims received on a certain day are for vandalism damages?

19. Passengers are making reservations for a particular flight on a small commuter plane 24 hours a day at a Poisson rate of 3 reservations per 8 hours. If 24 seats are available for the flight, what is the probability that by the end of the second day all the plane seats are reserved?

20. An insurance company claims that only 70% of the drivers regularly use seat belts. In a statistical survey, it was found that out of 20 randomly selected drivers, 12 regularly used seat belts. Is this sufficient evidence to conclude that the insurance company's claim is false?

21. From the set $\{x: 0 \leq x \leq 1\}$ numbers are selected at random and independently and rounded to three decimal places. What is the probability that 0.345 is obtained (a) for the first time on the 1000th selections; (b) for the third time on the 3000th selections?

22. The probability that a child of a certain family inherits a certain disease is 0.23 independently of other children inheriting the disease. If the family has five children and the disease is detected in one child, what is the probability that exactly two more children have the disease as well?

23. A bowl contains w white and b blue chips. Chips are drawn at random and with replacement until a blue chip is drawn. What is the probability that (a) exactly n draws are required; (b) at least n draws are required?

24. Experience shows that 75% of certain kinds of seeds germinate when planted under normal conditions. Determine the minimum number of seeds to be planted so that the chance of at least five of them germinating is more than 90%.

25. Suppose that n babies were born at a county hospital last week. Also suppose that the probability of a baby having blonde hair is p. If k of these n babies are blondes, what is the probability that the ith baby born is blonde?

26. A farmer, plagued by insects, seeks to attract birds to his property by distributing seeds over a wide area. Let λ be the average number of seeds per unit area, and suppose that the seeds are distributed in a way that the probability of having any number of seeds in a given area depends only on the size of the area and not on its location. Argue that X, the number of seeds that fall on the area A, is approximately a Poisson random variable with parameter λA.

27. Show that if all three of n, N, and $D \to \infty$ so that $n/N \to 0$, D/N converges to a small number, and $nD/N \to \lambda$, then for all x,

$$\frac{\binom{D}{x}\binom{N-D}{n-x}}{\binom{N}{n}} \to \frac{e^{-\lambda}\lambda^x}{x!}.$$

This formula shows that the Poisson distribution is the limit of the hypergeometric distribution.

Chapter 6

Continuous Random Variables

6.1 PROBABILITY DENSITY FUNCTIONS

As discussed in Section 4.2, the distribution function of a random variable X is a function F from $(-\infty, +\infty)$ to $\mathbf{R}$ defined by $F(t) = P(X \leq t)$. From the definition of F we deduced that it is nondecreasing, right continuous, and satisfies $\lim_{t \to \infty} F(t) = 1$ and $\lim_{t \to -\infty} F(t) = 0$. Furthermore, we showed that, for discrete random variables, distributions are step functions. We also proved that if X is a discrete random variable with set of possible values $\{x_1, x_2, \ldots\}$, probability mass function p, and distribution function F, then F has jump discontinuities at $x_1, x_2, \ldots$, where the magnitude of the jump at x_i is $p(x_i)$ and for $x_{n-1} \leq t < x_n$,

$$F(t) = P(X \leq t) = \sum_{i=1}^{n-1} p(x_i). \tag{6.1}$$

In the case of discrete random variables a very small change in t may cause relatively large changes in the values of F. For example, if t changes from $x_n - \varepsilon$ to x_n, $\varepsilon > 0$ being arbitrarily small, then $F(x)$ changes from $\sum_{i=1}^{n-1} p(x_i)$ to $\sum_{i=1}^{n} p(x_i)$, a change of magnitude $p(x_n)$, which might be large. In cases such as the lifetime of a random light bulb, the arrival time of a train at a station, and the weight of a random watermelon grown in a certain field, where the set of possible values of X is uncountable, small changes in x produce correspondingly small changes in the distribution of X. In such cases we expect that F, the distribution function of X, will be a continuous function. Random variables that have continuous distributions can be studied under general conditions. However, for practical reasons and mathematical simplicity, we restrict ourselves to a class of random variables that are called **absolutely continuous** and are defined as follows:

Definition *Let X be a random variable. Suppose that there exists a nonnegative real-valued function $f : \mathbf{R} \to [0, \infty)$ such that for any subset of real numbers A that*

231

can be constructed from intervals by a countable number of set operations,

$$P(X \in A) = \int_A f(x)\, dx. \tag{6.2}$$

*Then X is called **absolutely continuous** or, in this book, for simplicity, **continuous**. Therefore, whenever we say that X is continuous, we mean that it is absolutely continuous and hence satisfies (6.2). The function f is called the **probability density function**, or simply the **density function** of X.*

Let f be the density function of a random variable X with distribution function F. Some immediate properties of f are as follows:

(a) $F(t) = \int_{-\infty}^{t} f(x)\, dx.$

Let $A = (-\infty, t]$; then using (6.2), we can write

$$F(t) = P(X \le t) = P(X \in A) = \int_A f(x)\, dx = \int_{-\infty}^{t} f(x)\, dx.$$

Comparing (a) with (6.1), we see that f, the probability density function of a continuous random variable X, can be regarded as an extension of the idea of p, the probability mass function of a discrete random variable. As we study f further, the resemblance between f and p becomes clearer. For example, recall that if X is a discrete random variable with the set of possible values A and probability mass function p, then $\sum_{x \in A} p(x) = 1$. In the continuous case, the relation analogous to this is the following:

(b) $\int_{-\infty}^{\infty} f(x)\, dx = 1.$

This is true by (a) and the fact that $F(\infty) = 1$ $\big[$i.e., $\lim_{t \to \infty} F(t) = 1\big]$. It also follows from (6.2) with $A = \mathbf{R}$. Because of this property, in general, if a function $g : \mathbf{R} \to [0, \infty)$ satisfies $\int_{-\infty}^{\infty} g(x)\, dx = 1$, we say that g is a probability density function, or simply a density function.

(c) *If f is continuous, then $F'(x) = f(x)$.* This follows from (a) and the fundamental theorem of calculus: $\dfrac{d}{dt} \displaystyle\int_a^t f(x)\, dx = f(t)$. Note that even if f is not continuous, still $F'(x) = f(x)$ for every x at which f is continuous. *Therefore, the distribution functions of continuous random variables are integrals of their derivatives.*

(d) *For real numbers $a \le b$, $P(a \le X \le b) = \int_a^b f(t)\, dt.$*

Let $A = [a, b]$ in (6.2); then

$$P(a \le X \le b) = P(X \in A) = \int_A f(x)\, dx = \int_a^b f(x)\, dx.$$

Property (d) states that the probability of X being between a and b is equal to the area under the graph of f from a to b. Letting $a = b$ in (d), we obtain

$$P(X = a) = \int_a^a f(x)\,dx = 0.$$

This means that for any real number a, $P(X = a) = 0$. That is, the probability that a continuous random variable assumes a certain value is 0. From $P(X = a) = 0$, $\forall a \in \mathbf{R}$, we have that the probability of X lying in an interval does not depend on the endpoints of the interval. Therefore,

(e) $P(a < X < b) = P(a \le X < b) = P(a < X \le b) = P(a \le X \le b) = \int_a^b f(t)\,dt.$

Another implication of $P(X = a) = 0$, $\forall a \in \mathbf{R}$, is that *the value of the density function f at no point represents a probability.* The probabilistic significance of f is that its integral over any subset of real numbers B gives the probability that X lies in B. In particular,

> The area over an interval I under the graph of f represents the probability that the random variable X will belong to I. The area under f to the left of a given point t is $F(t)$, the value of the distribution function of X at t.

As an example, let X be a continuous random variable with probability density function f the graph of which is sketched in Figure 6.1. Then the shaded area under f is the probability that X is between a and b.

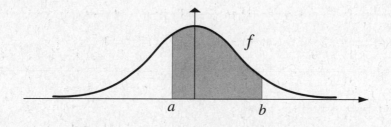

Figure 6.1   The shaded area under f is the probability that $X \in I = (a, b)$.

Intuitively, $f(a)$ is a measure that determines how likely it is for X to be close to a. To see this, note that if $\varepsilon > 0$ is very small, then $P(a - \varepsilon < X < a + \varepsilon)$ is the probability that X is close to a. Now

$$P(a - \varepsilon < X < a + \varepsilon) = \int_{a-\varepsilon}^{a+\varepsilon} f(t)\,dt$$

is the area under $f(t)$ from $a - \varepsilon$ to $a + \varepsilon$. This area is almost equal to the area of a rectangle with sides of lengths $(a + \varepsilon) - (a - \varepsilon) = 2\varepsilon$ and $f(a)$. Thus

$$P(a - \varepsilon < X < a + \varepsilon) = \int_{a-\varepsilon}^{a+\varepsilon} f(t)\, dt \approx 2\varepsilon f(a). \tag{6.3}$$

Similarly, for any other real number b (in the domain of f),

$$P(b - \varepsilon < X < b + \varepsilon) \approx 2\varepsilon f(b). \tag{6.4}$$

Relations (6.3) and (6.4) show that for small fixed $\varepsilon > 0$, if $f(a) < f(b)$, then

$$P(a - \varepsilon < X < a + \varepsilon) < P(b - \varepsilon < X < b + \varepsilon).$$

That is, if $f(a) < f(b)$, the probability that X is close to b is higher than the probability that X is close to a. Thus the larger $f(x)$, the higher the probability will be that X is close to x.

Example 6.1   Experience has shown that while walking in a certain park, the time X, in minutes, between seeing two people smoking has a density function of the form

$$f(x) = \begin{cases} \lambda x e^{-x} & x > 0 \\ 0 & \text{otherwise.} \end{cases}$$

(a)   Calculate the value of λ.

(b)   Find the probability distribution function of X.

(c)   What is the probability that Jeff, who has just seen a person smoking, will see another person smoking in 2 to 5 minutes? In at least 7 minutes?

Solution:

(a)   To determine λ, we use the property $\int_{-\infty}^{\infty} f(x)\, dx = 1$:

$$\int_{-\infty}^{\infty} f(x)\, dx = \int_{0}^{\infty} \lambda x e^{-x}\, dx = \lambda \int_{0}^{\infty} x e^{-x}\, dx.$$

Now, by integration by parts,

$$\int x e^{-x}\, dx = -(x + 1)e^{-x}.$$

Thus

$$\lambda \Big[-(x+1)e^{-x} \Big]_{0}^{\infty} = 1. \tag{6.5}$$

But as $x \to \infty$, using l'Hôpital's rule, we get

$$\lim_{x \to \infty} (x + 1)e^{-x} = \lim_{x \to \infty} \frac{x + 1}{e^x} = \lim_{x \to \infty} \frac{1}{e^x} = 0.$$

Therefore, (6.5) implies that $\lambda\big[0 - (-1)\big] = 1$ or $\lambda = 1$.

(b) To find F, the distribution function of X, note that $F(t) = 0$ if $t < 0$. For $t \geq 0$,

$$F(t) = \int_{-\infty}^{t} f(x)\,dx = \left[-(x+1)e^{-x}\right]_0^t = -(t+1)e^{-t} + 1.$$

Thus

$$F(t) = \begin{cases} 0 & \text{if } t < 0 \\ -(t+1)e^{-t} + 1 & \text{if } t \geq 0. \end{cases}$$

(c) The desired probabilities $P(2 < X < 5)$ and $P(X \geq 7)$ are calculated as follows:

$$P(2 < X < 5) = P(2 < X \leq 5) = P(X \leq 5) - P(X \leq 2) = F(5) - F(2)$$
$$= (1 - 6e^{-5}) - (1 - 3e^{-2}) = 3e^{-2} - 6e^{-5} \approx 0.37.$$
$$P(X \geq 7) = 1 - P(X < 7) = 1 - P(X \leq 7) = 1 - F(7)$$
$$= 8e^{-7} \approx 0.007.$$

Note: To calculate these quantities, we can use $P(2 < X < 5) = \int_2^5 f(t)\,dt$ and $P(X \geq 7) = \int_7^{\infty} f(t)\,dt$ as well.   ◆

Example 6.2

(a) Sketch the graph of the function

$$f(x) = \begin{cases} \dfrac{1}{2} - \dfrac{1}{4}|x - 3| & 1 \leq x \leq 5 \\ 0 & \text{otherwise,} \end{cases}$$

and show that it is the probability density function of a random variable X.

(b) Find F, the distribution function of X, and show that it is continuous.

(c) Sketch the graph of F.

Solution:

(a) Note that

$$f(x) = \begin{cases} 0 & x < 1 \\ \dfrac{1}{2} + \dfrac{1}{4}(x - 3) & 1 \leq x < 3 \\ \dfrac{1}{2} - \dfrac{1}{4}(x - 3) & 3 \leq x < 5 \\ 0 & x \geq 5. \end{cases}$$

Therefore, the graph of f is as shown in Figure 6.2.

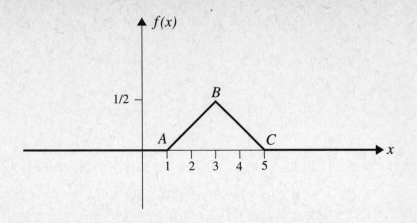

Figure 6.2   Density function of Example 6.2.

Now since $f(x) \geq 0$ and the area under f from 1 to 5, being the area of the triangle ABC (as seen from Figure 6.2) is $(1/2)(4 \times 1/2) = 1$, f is a density function of some random variable X.

(b)   To calculate the distribution function of X, we use the formula $F(t) = \int_{-\infty}^{t} f(x)\, dx$. For $t < 1$,

$$F(t) = \int_{-\infty}^{t} 0\, dx = 0;$$

for $1 \leq t < 3$,

$$F(t) = \int_{-\infty}^{t} f(x)\, dx = \int_{1}^{t} \left[\frac{1}{2} + \frac{1}{4}(x-3)\right] dx = \frac{1}{8}t^2 - \frac{1}{4}t + \frac{1}{8};$$

for $3 \leq t < 5$,

$$F(t) = \int_{-\infty}^{t} f(x)\, dx = \int_{1}^{3} \left[\frac{1}{2} + \frac{1}{4}(x-3)\right] dx + \int_{3}^{t} \left[\frac{1}{2} - \frac{1}{4}(x-3)\right] dx$$

$$= \frac{1}{2} + \left(-\frac{1}{8}t^2 + \frac{5}{4}t - \frac{21}{8}\right) = -\frac{1}{8}t^2 + \frac{5}{4}t - \frac{17}{8};$$

and for $t \geq 5$,

$$F(t) = \int_{-\infty}^{t} f(x)\, dx = \int_{1}^{3} \left[\frac{1}{2} + \frac{1}{4}(x-3)\right] dx + \int_{3}^{5} \left[\frac{1}{2} - \frac{1}{4}(x-3)\right] dx$$

$$= \frac{1}{2} + \frac{1}{2} = 1.$$

Therefore,

$$F(t) = \begin{cases} 0 & t < 1 \\ \dfrac{1}{8}t^2 - \dfrac{1}{4}t + \dfrac{1}{8} & 1 \le t < 3 \\ -\dfrac{1}{8}t^2 + \dfrac{5}{4}t - \dfrac{17}{8} & 3 \le t < 5 \\ 1 & t \ge 5. \end{cases}$$

F is continuous because

$$\lim_{t \to 1-} F(t) = 0 = F(1),$$

$$\lim_{t \to 3-} F(t) = \frac{1}{8}(3)^2 - \frac{1}{4}(3) + \frac{1}{8} = \frac{1}{2} = -\frac{1}{8}(3)^2 + \frac{5}{4}(3) - \frac{17}{8} = F(3),$$

$$\lim_{t \to 5-} F(t) = -\frac{1}{8}(5)^2 + \frac{5}{4}(5) - \frac{17}{8} = 1 = F(5).$$

(c)   The graph of F is shown in Figure 6.3.   ◆

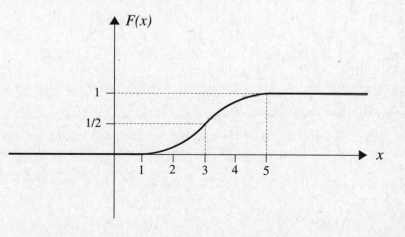

Figure 6.3   Distribution function of Example 6.2.

Remark 6.1   It is important to realize that there are random variables that are neither continuous nor discrete. Examples 4.7 and 4.10 give two such random variables.   ◆

EXERCISES

1. When a certain car breaks down, the time that it takes to fix it (in hours) is a random variable with the density function

$$f(x) = \begin{cases} ce^{-3x} & \text{if } 0 \le x < \infty \\ 0 & \text{otherwise.} \end{cases}$$

   (a) Calculate the value of c.

   (b) Find the probability that when this car breaks down, it takes at most 30 minutes to fix it.

2. The distribution function for the duration of a certain soap opera (in tens of hours) is

$$F(x) = \begin{cases} 1 - \dfrac{16}{x^2} & x \ge 4 \\ 0 & x < 4. \end{cases}$$

   (a) Calculate f, the probability density function of the soap opera.

   (b) Sketch the graphs of F and f.

   (c) What is the probability that the soap opera takes at most 50 hours? At least 60 hours? Between 50 and 70 hours? Between 10 and 35 hours?

3. The time it takes for a student to finish an aptitude test (in hours) has a density function of the form

$$f(x) = \begin{cases} c(x-1)(2-x) & \text{if } 1 < x < 2 \\ 0 & \text{elsewhere.} \end{cases}$$

   (a) Determine the constant c.

   (b) Calculate the distribution function of the time it takes for a randomly selected student to finish the aptitude test.

   (c) What is the probability that a student will finish the aptitude test in less than 75 minutes? Between $1\frac{1}{2}$ and 2 hours?

4. The lifetime of a tire selected randomly from a used tire shop is $10,000X$ miles, where X is a random variable with the density function

$$f(x) = \begin{cases} 2/x^2 & \text{if } 1 < x < 2 \\ 0 & \text{elsewhere.} \end{cases}$$

   (a) What percentage of the tires of this shop last fewer than 15,000 miles?

   (b) What percentage of those having lifetimes fewer than 15,000 miles last between 10,000 and 12,500 miles?

5. The probability density function of a random variable X is given by

$$f(x) = \begin{cases} \dfrac{c}{\sqrt{1-x^2}} & \text{if } -1 < x < 1 \\ 0 & \text{elsewhere.} \end{cases}$$

   (a) Calculate the value of c.

   (b) Find the probability distribution function of X.

6. Let X be a continuous random variable with density and distribution functions f and F, respectively. Assuming that $\alpha \in \mathbf{R}$ is a point at which $P(X \le \alpha) < 1$, prove that

$$h(x) = \begin{cases} \dfrac{f(x)}{1 - F(\alpha)} & \text{if } x \ge \alpha \\ 0 & \text{if } x < \alpha \end{cases}$$

is also a probability density function.

7. Let X be a continuous random variable with density function f. We say that X is **symmetric about** α if for all x,

$$P(X \ge \alpha + x) = P(X \le \alpha - x).$$

   (a) Prove that X is symmetric about α if and only if for all x, we have $f(\alpha - x) = f(\alpha + x)$.

   (b) Let X be a continuous random variable with probability density function

$$f(x) = \frac{1}{\sqrt{2\pi}} e^{-(x-3)^2/2}, \qquad x \in \mathbf{R},$$

   and Y be a continuous random variable with probability density function

$$g(x) = \frac{1}{\pi \left[1 + (x-1)^2 \right]}, \qquad x \in \mathbf{R}.$$

   Find the points about which X and Y are symmetric.

8. Suppose that the loss in a certain investment, in thousands of dollars, is a continuous random variable X that has a density function of the form

$$f(x) = \begin{cases} k(2x - 3x^2) & -1 < x < 0 \\ 0 & \text{elsewhere.} \end{cases}$$

   (a) Calculate the value of k.

   (b) Find the probability that the loss is at most $500.

9. Let X denote the lifetime of a radio, in years, manufactured by a certain company. The density function of X is given by

$$f(x) = \begin{cases} \dfrac{1}{15} e^{-x/15} & \text{if } 0 \le x < \infty \\ 0 & \text{elsewhere.} \end{cases}$$

What is the probability that, of eight such radios, at least four last more than 15 years?

10. Prove that if f and g are two probability density functions, then for $\alpha \ge 0$, $\beta \ge 0$, and $\alpha + \beta = 1$, $\alpha f + \beta g$ is also a probability density function.

11. The distribution function of a random variable X is given by

$$F(x) = \alpha + \beta \arctan \frac{x}{2}, \qquad -\infty < x < \infty.$$

Determine the constants α and β and the density function of X.

6.2   DENSITY FUNCTION OF A FUNCTION OF A RANDOM VARIABLE

In the theory and applications of probability, we are often faced with situations in which f, the density function of a random variable X, is known but the density function $h(X)$, for some function of X, is needed. In such cases there are two common methods for the evaluation of the density function of $h(X)$ that we will explain in this section. One method is to find the density of $h(X)$ by calculating its distribution function. Another method is to find it directly from the density function of X. We now explain the first method, called **the method of distribution functions**. To calculate the probability density function of $h(X)$, we calculate the distribution function of $h(X)$, $P\big(h(X) \le t\big)$, by finding a set A for which $h(X) \le t$ if and only if $X \in A$ and then computing $P(X \in A)$ from the distribution function of X. When $P\big(h(X) \le t\big)$ is found, the density function of $h(X)$, if it exists, is obtained by differentiation. Some examples follow.

Example 6.3   Let X be a continuous random variable with the probability density function

$$f(x) = \begin{cases} 2/x^2 & \text{if } 1 < x < 2 \\ 0 & \text{elsewhere.} \end{cases}$$

Find the distribution and the density functions of $Y = X^2$.

Solution: Let G and g be the distribution function and the density function of Y, respectively. By definition,

$$G(t) = P(Y \le t) = P(X^2 \le t) = P(-\sqrt{t} \le X \le \sqrt{t})$$

$$= \begin{cases} 0 & t < 1 \\ P(1 \le X \le \sqrt{t}) & 1 \le t < 4 \\ 1 & t \ge 4. \end{cases}$$

Now

$$P(1 \le X \le \sqrt{t}) = \int_1^{\sqrt{t}} \frac{2}{x^2}\,dx = \left[-\frac{2}{x}\right]_1^{\sqrt{t}} = 2 - \frac{2}{\sqrt{t}}.$$

Therefore,

$$G(t) = \begin{cases} 0 & t < 1 \\ 2 - \dfrac{2}{\sqrt{t}} & 1 \le t < 4 \\ 1 & t \ge 4. \end{cases}$$

The density function of Y, g, is found by differentiation of G:

$$g(t) = G'(t) = \begin{cases} \dfrac{1}{t\sqrt{t}} & \text{if } 1 \le t \le 4 \\ 0 & \text{elsewhere.} \end{cases} \quad \blacklozenge$$

Example 6.4   Let X be a continuous random variable with distribution function F and probability density function f. In terms of f, find the distribution and the density functions of $Y = X^3$.

Solution: Let G and g be the distribution function and the density function of Y. Then

$$G(t) = P(Y \le t) = P(X^3 \le t) = P(X \le \sqrt[3]{t}) = F(\sqrt[3]{t}).$$

Hence

$$g(t) = G'(t) = \frac{1}{3\sqrt[3]{t^2}} F'(\sqrt[3]{t}) = \frac{1}{3\sqrt[3]{t^2}} f(\sqrt[3]{t}),$$

where g is defined at all points $t \neq 0$ at which f is defined.   ◆

Example 6.5   The error of a measurement has the density function

$$f(x) = \begin{cases} 1/2 & \text{if } -1 < x < 1 \\ 0 & \text{otherwise.} \end{cases}$$

Find the distribution and the density functions of the magnitude of the error.

Solution: Let X be the error of the measurement. We want to find G, the distribution, and g, the density functions of $|X|$, the magnitude of the error. By definition,

$$G(t) = P(|X| \leq t) = P(-t \leq X \leq t) = \begin{cases} 0 & t < 0 \\ \int_{-t}^{t} \frac{1}{2} \, dx = t & 0 \leq t < 1 \\ 1 & t \geq 1. \end{cases}$$

The probability density function of Y, g, is obtained by differentiation of G:

$$g(t) = G'(t) = \begin{cases} 1 & \text{if } 0 \leq t \leq 1 \\ 0 & \text{elsewhere.} \end{cases}   ◆$$

In Examples 6.3, 6.4, and 6.5, the method used to find the density function of $h(X)$, a function of the random variable X, is to determine the distribution of $h(X)$ and then differentiate it. It is also possible to find the density function of $h(X)$ without obtaining its distribution function. We now explain this method, called the **method of transformations**.

Recall that a real-valued function $h \colon \mathbf{R} \to \mathbf{R}$ defined by $y = h(x)$ is *invertible* if and only if when solving $y = h(x)$ for x, we find that $x = h^{-1}(y)$ is itself a function. That is, $y = h(x)$ if and only if $x = h^{-1}(y)$, where $x = h^{-1}(y)$ is a function. For example, the function $y = h(x) = x^3$ is invertible because if we solve it for x, we obtain $x = h^{-1}(y) = \sqrt[3]{y}$, which is itself a function. The function $y = h(x) = x^2$ is not invertible if its domain is $(-\infty, \infty)$. When solving for x we obtain $x = \pm\sqrt{y}$, which is not a function. The function $y = x^2$ will be invertible if we restrict its domain to $[0, \infty)$, the set of nonnegative numbers. This follows since solving $y = x^2$ for x gives $x = \sqrt{y}$, which is a function. Other examples of invertible functions are $h(x) = e^x$, $x \in \mathbf{R}$; $h(x) = \ln x$, $x \in (0, \infty)$; and $h(x) = \sin x$, $x \in (0, \pi/2)$. Examples of functions that are not invertible are $h(x) = \sin x$, $x \in \mathbf{R}$; $h(x) = x^2 + 5$, $x \in \mathbf{R}$; and $h(x) = \exp(x^2)$, $x \in \mathbf{R}$.

In calculus, we have seen that a continuous function is invertible if and only if it is either strictly increasing or strictly decreasing. If it is strictly increasing, its inverse is also strictly increasing. If it is strictly decreasing, its inverse is strictly decreasing as well. We use these to prove the next theorem, which enables us to find the density function of $h(X)$ mainly by finding the inverse of h. For this application, all we need to know is how to find the inverse of a function, if it exists. Studying different criteria for invertibility does not serve our purpose.

Theorem 6.1   (Method of Transformations)   *Let X be a continuous random variable with density function f_X and the set of possible values A. For the invertible function $h: A \to \mathbf{R}$, let $Y = h(X)$ be a random variable with the set of possible values $B = h(A) = \{h(a): a \in A\}$. Suppose that the inverse of $y = h(x)$ is the function $x = h^{-1}(y)$, which is differentiable for all values of $y \in B$. Then f_Y, the density function of Y, is given by*

$$f_Y(y) = f_X\big(h^{-1}(y)\big)\big|(h^{-1})'(y)\big|, \qquad y \in B.$$

Proof: Let F_X and F_Y be distribution functions of X and $Y = h(X)$, respectively. Differentiability of h^{-1} implies that it is continuous. Since a continuous invertible function is strictly monotone, h^{-1} is either strictly increasing or strictly decreasing. If it is strictly increasing, $(h^{-1})'(y) > 0$, and hence $(h^{-1})'(y) = \big|(h^{-1})'(y)\big|$. Moreover, in this case h is also strictly increasing, so

$$F_Y(y) = P\big(h(X) \le y\big) = P\big(X \le h^{-1}(y)\big) = F_X\big(h^{-1}(y)\big).$$

Differentiating this by chain rule, we obtain

$$F_Y'(y) = (h^{-1})'(y)F_X'\big(h^{-1}(y)\big) = \big|(h^{-1})'(y)\big|f_X\big(h^{-1}(y)\big),$$

which gives the theorem. If h^{-1} is strictly decreasing, $(h^{-1})'(y) < 0$ and hence $\big|(h^{-1})'(y)\big| = -(h^{-1})'(y)$. In this case h is also strictly decreasing and we get

$$F_Y(y) = P\big(h(X) \le y\big) = P\big(X \ge h^{-1}(y)\big) = 1 - F_X\big(h^{-1}(y)\big).$$

Differentiating this by chain rule, we find that

$$F_Y'(y) = -(h^{-1})'(y)F_X'\big(h^{-1}(y)\big) = \big|(h^{-1})'(y)\big|f_X\big(h^{-1}(y)\big),$$

showing that the theorem is valid in this case as well.   ♦

Example 6.6   Let X be a random variable with the density function

$$f_X(x) = \begin{cases} 2e^{-2x} & \text{if } x > 0 \\ 0 & \text{otherwise.} \end{cases}$$

Using the method of transformations, find the probability density function of $Y = \sqrt{X}$.

Solution: The set of possible values of X is $A = (0, \infty)$. Let $h\colon (0, \infty) \to \mathbf{R}$ be defined by $h(x) = \sqrt{x}$. We want to find the density function of $Y = h(X) = \sqrt{X}$. The set of possible values of $h(X)$ is $B = \{h(a)\colon a \in A\} = (0, \infty)$. The function h is invertible with the inverse $x = h^{-1}(y) = y^2$, which is differentiable, and its derivative is $(h^{-1})'(y) = 2y$. Therefore, by Theorem 6.1,

$$f_Y(y) = f_X\big(h^{-1}(y)\big)\big|(h^{-1})'(y)\big| = 2e^{-2y^2}|2y|, \qquad y \in B = (0, \infty).$$

Since $y \in (0, \infty)$, $|2y| = 2y$ and we find that

$$f_Y(y) = \begin{cases} 4ye^{-2y^2} & \text{if } y > 0 \\ 0 & \text{otherwise} \end{cases}$$

is the density function of $Y = \sqrt{X}$. ◆

Example 6.7   Let X be a continuous random variable with the probability density function

$$f_X(x) = \begin{cases} 4x^3 & \text{if } 0 < x < 1 \\ 0 & \text{otherwise.} \end{cases}$$

Using the method of transformations, find the probability density function of $Y = 1 - 3X^2$.

Solution: The set of possible values of X is $A = (0, 1)$. Let $h\colon (0, 1) \to \mathbf{R}$ be defined by $h(x) = 1 - 3x^2$. We want to find the density function of $Y = h(X) = 1 - 3X^2$. The set of possible values of $h(X)$ is $B = \{h(a)\colon a \in A\} = (-2, 1)$. Since the domain of the function h is $(0, 1)$, h is invertible, and its inverse is found by solving $1 - 3x^2 = y$ for x, which gives $x = \sqrt{(1-y)/3}$. Therefore, the inverse of h is $x = h^{-1}(y) = \sqrt{(1-y)/3}$, which is differentiable, and its derivative is $(h^{-1})'(y) = -1/\big(2\sqrt{3(1-y)}\big)$. Using Theorem 6.1, we find that

$$f_Y(y) = f_X\big(h^{-1}(y)\big)\big|(h^{-1})'(y)\big| = 4\Big(\sqrt{\frac{1-y}{3}}\Big)^3\Big| -\frac{1}{2\sqrt{3(1-y)}}\Big| = \frac{2}{9}(1-y)$$

is the density function of Y when $y \in (-2, 1)$. ◆

EXERCISES

A

1. Let X be a continuous random variable with the density function

$$f(x) = \begin{cases} 1/4 & \text{if } x \in (-2, 2) \\ 0 & \text{otherwise.} \end{cases}$$

Using the method of distribution functions, find the probability density functions of $Y = X^3$ and $Z = X^4$.

2. Let X be a continuous random variable with distribution function F and density function f. Calculate the density function of the random variable $Y = e^X$.

3. Let the density function of X be

$$f(x) = \begin{cases} e^{-x} & \text{if } x > 0 \\ 0 & \text{elsewhere.} \end{cases}$$

Using the method of transformations, find the density functions of $Y = X\sqrt{X}$ and $Z = e^{-X}$.

4. Let X be a continuous random variable with the density function

$$f(x) = \begin{cases} 3e^{-3x} & \text{if } x > 0 \\ 0 & \text{otherwise.} \end{cases}$$

Using the method of transformations, find the probability density function of $Y = \log_2 X$.

5. Let the probability density function of X be

$$f(x) = \begin{cases} \lambda e^{-\lambda x} & \text{if } x \geq 0 \\ 0 & \text{otherwise,} \end{cases}$$

for some $\lambda > 0$. Using the method of distribution functions, calculate the probability density function of $Y = \sqrt[3]{X^2}$.

6. Let f be the probability density function of a random variable X. In terms of f, calculate the probability density function of X^2.

7. Let X be a random variable with the density function

$$f(x) = \frac{1}{\pi(1 + x^2)}, \qquad -\infty < x < \infty.$$

(X is called a **Cauchy random variable.**) Find the density function of $Z = \arctan X$.

B

8. Let X be a random variable with the probability density function given by

$$f(x) = \begin{cases} e^{-x} & \text{if } x \geq 0 \\ 0 & \text{elsewhere.} \end{cases}$$

Let

$$Y = \begin{cases} X & \text{if } X \leq 1 \\ 1/X & \text{if } X > 1. \end{cases}$$

Find the probability density function of Y.   ∎

6.3   EXPECTATIONS AND VARIANCES

Expectations of Continuous Random Variables

Let X be a continuous random variable with probability density and distribution functions f and F, respectively. To define $E(X)$, the average or expected value of X, first suppose that X only takes values from the interval $[a, b]$, and divide $[a, b]$ into n subintervals of equal lengths. Let $h = (b-a)/n, x_0 = a, x_1 = a+h, x_2 = a+2h, \ldots, x_n = a+nh = b.$ Then $a = x_0 < x_1 < x_2 < \cdots < x_n = b$ is a partition of $[a, b]$. Since F is continuous, let us assume that it is differentiable on (a, b). By the mean-value theorem (of calculus), there exists $t_i \in (x_{i-1}, x_i)$ such that

$$F(x_i) - F(x_{i-1}) = F'(t_i)(x_i - x_{i-1}), \qquad 1 \leq i \leq n,$$

or, equivalently,

$$P(x_{i-1} < X \leq x_i) = f(t_i)h, \qquad 1 \leq i \leq n. \tag{6.6}$$

If n is sufficiently large ($n \to \infty$), then h, the width of the intervals, is sufficiently small ($h \to 0$), and $f(x)$ does not vary appreciably over any subinterval of $(x_{i-1}, x_i]$,

$1 \leq i \leq n$. Thus t_i is an approximate value of X in the interval $(x_{i-1}, x_i]$. Now $\sum_{i=1}^{n} t_i P(x_{i-1} < X \leq x_i)$ finds the product of an approximate value of X when it is in $(x_{i-1}, x_i]$ and the probability that it is in $(x_{i-1}, x_i]$, and then sums over all these intervals. From the concept of expectation of a discrete random variable, it is clear that, as the lengths of these intervals get smaller and smaller, $\sum_{i=1}^{n} t_i P(x_{i-1} < X \leq x_i)$ gets closer and closer to the "average" value of X. So it is desirable to define $E(X)$ as

$$\lim_{n \to \infty} \sum_{i=1}^{n} t_i P(x_{i-1} < X \leq x_i).$$

But by (6.6) this is the same as $\lim_{n \to \infty} \sum_{i=1}^{n} t_i f(t_i) h$, where this limit is equal to $\int_a^b x f(x)\, dx$ as known from calculus. If X is not restricted to an interval $[a, b]$, a definition for $E(X)$ is motivated in the same way, but at the end the limit is taken as $a \to -\infty$ and $b \to \infty$.

Definition   *If X is a continuous random variable with probability density function f, the expected value of X is defined by*

$$E(X) = \int_{-\infty}^{\infty} x f(x)\, dx.$$

The expected value of X is also called the **mean**, or **mathematical expectation**, or simply the **expectation** of X, and as in the discrete case, sometimes it is denoted by EX, $E[X]$, μ, or μ_X.

To get a geometric feeling for mathematical expectation, consider a piece of cardboard of uniform density on which the graph of the density function f of a random variable X is drawn. Suppose that the cardboard is cut along the graph of f and we are asked to balance it on a given edge perpendicular to the x-axis. Then, to have it in equilibrium, we must balance the cardboard on the given edge at the point $x = E(X)$ on the x-axis.

Example 6.8   In a group of adult males, the difference between the uric acid value and 6, the standard value, is a random variable X with the following probability density function:

$$f(x) = \begin{cases} \dfrac{27}{490}(3x^2 - 2x) & \text{if } 2/3 < x < 3 \\[2mm] 0 & \text{elsewhere.} \end{cases}$$

Calculate the mean of these differences for the group.

Solution: By definition,

$$E(X) = \int_{-\infty}^{\infty} x f(x)\, dx = \int_{2/3}^{3} \frac{27}{490}(3x^3 - 2x^2)\, dx$$

$$= \frac{27}{490}\left[\frac{3}{4}x^4 - \frac{2}{3}x^3 \right]_{2/3}^{3} = \frac{283}{120} = 2.36. \quad \blacklozenge$$

Remark 6.2 If X is a continuous random variable with density function f, X is said to have a **finite expected value** if

$$\int_{-\infty}^{\infty} |x| f(x)\, dx < \infty;$$

that is, X has a finite expected value if the integral of $x f(x)$ converges absolutely. Otherwise, we say that the expected value of X is not finite. We now justify why the absolute convergence of the integral is required. Note that

$$E(X) = \int_{-\infty}^{\infty} x f(x)\, dx = \int_{-\infty}^{0} x f(x)\, dx + \int_{0}^{\infty} x f(x)\, dx$$

$$= -\int_{-\infty}^{0} (-x) f(x)\, dx + \int_{0}^{\infty} x f(x)\, dx,$$

where $\int_{-\infty}^{0}(-x) f(x)\, dx \geq 0$ and $\int_{0}^{\infty} x f(x)\, dx \geq 0$. Thus $E(X)$ is well defined if $\int_{-\infty}^{0}(-x) f(x)\, dx$ and $\int_{0}^{\infty} x f(x)\, dx$ are not both ∞. Moreover, $E(X) < \infty$ if neither of these two integrals is $+\infty$. Hence a necessary and sufficient condition for $E(X)$ to exist and to be finite is that $\int_{-\infty}^{0}(-x) f(x)\, dx < \infty$ and $\int_{0}^{\infty} x f(x)\, dx < \infty$. Since both of these integrals are finite if and only if

$$\int_{-\infty}^{\infty} |x| f(x)\, dx = \int_{-\infty}^{0} (-x) f(x)\, dx + \int_{0}^{\infty} x f(x)\, dx$$

is finite, we have that $E(X)$ is well defined and finite if and only if the integral $\int_{-\infty}^{\infty} x f(x)\, dx$ is absolutely convergent. ♦

Example 6.9 A random variable X with density function

$$f(x) = \frac{c}{1 + x^2}, \qquad -\infty < x < \infty,$$

is called a **Cauchy random variable**.

(a) Find c.

(b) Show that $E(X)$ does not exist.

Solution:

(a) Since f is a density function $\int_{-\infty}^{\infty} f(x)\, dx = 1$. Thus

$$\int_{-\infty}^{\infty} \frac{c}{1 + x^2}\, dx = c \int_{-\infty}^{\infty} \frac{dx}{1 + x^2} = 1.$$

Now

$$\int \frac{dx}{1+x^2} = \arctan x.$$

Since the range of $\arctan x$ is $(-\pi/2, +\pi/2)$, we get

$$1 = c \int_{-\infty}^{\infty} \frac{dx}{1+x^2} = c \Big[\arctan x \Big]_{-\infty}^{\infty} = c\Big[\frac{\pi}{2} - \Big(-\frac{\pi}{2}\Big)\Big] = c\pi.$$

Thus $c = 1/\pi$.

(b)   To show that $E(X)$ does not exist, note that

$$\int_{-\infty}^{\infty} |x| f(x)\,dx = \int_{-\infty}^{\infty} \frac{|x|\,dx}{\pi(1+x^2)} = 2 \int_0^{\infty} \frac{x\,dx}{\pi(1+x^2)}$$

$$= \frac{1}{\pi}\Big[\ln(1+x^2)\Big]_0^{\infty} = \infty. \quad \blacklozenge$$

Remark 6.3   In this book, unless otherwise specified, it is implicitly assumed that the expectation of a random variable is finite. $\blacklozenge$

The following theorem directly relates the distribution function of a random variable to its expectation. It enables us to find the expected value of a continuous random variable without calculating its probability density function. It also has important theoretical applications.

Theorem 6.2   *For any continuous random variable X with probability distribution function F and density function f,*

$$E(X) = \int_0^{\infty} \big[1 - F(t)\big]\,dt - \int_0^{\infty} F(-t)\,dt.$$

Proof: Note that

$$E(X) = \int_{-\infty}^{\infty} x f(x)\,dx = \int_{-\infty}^0 x f(x)\,dx + \int_0^{\infty} x f(x)\,dx$$

$$= -\int_{-\infty}^0 \Big(\int_0^{-x} dt\Big) f(x)\,dx + \int_0^{\infty} \Big(\int_0^x dt\Big) f(x)\,dx$$

$$= -\int_0^{\infty} \Big(\int_{-\infty}^{-t} f(x)\,dx\Big) dt + \int_0^{\infty} \Big(\int_t^{\infty} f(x)\,dx\Big) dt,$$

where the last equality is obtained by changing the order of integration. The theorem follows since $\int_{-\infty}^{-t} f(x)\,dx = F(-t)$ and $\int_t^{\infty} f(x)\,dx = P(X > t) = 1 - F(t)$. $\blacklozenge$

Remark 6.4   In the proof of this theorem we assumed that the random variable X is continuous. Even without this condition the theorem is still valid. Also note that, since $1 - F(t) = P(X > t)$, this theorem may be stated as follows.

For any random variable X,

$$E(X) = \int_0^\infty P(X > t)\, dt - \int_0^\infty P(X \le -t)\, dt.$$

In particular, if X is nonnegative, that is, $P(X < 0) = 0$, this theorem states that

$$E(X) = \int_0^\infty \left[1 - F(t)\right] dt = \int_0^\infty P(X > t)\, dt. \quad \blacklozenge$$

As an important application of Theorem 6.2, we now prove the law of the unconscious statistician, Theorem 4.2, for continuous random variables.

Theorem 6.3   *Let X be a continuous random variable with probability density function $f(x)$; then for any function $h: \mathbf{R} \to \mathbf{R}$,*

$$E\big[h(X)\big] = \int_{-\infty}^\infty h(x) f(x)\, dx.$$

Proof: Let

$$h^{-1}(t, \infty) = \big\{x: h(x) \in (t, \infty)\big\} = \big\{x: h(x) > t\big\}$$

with similar representation for $h^{-1}(-\infty, -t)$. Notice that we are not claiming that h has an inverse function. We are simply considering the set $\big\{x: h(x) \in (t, \infty)\big\}$, which is called the *inverse image* of (t, ∞) and is denoted by $h^{-1}(t, \infty)$. By Theorem 6.2,

$$E\big[h(X)\big] = \int_0^\infty P\big(h(X) > t\big)\, dt - \int_0^\infty P\big(h(X) \le -t\big)\, dt$$

$$= \int_0^\infty P\big(X \in h^{-1}(t, \infty)\big)\, dt - \int_0^\infty P\big(X \in h^{-1}(-\infty, -t]\big)\, dt$$

$$= \int_0^\infty \left(\int_{\{x:\, x \in h^{-1}(t, \infty)\}} f(x)\, dx\right) dt$$

$$- \int_0^\infty \left(\int_{\{x:\, x \in h^{-1}(-\infty, -t]\}} f(x)\, dx\right) dt$$

$$= \int_0^\infty \left(\int_{\{x:\, h(x) > t\}} f(x)\, dx\right) dt - \int_0^\infty \left(\int_{\{x:\, h(x) \le -t\}} f(x)\, dx\right) dt.$$

Now we change the order of integration for both of these double integrals. Since

$$\big\{(t, x): 0 < t < \infty,\ h(x) > t\big\} = \big\{(t, x): h(x) > 0,\ 0 < t < h(x)\big\},$$

and

$$\big\{(t, x): 0 < t < \infty,\ h(x) \le -t\big\} = \big\{(t, x): h(x) < 0,\ 0 < t \le -h(x)\big\},$$

we get

$$E[h(X)] = \int_{\{x\,:\,h(x)\,>\,0\}} \left(\int_0^{h(x)} dt \right) f(x)\, dx$$

$$- \int_{\{x\,:\,h(x)\,<\,0\}} \left(\int_0^{-h(x)} dt \right) f(x)\, dx$$

$$= \int_{\{x\,:\,h(x)\,>\,0\}} h(x) f(x)\, dx + \int_{\{x\,:\,h(x)\,<\,0\}} h(x) f(x)\, dx$$

$$= \int_{-\infty}^{\infty} h(x) f(x)\, dx.$$

Note that the last equality follows because $\displaystyle\int_{\{x\,:\,h(x)\,=\,0\}} h(x) f(x)\, dx = 0.$ ◆

Corollary   *Let X be a continuous random variable with probability density function $f(x)$. Let $h_1, h_2, \ldots, h_n$ be real-valued functions, and $\alpha_1, \alpha_2, \ldots, \alpha_n$ be real numbers. Then*

$$E\big[\alpha_1 h_1(X) + \alpha_2 h_2(X) + \cdots + \alpha_n h_n(X)\big]$$
$$= \alpha_1 E\big[h_1(X)\big] + \alpha_2 E\big[h_2(X)\big] + \cdots + \alpha_n E\big[h_n(X)\big].$$

Proof: In the discrete case, in the proof of the corollary of Theorem 4.2, replace $\sum$ by $\int$, and $p(x)$ by $f(x)\, dx$.   ◆

Just as in the discrete case, by this corollary we can write that, for example,

$$E(3X^4 + \cos X + 3e^X + 7) = 3E(X^4) + E(\cos X) + 3E(e^X) + 7.$$

Moreover, this corollary implies that if α and β are constants, then

$$E(\alpha X + \beta) = \alpha E(X) + \beta.$$

Example 6.10   A point X is selected from the interval $(0, \pi/4)$ randomly. Calculate $E(\cos 2X)$ and $E(\cos^2 X)$.

Solution: First we calculate the distribution function of X. Clearly,

$$F(t) = P(X \le t) = \begin{cases} 0 & t < 0 \\[2mm] \dfrac{t - 0}{\dfrac{\pi}{4} - 0} = \dfrac{4t}{\pi} & 0 \le t < \dfrac{\pi}{4} \\[4mm] 1 & t \ge \dfrac{\pi}{4}. \end{cases}$$

Thus f, the probability density function of X, is

$$f(t) = \begin{cases} \dfrac{4}{\pi} & 0 < t < \dfrac{\pi}{4} \\[2ex] 0 & \text{otherwise.} \end{cases}$$

Now, by Theorem 6.3,

$$E(\cos 2X) = \int_0^{\pi/4} \frac{4}{\pi} \cos 2x \, dx = \left[\frac{2}{\pi} \sin 2x\right]_0^{\pi/4} = \frac{2}{\pi}.$$

To calculate $E(\cos^2 X)$, note that $\cos^2 X = (1 + \cos 2X)/2$. So by the corollary of Theorem 6.3,

$$E(\cos^2 X) = E\left(\frac{1}{2} + \frac{1}{2}\cos 2X\right) = \frac{1}{2} + \frac{1}{2}E(\cos 2X) = \frac{1}{2} + \frac{1}{2} \cdot \frac{2}{\pi} = \frac{1}{2} + \frac{1}{\pi}. \quad \blacklozenge$$

Variances of Continuous Random Variables

Definition   *If X is a continuous random variable with $E(X) = \mu$, then $\mathrm{Var}(X)$ and σ_X, called the **variance** and **standard deviation** of X, respectively, are defined by*

$$\mathrm{Var}(X) = E\big[(X - \mu)^2\big],$$

$$\sigma_X = \sqrt{E\big[(X - \mu)^2\big]}.$$

Therefore, if f is the density function of X, then by Theorem 6.3,

$$\mathbf{Var}(X) = E\big[(X - \mu)^2\big] = \int_{-\infty}^{\infty} (x - \mu)^2 f(x) \, dx.$$

Also, as before, we have the following important relations whose proofs are analogous to those in the discrete case.

$$\mathbf{Var}(X) = E(X^2) - \big[E(X)\big]^2,$$

$$\mathbf{Var}(aX + b) = a^2 \mathbf{Var}(X), \quad \sigma_{aX+b} = |a|\sigma_X, \qquad a \text{ and } b \text{ being constants.}$$

Example 6.11   The time elapsed, in minutes, between the placement of an order of pizza and its delivery is random with the density function

$$f(x) = \begin{cases} 1/15 & \text{if } 25 < x < 40 \\[2ex] 0 & \text{otherwise.} \end{cases}$$

(a) Determine the mean and standard deviation of the time it takes for the pizza shop to deliver pizza.

(b) Suppose that it takes 12 minutes for the pizza shop to bake pizza. Determine the mean and standard deviation of the time it takes for the delivery person to deliver pizza.

Solution:

(a) Let the time elapsed between the placement of an order and its delivery be X minutes. Then

$$E(X) = \int_{25}^{40} x \frac{1}{15} dx = 32.5,$$

$$E(X^2) = \int_{25}^{40} x^2 \frac{1}{15} dx = 1075.$$

Therefore, $\text{Var}(X) = 1075 - (32.5)^2 = 18.75$, and hence $\sigma_X = \sqrt{\text{Var}(X)} = 4.33$.

(b) The time it takes for the delivery person to deliver pizza is $X - 12$. Therefore, the desired quantities are

$$E(X - 12) = E(X) - 12 = 32.5 - 12 = 20.5$$
$$\sigma_{X-12} = |1|\sigma_X = \sigma_X = 4.33. \quad \blacklozenge$$

Remark 6.5   For a continuous random variable X, the moments, absolute moments, moments about a constant c, and central moments are all defined in a manner similar to those of Section 4.5 (the discrete case). $\blacklozenge$

★ **Remark 6.6**   In Remark 4.2, we showed that, for a discrete random variable X and positive integer n, if $E(X^{n+1})$ exists, then $E(X^n)$ also exists. That is, the existence of higher moments of a discrete random variable implies the existence of its lower moments. In particular, the existence of $E(X^2)$ implies the existence of $E(X)$ and, hence, $\text{Var}(X)$. These facts are also true for continuous random variables, and their proofs are similar. Let X be a continuous random variable with probability density function $f(x)$. Let n be a positive integer. We will show that if $E(X^{n+1})$ exists, then $E(X^n)$ also exists. Clearly,

$$\int_{-1}^{1} |x|^n f(x) \, dx \le \int_{-1}^{1} f(x) \, dx \le \int_{-\infty}^{\infty} f(x) \, dx = 1;$$

$$\int_{|x|>1} |x|^n f(x) \, dx \le \int_{|x|>1} |x|^{n+1} f(x) \, dx \le \int_{-\infty}^{\infty} |x|^{n+1} f(x) \, dx.$$

These inequalities yield

$$\int_{-\infty}^{\infty} |x|^n f(x)\,dx = \int_{-1}^{1} |x|^n f(x)\,dx + \int_{|x|>1} |x|^n f(x)\,dx \le 1 + \int_{-\infty}^{\infty} |x|^{n+1} f(x)\,dx.$$

Therefore, $\displaystyle\int_{-\infty}^{\infty} |x|^{n+1} f(x)\,dx < \infty$ implies that $\displaystyle\int_{-\infty}^{\infty} |x|^n f(x)\,dx < \infty$. By Remark 6.2, this shows that the existence of $E(X^{n+1})$ implies the existence of $E(X^n)$. $\blacklozenge$

EXERCISES

A

1. The distribution function for the duration of a certain soap opera (in tens of hours) is

$$F(x) = \begin{cases} 1 - \dfrac{16}{x^2} & \text{if } x \ge 4 \\[2mm] 0 & \text{if } x < 4. \end{cases}$$

   (a) Find $E(X)$.

   (b) Show that $\text{Var}(X)$ does not exist.

2. The time it takes for a student to finish an aptitude test (in hours) has the density function

$$f(x) = \begin{cases} 6(x-1)(2-x) & \text{if } 1 < x < 2 \\ 0 & \text{otherwise.} \end{cases}$$

   Determine the mean and standard deviation of the time it takes for a randomly selected student to finish the aptitude test.

3. The mean and standard deviation of the lifetime of a car muffler manufactured by company A are 5 and 2 years, respectively. These quantities for car mufflers manufactured by company B are, respectively, 4 years and 18 months. Brian buys one muffler from company A and one from company B. That of company A lasts 4 years and 3 months, and that of company B lasts 3 years and 9 months. Determine which of these mufflers has performed relatively better.
   Hint: Find the standardized lifetimes of the mufflers and compare (see Section 4.6).

4. A random variable X has the density function

$$f(x) = \begin{cases} 3e^{-3x} & \text{if } 0 \le x < \infty \\ 0 & \text{otherwise.} \end{cases}$$

Calculate $E(e^X)$.

5. Find the expected value of a random variable X with the density function

$$f(x) = \begin{cases} \dfrac{1}{\pi\sqrt{1 - x^2}} & \text{if } -1 < x < 1 \\ 0 & \text{otherwise.} \end{cases}$$

6. Let Y be a continuous random variable with probability distribution function

$$F(y) = \begin{cases} e^{-k(\alpha - y)/A} & -\infty < y \le \alpha \\ 1 & y > \alpha, \end{cases}$$

where A, k, and α are positive constants. (Such distribution functions arise in the study of local computer network performance.) Find $E(Y)$.

7. Let the probability density function of tomorrow's Celsius temperature be h. In terms of h, calculate the corresponding probability density function and its expectation for Fahrenheit temperature.
Hint: Let C and F be tomorrow's temperature in Celsius and Fahrenheit, respectively. Then $F = 1.8C + 32$.

8. Let X be a continuous random variable with probability density function

$$f(x) = \begin{cases} 2/x^2 & \text{if } 1 < x < 2 \\ 0 & \text{elsewhere.} \end{cases}$$

Find $E(\ln X)$.

9. A right triangle has a hypotenuse of length 9. If the probability density function of one side's length is given by

$$f(x) = \begin{cases} x/6 & \text{if } 2 < x < 4 \\ 0 & \text{otherwise,} \end{cases}$$

what is the expected value of the length of the other side?

10. Let X be a random variable with probability density function

$$f(x) = \frac{1}{2}e^{-|x|}, \qquad -\infty < x < \infty.$$

Calculate Var(X).

B

11. Let X be a random variable with the probability density function

$$f(x) = \frac{1}{\pi(1+x^2)}, \qquad -\infty < x < \infty.$$

Prove that $E(|X|^\alpha)$ converges if $0 < \alpha < 1$ and diverges if $\alpha \geq 1$.

12. Suppose that X, the interarrival time between two customers entering a certain postoffice, satisfies

$$P(X > t) = \alpha e^{-\lambda t} + \beta e^{-\mu t}, \qquad t \geq 0,$$

where $\alpha + \beta = 1$, $\alpha \geq 0$, $\beta \geq 0$, $\lambda > 0$, $\mu > 0$. Calculate the expected value of X.

Hint: For a fast calculation, use Remark 6.4.

13. For $n \geq 1$, let X_n be a continuous random variable with the probability density function

$$f_n(x) = \begin{cases} \dfrac{c_n}{x^{n+1}} & \text{if } x \geq c_n \\ 0 & \text{otherwise.} \end{cases}$$

X_n's are called **Pareto random variables** and are used to study income distributions.

  (a) Calculate $c_n, n \geq 1$.

  (b) Find $E(X_n), n \geq 1$.

  (c) Determine the density function of $Z_n = \ln X_n, n \geq 1$.

  (d) For what values of m does $E(X_n^{m+1})$ exist?

14. Let X be a continuous random variable with the probability density function

$$f(x) = \begin{cases} \dfrac{1}{\pi} x \sin x & \text{if } 0 < x < \pi \\ 0 & \text{otherwise.} \end{cases}$$

Prove that

$$E(X^{n+1}) + (n+1)(n+2)E(X^{n-1}) = \pi^{n+1}.$$

15. Let X be a continuous random variable with density function f. A number t is said to be the **median** of X if

$$P(X \leq t) = P(X \geq t) = \frac{1}{2}.$$

By Exercise 7, Section 6.1, X is symmetric about α if and only if for all x we have $f(\alpha - x) = f(\alpha + x)$. Show that if X is symmetric about α, then

$$E(X) = \text{Median}(X) = \alpha.$$

16. Let X be a continuous random variable with probability density function $f(x)$. Determine the value of y for which $E(|X - y|)$ is minimum.

17. Let X be a nonnegative random variable with distribution function F. Define

$$I(t) = \begin{cases} 1 & \text{if } X > t \\ 0 & \text{otherwise.} \end{cases}$$

(a) Prove that $\int_0^\infty I(t)\,dt = X$.

(b) By calculating the expected value of both sides of part (a), prove that

$$E(X) = \int_0^\infty \big[1 - F(t)\big]\,dt.$$

This is a special case of Theorem 6.2.

(c) For $r > 0$, use part (b) to prove that

$$E(X^r) = r \int_0^\infty t^{r-1}\big[1 - F(t)\big]\,dt.$$

18. Let X be a continuous random variable. Prove that

$$\sum_{n=1}^\infty P\big(|X| \ge n\big) \le E\big(|X|\big) \le 1 + \sum_{n=1}^\infty P\big(|X| \ge n\big).$$

These important inequalities show that $E(|X|) < \infty$ if and only if the series $\sum_{n=1}^\infty P(|X| \ge n)$ converges.

Hint: By Exercise 17,

$$E\big(|X|\big) = \int_0^\infty P\big(|X| > t\big)\,dt = \sum_{n=0}^\infty \int_n^{n+1} P\big(|X| > t\big)\,dt.$$

Note that on the interval $[n, n+1)$,

$$P\big(|X| \ge n + 1\big) < P\big(|X| > t\big) \le P\big(|X| \ge n\big).$$

19. Let X be the random variable introduced in Exercise 12. Applying the results of Exercise 17, calculate $\text{Var}(X)$.

20. Suppose that X is the lifetime of a randomly selected fan used in certain types of diesel engines. Let Y be a randomly selected competing fan for the same type of diesel engines manufactured by another company. To compare the lifetimes X and Y, it is not sufficient to compare $E(X)$ and $E(Y)$. For example, $E(X) > E(Y)$ does not necessarily imply that the first manufacture's fan outlives the second manufacture's fan. Knowing $\text{Var}(X)$ and $\text{Var}(Y)$ will help, but variance is also a crude measure. One of the best tools for comparing random variables in such situations is *stochastic comparison*. let X and Y be two random variables. We say that X is **stochastically larger** than Y, denoted by $X \geq_{st} Y$, if for all t,

$$P(X > t) \geq P(Y > t).$$

Show that if $X \geq_{st} Y$, then $E(X) \geq E(Y)$, but not conversely.
Hint: Use Theorem 6.2.

21. Let X be a continuous random variable with probability density function f. Show that if $E(X)$ exists; that is, if $\int_{-\infty}^{\infty} |x| f(x) \, dx < \infty$, then

$$\lim_{x \to -\infty} x P(X \leq x) = \lim_{x \to \infty} x P(X > x) = 0.$$

∎

REVIEW PROBLEMS

---

1. Let X be a random number from $(0, 1)$. Find the probability density function of $Y = 1/X$.

2. Let X be a continuous random variable with the probability density function

$$f(x) = \begin{cases} 2/x^3 & \text{if } x > 1 \\ 0 & \text{otherwise.} \end{cases}$$

Find $E(X)$ and $\text{Var}(X)$ if they exist.

3. Let X be a continuous random variable with density function

$$f(x) = 6x(1 - x), \qquad 0 < x < 1.$$

What is the probability that X is within two standard deviations of the mean?

4. Let X be a random variable with density function

$$f(x) = \frac{e^{-|x|}}{2}, \qquad -\infty < x < \infty.$$

Find $P(-2 < X < 1)$.

5. Does there exist a constant c for which the following is a density function?

$$f(x) = \begin{cases} \dfrac{c}{1+x} & \text{if } x > 0 \\ 0 & \text{otherwise.} \end{cases}$$

6. Let X be a random variable with density function

$$f(x) = \begin{cases} 4x^3/15 & 1 \le x \le 2 \\ 0 & \text{otherwise.} \end{cases}$$

Find the density functions of $Y = e^X$, $Z = X^2$, and $W = (X-1)^2$.

7. The probability density function of a continuous random variable X is

$$f(x) = \begin{cases} 30x^2(1-x)^2 & \text{if } 0 < x < 1 \\ 0 & \text{otherwise.} \end{cases}$$

Find the probability density function of $Y = X^4$.

8. Let F, the distribution of a random variable X, be defined by

$$F(x) = \begin{cases} 0 & x < -1 \\ \dfrac{1}{2} + \dfrac{\arcsin x}{\pi} & -1 \le x < 1 \\ 1 & x \ge 1, \end{cases}$$

where $\arcsin x$ lies between $-\pi/2$ and $\pi/2$. Find f, the probability density function of X and $E(X)$.

9. Prove or disprove: If $\sum_{i=1}^{n} \alpha_i = 1$, $\alpha_i \ge 0$, $\forall i$, and $\{f_i\}_{i=1}^{n}$ is a sequence of density functions, then $\sum_{i=1}^{n} \alpha_i f_i$ is a probability density function.

10. Let X be a continuous random variable with set of possible values $\{x : 0 < x < \alpha\}$ (where $\alpha < \infty$), distribution function F, and density function f. Using integration by parts, prove the following special case of Theorem 6.2.

$$E(X) = \int_0^\alpha \left[1 - F(t)\right] dt.$$

11. The lifetime (in hours) of a light bulb manufactured by a certain company is a random variable with probability density function

$$f(x) = \begin{cases} 0 & \text{if } x \le 500 \\ \dfrac{5 \times 10^5}{x^3} & \text{if } x > 500. \end{cases}$$

Suppose that, for all nonnegative real numbers a and b, the event that any light bulb lasts at least a hours is independent of the event that any other light bulb lasts at least b hours. Find the probability that, of six such light bulbs selected at random, exactly two last over 1000 hours.

12. Let X be a continuous random variable with distribution function F and density function f. Find the distribution function and the density function of $Y = |X|$. ∎

Chapter 7

Special Continuous Distributions

In this chapter we study some examples of continuous random variables. These random variables appear frequently in theory and applications of probability, statistics, and branches of science and engineering.

7.1 UNIFORM RANDOM VARIABLES

In Sections 1.6 and 1.7 we explained that in random selection of a point from an interval (a, b), the probability of the occurrence of any particular point is zero. As a result, we stated that if $[\alpha, \beta] \subseteq (a, b)$, the events that the point falls in $[\alpha, \beta]$, (α, β), $[\alpha, \beta)$, and $(\alpha, \beta]$ are all equiprobable. Moreover, we said that a point is randomly selected from an interval (a, b) if any two of its subintervals that have the same length are equally likely to include the point. We also mentioned that the probability associated with the event that the subinterval (α, β) includes the point is defined to be $(\beta - \alpha)/(b - a)$. Applications of these facts have been discussed throughout the book. Therefore, their significance should be clear by now. In particular, in Chapter 13 we show that the core of computer simulations is selection of random points from intervals. In this section we introduce the concept of a uniform random variable. Then we study its properties and applications. As we will see now, uniform random variables are directly related to random selection of points from intervals.

Suppose that X is the value of the random point selected from an interval (a, b). Then X is called a **uniform random variable** over (a, b). Let F and f be probability distribution and density functions of X, respectively. Clearly,

$$F(t) = \begin{cases} 0 & t < a \\ \dfrac{t - a}{b - a} & a \leq t < b \\ 1 & t \geq b. \end{cases}$$

Therefore,

$$f(t) = F'(t) = \begin{cases} \dfrac{1}{b-a} & \text{if } a < t < b \\[2mm] 0 & \text{otherwise.} \end{cases} \tag{7.1}$$

Definition   *A random variable X is said to be* **uniformly distributed** *over an interval* (a, b) *if its probability density function is given by* (7.1).

Another way of reaching this definition is to note that $f(x)$ is a measure that determines how likely it is for X to be close to x. Since for all $x \in (a, b)$ the probability that X is close to x is the same, f should be a nonzero constant on (a, b); zero, elsewhere. Therefore,

$$f(x) = \begin{cases} c & \text{if } a < x < b \\[2mm] 0 & \text{elsewhere.} \end{cases}$$

Now $\int_a^b f(x)\, dx = 1$ implies that $\int_a^b c\, dx = c(b - a) = 1$. Thus $c = 1/(b - a)$.

Figure 7.1 represents the graphs of f and F, the density function and the distribution function of a uniform random variable over the interval (a, b).

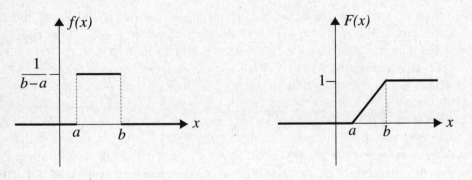

Figure 7.1   Density and distribution functions of a uniform random variable.

In random selections of a large number of points from (a, b), we expect that the average of the values of the points will be approximately $(a + b)/2$, the midpoint of (a, b). This can be shown by calculating $E(X)$, the expected value of a uniform random variable X over (a, b).

$$E(X) = \int_a^b x\, \frac{1}{b-a}\, dx = \frac{1}{b-a}\left[\frac{1}{2}x^2\right]_a^b = \frac{1}{b-a}\left(\frac{1}{2}b^2 - \frac{1}{2}a^2\right)$$

$$= \frac{(b-a)(b+a)}{2(b-a)} = \frac{a+b}{2}.$$

To find Var(X), note that

$$E(X^2) = \int_a^b x^2 \frac{1}{b-a} dx = \frac{1}{3} \frac{b^3 - a^3}{b-a} = \frac{1}{3}(a^2 + ab + b^2).$$

Hence

$$\text{Var}(X) = E(X^2) - \left[E(X)\right]^2 = \frac{1}{3}(a^2 + ab + b^2) - \left(\frac{a+b}{2}\right)^2 = \frac{(b-a)^2}{12}.$$

We have shown that

If X is uniformly distributed over an interval (a, b), then

$$E(X) = \frac{a+b}{2}, \qquad \text{Var}(X) = \frac{(b-a)^2}{12}, \qquad \sigma_X = \frac{b-a}{\sqrt{12}}.$$

It is interesting to note that the expected value and the variance of a randomly selected integer Y from the set $\{1, 2, 3, \dots, N\}$ are very similar to $E(X)$ and Var(X) obtained above. By Exercise 5 of Section 4.5, $E(Y) = (1 + N)/2$ and Var$(Y) = (N^2 - 1)/12$.

Example 7.1   Starting at 5:00 A.M., every half hour there is a flight from San Francisco airport to Los Angeles International airport. Suppose that none of these planes is completely sold out and that they always have room for passengers. A person who wants to fly to L.A. arrives at the airport at a random time between 8:45 A.M. and 9:45 A.M. Find the probability that she waits (a) at most 10 minutes; (b) at least 15 minutes.

Solution: Let the passenger arrive at the airport X minutes past 8:45. Then X is a uniform random variable over the interval $(0, 60)$. Hence the density function of X is given by

$$f(x) = \begin{cases} 1/60 & \text{if } 0 < x < 60 \\ 0 & \text{elsewhere.} \end{cases}$$

Now the passenger waits at most 10 minutes if she arrives between 8:50 and 9:00 or 9:20 and 9:30; that is, if $5 < X < 15$ or $35 < X < 45$. So the answer to (a) is

$$P(5 < X < 15) + P(35 < X < 45) = \int_5^{15} \frac{1}{60} dx + \int_{35}^{45} \frac{1}{60} dx = \frac{1}{3}.$$

The passenger waits at least 15 minutes if she arrives between 9:00 and 9:15 or 9:30 and 9:45; that is, if $15 < X < 30$ or $45 < X < 60$. Thus the answer to (b) is

$$P(15 < X < 30) + P(45 < X < 60) = \int_{15}^{30} \frac{1}{60} dx + \int_{45}^{60} \frac{1}{60} dx = \frac{1}{2}. \quad \blacklozenge$$

Example 7.2   A person arrives at a bus station every day at 7:00 A.M. If a bus arrives at a random time between 7:00 A.M. and 7:30 A.M., what is the average time spent waiting?

Solution: If the bus arrives X minutes past 7:00 A.M., then X is a uniform random variable over the interval $(0, 30)$. Hence the average waiting time is

$$E(X) = \frac{0 + 30}{2} = 15 \text{ minutes.} \quad \blacklozenge$$

The uniform distribution is often used to solve elementary problems in **geometric probability**. In Chapter 8 we solve several such problems. Here we discuss only a famous problem introduced by the French mathematician Joseph Bertrand (1822–1900) in 1889. It is called **Bertrand's paradox**. Bertrand seriously doubted that probability could be defined on infinite sample spaces. To make his point, he posed this problem:

Example 7.3   What is the probability that a *random chord* of a circle is longer than a side of an equilateral triangle inscribed into the circle?

Solution: Since the exact meaning of a *random chord* is not given, we cannot solve this problem as stated. We interpret the expression *random chord* in three different ways and solve the problem in each case. Let the center of the circle be at C and its radius be r.

First interpretation: To draw a random chord, by considerations of symmetry, first choose a random point A on the circle and connect it to C, the center; then choose a random number d from $(0, r)$ and place M on AC so that $\overline{CM} = d$. Finally, from M draw a chord perpendicular to the radius AC (see Figure 7.2). Since $d < r/2$ if and only if the chord is longer than a side of an equilateral triangle inscribed into the circle and d is uniformly distributed over $(0, r)$, the desired probability is

$$P\left(d < \frac{r}{2}\right) = \frac{r/2}{r} = \frac{1}{2}.$$

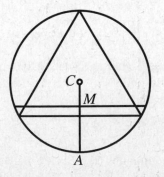

Figure 7.2   First interpretation.

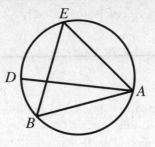

Figure 7.3   Second interpretation.

Second interpretation: To draw a random chord, by considerations of symmetry, first choose a random point A and then another random point D on the circle and connect AD. Let B and E be the points on the circle that make ABE an equilateral triangle. The random chord AD is longer than a side of ABE if and only if D lies on the arc BE. Since the length of the arc BE is one-third of the length of the whole circle and D is a random point on the circle, D lies on the arc BE with probability $1/3$. Thus the desired probability is $1/3$ as well (see Figure 7.3).

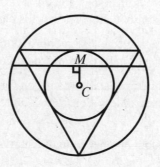

Figure 7.4   Third interpretation.

Third interpretation: Since a chord is perpendicular to the radius connecting its midpoint to the center of the circle, every chord is uniquely determined by its midpoint. To draw a random chord, choose a random point M inside the circle, connect it to C, and draw a chord perpendicular to MC from M. It is clear that the chord is longer than a side of the equilateral triangle inscribed into the circle if and only if its midpoint M lies inside the circle centered at C with radius $r/2$ (see Figure 7.4). Since each choice of M uniquely determines one choice of a chord, the desired probability is the area of the small circle divided by the area of the original circle (a fact discussed in detail in Section 8.1). It is equal to

$$\frac{\pi(r/2)^2}{\pi r^2} = \frac{1}{4}.$$

As we showed, three different interpretations of random chord resulted in three different answers. Because of this, Bertrand's problem was once considered a paradox. At that time, one did not pay attention to the fact that the three interpretations correspond to three different experiments concerning the selection of a random chord. In this process we are dealing with three different probability functions defined on the same set of events.

◆

EXERCISES

A

1. It takes a professor a random time between 20 and 27 minutes to walk from his home to school every day. If he has a class at 9:00 A.M. and he leaves home at 8:37 A.M., find the probability that he reaches his class on time.

2. Suppose that 15 points are selected at random and independently from the interval $(0, 1)$. How many of them can be expected to be greater than $3/4$?

3. The time at which a bus arrives at a station is uniform over an interval (a, b) with mean 2:00 P.M. and standard deviation $\sqrt{12}$ minutes. Determine the values of a and b.

4. Suppose that b is a random number from the interval $(-3, 3)$. What is the probability that the quadratic equation $x^2 + bx + 1 = 0$ has at least one real root?

5. The radius of a sphere is a random number between 2 and 4. What is the expected value of its volume? What is the probability that its volume is at most 36π?

6. A point is selected at random on a line segment of length ℓ. What is the probability that the longer segment is at least twice as long as the shorter segment?

7. A point is selected at random on a line segment of length ℓ. What is the probability that none of the two segments is smaller than $\ell/3$?

8. From the class of all triangles one is selected at random. What is the probability that it is obtuse?
   Hint: The largest angle of a triangle is less than 180 degrees but greater than or equal to 60 degrees.

9. A farmer who has two pieces of lumber of lengths a and b $(a < b)$ decides to build a pen in the shape of a triangle for his chickens. He sends his foolish son out to cut the longer piece and the boy, without taking any thought as to the ultimate purpose, makes a cut on the lumber of length b, at a point selected randomly. What are the chances that the two resulting pieces and the piece of length a can be used to form a triangular pen?

    Hint: Three segments form a triangle if and only if the length of any one of them is less than the sum of the lengths of the remaining two.

10. Let θ be a random number between $-\pi/2$ and $\pi/2$. Find the probability density function of $X = \tan \theta$.

11. Let X be a random number from $[0, 1]$. Find the probability mass function of $[nX]$, the greatest integer less than or equal to nX.

B

12. Let X be a random number from $(0, 1)$. Find the density functions of (a) $Y = -\ln(1 - X)$, and (b) $Z = X^n$.

13. Let X be a uniform random variable over the interval $(0, 1 + \theta)$, where $0 < \theta < 1$ is a given parameter. Find a function of X, say $g(X)$, so that $E[g(X)] = \theta^2$.

14. Let X be a continuous random variable with distribution function F. Prove that $F(X)$ is uniformly distributed over $(0, 1)$.

15. Let g be a nonnegative real-valued function on $\mathbf{R}$ that satisfies the relation $\int_{-\infty}^{\infty} g(t)\, dt = 1$. Show that if, for a random variable X, the random variable $Y = \int_{-\infty}^{X} g(t)\, dt$ is uniform, then g is the density function of X.

16. The sample space of an experiment is $S = (0, 1)$, and for every subset A of S, $P(A) = \int_A dx$. Let X be a random variable defined on S by $X(\omega) = 5\omega - 1$. Prove that X is a uniform random variable over the interval $(-1, 4)$.

17. Let Y be a random number from $(0, 1)$. Let X be the second digit of $\sqrt{Y}$. Prove that for $n = 0, 1, 2, \ldots, 9$, $P(X = n)$ increases as n increases. This is remarkable because it shows that $P(X = n)$, $n = 1, 2, 3, \ldots$ is not constant. That is, Y is uniform but X is not.

■

7.2   NORMAL RANDOM VARIABLES

In search of formulas to approximate binomial probabilities, Poisson was not alone. Other mathematicians had also realized the importance of such investigations. In 1718, before Poisson, De Moivre had discovered the following approximation, which is completely different from Poisson's.

De Moivre's Theorem   *Let X be a binomial random variable with parameters n and 1/2. Then for any numbers a and b, a < b,*

$$\lim_{n\to\infty} P\left(a < \frac{X - (1/2)n}{(1/2)\sqrt{n}} < b\right) = \frac{1}{\sqrt{2\pi}} \int_a^b e^{-t^2/2}\, dt.$$

Note that in this formula $(1/2)n = E(X)$ and $(1/2)\sqrt{n} = \sigma_X$.

De Moivre's theorem was appreciated by Laplace, who recognized its importance. In 1812 he generalized it to binomial random variables with parameters n and p. He showed the following theorem, now called the De Moivre-Laplace theorem.

Theorem 7.1   (De Moivre-Laplace Theorem)   *Let X be a binomial random variable with parameters n and p. Then for any numbers a and b, a < b,*

$$\lim_{n\to\infty} P\left(a < \frac{X - np}{\sqrt{np(1 - p)}} < b\right) = \frac{1}{\sqrt{2\pi}} \int_a^b e^{-t^2/2}\, dt.$$

Note that np and $\sqrt{np(1 - p)}$ appearing in this formula are, respectively, $E(X)$ and σ_X.

Poisson's approximation, as discussed, is good when n is large, p is relatively small, and np is appreciable. The De Moivre-Laplace formula yields excellent approximations for values of n and p for which $np(1 - p) \geq 10$.

To find his approximation, Poisson first considered a binomial random variable X with parameters (n, p). Then he showed that if $\lambda = np$, $P(X = i)$ is approximately $(e^{-\lambda}\lambda^i)/i!$ for large n. Finally, he proved that $(e^{-\lambda}\lambda^i)/i!$ itself is a probability mass function. The same procedure can be followed for the De Moivre-Laplace approximation as well. By this theorem, if X is a binomial random variable with parameters (n, p), the sequence of probabilities

$$P\left(\frac{X - np}{\sqrt{np(1 - p)}} \leq t\right), \qquad n = 1, 2, 3, 4, \ldots,$$

converges to $\dfrac{1}{\sqrt{2\pi}} \displaystyle\int_{-\infty}^t e^{-x^2/2}\, dx$, where the function $\Phi(t) = \dfrac{1}{\sqrt{2\pi}} \displaystyle\int_{-\infty}^t e^{-x^2/2}\, dx$ is a distribution function itself. To prove that Φ is a distribution function, note that it is increasing, continuous, and $\Phi(-\infty) = 0$. The proof of $\Phi(\infty) = 1$ is tricky. We use the following ingenious technique introduced by Gauss to show it. Let

$$I = \int_{-\infty}^{\infty} e^{-x^2/2}\, dx.$$

Then

$$I^2 = \left(\int_{-\infty}^{\infty} e^{-x^2/2}\, dx\right)\left(\int_{-\infty}^{\infty} e^{-y^2/2}\, dy\right) = \int_{-\infty}^{\infty}\int_{-\infty}^{\infty} e^{-(x^2+y^2)/2}\, dx\, dy.$$

To evaluate this double integral, we change the variables to polar coordinates. That is, we let $x = r \cos \theta$, $y = r \sin \theta$. We get $dx\, dy = r\, d\theta\, dr$ and

$$I^2 = \int_0^\infty \int_0^{2\pi} e^{-r^2/2} r\, d\theta\, dr = \int_0^\infty e^{-r^2/2} r \left(\int_0^{2\pi} d\theta \right) dr$$

$$= 2\pi \int_0^\infty r e^{-r^2/2}\, dr = 2\pi \left[-e^{-r^2/2} \right]_0^\infty = 2\pi.$$

Thus

$$I = \int_{-\infty}^\infty e^{-x^2/2}\, dx = \sqrt{2\pi},$$

and hence

$$\Phi(\infty) = \frac{1}{\sqrt{2\pi}} \int_{-\infty}^\infty e^{-x^2/2}\, dx = 1.$$

Therefore, Φ is a distribution function.

Definition   *A random variable X is called **standard normal** if its distribution function is Φ, that is, if*

$$P(X \le t) = \Phi(t) \equiv \frac{1}{\sqrt{2\pi}} \int_{-\infty}^t e^{-x^2/2}\, dx.$$

By the fundamental theorem of calculus, f, the density function of a standard normal random variable, is given by

$$f(x) = \Phi'(x) = \frac{1}{\sqrt{2\pi}} e^{-x^2/2}.$$

The standard normal density function is a bell-shaped curve that is symmetric about the y-axis (see Figure 7.5).

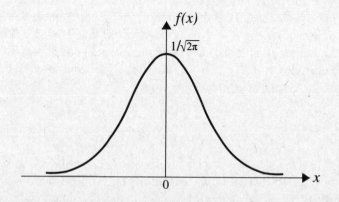

Figure 7.5   Graph of the standard normal density function.

Since Φ is the distribution function of the standard normal random variable, $\Phi(t)$ is the area under this curve from $-\infty$ to t. Because $\Phi(\infty) = 1$ and the curve is symmetric about the y-axis, $\Phi(0) = 1/2$. Moreover,

$$\Phi(-t) = 1 - \Phi(t).$$

To see this, note that

$$\Phi(-t) = \frac{1}{\sqrt{2\pi}} \int_{-\infty}^{-t} e^{-x^2/2}\, dx.$$

Substituting $u = -x$, we obtain

$$\Phi(-t) = \frac{-1}{\sqrt{2\pi}} \int_{\infty}^{t} e^{-u^2/2}\, du = \frac{1}{\sqrt{2\pi}} \int_{t}^{\infty} e^{-u^2/2}\, du$$

$$= \frac{1}{\sqrt{2\pi}} \int_{-\infty}^{\infty} e^{-u^2/2}\, du - \frac{1}{\sqrt{2\pi}} \int_{-\infty}^{t} e^{-u^2/2}\, du$$

$$= 1 - \Phi(t).$$

Correction for Continuity

Thus far we have shown that the De Moivre-Laplace theorem approximates the distribution of a discrete random variable by that of a continuous one. But we have not demonstrated how this approximation works in practice. To do so, first we explain how, in general, a probability based on a discrete random variable is approximated by a probability based on a continuous one. Let X be a discrete random variable with probability mass function $p(x)$, and suppose that we want to find $P(i \le X \le j)$, $i < j$. Consider the **histogram** of X, as sketched in Figure 7.6 from i to j. In that figure the base of each rectangle equals 1, and the height (and therefore the area) of the rectangle with the base midpoint k is $p(k)$, $i \le k \le j$. Thus the sum of the areas of all rectangles is $\sum_{k=i}^{j} p(k)$, which is the exact value of $P(i \le X \le j)$. Now suppose that $f(x)$, the density function of a continuous random variable, sketched in Figure 7.7, is a good approximation to $p(x)$. Then, as this figure shows, $P(i \le X \le j)$, the sum of the areas of all rectangles of the figure, is approximately the area under $f(x)$ from $i - 1/2$ to $j + 1/2$ rather than from i to j. That is,

$$P(i \le X \le j) \approx \int_{i-1/2}^{j+1/2} f(x)\, dx.$$

This adjustment is called **correction for continuity** and is necessary for approximation of the distribution of a discrete random variable with that of a continuous one. Similarly, the following corrections for continuity are made to calculate the given probabilities.

$$P(X = k) \approx \int_{k-1/2}^{k+1/2} f(x)\,dx,$$

$$P(X \geq i) \approx \int_{i-1/2}^{\infty} f(x)\,dx,$$

$$P(X \leq j) \approx \int_{-\infty}^{j+1/2} f(x)\,dx.$$

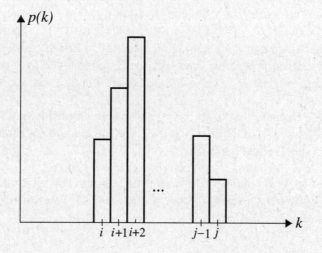
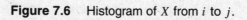

Figure 7.6   Histogram of X from i to j.

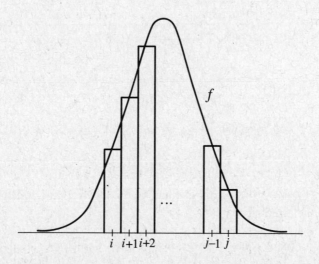

Figure 7.7   Histogram of X and the density function f.

In real-world problems, or even sometimes in theoretical ones, to apply the De Moivre-Laplace theorem, we need to calculate the numerical values of $\int_a^b e^{-x^2/2}\, dx$ for some real numbers a and b. Since $e^{-x^2/2}$ has no antiderivative in terms of elementary functions, such integrals are approximated by numerical techniques. Tables for such approximations have been available since 1799. Nowadays, most of the scientific calculators give excellent approximate values for these integrals. Table 1 of the Appendix gives the values of $\Phi(x) = (1/\sqrt{2\pi})\int_{-\infty}^x e^{-y^2/2}\, dy$ for $x = -3.89$ to $x = 0$. For $x \le -3.90$, $\Phi(x) \approx 0$. Table 2 of the Appendix gives the values of $\Phi(x)$ for $x = 0$ to $x = 3.89$. For $x > 3.89$, $\Phi(x) \approx 1$. Note that, using the relation $\Phi(x) = 1 - \Phi(-x)$, we can also use Table 1 to find $\Phi(x)$ for $x > 0$ and Table 2 to find $\Phi(x)$ for $x < 0$. Therefore, one of the tables 1 or 2 is sufficient to find $\Phi(x)$ for all values of x. However, for convenience, we have included a table for negative values of x and a separate table for positive values of x. The following example shows how the De Moivre-Laplace theorem is applied.

Example 7.4   Suppose that of all the clouds that are seeded with silver iodide, 58% show splendid growth. If 60 clouds are seeded with silver iodide, what is the probability that exactly 35 show splendid growth?

Solution: Let X be the number of clouds that show splendid growth. Then $E(X) = 60(0.58) = 34.80$ and $\sigma_X = \sqrt{60(0.58)(1 - 0.58)} = 3.82$. By correction for continuity and De Moivre-Laplace theorem,

$$P(X = 35) \approx P(34.5 < X < 35.5)$$

$$= P\left(\frac{34.5 - 34.80}{3.82} < \frac{X - 34.80}{3.82} < \frac{35.5 - 34.80}{3.82}\right)$$

$$= P\left(-0.08 < \frac{X - 34.8}{3.82} < 0.18\right) \approx \frac{1}{\sqrt{2\pi}}\int_{-0.08}^{0.18} e^{-x^2/2}\, dx$$

$$= \frac{1}{\sqrt{2\pi}}\int_{-\infty}^{0.18} e^{-x^2/2}\, dx - \frac{1}{\sqrt{2\pi}}\int_{-\infty}^{-0.08} e^{-x^2/2}\, dx$$

$$= \Phi(0.18) - \Phi(-0.08) = 0.5714 - 0.4681 = 0.1033.$$

The exact value of $P(X = 35)$ is $\binom{60}{35}(0.58)^{35}(0.42)^{25}$, which up to four decimal points equals 0.1039. Therefore, the answer obtained by the De Moivre-Laplace approximation is very close to the actual probability.   ◆

★ **Remark 7.1**   Using today's powerful scientific calculators, calculating the numerical value of $\Phi(x) = (1/\sqrt{2\pi})\int_{-\infty}^x e^{-y^2/2}\, dy$ up to several decimal points of accuracy is almost as easy as addition or subtraction. However, both for theoretical and computational reasons, approximating $\Phi(x)$ by simpler functions has been the subject of many studies. One such approximation was given by A. K. Shah in his paper "A Simpler Approximation for Areas Under the Standard Normal Curve" (*The American Statistician*, **39**, 80, 1985). According to this approximation, for $0 \leq x \leq 2.2$,

$$\Phi(x) \approx 0.5 + \frac{x(4.4 - x)}{10},$$

where the error of approximation is at most 0.005. That is,

$$\left| \Phi(x) - 0.5 - \frac{x(4.4 - x)}{10} \right| \leq 0.005. \quad \blacklozenge$$

Now that the importance of the standard normal random variable is clarified, we will calculate its expectation and variance. The graph of the standard normal density function (Figure 7.5) suggests that its expectation is zero. To prove this, let X be a standard normal random variable. Then

$$E(X) = \frac{1}{\sqrt{2\pi}} \int_{-\infty}^{\infty} x e^{-x^2/2} \, dx = 0,$$

because the integrand, $x e^{-x^2/2}$, is a finite odd function, and the integral is taken from $-\infty$ to $+\infty$. To calculate Var(X), note that

$$E(X^2) = \frac{1}{\sqrt{2\pi}} \int_{-\infty}^{\infty} x^2 e^{-x^2/2} \, dx.$$

Using integration by parts, we get (let $u = x$, $dv = x e^{-x^2/2} \, dx$)

$$\int_{-\infty}^{\infty} x x e^{-x^2/2} \, dx = \left[-x e^{-x^2/2} \right]_{-\infty}^{\infty} + \int_{-\infty}^{\infty} e^{-x^2/2} \, dx = 0 + \sqrt{2\pi} = \sqrt{2\pi}.$$

Therefore, $E(X^2) = 1$, Var$(X) = E(X^2) - \left[E(X) \right]^2 = 1$, and $\sigma_X = \sqrt{\text{Var}(X)} = 1$. We have shown that

The expected value of a standard normal random variable is 0. Its standard deviation is 1.

Although some natural phenomena obey a standard normal distribution, when it comes to the analysis of data, due to the lack of parameters in the standard normal distribution, it cannot be used. To overcome this difficulty, mathematicians generalized the standard normal distribution by introducing the following density function.

Definition   *A random variable X is called **normal**, with parameters μ and σ, if its density function is given by*

$$f(x) = \frac{1}{\sigma\sqrt{2\pi}} \exp\left[\frac{-(x - \mu)^2}{2\sigma^2} \right], \qquad -\infty < x < \infty.$$

f is a density function because by the change of variable $y = (x - \mu)/\sigma$:

$$\int_{-\infty}^{\infty} \frac{1}{\sigma\sqrt{2\pi}} \exp\left[\frac{-(x - \mu)^2}{2\sigma^2}\right] dx = \frac{1}{\sqrt{2\pi}} \int_{-\infty}^{\infty} e^{-y^2/2} \, dy = \Phi(\infty) = 1.$$

If X is a normal random variable with parameters μ and σ, we write $X \sim N(\mu, \sigma^2)$.

One of the first applications of $N(\mu, \sigma^2)$ was given by Gauss in 1809. Gauss used $N(\mu, \sigma^2)$ to model the errors of observations in astronomy. For this reason, the normal distribution is sometimes called **the Gaussian distribution.** Larsen and Marx, in their *An Introduction to Mathematical Statistics and Its Applications* (Prentice Hall, Upper Saddle River, N.J., 1986), explain that $N(\mu, \sigma^2)$ was popularized by the Belgian scholar Lambert Quetelet (1796–1874), who used it successfully in data analysis in many situations.[†] The chest measurement of Scottish soldiers is one of the studies of $N(\mu, \sigma^2)$ by Quetelet. He measured X_i, $i = 1, 2, 3, \ldots, 5738$, the respective sizes of the chests of 5738 Scottish soldiers, and, using statistical methods, found that their average and standard deviation were 39.8 and 2.05 inches, respectively. Then he counted the number of soldiers who had a chest of size i, $i = 33, 34, \ldots, 48$, and calculated the relative frequencies of sizes 33 through 48. He showed that for $i = 33, \ldots, 48$ the relative frequency of the size i is very close to $P(i - 1/2 < X < i + 1/2)$, where X is $N(\mu, \sigma^2)$ with $\mu = 39.8$ and $\sigma = 2.05$. Hence he concluded that the measure of the chest of a Scottish soldier is approximately a normal random variable with parameters 39.8 and 2.05.

The following lemma shows that, by a simple change of variable, $N(\mu, \sigma^2)$ can be transformed to $N(0, 1)$.

Lemma 7.1   *If $X \sim N(\mu, \sigma^2)$, then $Z = (X - \mu)/\sigma$ is $N(0, 1)$. That is, if $X \sim N(\mu, \sigma^2)$, the standardized X is $N(0, 1)$.*

Proof: We show that the distribution function of Z is $(1/\sqrt{2\pi}) \int_{-\infty}^{x} e^{-y^2/2} \, dy$. Note that

$$P(Z \le x) = P\left(\frac{X - \mu}{\sigma} \le x\right) = P(X \le \sigma x + \mu)$$

$$= \frac{1}{\sigma\sqrt{2\pi}} \int_{-\infty}^{\sigma x + \mu} \exp\left[\frac{-(t - \mu)^2}{2\sigma^2}\right] dt.$$

Let $y = (t - \mu)/\sigma$; then $dt = \sigma \, dy$ and we get

$$P(Z \le x) = \frac{1}{\sigma\sqrt{2\pi}} \int_{-\infty}^{x} e^{-y^2/2} \sigma \, dy = \frac{1}{\sqrt{2\pi}} \int_{-\infty}^{x} e^{-y^2/2} \, dy. \quad \blacklozenge$$

[†]An excellent treatise on the work of Quetelet concerning the applications of probability to the measurement of uncertainty in the social sciences can be found in Chapter 5 of *The History of Statistics* by Stephen M. Stigler, The Belknap Press of Harvard University Press, 1986.

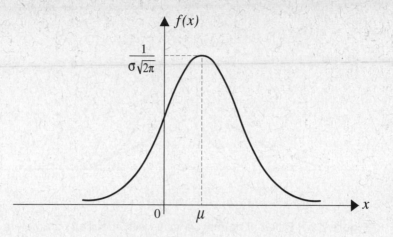

Figure 7.8   Density of $N(\mu, \sigma^2)$.

Let $X \sim N(\mu, \sigma^2)$. Quetelet's analysis of the sizes of the chests of Scottish soldiers is an indication that μ and σ are the expected value and the standard deviation of X, respectively. To prove this, note that the random variable $Z = (X - \mu)/\sigma$ is $N(0, 1)$ and $X = \sigma Z + \mu$. Hence $E(X) = E(\sigma Z + \mu) = \sigma E(Z) + \mu = \mu$ and $\text{Var}(X) = \text{Var}(\sigma Z + \mu) = \sigma^2 \text{Var}(Z) = \sigma^2$. Therefore,

> The parameters μ and σ that appear in the formula of the density function of X are its expected value and standard deviation, respectively.

The graph of $f(x) = \dfrac{1}{\sigma\sqrt{2\pi}} \exp\left[\dfrac{-(t - \mu)^2}{2\sigma^2}\right]$ is a bell-shaped curve symmetric about $x = \mu$ with the maximum at $(\mu, 1/\sigma\sqrt{2\pi})$ and inflection points at $\mu \pm \sigma$ (see Figures 7.8 and 7.9).

The transformation $Z = (X - \mu)/\sigma$ enables us to use Tables 1 and 2 of the Appendix to calculate the probabilities concerning X. Some examples follow.

Example 7.5   Suppose that a Scottish soldier's chest size is normally distributed with mean 39.8 and standard deviation 2.05 inches, respectively. What is the probability that of 20 randomly selected Scottish soldiers, five have a chest of at least 40 inches?

Solution: Let p be the probability that a randomly selected Scottish soldier has a chest of 40 or more inches. If X is the normal random variable with mean 39.8 and standard deviation 2.05, then

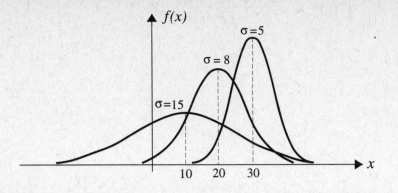

Figure 7.9   Different normal densities with specified parameters.

$$p = P(X \geq 40) = P\left(\frac{X - 39.8}{2.05} \geq \frac{40 - 39.8}{2.05}\right) = P\left(\frac{X - 39.8}{2.05} \geq 0.10\right)$$
$$= P(Z \geq 0.10) = 1 - \Phi(0.1) \approx 1 - 0.5398 \approx 0.46.$$

Therefore, the probability that of 20 randomly selected Scottish soldiers, five have a chest of at least 40 inches is

$$\binom{20}{5}(0.46)^5(0.54)^{15} \approx 0.03. \quad \blacklozenge$$

Example 7.6   Let X, the grade of a randomly selected student in a test of a probability course, be a normal random variable. A professor is said to grade such a test *on the curve* if he finds the average μ and the standard deviation σ of the grades and then assigns letter grades according to the following table.

Range of the grade	$X \geq \mu + \sigma$	$\mu \leq X < \mu + \sigma$	$\mu - \sigma \leq X < \mu$	$\mu - 2\sigma \leq X < \mu - \sigma$	$X < \mu - 2\sigma$
Letter grade	A	B	C	D	F

Suppose that the professor of the probability course grades the test on the curve. Determine the percentage of the students who will get A, B, C, D, and F, respectively.

Solution: By the fact that $(X - \mu)/\sigma$ is standard normal,

$$P(X \geq \mu + \sigma) = P\left(\frac{X - \mu}{\sigma} \geq 1\right) = 1 - \Phi(1) \approx 0.1587,$$

$$P(\mu \leq X < \mu + \sigma) = P\left(0 \leq \frac{X - \mu}{\sigma} < 1\right) = \Phi(1) - \Phi(0) \approx 0.3413,$$

$$P(\mu - \sigma \leq X < \mu) = P\left(-1 \leq \frac{X - \mu}{\sigma} < 0\right) = \Phi(0) - \Phi(-1)$$

$$= 0.5 - 0.1587 \approx 0.3413,$$

$$P(\mu - 2\sigma \leq X < \mu - \sigma) = P\left(-2 \leq \frac{X - \mu}{\sigma} < -1\right) = \Phi(-1) - \Phi(-2)$$

$$= 0.1587 - 0.0228 \approx 0.1359,$$

$$P(X < \mu - 2\sigma) = P\left(\frac{X - \mu}{\sigma} < -2\right) = \Phi(-2) \approx 0.0228.$$

Therefore, approximately 16% should get A, 34% B, 34% C, 14% D, and 2% F. If an instructor grades a test on the curve, instead of calculating μ and σ, he or she may assign A to the top 16%, B to the next 34%, and so on. ◆

Example 7.7 The scores on an achievement test given to 100,000 students are normally distributed with mean 500 and standard deviation 100. What should the score of a student be to place him among the top 10% of all students?

Solution: Letting X be a normal random variable with mean 500 and standard deviation 100, we must find x so that $P(X \geq x) = 0.10$ or $P(X < x) = 0.90$. This gives

$$P\left(\frac{X - 500}{100} < \frac{x - 500}{100}\right) = 0.90.$$

Thus

$$\Phi\left(\frac{x - 500}{100}\right) = 0.90.$$

From Table 2 of the Appendix, we have that $\Phi(1.28) \approx 0.8997$, implying that $(x - 500)/100 \approx 1.28$. This gives $x \approx 628$; therefore, a student should earn 628 or more to be among the top 10% of the students. ◆

Example 7.8   (Investment) The annual rate of return for a share of a specific stock is a normal random variable with mean 10% and standard deviation 12%. Ms. Couture buys 100 shares of the stock at a price of $60 per share. What is the probability that after a year her net profit from that investment is at least $750? Ignore transaction costs and assume that there is no annual dividend.

Solution: Let r be the rate of return of this stock. The random variable r is normal with $\mu = 0.10$ and $\sigma = 0.12$. Let X be the price of the total shares of the stock that Ms. Couture buys this year. We are given that $X = 6000$. Let Y be the total value of the shares next year. The desired probability is

$$P(Y - X \geq 750) = P\left(\frac{Y - X}{X} \geq \frac{750}{X}\right) = P\left(r \geq \frac{750}{6000}\right)$$

$$= P(r \geq 0.125) = P\left(Z \geq \frac{0.125 - 0.10}{0.12}\right)$$

$$= P(Z \geq 0.21) = 1 - P(Z < 0.21)$$

$$= 1 - \Phi(0.21) = 1 - 0.5832 = 0.4168. \quad \blacklozenge$$

It is important to note that a large family of random variables in mathematics and statistics have distributions that are either normal or approximately normal. One of the reasons for this is the celebrated central limit theorem which we will study in Chapter 11. To express that theorem, we will use Richard von Mises' words from his *Probability, Statistics and the Truth* (page 171, Dover Publications, 1981):

> The normal curve represents the distribution in all cases where a final collective is formed by combination of a very large number of initial collectives, the attribute in the final collective being the sum of the results in the initial collectives. The original collectives are not necessarily simple alternatives as they are in Bernoulli's problem. It is not even necessary for them to have the same attributes or the same distributions. The only conditions are that a very great number of collectives are combined and that the attributes are mixed in such a way that the final attribute is the sum of all the original ones. Under these conditions the final distribution is always represented by a normal curve.

For example, consider the annual profit of an insurance company that sells thousands of policies each year. This is a normal random variable since the total profit is the sum of profits (losses being treated as negative profits) obtained from all policies sold. Some policies might be very complicated (i.e., those sold to the banks or oil companies). Moreover, the profit from each policy is itself a random variable which depends on many factors: the premium, risk involved, damages to be paid, quit claims, profit or loss from investments, the level of competition in the industry, and so on. To show the wide range of applications of normal random variables, in the next example, we will present, without proof, a celebrated formula in finance which is constructed based on normal distribution function.

★ **Example 7.9**   (**Black-Scholes Call Option Formula**)   In natural processes, the changes that take place are often the result of a sum of many insignificant random factors. Although the effect of each of these factors, separately, may be ignored, the effect of their sum in general may not. Therefore, the study of distributions of sums of a large number of independent random variables is essential and is the subject of

Chapter 11. In Section 11.5, we will show that for an immense number of random phenomena, the distribution functions of such sums are approximately normal. In finance, random fluctuations of prices of shares of stocks, or other assets, are a result of an enormity of political, social, economical, and financial factors independently assessed by a large number of traders. For that reason financial scholars and consultants use normal distributions as models for prices of assets in general, and shares of stocks in particular. That is why the *Black-Scholes call option formula*, presented next, is based on normal distribution function.

In finance, a contract that entitles a person to purchase a commodity is called **call option**. To explain the concept of call option, we will present an example. Suppose that Misha purchases, for a price of $7 per share, the **option** of buying a block of 100 shares in a company at $80 per share during, say, the next six months. That is, Misha pays $700 to a *seller* and buys the *right* of purchasing 100 shares in the company at any time before or including the agreed expiration day at $80 per share. Misha is not obliged to purchase any number of the 100 shares he simply has the option to purchase. However, the premium paid to buy call options is not recovered whether or not he buys any number of the shares. In the language of finance, Misha is the **option holder**, the seller is the **writer**, and the predetermined price per share of $80, which is valid until the expiration day is called **strike price**. If Misha actually does buy a block of at least one share in the company, then we say that he **exercises** the option.

Suppose that sometime before the expiration day, the price of the shares in the company rise to, say, $95 per share. If at that time Misha purchases 100 shares for $80 per share, and sells them immediately for $95, he makes a profit of $9500 − $8000 − $700 = $800. On the other hand, if the price does not rise by the expiration day, Misha will lose $700 for having purchased the call option to buy the block of 100 shares. As for the writer, in the latter case, she makes a profit of $700, whereas in the former case, she will end up selling the block of 100 shares, which is worth $9500, for only $8000, losing an opportunity to make $1500 − $700 = $800 more.

From this example, it should be evident that determining a fair price for an option is a major problem in *option theory*. The question is whether or not the right to buy a share for a price of $7 is fair to both Misha and the writer. Until 1973, it was widely believed that the call option should be determined based on the principle assumption that the average return from the stock should be lower than the average return from the option itself. However, in 1973, Fisher Black, an applied mathematician and a financial consultant, and MIT professor Myron Scholes presented a paper with a new approach. They argued, among other points, that all risk will be eliminated if option holders or writers chose two lines of investments with identical returns, buy one and sell the other simultaneously on the expiration day. Based on such assumptions, in terms of normal distribution function, they derived formulas for call options that are extremely popular. To state a version of Black-Scholes' widely used call option formula, note that, since periods remaining to expiration are usually short, those stocks with higher standard deviation will rise above strick price with higher probability. Therefore, they should have higher option values.

Interest rate also has an effect on the value of an option. Usually, as interest rate increases, on average, the growth rate of stock price also increases. This should in turn increase the prices of options. Since dividends essentially reduce the stock price, they have negative effect on the values of the call option. For this reason, in the next formula we assume that no dividends are paid before or on the expiration day.

We are now ready to state the *Black-Scholes call option formula* for *European options*, in which, unlike *American options*, the option holder can only exercise on the expiration day. The appearance of the normal distribution function in the following formula has something to do with the fact that the number of upward movements in price of a share of a stock is binomial, and the limiting behavior of binomial is normal.

Black-Scholes Call Option Formula: *In a European call option with standard deviation σ, let K be the strike price, S be the current price, and τ denote the time period over which the option is valid. Suppose that interest is compounded continuously at a rate r. Then the Black-Scholes formula for the value of call option is given by*

$$C = S\Phi(\alpha) - Ke^{-r\tau}\Phi(\alpha - \sigma\sqrt{\tau}),$$

where

$$\alpha = \frac{\ln(S/K) + (r + \sigma^2/2)\tau}{\sigma\sqrt{\tau}},$$

and, as usual, Φ is the distribution function of a standard normal random variable.

To find the Black-Scholes call option value, per share, for the block of 100 shares of the stock Misha is interested in, suppose that the current price of the stock, per share, is $85 with standard deviation of 18% per year, and the option is valid for 6 months. The strike price is $80 and the interest rate is 12%. Therefore, $S = 85$, $K = 80$, $\sigma = 0.18$, $r = 0.12$, and $\tau = 6/12 = 1/2$. Hence

$$\alpha = \frac{\ln(85/80) + (0.12 + 0.18^2/2)(1/2)}{0.18\sqrt{1/2}} = 1.01,$$

and thus the value of the call option is

$$\begin{aligned}
C &= 85\Phi(1.01) - 80e^{-(0.12)(1/2)}\Phi\big(1.01 - 0.18\sqrt{1/2}\big) \\
&= 85\Phi(1.01) - 75.34\Phi(0.88) \\
&= 85(0.8438) - 75.34(0.8106) = 10.65. \quad \blacklozenge
\end{aligned}$$

EXERCISES

A

1. Suppose that 90% of the patients with a certain disease can be cured with a certain drug. What is the approximate probability that, of 50 such patients, at least 45 can be cured with the drug?

2. A small college has 1095 students. What is the approximate probability that more than five students were born on Christmas day? Assume that the birthrates are constant throughout the year and that each year has 365 days.

3. Let $\Psi(x) = 2\Phi(x) - 1$. The function Ψ is called the **positive normal distribution.** Prove that if Z is standard normal, then $|Z|$ is positive normal.

4. Let Z be a standard normal random variable and α be a given constant. Find the real number x that maximizes $P(x < Z < x + \alpha)$.

5. Let X be a standard normal random variable. Calculate $E(X \cos X)$, $E(\sin X)$, and $E\left(\dfrac{X}{1 + X^2}\right)$.

6. The ages of subscribers to a certain newspaper are normally distributed with mean 35.5 years and standard deviation 4.8. What is the probability that the age of a random subscriber is (a) more than 35.5 years; (b) between 30 and 40 years?

7. The grades for a certain exam are normally distributed with mean 67 and variance 64. What percent of students get A(≥ 90), B($80 - 90$), C($70 - 80$), D($60 - 70$), and F(< 60)?

8. Suppose that the distribution of the diastolic blood pressure is normal for a randomly selected person in a certain population is normal with mean 80 mm Hg and standard deviation 7 mm Hg. If people with diastolic blood pressures 95 or above are considered hypertensive and people with diastolic blood pressures above 89 and below 95 are considered to have mild hypertension, what percent of that population have mild hypertension and what percent are hypertensive? Assume in that population no one has abnormal systolic blood pressure.

9. The length of an aluminum-coated steel sheet manufactured by a certain factory is approximately normal with mean 75 centimeters and standard deviation 1 centimeter. Find the probability that a randomly selected sheet manufactured by this factory is between 74.5 and 75.8 centimeters.

10. Suppose that the IQ of a randomly selected student from a university is normal with mean 110 and standard deviation 20. Determine the interval of values that is centered at the mean and includes 50% of the IQ's of the students at that university.

11. The amount of cereal in a box is normal with mean 16.5 ounces. If the packager is required to fill at least 90% of the cereal boxes with 16 or more ounces of cereal, what is the largest standard deviation for the amount of cereal in a box?

12. Suppose that the scores on a certain manual dexterity test are normal with mean 12 and standard deviation 3. If eight randomly selected individuals take the test, what is the probability that none will make a score less than 14?

13. A number t is said to be the **median** of a continuous random variable X if

$$P(X \le t) = P(X \ge t) = 1/2.$$

Calculate the median of the normal random variable with parameters μ and σ^2.

14. Let $X \sim N(\mu, \sigma^2)$. Prove that $P(|X - \mu| > k\sigma)$ does not depend on μ or σ.

15. Suppose that lifetimes of light bulbs produced by a certain company are normal random variables with mean 1000 hours and standard deviation 100 hours. Is this company correct when it claims that 95% of its light bulbs last at least 900 hours?

16. Suppose that lifetimes of light bulbs produced by a certain company are normal random variables with mean 1000 hours and standard deviation 100 hours. Suppose that lifetimes of light bulbs produced by a second company are normal random variables with mean 900 hours and standard deviation 150 hours. Howard buys one light bulb manufactured by the first company and one by the second company. What is the probability that at least one of them lasts 980 or more hours?

17. (Investment)   The annual rate of return for a share of a specific stock is a normal random variable with mean 0.12 and standard deviation of 0.06. The current price of the stock is \$35 per share. Mrs. Lovotti would like to purchase enough shares of this stock to make at least \$1000 profit with a probability of at least 90% in one year. Find the minimum number of shares that she should buy. Ignore transaction costs and assume that there are no annual dividends.

18. Find the expected value and the variance of a random variable with the probability density function

$$f(x) = \sqrt{\frac{2}{\pi}} e^{-2(x-1)^2}.$$

19. Let $X \sim N(\mu, \sigma^2)$. Find the probability distribution function of $|X - \mu|$ and its expected value.

20. Determine the value(s) of k for which the following is the probability density function of a normal random variable.

$$f(x) = \sqrt{k} e^{-k^2 x^2 - 2kx - 1}, \qquad -\infty < x < \infty.$$

21. The viscosity of a brand of motor oil is normal with mean 37 and standard deviation 10. What is the lowest possible viscosity for a specimen that has viscosity higher than at least 90% of that brand of motor oil?

22. In a certain town the length of residence of a family in a home is normal with mean 80 months and variance 900. What is the probability that of 12 independent families, living on a certain street of that town, at least three will have lived there more than eight years?

23. Let $\alpha \in (-\infty, \infty)$ and $Z \sim N(0, 1)$; find $E(e^{\alpha Z})$.

24. Let $X \sim N(0, \sigma^2)$. Calculate the density function of $Y = X^2$.

25. Let $X \sim N(\mu, \sigma^2)$. Calculate the density function of $Y = e^X$.

26. Let $X \sim N(0, 1)$. Calculate the density function of $Y = \sqrt{|X|}$.

B

27. Suppose that the odds are 1 to 5000 in favor of a customer of a particular bookstore buying a certain fiction bestseller . If 800 customers enter the store every day, how many copies of that bestseller should the store stock every month so that, with a probability of more than 98%, it does not run out of this book? For simplicity, assume that a month is 30 days.

28. Every day a factory produces 5000 light bulbs, of which 2500 are type I and 2500 are type II. If a sample of 40 light bulbs is selected at random to be examined for defects, what is the approximate probability that this sample contains at least 18 light bulbs of each type?

29. To examine the accuracy of an algorithm that selects random numbers from the set $\{1, 2, \ldots, 40\}$, 100,000 numbers are selected and there are 3500 ones. Given that the expected number of ones is 2500, is it fair to say that this algorithm is not accurate?

30. Prove that for some constant k, $f(x) = ka^{-x^2}, a \in (0, \infty)$, is a normal probability density function.

31. **(a)** Prove that for all $x > 0$,

$$\frac{1}{x\sqrt{2\pi}}\left(1 - \frac{1}{x^2}\right)e^{-x^2/2} < 1 - \Phi(x) < \frac{1}{x\sqrt{2\pi}}e^{-x^2/2}.$$

Hint: Integrate the following inequalities:

$$(1 - 3y^{-4})e^{-y^2/2} < e^{-y^2/2} < (1 + y^{-2})e^{-y^2/2}.$$

(b) Use part (a) to prove that $1 - \Phi(x) \sim \dfrac{1}{x\sqrt{2\pi}}e^{-x^2/2}$. That is, as $x \to \infty$, the ratio of the two sides approaches 1.

32. Let Z be a standard normal random variable. Show that for $x > 0$,

$$\lim_{t \to \infty} P\left(Z > t + \frac{x}{t} \mid Z \geq t\right) = e^{-x}.$$

Hint: Use part (b) of Exercise 31.

33. The amount of soft drink in a bottle is a normal random variable. Suppose that in 7% of the bottles containing this soft drink there are less than 15.5 ounces, and in 10% of them there are more than 16.3 ounces. What are the mean and standard deviation of the amount of soft drink in a randomly selected bottle?

34. At an archaeological site 130 skeletons are found and their heights are measured and found to be approximately normal with mean 172 centimeters and variance 81 centimeters. At a nearby site, five skeletons are discovered and it is found that the heights of exactly three of them are above 185 centimeters. Based on this information is it reasonable to assume that the second group of skeletons belongs to the same family as the first group of skeletons?

35. In a forest, the number of trees that grow in a region of area R has a Poisson distribution with mean λR, where λ is a positive real number. Find the expected value of the distance from a certain tree to its nearest neighbor.

36. Let $I = \int_0^\infty e^{-x^2/2}\, dx$; then

$$I^2 = \int_0^\infty \left[\int_0^\infty e^{-(x^2+y^2)/2}\, dy\right] dx.$$

Let $y/x = s$ and change the order of integration to show that $I^2 = \pi/2$. This gives an alternative proof of the fact that Φ is a distribution function. The advantage of this method is that it avoids polar coordinates. ∎

7.3   EXPONENTIAL RANDOM VARIABLES

Let $\{N(t): t \geq 0\}$ be a Poisson process. Then, as discussed in Section 5.2, $N(t)$ is the number of "events" that have occurred at or prior to time t. Let X_1 be the time of the first event, X_2 be the elapsed time between the first and the second events, X_3 be the elapsed time between the second and third events, and so on. The sequence of random variables $\{X_1, X_2, X_3, \ldots\}$ is called the **sequence of interarrival times** of the Poisson process $\{N(t): t \geq 0\}$. Let $\lambda = E[N(1)]$; then

$$P(N(t) = n) = \frac{e^{-\lambda t}(\lambda t)^n}{n!}.$$

This enables us to calculate the probability distribution functions of the random variables X_i, $i \geq 1$. For $t \geq 0$,

$$P(X_1 > t) = P\big(N(t) = 0\big) = e^{-\lambda t}.$$

Therefore,

$$P(X_1 \leq t) = 1 - P(X_1 > t) = 1 - e^{-\lambda t}.$$

Since a Poisson process is stationary and possesses independent increments, at any time t, the process probabilistically starts all over again. Hence the interarrival time of any two consecutive events has the same distribution as X_1; that is, the sequence $\{X_1, X_2, X_3, \dots\}$ is identically distributed. Therefore, for all $n \geq 1$,

$$P(X_n \leq t) = P(X_1 \leq t) = \begin{cases} 1 - e^{-\lambda t} & t \geq 0 \\ 0 & t < 0. \end{cases}$$

Let

$$F(t) = \begin{cases} 1 - e^{-\lambda t} & t \geq 0 \\ 0 & t < 0 \end{cases}$$

for some $\lambda > 0$. Then F is the distribution function of X_n for all $n \geq 1$. It is called **exponential distribution** and is one of the most important distributions of pure and applied probability. Since

$$f(t) = F'(t) = \begin{cases} \lambda e^{-\lambda t} & t \geq 0 \\ 0 & t < 0 \end{cases} \tag{7.2}$$

and

$$\int_0^\infty \lambda e^{-\lambda t}\, dt = \lim_{b \to \infty} \int_0^b \lambda e^{-\lambda t}\, dt = \lim_{b \to \infty} \Big[-e^{-\lambda t} \Big]_0^b = \lim_{b \to \infty} (1 - e^{-\lambda b}) = 1,$$

f is a density function.

Definition   *A continuous random variable X is called **exponential** with parameter $\lambda > 0$ if its density function is given by (7.2).*

Because the interarrival times of a Poisson process are exponential, the following are examples of random variables that might be exponential.

1.   The interarrival time between two customers at a post office

2.   The duration of Jim's next telephone call

3. The time between two consecutive earthquakes in California

4. The time between two accidents at an intersection

5. The time until the next baby is born in a hospital

6. The time until the next crime in a certain town

7. The time to failure of the next fiber segment in a large group of such segments when all of them are initially fault free

8. The time interval between the observation of two consecutive shooting stars on a summer evening

9. The time between two consecutive fish caught by a fisherman from a large lake with lots of fish

From Section 5.2 we know that λ is the average number of the events in one time unit; that is, $E[N(1)] = \lambda$. Therefore, we should expect an average time $1/\lambda$ between two consecutive events. To prove this, let X be an exponential random variable with parameter λ; then

$$E(X) = \int_{-\infty}^{\infty} x f(x)\, dx = \int_{0}^{\infty} x(\lambda e^{-\lambda x})\, dx.$$

Using integration by parts with $u = x$ and $dv = \lambda e^{-\lambda x}\, dx$, we obtain

$$E(X) = \left[-xe^{-\lambda x} \right]_{0}^{\infty} + \int_{0}^{\infty} e^{-\lambda x}\, dx = 0 - \left[\frac{1}{\lambda} e^{-\lambda x} \right]_{0}^{\infty} = \frac{1}{\lambda}.$$

A similar calculation shows that

$$E(X^2) = \int_{-\infty}^{\infty} x^2 f(x)\, dx = \int_{0}^{\infty} x^2(\lambda e^{-\lambda x})\, dx = \frac{2}{\lambda^2}.$$

Hence

$$\mathrm{Var}(X) = E(X^2) - \left[E(X) \right]^2 = \frac{2}{\lambda^2} - \frac{1}{\lambda^2} = \frac{1}{\lambda^2},$$

and therefore $\sigma_X = 1/\lambda$. We have shown that

For an exponential random variable with parameter λ,

$$E(X) = \sigma_X = \frac{1}{\lambda}, \qquad \mathrm{Var}(X) = \frac{1}{\lambda^2}.$$

Figures 7.10 and 7.11 represent the graphs of the exponential density and exponential distribution functions, respectively.

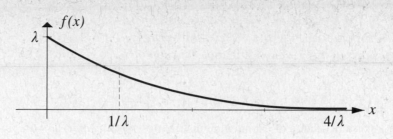

Figure 7.10   Exponential density function with parameter λ.

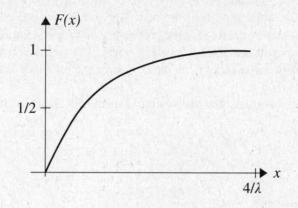

Figure 7.11   Exponential distribution function.

Example 7.10   Suppose that every three months , on average, an earthquake occurs in California. What is the probability that the next earthquake occurs after three but before seven months?

Solution: Let X be the time (in months) until the next earthquake; it can be assumed that X is an exponential random variable with $1/\lambda = 3$ or $\lambda = 1/3$. To calculate $P(3 < X < 7)$, note that since F, the distribution function of X, is given by

$$F(t) = P(X \le t) = 1 - e^{-t/3} \quad \text{for } t > 0,$$

we can write

$$P(3 < X < 7) = F(7) - F(3) = (1 - e^{-7/3}) - (1 - e^{-1}) \approx 0.27. \quad \blacklozenge$$

Example 7.11   At an intersection there are two accidents per day, on average. What is the probability that after the next accident there will be no accidents at all for the next two days?

Solution: Let X be the time (in days) between the next two accidents. It can be assumed that X is exponential with parameter λ, satisfying $1/\lambda = 1/2$, so that $\lambda = 2$. To find $P(X > 2)$, note that F, the distribution function of X, is given by $F(t) = 1 - e^{-2t}$, $t > 0$. Hence

$$P(X > 2) = 1 - P(X \leq 2) = 1 - F(2) = e^{-4} \approx 0.02. \quad \blacklozenge$$

An important feature of exponential distribution is its *memoryless property.* A non-negative random variable X is called **memoryless** if, for all $s, t \geq 0$,

$$P(X > s + t \mid X > t) = P(X > s). \tag{7.3}$$

If, for example, X is the lifetime of some type of instrument, then (7.3) means that there is no deterioration with age of the instrument. The probability that a new instrument will last more than s years is the same as the probability that a used instrument that has lasted more than t years will last at least another s years. In other words, the probability that such an instrument will deteriorate in the next s years does not depend on the age of the instrument.

To show that an exponential distribution is memoryless, note that (7.3) is equivalent to

$$\frac{P(X > s + t, X > t)}{P(X > t)} = P(X > s)$$

and

$$P(X > s + t) = P(X > s)P(X > t). \tag{7.4}$$

Now since

$$P(X > s + t) = 1 - [1 - e^{-\lambda(s+t)}] = e^{-\lambda(s+t)},$$
$$P(X > s) = 1 - (1 - e^{-\lambda s}) = e^{-\lambda s},$$

and

$$P(X > t) = 1 - (1 - e^{-\lambda t}) = e^{-\lambda t},$$

we have that (7.4) follows. Hence X is memoryless. It can be shown that exponential is the only continuous distribution which possesses a memoryless property (see Exercise 15).

Example 7.12   The lifetime of a TV tube (in years) is an exponential random variable with mean 10. If Jim bought his TV set 10 years ago, what is the probability that its tube will last another 10 years?

Solution: Let X be the lifetime of the tube. Since X is an exponential random variable, there is no deterioration with age of the tube. Hence

$$P(X > 20 \mid X > 10) = P(X > 10) = 1 - [1 - e^{(-1/10)10}] \approx 0.37. \quad \blacklozenge$$

Example 7.13 Suppose that, on average, two earthquakes occur in San Francisco and two in Los Angeles every year. If the last earthquake in San Francisco occurred 10 months ago and the last earthquake in Los Angeles occurred two months ago, what is the probability that the next earthquake in San Francisco occurs after the next earthquake in Los Angeles?

Solution: It can be assumed that the number of earthquakes in San Francisco and Los Angeles are both Poisson processes with common rate $\lambda = 2$. Hence the times between two consecutive earthquakes in Los Angeles and two consecutive earthquakes in San Francisco are both exponentially distributed with the common mean $1/\lambda = 1/2$. Because of the memoryless property of the exponential distribution, it does not matter when the last earthquakes in San Francisco and Los Angeles have occurred. The times between now and the next earthquake in San Francisco and the next earthquake in Los Angeles both have the same distribution. Since these time periods are exponentially distributed with the same parameter, by symmetry, the probability that the next earthquake in San Francisco occurs after that in Los Angeles is 1/2. ♦

Relationship between Exponential and Geometric: Recall that if a Bernoulli trial is performed successively and independently, then the number of trials until the *first success* occurs is geometric. Furthermore, the number of trials *between two consecutive successes* is also geometric. Sometimes exponential is considered to be the continuous analog of geometric because, for a Poisson process, the time it will take until the *first event* occurs is exponential, and the time between *two consecutive events* is also exponential. Moreover, exponential is the only memoryless continuous distribution, and geometric is the only memoryless discrete distribution. It is also interesting to know that if X is an exponential random variable, then $[X]$, the integer part of X; i.e, the greatest integer less than or equal to X, is geometric (see Exercise 14). ♦

Remark 7.2 In this section, we showed that if $\{N(t): t \geq 0\}$ is a Poisson process with rate λ, then the interarrival times of the process form an independent sequence of identically distributed exponential random variables with mean $1/\lambda$. Using the tools of an area of probability called *renewal theory*, we can prove that the converse of this fact is also true:

> If, for some process, $N(t)$ is the number of "events" occurring in $[0, t]$, and if the times between consecutive events form a sequence of independent and identically distributed exponential random variables with mean $1/\lambda$, then $\{N(t): t \geq 0\}$ is a Poisson process with rate λ. ♦

EXERCISES

A

1. Customers arrive at a postoffice at a Poisson rate of three per minute. What is the probability that the next customer does not arrive during the next 3 minutes?

2. Find the median of an exponential random variable with rate λ. Recall that for a continuous distribution F, the median $Q_{0.5}$ is the point at which $F(Q_{0.5}) = 1/2$.

3. Let X be an exponential random variable with mean 1. Find the probability density function of $Y = -\ln X$.

4. The time between the first and second heart attacks for a certain group of people is an exponential random variable. If 50% of those who have had a heart attack will have another one within the next five years, what is the probability that a person who had one heart attack five years ago will not have another one in the next five years?

5. Guests arrive at a hotel, in accordance with a Poisson process, at a rate of five per hour. Suppose that for the last 10 minutes no guest has arrived. What is the probability that (a) the next one will arrive in less than 2 minutes; (b) from the arrival of the tenth to the arrival of the eleventh guest takes no more than 2 minutes?

6. Let X be an exponential random variable with parameter λ. Find

$$P\big(|X - E(X)| \ge 2\sigma_X\big).$$

7. Suppose that, at an Italian restaurant, the time, in minutes, between two customers ordering pizza is exponential with parameter λ. What is the probability that (a) no customer orders pizza during the next t minutes; (b) the next pizza order is placed in at least t minutes but no later than s minutes ($t < s$)?

8. Suppose that the time it takes for a novice secretary to type a document is exponential with mean 1 hour. If at the beginning of a certain eight-hour working day the secretary receives 12 documents to type, what is the probability that she will finish them all by the end of the day?

9. The profit is $350 for each computer assembled by a certain person. Suppose that the assembler guarantees his computers for one year and the time between two failures of a computer is exponential with mean 18 months. If it costs the assembler $40 to repair a failed computer, what is the expected profit per computer?
   Hint: Let $N(t)$ be the number of times that the computer fails in $[0, t]$. Then $\{N(t): t \ge 0\}$ is a Poisson process with parameter $\lambda = 1/18$.

10. Mr. Jones is waiting to make a phone call at a train station. There are two public telephone booths next to each other, occupied by two persons, say A and B. If the duration of each telephone call is an exponential random variable with $\lambda = 1/8$, what is the probability that among Mr. Jones, A, and B, Mr. Jones will not be the last to finish his call?

11. In a factory, a certain machine operates for a period which is exponentially distributed with parameter λ. Then it breaks down and will be in repair shop for a period, which is also exponentially distributed with mean $1/\lambda$. The operating and the repair times are independent. For this machine, we say that a change of "state" occurs each time that it breaks down, or each time that it is fixed. In a time interval of length t, find the probability mass function of the number of times a change of state occurs.

B

12. In data communication, messages are usually combinations of characters, and each character consists of a number of bits. A bit is the smallest unit of information and is either 1 or 0. Suppose that L, the length of a character (in bits) is a geometric random variable with parameter p. If a sender emits messages at the rate of 1000 bits per second, what is the distribution of T, the time it takes the sender to emit a character?

13. The random variable X is called **double exponentially distributed** if its density function is given by

$$f(x) = ce^{-|x|}, \quad -\infty < x < +\infty.$$

(a) Find the value of c.

(b) Prove that $E(X^{2n}) = (2n)!$ and $E(X^{2n+1}) = 0$.

14. Let X, the lifetime (in years) of a radio tube, be exponentially distributed with mean $1/\lambda$. Prove that $[X]$, the integer part of X, which is the complete number of years that the tube works, is a geometric random variable.

15. Prove that if X is a positive, continuous, memoryless random variable with distribution function F, then $F(t) = 1 - e^{-\lambda t}$ for some $\lambda > 0$. This shows that the exponential is the only distribution on $(0, \infty)$ with the memoryless property.

∎

7.4   GAMMA DISTRIBUTIONS

Let $\{N(t): t \geq 0\}$ be a Poisson process, X_1 be the time of the first event, and for $n \geq 2$, let X_n be the time between the $(n-1)$st and nth events. As we explained in Section 7.3, $\{X_1, X_2, \dots\}$ is a sequence of identically distributed exponential random variables with mean $1/\lambda$, where λ is the rate of $\{N(t): t \geq 0\}$. For this Poisson process let X be the time of the nth event. Then X is said to have a **gamma distribution** with parameters (n, λ). Therefore, exponential is the time we will wait for the first event to occur, and gamma is the time we will wait for the nth event to occur. Clearly, a gamma distribution with parameters $(1, \lambda)$ is identical with an exponential distribution with parameter λ.

Let X be a gamma random variable with parameters (n, λ). To find f, the density function of X, note that $\{X \leq t\}$ occurs if the time of the nth event is in $[0, t]$, that is, if the number of events occurring in $[0, t]$ is at least n. Hence F, the distribution function of X, is given by

$$F(t) = P(X \leq t) = P\big(N(t) \geq n\big) = \sum_{i=n}^{\infty} \frac{e^{-\lambda t}(\lambda t)^i}{i!}.$$

Differentiating F, the density function f is obtained:

$$f(t) = \sum_{i=n}^{\infty} \left[-\lambda e^{-\lambda t}\frac{(\lambda t)^i}{i!} + \lambda e^{-\lambda t}\frac{(\lambda t)^{i-1}}{(i-1)!} \right]$$

$$= \sum_{i=n}^{\infty} -\lambda e^{-\lambda t}\frac{(\lambda t)^i}{i!} + \left[\lambda e^{-\lambda t}\frac{(\lambda t)^{n-1}}{(n-1)!} + \sum_{i=n+1}^{\infty} \lambda e^{-\lambda t}\frac{(\lambda t)^{i-1}}{(i-1)!} \right]$$

$$= \left[-\sum_{i=n}^{\infty} \lambda e^{-\lambda t}\frac{(\lambda t)^i}{i!} \right] + \lambda e^{-\lambda t}\frac{(\lambda t)^{n-1}}{(n-1)!} + \left[\sum_{i=n}^{\infty} \lambda e^{-\lambda t}\frac{(\lambda t)^i}{i!} \right]$$

$$= \lambda e^{-\lambda t}\frac{(\lambda t)^{n-1}}{(n-1)!}.$$

The density function

$$f(x) = \begin{cases} \lambda e^{-\lambda x}\dfrac{(\lambda x)^{n-1}}{(n-1)!} & \text{if } x \geq 0 \\[2mm] 0 & \text{elsewhere} \end{cases}$$

is called the **gamma** (or **n-Erlang**) **density** with parameters (n, λ).

Now we extend the definition of the gamma density from parameters (n, λ) to (r, λ), where $r > 0$ is not necessarily a positive integer. As we shall see later, this extension has useful applications in probability and statistics. In the formula of the gamma density function, the term $(n-1)!$ is defined only for positive integers. So the only obstacle in such an extension is to find a function of r that has the basic property of the factorial function, namely, $n! = n \cdot (n - 1)!$, and coincides with $(n - 1)!$ when n is a positive integer. The function with these properties is $\Gamma: (0, \infty) \to \mathbf{R}$ defined by

$$\Gamma(r) = \int_0^\infty t^{r-1} e^{-t} \, dt.$$

The property analogous to $n! = n \cdot (n - 1)!$ is

$$\Gamma(r + 1) = r\Gamma(r), \quad r > 1,$$

which is obtained by an integration by parts applied to $\Gamma(r + 1)$ with $u = t^r$ and $dv \doteq e^{-t} \, dt$:

$$\Gamma(r + 1) = \int_0^\infty t^r e^{-t} \, dt = \left[-t^r e^{-t} \right]_0^\infty + r \int_0^\infty t^{r-1} e^{-t} \, dt$$

$$= r \int_0^\infty t^{r-1} e^{-t} \, dt = r\Gamma(r).$$

To show that $\Gamma(n)$ coincides with $(n - 1)!$ when n is a positive integer, note that

$$\Gamma(1) = \int_0^\infty e^{-t} \, dt \doteq 1.$$

Therefore,

$$\Gamma(2) = (2 - 1)\Gamma(2 - 1) = 1 = 1!,$$
$$\Gamma(3) = (3 - 1)\Gamma(3 - 1) = 2 \cdot 1 = 2!,$$
$$\Gamma(4) = (4 - 1)\Gamma(4 - 1) = 3 \cdot 2 \cdot 1 = 3!.$$

Repetition of this process or a simple induction implies that

$$\Gamma(n + 1) = n!.$$

Hence $\Gamma(r + 1)$ is the natural generalization of $n!$ for a noninteger $r > 0$. This motivates the following definition.

Definition   *A random variable X with probability density function*

$$f(x) = \begin{cases} \dfrac{\lambda e^{-\lambda x} (\lambda x)^{r-1}}{\Gamma(r)} & \text{if } x \geq 0 \\[2mm] 0 & \text{elsewhere} \end{cases}$$

*is said to have a **gamma** distribution with parameters (r, λ), $\lambda > 0$, $r > 0$.*

Figures 7.12 and 7.13 demonstrate the shape of the gamma density function for several values of r and λ.

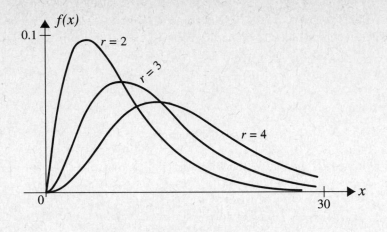

Figure 7.12   Gamma densities for $\lambda = 1/4$.

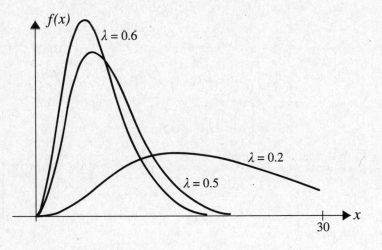

Figure 7.13   Gamma densities for $r = 4$.

Example 7.14   Suppose that, on average, the number of β-particles emitted from a radioactive substance is four every second. What is the probability that it takes at least 2 seconds before the next two β-particles are emitted?

Solution: Let $N(t)$ denote the number of β-particles emitted from a radioactive substance in $[0, t]$. It is reasonable to assume that $\big\{ N(t) \colon t \geq 0 \big\}$ is a Poisson process. Let 1 second be the time unit; then $\lambda = E\big[N(1)\big] = 4$. X, the time between now and when the second β-particle is emitted, has a gamma distribution with parameters $(2, 4)$. Therefore,

$$P(X \geq 2) = \int_2^\infty \frac{4e^{-4x}(4x)^{2-1}}{\Gamma(2)} \, dx = \int_2^\infty 16xe^{-4x} \, dx$$

$$= \left[-4xe^{-4x} \right]_2^\infty - \int_2^\infty -4e^{-4x} \, dx = 8e^{-8} + e^{-8} \approx 0.003.$$

Note that an alternative solution to this problem is

$$P(X \geq 2) = P\big(N(2) \leq 1\big) = P\big(N(2) = 0\big) + P\big(N(2) = 1\big)$$

$$= \frac{e^{-8}(8)^0}{0!} + \frac{e^{-8}(8)^1}{1!} = 9e^{-8} \approx 0.003. \quad \blacklozenge$$

Let X be a gamma random variable with parameters (r, λ). To find $E(X)$ and $\mathrm{Var}(X)$, note that for all $n \geq 0$,

$$E(X^n) = \int_0^\infty x^n \frac{\lambda e^{-\lambda x}(\lambda x)^{r-1}}{\Gamma(r)} \, dx = \frac{\lambda^r}{\Gamma(r)} \int_0^\infty x^{n+r-1} e^{-\lambda x} \, dx.$$

Let $t = \lambda x$; then $dt = \lambda \, dx$, so

$$E(X^n) = \frac{\lambda^r}{\Gamma(r)} \int_0^\infty \frac{t^{n+r-1}}{\lambda^{n+r-1}} e^{-t} \frac{1}{\lambda} dt = \frac{\lambda^r}{\Gamma(r)\lambda^{n+r}} \int_0^\infty t^{n+r-1} e^{-t} \, dt = \frac{\Gamma(n+r)}{\Gamma(r)\lambda^n}.$$

For $n = 1$, this gives

$$E(X) = \frac{\Gamma(r+1)}{\Gamma(r)\lambda} = \frac{r\Gamma(r)}{\lambda\Gamma(r)} = \frac{r}{\lambda}.$$

For $n = 2$,

$$E(X^2) = \frac{\Gamma(r+2)}{\Gamma(r)\lambda^2} = \frac{(r+1)\Gamma(r+1)}{\lambda^2\Gamma(r)} = \frac{(r+1)r\Gamma(r)}{\lambda^2\Gamma(r)} = \frac{r^2+r}{\lambda^2}.$$

Thus

$$\mathrm{Var}(X) = \frac{r^2+r}{\lambda^2} - \left(\frac{r}{\lambda}\right)^2 = \frac{r}{\lambda^2}.$$

We have shown that

For a gamma random variable with parameters r and λ,

$$E(X) = \frac{r}{\lambda}, \qquad \mathrm{Var}(X) = \frac{r}{\lambda^2}, \qquad \sigma_X = \frac{\sqrt{r}}{\lambda}.$$

Example 7.15   There are 100 questions in a test. Suppose that, for all $s > 0$ and $t > 0$, the event that it takes t minutes to answer one question is independent of the event that it takes s minutes to answer another one. If the time that it takes to answer a question is exponential with mean 1/2, find the distribution, the average time, and the standard deviation of the time it takes to do the entire test.

Solution: Let X be the time to answer a question and $N(t)$ the number of questions answered by time t. Then $\{N(t): t \geq 0\}$ is a Poisson process at the rate of $\lambda = 1/E(X) = 2$ per minute. Therefore, the time that it takes to complete all the questions is gamma with parameters $(100, 2)$. The average time to finish the test is $r/\lambda = 100/2 = 50$ minutes with standard deviation $\sqrt{r/\lambda^2} = \sqrt{100/4} = 5$.   ◆

Relationship between Gamma and Negative Binomial: Recall that if a Bernoulli trial is performed successively and independently, then the number of trials until the rth success occurs is negative binomial. Sometimes gamma is viewed as the continuous analog of negative binomial, one reason being that, for a Poisson process, the time it will take until the rth event occurs is gamma. There are more serious relationships between the two distributions that we will not discuss. For example, in some sense, in limit, negative binomial's distribution approaches gamma distribution.   ◆

EXERCISES

A

1.   Show that the gamma density function with parameters (r, λ) has a unique maximum at $(r - 1)/\lambda$.

2.   Let X be a gamma random variable with parameters (r, λ). Find the distribution function of cX, where c is a positive constant.

3.   In a hospital, babies are born at a Poisson rate of 12 per day. What is the probability that it takes at least seven hours before the next three babies are born?

4.   Let f be the density function of a gamma random variable X, with parameters (r, λ). Prove that $\int_{-\infty}^{\infty} f(x)\, dx = 1$.

5.   Customers arrive at a restaurant at a Poisson rate of 12 per hour. If the restaurant makes a profit only after 30 customers have arrived, what is the expected length of time until the restaurant starts to make profit?

6.   A manufacturer produces light bulbs at a Poisson rate of 200 per hour. The probability that a light bulb is defective is 0.015. During production, the light bulbs are tested one by one, and the defective ones are put in a special can that holds up to a maximum of 25 light bulbs. On average, how long does it take until the can is filled?

B

7. For $n = 0, 1, 2, 3, \ldots$, calculate $\Gamma(n + 1/2)$.

8. **(a)** Let Z be a standard normal random variable. Show that the random variable $Y = Z^2$ is gamma and find its parameters.

   (b) Let X be a normal random variable with mean μ and standard deviation σ. Find the distribution function of $W = \left(\dfrac{X - \mu}{\sigma}\right)^2$.

9. Howard enters a bank that has n tellers. All the tellers are busy serving customers, and there is exactly one queue being served by all tellers, with one customer ahead of Howard waiting to be served. If the service time of a customer is exponential with parameter λ, find the distribution of the waiting time for Howard in the queue.

10. In data communication, messages are usually combinations of characters, and each character consists of a number of bits. A bit is the smallest unit of information and is either 1 or 0. Suppose that the length of a character (in bits) is a geometric random variable with parameter p. Suppose that a sender emits messages at the rate of 1000 bits per second. What is the distribution of T, the time it takes the sender to emit a message combined of k characters of independent lengths?
Hint: Let $N(t)$ be the number of characters emitted at or prior to t. First argue that $\{N(t) : t \geq 0\}$ is a Poisson process and find its parameter. ∎

7.5   BETA DISTRIBUTIONS

A random variable X is called **beta** with parameters $(\alpha, \beta), \alpha > 0, \beta > 0$ if f, its density function, is given by

$$f(x) = \begin{cases} \dfrac{1}{B(\alpha, \beta)} x^{\alpha-1}(1 - x)^{\beta-1} & \text{if } 0 < x < 1 \\ \\ 0 & \text{otherwise,} \end{cases}$$

where

$$B(\alpha, \beta) = \int_0^1 x^{\alpha-1}(1 - x)^{\beta-1}\, dx.$$

$B(\alpha, \beta)$ is related to the gamma function by the relation

$$B(\alpha, \beta) = \frac{\Gamma(\alpha)\Gamma(\beta)}{\Gamma(\alpha + \beta)}.$$

Beta density occurs in a natural way in the study of the median of a sample of random points from $(0, 1)$. Let $X_{(1)}$ be the smallest of these numbers, $X_{(2)}$ be the second smallest,

$\ldots$, $X_{(i)}$ be the ith smallest, $\ldots$, and $X_{(n)}$ be the largest of these numbers. If $n = 2k+1$ is odd, $X_{(k+1)}$ is called the **median** of these n random numbers, whereas if $n = 2k$ is even, $[X_{(k)} + X_{(k+1)}]/2$ is called the **median.** It can be shown that the median of $(2n+1)$ random numbers from the interval $(0, 1)$ is a *beta* random variable with parameters $(n + 1, n + 1)$. As Figures 7.14 and 7.15 show, by changing the values of parameters α and β, beta densities cover a wide range of different shapes. If $\alpha = \beta$, the median is $x = 1/2$, and the density of the beta random variable is symmetric about the median. In particular, if $\alpha = \beta = 1$, the uniform density over the interval $(0, 1)$ is obtained.

Beta distributions are often appropriate models for random variables that vary between two finite limits—an upper and a lower. For this reason the following are examples of random variables that might be beta.

1. The fraction of people in a community who use a certain product in a given period of time

2. The percentage of total farm acreage that produces healthy watermelons

3. The distance from one end of a tree to that point where it breaks in a severe storm

In these three instances the random variables are restricted between 0 and 1, 0 and 100, and 0 and the length of the tree, respectively.

To find the expected value and the variance of a beta random variable X, with parameters (α, β), note that for $n \geq 1$,

$$E(X^n) = \frac{1}{B(\alpha, \beta)} \int_0^1 x^{\alpha+n-1}(1-x)^{\beta-1}\, dx = \frac{B(\alpha+n, \beta)}{B(\alpha, \beta)} = \frac{\Gamma(\alpha+n)\Gamma(\alpha+\beta)}{\Gamma(\alpha)\Gamma(\alpha+\beta+n)}.$$

Letting $n = 1$ and $n = 2$ in this relation, we find that

$$E(X) = \frac{\Gamma(\alpha+1)\Gamma(\alpha+\beta)}{\Gamma(\alpha)\Gamma(\alpha+\beta+1)} = \frac{\alpha\Gamma(\alpha)\Gamma(\alpha+\beta)}{\Gamma(\alpha)\,(\alpha+\beta)\Gamma(\alpha+\beta)} = \frac{\alpha}{\alpha+\beta},$$

$$E(X^2) = \frac{\Gamma(\alpha+2)\Gamma(\alpha+\beta)}{\Gamma(\alpha)\Gamma(\alpha+\beta+2)} = \frac{(\alpha+1)\alpha}{(\alpha+\beta+1)(\alpha+\beta)}.$$

Thus

$$\text{Var}(X) = E(X^2) - \big[E(X)\big]^2 = \frac{\alpha\beta}{(\alpha+\beta+1)(\alpha+\beta)^2}.$$

We have established the following formulas:

For a beta random variable with parameters (α, β),

$$E(X) = \frac{\alpha}{\alpha+\beta}, \qquad \text{Var}(X) = \frac{\alpha\beta}{(\alpha+\beta+1)(\alpha+\beta)^2}.$$

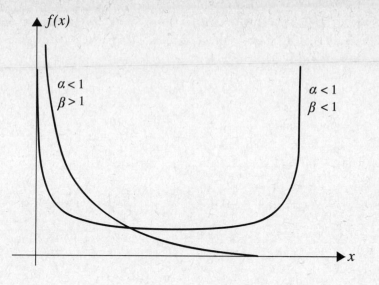

Figure 7.14   Beta densities for $\alpha < 1$, $\beta < 1$ and $\alpha < 1$, $\beta > 1$.

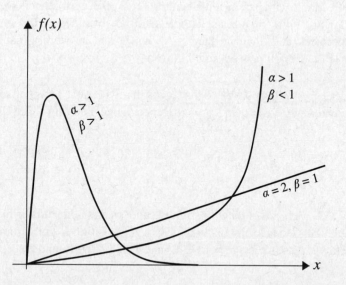

Figure 7.15   Beta densities for the indicated values of α and β.

There is also an interesting relation between beta and binomial distributions that the following theorem explains.

Theorem 7.2   *Let α and β be positive integers, X be a beta random variable with parameters α and β, and Y be a binomial random variable with parameters $\alpha + \beta - 1$ and p, $0 < p < 1$. Then*

$$P(X \le p) = P(Y \ge \alpha).$$

Proof: From the definition,

$$
\begin{aligned}
P(X \le p) &= \frac{(\alpha + \beta - 1)!}{(\alpha - 1)! \, (\beta - 1)!} \int_0^p x^{\alpha-1}(1-x)^{\beta-1} \, dx \\
&= (\alpha + \beta - 1) \binom{\alpha + \beta - 2}{\alpha - 1} \int_0^p x^{\alpha-1}(1-x)^{\beta-1} \, dx \\
&= \sum_{i=\alpha}^{\alpha+\beta-1} \binom{\alpha + \beta - 1}{i} p^i (1-p)^{\alpha+\beta-1-i} \\
&= P(Y \ge \alpha),
\end{aligned}
$$

where the third equality follows from using integration by parts $(\alpha - 1)$ times.   ◆

Example 7.16   The proportion of the stocks that will increase in value tomorrow is a beta random variable with parameters α and β. Today, these parameters were determined to be $\alpha = 5$ and $\beta = 4$, by a financial analyst, after she assessed the current political, social, economical, and financial factors. What is the probability that tomorrow the values of at least 70% of the stocks will move up?

Solution: Let X be the proportion of the stocks that will move up in value tomorrow; X is beta with parameters $\alpha = 5$ and $\beta = 4$. Hence the desired probability is

$$P(X \ge 0.70) = \int_{0.70}^1 \frac{1}{B(5, 4)} x^4 (1-x)^3 \, dx = 280 \int_{0.70}^1 x^4 (1-x)^3 \, dx = 0.194. \quad ◆$$

Example 7.17   A beam of length ℓ, rigidly supported at both ends, is hit suddenly at a random point. This leads to a break in the beam at a position X units from the right end. If X/ℓ is beta with parameters $\alpha = \beta = 3$, find $E(X)$, $\text{Var}(X)$, and $P(\ell/5 < X < \ell/4)$.

Solution: Since

$$E(X/\ell) = \frac{\alpha}{\alpha + \beta} = \frac{3}{6} = \frac{1}{2},$$

$$\text{Var}(X/\ell) = \frac{\alpha\beta}{(\alpha + \beta + 1)(\alpha + \beta)^2} = \frac{9}{7(6)^2} = \frac{1}{28},$$

we have that $E(X) = \ell/2$, $\text{Var}(X) = \ell^2/28$, and

$$P(\ell/5 < X < \ell/4) = P(1/5 < X/\ell < 1/4) = \int_{1/5}^{1/4} \frac{1}{B(3,3)} x^2(1-x)^2 \, dx$$

$$= 30 \int_{1/5}^{1/4} x^2(1-x)^2 \, dx = \frac{72953}{1,600,000} = 0.046. \quad \blacklozenge$$

Example 7.18 Suppose that all 25 passengers of a commuter plane, departing at 2:00 P.M., arrive at random times between 12:45 and 1:45 P.M. Find the probability that the median of the arrival times of the passengers is at or prior to 1:12 P.M.

Solution: Let the arrival time of a randomly selected passenger be W hours past 12:45. Then W is a uniform random variable over the interval $(0, 1)$. Let X be the median of the arrival times of the 25 passengers. Since the median of $(2n + 1)$ random numbers from $(0, 1)$ is beta with parameters $(n+1, n+1)$, we have that X is beta with parameters $(13, 13)$. Since the length of the time interval from 12:45 to 1:12 is 27 minutes, the desired quantity is $P(X \le 27/60)$. To calculate this, let Y be a binomial random variable with parameters 25 and $p = 27/60 = 0.45$. Then, by Theorem 7.2,

$$P(X \le 0.45) = P(Y \ge 13) = \sum_{i=13}^{25} \binom{25}{i} (0.45)^i (0.55)^{25-i} \approx 0.306. \quad \blacklozenge$$

EXERCISES

A

1. Is the following the probability density function of some beta random variable X? If so, find $E(X)$ and $\text{Var}(X)$.

$$f(x) = \begin{cases} 12x(1-x)^2 & 0 < x < 1 \\ 0 & \text{otherwise.} \end{cases}$$

2. Is the following a probability density function? Why or why not?

$$f(x) = \begin{cases} 120x^2(1-x)^4 & 0 < x < 1 \\ 0 & \text{otherwise.} \end{cases}$$

3. For what value of c is the following a probability density function of some random variable X? Find $E(X)$ and $\text{Var}(X)$.

$$f(x) = \begin{cases} cx^4(1-x)^5 & 0 < x < 1 \\ 0 & \text{otherwise.} \end{cases}$$

4. Suppose that new blood pressure medicines introduced are effective on $100p\%$ of the patients, where p is a beta random variable with parameters $\alpha = 20$ and $\beta = 13$. What is the probability that a new blood pressure medicine is effective on at least 60% of the hypertensive population?

5. The proportion of resistors a procurement office of an engineering firm orders every month, from a specific vendor, is a beta random variable with mean 1/3 and variance 1/18. What is the probability that next month, the procurement office orders at least 7/12th of its purchase from this vendor?

6. At a certain university, the fraction of students who get a C in any section of a certain course is uniform over $(0, 1)$. Find the probability that the median of these fractions for the 13 sections of the course that are offered next semester is at least 0.40.

7. Suppose that while daydreaming, the fraction X of the time that one commits brave deeds is beta with parameters $(5, 21)$. What is the probability that next time Jeff is daydreaming, he commits brave deeds at least 1/4 of the time?

8. For complicated projects such as construction of spacecrafts, project managers estimate two quantities, a and b, the minimum and maximum lengths of time it will take for a project to be completed, respectively. In estimating b, they consider all of the possible complications that might delay the completion date. Experience shows that the actual length of time it takes for a project to be completed is a random variable defined by

$$Y = a + (b-a)X,$$

where X is beta with parameters α and β ($\alpha > 0$, $\beta > 0$) that can be determined for the project.

  (a) Find $E(Y)$ and $\text{Var}(Y)$.

  (b) Find the probability density function of Y.

  (c) Suppose that it takes Y years to complete a specific project, where $Y = 2 + 4X$, and X is beta with parameters $\alpha = 2$ and $\beta = 3$. What is the probability that it takes less than 3 years to complete the project?

B

9. Under what conditions and about which point(s) is the probability density function of a beta random variable symmetric?

10. For $\alpha, \beta > 0$, show that

$$B(\alpha, \beta) = 2 \int_0^\infty t^{2\alpha-1}(1+t^2)^{-(\alpha+\beta)}\, dt.$$

Hint: Make the substitution $x = t^2/(1+t^2)$ in

$$B(\alpha, \beta) = \int_0^1 x^{\alpha-1}(1-x)^{\beta-1}\, dx.$$

11. Prove that

$$B(\alpha, \beta) = \frac{\Gamma(\alpha)\Gamma(\beta)}{\Gamma(\alpha + \beta)}.$$

12. For an integer $n \geq 3$, let X be a random variable with the probability density function

$$f(x) = \frac{\Gamma\left(\dfrac{n+1}{2}\right)}{\sqrt{n\pi}\,\Gamma\left(\dfrac{n}{2}\right)}\left(1 + \frac{x^2}{n}\right)^{-(n+1)/2}, \quad -\infty < x < \infty.$$

Such random variables have significant applications in statistics. They are called **t-distributed with n degrees of freedom.** Using the previous two exercises, find $E(X)$ and $\text{Var}(X)$.

★ 7.6   SURVIVAL ANALYSIS AND HAZARD FUNCTIONS

In this section, we will study the risk or rate of failure, per unit of time, of lifetimes that have already survived a certain length of time. The term *lifetime* is broad and applies to appropriate quantities in various models in science, engineering, and business. Depending on the context of a study, by a *lifetime*, we might mean the lifetime of a machine, an electrical component, a living organism, an organism that is under an experimental medical treatment, a financial contract such as a mortgage, the waiting time until a customer arrives at a bank, or the time it takes until a customer is served at a post office. The failure for a living organism is its death. For a mortgage, it is when the mortgage is paid off. For the time until a customer arrives at a bank, failure occurs when a customer enter the bank.

In the following discussion, for convenience, we will talk about the lifetime of a system. However, our definitions and results are general and apply to other examples such as the ones just stated. Let the lifetime of a system be X, where X is a nonnegative, continuous random variable with probability distribution function F and probability density function f. The function defined by

$$\bar{F}(t) = 1 - F(t) = P(X > t)$$

is called the **survival function** of X. For $t > 0$, $\bar{F}(t)$ is the probability that the system has already survived at least t units of time. Furthermore, the following relation, shown in Remark 6.4, enables us to calculate $E(X)$, the expected value of the lifetime of the system, using $\bar{F}(t)$:

$$E(X) = \int_0^\infty \bar{F}(t)\, dt.$$

The probability that a system, which has survived at least t units of time, fails on the time interval $(t, t + \Delta_t]$ is given by

$$P(X \le t + \Delta_t \mid X > t) = \frac{P(t < X \le t + \Delta_t)}{P(X > t)} = \frac{F(t + \Delta_t) - F(t)}{\bar{F}(t)}.$$

To find the instantaneous failure rate of a system of age t at time t, note that during the interval $(t, t + \Delta_t]$, the system fails at a rate of

$$\frac{1}{\Delta_t} P(X \le t + \Delta_t \mid X > t) = \frac{1}{\bar{F}(t)} \cdot \frac{F(t + \Delta t) - F(t)}{\Delta_t}$$

per unit of time. As $\Delta_t \to 0$, this quantity approaches the instantaneous failure rate of the system at time t, given that it has already survived t units of time. Let

$$\lambda(t) = \lim_{\Delta_t \to 0} \frac{1}{\bar{F}(t)} \cdot \frac{F(t + \Delta_t) - F(t)}{\Delta_t}$$

$$= \frac{1}{\bar{F}(t)} \cdot \lim_{\Delta_t \to 0} \frac{F(t + \Delta_t) - F(t)}{\Delta_t}$$

$$= \frac{F'(t)}{\bar{F}(t)} = \frac{f(t)}{\bar{F}t)}.$$

Then $\lambda(t)$ is called the **hazard function** of the random variable X. It is the instantaneous failure rate at t, per unit of time, given that the system has already survived until time t. Note that $\lambda(t) \ge 0$, but it *is not* a probability density function.

Remark 7.3   An alternate term used for $\bar{F}(t)$, the survival function of X, is the *reliability function*. Other terms used for $\lambda(t)$, the hazard function, are *hazard rate, failure rate* *function, failure rate, intensity rate*, and *conditional failure rate*; sometimes actuarial scientists call it *force of mortality.*   ◆

We know that if $\Delta_t > 0$ is very small,

$$P(t < X \le t + \Delta_t) = \int_t^{t+\Delta_t} f(x)\, dx$$

is the area under f from t to $t + \Delta_t$. This area is almost equal to the area of a rectangle with sides of length Δ_t and $f(t)$. Thus

$$P(t < X \le t + \Delta_t) \approx f(t)\Delta_t.$$

The smaller Δ_t, the closer $f(t)\Delta_t$ is to the probability that the system fails in $(t, t+\Delta_t]$. This approximation implies that, for infinitesimal $\Delta_t > 0$,

$$\lambda(t)\Delta_t = \frac{f(t)\Delta_t}{\bar{F}(t)} \approx \frac{P(t < X \le t + \Delta_t)}{P(X > t)} = P(X \le t + \Delta_t \mid X > t).$$

Therefore,

For very small values of $\Delta_t > 0$, the quantity $\lambda(t)\Delta_t$ is approximately the conditional probability that the system fails in $(t, t+\Delta_t)$, given that it has lasted at least until t. That is,

$$P(X \le t + \Delta_t \mid X > t) \approx \lambda(t)\Delta_t. \tag{7.5}$$

Example 7.19   Dr. Hirsch has informed one of his employees, Dr. Kizanis, that she will receive her next year's employment contract at a random time between 10:00 A.M. and 3:00 P.M. Suppose that Dr. Kizanis will receive her contract X minutes past 10:00 A.M. Then X is a uniform random variable over the interval $(0, 300)$. Hence the probability density function of X is given by

$$f(t) = \begin{cases} 1/300 & \text{if } 0 < t < 300 \\ 0 & \text{otherwise.} \end{cases}$$

Straightforward calculations show that the survival function of X is given by

$$\bar{F}(t) = P(X > t) = \begin{cases} 1 & \text{if } t < 0 \\ \dfrac{300 - t}{300} & \text{if } 0 \le t < 300 \\ 0 & \text{if } t \ge 300. \end{cases}$$

The hazard function, $\lambda(t) = f(t)/\bar{F}(t)$, is defined only for $t < 300$. It is unbounded for $t \geq 300$. We have

$$\lambda(t) = \begin{cases} 0 & \text{if } t < 0 \\ \dfrac{1}{300 - t} & \text{if } 0 \leq t < 300. \end{cases}$$

Since $\lambda(0) = 0.003333$, at 10:00 A.M. the instantaneous arrival rate of the contract (the failure rate in this context) is 0.003333 per minute. At noon this rate will increase to $\lambda(120) = 0.0056$, at 2:00 P.M. to $\lambda(240) = 0.017$, at 2:59 P.M. it reaches $\lambda(299) = 1$. One second before 3:00 P.M., the instantaneous arrival of the contract is, approximately, $\lambda(299.983) = 58.82$. This shows that if Dr. Kizanis has not received her contract by one second before 3:00 P.M., the instantaneous arrival rate at that time is very high. That rate approaches ∞, as the time approaches 3:00 P.M. To translate all these into probabilities, let $\Delta_t = 1/60$. Then $f(t)\Delta_t$ is approximately the probability that the contract will arrive within one second after t, whereas $\lambda(t)\Delta_t$ is approximately the probability that the contract will arrive within one second after t, *given* that it has not yet arrived by time t. The following table shows these probabilities at the indicated times:

t	0	120	240	299	299.983
$f(t)\Delta_t$	0.000056	0.000056	0.000056	0.000056	0.000056
$\lambda(t)\Delta_t$	0.000056	0.000093	0.000283	0.0167	0.98

The fact that $f(t)\Delta_t$ is constant is expected because X is uniformly distributed over $(0, 300)$, and all of the intervals under consideration are subintervals of $(0, 300)$ of equal lengths. ♦

Let X be a nonnegative continuous random variable with probability distribution function F, probability density function f, survival function $\bar{F}$, and hazard function $\lambda(t)$. We will now calculate $\bar{F}$ and f in terms of $\lambda(t)$. The formulas obtained are very useful in various applications. Let $G(t) = -\ln\left[\bar{F}(t)\right]$. Then

$$G'(t) = \frac{f(t)}{\bar{F}(t)} = \lambda(t).$$

Consequently,

$$\int_0^t G'(u)\,du = \int_0^t \lambda(u)\,du.$$

Hence

$$G(t) - G(0) = \int_0^t \lambda(u)\,du.$$

Since X is nonnegative, $G(0) = -\ln\left[1 - F(0)\right] = -\ln 1 = 0$. So

$$-\ln\left[\bar{F}(t)\right] - 0 = G(t) - G(0) = \int_0^t \lambda(u)\,du$$

implies that

$$\bar{F}(t) = \exp\left(-\int_0^t \lambda(u)\, du\right). \tag{7.6}$$

By this equation,

$$F(t) = 1 - \exp\left[-\int_0^t \lambda(u)\, du\right].$$

Differentiating both sides of this relation, with respect to t, yields

$$f(t) = \lambda(t) \exp\left[-\int_0^t \lambda(u)\, du\right]. \tag{7.7}$$

This demonstrates that the hazard function *uniquely* determines the probability density function.

In reliability theory, a branch of engineering, it is often observed that $\lambda(t)$, the hazard function of the lifetime of a manufactured machine, is initially large due to undetected defective components during testing. Later, $\lambda(t)$ will decrease and remains more or less the same until a time when it increases again due to aging, which makes worn-out components more likely to fail. A random variable X is said to have **an increasing failure rate** if $\lambda(t)$ is increasing. It is said to have **a decreasing failure rate** if $\lambda(t)$ is decreasing. In the next example, we will show that if $\lambda(t)$ is neither increasing nor decreasing; that is, if $\lambda(t)$ is a constant, then X is an exponential random variable. In such a case, the fact that aging does not change the failure rate is consistent with the memoryless property of exponential random variables. To summarize, the lifetime of a manufactured machine is likely to have a decreasing failure rate in the beginning, a constant failure rate later on, and an increasing failure rate due to wearing out after an aging process. Similarly, lifetimes of living organisms, after a certain age, have increasing failure rates. However, for a newborn baby, the longer he or she survives, the chances of surviving is higher. This means that as t increases, $\lambda(t)$ decreases. That is, $\lambda(t)$ is decreasing.

Example 7.20   In this example, we will prove the following important theorem:

Let $\lambda(t)$, the hazard function of a continuous, nonnegative random variable X, be a constant λ. Then X is an exponential random variable with parameter λ.

To show this theorem, let f be the probability density function of X. By (7.7),

$$f(t) = \lambda \exp\left(-\int_0^t \lambda\, du\right) = \lambda e^{-\lambda t}.$$

This is the density function of an exponential random variable with parameter λ. Thus a random variable with constant hazard function is exponential. As mentioned previously, this result is not surprising. The fact that an exponential random variable is memoryless implies that age has no effect on the distribution of the remaining lifetimes of exponentially distributed random variables. For systems that have exponential lifetime distribution, failures are not due to aging and wearing out. In fact, such systems do not wear out at all. Failures occur abruptly. There is no transition from the previous "state" of the system and no preparation for, or gradual approach to, failure. ◆

EXERCISES

1. Experience shows that the failure rate of a certain electrical component is a linear function. Suppose that after two full days of operation, the failure rate is 10% per hour and after three full days of operation, it is 15% per hour.

    (a) Find the probability that the component operates for at least 30 hours.

    (b) Suppose that the component has been operating for 30 hours. What is the probability that it fails within the next hour?

2. One of the most popular distributions used to model the lifetimes of electric components is the **Weibull distribution**, whose probability density function is given by

$$f(t) = \alpha t^{\alpha-1} e^{-t^{\alpha}}, \quad t > 0, \ \alpha > 0.$$

Determine for which values of α the hazard function of a Weibull random variable is increasing, for which values it is decreasing, and for which values it is constant. ∎

REVIEW PROBLEMS

1. For a restaurant, the time it takes to deliver pizza (in minutes) is uniform over the interval $(25, 37)$. Determine the proportion of deliveries that are made in less than half an hour.

2. It is known that the weight of a random woman from a community is normal with mean 130 pounds and standard deviation 20. Of the women in that community who weigh above 140 pounds, what percent weigh over 170 pounds?

3. One thousand random digits are generated. What is the probability that digit 5 is generated at most 93 times?

4. Let X, the lifetime of a light bulb, be an exponential random variable with parameter λ. Is it possible that X satisfies the following relation?

$$P(X \le 2) = 2P(2 < X \le 3).$$

If so, for what value of λ?

5. The time that it takes for a computer system to fail is exponential with mean 1700 hours. If a lab has 20 such computer systems, what is the probability that at least two fail before 1700 hours of use?

6. Let X be a uniform random variable over the interval $(0, 1)$. Calculate $E(-\ln X)$.

7. Suppose that the diameter of a randomly selected disk produced by a certain manufacturer is normal with mean 4 inches and standard deviation 1 inch. Find the distribution function of the diameter of a randomly chosen disk, in centimeters.

8. Let X be an exponential random variable with parameter λ. Prove that

$$P(\alpha \le X \le \alpha + \beta) \le P(0 \le X \le \beta).$$

9. The time that it takes for a calculus student to answer all the questions on a certain exam is an exponential random variable with mean 1 hour and 15 minutes. If all 10 students of a calculus class are taking that exam, what is the probability that at least one of them completes it in less than one hour?

10. Determine the value(s) of k for which the following is a density function.

$$f(x) = ke^{-x^2+3x+2}, \qquad -\infty < x < \infty.$$

11. The grades of students in a calculus-based probability course are normal with mean 72 and standard deviation 7. If 90, 80, 70, and 60 are the respective lowest, A, B, C, and D, what percent of students in this course get A's, B's, C's, D's, and F's?

12.. The number of minutes that a train from Milan to Rome is late is an exponential random variable X with parameter λ. Find $P(X > E(X))$.

13. In a measurement, a number is rounded off to the nearest k decimal places. Let X be the rounding error. Determine the probability distribution function of X and its parameters.

14. Suppose that the weights of passengers taking an elevator in a certain building are normal with mean 175 pounds and standard deviation 22. What is the minimum weight for a passenger who outweighs at least 90% of the other passengers?

15. The breaking strength of a certain type of yarn produced by a certain vendor is normal with mean 95 and standard deviation 11. What is the probability that, in a random sample of size 10 from the stock of this vendor, the breaking strengths of at least two are over 100?

16. The number of phone calls to a specific exchange is a Poisson process with rate 23 per hour. Calculate the probability that the time until the 91st call is at least 4 hours.

17. Let X be a uniform random variable over the interval $(1 - \theta, 1 + \theta)$, where $0 < \theta < 1$ is a given parameter. Find a function of X, say $g(X)$, so that $E[g(X)] = \theta^2$.

18. A beam of length ℓ is rigidly supported at both ends. Experience shows that whenever the beam is hit at a random point, it breaks at a position X units from the right end, where X/ℓ is a beta random variable. If $E(X) = 3\ell/7$ and $\text{Var}(X) = 3\ell^2/98$, find $P(\ell/7 < X < \ell/3)$.

■

Chapter 8

Bivariate Distributions

8.1 JOINT DISTRIBUTIONS OF TWO RANDOM VARIABLES

Joint Probability Mass Functions

Thus far we have studied probability mass functions of single discrete random variables and probability density functions of single continuous random variables. We now consider two or more random variables that are defined simultaneously on the same sample space. In this section we consider such cases with two variables. Cases of three or more variables are studied in Chapter 9.

Definition *Let X and Y be two discrete random variables defined on the same sample space. Let the sets of possible values of X and Y be A and B, respectively. The function*

$$p(x, y) = P(X = x, Y = y)$$

*is called the **joint probability mass function** of X and Y.*

Note that $p(x, y) \geq 0$. If $x \notin A$ or $y \notin B$, then $p(x, y) = 0$. Also,

$$\sum_{x \in A} \sum_{y \in B} p(x, y) = 1. \qquad (8.1)$$

Let X and Y have joint probability mass function $p(x, y)$. Let p_X be the probability mass function of X. Then

$$p_X(x) = P(X = x) = P(X = x, Y \in B)$$

$$= \sum_{y \in B} P(X = x, Y = y) = \sum_{y \in B} p(x, y).$$

Similarly, p_Y, the probability mass function of Y, is given by

$$p_Y(y) = \sum_{x \in A} p(x, y).$$

These relations motivate the following definition.

Definition   *Let X and Y have joint probability mass function $p(x, y)$. Let A be the set of possible values of X and B be the set of possible values of Y. Then the functions $p_X(x) = \sum_{y \in B} p(x, y)$ and $p_Y(y) = \sum_{x \in A} p(x, y)$ are called, respectively, the **marginal probability mass functions** of X and Y.*

Example 8.1   A small college has 90 male and 30 female professors. An ad hoc committee of five is selected at random to write the vision and mission of the college. Let X and Y be the number of men and women on this committee, respectively.

(a)   Find the joint probability mass function of X and Y.

(b)   Find p_X and p_Y, the marginal probability mass functions of X and Y.

Solution:

(a)   The set of possible values for both X and Y is $\{0, 1, 2, 3, 4, 5\}$. The joint probability mass function of X and Y, $p(x, y)$, is given by

$$
p(x, y) = \begin{cases} \dfrac{\dbinom{90}{x}\dbinom{30}{y}}{\dbinom{120}{5}} & \text{if } x, y \in \{0, 1, 2, 3, 4, 5\}, \quad x + y = 5 \\[4mm] 0 & \text{otherwise.} \end{cases}
$$

(b)   To find p_X and p_Y, the marginal probability mass functions of X and Y, respectively, note that $p_X(x) = \sum_{y=0}^{5} p(x, y)$, $p_Y(y) = \sum_{x=0}^{5} p(x, y)$. Since $p(x, y) = 0$ if $x + y \neq 5$, $\sum_{y=0}^{5} p(x, y) = p(x, 5 - x)$ and $\sum_{x=0}^{5} p(x, y) = p(5 - y, y)$. Therefore,

$$
p_X(x) = \frac{\dbinom{90}{x}\dbinom{30}{5-x}}{\dbinom{120}{5}}, \qquad x \in \{0, 1, 2, 3, 4, 5\},
$$

$$
p_Y(y) = \frac{\dbinom{90}{5-y}\dbinom{30}{y}}{\dbinom{120}{5}}, \qquad y \in \{0, 1, 2, 3, 4, 5\}.
$$

Note that, as expected, p_X and p_Y are hypergeometric. ◆

Example 8.2   Roll a balanced die and let the outcome be X. Then toss a fair coin X times and let Y denote the number of tails. What is the joint probability mass function of X and Y and the marginal probability mass functions of X and Y?

Solution: Let $p(x, y)$ be the joint probability mass function of X and Y. Clearly, $X \in \{1, 2, 3, 4, 5, 6\}$ and $Y \in \{0, 1, 2, 3, 4, 5, 6\}$. Now if $X = 1$, then $Y = 0$ or 1; we have

$$p(1, 0) = P(X = 1, Y = 0) = P(X = 1)P(Y = 0 \mid X = 1) = \frac{1}{6} \cdot \frac{1}{2} = \frac{1}{12},$$

$$p(1, 1) = P(X = 1, Y = 1) = P(X = 1)P(Y = 1 \mid X = 1) = \frac{1}{6} \cdot \frac{1}{2} = \frac{1}{12}.$$

If $X = 2$, then $y = 0, 1$, or 2, where

$$p(2, 0) = P(X = 2, Y = 0) = P(X = 2)P(Y = 0 \mid X = 2) = \frac{1}{6} \cdot \frac{1}{4} = \frac{1}{24}.$$

Similarly, $p(2, 1) = 1/12$, $p(2, 2) = 1/24$. If $X = 3$, then $y = 0, 1, 2$, or 3, and

$$p(3, 0) = P(X = 3, Y = 0) = P(X = 3)P(Y = 0 \mid X = 3)$$
$$= \frac{1}{6}\binom{3}{0}\left(\frac{1}{2}\right)^{0}\left(\frac{1}{2}\right)^{3} = \frac{1}{48},$$

$$p(3, 1) = P(X = 3, Y = 1) = P(X = 3)P(Y = 1 \mid X = 3)$$
$$= \frac{1}{6}\binom{3}{1}\left(\frac{1}{2}\right)^{2}\left(\frac{1}{2}\right)^{1} = \frac{3}{48}.$$

Similarly, $p(3, 2) = 3/48$, $p(3, 3) = 1/48$. Similar calculations will yield the following table for $p(x, y)$.

x	0	1	2	3	4	5	6	$p_X(x)$
1	1/12	1/12	0	0	0	0	0	1/6
2	1/24	2/24	1/24	0	0	0	0	1/6
3	1/48	3/48	3/48	1/48	0	0	0	1/6
4	1/96	4/96	6/96	4/96	1/96	0	0	1/6
5	1/192	5/192	10/192	10/192	5/192	1/192	0	1/6
6	1/384	6/384	15/384	20/384	15/384	6/384	1/384	1/6
$p_Y(y)$	63/384	120/384	99/384	64/384	29/384	8/384	1/384	

(The column header "y" spans columns 0 through 6.)

Note that $p_X(x) = P(X = x)$ and $p_Y(y) = P(Y = y)$, the probability mass functions of X and Y, are obtained by summing up the rows and the columns of this table, respectively. ♦

Let X and Y be discrete random variables with joint probability mass function $p(x, y)$. Let the sets of possible values of X and Y be A and B, respectively. To find $E(X)$ and $E(Y)$, first we calculate p_X and p_Y, the marginal probability mass functions of X and Y, respectively. Then we will use the following formulas.

$$E(X) = \sum_{x \in A} x p_X(x); \qquad E(Y) = \sum_{y \in B} y p_Y(y).$$

Example 8.3   Let the joint probability mass function of random variables X and Y be given by

$$p(x, y) = \begin{cases} \dfrac{1}{70} x(x + y) & \text{if } x = 1, 2, 3, \quad y = 3, 4 \\ 0 & \text{elsewhere.} \end{cases}$$

Find $E(X)$ and $E(Y)$.

Solution: To find $E(X)$ and $E(Y)$, first we need to calculate $p_X(x)$ and $p_Y(y)$, the marginal probability mass functions of X and Y, respectively. Note that

$$p_X(x) = p(x, 3) + p(x, 4)$$

$$= \frac{1}{70} x(x + 3) + \frac{1}{70} x(x + 4)$$

$$= \frac{1}{35} x^2 + \frac{1}{10} x, \qquad x = 1, 2, 3;$$

$$p_Y(y) = p(1, y) + p(2, y) + p(3, y)$$

$$= \frac{1}{70}(1 + y) + \frac{2}{70}(2 + y) + \frac{3}{70}(3 + y)$$

$$= \frac{1}{5} + \frac{3}{35} y, \qquad y = 3, 4.$$

Therefore,

$$E(X) = \sum_{x=1}^{3} x p_X(x) = \sum_{x=1}^{3} x\left(\frac{1}{35} x^2 + \frac{1}{10} x\right) = \frac{17}{7} \approx 2.43;$$

$$E(Y) = \sum_{y=3}^{4} y p_Y(y) = \sum_{y=3}^{4} y\left(\frac{1}{5} + \frac{3}{35} y\right) = \frac{124}{35} \approx 3.54. \quad \blacklozenge$$

We will now state the following generalization of Theorem 4.2 from one dimension to two. As an immediate application of this important generalization, we will show that the expected value of the sum of two random variables is equal to the sum of their expected values. We will prove a generalization of this theorem in Chapter 10 and discuss some of its many applications in that chapter.

Theorem 8.1   *Let $p(x, y)$ be the joint probability mass function of discrete random variables X and Y. Let A and B be the set of possible values of X and Y, respectively. If h is a function of two variables from $\mathbf{R}^2$ to $\mathbf{R}$, then $h(X, Y)$ is a discrete random variable with the expected value given by*

$$E[h(X, Y)] = \sum_{x \in A} \sum_{y \in B} h(x, y) p(x, y),$$

provided that the sum is absolutely convergent.

Corollary   *For discrete random variables X and Y,*

$$E(X + Y) = E(X) + E(Y).$$

Proof: In Theorem 8.1 let $h(x, y) = x + y$. Then

$$E(X + Y) = \sum_{x \in A} \sum_{y \in B} (x + y) p(x, y)$$

$$= \sum_{x \in A} \sum_{y \in B} x p(x, y) + \sum_{x \in A} \sum_{y \in B} y p(x, y)$$

$$= E(X) + E(Y). \quad \blacklozenge$$

Joint Probability Density Functions

To define the concept of *joint probability density function* of two continuous random variables X and Y, recall from Section 6.1 that a single random variable X is called continuous if there exists a nonnegative real-valued function $f: \mathbf{R} \to [0, \infty)$ such that for any subset A of real numbers that can be constructed from intervals by a countable number of set operations,

$$P(X \in A) = \int_A f(x)\, dx.$$

This definition is generalized in the following obvious way:

Definition   *Two random variables X and Y, defined on the same sample space, have a continuous joint distribution if there exists a nonnegative function of two variables, $f(x, y)$ on $\mathbf{R} \times \mathbf{R}$, such that for any region R in the xy-plane that can be formed from rectangles by a countable number of set operations,*

$$P((X, Y) \in R) = \iint_R f(x, y)\, dx\, dy. \tag{8.2}$$

*The function $f(x, y)$ is called the **joint probability density function** of X and Y.*

Note that if in (8.2) the region R is a plane curve, then $P\big((X, Y) \in R\big) = 0$. That is, the probability that (X, Y) lies on any curve (in particular, a circle, an ellipse, a straight line, etc.) is 0.

Let $R = \big\{(x, y): x \in A, \, y \in B\big\}$, where A and B are *any* subsets of real numbers that can be constructed from intervals by a countable number of set operations. Then (8.2) gives

$$P(X \in A, Y \in B) = \int_B \int_A f(x, y) \, dx \, dy. \tag{8.3}$$

Letting $A = (-\infty, \infty)$, $B = (-\infty, \infty)$, (8.3) implies the relation

$$\int_{-\infty}^{\infty} \int_{-\infty}^{\infty} f(x, y) \, dx \, dy = 1,$$

which is the continuous analog of (8.1). The relation (8.3) also implies that

$$P(X = a, Y = b) = \int_b^b \int_a^a f(x, y) \, dx \, dy = 0.$$

Hence, for $a < b$ and $c < d$,

$$P(a < X \le b, \, c \le Y \le d) = P(a < X < b, \, c < Y < d)$$
$$= P(a \le X < b, \, c \le Y < d) = \cdots$$
$$= \int_c^d \left(\int_a^b f(x, y) \, dx \right) dy.$$

Because, for real numbers a and b, $P(X = a, Y = b) = 0$, in general, $f(a, b)$ is not equal to $P(X = a, Y = b)$. At no point do the values of f represent probabilities. Intuitively, $f(a, b)$ is a measure that determines how likely it is that X is close to a and Y is close to b. To see this, note that if ε and δ are very small positive numbers, then $P(a - \varepsilon < X < a + \varepsilon, \, b - \delta < Y < b + \delta)$ is the probability that X is close to a and Y is close to b. Now

$$P(a - \varepsilon < X < a + \varepsilon, \, b - \delta < Y < b + \delta) = \int_{b-\delta}^{b+\delta} \int_{a-\varepsilon}^{a+\varepsilon} f(x, y) \, dx \, dy$$

is the volume under the surface $z = f(x, y)$, above the region $(a - \varepsilon, a + \varepsilon) \times (b - \delta, b + \delta)$. For infinitesimal ε and δ, this volume is approximately equal to the volume of a rectangular parallelepiped with sides of lengths 2ε, 2δ, and height $f(a, b)$ [i.e., $(2\varepsilon)(2\delta)f(a, b) = 4\varepsilon\delta f(a, b)$]. Therefore,

$$P(a - \varepsilon < X < a + \varepsilon, \, b - \delta < Y < b + \delta) \approx 4\varepsilon\delta f(a, b).$$

Hence for fixed small values of ϵ and δ, we observe that a larger value of $f(a, b)$ makes $P(a - \varepsilon < X < a + \varepsilon, \, b - \delta < Y < b + \delta)$ larger. That is, the larger the value of $f(a, b)$, the higher the probability that X and Y are close to a and b, respectively.

Let X and Y have joint probability density function $f(x, y)$. Let f_Y be the probability density function of Y. To find f_Y in terms of f, note that, on the one hand, for any subset B of $\mathbf{R}$,

$$P(Y \in B) = \int_B f_Y(y)\, dy, \tag{8.4}$$

and, on the other hand, using (8.3),

$$P(Y \in B) = P\big(X \in (-\infty, \infty), Y \in B\big) = \int_B \left(\int_{-\infty}^{\infty} f(x, y)\, dx \right) dy.$$

Comparing this with (8.4), we can write

$$f_Y(y) = \int_{-\infty}^{\infty} f(x, y)\, dx. \tag{8.5}$$

Similarly,

$$f_X(x) = \int_{-\infty}^{\infty} f(x, y)\, dy. \tag{8.6}$$

Therefore, it is reasonable to make the following definition:

Definition   *Let X and Y have joint probability density function $f(x, y)$; then the functions f_X and f_Y, given by (8.6) and (8.5), are called, respectively, the **marginal probability density functions** of X and Y.*

Note that while from the joint probability density function of X and Y we can find the marginal probability density functions of X and Y, it is not possible, in general, to find the joint probability density function of two random variables from their marginals. This is because the marginal probability density function of a random variable X gives information about X without looking at the possible values of other random variables. However, sometimes with more information the marginal probability density functions enable us to find the joint probability density function of two random variables. An example of such a case, studied in Section 8.2, is when X and Y are independent random variables.

Let X and Y be two random variables (discrete, continuous, or mixed). The **joint probability distribution function**, or *joint cumulative probability distribution function*, or simply the *joint distribution of X and Y*, is defined by

$$F(t, u) = P(X \le t, Y \le u)$$

for all $-\infty < t, u < \infty$. The **marginal probability distribution function of X**, F_X, can be found from F as follows:

$$F_X(t) = P(X \le t) = P(X \le t, Y < \infty) = P\big(\lim_{n \to \infty} \{X \le t, Y \le n\}\big)$$

$$= \lim_{n \to \infty} P(X \le t, Y \le n) = \lim_{n \to \infty} F(t, n) \equiv F(t, \infty).$$

To justify the fourth equality, note that the sequence of events $\{X \le t, Y \le n\}$, $n \ge 1$, is an increasing sequence. Therefore, by continuity of probability function (Theorem 1.8),

$$P\left(\lim_{n \to \infty} \{X \le t, Y \le n\}\right) = \lim_{n \to \infty} P(X \le t, Y \le n).$$

Similarly, F_Y, the **marginal probability distribution function of Y**, is

$$F_Y(u) = P(Y \le u) = \lim_{n \to \infty} F(n, u) \equiv F(\infty, u).$$

Now suppose that the joint probability density function of X and Y is $f(x, y)$. Then

$$F(x, y) = P(X \le x, Y \le y) = P\big(X \in (-\infty, x], Y \in (-\infty, y]\big)$$

$$= \int_{-\infty}^{y} \int_{-\infty}^{x} f(t, u)\, dt\, du. \tag{8.7}$$

Assuming that the partial derivatives of F exist, by differentiation of (8.7), we get

$$f(x, y) = \frac{\partial^2}{\partial x\, \partial y} F(x, y).$$

Moreover, from (8.7), we obtain

$$F_X(x) = F(x, \infty) = \int_{-\infty}^{\infty} \left(\int_{-\infty}^{x} f(t, u)\, dt \right) du$$

$$= \int_{-\infty}^{x} \left(\int_{-\infty}^{\infty} f(t, u)\, du \right) dt = \int_{-\infty}^{x} f_X(t)\, dt,$$

and, similarly,

$$F_Y(y) = \int_{-\infty}^{y} f_Y(u)\, du.$$

These relations show that if X and Y have joint probability density function $f(x, y)$, then X and Y are continuous random variables with density functions f_X and f_Y, and distribution functions F_X and F_Y, respectively. Therefore, $F_X'(x) = f_X(x)$ and $F_Y'(y) = f_Y(y)$. We also have

$$E(X) = \int_{-\infty}^{\infty} x f_X(x)\, dx; \qquad E(Y) = \int_{-\infty}^{\infty} y f_Y(y)\, dy.$$

Example 8.4   The joint probability density function of random variables X and Y is given by

$$f(x, y) = \begin{cases} \lambda x y^2 & 0 \le x \le y \le 1 \\ 0 & \text{otherwise.} \end{cases}$$

(a) Determine the value of λ.

(b) Find the marginal probability density functions of X and Y.

(c) Calculate $E(X)$ and $E(Y)$.

Solution:

(a) To find λ, note that $\int_{-\infty}^{\infty} \int_{-\infty}^{\infty} f(x, y)\, dx\, dy = 1$ gives

$$\int_0^1 \left(\int_x^1 \lambda x y^2 \, dy \right) dx = 1.$$

Therefore,

$$1 = \int_0^1 \left(\int_x^1 y^2 \, dy \right) \lambda x \, dx = \int_0^1 \left[\frac{1}{3} y^3 \right]_x^1 \lambda x \, dx$$

$$= \lambda \int_0^1 \left(\frac{1}{3} - \frac{1}{3} x^3 \right) x \, dx = \frac{\lambda}{3} \int_0^1 (1 - x^3) x \, dx = \frac{\lambda}{3} \int_0^1 (x - x^4) \, dx$$

$$= \frac{\lambda}{3} \left[\frac{1}{2} x^2 - \frac{1}{5} x^5 \right]_0^1 = \frac{\lambda}{10},$$

and hence $\lambda = 10$.

(b) To find f_X and f_Y, the respective marginal probability density functions of X and Y, we use (8.6) and (8.5):

$$f_X(x) = \int_{-\infty}^{\infty} f(x, y) \, dy = \int_x^1 10 x y^2 \, dy = \left[\frac{10}{3} x y^3 \right]_x^1$$

$$= \frac{10}{3} x (1 - x^3), \qquad 0 \le x \le 1;$$

$$f_Y(y) = \int_{-\infty}^{\infty} f(x, y) \, dx = \int_0^y 10 x y^2 \, dx = \left[5 x^2 y^2 \right]_0^y$$

$$= 5 y^4, \qquad 0 \le y \le 1.$$

(c) To find $E(X)$ and $E(Y)$, we use the results obtained in part (b):

$$E(X) = \int_0^1 x \cdot \frac{10}{3} x (1 - x^3) \, dx = \frac{5}{9};$$

$$E(Y) = \int_0^1 y \cdot 5 y^4 \, dy = \frac{5}{6}. \quad \blacklozenge$$

Example 8.5   For $\lambda > 0$, let

$$F(x, y) = \begin{cases} 1 - \lambda e^{-\lambda(x+y)} & \text{if } x > 0, \quad y > 0 \\ 0 & \text{otherwise.} \end{cases}$$

Determine if F is the joint probability distribution function of two random variables X and Y.

Solution: If F is the joint probability distribution function of two random variables X and Y, then $\dfrac{\partial^2}{\partial x\, \partial y} F(x, y)$ is the joint probability density function of X and Y. But

$$\frac{\partial^2}{\partial x\, \partial y} F(x, y) = \begin{cases} -\lambda^3 e^{-\lambda(x+y)} & \text{if } x > 0, \quad y > 0 \\ 0 & \text{otherwise.} \end{cases}$$

Since $\dfrac{\partial^2}{\partial x\, \partial y} F(x, y) < 0$, it cannot be a joint probability density function. Therefore, F is not a joint probability distribution function.   ◆

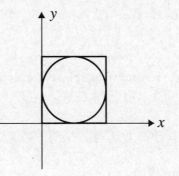

Figure 8.1   Geometric model of Example 8.6.

Example 8.6   A circle of radius 1 is inscribed in a square with sides of length 2. A point is selected at random from the square. What is the probability that it is inside the circle? Note that by a point being selected at random from the square we mean that the point is selected in a way that all the subsets of equal areas of the square are equally likely to contain the point.

Solution: Let the square and the circle be situated in the coordinate system as shown in Figure 8.1. Let the coordinates of the point selected at random be (X, Y); then X and Y are random variables. By definition, regions inside the square with equal areas are equally likely to contain (X, Y). Hence, for all (a, b) inside the square, the probability that (X, Y) is "close" to (a, b) is the same. Let $f(x, y)$ be the joint probability density function of X and Y. Since $f(x, y)$ is a measure that determines how likely it is that X is close to x and Y is close to y, $f(x, y)$ must be constant for the points inside the square, and 0 elsewhere. Therefore, for some $c > 0$,

$$f(x, y) = \begin{cases} c & \text{if } 0 < x < 2, \quad 0 < y < 2 \\ 0 & \text{otherwise,} \end{cases}$$

where $\int_{-\infty}^{\infty} \int_{-\infty}^{\infty} f(x, y) \, dx \, dy = 1$ gives

$$\int_0^2 \int_0^2 c \, dx \, dy = 1,$$

implying that $c = 1/4$. Now let R be the region inside the circle. Then the desired probability is

$$\iint_R f(x, y) \, dx \, dy = \iint_R \frac{1}{4} \, dx \, dy = \frac{\displaystyle\iint_R dx \, dy}{4}.$$

Note that $\displaystyle\iint_R dx \, dy$ is the area of the circle, and 4 is the area of the square. Thus the desired probability is

$$\frac{\text{area of the circle}}{\text{area of the square}} = \frac{\pi(1)^2}{4} = \frac{\pi}{4}. \quad \blacklozenge$$

What we showed in Example 8.6 is true, in general. Let S be a bounded region in the Euclidean plane and suppose that R is a region inside S. Fix a coordinate system, and let the coordinates of a point selected at random from S be (X, Y). By an argument similar to that of Example 8.6, we have that, for some $c > 0$, the joint probability density function of X and Y, $f(x, y)$, is given by

$$f(x, y) = \begin{cases} c & \text{if } (x, y) \in S \\ 0 & \text{otherwise,} \end{cases}$$

where

$$\iint_S f(x, y) \, dx \, dy = 1.$$

This gives $c \displaystyle\iint_S dx\, dy = 1$, or, equivalently, $c \times \text{area}(S) = 1$. Therefore, $c = 1/\text{area}(S)$

and hence

$$f(x, y) = \begin{cases} \dfrac{1}{\text{area}(S)} & \text{if } (x, y) \in S \\[2ex] 0 & \text{otherwise.} \end{cases}$$

Thus

$$P\big((X, Y) \in R\big) = \iint_R f(x, y)\, dx\, dy = \frac{1}{\text{area}(S)} \iint_R dx\, dy = \frac{\text{area}(R)}{\text{area}(S)}.$$

Based on these observations, we make the following definition.

Definition   *Let S be a subset of the plane with area A(S). A point is said to be* ***randomly selected*** *from S if for any subset R of S with area A(R), the probability that R contains the point is A(R)/A(S).*

This definition is essential in the field of **geometric probability**. By the following examples, we will show how it can help to solve problems readily.

Example 8.7   A man invites his fiancée to a fine hotel for a Sunday brunch. They decide to meet in the lobby of the hotel between 11:30 A.M. and 12 noon. If they arrive at random times during this period, what is the probability that they will meet within 10 minutes?

Solution: Let X and Y be the minutes past 11:30 A.M. that the man and his fiancée arrive at the lobby, respectively. Let

$$S = \big\{(x, y): 0 \le x \le 30,\ 0 \le y \le 30\big\}, \quad \text{and} \quad R = \big\{(x, y) \in S: |x - y| \le 10\big\}.$$

Then the desired probability, $P\big(|X - Y| \le 10\big)$, is given by

$$P\big(|X - Y| \le 10\big) = \frac{\text{area of } R}{\text{area of } S} = \frac{\text{area of } R}{30 \times 30} = \frac{\text{area}(R)}{900}.$$

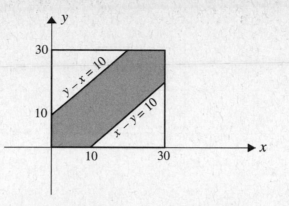

Figure 8.2   Geometric model of Example 8.7.

$R = \{(x, y) \in S : x - y \leq 10$ and $y - x \leq 10\}$ is the shaded region of Figure 8.2, and its area is the area of the square minus the areas of the two unshaded triangles: $(30)(30) - 2(1/2 \times 20 \times 20) = 500$. Hence the desired probability is $500/900 = 5/9$.

$\blacklozenge$

Example 8.8   A farmer decides to build a pen in the shape of a triangle for his chickens. He sends his son out to cut the lumber and the boy, without taking any thought as to the ultimate purpose, makes two cuts at two points selected at random. What are the chances that the resulting three pieces of lumber can be used to form a triangular pen?

Solution: Suppose that the length of the lumber is ℓ. Let A and B be the random points placed on the lumber; let the distances of A and B from the left end of the lumber be denoted by X and Y, respectively. If $X < Y$, the lumber is divided into three parts of lengths X, $Y - X$, and $\ell - Y$; otherwise, it is divided into three parts of lengths Y, $X - Y$, and $\ell - X$. Because of symmetry, we calculate the probability that $X < Y$, and X, $Y - X$, and $\ell - Y$ form a triangle. Then we multiply the result by 2 to obtain the desired probability. We know that three segments form a triangle if and only if the length of any one of them is less than the sum of the lengths of the remaining two. Therefore, we must have

$$X < Y$$
$$X < (Y - X) + (\ell - Y)$$
$$(Y - X) < X + (\ell - Y)$$
$$\ell - Y < X + (Y - X),$$

or, equivalently,

$$X < Y, \quad X < \frac{\ell}{2}, \quad Y < X + \frac{\ell}{2}, \quad \text{and} \quad Y > \frac{\ell}{2}.$$

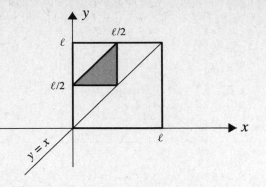

Figure 8.3   Geometric model of Example 8.8.

Since (X, Y) is a random point from the square $(0, \ell) \times (0, \ell)$, the probability that it satisfies these inequalities is the area of R, where

$$R = \left\{ (x, y) : x < y, \ x < \frac{\ell}{2}, \ y < x + \frac{\ell}{2}, \ y > \frac{\ell}{2} \right\},$$

divided by ℓ^2, the area of the square. Note that R is the shaded region of Figure 8.3 and its area is 1/8 of the area of the square. Thus the probability that (X, Y) lies in R is 1/8 and hence the desired probability is 1/4.   ◆

Before closing this section, we state the continuous analog of Theorem 8.1, which is a generalization of Theorem 6.3 from one dimension to two. As an immediate application of this important generalization, we will show the continuous version of the following theorem that we proved above for discrete random variables: The expected value of the sum of two random variables is equal to the sum of their expected values.

Theorem 8.2   *Let $f(x, y)$ be the joint probability density function of random variables X and Y. If h is a function of two variables from $\mathbf{R}^2$ to $\mathbf{R}$, then $h(X, Y)$ is a random variable with the expected value given by*

$$E\big[h(X, Y)\big] = \int_{-\infty}^{\infty} \int_{-\infty}^{\infty} h(x, y) f(x, y) \, dx \, dy,$$

provided that the integral is absolutely convergent.

Corollary   *For random variables X and Y,*

$$E(X + Y) = E(X) + E(Y).$$

Proof: In Theorem 8.2 let $h(x, y) = x + y$. Then

$$E(X + Y) = \int_{-\infty}^{\infty} \int_{-\infty}^{\infty} (x + y) f(x, y) \, dx \, dy$$

$$= \int_{-\infty}^{\infty} \int_{-\infty}^{\infty} x f(x, y) \, dx \, dy + \int_{-\infty}^{\infty} \int_{-\infty}^{\infty} y f(x, y) \, dx \, dy$$

$$= E(X) + E(Y). \quad \blacklozenge$$

Example 8.9 Let X and Y have joint probability density function

$$f(x, y) = \begin{cases} \dfrac{3}{2}(x^2 + y^2) & \text{if } 0 < x < 1, \quad 0 < y < 1 \\ \\ 0 & \text{otherwise.} \end{cases}$$

Find $E(X^2 + Y^2)$.

Solution: By Theorem 8.2,

$$E(X^2 + Y^2) = \int_{-\infty}^{\infty} \int_{-\infty}^{\infty} (x^2 + y^2) f(x, y) \, dx \, dy = \int_{0}^{1} \int_{0}^{1} \frac{3}{2}(x^2 + y^2)^2 \, dx \, dy$$

$$= \frac{3}{2} \int_{0}^{1} \int_{0}^{1} (x^4 + 2x^2 y^2 + y^4) \, dx \, dy = \frac{14}{15}. \quad \blacklozenge$$

EXERCISES

A

1. Let the joint probability mass function of discrete random variables X and Y be given by

$$p(x, y) = \begin{cases} k\left(\dfrac{x}{y}\right) & \text{if } x = 1, 2, \quad y = 1, 2 \\ \\ 0 & \text{otherwise.} \end{cases}$$

Determine (a) the value of the constant k, (b) the marginal probability mass functions of X and Y, (c) $P(X > 1 \mid Y = 1)$, (d) $E(X)$ and $E(Y)$.

2. Let the joint probability mass function of discrete random variables X and Y be given by

$$p(x, y) = \begin{cases} c(x + y) & \text{if } x = 1, 2, 3, \quad y = 1, 2 \\ 0 & \text{otherwise.} \end{cases}$$

Determine (a) the value of the constant c, (b) the marginal probability mass functions of X and Y, (c) $P(X \geq 2 \mid Y = 1)$, (d) $E(X)$ and $E(Y)$.

3. Let the joint probability mass function of discrete random variables X and Y be given by

$$p(x, y) = \begin{cases} k(x^2 + y^2) & \text{if } (x, y) = (1, 1), (1, 3), (2, 3) \\ 0 & \text{otherwise.} \end{cases}$$

Determine (a) the value of the constant k, (b) the marginal probability mass functions of X and Y, and (c) $E(X)$ and $E(Y)$.

4. Let the joint probability mass function of discrete random variables X and Y be given by

$$p(x, y) = \begin{cases} \dfrac{1}{25}(x^2 + y^2) & \text{if } x = 1, 2, \quad y = 0, 1, 2 \\ 0 & \text{otherwise.} \end{cases}$$

Find $P(X > Y)$, $P(X + Y \leq 2)$, and $P(X + Y = 2)$.

5. Thieves stole four animals at random from a farm that had seven sheep, eight goats, and five burros. Calculate the joint probability mass function of the number of sheep and goats stolen.

6. Two dice are rolled. The sum of the outcomes is denoted by X and the absolute value of their difference by Y. Calculate the joint probability mass function of X and Y and the marginal probability mass functions of X and Y.

7. In a community 30% of the adults are Republicans, 50% are Democrats, and the rest are independent. For a randomly selected person, let

$$X = \begin{cases} 1 & \text{if he or she is a Republican} \\ 0 & \text{otherwise,} \end{cases}$$

$$Y = \begin{cases} 1 & \text{if he or she is a Democrat} \\ 0 & \text{otherwise.} \end{cases}$$

Calculate the joint probability mass function of X and Y.

8. From an ordinary deck of 52 cards, seven cards are drawn at random and without replacement. Let X and Y be the number of hearts and the number of spades drawn, respectively.

(a) Find the joint probability mass function of X and Y.

(b) Calculate $P(X \geq Y)$.

9. Let the joint probability density function of random variables X and Y be given by

$$f(x, y) = \begin{cases} 2 & \text{if } 0 \leq y \leq x \leq 1 \\ 0 & \text{elsewhere.} \end{cases}$$

(a) Calculate the marginal probability density functions of X and Y, respectively.

(b) Find $E(X)$ and $E(Y)$.

(c) Calculate $P(X < 1/2)$, $P(X < 2Y)$, and $P(X = Y)$.

10. Let the joint probability density function of random variables X and Y be given by

$$f(x, y) = \begin{cases} 8xy & \text{if } 0 \leq y \leq x \leq 1 \\ 0 & \text{elsewhere.} \end{cases}$$

(a) Calculate the marginal probability density functions of X and Y, respectively.

(b) Calculate $E(X)$ and $E(Y)$.

11. Let the joint probability density function of random variables X and Y be given by

$$f(x, y) = \begin{cases} \dfrac{1}{2}ye^{-x} & \text{if } x > 0, \quad 0 < y < 2 \\ 0 & \text{elsewhere.} \end{cases}$$

Find the marginal probability density functions of X and Y.

12. Let X and Y have the joint probability density function

$$f(x, y) = \begin{cases} 1 & \text{if } 0 \leq x \leq 1, \quad 0 \leq y \leq 1 \\ 0 & \text{elsewhere.} \end{cases}$$

Calculate $P(X+Y \leq 1/2)$, $P(X-Y \leq 1/2)$, $P(XY \leq 1/4)$, and $P(X^2+Y^2 \leq 1)$.

13. Let R be the bounded region between $y = x$ and $y = x^2$. A random point (X, Y) is selected from R.

   (a) Find the joint probability density function of X and Y.

   (b) Calculate the marginal probability density functions of X and Y.

   (c) Find $E(X)$ and $E(Y)$.

14. A man invites his fiancée to an elegant hotel for a Sunday brunch. They decide to meet in the lobby of the hotel between 11:30 A.M. and 12 noon. If they arrive at random times during this period, what is the probability that the first to arrive has to wait at least 12 minutes?

15. A farmer makes cuts at two points selected at random on a piece of lumber of length ℓ. What is the expected value of the length of the middle piece?

16. On a line segment AB of length ℓ, two points C and D are placed at random and independently. What is the probability that C is closer to D than to A?

17. Two points X and Y are selected at random and independently from the interval $(0, 1)$. Calculate $P(Y \leq X \text{ and } X^2 + Y^2 \leq 1)$.

B

18. Let X and Y be random variables with finite expectations. Show that if $P(X \leq Y) = 1$, then $E(X) \leq E(Y)$.

19. Suppose that h is the probability density function of a continuous random variable. Let the joint probability density function of two random variables X and Y be given by

$$f(x, y) = h(x)h(y), \qquad x \in \mathbf{R}, \quad y \in \mathbf{R}.$$

Prove that $P(X \geq Y) = 1/2$.

20. Let g and h be two probability density functions with probability distribution functions G and H, respectively. Show that for $-1 \leq \alpha \leq 1$, the function

$$f(x, y) = g(x)h(y)\big(1 + \alpha[2G(x) - 1][2H(y) - 1]\big)$$

is a joint probability density function of two random variables. Moreover, prove that g and h are the marginal probability density functions of f.
Note: This exercise gives an infinite family of joint probability density functions all with the same marginal probability density function of X and of Y.

21. Three points M, N, and L are placed on a circle at random and independently. What is the probability that MNL is an acute angle?

22. Two numbers x and y are selected at random from the interval $(0, 1)$. For $i = 0, 1, 2$, determine the probability that the integer nearest to $x + y$ is i.
Note: This problem was given by Hilton and Pedersen in the paper "A Role for Untraditional Geometry in the Curriculum," published in the December 1989 issue of *Kolloquium Mathematik-Didaktik der Universität Bayreuth*. It is not hard to solve, but it has interesting consequences in arithmetic.

23. A farmer who has two pieces of lumber of lengths a and b ($a < b$) decides to build a pen in the shape of a triangle for his chickens. He sends his son out to cut the lumber and the boy, without taking any thought as to the ultimate purpose, makes two cuts, one on each piece, at randomly selected points. He then chooses three of the resulting pieces at random and take them to his father. What are the chances that they can be used to form a triangular pen?

24. Two points are placed on a segment of length ℓ independently and at random to divide the line into three parts. What is the probability that the length of none of the three parts exceeds a given value α, $\ell/3 \leq \alpha \leq \ell$?

25. A point is selected at random and uniformly from the region

$$R = \{(x, y) \colon |x| + |y| \leq 1\}.$$

Find the probability density function of the x-coordinate of the point selected at random.

26. Let X and Y be continuous random variables with joint probability density function $f(x, y)$. Let $Z = Y/X$, $X \neq 0$. Prove that the probability density function of Z is given by

$$f_Z(z) = \int_{-\infty}^{\infty} |x| f(x, xz) \, dx.$$

27. Consider a disk centered at O with radius R. Suppose that $n \geq 3$ points P_1, P_2, $\ldots$, P_n are independently placed at random inside the disk. Find the probability that all these points are contained in a closed semicircular disk.
Hint: For each $1 \leq i \leq n$, let A_i be the endpoint of the *radius* through P_i and let B_i be the corresponding antipodal point. Let D_i be the closed semicircular disk $O A_i B_i$ with negatively (clockwise) oriented boundary. Note that there is at most one D_i, $1 \leq i \leq n$, that contains all the P_i's.

28. For $\alpha > 0$, $\beta > 0$, and $\gamma > 0$, the following function is called the **bivariate Dirichlet probability density function**

$$f(x, y) = \frac{\Gamma(\alpha + \beta + \gamma)}{\Gamma(\alpha)\Gamma(\beta)\Gamma(\gamma)} x^{\alpha-1} y^{\beta-1} (1 - x - y)^{\gamma-1}$$

if $x \geq 0$, $y \geq 0$, and $x + y \leq 1$; $f(x, y) = 0$, otherwise. Prove that f_X, the marginal probability density function of X, is beta with parameters $(\alpha, \beta + \gamma)$;

and f_Y is beta with the parameters $(\beta, \alpha + \gamma)$.

Hint: Note that

$$\frac{\Gamma(\alpha + \beta + \gamma)}{\Gamma(\alpha)\Gamma(\beta)\Gamma(\gamma)} = \left[\frac{\Gamma(\alpha + \beta + \gamma)}{\Gamma(\alpha)\Gamma(\beta + \gamma)}\right]\left[\frac{\Gamma(\beta + \gamma)}{\Gamma(\beta)\Gamma(\gamma)}\right] = \frac{1}{B(\alpha, \beta + \gamma)}\frac{1}{B(\beta, \gamma)}.$$

29. As Liu Wen from Hebei University of Technology in Tianjin, China, has noted in the April 2001 issue of *The American Mathematical Monthly,* in some reputable probability and statistics texts it has been asserted that "if a two-dimensional distribution function $F(x, y)$ has a continuous density of $f(x, y)$, then

$$f(x, y) = \frac{\partial^2 F(x, y)}{\partial x\, \partial y}. \tag{8.8}$$

Furthermore, some intermediate textbooks in probability and statistics even assert that at a point of continuity for $f(x, y)$, $F(x, y)$ is twice differentiable, and (8.8) holds at that point." Let the joint probability density function of random variables X and Y be given by

$$f(x, y) = \begin{cases} 1 + 2xe^y & \text{if } y \geq 0,\ (-1/2)e^{-y} \leq x \leq 0 \\[2mm] 1 - 2xe^y & \text{if } y \geq 0,\ 0 \leq x \leq (1/2)e^{-y} \\[2mm] 1 + 2xe^{-y} & \text{if } y \leq 0,\ (-1/2)e^y \leq x \leq 0 \\[2mm] 1 - 2xe^{-y} & \text{if } y \leq 0,\ 0 \leq x \leq (1/2)e^y \\[2mm] 0 & \text{otherwise.} \end{cases}$$

Show that even though f is continuous everywhere, the partial derivatives of its distribution function F do not exist at $(0, 0)$. This counterexample is constructed based on a general example given by Liu Wen in the aforementioned paper. ∎

8.2 INDEPENDENT RANDOM VARIABLES

Two random variables X and Y are called **independent** if, for arbitrary subsets A and B of real numbers, the events $\{X \in A\}$ and $\{Y \in B\}$ are independent, that is, if

$$P(X \in A, Y \in B) = P(X \in A)P(Y \in B). \tag{8.9}$$

Using the axioms of probability, we can prove that X and Y are independent if and only if for any two real numbers a and b,

$$P(X \leq a, Y \leq b) = P(X \leq a)P(Y \leq b). \tag{8.10}$$

Hence (8.9) and (8.10) are equivalent.

Relation (8.10) states that X and Y are independent random variables if and only if their joint probability distribution function is the product of their marginal distribution functions. The following theorem states this fact.

Theorem 8.3   *Let X and Y be two random variables defined on the same sample space. If F is the joint probability distribution function of X and Y, then X and Y are independent if and only if for all real numbers t and u,*

$$F(t, u) = F_X(t)F_Y(u).$$

Independence of Discrete Random Variables

If X and Y are discrete random variables with sets of possible values E and F, respectively, and joint probability mass function $p(x, y)$, then the definition of independence is satisfied if for all $x \in E$ and $y \in F$, the events $\{X = x\}$ and $\{Y = y\}$ are independent; that is,

$$P(X = x, Y = y) = P(X = x)P(Y = y). \tag{8.11}$$

Relation (8.11) shows that X and Y are independent if and only if their joint probability mass function is the product of the marginal probability mass functions of X and Y. Therefore, we have the following theorem.

Theorem 8.4   *Let X and Y be two discrete random variables defined on the same sample space. If $p(x, y)$ is the joint probability mass function of X and Y, then X and Y are independent if and only if for all real numbers x and y,*

$$p(x, y) = p_X(x)p_Y(y).$$

Let X and Y be discrete *independent* random variables with sets of possible values A and B, respectively. Then (8.11) implies that for all $x \in A$ and $y \in B$,

$$P(X = x \mid Y = y) = P(X = x)$$

and

$$P(Y = y \mid X = x) = P(Y = y).$$

Hence X and Y are *independent if knowing the value of one of them does not change the probability mass function of the other.*

Example 8.10 Suppose that 4% of the bicycle fenders, produced by a stamping machine from the strips of steel, need smoothing. What is the probability that, of the next 13 bicycle fenders stamped by this machine, two need smoothing and, of the next 20, three need smoothing?

Solution: Let X be the number of bicycle fenders among the first 13 that need smoothing. Let Y be the number of those among the next 7 that need smoothing. We want to calculate $P(X = 2, Y = 1)$. Since X and Y are independent binomial random variables with parameters $(13, 0.04)$ and $(7, 0.04)$, respectively, we can write

$$P(X = 2, Y = 1) = P(X = 2)P(Y = 1)$$

$$= \binom{13}{2}(0.04)^2(0.96)^{11}\binom{7}{1}(0.04)^1(0.96)^6 \approx 0.0175. \quad \blacklozenge$$

The idea of Example 8.10 can be stated in general terms: Suppose that $n + m$ independent Bernoulli trials, each with parameter p, are performed. If X and Y are the number of successes in the first n and in the last m trials, then X and Y are binomial random variables with parameters (n, p) and (m, p), respectively. Furthermore, they are independent because knowing the number of successes in the first n trials does not change the probability mass function of the number of successes in the last m trials. Hence

$$P(X = i, Y = j) = P(X = i)P(Y = j)$$

$$= \binom{n}{i}p^i(1 - p)^{n-i}\binom{m}{j}p^j(1 - p)^{m-j}$$

$$= \binom{n}{i}\binom{m}{j}p^{i+j}(1 - p)^{(n+m)-(i+j)}.$$

We now prove that functions of independent random variables are also independent.

Theorem 8.5 *Let X and Y be independent random variables and $g: \mathbf{R} \to \mathbf{R}$ and $h: \mathbf{R} \to \mathbf{R}$ be real-valued functions; then $g(X)$ and $h(Y)$ are also independent random variables.*

Proof: To show that $g(X)$ and $h(Y)$ are independent, by (8.10) it suffices to prove that, for any two real numbers a and b,

$$P\big(g(X) \leq a, h(Y) \leq b\big) = P\big(g(X) \leq a\big)P\big(h(Y) \leq b\big).$$

Let $A = \{x: g(x) \leq a\}$ and $B = \{y: h(y) \leq b\}$. Clearly, $x \in A$ if and only if $g(x) \leq a$, and $y \in B$ if and only if $h(y) \leq b$. Therefore,

$$P\big(g(X) \leq a, h(Y) \leq b\big) = P(X \in A, Y \in B) = P(X \in A)P(Y \in B)$$

$$= P\big(g(X) \leq a\big)P\big(h(Y) \leq b\big). \quad \blacklozenge$$

By this theorem, if X and Y are independent random variables, then sets such as $\{X^2, Y\}$, $\{\sin X, e^Y\}$, $\{X^2 - 2X, Y^3 + 3Y\}$ are sets of independent random variables.

Another important property of independent random variables is that the expected value of their product is equal to the product of their expected values. To prove this, we use Theorem 8.1.

Theorem 8.6   *Let X and Y be independent random variables. Then for all real-valued functions $g: \mathbf{R} \to \mathbf{R}$ and $h: \mathbf{R} \to \mathbf{R}$,*

$$E[g(X)h(Y)] = E[g(X)]E[h(Y)],$$

where, as usual, we assume that $E[g(X)]$ and $E[h(Y)]$ are finite.

Solution: Let A be the set of possible values of X, and B be the set of possible values for Y. Let $p(x, y)$ be the joint probability mass function of X and Y. Then

$$E[g(X)h(Y)] = \sum_{x \in A} \sum_{y \in B} g(x)h(y)p(x, y) = \sum_{x \in A} \sum_{y \in B} g(x)h(y)p_X(x)p_Y(y)$$

$$= \sum_{x \in A} \left[g(x)p_X(x) \sum_{y \in B} h(y)p_Y(y) \right] = \sum_{x \in A} g(x)p_X(x)E[h(Y)]$$

$$= E[h(Y)] \sum_{x \in A} g(x)p_X(x) = E[h(Y)]E[g(X)]. \quad \blacklozenge$$

By this theorem, if X and Y are independent, then $E(XY) = E(X)E(Y)$, and relations such as the following are valid.

$$E(X^2|Y|) = E(X^2)E(|Y|),$$

$$E[(\sin X)e^Y] = E(\sin X)E(e^Y),$$

$$E[(X^2 - 2X)(Y^3 + 3Y)] = E(X^2 - 2X)E(Y^3 + 3Y).$$

However, the converse of Theorem 8.6 is not necessarily true. That is, two random variables X and Y might be dependent while $E(XY) = E(X)E(Y)$. Here is an example:

Example 8.11   Let X be a random variable with the set of possible values $\{-1, 0, 1\}$ and probability mass function $p(-1) = p(0) = p(1) = 1/3$. Letting $Y = X^2$, we have

$$E(X) = -1 \cdot \frac{1}{3} + 0 \cdot \frac{1}{3} + 1 \cdot \frac{1}{3} = 0,$$

$$E(Y) = E(X^2) = (-1)^2 \cdot \frac{1}{3} + 0^2 \cdot \frac{1}{3} + (1)^2 \cdot \frac{1}{3} = \frac{2}{3},$$

$$E(XY) = E(X^3) = (-1)^3 \cdot \frac{1}{3} + 0^3 \cdot \frac{1}{3} + (1)^3 \cdot \frac{1}{3} = 0.$$

Thus $E(XY) = E(X)E(Y)$ while, clearly, X and Y are dependent. $\blacklozenge$

Independence of Continuous Random Variables

For jointly continuous *independent* random variables X and Y having joint probability distribution function $F(x, y)$ and joint probability density function $f(x, y)$, Theorem 8.3 implies that, for any two real numbers x and y,

$$F(x, y) = F_X(x)F_Y(y). \tag{8.12}$$

Differentiating (8.12) with respect to x, we get

$$\frac{\partial F}{\partial x} = f_X(x)F_Y(y),$$

which upon differentiation with respect to y yields

$$\frac{\partial^2 F}{\partial y \partial x} = f_X(x)f_Y(y),$$

or, equivalently,

$$f(x, y) = f_X(x)f_Y(y). \tag{8.13}$$

We leave it as an exercise that, if (8.13) is valid, then X and Y are independent. We have the following theorem:

Theorem 8.7   *Let X and Y be jointly continuous random variables with joint proba-bility density function $f(x, y)$. Then X and Y are independent if and only if $f(x, y)$ is the product of their marginal densities $f_X(x)$ and $f_Y(y)$.*

Example 8.12   Stores A and B, which belong to the same owner, are located in two different towns. If the probability density function of the weekly profit of each store, in thousands of dollars, is given by

$$f(x) = \begin{cases} x/4 & \text{if } 1 < x < 3 \\ 0 & \text{otherwise,} \end{cases}$$

and the profit of one store is independent of the other, what is the probability that next week one store makes at least \$500 more than the other store?

Solution: Let X and Y denote next week's profits of A and B, respectively. The desired probability is $P(X > Y + 1/2) + P(Y > X + 1/2)$. Since X and Y have the same probability density function, by symmetry, this sum equals $2P(X > Y + 1/2)$. To calculate this, we need to know $f(x, y)$, the joint probability density function of X and Y. Since X and Y are independent,

$$f(x, y) = f_X(x)f_Y(y),$$

where

$$f_X(x) = \begin{cases} x/4 & \text{if } 1 < x < 3 \\ 0 & \text{otherwise,} \end{cases} \qquad f_Y(y) = \begin{cases} y/4 & \text{if } 1 < y < 3 \\ 0 & \text{otherwise.} \end{cases}$$

Thus

$$f(x, y) = \begin{cases} xy/16 & \text{if } 1 < x < 3,\ 1 < y < 3 \\ 0 & \text{otherwise.} \end{cases}$$

The desired probability, as seen from Figure 8.4, is therefore equal to

$$2P\left(X > Y + \frac{1}{2}\right) = 2P\left((X, Y) \in \left\{(x, y): \frac{3}{2} < x < 3,\ 1 < y < x - \frac{1}{2}\right\}\right)$$

$$= 2\int_{3/2}^{3} \left(\int_{1}^{x-1/2} \frac{xy}{16} dy\right) dx = \frac{1}{8}\int_{3/2}^{3} \left[\frac{xy^2}{2}\right]_{1}^{x-1/2} dx$$

$$= \frac{1}{16}\int_{3/2}^{3} x\left[\left(x - \frac{1}{2}\right)^2 - 1\right] dx$$

$$= \frac{1}{16}\int_{3/2}^{3} \left(x^3 - x^2 - \frac{3}{4}x\right) dx$$

$$= \frac{1}{16}\left[\frac{1}{4}x^4 - \frac{1}{3}x^3 - \frac{3}{8}x^2\right]_{3/2}^{3} = \frac{549}{1024} \approx 0.54. \quad \blacklozenge$$

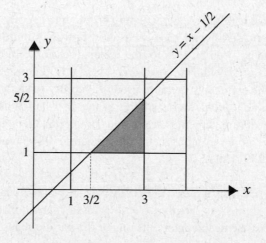

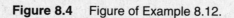

Figure 8.4   Figure of Example 8.12.

Example 8.13   A point is selected at random from the rectangle

$$R = \{(x, y) \in \mathbf{R}^2 : 0 < x < a, \ 0 < y < b\}.$$

Let X be the x-coordinate and Y be the y-coordinate of the point selected. Determine if X and Y are independent random variables.

Solution: From Section 8.1 we know that $f(x, y)$, the joint probability density function of X and Y, is given by

$$f(x, y) = \begin{cases} \dfrac{1}{\text{area}(R)} = \dfrac{1}{ab} & \text{if } (x, y) \in R \\ \\ 0 & \text{elsewhere.} \end{cases}$$

Now the marginal density functions of X and Y, f_X and f_Y, are given by

$$f_X(x) = \int_0^b \frac{1}{ab}dy = \frac{1}{a}, \qquad x \in (0, a),$$

$$f_Y(y) = \int_0^a \frac{1}{ab}dx = \frac{1}{b}, \qquad y \in (0, b).$$

Therefore, $f(x, y) = f_X(x)f_Y(y)$, $\forall x, y \in \mathbf{R}$, and hence X and Y are independent.
◆

We now explain one of the most interesting problems of geometric probability, **Buffon's needle problem.** In Chapter 13, we will show how the solution of this problem and the Monte Carlo method can be used to find estimations for π by simulation. Georges Louis Buffon (1707–1784), who proposed and solved this famous problem, was a French naturalist who, in the eighteenth century, used probability to study natural phenomena. In addition to the needle problem, his studies of the distribution and expectation of the remaining lifetimes of human beings are famous among mathematicians. These works, together with many more, are published in his gigantic 44-volume, *Histoire Naturelle* (Natural History; 1749–1804).

Example 8.14   (Buffon's Needle Problem)   A plane is ruled with parallel lines a distance d apart. A needle of length ℓ, $\ell < d$, is tossed at random onto the plane. What is the probability that the needle intersects one of the parallel lines?

Solution: Let X denote the distance from the center of the needle to the closest line, and let Θ denote the angle between the needle and the line. The position of the needle is completely determined by the coordinates X and Θ.

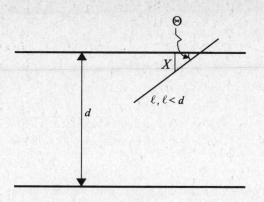

Figure 8.5   Buffon's needle problem.

As Figure 8.5 shows, the needle intersects the line if and only if the length of the hypotenuse of the triangle, $X / \sin \Theta$, is less than $\ell/2$. Since X and Θ are independent uniform random variables over the intervals $(0, d/2)$ and $(0, \pi)$, respectively, the probability that the needle intersects a line is $\iint_R f(x, \theta) \, dx \, d\theta$, where $f(x, \theta)$ is the joint probability density function of X and Θ and is given by

$$f(x, \theta) = f_X(x) f_\Theta(\theta) = \begin{cases} \dfrac{2}{\pi d} & \text{if } 0 \leq \theta \leq \pi, \ 0 \leq x \leq \dfrac{d}{2}. \\ 0 & \text{elsewhere} \end{cases}$$

and

$$R = \left\{ (x, \theta) : \frac{x}{\sin \theta} < \frac{\ell}{2} \right\} = \left\{ (x, \theta) : x < \frac{\ell}{2} \sin \theta \right\}.$$

Thus the desired probability is

$$\iint_R f(x, \theta) \, dx \, d\theta = \int_0^\pi \int_0^{(\ell/2) \sin \theta} \frac{2}{\pi d} \, dx \, d\theta = \frac{\ell}{\pi d} \int_0^\pi \sin \theta \, d\theta = \frac{2 \ell}{\pi d}. \quad \blacklozenge$$

Example 8.15   Prove that two random variables X and Y with the following joint probability density function are not independent.

$$f(x, y) = \begin{cases} 8xy & 0 \leq x \leq y \leq 1 \\ 0 & \text{otherwise.} \end{cases}$$

Solution: To check the validity of (8.13), we first calculate f_X and f_Y.

$$f_X(x) = \int_x^1 8xy\,dy = 4x(1 - x^2), \qquad\qquad 0 \le x \le 1,$$

$$f_Y(y) = \int_0^y 8xy\,dx = 4y^3, \qquad\qquad 0 \le y \le 1.$$

Now since $f(x, y) \ne f_X(x) f_Y(y)$, X and Y are dependent. This is expected because of the range $0 \le x \le y \le 1$, which is not a Cartesian product of one-dimensional regions.
◆

The results of Theorems 8.5 and 8.6 are valid for continuous random variables as well:

> Let X and Y be independent continuous random variables and $g: \mathbf{R} \to \mathbf{R}$ and $h: \mathbf{R} \to \mathbf{R}$ be real-valued functions; then $g(X)$ and $h(Y)$ are also independent random variables.

The proof of this fact is identical to the proof of Theorem 8.5. The continuous analog of Theorem 8.6 and its proof are as follows:

> Let X and Y be independent continuous random variables. Then for all real-valued functions $g: \mathbf{R} \to \mathbf{R}$ and $h: \mathbf{R} \to \mathbf{R}$,
>
> $$E[g(X)h(Y)] = E[g(X)]E[h(Y)],$$
>
> where, as usual, we assume that $E[g(X)]$ and $E[h(Y)]$ are finite.

Proof: Let $f(x, y)$ be the joint probability density function of X and Y. Then

$$
\begin{aligned}
E[g(X)h(Y)] &= \int_{-\infty}^{\infty} \int_{-\infty}^{\infty} g(x)h(y)f(x, y)\,dx\,dy \\
&= \int_{-\infty}^{\infty} \int_{-\infty}^{\infty} g(x)h(y)f_X(x)f_Y(y)\,dx\,dy \\
&= \int_{-\infty}^{\infty} h(y)f_Y(y)\left(\int_{-\infty}^{\infty} g(x)f_X(x)\,dx \right) dy \\
&= \left(\int_{-\infty}^{\infty} g(x)f_X(x)\,dx \right)\left(\int_{-\infty}^{\infty} h(y)f_Y(y)\,dy \right) \\
&= E[g(X)]E[h(Y)]. \quad ◆
\end{aligned}
$$

Again, by this theorem, if X and Y are independent, then $E(XY) = E(X)E(Y)$. As we know from the discrete case, the converse of this fact is not *necessarily* true. (See Example 8.11.)

EXERCISES

A

1. Let the joint probability mass function of random variables X and Y be given by

$$p(x, y) = \begin{cases} \dfrac{1}{25}(x^2 + y^2) & \text{if } x = 1, 2, \quad y = 0, 1, 2 \\ \\ 0 & \text{elsewhere.} \end{cases}$$

Are X and Y independent? Why or why not?

2. Let the joint probability mass function of random variables X and Y be given by

$$p(x, y) = \begin{cases} \dfrac{1}{7}x^2 y & \text{if } (x, y) = (1, 1), (1, 2), (2, 1) \\ \\ 0 & \text{elsewhere.} \end{cases}$$

Are X and Y independent? Why or why not?

3. Let X and Y be independent random variables each having the probability mass function

$$p(x) = \frac{1}{2}\left(\frac{2}{3}\right)^x, \qquad x = 1, 2, 3, \dots .$$

Find $P(X = 1, Y = 3)$ and $P(X + Y = 3)$.

4. From an ordinary deck of 52 cards, eight cards are drawn at random and without replacement. Let X and Y be the number of clubs and spades, respectively. Are X and Y independent?

5. What is the probability that there are exactly two girls among the first seven and exactly four girls among the first 15 babies born in a hospital in a given week? Assume that the events that a child born is a girl or is a boy are equiprobable.

6. Let X and Y be two independent random variables with distribution functions F and G, respectively. Find the distribution functions of $\max(X, Y)$ and $\min(X, Y)$.

7. A fair coin is tossed n times by Adam and n times by Andrew. What is the probability that they get the same number of heads?

8. The joint probability mass function $p(x, y)$ of the random variables X and Y is given by the following table. Determine if X and Y are independent.

		y		
x	0	1	2	3
0	0.1681	0.1804	0.0574	0.0041
1	0.1804	0.1936	0.0616	0.0044
2	0.0574	0.0616	0.0196	0.0014
3	0.0041	0.0044	0.0014	0.0001

9. Let the joint probability density function of random variables X and Y be given by

$$f(x, y) = \begin{cases} 2 & \text{if } 0 \le y \le x \le 1 \\ 0 & \text{elsewhere.} \end{cases}$$

Are X and Y independent? Why or why not?

10. Suppose that the amount of cholesterol in a certain type of sandwich is $100X$ milligrams, where X is a random variable with the following density function:

$$f(x) = \begin{cases} \dfrac{2x + 3}{18} & \text{if } 2 < x < 4 \\ 0 & \text{otherwise.} \end{cases}$$

Find the probability that two such sandwiches made independently have the same amount of cholesterol.

11. Let the joint probability density function of random variables X and Y be given by

$$f(x, y) = \begin{cases} x^2 e^{-x(y+1)} & \text{if } x \ge 0, \quad y \ge 0 \\ 0 & \text{elsewhere.} \end{cases}$$

Are X and Y independent? Why or why not?

12. Let the joint probability density function of X and Y be given by

$$f(x, y) = \begin{cases} 8xy & \text{if } 0 \le x < y \le 1 \\ 0 & \text{otherwise.} \end{cases}$$

Determine if $E(XY) = E(X)E(Y)$.

13. Let the joint probability density function of X and Y be given by

$$f(x, y) = \begin{cases} 2e^{-(x+2y)} & \text{if } x \geq 0, \quad y \geq 0 \\ 0 & \text{otherwise.} \end{cases}$$

Find $E(X^2Y)$.

14. Let X and Y be two independent random variables with the same probability density function given by

$$f(x) = \begin{cases} e^{-x} & \text{if } 0 < x < \infty \\ 0 & \text{elsewhere.} \end{cases}$$

Show that g, the probability density function of X/Y, is given by

$$g(t) = \begin{cases} \dfrac{1}{(1+t)^2} & \text{if } 0 < t < \infty \\ 0 & t \leq 0. \end{cases}$$

15. Let X and Y be independent exponential random variables both with mean 1. Find $E[\max(X, Y)]$.

16. Let X and Y be independent random points from the interval $(-1, 1)$. Find $E[\max(X, Y)]$.

17. Let X and Y be independent random points from the interval $(0, 1)$. Find the probability density function of the random variable XY.

18. A point is selected at random from the disk

$$R = \{(x, y) \in \mathbf{R}^2 : x^2 + y^2 \leq 1\}.$$

Let X be the x-coordinate and Y be the y-coordinate of the point selected. Determine if X and Y are independent random variables.

19. Six brothers and sisters who are all either under 10 or in their early teens are having dinner with their parents and four grandparents. Their mother unintentionally feeds the entire family (including herself) a type of poisonous mushrooms that makes 20% of the adults and 30% of the children sick. What is the probability that more adults than children get sick?

20. The lifetimes of mufflers manufactured by company A are random with the following density function:

$$f(x) = \begin{cases} \dfrac{1}{6}e^{-x/6} & \text{if } x > 0 \\ 0 & \text{elsewhere.} \end{cases}$$

The lifetimes of mufflers manufactured by company B are random with the following density function:

$$g(y) = \begin{cases} \dfrac{2}{11}e^{-2y/11} & \text{if } y > 0 \\ \\ 0 & \text{elsewhere.} \end{cases}$$

Elizabeth buys two mufflers, one from company A and the other one from company B and installs them on her cars at the same time. What is the probability that the muffler of company B outlasts that of company A?

B

21. Let E be an event; the random variable I_E, defined as follows, is called the **indicator** of E:

$$I_E = \begin{cases} 1 & \text{if } E \text{ occurs} \\ 0 & \text{otherwise.} \end{cases}$$

Show that A and B are independent events if and only if I_A and I_B are independent random variables.

22. Let B and C be two independent random variables both having the following density function:

$$f(x) = \begin{cases} \dfrac{3x^2}{26} & \text{if } 1 < x < 3 \\ \\ 0 & \text{otherwise.} \end{cases}$$

What is the probability that the quadratic equation $X^2 + BX + C = 0$ has two real roots?

23. Let the joint probability density function of two random variables X and Y satisfy

$$f(x, y) = g(x)h(y), \qquad -\infty < x < \infty, \quad -\infty < y < \infty,$$

where g and h are two functions from $\mathbf{R}$ to $\mathbf{R}$. Show that X and Y are independent.

24. Let X and Y be two independent random points from the interval $(0, 1)$. Calculate the probability distribution function and the probability density function of $\max(X, Y)/\min(X, Y)$.

25. Suppose that X and Y are independent, identically distributed exponential random variables with mean $1/\lambda$. Prove that $X/(X + Y)$ is uniform over $(0, 1)$.

26. Let $f(x, y)$ be the joint probability density function of two continuous random variables; f is called **circularly symmetrical** if it is a function of $\sqrt{x^2 + y^2}$, the distance of (x, y) from the origin; that is, if there exists a function φ so that $f(x, y) = \varphi(\sqrt{x^2 + y^2})$. Prove that if X and Y are independent random variables, their joint probability density function is circularly symmetrical if and only if they are both normal with mean 0 and equal variance.

Hint: Suppose that f is circularly symmetrical; then

$$f_X(x) f_Y(y) = \varphi(\sqrt{x^2 + y^2}).$$

Differentiating this relation with respect to x yields

$$\frac{\varphi'(\sqrt{x^2 + y^2})}{\varphi(\sqrt{x^2 + y^2})\sqrt{x^2 + y^2}} = \frac{f_X'(x)}{x f_X(x)}.$$

This implies that both sides are constants, so that

$$\frac{f_X'(x)}{x f_X(x)} = k$$

for some constant k. Solve this and use the fact that f_X is a probability density function to show that f_X is normal. Repeat the same procedure for f_Y. ∎

8.3   CONDITIONAL DISTRIBUTIONS

Let X and Y be two discrete or two continuous random variables. In this chapter we study the distribution function and the expected value of the random variable X given that $Y = y$.

Conditional Distributions: Discrete Case

Let X be a discrete random variable with set of possible values A, and let Y be a discrete random variable with set of possible values B. Let $p(x, y)$ be the joint probability mass function of X and Y, and let p_X and p_Y be the marginal probability mass functions of X and Y. When no information is given about the value of Y,

$$p_X(x) = P(X = x) = \sum_{y \in B} P(X = x, Y = y) = \sum_{y \in B} p(x, y) \tag{8.14}$$

is used to calculate the probabilities of events concerning X. However, if the value of Y is known, then instead of $p_X(x)$, the **conditional probability mass function of X given that $Y = y$** is used. This function, denoted by $p_{X|Y}(x|y)$, is defined as follows:

$$p_{X|Y}(x|y) = P(X = x \mid Y = y) = \frac{P(X = x, Y = y)}{P(Y = y)} = \frac{p(x, y)}{p_Y(y)},$$

where $x \in A$, $y \in B$, and $p_Y(y) > 0$. Hence, **for $x \in A$, $y \in B$, and $p_Y(y) > 0$,**

$$p_{X|Y}(x|y) = \frac{p(x, y)}{p_Y(y)}. \tag{8.15}$$

Note that

$$\sum_{x \in A} p_{X|Y}(x|y) = \sum_{x \in A} \frac{p(x, y)}{p_Y(y)} = \frac{1}{p_Y(y)} \sum_{x \in A} p(x, y) = \frac{1}{p_Y(y)} p_Y(y) = 1.$$

Hence for any fixed $y \in B$, $p_{X|Y}(x|y)$ is itself a probability mass function with the set of possible values A. If X and Y are independent, $p_{X|Y}$ coincides with p_X because

$$p_{X|Y}(x|y) = \frac{p(x, y)}{p_Y(y)} = \frac{P(X = x, Y = y)}{P(Y = y)} = \frac{P(X = x)P(Y = y)}{P(Y = y)}$$

$$= P(X = x) = p_X(x).$$

Similar to $p_{X|Y}(x|y)$, the **conditional distribution function of X, given that $Y = y$** is defined as follows:

$$F_{X|Y}(x|y) = P(X \le x \mid Y = y) = \sum_{t \le x} P(X = t \mid Y = y) = \sum_{t \le x} p_{X|Y}(t|y).$$

Example 8.16   Let the joint probability mass function of X and Y be given by

$$p(x, y) = \begin{cases} \dfrac{1}{15}(x + y) & \text{if } x = 0, 1, 2, \ y = 1, 2 \\ 0 & \text{otherwise.} \end{cases}$$

Find $p_{X|Y}(x|y)$ and $P(X = 0 \mid Y = 2)$.

Solution: To use $p_{X|Y}(x|y) = p(x, y)/p_Y(y)$, we must first calculate $p_Y(y)$.

$$p_Y(y) = \sum_{x=0}^{2} p(x, y) = \frac{1 + y}{5}.$$

Therefore,

$$p_{X|Y}(x|y) = \frac{(x + y)/15}{(1 + y)/5} = \frac{x + y}{3(1 + y)}, \quad x = 0, 1, 2 \text{ when } y = 1 \text{ or } 2.$$

In particular, $P(X = 0 \mid Y = 2) = \dfrac{0 + 2}{3(1 + 2)} = \dfrac{2}{9}$.   ◆

Example 8.17   Let $N(t)$ be the number of males who enter a certain post office at or prior to time t. Let $M(t)$ be the number of females who enter a certain post office at or prior to t. Suppose that $\{N(t): t \geq 0\}$ and $\{M(t): t \geq 0\}$ are independent Poisson processes with rates λ and μ, respectively. So for all $t > 0$ and $s > 0$, $N(t)$ is independent of $M(s)$. If at some instant t, $N(t) + M(t) = n$, what is the conditional probability mass function of $N(t)$?

Solution: For simplicity, let $K(t) = N(t) + M(t)$ and $p(x, n)$ be the joint probability mass function of $N(t)$ and $K(t)$. Then $p_{N(t)|K(t)}(x|n)$, the desired probability mass function, is found as follows:

$$p_{N(t)|K(t)}(x|n) = \frac{p(x, n)}{p_{K(t)}(n)} = \frac{P\big(N(t) = x, K(t) = n\big)}{P\big(K(t) = n\big)}$$

$$= \frac{P\big(N(t) = x, M(t) = n - x\big)}{P\big(K(t) = n\big)}.$$

Since $N(t)$ and $M(t)$ are independent random variables and $K(t) = N(t) + M(t)$ is a Poisson random variable with rate $\lambda t + \mu t$ (accept this for now; we will prove it in Theorem 11.5),

$$p_{N(t)|K(t)}(x|n) = \frac{P\big(N(t) = x\big)P\big(M(t) = n - x\big)}{P\big(K(t) = n\big)} = \frac{\dfrac{e^{-\lambda t}(\lambda t)^x}{x!}\dfrac{e^{-\mu t}(\mu t)^{n-x}}{(n - x)!}}{\dfrac{e^{-(\lambda t + \mu t)}(\lambda t + \mu t)^n}{n!}}$$

$$= \frac{n!}{x!\,(n - x)!}\frac{\lambda^x \mu^{n-x}}{(\lambda + \mu)^n} = \binom{n}{x}\Big(\frac{\lambda}{\lambda + \mu}\Big)^x\Big(\frac{\mu}{\lambda + \mu}\Big)^{n-x}$$

$$= \binom{n}{x}\Big(\frac{\lambda}{\lambda + \mu}\Big)^x\Big(1 - \frac{\lambda}{\lambda + \mu}\Big)^{n-x}, \qquad x = 0, 1, \dots, n.$$

This shows that the conditional probability mass function of the number of males who have entered the post office at or prior to t, given that altogether n persons have entered the post office during this period, is binomial with parameters n and $\lambda/(\lambda + \mu)$.   ◆

We now generalize the concept of mathematical expectation to the conditional case. Let X and Y be discrete random variables, and let the set of possible values of X be A. The **conditional expectation of the random variable X given that $Y = y$** is as follows:

$$E(X \mid Y = y) = \sum_{x \in A} x P(X = x \mid Y = y) = \sum_{x \in A} x p_{X|Y}(x|y),$$

where $p_Y(y) > 0$. Hence, by definition, for $p_Y(y) > 0$,

$$E(X \mid Y = y) = \sum_{x \in A} x p_{X|Y}(x|y). \tag{8.16}$$

Note that conditional expectations are simply ordinary expectations computed relative to conditional distributions. For this reason, they satisfy the same properties that ordinary expectations do. For example, if h is an ordinary function from $\mathbf{R}$ to $\mathbf{R}$, then for the discrete random variables X and Y with set of possible values A for X, the expected value of $h(X)$ is obtained from

$$E\big[h(X) \mid Y = y\big] = \sum_{x \in A} h(x) p_{X|Y}(x|y). \qquad (8.17)$$

Example 8.18   Calculate the expected number of aces in a randomly selected poker hand that is found to have exactly two jacks.

Solution: Let X and Y be the number of aces and jacks in a random poker hand, respectively. Then

$$E(X \mid Y = 2) = \sum_{x=0}^{3} x p_{X|Y}(x|2) = \sum_{x=0}^{3} x \frac{p(x, 2)}{p_Y(2)}$$

$$= \sum_{x=0}^{3} x \frac{\dfrac{\dbinom{4}{2}\dbinom{4}{x}\dbinom{44}{3-x}}{\dbinom{52}{5}}}{\dfrac{\dbinom{4}{2}\dbinom{48}{3}}{\dbinom{52}{5}}} = \sum_{x=0}^{3} x \frac{\dbinom{4}{x}\dbinom{44}{3-x}}{\dbinom{48}{3}} \approx 0.25. \quad \blacklozenge$$

Example 8.19   While rolling a balanced die successively, the first 6 occurred on the third roll. What is the expected number of rolls until the first 1?

Solution: Let X and Y be the number of rolls until the first 1 and the first 6, respectively. The required quantity, $E(X \mid Y = 3)$, is calculated from

$$E(X \mid Y = 3) = \sum_{x=1}^{\infty} x p_{X|Y}(x|3) = \sum_{x=1}^{\infty} x \frac{p(x, 3)}{p_Y(3)}.$$

Clearly,

$$p_Y(3) = \Big(\frac{5}{6}\Big)^2\Big(\frac{1}{6}\Big) = \frac{25}{216}, \qquad p(1, 3) = \Big(\frac{1}{6}\Big)\Big(\frac{5}{6}\Big)\Big(\frac{1}{6}\Big) = \frac{5}{216},$$

$$p(2, 3) = \Big(\frac{4}{6}\Big)\Big(\frac{1}{6}\Big)\Big(\frac{1}{6}\Big) = \frac{4}{216}, \qquad p(3, 3) = 0,$$

and for $x > 3$,

$$p(x, 3) = \left(\frac{4}{6}\right)^2 \left(\frac{1}{6}\right) \left(\frac{5}{6}\right)^{x-4} \left(\frac{1}{6}\right) = \frac{1}{81} \left(\frac{5}{6}\right)^{x-4}.$$

Therefore,

$$E(X \mid Y = 3) = \sum_{x=1}^{\infty} x \frac{p(x, 3)}{p_Y(3)} = \frac{216}{25} \sum_{x=1}^{\infty} x p(x, 3)$$

$$= \frac{216}{25} \left[1 \cdot \frac{5}{216} + 2 \cdot \frac{4}{216} + \sum_{x=4}^{\infty} x \cdot \frac{1}{81} \left(\frac{5}{6}\right)^{x-4} \right]$$

$$= \frac{13}{25} + \frac{8}{75} \sum_{x=4}^{\infty} x \left(\frac{5}{6}\right)^{x-4}$$

$$= \frac{13}{25} + \frac{8}{75} \sum_{x=4}^{\infty} (x - 4 + 4) \left(\frac{5}{6}\right)^{x-4}$$

$$= \frac{13}{25} + \frac{8}{75} \left[\sum_{x=4}^{\infty} (x - 4) \left(\frac{5}{6}\right)^{x-4} + 4 \sum_{x=4}^{\infty} \left(\frac{5}{6}\right)^{x-4} \right]$$

$$= \frac{13}{25} + \frac{8}{75} \left[\sum_{x=0}^{\infty} x \left(\frac{5}{6}\right)^{x} + 4 \sum_{x=0}^{\infty} \left(\frac{5}{6}\right)^{x} \right]$$

$$= \frac{13}{25} + \frac{8}{75} \left[\frac{5/6}{(1 - 5/6)^2} + 4 \left(\frac{1}{1 - 5/6}\right) \right] = 6.28. \quad \blacklozenge$$

Example 8.20   (The Box Problem: To Switch or Not to Switch)   Suppose that
there are two identical closed boxes, one containing twice as much money as the other.
David is asked to choose one of these boxes for himself. He picks a box at random and
opens it. If, after observing the content of the box, he has the option to exchange it for
the other box, should he exchange or not?

Solution: Suppose that David finds $\$x$ in the box. Then the other box has $\$(2x)$ with
probability 1/2 and $\$(x/2)$ with probability 1/2. Therefore, the expected value of the
dollar amount in the other box is

$$\frac{x}{2} \cdot \frac{1}{2} + 2x \cdot \frac{1}{2} = 1.25x.$$

This implies that David should switch no matter how much he finds in the box he picks
at random. But as Steven J. Brams and D. Marc Kilgour observe in their paper "The Box
Problem: To Switch or Not to Switch" (*Mathematics Magazine*, February 1995, Volume
68, Number 1),

It seems paradoxical that it would always be better to switch to the second box, no matter how much money you found in the first one.

They go on to explain,

What is needed to determine whether or not switching is worthwhile is some prior notion of the amount of money to be found in each box. For particular, knowing the prior probability distribution would enable you to calculate whether the amount found in the first box is less than the expected value of what is in the other box—and, therefore, whether a switch is profitable.

Let L be the event that David picked the box with the larger amount. Let X be the dollar amount David finds in the box he picks. According to a general exchange condition, discovered by Brams and Kilgour, on the average, an exchange is profitable if and only if for the observed value x,

$$P(L \mid X = x) < 2/3.$$

That is, David should switch if and only if the conditional probability that he picked the box with the larger amount given that he found $\$x$ is less than 2/3. To show this, let Y be the amount in the box David did not pick, and let S be the event that David picked the box with the smaller amount. Then

$$E(Y \mid X = x) = \frac{x}{2} \cdot P(L \mid X = x) + 2x \cdot P(S \mid X = x).$$

On average, exchange is profitable if and only if $E(Y \mid X = x) > x$, or, equivalently, if and only if

$$\frac{x}{2} \cdot P(L \mid X = x) + 2x \cdot P(S \mid X = x) > x.$$

Given that $P(S \mid X = x) = 1 - P(L \mid X = x)$, this relation yields

$$\frac{1}{2} P(L \mid X = x) + 2\big[1 - P(L \mid X = x)\big] > 1,$$

which implies that $P(L \mid X = x) < 2/3$.

As an example, suppose that the box with the larger amount contains $\$1$, $\$2$, $\$4$, or $\$8$ with equal probabilities. Then, for $x = 1/2, 1, 2$, and 4, David should switch , for $x = 8$ he should not. This is because, for example, by Bayes' formula (Theorem 3.5),

$$P(L \mid X = 2) = \frac{P(X = 2 \mid L)P(L)}{P(X = 2 \mid L)P(L) + P(X = 2 \mid S)P(S)}$$

$$= \frac{\dfrac{1}{4} \times \dfrac{1}{2}}{\dfrac{1}{4} \times \dfrac{1}{2} + \dfrac{1}{4} \times \dfrac{1}{2}} = \frac{1}{2} < \frac{2}{3},$$

whereas $P(L \mid X = 8) = 1 > 2/3$. ◆

Conditional Distributions: Continuous Case

Now let X and Y be two continuous random variables with the joint probability density function $f(x, y)$. Again, when no information is given about the value of Y, $f_X(x) = \int_{-\infty}^{\infty} f(x, y)\, dy$ is used to calculate the probabilities of events concerning X. However, when the value of Y is known, to find such probabilities, $f_{X|Y}(x|y)$, the **conditional probability density function of X given that $Y = y$** is used. Similar to the discrete case, $f_{X|Y}(x|y)$ is defined as follows:

$$f_{X|Y}(x|y) = \frac{f(x, y)}{f_Y(y)}, \tag{8.18}$$

provided that $f_Y(y) > 0$. Note that

$$\int_{-\infty}^{\infty} f_{X|Y}(x|y)\, dx = \int_{-\infty}^{\infty} \frac{f(x, y)}{f_Y(y)}\, dx = \frac{1}{f_Y(y)} \int_{-\infty}^{\infty} f(x, y)\, dx = \frac{1}{f_Y(y)} f_Y(y) = 1,$$

showing that for a fixed y, $f_{X|Y}(x|y)$ is itself a probability density function. If X and Y are independent, then $f_{X|Y}$ coincides with f_X because

$$f_{X|Y}(x|y) = \frac{f(x, y)}{f_Y(y)} = \frac{f_X(x) f_Y(y)}{f_Y(y)} = f_X(x).$$

Similarly, the **conditional probability density function of Y given that $X = x$** is defined by

$$f_{Y|X}(y|x) = \frac{f(x, y)}{f_X(x)}, \tag{8.19}$$

provided that $f_X(x) > 0$. Also, as we expect, $F_{X|Y}(x|y)$, the **conditional probability distribution function of X given that $Y = y$** is defined as follows:

$$F_{X|Y}(x|y) = P(X \leq x \mid Y = y) = \int_{-\infty}^{x} f_{X|Y}(t|y)\, dt.$$

Therefore,

$$\frac{d}{dt} F_{X|Y}(x|y) = f_{X|Y}(x|y). \tag{8.20}$$

Example 8.21   Let X and Y be continuous random variables with joint probability density function

$$f(x, y) = \begin{cases} \dfrac{3}{2}(x^2 + y^2) & \text{if } 0 < x < 1, \quad 0 < y < 1 \\[2mm] 0 & \text{otherwise.} \end{cases}$$

Find $f_{X|Y}(x|y)$.

Solution: By definition, $f_{X|Y}(x|y) = \dfrac{f(x, y)}{f_Y(y)}$, where

$$f_Y(y) = \int_{-\infty}^{\infty} f(x, y)\, dx = \int_0^1 \frac{3}{2}(x^2 + y^2)\, dx = \frac{3}{2}y^2 + \frac{1}{2}.$$

Thus

$$f_{X|Y}(x|y) = \frac{3/2(x^2 + y^2)}{(3/2)y^2 + 1/2} = \frac{3(x^2 + y^2)}{3y^2 + 1}$$

for $0 < x < 1$ and $0 < y < 1$. Everywhere else, $f_{X|Y}(x|y) = 0$. ◆

Example 8.22 First, a point Y is selected at random from the interval $(0, 1)$. Then another point X is chosen at random from the interval $(0, Y)$. Find the probability density function of X.

Solution: Let $f(x, y)$ be the joint probability density function of X and Y. Then

$$f_X(x) = \int_{-\infty}^{\infty} f(x, y)\, dy,$$

where from $f_{X|Y}(x|y) = \dfrac{f(x, y)}{f_Y(y)}$, we obtain

$$f(x, y) = f_{X|Y}(x|y) f_Y(y).$$

Therefore,

$$f_X(x) = \int_{-\infty}^{\infty} f_{X|Y}(x|y) f_Y(y)\, dy.$$

Since Y is uniformly distributed over $(0, 1)$,

$$f_Y(y) = \begin{cases} 1 & \text{if } 0 < y < 1 \\ 0 & \text{elsewhere.} \end{cases}$$

Since given $Y = y$, X is uniformly distributed over $(0, y)$,

$$f_{X|Y}(x|y) = \begin{cases} 1/y & \text{if } 0 < y < 1, \quad 0 < x < y \\ 0 & \text{elsewhere.} \end{cases}$$

Thus

$$f_X(x) = \int_{-\infty}^{\infty} f_{X|Y}(x|y) f_Y(y)\, dy = \int_x^1 \frac{dy}{y} = \ln 1 - \ln x = -\ln x.$$

Therefore,

$$f_X(x) = \begin{cases} -\ln x & \text{if } 0 < x < 1 \\ 0 & \text{elsewhere.} \end{cases} \quad \blacklozenge$$

Example 8.23   Let the conditional probability density function of X, given that $Y = y$, be

$$f_{X|Y}(x|y) = \frac{x+y}{1+y}e^{-x}, \qquad 0 < x < \infty, \quad 0 < y < \infty.$$

Find $P(X < 1 \mid Y = 2)$.

Solution: The probability density function of X given $Y = 2$ is

$$f_{X|Y}(x|2) = \frac{x+2}{3}e^{-x}, \qquad 0 < x < \infty.$$

Therefore,

$$P(X < 1 \mid Y = 2) = \int_0^1 \frac{x+2}{3}e^{-x}\,dx = \frac{1}{3}\left[\int_0^1 xe^{-x}\,dx + \int_0^1 2e^{-x}\,dx\right]$$

$$= \frac{1}{3}\left[-xe^{-x} - e^{-x}\right]_0^1 - \frac{2}{3}\left[e^{-x}\right]_0^1 = 1 - \frac{4}{3}e^{-1} \approx 0.509.$$

Note that while we calculated $P(X < 1 \mid Y = 2)$, it is not possible to calculate probabilities such as $P(X < 1 \mid 2 < Y < 3)$ using $f_{X|Y}(x|y)$. This is because the probability density function of X, given that $2 < Y < 3$, cannot be found from the conditional probability density function of X, given $Y = y$. $\blacklozenge$

Similar to the case where X and Y are discrete, for continuous random variables X and Y with joint probability density function $f(x, y)$, the **conditional expectation of X given that $Y = y$** is as follows:

$$E(X \mid Y = y) = \int_{-\infty}^{\infty} x f_{X|Y}(x|y)\,dx, \tag{8.21}$$

where $f_Y(y) > 0$.

As explained in the discrete case, conditional expectations are simply ordinary expectations computed relative to conditional distributions. For this reason, they satisfy the same properties that ordinary expectations do. For example, if h is an ordinary function from $\mathbf{R}$ to $\mathbf{R}$, then, for continuous random variables X and Y, with joint probability density function $f(x, y)$,

$$E\big[h(X) \mid Y = y\big] = \int_{-\infty}^{\infty} h(x) f_{X|Y}(x|y)\,dx. \tag{8.22}$$

In particular, this implies that the conditional variance of X given that $Y = y$ is given by

$$\sigma^2_{X|Y=y} = \int_{-\infty}^{\infty} \left[x - E(X \mid Y = y)\right]^2 f_{X|Y}(x|y)\, dx. \tag{8.23}$$

Example 8.24 Let X and Y be continuous random variables with joint probability density function

$$f(x, y) = \begin{cases} e^{-y} & \text{if } y > 0, \ \ 0 < x < 1 \\ 0 & \text{elsewhere.} \end{cases}$$

Find $E(X \mid Y = 2)$.

Solution: From the definition,

$$E(X \mid Y = 2) = \int_{-\infty}^{\infty} x f_{X|Y}(x|2)\, dx = \int_{0}^{1} x \frac{f(x, 2)}{f_Y(2)}\, dx = \int_{0}^{1} x \frac{e^{-2}}{f_Y(2)}\, dx.$$

But $f_Y(2) = \int_0^1 f(x, 2)\, dx = \int_0^1 e^{-2}\, dx = e^{-2}$; therefore,

$$E(X \mid Y = 2) = \int_{0}^{1} x \frac{e^{-2}}{e^{-2}}\, dx = \frac{1}{2}. \quad \blacklozenge$$

Example 8.25   The lifetimes of batteries manufactured by a certain company are identically distributed with probability distribution and probability density functions F and f, respectively. In terms of F, f, and s, find the expected value of the lifetime of an s-hour-old battery.

Solution: Let X be the lifetime of the s-hour-old battery. We want to calculate $E(X \mid X > s)$. Let

$$F_{X|X>s}(t) = P(X \le t \mid X > s),$$

and $f_{X|X>s}(t) = F'_{X|X>s}(t)$. Then

$$E(X \mid X > s) = \int_{0}^{\infty} t f_{X|X>s}(t)\, dt.$$

Now

$$F_{X|X>s}(t) = P(X \le t \mid X > s) = \frac{P(X \le t, X > s)}{P(X > s)}$$

$$= \begin{cases} 0 & \text{if } t \le s \\[2mm] \dfrac{P(s < X \le t)}{P(X > s)} & \text{if } t > s. \end{cases}$$

Therefore,

$$F_{X|X>s}(t) = \begin{cases} 0 & \text{if } t \leq s \\ \dfrac{F(t) - F(s)}{1 - F(s)} & \text{if } t > s. \end{cases}$$

Differentiating $F_{X|X>s}(t)$ with respect to t, we obtain

$$f_{X|X>s}(t) = \begin{cases} 0 & \text{if } t \leq s \\ \dfrac{f(t)}{1 - F(s)} & \text{if } t > s. \end{cases}$$

This yields

$$E(X \mid X > s) = \int_0^\infty t f_{X|X>s}(t)\, dt = \frac{1}{1 - F(s)} \int_s^\infty t f(t)\, dt. \quad \blacklozenge$$

★ **Remark 8.1**   Suppose that, in Example 8.25, a battery manufactured by the company is installed at time 0 and begins to operate. If at time s an inspector finds the battery dead, then the expected lifetime of the dead battery is $E(X \mid X < s)$. Similar to the calculations in that example, we can show that

$$E(X \mid X < s) = \frac{1}{F(s)} \int_0^s t f(t)\, dt.$$

(See Exercise 21.)   ♦

EXERCISES

A

1. Let the joint probability mass function of discrete random variables X and Y be given by

$$p(x, y) = \begin{cases} \dfrac{1}{25}(x^2 + y^2) & \text{if } x = 1, 2, \quad y = 0, 1, 2 \\ 0 & \text{otherwise.} \end{cases}$$

Find $p_{X|Y}(x|y)$, $P(X = 2 \mid Y = 1)$, and $E(X \mid Y = 1)$.

2. Let the joint probability density function of continuous random variables X and Y be given by

$$f(x, y) = \begin{cases} 2 & \text{if } 0 < x < y < 1 \\ 0 & \text{elsewhere.} \end{cases}$$

Find $f_{X|Y}(x|y)$.

3. An unbiased coin is flipped until the sixth head is obtained. If the third head occurs on the fifth flip, what is the probability mass function of the number of flips?

4. Let the conditional probability density function of X given that $Y = y$ be given by

$$f_{X|Y}(x|y) = \frac{3(x^2 + y^2)}{3y^2 + 1}, \qquad 0 < x < 1, \quad 0 < y < 1.$$

Find $P(1/4 < X < 1/2 \mid Y = 3/4)$.

5. Let X and Y be independent discrete random variables. Prove that for all y, $E(X \mid Y = y) = E(X)$. Do the same for continuous random variables X and Y.

6. Let X and Y be continuous random variables with joint probability density function

$$f(x, y) = \begin{cases} x + y & \text{if } 0 \le x \le 1, \quad 0 \le y \le 1 \\ 0 & \text{elsewhere.} \end{cases}$$

Calculate $f_{X|Y}(x|y)$.

7. Let X and Y be continuous random variables with joint probability density function given by

$$f(x, y) = \begin{cases} e^{-x(y+1)} & \text{if } x \ge 0, \quad 0 \le y \le e - 1 \\ 0 & \text{elsewhere.} \end{cases}$$

Calculate $E(X \mid Y = y)$.

8. First a point Y is selected at random from the interval $(0, 1)$. Then another point X is selected at random from the interval $(Y, 1)$. Find the probability density function of X.

9. Let (X, Y) be a random point from a unit disk centered at the origin. Find $P(0 \le X \le 4/11 \mid Y = 4/5)$.

10. The joint probability density function of X and Y is given by

$$f(x, y) = \begin{cases} c e^{-x} & \text{if } x \ge 0, \quad |y| < x \\ 0 & \text{otherwise.} \end{cases}$$

    (a) Determine the constant c.

    (b) Find $f_{X|Y}(x|y)$ and $f_{Y|X}(y|x)$.

    (c) Calculate $E(Y \mid X = x)$ and $\text{Var}(Y \mid X = x)$.

11. Leon leaves his office every day at a random time between 4:30 P.M. and 5:00 P.M. If he leaves t minutes past 4:30, the time it will take him to reach home is a random number between 20 and $20 + (2t)/3$ minutes. Let Y be the number of minutes past 4:30 that Leon leaves his office tomorrow and X be the number of minutes it takes him to reach home. Find the joint probability density function of X and Y.

12. Show that if $\{N(t): t \geq 0\}$ is a Poisson process, the conditional distribution of the first arrival time given $N(t) = 1$ is uniform on $(0, t)$.

13. In a sequence of independent Bernoulli trials, let X be the number of successes in the first m trials and Y be the number of successes in the first n trials, $m < n$. Show that the conditional distribution of X, given $Y = y$, is hypergeometric. Also, find the conditional distribution of Y given $X = x$.

B

14. A point is selected at random and uniformly from the region

$$R = \{(x, y): |x| + |y| \leq 1\}.$$

Find the conditional probability density function of X given $Y = y$.

15. Let $\{N(t): t \geq 0\}$ be a Poisson process. For $s < t$ show that the conditional distribution of $N(s)$ given $N(t) = n$ is binomial with parameters n and $p = s/t$. Also find the conditional distribution of $N(t)$ given $N(s) = k$.

16. Cards are drawn from an ordinary deck of 52, one at a time, randomly and with replacement. Let X and Y denote the number of draws until the first ace and the first king are drawn, respectively. Find $E(X \mid Y = 5)$.

17. A box contains 10 red and 12 blue chips. Suppose that 18 chips are drawn, one by one, at random and with replacement. If it is known that 10 of them are blue, show that the expected number of blue chips in the first nine draws is five.

18. Let X and Y be continuous random variables with joint probability density function

$$f(x, y) = \begin{cases} n(n-1)(y-x)^{n-2} & \text{if } 0 \leq x \leq y \leq 1 \\ 0 & \text{otherwise.} \end{cases}$$

Find the conditional expectation of Y given that $X = x$.

19. A point (X, Y) is selected randomly from the triangle with vertices $(0, 0)$, $(0, 1)$, and $(1, 0)$.

    (a)   Find the joint probability density function of X and Y.

    (b)   Calculate $f_{X|Y}(x|y)$.

    (c)   Evaluate $E(X \mid Y = y)$.

20.   Let X and Y be discrete random variables with joint probability mass function

$$p(x, y) = \frac{1}{e^2 y! \, (x - y)!}, \qquad x = 0, 1, 2, \dots, \qquad y = 0, 1, 2, \dots, x,$$

$p(x, y) = 0$, elsewhere. Find $E(Y \mid X = x)$.

21.   The lifetimes of batteries manufactured by a certain company are identically distributed with probability distribution and density functions F and f, respectively. Suppose that a battery manufactured by this company is installed at time 0 and begins to operate. If at time s an inspector finds the battery dead, in terms of F, f, and s, find the expected lifetime of the dead battery.

                                                                         ■

8.4   TRANSFORMATIONS OF TWO RANDOM VARIABLES

In our preceding discussions of random variables, cases have arisen where we have calculated the distribution and the density functions of a function of a random variable X. Functions such as X^2, e^X, $\cos X$, $X^3 + 1$, and so on. In particular, in Section 6.2 we explained how, in general, density functions and distribution functions of such functions can be obtained. In this section we demonstrate a method for finding the joint density function of functions of two random variables. The key is the following, which is the analog of the change of variable theorem for functions of several variables.

Theorem 8.8   *Let X and Y be continuous random variables with joint probability density function $f(x, y)$. Let h_1 and h_2 be real-valued functions of two variables, $U = h_1(X, Y)$ and $V = h_2(X, Y)$. Suppose that*

    (a)   *$u = h_1(x, y)$ and $v = h_2(x, y)$ defines a one-to-one transformation of a set R in the xy-plane onto a set Q in the uv-plane. That is, for $(u, v) \in Q$, the system of two equations in two unknowns,*

$$\begin{cases} h_1(x, y) = u \\ h_2(x, y) = v, \end{cases} \tag{8.24}$$

has a unique solution $x = w_1(u, v)$ and $y = w_2(u, v)$ for x and y, in terms of u and v; and

(b)   *the functions w_1 and w_2 have continuous partial derivatives, and the Jacobian of the transformation $x = w_1(u, v)$ and $y = w_2(u, v)$ is nonzero at all points $(u, v) \in Q$; that is, the following 2×2 determinant is nonzero on Q:*

$$J = \begin{vmatrix} \dfrac{\partial w_1}{\partial u} & \dfrac{\partial w_1}{\partial v} \\[2ex] \dfrac{\partial w_2}{\partial u} & \dfrac{\partial w_2}{\partial v} \end{vmatrix} = \frac{\partial w_1}{\partial u}\frac{\partial w_2}{\partial v} - \frac{\partial w_1}{\partial v}\frac{\partial w_2}{\partial u} \neq 0.$$

Then the random variables U and V are jointly continuous with the joint probability density function $g(u, v)$ given by

$$g(u, v) = \begin{cases} f\big(w_1(u, v), w_2(u, v)\big)|J| & (u, v) \in Q \\ 0 & \text{elsewhere.} \end{cases} \tag{8.25}$$

Theorem 8.8 is a result of the change of a variable theorem in double integrals. To see this, let B be a subset of Q in the uv-plane. Suppose that in the xy-plane, $A \subseteq R$ is the set that is transformed to B by the one-to-one transformation (8.24). Clearly, the events $(U, V) \in B$ and $(X, Y) \in A$ are equiprobable. Therefore,

$$P\big((U, V) \in B\big) = P\big((X, Y) \in A\big) = \iint\limits_{A} f(x, y)\, dx\, dy.$$

Using the change of variable formula for double integrals, we have

$$\iint\limits_{A} f(x, y)\, dx\, dy = \iint\limits_{B} f\big(w_1(u, v), w_2(u, v)\big)|J|\, du\, dv.$$

Hence

$$P\big((U, V) \in B\big) = \iint\limits_{B} f\big(w_1(u, v), w_2(u, v)\big)|J|\, du\, dv.$$

Since this is true for all subsets B of Q, it shows that $g(u, v)$, the joint density of U and V, is given by (8.25).

Example 8.26   Let X and Y be positive independent random variables with the identical probability density function e^{-x} for $x > 0$. Find the joint probability density function of $U = X + Y$ and $V = X/Y$.

Solution: Let $f(x, y)$ be the joint probability density function of X and Y. Then

$$f_X(x) = \begin{cases} e^{-x} & \text{if } x > 0 \\ 0 & \text{if } x \le 0, \end{cases}$$

$$f_Y(y) = \begin{cases} e^{-y} & \text{if } y > 0 \\ 0 & \text{if } y \le 0. \end{cases}$$

Therefore,

$$f(x, y) = f_X(x) f_Y(y) = \begin{cases} e^{-(x+y)} & \text{if } x > 0 \text{ and } y > 0 \\ 0 & \text{elsewhere.} \end{cases}$$

Let $h_1(x, y) = x + y$ and $h_2(x, y) = x/y$. Then the system of equations

$$\begin{cases} x + y = u \\ \dfrac{x}{y} = v \end{cases}$$

has the unique solution $x = (uv)/(v + 1)$, $y = u/(v + 1)$, and

$$J = \begin{vmatrix} \dfrac{v}{v + 1} & \dfrac{u}{(v + 1)^2} \\ \dfrac{1}{v + 1} & -\dfrac{u}{(v + 1)^2} \end{vmatrix} = -\dfrac{u}{(v + 1)^2} \neq 0,$$

since $x > 0$ and $y > 0$ imply that $u > 0$ and $v > 0$; that is,

$$Q = \big\{ (u, v) : u > 0 \text{ and } v > 0 \big\}.$$

Hence, by Theorem 8.8, $g(u, v)$, the joint probability density function of U and V, is given by

$$g(u, v) = e^{-u} \dfrac{u}{(v + 1)^2}, \qquad u > 0 \text{ and } v > 0. \quad \blacklozenge$$

The following example proves a well-known theorem called **Box-Muller's theorem**, which, as we explain in Section 13.4, is used to simulate normal random variables (see Theorem 13.3).

Example 8.27   Let X and Y be two independent uniform random variables over $(0, 1)$; show that the random variables $U = \cos(2\pi X)\sqrt{-2\ln Y}$ and $V = \sin(2\pi X)\sqrt{-2\ln Y}$ are independent standard normal random variables.

Solution: Let $h_1(x, y) = \cos(2\pi x)\sqrt{-2\ln y}$ and $h_2(x, y) = \sin(2\pi x)\sqrt{-2\ln y}$. Then the system of equations

$$\begin{cases} \cos(2\pi x)\sqrt{-2\ln y} = u \\ \sin(2\pi x)\sqrt{-2\ln y} = v \end{cases}$$

defines a one-to-one transformation of the set

$$R = \{(x, y): 0 < x < 1, \; 0 < y < 1\}$$

onto

$$Q = \{(u, v): \; -\infty < u < \infty, \; -\infty < v < \infty\};$$

hence it can be solved uniquely in terms of x and y. To see this, square both sides of these equations and sum them up. We obtain $-2\ln y = u^2 + v^2$, which gives $y = \exp\left[-(u^2 + v^2)/2\right]$. Putting $-2\ln y = u^2 + v^2$ back into these equations, we get

$$\cos 2\pi x = \frac{u}{\sqrt{u^2 + v^2}} \quad \text{and} \quad \sin 2\pi x = \frac{v}{\sqrt{u^2 + v^2}},$$

which enable us to determine the unique value of x. For example, if $u > 0$ and $v > 0$, then $2\pi x$ is uniquely determined in the first quadrant from $2\pi x = \arccos(u/\sqrt{u^2 + v^2})$. Hence the first condition of Theorem 8.8 is satisfied. To check the second condition, note that for $u > 0$ and $v > 0$,

$$w_1(u, v) = \frac{1}{2\pi}\arccos\left(\frac{u}{\sqrt{u^2 + v^2}}\right),$$

$$w_2(u, v) = \exp\left[-(u^2 + v^2)/2\right].$$

Hence

$$J = \begin{vmatrix} \dfrac{\partial w_1}{\partial u} & \dfrac{\partial w_1}{\partial v} \\[2mm] \dfrac{\partial w_2}{\partial u} & \dfrac{\partial w_2}{\partial v} \end{vmatrix} = \begin{vmatrix} \dfrac{-v}{2\pi(u^2 + v^2)} & \dfrac{u}{2\pi(u^2 + v^2)} \\[2mm] -u\exp\left[-(u^2 + v^2)/2\right] & -v\exp\left[-(u^2 + v^2)/2\right] \end{vmatrix}$$

$$= \frac{1}{2\pi}\exp\left[-(u^2 + v^2)/2\right] \neq 0.$$

Now X and Y being two independent uniform random variables over $(0, 1)$ imply that f, their joint probability density function is

$$f(x, y) = f_X(x)f_Y(y) = \begin{cases} 1 & \text{if } 0 < x < 1, \; 0 < y < 1 \\ 0 & \text{elsewhere.} \end{cases}$$

Hence, by Theorem 8.8, $g(u, v)$, the joint probability density function of U and V is given by

$$g(u, v) = \frac{1}{2\pi} \exp\left[- (u^2 + v^2)/2\right], \quad -\infty < u < \infty, \ -\infty < v < \infty.$$

The probability density function of U is calculated as follows:

$$g_U(u) = \int_{-\infty}^{\infty} \frac{1}{2\pi} \exp\left(- \frac{u^2 + v^2}{2}\right) dv = \frac{1}{2\pi} \exp\left(\frac{-u^2}{2}\right) \int_{-\infty}^{\infty} \exp\left(\frac{-v^2}{2}\right) dv,$$

where $\int_{-\infty}^{\infty} (1/\sqrt{2\pi}) \exp(-v^2/2) \, dv = 1$ implies that $\int_{-\infty}^{\infty} \exp(-v^2/2) \, dv = \sqrt{2\pi}$. Therefore,

$$g_U(u) = \frac{1}{\sqrt{2\pi}} \exp\left(\frac{-u^2}{2}\right),$$

which shows that U is standard normal. Similarly,

$$g_V(v) = \frac{1}{\sqrt{2\pi}} \exp\left(\frac{-v^2}{2}\right).$$

Since $g(u, v) = g_U(u)g_V(v)$, U and V are independent standard normal random variables. ◆

As an application of Theorem 8.8, we now prove the following theorem, an excellent resource for calculation of density and distribution functions of sums of continuous *independent* random variables.

Theorem 8.9   (Convolution Theorem)   *Let X and Y be continuous independent random variables with probability density functions f_1 and f_2 and probability distribution functions F_1 and F_2, respectively. Then g and G, the probability density and distribution functions of $X + Y$, respectively, are given by*

$$g(t) = \int_{-\infty}^{\infty} f_1(x) f_2(t - x) \, dx,$$

$$G(t) = \int_{-\infty}^{\infty} f_1(x) F_2(t - x) \, dx.$$

Proof: Let $f(x, y)$ be the joint probability density function of X and Y. Then $f(x, y) = f_1(x) f_2(y)$. Let $U = X + Y$, $V = X$, $h_1(x, y) = x + y$, and $h_2(x, y) = x$. Then the system of equations

$$\begin{cases} x + y = u \\ x = v \end{cases}$$

has the unique solution $x = v$, $y = u - v$, and

$$J = \begin{vmatrix} \dfrac{\partial x}{\partial u} & \dfrac{\partial x}{\partial v} \\[2ex] \dfrac{\partial y}{\partial u} & \dfrac{\partial y}{\partial v} \end{vmatrix} = \begin{vmatrix} 0 & 1 \\ 1 & -1 \end{vmatrix} = -1 \neq 0.$$

Hence, by Theorem 8.8, the joint probability density function of U and V, $\psi(u, v)$, is given by

$$\psi(u, v) = f_1(v) f_2(u - v)|J| = f_1(v) f_2(u - v).$$

Therefore, the marginal probability density function of $U = X + Y$ is

$$g(u) = \int_{-\infty}^{\infty} \psi(u, v)\, dv = \int_{-\infty}^{\infty} f_1(v) f_2(u - v)\, dv,$$

which is the same as

$$g(t) = \int_{-\infty}^{\infty} f_1(x) f_2(t - x)\, dx.$$

To find $G(t)$, the distribution function of $X + Y$, note that

$$G(t) = \int_{-\infty}^{t} g(u)\, du = \int_{-\infty}^{t} \left(\int_{-\infty}^{\infty} f_1(x) f_2(u - x)\, dx \right) du$$

$$= \int_{-\infty}^{\infty} \left(\int_{-\infty}^{t} f_2(u - x)\, du \right) f_1(x)\, dx$$

$$= \int_{-\infty}^{\infty} F_2(t - x) f_1(x)\, dx,$$

where, letting $s = u - x$, the last equality follows from

$$\int_{-\infty}^{t} f_2(u - x)\, du = \int_{-\infty}^{t-x} f_2(s)\, ds = F_2(t - x). \quad \blacklozenge$$

Note that by symmetry we can also write

$$g(t) = \int_{-\infty}^{\infty} f_2(y) f_1(t - y)\, dy,$$

$$G(t) = \int_{-\infty}^{\infty} f_2(y) F_1(t - y)\, dy.$$

Definition   *Let f_1 and f_2 be two probability density functions. Then the function $g(t)$, defined by*

$$g(t) = \int_{-\infty}^{\infty} f_1(x) f_2(t - x)\, dx,$$

is called the **convolution** *of f_1 and f_2.*

Theorem 8.8 shows that

> If X and Y are independent continuous random variables, the probability density function of $X + Y$ is the convolution of the probability density functions of X and Y.

Theorem 8.9 is also valid for discrete random variables. Let p_X and p_Y be probability mass functions of two discrete random variables X and Y. Then the function

$$p(z) = \sum_x p_X(x) p_Y(z - x)$$

is called the **convolution** of p_X and p_Y. It is readily seen that if X and Y are independent, the probability mass function of $X + Y$ is the convolution of the probability mass functions of X and Y:

$$P(X + Y = z) = \sum_x P(X = x, Y = z - x) = \sum_x P(X = x) P(Y = z - x)$$

$$= \sum_x p_X(x) p_Y(z - x).$$

Exercise 5, a famous example given by W. J. Hall, shows that the converse of Theorem 8.9 is not valid. That is, it may happen that the probability mass function of two dependent random variables X and Y is the convolution of the probability mass functions of X and Y.

Example 8.28 Let X and Y be independent exponential random variables, each with parameter λ. Find the distribution function of $X + Y$.

Solution: Let f_1 and f_2 be the probability density functions of X and Y, respectively. Then

$$f_1(x) = f_2(x) = \begin{cases} \lambda e^{-\lambda x} & \text{if } x \geq 0 \\ 0 & \text{otherwise.} \end{cases}$$

Hence

$$f_2(t - x) = \begin{cases} \lambda e^{-\lambda(t-x)} & \text{if } x \leq t \\ 0 & \text{otherwise.} \end{cases}$$

By convolution theorem, h, the probability density function of $X + Y$, is given by

$$h(t) = \int_{-\infty}^{\infty} f_2(t - x) f_1(x) \, dx = \int_0^t \lambda e^{-\lambda(t-x)} \cdot \lambda e^{-\lambda x} dx = \lambda^2 t e^{-\lambda t}.$$

This is the density function of a gamma random variable with parameters 2 and λ. Hence $X + Y$ is gamma with parameters 2 and λ. We will study a generalization of this important theorem in Section 11.2.   ◆

EXERCISES

A

1. Let X and Y be independent random numbers from the interval $(0, 1)$. Find the joint probability density function of $U = -2 \ln X$ and $V = -2 \ln Y$.

2. Let X and Y be two positive independent continuous random variables with the probability density functions $f_1(x)$ and $f_2(y)$, respectively. Find the probability density function of $U = X/Y$.
    Hint: Let $V = X$; find the joint probability density function of U and V. Then calculate the marginal probability density function of U.

3. Let $X \sim N(0, 1)$ and $Y \sim N(0, 1)$ be independent random variables. Find the joint probability density function of $R = \sqrt{X^2 + Y^2}$ and $\Theta = \arctan(Y/X)$. Show that R and Θ are independent. Note that (R, Θ) is the polar coordinate representation of (X, Y).

4. From the interval $(0, 1)$, two random numbers are selected independently. Show that the probability density function of their sum is given by

$$g(t) = \begin{cases} t & \text{if } 0 \le t < 1 \\ 2 - t & \text{if } 1 \le t < 2 \\ 0 & \text{otherwise.} \end{cases}$$

5. Let $-1/9 < c < 1/9$ be a constant. Let $p(x, y)$, the joint probability mass function of the random variables X and Y, be given by the following table:

		y	
x	-1	0	1
-1	$1/9$	$1/9 - c$	$1/9 + c$
0	$1/9 + c$	$1/9$	$1/9 - c$
1	$1/9 - c$	$1/9 + c$	$1/9$

    (a)   Show that the probability mass function of $X + Y$ is the convolution function of the probability mass functions of X and Y for all c.

    (b)   Show that X and Y are independent if and only if $c = 0$.

B

6. Let X and Y be independent random variables with common probability density function

$$f(x) = \begin{cases} \dfrac{1}{x^2} & \text{if } x \geq 1 \\ 0 & \text{elsewhere.} \end{cases}$$

Calculate the joint probability density function of $U = X/Y$ and $V = XY$.

7. Let X and Y be independent random variables with common probability density function

$$f(x) = \begin{cases} e^{-x} & \text{if } x > 0 \\ 0 & \text{elsewhere.} \end{cases}$$

Find the joint probability density function of $U = X + Y$ and $V = e^X$.

8. Prove that if X and Y are independent standard normal random variables, then $X + Y$ and $X - Y$ are independent random variables. This is a special case of the following important theorem.

> Let X and Y be independent random variables with a common distribution F. The random variables $X + Y$ and $X - Y$ are independent if and only if F is a normal distribution function.

9. Let X and Y be independent (strictly positive) gamma random variables with parameters (r_1, λ) and (r_2, λ), respectively. Define $U = X + Y$ and $V = X/(X + Y)$.

   (a) Find the joint probability density function of U and V.

   (b) Prove that U and V are independent.

   (c) Show that U is gamma and V is beta.

10. Let X and Y be independent (strictly positive) exponential random variables each with parameter λ. Are the random variables $X + Y$ and X/Y independent?

■

REVIEW PROBLEMS

1. The joint probability mass function of X and Y is given by the following table.

		x	
y	1	2	3
2	0.05	0.25	0.15
4	0.14	0.10	0.17
6	0.10	0.02	0.02

   (a) Find $P(XY \leq 6)$.

   (b) Find $E(X)$ and $E(Y)$.

2. A fair die is tossed twice. The sum of the outcomes is denoted by X and the largest value by Y. (a) Calculate the joint probability mass function of X and Y; (b) find the marginal probability mass functions of X and Y; (c) find $E(X)$ and $E(Y)$.

3. Calculate the probability mass function of the number of spades in a random bridge hand that includes exactly four hearts.

4. Suppose that three cards are drawn at random from an ordinary deck of 52 cards. If X and Y are the numbers of diamonds and clubs, respectively, calculate the joint probability mass function of X and Y.

5. Calculate the probability mass function of the number of spades in a random bridge hand that includes exactly four hearts and three clubs.

6. Let the joint probability density function of X and Y be given by

$$f(x, y) = \begin{cases} \dfrac{c}{x} & \text{if } 0 < y < x, \ 0 < x < 2 \\ 0 & \text{elsewhere.} \end{cases}$$

   (a) Determine the value of c.

   (b) Find the marginal probability density functions of X and Y.

7. Let X and Y have the joint probability density function below. Determine if $E(XY) = E(X)E(Y)$.

$$f(x, y) = \begin{cases} \dfrac{3}{4}x^2 y + \dfrac{1}{4}y & \text{if } 0 < x < 1 \text{ and } 0 < y < 2 \\ \\ 0 & \text{elsewhere.} \end{cases}$$

8. Prove that the following cannot be the joint probability distribution function of two random variables X and Y.

$$F(x, y) = \begin{cases} 1 & \text{if } x + y \geq 1 \\ 0 & \text{if } x + y < 1. \end{cases}$$

9. Three concentric circles of radii r_1, r_2, and r_3, $r_1 > r_2 > r_3$, are the boundaries of the regions that form a circular target. If a person fires a shot at random at the target, what is the probability that it lands in the middle region?

10. A fair coin is flipped 20 times. If the total number of heads is 12, what is the expected number of heads in the first 10 flips?

11. Let the joint probability distribution function of the lifetimes of two brands of lightbulb be given by

$$F(x, y) = \begin{cases} (1 - e^{-x^2})(1 - e^{-y^2}) & \text{if } x > 0, \ y > 0 \\ 0 & \text{otherwise.} \end{cases}$$

Find the probability that one lightbulb lasts more than twice as long as the other.

12. For $\Omega = \{(x, y): 0 < x + y < 1, \ 0 < x < 1, 0 < y < 1\}$, a region in the plane, let

$$f(x, y) = \begin{cases} 3(x + y) & \text{if } (x, y) \in \Omega \\ 0 & \text{otherwise} \end{cases}$$

be the joint probability density function of the random variables X and Y. Find the marginal probability density functions of X and Y, and $P(X + Y > 1/2)$.

13. Let X and Y be continuous random variables with the joint probability density function

$$f(x, y) = \begin{cases} e^{-y} & \text{if } y > 0, \ 0 < x < 1 \\ 0 & \text{elsewhere.} \end{cases}$$

Find $E(X^n \mid Y = y), n \geq 1$.

14. From an ordinary deck of 52 cards, cards are drawn successively and with replacement. Let X and Y denote the number of spades in the first 10 cards and in the second 15 cards, respectively. Calculate the joint probability mass function of X and Y.

15. Let the joint probability density function of X and Y be given by

$$f(x, y) = \begin{cases} cx(1 - x) & \text{if } 0 \le x \le y \le 1 \\ 0 & \text{otherwise.} \end{cases}$$

   (a) Determine the value of c.

   (b) Determine if X and Y are independent.

16. A point is selected at random from the bounded region between the curves $y = x^2 - 1$ and $y = 1 - x^2$. Let X be the x-coordinate, and let Y be the y-coordinate of the point selected. Determine if X and Y are independent.

17. Let X and Y be two independent uniformly distributed random variables over the intervals $(0, 1)$ and $(0, 2)$, respectively. Find the probability density function of X/Y.

18. If F is the probability distribution function of a random variable X, is $G(x, y) = F(x) + F(y)$ a joint probability distribution function?

19. A bar of length ℓ is broken into three pieces at two random spots. What is the probability that the length of at least one piece is less than $\ell/20$?

20. There are prizes in 10% of the boxes of a certain type of cereal. Let X be the number of boxes of such cereal that Kim should buy to find a prize. Let Y be the number of additional boxes of such cereal that she should purchase to find another prize. Calculate the joint probability mass function of X and Y.

21. Let the joint probability density function of random variables X and Y be given by

$$f(x, y) = \begin{cases} 1 & \text{if } |y| < x, \quad 0 < x < 1 \\ 0 & \text{otherwise.} \end{cases}$$

   Show that $E(Y \mid X = x)$ is a linear function of x while $E(X \mid Y = y)$ is not a linear function of y.

22. **(The Wallet Paradox)**   Consider the following "paradox" given by Martin Gardner in his book *Aha! Gotcha* (W. H. Freeman and Company, New York, 1981).

   Each of two persons places his wallet on the table. Whoever has the smallest amount of money in his wallet, wins all the money in the other wallet. Each of the players reason as follows: "I may lose what I have but I may also win more than I have. So the game is to my advantage."

As Kent G. Merryfield, Ngo Viet, and Saleem Watson have observed in their paper "The Wallet Paradox" in the August–September 1997 issue of the *American Mathematical Monthly*,

> Paradoxically, it seems that the game is to the advantage of both play-
> ers. . . . However, the inference that "the game is to my advantage" is
> the source of the apparent paradox, because it does not take into ac-
> count the *probabilities* of winning or losing. In other words, if the game
> is played many times, how often does a player win? How often does
> he lose? And by how much?

Following the analysis of Kent G. Merryfield, Ngo Viet, and Saleem Watson, let X and Y be the amount of money in the wallets of players A and B, respectively. Let W_A and W_B be the amount of money that player A and B will win, respectively. $W_A(X, Y) = -W_B(X, Y)$ and

$$W_A(X, Y) = \begin{cases} -X & \text{if } X > Y \\ Y & \text{if } X < Y \\ 0 & \text{if } X = Y. \end{cases}$$

Suppose that the distribution function of the money in each player's wallet is the same; that is, X and Y are independent, identically distributed random variables on some interval $[a, b]$ or $[a, \infty), 0 \le a < b < \infty$. Show that

$$E(W_A) = E(W_B) = 0.$$

Chapter 9

Multivariate Distributions

9.1 JOINT DISTRIBUTIONS OF $n > 2$ RANDOM VARIABLES

Joint Probability Mass Functions

The following definition generalizes the concept of joint probability mass function of two discrete random variables to $n > 2$ discrete random variables.

Definition *Let $X_1, X_2, \ldots, X_n$ be discrete random variables defined on the same sample space, with sets of possible values $A_1, A_2, \ldots, A_n$, respectively. The function*

$$p(x_1, x_2, \ldots, x_n) = P(X_1 = x_1, X_2 = x_2, \ldots, X_n = x_n)$$

*is called the **joint probability mass function** of $X_1, X_2, \ldots, X_n$.*

Note that

(a) $p(x_1, x_2, \ldots, x_n) \geq 0$.

(b) If for some i, $1 \leq i \leq n$, $x_i \notin A_i$, then $p(x_1, x_2, \ldots, x_n) = 0$.

(c) $\sum_{x_i \in A_i,\ 1 \leq i \leq n} p(x_1, x_2, \ldots, x_n) = 1$.

Moreover, if the joint probability mass function of random variables $X_1, X_2, \ldots, X_n$, $p(x_1, x_2, \ldots, x_n)$, is given, then for $1 \leq i \leq n$, the **marginal probability mass function** of X_i, p_{X_i}, can be found from $p(x_1, x_2, \ldots, x_n)$ by

$$p_{X_i}(x_i) = P(X_i = x_i) = P(X_i = x_i;\ X_j \in A_j,\ 1 \leq j \leq n, j \neq i)$$

$$= \sum_{x_j \in A_j,\ j \neq i} p(x_1, x_2, \ldots, x_n). \tag{9.1}$$

More generally, to find the **joint probability mass function marginalized** over a given set of k of these random variables, we sum up $p(x_1, x_2, \ldots, x_n)$ over all possible values of the remaining $n - k$ random variables. For example, if $p(x, y, z)$ denotes the joint probability mass function of random variables $X, Y,$ and $Z,$ then

$$p_{X,Y}(x, y) = \sum_z p(x, y, z)$$

is the joint probability mass function marginalized over X and $Y,$ whereas

$$p_{Y,Z}(y, z) = \sum_x p(x, y, z)$$

is the joint probability mass function marginalized over Y and $Z.$

Example 9.1   Dr. Shams has 23 hypertensive patients, of whom five do not use any medicine but try to lower their blood pressures by self-help: dieting, exercise, not smoking, relaxation, and so on. Of the remaining 18 patients, 10 use beta blockers and 8 use diuretics. A random sample of seven of all these patients is selected. Let $X, Y,$ and Z be the number of the patients in the sample trying to lower their blood pressures by self-help, beta blockers, and diuretics, respectively. Find the joint probability mass function and the marginal probability mass functions of $X, Y,$ and $Z.$

Solution: Let $p(x, y, z)$ be the joint probability mass function of $X, Y,$ and $Z.$ Then for $0 \le x \le 5,\ 0 \le y \le 7,\ 0 \le z \le 7,\ x + y + z = 7,$

$$p(x, y, z) = \frac{\binom{5}{x}\binom{10}{y}\binom{8}{z}}{\binom{23}{7}};$$

$p(x, y, z) = 0,$ otherwise. The marginal probability mass function of $X,\ p_X,$ is obtained as follows:

$$p_X(x) = \sum_{\substack{x+y+z=7 \\ 0\le y\le 7 \\ 0\le z\le 7}} p(x, y, z) = \sum_{y=0}^{7-x} \frac{\binom{5}{x}\binom{10}{y}\binom{8}{7-x-y}}{\binom{23}{7}}$$

$$= \frac{\binom{5}{x}}{\binom{23}{7}} \sum_{y=0}^{7-x} \binom{10}{y}\binom{8}{7-x-y}.$$

Now $\displaystyle\sum_{y=0}^{7-x} \binom{10}{y}\binom{8}{7-x-y}$ is the total number of the ways we can choose $7 - x$ patients from 18 (in Exercise 41 of Section 2.4, let $m = 10,\ n = 8,$ and $r = 7 - x$); hence it is equal to $\dbinom{18}{7 - x}$. Therefore, as expected,

$$p_X(x) = \frac{\dbinom{5}{x}\dbinom{18}{7 - x}}{\dbinom{23}{7}}, \qquad x = 0, 1, 2, 3, 4, 5.$$

Similarly,

$$p_Y(y) = \frac{\dbinom{10}{y}\dbinom{13}{7 - y}}{\dbinom{23}{7}}, \qquad y = 0, 1, 2, \ldots, 7,$$

$$p_Z(z) = \frac{\dbinom{8}{z}\dbinom{15}{7 - z}}{\dbinom{23}{7}}, \qquad z = 0, 1, 2, \ldots, 7. \quad \blacklozenge$$

Remark 9.1   A joint probability mass function, such as $p(x, y, z)$ of Example 9.1, is called *multivariate hypergeometric*. In general, suppose that a box contains n_1 marbles of type 1, n_2 marbles of type 2, ..., and n_r marbles of type r. If n marbles are drawn at random and X_i $(i = 1, 2, \ldots, r)$ is the number of the marbles of type i drawn, the joint probability mass function of $X_1, X_2, \ldots, X_r$ is called **multivariate hypergeometric** and is given by

$$p(x_1, x_2, \ldots, x_r) = \frac{\dbinom{n_1}{x_1}\dbinom{n_2}{x_2} \cdots \dbinom{n_r}{x_r}}{\dbinom{n_1 + n_2 + \cdots + n_r}{n}},$$

where $x_1 + x_2 + \cdots + x_r = n$. $\quad \blacklozenge$

We now extend the definition of the joint probability distribution from 2 to $n > 2$ random variables. Let $X_1, X_2, \ldots, X_n$ be n random variables (discrete, continuous, or mixed). Then the **joint probability distribution function** of $X_1, X_2, \ldots, X_n$ is defined by

$$F(t_1, t_2, \ldots, t_n) = P(X_1 \leq t_1, X_2 \leq t_2, \ldots, X_n \leq t_n) \tag{9.2}$$

for all $-\infty < t_i < +\infty$, $i = 1, 2, \ldots, n$.

The **marginal probability distribution function** of X_i, $1 \le i \le n$, can be found from F as follows:

$$F_{X_i}(t_i) = P(X_i \le t_i)$$

$$= P(X_1 < \infty, \ldots, X_{i-1} < \infty, X_i \le t_i, X_{i+1} < \infty, \ldots, X_n < \infty)$$

$$= \lim_{\substack{t_j \to \infty \\ 1 \le j \le n, \, j \ne i}} F(t_1, t_2, \ldots, t_n). \tag{9.3}$$

More generally, to find the **joint probability distribution function marginalized** over a given set of k of these random variables, we calculate the limit of $F(t_1, t_2, \ldots, t_n)$ as $t_j \to \infty$, for every j that belongs to one of the remaining $n - k$ variables. For example, if $F(x, y, z, t)$ denotes the joint probability density function of random variables X, Y, Z, and T, then the joint probability distribution function marginalized over Y and T is given by

$$F_{Y,T}(y, t) = \lim_{x, z \to \infty} F(x, y, z, t).$$

Just as the joint probability distribution function of two random variables, we have that F, the joint probability distribution of n random variables, satisfies the following:

(a) F is nondecreasing in each argument.

(b) F is right continuous in each argument.

(c) $F(t_1, t_2, \ldots, t_{i-1}, -\infty, t_{i+1}, \ldots, t_n) = 0$ for $i = 1, 2, \ldots, n$.

(d) $F(\infty, \infty, \ldots, \infty) = 1$.

We now generalize the concept of independence of random variables from two to any number of random variables.

Suppose that $X_1, X_2, \ldots, X_n$ are random variables (discrete, continuous, or mixed) on a sample space. We say that they are **independent** if, for arbitrary subsets $A_1, A_2, \ldots, A_n$ of real numbers,

$$P(X_1 \in A_1, X_2 \in A_2, \ldots, X_n \in A_n) = P(X_1 \in A_1)P(X_2 \in A_2) \cdots P(X_n \in A_n).$$

Similar to the case of two random variables, $X_1, X_2, \ldots, X_n$ are independent if and only if, for any $x_i \in \mathbf{R}$, $i = 1, 2, \ldots, n$,

$$P(X_1 \le x_1, X_2 \le x_2, \ldots, X_n \le x_n) = P(X_1 \le x_1)P(X_2 \le x_2) \cdots P(X_n \le x_n).$$

That is, $X_1, X_2, \ldots, X_n$ **are independent if and only if**

$$F(x_1, x_2, \ldots, x_n) = F_{X_1}(x_1)F_{X_2}(x_2) \cdots F_{X_n}(x_n).$$

If $X_1, X_2, \ldots, X_n$ are discrete, the definition of independence reduces to the following condition:

$$P(X_1 = x_1, \ldots, X_n = x_n) = P(X_1 = x_1) \cdots P(X_n = x_n) \qquad (9.4)$$

for any set of points, x_i, $i = 1, 2, \ldots, n$.

Let X, Y, and Z be independent random variables. Then, by definition, for arbitrary subsets A_1, A_2, and A_3 of $\mathbf{R}$,

$$P(X \in A_1, Y \in A_2, Z \in A_3) = P(X \in A_1)P(Y \in A_2)P(Z \in A_3). \qquad (9.5)$$

Now, if in (9.5) we let $A_2 = \mathbf{R}$, then since the event $Y \in \mathbf{R}$ is certain and has probability 1, we get

$$P(X \in A_1, Z \in A_3) = P(X \in A_1)P(Z \in A_3).$$

This shows that X and Z are independent random variables. In the same way it can be shown that $\{X, Y\}$, and $\{Y, Z\}$ are also independent sets. Similarly, if $\{X_1, X_2, \ldots, X_n\}$ is a sequence of independent random variables, its subsets are also independent sets of random variables. This observation motivates the following definition for the concept of independence of any collection of random variables.

Definition   *A collection of random variables is called **independent** if all of its finite subcollections are independent.*

Sometimes independence of a collection of random variables is self-evident and requires no checking. For example, let E be an experiment with the sample space S. Let X be a random variable defined on S. If the experiment E is repeated n times independently and on the ith experiment X is called X_i, the sequence $\{X_1, X_2, \ldots, X_n\}$ is an independent sequence of random variables.

It is also important to know that the result of Theorem 8.5 is true for any number of random variables:

If $\{X_1, X_2, \ldots\}$ is a sequence of independent random variables and for $i = 1, 2, \ldots$, $g_i : \mathbf{R} \to \mathbf{R}$ is a real-valued function, then the sequence $\{g_1(X_1), g_2(X_2), \ldots\}$ is also an independent sequence of random variables.

The following theorem is the generalization of Theorem 8.4. It follows from (9.4).

Theorem 9.1   *Let X_1, X_2, $\ldots$, X_n be jointly discrete random variables with the joint probability mass function $p(x_1, x_2, \ldots, x_n)$. Then X_1, X_2, $\ldots$, X_n are independent if and only if $p(x_1, x_2, \ldots, x_n)$ is the product of their marginal densities $p_{X_1}(x_1)$, $p_{X_2}(x_2)$, $\ldots$, $p_{X_n}(x_n)$.*

The following is a generalization of Theorem 8.1 from dimension 2 to n.

Theorem 9.2   *Let $p(x_1, x_2, \ldots, x_n)$ be the joint probability mass function of discrete random variables $X_1, X_2, \ldots, X_n$. For $1 \le i \le n$, let A_i be the set of possible values of X_i. If h is a function of n variables from $\mathbf{R}^n$ to $\mathbf{R}$, then $Y = h(X_1, X_2, \ldots, X_n)$ is a discrete random variable with expected value given by*

$$E(Y) = \sum_{x_n \in A_n} \cdots \sum_{x_1 \in A_1} h(x_1, x_2, \ldots, x_n) p(x_1, x_2, \ldots, x_n)$$

provided that the sum is finite.

Using Theorems 9.1 and 9.2, an almost identical proof to that of Theorem 8.6 implies that

> The expected value of the product of several independent discrete random variables is equal to the product of their expected values.

Example 9.2   Let

$$p(x, y, z) = k(x^2 + y^2 + yz), \quad x = 0, 1, 2; \ y = 2, 3; \ z = 3, 4.$$

(a)   For what value of k is $p(x, y, z)$ a joint probability mass function?

(b)   Suppose that, for the value of k found in part (a), $p(x, y, z)$ is the joint probability mass function of random variables X, Y, and Z. Find $P_{Y,Z}(y, z)$ and $p_Z(z)$.

(c)   Find $E(XZ)$.

Solution:

(a)   We must have

$$k \sum_{x=0}^{2} \sum_{y=2}^{3} \sum_{z=3}^{4} (x^2 + y^2 + yz) = 1.$$

This implies that $k = 1/203$.

(b)
$$P_{Y,Z}(y, z) = \frac{1}{203} \sum_{x=0}^{2} (x^2 + y^2 + yz)$$

$$= \frac{1}{203}(3y^2 + 3yz + 5), \quad y = 2, 3; \ z = 3, 4.$$

$$p_Z(z) = \frac{1}{203} \sum_{x=0}^{2} \sum_{y=2}^{3} (x^2 + y^2 + yz)$$

$$= \frac{15}{203}z + \frac{7}{29}, \quad z = 3, 4.$$

An alternate way to find $p_Z(z)$ is to use the result obtained in part (b):

$$p_Z(z) = \sum_{y=2}^{3} p_{Y,Z}(y, z) = \frac{1}{203} \sum_{y=2}^{3} (3y^2 + 3yz + 5)$$

$$= \frac{15}{203} z + \frac{7}{29}, \quad z = 3, 4.$$

(c) By Theorem 9.2,

$$E(XZ) = \frac{1}{203} \sum_{x=0}^{2} \sum_{y=2}^{3} \sum_{z=3}^{4} xz(x^2 + y^2 + yz) = \frac{774}{203} \approx 3.81. \quad \blacklozenge$$

★ Example 9.3 (Reliability of Systems)[†] Consider a system consisting of n components denoted by $1, 2, \ldots, n$. Suppose that component i, $1 \le i \le n$, is either functioning or not functioning, with no other performance capabilities. Let X_i be a Bernoulli random variable defined by

$$X_i = \begin{cases} 1 & \text{if the component } i \text{ is functioning,} \\ 0 & \text{if the component } i \text{ is not functioning.} \end{cases}$$

The random variable X_i determines the performance mode of the ith component of the system. Suppose that

$$p_i = P(X_i = 1) = 1 - P(X_i = 0).$$

The value p_i, the probability that the ith component functions, is called the **reliability** of the ith component. Throughout we assume that the components function *independently* of each other. That is, $\{X_1, X_2, \ldots, X_n\}$ is an independent set of random variables.

Suppose that the system itself also has only two performance capabilities, functioning and not functioning. Let X be a Bernoulli random variable defined by

$$X = \begin{cases} 1 & \text{if the system is functioning,} \\ 0 & \text{if the system is not functioning.} \end{cases}$$

The random variable X determines the performance mode of the system. Suppose that

$$r = P(X = 1) = 1 - P(X = 0).$$

The value r, the probability that the system functions, is called the **reliability** of the system. We consider only systems for which it is possible to determine r from knowledge of the performance modes of the components.

[†]If this example is skipped, then all exercises and examples in this and future chapters marked "(Reliability of Systems)" should be skipped as well.

As an example, a system is called **series system** when it functions if and only if all n of its components function. For such a system,

$$X = \min(X_1, X_2, \ldots, X_n) = X_1 X_2 \cdots X_n.$$

This is because, if at least one component does not function, then $X_i = 0$ for at least one i, which implies that $X = 0$, indicating that the system does not function. Similarly, if all n components of the system function, then $X_i = 1$ for $1 \leq i \leq n$, implying that $X = 1$, which indicates that the system functions. For a series system, the preceding relationship, expressing X in terms of $X_1, X_2, \ldots, X_n$, enables us to find the reliability of the system:

$$\begin{aligned} r = P(X = 1) &= P\big(\min(X_1, X_2, \ldots X_n) = 1\big) \\ &= P(X_1 = 1, X_2 = 1, \ldots, X_n = 1) \\ &= P(X_1 = 1)P(X_2 = 1) \cdots P(X_n = 1) = \prod_{i=1}^{n} p_i. \end{aligned}$$

Figure 9.1   Geometric representation for a series system.

Geometrically, a series system is often represented by the diagram of Figure 9.1. The reason for this representation is that, for a signal fed in at the input to be transmitted to the output, all components must be functioning. Otherwise, the signal cannot go all the way through.

As another example, consider a **parallel system**. Such a system functions if and only if at least one of its n components functions. For a parallel system,

$$X = \max(X_1, X_2, \ldots, X_n) = 1 - (1 - X_1)(1 - X_2) \cdots (1 - X_n).$$

This is because, if at least one component functions, then for some i, $X_i = 1$, implying that $X = 1$, whereas if none of the components functions, then for all i, $X_i = 0$ implying that $X = 0$. The reliability of a parallel system is given by

$$\begin{aligned} r = P(X = 1) &= P\big(\max(X_1, X_2, \ldots, X_n) = 1\big) \\ &= 1 - P\big(\max(X_1, X_2, \ldots, X_n) = 0\big) \\ &= 1 - P(X_1 = 0, X_2 = 0, \ldots, X_n = 0) \\ &= 1 - P(X_1 = 0)P(X_2 = 0) \cdots P(X_n = 0) = 1 - \prod_{i=1}^{n}(1 - p_i). \end{aligned}$$

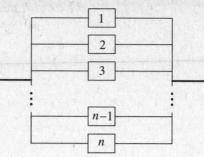

Figure 9.2   Geometric representation for a parallel system.

Geometrically, a parallel system is shown by the diagram of Figure 9.2. This representation shows that, for a signal fed in at the input, at least one component must function to transmit it to the output.

Calculation of r, the reliability of a given system, is not, in general, as easy as it was for series and parallel systems. One useful technique for evaluating r is to write X, if possible, in terms of $X_1, X_2, \ldots, X_n$, and then use the following simple relation:

$$r = P(X = 1) = 1 \cdot P(X = 1) + 0 \cdot P(X = 0) = E(X).$$

An example follows.

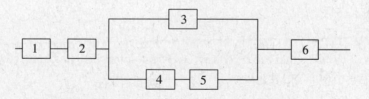

Figure 9.3   A combination of series and parallel systems.

Consider the system whose structure is as shown in Figure 9.3. This system is a combination of series and parallel systems. The fact that for a series system, $X = X_1 X_2 \cdots X_n$, and for a parallel system,

$$X = 1 - (1 - X_1)(1 - X_2) \cdots (1 - X_n),$$

enable us to find out immediately that, for this system,

$$X = X_1 X_2 \big[1 - (1 - X_3)(1 - X_4 X_5)\big] X_6 = X_1 X_2 X_6 (X_3 + X_4 X_5 - X_3 X_4 X_5).$$

Since for $1 \le i \le 6$,

$$E(X_i) = 1 \cdot P(X_i = 1) + 0 \cdot P(X_i = 0) = p_i,$$

and the X_i's are independent random variables, $r = E(X)$ yields

$$r = E(X) = E(X_1)E(X_2)E(X_6)\big[E(X_3) + E(X_4)E(X_5) - E(X_3)E(X_4)E(X_5)\big]$$
$$= p_1 p_2 p_6 (p_3 + p_4 p_5 - p_3 p_4 p_5). \quad \blacklozenge$$

Joint Probability Density Functions

We now define the concept of a jointly continuous density function for more than two random variables.

Definition   *Let $X_1, X_2, \ldots, X_n$ be continuous random variables defined on the same sample space. We say that $X_1, X_2, \ldots, X_n$ have a **continuous joint distribution** if there exists a nonnegative function of n variables, $f(x_1, x_2, \ldots, x_n)$, on $\mathbf{R} \times \mathbf{R} \times \cdots \times \mathbf{R} \equiv \mathbf{R}^n$ such that for any region R in $\mathbf{R}^n$ that can be formed from n-dimensional rectangles by a countable number of set operations,*

$$P\big((X_1, X_2, \ldots, X_n) \in R\big) = \underset{R}{\int \cdots \int} f(x_1, x_2, \ldots, x_n)\, dx_1\, dx_2 \cdots dx_n. \quad (9.6)$$

*The function $f(x_1, x_2, \ldots, x_n)$ is called the **joint probability density function** of $X_1, X_2, \ldots, X_n$.*

Let $R = \big\{(x_1, x_2, \ldots, x_n): x_i \in A_i,\ 1 \le i \le n\big\}$, where $A_i,\ 1 \le i \le n$, is any subset of real numbers that can be constructed from intervals by a countable number of set operations. Then (9.6) gives

$$P(X_1 \in A_1, X_2 \in A_2, \ldots, X_n \in A_n)$$
$$= \int_{A_n} \int_{A_{n-1}} \cdots \int_{A_1} f(x_1, x_2, \ldots, x_n)\, dx_1\, dx_2 \cdots dx_n.$$

Letting $A_i = (-\infty, +\infty),\ 1 \le i \le n$, this implies that

$$\int_{-\infty}^{+\infty} \cdots \int_{-\infty}^{+\infty} f(x_1, x_2, \ldots, x_n)\, dx_1\, dx_2 \cdots dx_n = 1.$$

Let f_{X_i} be the marginal probability density function of $X_i,\ 1 \le i \le n$. Then

$$f_{X_i}(x_i) = \underbrace{\int_{-\infty}^{+\infty} \cdots \int_{-\infty}^{+\infty}}_{n-1 \text{ terms}} f(x_1, x_2, \ldots, x_n)\, dx_1 \cdots dx_{i-1}\, dx_{i+1} \cdots dx_n. \quad (9.7)$$

Therefore, for instance,

$$f_{X_3}(x_3) = \underbrace{\int_{-\infty}^{+\infty} \cdots \int_{-\infty}^{+\infty}}_{n-1 \text{ terms}} f(x_1, x_2, \ldots, x_n) \, dx_1 \, dx_2 \, dx_4 \cdots dx_n.$$

More generally, to find the **joint probability density function marginalized** over a given set of k of these random variables, we integrate $f(x_1, x_2, \ldots, x_n)$ over all possible values of the remaining $n - k$ random variables. For example, if $f(x, y, z, t)$ denotes the joint probability density function of random variables X, Y, Z, and T, then

$$f_{Y,T}(y, t) = \int_{-\infty}^{+\infty} \int_{-\infty}^{+\infty} f(x, y, z, t) \, dx \, dz$$

is the joint probability density function marginalized over Y and T, whereas

$$f_{X,Z,T}(x, z, t) = \int_{-\infty}^{+\infty} f(x, y, z, t) \, dy$$

is the joint probability density function marginalized over X, Z, and T.

The following theorem is the generalization of Theorem 8.7 and the continuous analog of Theorem 9.1. Its proof is similar to the proof of Theorem 8.7.

Theorem 9.3   *Let $X_1, X_2, \ldots, X_n$ be jointly continuous random variables with the joint probability density function $f(x_1, x_2, \ldots, x_n)$. Then $X_1, X_2, \ldots, X_n$ are independent if and only if $f(x_1, x_2, \ldots, x_n)$ is the product of their marginal densities $f_{X_1}(x_1)$, $f_{X_2}(x_2)$, $\ldots, f_{X_n}(x_n)$.*

Let F be the joint probability distribution function of jointly continuous random variables $X_1, X_2, \ldots, X_n$, with the joint probability density function $f(x_1, x_2, \ldots, x_n)$, then

$$F(t_1, t_2, \ldots, t_n) = \int_{-\infty}^{t_n} \int_{-\infty}^{t_{n-1}} \cdots \int_{-\infty}^{t_1} f(x_1, x_2, \ldots, x_n) \, dx_1 \cdots dx_n, \qquad (9.8)$$

and

$$f(x_1, x_2, \ldots, x_n) = \frac{\partial^n F(x_1, x_2, \ldots, x_n)}{\partial x_1 \partial x_2 \cdots \partial x_n}. \qquad (9.9)$$

The following is the continuous analog of Theorem 9.2. It is also a generalization of Theorem 8.2 from dimension 2 to n.

Theorem 9.4   *Let $f(x_1, x_2, \ldots, x_n)$ be the joint probability density function of random variables $X_1, X_2, \ldots, X_n$. If h is a function of n variables from $\mathbf{R}^n$ to $\mathbf{R}$, then $Y = h(X_1, X_2, \ldots, X_n)$ is a random variable with expected value given by*

$$E(Y) = \int_{-\infty}^{\infty} \cdots \int_{-\infty}^{\infty} h(x_1, x_2, \ldots, x_n) f(x_1, x_2, \ldots, x_n) \, dx_1 \, dx_2 \cdots dx_n,$$

provided that the integral is absolutely convergent.

Using Theorems 9.3 and 9.4, an almost identical proof to that of Theorem 8.6 implies that

> The expected value of the product of several independent random variables is equal to the product of their expected values.

Example 9.4   A system has n components, the lifetime of each being an exponential random variable with parameter λ. Suppose that the lifetimes of the components are independent random variables, and the system fails as soon as any of its components fails. Find the probability density function of the time until the system fails.

Solution: Let $X_1, X_2, \ldots, X_n$ be the lifetimes of the components. Then $X_1, X_2, \ldots, X_n$ are independent random variables and for $i = 1, 2, \ldots, n$,

$$P(X_i \le t) = 1 - e^{-\lambda t}.$$

Letting X be the time until the system fails, we have $X = \min(X_1, X_2, \ldots, X_n)$. Therefore,

$$P(X > t) = P\big(\min(X_1, X_2, \ldots, X_n) > t\big) = P(X_1 > t, X_2 > t, \ldots, X_n > t)$$
$$= P(X_1 > t)P(X_2 > t) \cdots P(X_n > t) = (e^{-\lambda t})(e^{-\lambda t}) \cdots (e^{-\lambda t}) = e^{-n\lambda t}.$$

Let f be the probability density function of X; then

$$f(t) = \frac{d}{dt} P(X \le t) = \frac{d}{dt}(1 - e^{-n\lambda t}) = n\lambda e^{-n\lambda t}. \quad \blacklozenge$$

Example 9.5

(a)   Prove that the following is a joint probability density function.

$$f(x, y, z, t) = \begin{cases} \dfrac{1}{xyz} & \text{if } 0 < t \le z \le y \le x \le 1 \\ \\ 0 & \text{elsewhere.} \end{cases}$$

(b)   Suppose that f is the joint probability density function of random variables X, Y, Z, and T. Find $f_{Y,Z,T}(y, z, t)$, $f_{X,T}(x, t)$, and $f_Z(z)$.

Solution:

(a) Since $f(x, y, z, t) \geq 0$ and

$$\int_0^1 \int_0^x \int_0^y \int_0^z \frac{1}{xyz}\, dt\, dz\, dy\, dx = \int_0^1 \int_0^x \int_0^y \frac{1}{xy}\, dz\, dy\, dx$$

$$= \int_0^1 \int_0^x \frac{1}{x}\, dy\, dx = \int_0^1 dx = 1,$$

f is a joint probability density function.

(b) For $0 < t \leq z \leq y \leq 1$,

$$f_{Y,Z,T}(y, z, t) = \int_y^1 \frac{1}{xyz}\, dx = \frac{1}{yz} \ln x \Big|_y^1 = -\frac{\ln y}{yz}.$$

Therefore,

$$f_{Y,Z,T}(y, z, t) = \begin{cases} -\dfrac{\ln y}{yz} & \text{if } 0 < t \leq z \leq y \leq 1 \\[2mm] 0 & \text{elsewhere.} \end{cases}$$

To find $f_{X,T}(x, t)$, we have that for $0 < t \leq x \leq 1$,

$$f_{X,T}(x, t) = \int_t^x \int_t^y \frac{1}{xyz}\, dz\, dy = \int_t^x \left[\frac{\ln z}{xy}\right]_t^y dy$$

$$= \int_t^x \left(\frac{\ln y}{xy} - \frac{\ln t}{xy}\right) dy = \left[\frac{1}{2x}(\ln y)^2 - \frac{\ln t}{x} \ln y\right]_t^x$$

$$= \frac{1}{2x}(\ln x)^2 - \frac{1}{x}(\ln t)(\ln x) + \frac{1}{2x}(\ln t)^2 = \frac{1}{2x}(\ln x - \ln t)^2$$

$$= \frac{1}{2x} \ln^2 \frac{x}{t}.$$

Therefore,

$$f_{X,T}(x, t) = \begin{cases} \dfrac{1}{2x} \ln^2 \dfrac{x}{t} & \text{if } 0 < t \leq x \leq 1 \\[2mm] 0 & \text{otherwise.} \end{cases}$$

To find $f_Z(z)$, we have that for $0 < z \leq 1$,

$$f_Z(z) = \int_z^1 \int_z^x \int_0^z \frac{1}{xyz}\, dt\, dy\, dx = \int_z^1 \int_z^x \frac{1}{xy}\, dy\, dx$$

$$= \int_z^1 \left[\frac{1}{x}\ln y\right]_z^x dx = \int_z^1 \left(\frac{1}{x}\ln x - \frac{1}{x}\ln z\right) dx$$

$$= \left[\frac{1}{2}(\ln x)^2 - (\ln x)(\ln z)\right]_z^1 = \frac{1}{2}(\ln z)^2.$$

Thus

$$f_Z(z) = \begin{cases} \dfrac{1}{2}(\ln z)^2 & \text{if } 0 < z \le 1 \\[2ex] 0 & \text{otherwise.} \end{cases} \quad \blacklozenge$$

Random Sample

Definition: *We say that n random variables $X_1, X_2, \ldots, X_n$ form a **random sample** of size n, from a (continuous or discrete) distribution function F, if they are independent and, for $1 \le i \le n$, the distribution function of X_i is F. Therefore, elements of a random sample are independent and identically distributed.*

To explain this definition, suppose that the lifetime distribution of the light bulbs manufactured by a company is exponential with parameter λ. To estimate $1/\lambda$, the average lifetime of a light bulb, for some positive integer n, we choose n light bulbs at random and independently from those manufactured by the company. For $1 \le i \le n$, let X_i be the lifetime of the ith light bulb selected. Then $\{X_1, X_2, \ldots, X_n\}$ is a random sample of size n from the exponential distribution with parameter λ. That is, for $1 \le i \le n$, X_i's are independent, and X_i is exponential with parameter λ. Clearly, an estimation of $1/\lambda$ is the mean of the random sample $X_1, X_2, \ldots, X_n$ denoted by $\bar{X}$:

$$\bar{X} = \frac{X_1 + X_2 + \cdots + X_n}{n}.$$

Thus all we need to do is to measure the lifetime of each of the n light bulbs of the random sample and find the average of the observed values. In Sections 11.3 and 11.5, we will discuss methods to calculate n, the sample size, so that the error of estimation does not exceed a predetermined quantity.

EXERCISES

A

1. From an ordinary deck of 52 cards, 13 cards are selected at random. Calculate the joint probability mass function of the numbers of hearts, diamonds, clubs, and spades selected.

2. A jury of 12 people is randomly selected from a group of eight Afro-American, seven Hispanic, three Native American, and 20 white potential jurors. Let A, H, N, and W be the number of Afro-American, Hispanic, Native American, and white jurors selected, respectively. Calculate the joint probability mass function of A, H, N, W and the marginal probability mass function of A.

3. Let $p(x, y, z) = (xyz)/162$, $x = 4, 5$, $y = 1, 2, 3$, and $z = 1, 2$, be the joint probability mass function of the random variables X, Y, Z.

    (a) Calculate the joint marginal probability mass functions of X, Y; Y, Z; and X, Z.

    (b) Find $E(YZ)$.

4. Let the joint probability density function of X, Y, and Z be given by

    $$f(x, y, z) = \begin{cases} 6e^{-x-y-z} & \text{if } 0 < x < y < z < \infty \\ 0 & \text{elsewhere.} \end{cases}$$

    (a) Find the marginal joint probability density function of X, Y; X, Z; and Y, Z.

    (b) Find $E(X)$.

5. From the set of families with two children a *family* is selected at random. Let $X_1 = 1$ if the first child of the family is a girl; $X_2 = 1$ if the second child of the family is a girl; and $X_3 = 1$ if the family has exactly one boy. For $i = 1, 2, 3$, let $X_i = 0$ in other cases. Determine if X_1, X_2, and X_3 are independent. Assume that in a family the probability that a child is a girl is independent of the gender of the other children and is 1/2.

6. Let X, Y, and Z be jointly continuous with the following joint probability density function:

    $$f(x, y, z) = \begin{cases} x^2 e^{-x(1+y+z)} & \text{if } x, y, z > 0 \\ 0 & \text{otherwise.} \end{cases}$$

    Are X, Y, and Z independent? Are they pairwise independent?

7. Let the joint probability distribution function of X, Y, and Z be given by

$$F(x, y, z) = (1 - e^{-\lambda_1 x})(1 - e^{-\lambda_2 y})(1 - e^{-\lambda_3 z}), \quad x, y, z > 0,$$

where $\lambda_1, \lambda_2, \lambda_3 > 0$.

(a) Are X, Y, and Z independent?

(b) Find the joint probability density function of X, Y, and Z.

(c) Find $P(X < Y < Z)$.

8. **(a)** Show that the following is a joint probability density function.

$$f(x, y, z) = \begin{cases} -\dfrac{\ln x}{xy} & \text{if } 0 < z \le y \le x \le 1 \\ 0 & \text{otherwise.} \end{cases}$$

(b) Suppose that f is the joint probability density function of X, Y, and Z. Find $f_{X,Y}(x, y)$ and $f_Y(y)$.

9. Inside a circle of radius R, n points are selected at random and independently. Find the probability that the distance of the nearest point to the center is at least r.

10. A point is selected at random from the cube

$$\Omega = \big\{(x, y, z): -a \le x \le a, -a \le y \le a, -a \le z \le a\big\}.$$

What is the probability that it is inside the sphere inscribed in the cube?

11. Is the following a joint probability density function?

$$f(x_1, x_2, \dots, x_n) = \begin{cases} e^{-x_n} & \text{if } 0 < x_1 < x_2 < \cdots < x_n \\ 0 & \text{otherwise.} \end{cases}$$

12. Suppose that the lifetimes of radio transistors are independent exponential random variables with mean five years. Arnold buys a radio and decides to replace its transistor upon failure two times: once when the original transistor dies and once when the replacement dies. He stops using the radio when the second replacement of the transistor goes out of order. Assuming that Arnold repairs the radio if it fails for any other reason, find the probability that he uses the radio at least 15 years.

13. Let $X_1, X_2, \dots, X_n$ be independent exponential random variables with means $1/\lambda_1, 1/\lambda_2, \dots, 1/\lambda_n$, respectively. Find the probability distribution function of $X = \min(X_1, X_2, \dots, X_n)$.

14. **(Reliability of Systems)** Suppose that a system functions if and only if at least k $(1 \le k \le n)$ of its components function. Furthermore, suppose that $p_i = p$ for $1 \le i \le n$. Find the reliability of this system. (Such a system is said to be a **k-out-of-n system**.)

B

15. An item has n parts, each with an exponentially distributed lifetime with mean $1/\lambda$. If the failure of one part makes the item fail, what is the average lifetime of the item?
Hint: Use the result of Exercise 13.

16. Suppose that the lifetimes of a certain brand of transistor are identically distributed and independent random variables with probability distribution function F. These transistors are randomly selected, one at a time, and their lifetimes are measured. Let the Nth be the first transistor that will last longer than s hours. Let X_N be the lifetime of this transistor. Are N and X_N independent random variables?

17. **(Reliability of Systems)**   Consider the system whose structure is shown in Figure 9.4. Find the reliability of this system.

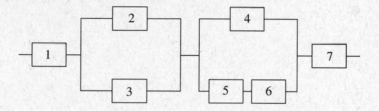

Figure 9.4   A diagram for the system of Exercise 17.

18. Let $X_1, X_2, \ldots, X_n$ be n independent random numbers from the interval $(0, 1)$. Find $E\left(\max_{1 \le i \le n} X_i\right)$ and $E\left(\min_{1 \le i \le n} X_i\right)$.

19. Let F be a probability distribution function. Prove that the functions F^n and $1 - (1 - F)^n$ are also probability distribution functions.
Hint: Let $X_1, X_2, \ldots, X_n$ be independent random variables each with the probability distribution function F. Find the probability distribution functions of the random variables $\max(X_1, X_2, \ldots, X_n)$ and $\min(X_1, X_2, \ldots, X_n)$.

20. Let $X_1, X_2, \ldots, X_n$ be n independent random numbers from $(0, 1)$, and $Y_n = n \cdot \min(X_1, X_2, \ldots, X_n)$. Prove that

$$\lim_{n \to \infty} P(Y_n > x) = e^{-x}, \qquad x \ge 0.$$

21. Suppose that h is the probability density function of a continuous random variable. Let the joint probability density function of X, Y, and Z be

$$f(x, y, z) = h(x)h(y)h(z), \qquad x, y, z \in \mathbf{R}.$$

Prove that $P(X < Y < Z) = 1/6$.

22. **(Reliability of Systems)**   To transfer water from point A to point B, a water-supply system with five water pumps located at the points 1, 2, 3, 4, and 5 is designed as in Figure 9.5. Suppose that whenever the system is turned on for water to flow from A to B, pump i, $i \leq 5$, functions with probability p_i independent of the other pumps. What is the probability that, at such a time, water reaches B?

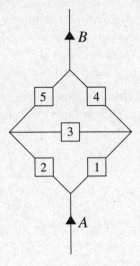

Figure 9.5   The water-supply system of Exercise 22.

23.   A point is selected at random from the pyramid

$$V = \{(x, y, z): x, y, z \geq 0, \ x + y + z \leq 1\}.$$

Letting (X, Y, Z) be its coordinates, determine if X, Y, and Z are independent. *Hint:*   Recall that the volume of a pyramid is $Bh/3$, where h is the height and B is the area of the base.

24.   **(Roots of Quadratic Equations)**   Three numbers A, B, and C are selected at random and independently from the interval $(0, 1)$. Determine the probability that the quadratic equation $Ax^2 + Bx + C = 0$ has real roots. In other words, what fraction of "all possible quadratic equations" with coefficients in $(0, 1)$ have real roots?

25.   **(Roots of Cubic Equations)**   Solve the following exercise posed by S. A. Patil and D. S. Hawkins, Tennessee Technological University, Cookeville, Tennessee, in *The College Mathematics Journal*, September 1992.

Let A, B, and C be independent random variables uniformly distributed on [0, 1]. What is the probability that all of the roots of the cubic equation $x^3 + Ax^2 + Bx + C = 0$ are real?

■

9.2   ORDER STATISTICS

Definition   *Let* $\{X_1, X_2, \ldots, X_n\}$ *be an independent set of identically distributed continuous random variables with the common density and distribution functions* f *and* F, *respectively. Let* $X_{(1)}$ *be the smallest value in* $\{X_1, X_2, \ldots, X_n\}$, $X_{(2)}$ *be the second smallest value,* $X_{(3)}$ *be the third smallest, and, in general,* $X_{(k)}$ $(1 \leq k \leq n)$ *be the kth smallest value in* $\{X_1, X_2, \ldots, X_n\}$. *Then* $X_{(k)}$ *is called the **kth order statistic**, and the set* $\{X_{(1)}, X_{(2)}, \ldots, X_{(n)}\}$ *is said to consist of the **order statistics** of* $\{X_1, X_2, \ldots, X_n\}$.

By this definition, for example, if at a sample point ω of the sample space, $X_1(\omega) = 8$, $X_2(\omega) = 2$, $X_3(\omega) = 5$, and $X_4(\omega) = 6$, then the order statistics of $\{X_1, X_2, X_3, X_4\}$ is $\{X_{(1)}, X_{(2)}, X_{(3)}, X_{(4)}\}$, where $X_{(1)}(\omega) = 2$, $X_{(2)}(\omega) = 5$, $X_{(3)}(\omega) = 6$, and $X_{(4)}(\omega) = 8$. Continuity of X_i's implies that $P(X_{(i)} = X_{(j)}) = 0$. Hence

$$P(X_{(1)} < X_{(2)} < X_{(3)} < \cdots < X_{(n)}) = 1.$$

Unlike X_i's, the random variables $X_{(i)}$'s are neither independent nor identically distributed.

There are many useful and practical applications of order statistics in different branches of pure and applied probability, as well as in estimation theory. To show how it arises, we present three examples.

Example 9.6   Suppose that customers arrive at a warehouse from n different locations. Let X_i, $1 \leq i \leq n$, be the time until the arrival of the next customer from location i; then $X_{(1)}$ is the arrival time of the next customer to the warehouse.   ◆

Example 9.7   Suppose that a machine consists of n components with the lifetimes $X_1, X_2, \ldots, X_n$, respectively, where X_i's are independent and identically distributed. Suppose that the machine remains operative unless k or more of its components fail. Then $X_{(k)}$, the kth order statistic of $\{X_1, X_2, \ldots, X_n\}$, is the time when the machine fails. Also, $X_{(1)}$ is the failure time of the first component.   ◆

Example 9.8   Let $X_1, X_2, \ldots, X_n$ be a random sample of size n from a population with continuous distribution F. Then the following important statistical concepts are expressed in terms of order statistics:

(i)   The sample range is $X_{(n)} - X_{(1)}$.

(ii)   The sample midrange is $[X_{(n)} + X_{(1)}]/2$.

(iii)   The sample median is

$$m = \begin{cases} X_{(i+1)} & \text{if } n = 2i + 1 \\ \dfrac{X_{(i)} + X_{(i+1)}}{2} & \text{if } n = 2i. \end{cases} \quad \blacklozenge$$

We will now determine the probability distribution and the probability density functions of $X_{(k)}$, the kth order statistic.

Theorem 9.5   *Let $\{X_{(1)}, X_{(2)}, \dots, X_{(n)}\}$ be the order statistics of the independent and identically distributed continuous random variables $X_1, X_2, \dots, X_n$ with the common probability distribution and probability density functions F and f, respectively. Then F_k and f_k, the probability distribution and probability density functions of $X_{(k)}$, respectively, are given by*

$$F_k(x) = \sum_{i=k}^{n} \binom{n}{i} [F(x)]^i [1 - F(x)]^{n-i}, \quad -\infty < x < \infty, \tag{9.10}$$

and

$$f_k(x) = \frac{n!}{(k-1)!\,(n-k)!} f(x) [F(x)]^{k-1} [1 - F(x)]^{n-k}, \quad -\infty < x < \infty. \tag{9.11}$$

Proof: Let $-\infty < x < \infty$. To calculate $P(X_{(k)} \leq x)$, note that $X_{(k)} \leq x$ if and only if at least k of the random variables $X_1, X_2, \dots, X_n$ are in $(-\infty, x]$. Thus

$$F_k(x) = P(X_{(k)} \leq x)$$

$$= \sum_{i=k}^{n} P\big(i \text{ of the random variables } X_1, X_2, \dots, X_n \text{ are in } (-\infty, x]\big)$$

$$= \sum_{i=k}^{n} \binom{n}{i} [F(x)]^i [1 - F(x)]^{n-i},$$

where the last equality follows because from the random variables $X_1, X_2, \dots, X_n$ the number of those that lie in $(-\infty, x]$ has binomial distribution with parameters (n, p), $p = F(x)$.

We will now obtain f_k by differentiating F_k:

$$f_k(x) = \sum_{i=k}^{n} \binom{n}{i} i f(x) [F(x)]^{i-1} [1 - F(x)]^{n-i}$$

$$- \sum_{i=k}^{n} \binom{n}{i} [F(x)]^i (n-i) f(x) [1 - F(x)]^{n-i-1}$$

$$= \sum_{i=k}^{n} \frac{n!}{(i-1)!(n-i)!} f(x) [F(x)]^{i-1} [1 - F(x)]^{n-i}$$

$$- \sum_{i=k}^{n} \frac{n!}{i!(n-i-1)!} f(x) [F(x)]^i [1 - F(x)]^{n-i-1}$$

$$= \sum_{i=k}^{n} \frac{n!}{(i-1)!(n-i)!} f(x) [F(x)]^{i-1} [1 - F(x)]^{n-i}$$

$$- \sum_{i=k+1}^{n} \frac{n!}{(i-1)!(n-i)!} f(x) [F(x)]^{i-1} [1 - F(x)]^{n-i}.$$

After cancellations, this gives (9.11).   ◆

Remark 9.2 Note that by (9.10) and (9.11), respectively, F_1 and f_1, the probability distribution and the probability density functions of $X_{(1)} = \min(X_1, X_2, \ldots, X_n)$, are found to be

$$F_1(x) = \sum_{i=1}^{n} \binom{n}{i} [F(x)]^i [1 - F(x)]^{n-i}$$

$$= \sum_{i=0}^{n} \binom{n}{i} [F(x)]^i [1 - F(x)]^{n-i} - [1 - F(x)]^n$$

$$= [F(x) + [1 - F(x)]]^n - [1 - F(x)]^n$$

$$= 1 - [1 - F(x)]^n, \qquad -\infty < x < \infty.$$

and

$$f_1(x) = nf(x) [1 - F(x)]^{n-1}, \qquad -\infty < x < \infty.$$

Also, F_n and f_n, the probability distribution and probability density functions of $X = \max(X_1, X_2, \ldots, X_n)$, respectively, are found to be

$$F_n(x) = [F(x)]^n, \qquad -\infty < x < \infty,$$

and

$$f_n(x) = nf(x) [F(x)]^{n-1}, \qquad -\infty < x < \infty.$$

These quantities can also be calculated directly (see, for example, Example 9.4 or Exercise 13, Section 9.1).   ◆

Example 9.9 Let $X_1, X_2, \ldots, X_{2n+1}$ be $2n + 1$ random numbers from $(0, 1)$. Then f and F, the respective probability density and distribution functions of X_i's, are given by

$$f(x) = \begin{cases} 1 & \text{if } 0 < x < 1 \\ 0 & \text{elsewhere,} \end{cases}$$

$$F(x) = \begin{cases} 0 & \text{if } x < 0 \\ x & \text{if } 0 \le x < 1 \\ 1 & \text{if } x \ge 1. \end{cases}$$

Using (9.11), the probability density function of $X_{(n+1)}$, the median of these numbers, is found to be

$$f_{n+1}(x) = \frac{(2n+1)!}{n!\,n!}\, x^n (1-x)^n, \qquad 0 < x < 1;$$

0, elsewhere. Since for the beta function, $B(n+1, n+1) = (n!\,n!)/(2n+1)!$, we have

$$f_{n+1}(x) = \frac{1}{B(n+1, n+1)}\, x^n (1-x)^n, \qquad 0 < x < 1;$$

0, elsewhere. Hence $X_{(n+1)}$ is beta with parameters $(n+1, n+1)$. In fact, it is straightforward to see that in this example, for $1 \le k \le 2n+1$, $X_{(k)}$ is beta with parameters k and $2n - k + 2$.   ◆

The following theorem gives the joint probability density function of $X_{(i)}$ and $X_{(j)}$. Its proof is similar to that of Theorem 9.5.

Theorem 9.6 *Let $\{X_{(1)}, X_{(2)}, \ldots, X_{(n)}\}$ be the order statistics of the independent and identically distributed continuous random variables $X_1, X_2, \ldots, X_n$ with the common probability density and probability distribution functions f and F, respectively. Then for $x < y$, $f_{ij}(x, y)$, the joint probability density function of $X_{(i)}$ and $X_{(j)}$ ($i < j$), is given by*

$$f_{ij}(x, y) =$$
$$\frac{n!}{(i-1)!\,(j-i-1)!\,(n-j)!}\, f(x) f(y) [F(x)]^{i-1} [F(y) - F(x)]^{j-i-1} [1 - F(y)]^{n-j}.$$

It is clear that for $x \ge y$, $f_{ij}(x, y) = 0$.

Another important quantity that is used often is the joint probability density function of $X_{(1)}, X_{(2)}, \ldots, X_{(n)}$. The following theorem, whose proof we skip, expresses the general form of this function.

Theorem 9.7   *Let* $\{X_{(1)}, X_{(2)}, \ldots, X_{(n)}\}$ *be the order statistics of the independent and identically distributed continuous random variables* $X_1, X_2, \ldots, X_n$ *with the common probability density and distribution functions* f *and* F, *respectively. Then* $f_{12\cdots n}$, *the joint probability density function of* $X_{(1)}, X_{(2)}, \ldots, X_{(n)}$, *is given by*

$$f_{12\cdots n}(x_1, x_2, \ldots, x_n) =$$

$$\begin{cases} n! f(x_1) f(x_2) \cdots f(x_n) & -\infty < x_1 < x_2 < \cdots < x_n < \infty \\ 0 & \text{otherwise.} \end{cases}$$

Justification: For small $\varepsilon > 0$,

$$P(x_1 - \varepsilon < X_{(1)} < x_1 + \varepsilon, \ldots, x_n - \varepsilon < X_{(n)} < x_n + \varepsilon)$$

$$= \int_{x_n - \varepsilon}^{x_n + \varepsilon} \int_{x_{n-1} - \varepsilon}^{x_{n-1} + \varepsilon} \cdots \int_{x_1 - \varepsilon}^{x_1 + \varepsilon} f_{12\cdots n}(x_1, x_2, \ldots, x_n) dx_1 \, dx_2 \cdots dx_n$$

$$\approx 2^n \varepsilon^n f_{12\cdots n}(x_1, x_2, \ldots, x_n). \tag{9.12}$$

Now for $x_1 < x_2 < \cdots < x_n$, let $\mathcal{P}$ be the set of all permutations of $\{x_1, x_2, \cdots, x_n\}$; then $\mathcal{P}$ has $n!$ elements and we can write

$$P(x_1 - \varepsilon < X_{(1)} < x_1 + \varepsilon, \ldots, x_n - \varepsilon < X_{(n)} < x_n + \varepsilon)$$

$$= \sum_{\{x_{i_1}, x_{i_2}, \ldots, x_{i_n}\} \in \mathcal{P}} P(x_{i_1} - \varepsilon < X_1 < x_{i_1} + \varepsilon, \ldots, x_{i_n} - \varepsilon < X_n < x_{i_n} + \varepsilon)$$

$$\approx \sum_{\{x_{i_1}, x_{i_2}, \ldots, x_{i_n}\} \in \mathcal{P}} 2^n \varepsilon^n f(x_{i_1}) f(x_{i_2}) \cdots f(x_{i_n}). \tag{9.13}$$

This is because $X_1, X_2, \ldots, X_n$ are independent, and hence their joint probability density function is the product of their marginal probability density functions. Putting (9.12) and (9.13) together, we obtain

$$\sum_{\{x_{i_1}, x_{i_2}, \ldots, x_{i_n}\} \in \mathcal{P}} f(x_{i_1}) f(x_{i_2}) \cdots f(x_{i_n}) = f_{12\cdots n}(x_1, x_2, \ldots, x_n). \tag{9.14}$$

But

$$f(x_{i_1}) f(x_{i_2}) \cdots f(x_{i_n}) = f(x_1) f(x_2) \cdots f(x_n).$$

Therefore,

$$\sum_{\{x_{i_1}, x_{i_2}, \ldots, x_{i_n}\} \in \mathcal{P}} f(x_{i_1}) f(x_{i_2}) \cdots f(x_{i_n}) =$$

$$\sum_{\{x_{i_1}, x_{i_2}, \ldots, x_{i_n}\} \in \mathcal{P}} f(x_1) f(x_2) \cdots f(x_n) = n! f(x_1) f(x_2) \cdots f(x_n). \tag{9.15}$$

Relations (9.14) and (9.15) imply that

$$f_{12\cdots n}(x_1, x_2, \ldots, x_n) = n! f(x_1) f(x_2) \cdots f(x_n). \quad \blacklozenge$$

Example 9.10   The distance between two towns, A and B, is 30 miles. If three gas stations are constructed independently at randomly selected locations between A and B, what is the probability that the distance between any two gas stations is at least 10 miles?

Solution: Let X_1, X_2, and X_3 be the locations at which the gas stations are constructed. The probability density function of X_1, X_2, and X_3 is given by

$$f(x) = \begin{cases} \dfrac{1}{30} & \text{if } 0 < x < 30 \\[2mm] 0 & \text{elsewhere.} \end{cases}$$

Therefore, by Theorem 9.7, f_{123}, the joint probability density function of the order statistics of X_1, X_2, and X_3, is as follows.

$$f_{123}(x_1, x_2, x_3) = 3!\left(\frac{1}{30}\right)^3, \qquad 0 < x_1 < x_2 < x_3 < 30.$$

Using this, we have that the desired probability is given by the following triple integral.

$$P(X_{(1)} + 10 < X_{(2)} \text{ and } X_{(2)} + 10 < X_{(3)})$$
$$= \int_0^{10} \int_{x_1+10}^{20} \int_{x_2+10}^{30} f_{123}(x_1, x_2, x_3) \, dx_3 \, dx_2 \, dx_1 = \frac{1}{27} \quad \blacklozenge$$

EXERCISES

A

1. Let X_1, X_2, X_3, and X_4 be four independently selected random numbers from $(0, 1)$. Find $P(1/4 < X_{(3)} < 1/2)$.

2. Two random points are selected from $(0, 1)$ independently. Find the probability that one of them is at least three times the other.

3. Let X_1, X_2, X_3, and X_4 be independent exponential random variables, each with parameter λ. Find $P(X_{(4)} \geq 3\lambda)$.

4. Let $X_1, X_2, X_3, \ldots, X_n$ be a sequence of nonnegative, identically distributed, and independent random variables. Let F be the probability distribution function of $X_i, 1 \leq i \leq n$. Prove that

$$E[X_{(n)}] = \int_0^\infty \left(1 - [F(x)]^n\right) dx.$$

Hint: Use Theorem 6.2.

5. Let $X_1, X_2, X_3, \ldots, X_m$ be a sequence of nonnegative, independent binomial random variables, each with parameters (n, p). Find the probability mass function of $X_{(i)}, 1 \leq i \leq m$.

6. Prove that G, the probability distribution function of $[X_{(1)} + X_{(n)}]/2$, the midrange of a random sample of size n from a population with continuous probability distribution function F and probability density function f, is given by

$$G(t) = n \int_{-\infty}^t [F(2t - x) - F(x)]^{n-1} f(x) dx.$$

Hint: Use Theorem 9.6 to find f_{1n}; then integrate over the region $x + y \leq 2t$ and $x \leq y$.

B

7. Let X_1 and X_2 be two independent exponential random variables each with parameter λ. Show that $X_{(1)}$ and $X_{(2)} - X_{(1)}$ are independent.

8. Let X_1 and X_2 be two independent $N(0, \sigma^2)$ random variables. Find $E[X_{(1)}]$.
   Hint: Let $f_{12}(x, y)$ be the joint probability density function of $X_{(1)}$ and $X_{(2)}$. The desired quantity is $\iint x f_{12}(x, y) dx dy$, where the integration is taken over an appropriate region.

9. Let $X_1, X_2, \ldots, X_n$ be a random sample of size n from a population with continuous probability distribution function F and probability density function f.

   (a) Calculate the probability density function of the sample range, $R = X_{(n)} - X_{(1)}$.

   (b) Use (a) to find the probability density function of the sample range of n random numbers from $(0, 1)$.

10. Let $X_1, X_2, \ldots, X_n$ be n independently randomly selected points from the interval $(0, \theta), \theta > 0$. Prove that
    $$E(R) = \frac{n-1}{n+1}\theta,$$
    where $R = X_{(n)} - X_{(1)}$ is the range of these points.
    Hint: Use part (a) of Exercise 9. Also compare this with Exercise 18, Section 9.1.

9.3   MULTINOMIAL DISTRIBUTIONS

Multinomial distribution is a generalization of a binomial. Suppose that, whenever an experiment is performed, one of the disjoint outcomes $A_1, A_2, \ldots, A_r$ will occur. Let $P(A_i) = p_i$, $1 \leq i \leq r$. Then $p_1 + p_2 + \cdots + p_r = 1$. If, in n independent performances of this experiment, X_i, $i = 1, 2, 3, \ldots, r$, denotes the number of times that A_i occurs, then $p(x_1, \ldots, x_r)$, the joint probability mass function of $X_1, X_2, \ldots, X_r$, is called **multinomial joint probability mass function**, and its distribution is said to be a **multinomial distribution**. For any set of nonnegative integers $\{x_1, x_2, \ldots, x_r\}$ with $x_1 + x_2 + \cdots + x_r = n$,

$$p(x_1, x_2, \ldots, x_r) = P(X_1 = x_1, X_2 = x_2, \ldots, X_r = x_r)$$

$$= \frac{n!}{x_1!\, x_2! \cdots x_r!}\, p_1^{x_1} p_2^{x_2} \cdots p_r^{x_r}. \tag{9.16}$$

To prove this relation, recall that, by Theorem 2.4, the number of distinguishable permutations of n objects of r different types where x_1 are alike, x_2 are alike, $\cdots$, x_r are alike $(n = x_1 + \cdots + x_r)$ is $n!/(x_1!\, x_2! \cdots x_r!)$. Hence there are $n!/(x_1!\, x_2! \cdots x_r!)$ sequences of $A_1, A_2, \ldots, A_r$ in which the number of A_i's is x_i, $i = 1, 2, \ldots, r$. Relation (9.16) follows since the probability of the occurrence of any of these sequences is $p_1^{x_1} p_2^{x_2} \cdots p_r^{x_r}$. By Theorem 2.6, $p(x_1, x_2, \ldots, x_r)$'s given by (9.16) are the terms in the expansion of $(p_1 + p_2 + \cdots + p_r)^n$. For this reason, the multinomial distribution sometimes is called the **polynomial distribution**. The following relation guarantees that $p(x_1, x_2, \ldots, x_r)$ is a joint probability mass function:

$$\sum_{x_1 + x_2 + \cdots + x_r = n} p(x_1, x_2, \ldots, x_r)$$

$$= \sum_{x_1 + x_2 + \cdots + x_r = n} \frac{n!}{x_1!\, x_2! \cdots x_r!}\, p_1^{x_1} p_2^{x_2} \cdots p_r^{x_r} = (p_1 + p_2 + \cdots + p_r)^n = 1.$$

Note that, for $r = 2$, the multinomial distribution coincides with the binomial distribution. This is because, for $r = 2$, the experiment has only two possible outcomes, A_1 and A_2. Hence it is a Bernoulli trial.

Example 9.11   In a certain town, at 8:00 P.M., 30% of the TV viewing audience watch the news, 25% watch a certain comedy, and the rest watch other programs. What is the probability that, in a statistical survey of seven randomly selected viewers, exactly three watch the news and at least two watch the comedy?

Solution: In the random sample, let X_1, X_2, and X_3 be the numbers of viewers who watch the news, the comedy, and other programs, respectively. Then the joint distribution of X_1, X_2, and X_3 is multinomial with $p_1 = 0.30$, $p_2 = 0.25$, and $p_3 = 0.45$. Therefore, for $i + j + k = 7$,

$$P(X_1 = i, X_2 = j, X_3 = k) = \frac{7!}{i!\, j!\, k!}(0.30)^i (0.25)^j (0.45)^k.$$

The desired probability equals

$$P(X_1 = 3, X_2 \geq 2) = P(X_1 = 3, X_2 = 2, X_3 = 2)$$
$$+ P(X_1 = 3, X_2 = 3, X_3 = 1) + P(X_1 = 3, X_2 = 4, X_3 = 0)$$
$$= \frac{7!}{3!\,2!\,2!}(0.30)^3(0.25)^2(0.45)^2 + \frac{7!}{3!\,3!\,1!}(0.30)^3(0.25)^3(0.45)^1$$
$$+ \frac{7!}{3!\,4!\,0!}(0.30)^3(0.25)^4(0.45)^0 \approx 0.103. \quad \blacklozenge$$

Example 9.12   A warehouse contains 500 TV sets, of which 25 are defective, 300 are in working condition but used, and the rest are brand new. What is the probability that, in a random sample of five TV sets from this warehouse, there are exactly one defective and exactly two brand new sets?

Solution: In the random sample, let X_1 be the number of defective TV sets, X_2 be the number of TV sets in working condition but used, and X_3 be the number of brand new sets. The desired probability is given by

$$P(X_1 = 1, X_2 = 2, X_3 = 2) = \frac{\binom{25}{1}\binom{300}{2}\binom{175}{2}}{\binom{500}{5}} \approx 0.067.$$

Note that, since the number of sets is large and the sample size is small, the joint distribution of X_1, X_2, and X_3 is *approximately* multinomial with $p_1 = 25/500 = 1/20$, $p_2 = 300/500 = 3/5$, and $p_3 = 175/500 = 7/20$. Hence

$$P(X_1 = 1, X_2 = 2, X_3 = 2) \approx \frac{5!}{1!\,2!\,2!}\left(\frac{1}{20}\right)^1\left(\frac{3}{5}\right)^2\left(\frac{7}{20}\right)^2 \approx 0.066. \quad \blacklozenge$$

Example 9.13   (Marginals of Multinomials)   Let $X_1, X_2, \ldots, X_r$ $(r \geq 4)$ have the joint multinomial probability mass function $p(x_1, x_2, \ldots, x_r)$ with parameters n and $p_1, p_2, \ldots, p_r$. Find the marginal probability mass functions p_{X_1} and p_{X_1, X_2, X_3}.

Solution: From the definition of marginal probability mass functions,

$$p_{X_1}(x_1) = \sum_{x_2+x_3+\cdots+x_r=n-x_1} \frac{n!}{x_1!\,x_2!\cdots x_r!}\, p_1^{x_1} p_2^{x_2} \cdots p_r^{x_r}$$

$$= \frac{n!}{x_1!\,(n-x_1)!}\, p_1^{x_1} \sum_{x_2+x_3+\cdots+x_r=n-x_1} \frac{(n-x_1)!}{x_2!\,x_3!\cdots x_r!}\, p_2^{x_2} p_3^{x_3} \cdots p_r^{x_r}$$

$$= \frac{n!}{x_1!\,(n-x_1)!}\, p_1^{x_1}\, (p_2 + p_3 + \cdots + p_r)^{n-x_1}$$

$$= \frac{n!}{x_1!\,(n-x_1)!}\, p_1^{x_1}\, (1 - p_1)^{n-x_1},$$

where the next-to-last equality follows from multinomial expansion (Theorem 2.6) and the last inequality follows from $p_1 + p_2 + \cdots + p_r = 1$. This shows that *the marginal probability mass function of X_1 is binomial with parameters n and p_1*. To find the joint probability mass function marginalized over X_1, X_2, and X_3, let $k = n - x_1 - x_2 - x_3$; then

$$p_{X_1,X_2,X_3}(x_1, x_2, x_3)$$

$$= \sum_{x_4 + \cdots + x_r = k} \frac{n!}{x_1! \, x_2! \, x_3! \, x_4! \cdots x_r!} \, p_1^{x_1} p_2^{x_2} p_3^{x_3} p_4^{x_4} \cdots p_r^{x_r}$$

$$= \frac{n!}{x_1! \, x_2! \, x_3! \, k!} \, p_1^{x_1} p_2^{x_2} p_3^{x_3} \sum_{x_4 + \cdots + x_r = k} \frac{k!}{x_4! \cdots x_r!} \, p_4^{x_4} \cdots p_r^{x_r}$$

$$= \frac{n!}{x_1! \, x_2! \, x_3! \, k!} \, p_1^{x_1} p_2^{x_2} p_3^{x_3} \, (p_4 + \cdots + p_r)^k$$

$$= \frac{n!}{x_1! \, x_2! \, x_3! \, (n - x_1 - x_2 - x_3)!} \, p_1^{x_1} p_2^{x_2} p_3^{x_3} \, (1 - p_1 - p_2 - p_3)^{n - x_1 - x_2 - x_3}.$$

This shows that *the joint probability mass function marginalized over X_1, X_2, and X_3 is multinomial with parameters n and p_1, p_2, p_3, $1 - p_1 - p_2 - p_3$.* ◆

Remark 9.3   The method of Example 9.13 can be extended to prove the following theorem:

Let the joint distribution of the random variables X_1, X_2, ..., X_r be multinomial with parameters n and $p_1, p_2, \ldots, p_r$. The joint probability mass function marginalized over a subset $X_{i_1}, X_{i_2}, \ldots, X_{i_k}$ of k ($k > 1$) of these r random variables is multinomial with parameters n and p_{i_1}, $p_{i_2}, \ldots, p_{i_k}, 1 - p_{i_1} - p_{i_2} - \cdots - p_{i_k}$. ◆

EXERCISES

A

1. Light bulbs manufactured by a certain factory last a random time between 400 and 1200 hours. What is the probability that, of eight such bulbs, three burn out before 550 hours, two burn out after 800 hours, and three burn out after 550 but before 800 hours?

2. An urn contains 100 chips of which 20 are blue, 30 are red, and 50 are green. We draw 20 chips at random and with replacement. Let B, R, and G be the number of blue, red, and green chips, respectively. Calculate the joint probability mass function of B, R, and G.

3. Suppose that each day the price of a stock moves up 1/8 of a point with probability 1/4, remains the same with probability 1/3, and moves down 1/8 of a point with probability 5/12. If the price fluctuations from one day to another are independent, what is the probability that after six days the stock has its original price?

4. At a certain college, 16% of the calculus students get A's, 34% B's, 34% C's, 14% D's, and 2% F's. What is the probability that, of 15 calculus students selected at random, five get B's, five C's, two D's, and at least two A's?

5. Suppose that 50% of the watermelons grown on a farm are classified as large, 30% as medium, and 20% as small. Joanna buys five watermelons at random from this farm. What is the probability that (a) at least two of them are large; (b) two of them are large, two are medium, and one is small; (c) exactly two of them are medium if it is given that at least two of them are large?

6. Suppose that the ages of 30% of the teachers of a country are over 50, 20% are between 40 and 50, and 50% are below 40. In a random committee of 10 teachers from this country, two are above 50. What is the probability mass function of those who are below 40?

7. **(Genetics)**   As we know, in humans, for blood type, there are three alleles A, B, and O. The alleles A and B are codominant to each other and dominant to O. A man of genotype AB marries a woman of genotype BO. If they have six children, what is the probability that three will have type B blood, two will have type A blood, and one will have type AB blood?

8. **(Genetics)**   Let p and q be positive numbers with $p + q = 1$. For a gene with dominant allele A and recessive allele a, let p^2, $2pq$, and q^2 be the probabilities that a randomly selected person from a population has genotype AA, Aa, and aa, respectively. A group of six members of the population is selected randomly. Determine the value of p that maximizes the probability of the event of obtaining two AA's, two Aa's, and two aa's.

B

9. Customers enter a department store at the rate of three per minute, in accordance with a Poisson process. If 30% of them buy nothing, 20% pay cash, 40% use charge cards, and 10% write personal checks, what is the probability that in five operating minutes of the store, five customers use charge cards, two write personal checks, and three pay cash?

REVIEW PROBLEMS

1. An urn contains 100 chips of which 20 are blue, 30 are red, and 50 are green. Suppose that 20 chips are drawn at random and without replacement. Let B, R, and G be the number of blue, red, and green chips, respectively. Calculate the joint probability mass function of B, R, and G.

2. Let X be the smallest number obtained in rolling a balanced die n times. Calculate the probability distribution function and the probability mass function of X.

3. Suppose that n points are selected at random and independently inside the cube

$$\Omega = \{(x, y, z): -a \le x \le a, -a \le y \le a, -a \le z \le a\}.$$

Find the probability that the distance of the nearest point to the center is at least r $(r < a)$.

4. The joint probability density function of random variables X, Y, and Z is given by

$$f(x, y, z) = \begin{cases} c(x + y + 2z) & \text{if } 0 \le x, y, z \le 1 \\ 0 & \text{otherwise.} \end{cases}$$

   (a) Determine the value of c.

   (b) Find $P(X < 1/3 \mid Y < 1/2, \; Z < 1/4)$.

5. A fair die is tossed 18 times. What is the probability that each face appears three times?

6. Alvie, a marksman, fires seven independent shots at a target. Suppose that the probabilities that he hits the bull's-eye, he hits the target but not the bull's-eye, and he misses the target are 0.4, 0.35, and 0.25, respectively. What is the probability that he hits the bull's-eye three times, the target but not the bull's-eye two times, and misses the target two times?

7. A system consists of n components whose lifetimes form an independent sequence of random variables. In order for the system to function, all components must function. Let $F_1, F_2, \ldots, F_n$ be the distribution functions of the lifetimes of the components of the system. In terms of $F_1, F_2, \ldots, F_n$, find the survival function of the lifetime of the system.

8. A system consists of n components whose lifetimes form an independent sequence of random variables. Suppose that the system functions as long as at least one of its components functions. Let $F_1, F_2, \ldots, F_n$ be the distribution functions of the lifetimes of the components of the system. In terms of $F_1, F_2, \ldots, F_n$, find the survival function of the lifetime of the system.

9. A bar of length ℓ is broken into three pieces at two random spots. What is the probability that the length of at least one piece is less than $\ell/20$?

10. Let X_1, X_2, and X_3 be independent random variables from $(0, 1)$. Find the probability density function and the expected value of the midrange of these random variables $[X_{(1)} + X_{(3)}]/2$.

■

Chapter 10

More Expectations and Variances

10.1 EXPECTED VALUES OF SUMS OF RANDOM VARIABLES

As we have seen, for a discrete random variable X with set of possible values A and probability mass function p, $E(X)$ is defined by $\sum_{x \in A} x p(x)$. For a continuous random variable X with probability density function f, the same quantity, $E(X)$, is defined by $\int_{-\infty}^{\infty} x f(x)\, dx$. Recall that for a random variable X, the expected value, $E(X)$, might not exist (see Examples 4.18 and 4.19 and Exercise 11, Section 6.3). In the following discussion we always assume that the expected value of a random variable exists.

To begin, we prove the linearity property of expectation for continuous random variables.

Theorem 10.1 *For random variables $X_1, X_2, \ldots, X_n$ defined on the same sample space,*

$$E\left(\sum_{i=1}^{n} \alpha_i X_i \right) = \sum_{i=1}^{n} \alpha_i E(X_i).$$

Proof: For convenience, we prove this for continuous random variables. For discrete random variables the proof is similar. Let $f(x_1, x_2, \ldots, x_n)$ be the joint probability density function of $X_1, X_2, \ldots, X_n$; then, by Theorem 9.4,

$$E\left(\sum_{i=1}^{n} \alpha_i X_i \right) = \int_{-\infty}^{\infty} \int_{-\infty}^{\infty} \cdots \int_{-\infty}^{\infty} \left(\sum_{i=1}^{n} \alpha_i x_i \right) f(x_1, x_2, \ldots, x_n)\, dx_1\, dx_2 \cdots dx_n$$

$$= \sum_{i=1}^{n} \alpha_i \int_{-\infty}^{\infty} \int_{-\infty}^{\infty} \cdots \int_{-\infty}^{\infty} x_i f(x_1, x_2, \ldots, x_n)\, dx_1\, dx_2 \cdots dx_n$$

$$= \sum_{i=1}^{n} \alpha_i E(X_i). \quad \blacklozenge$$

In Theorem 10.1, letting $\alpha_i = 1, 1 \leq i \leq n$, we have the following important corollary.

Corollary   *Let $X_1, X_2, \ldots, X_n$ be random variables on the same sample space. Then*

$$E(X_1 + X_2 + \cdots + X_n) = E(X_1) + E(X_2) + \cdots + E(X_n).$$

Example 10.1   A die is rolled 15 times. What is the expected value of the sum of the outcomes?

Solution: Let X be the sum of the outcomes, and for $i = 1, 2, \ldots, 15$; let X_i be the outcome of the ith roll. Then $X = X_1 + X_2 + \cdots + X_{15}$. Thus

$$E(X) = E(X_1) + E(X_2) + \cdots + E(X_{15}).$$

Now since for all values of i, X_i is 1, 2, 3, 4, 5, and 6, with probability of 1/6 for all six values,

$$E(X_i) = 1 \cdot \frac{1}{6} + 2 \cdot \frac{1}{6} + 3 \cdot \frac{1}{6} + 4 \cdot \frac{1}{6} + 5 \cdot \frac{1}{6} + 6 \cdot \frac{1}{6} = \frac{7}{2}.$$

Hence $E(X) = 15(7/2) = 52.5.$ ◆

Example 10.1 and the following examples show the power of Theorem 10.1 and its corollary. Note that in Example 10.1, while the formula

$$E(X_1 + X_2 + \cdots + X_n) = E(X_1) + E(X_2) + \cdots + E(X_n)$$

enables us to compute $E(X)$ readily, computing $E(X)$ directly from probability mass function of X is not easy. This is because calculation of the probability mass function of X is time-consuming and cumbersome.

Example 10.2   A well-shuffled ordinary deck of 52 cards is divided randomly into four piles of 13 each. Counting jack, queen, and king as 11, 12, and 13, respectively, we say that a match occurs in a pile if the jth card is j. What is the expected value of the total number of matches in all four piles?

Solution: Let $X_i, i = 1, 2, 3, 4$ be the number of matches in the ith pile. $X = X_1 + X_2 + X_3 + X_4$ is the total number of matches, and $E(X) = E(X_1) + E(X_2) + E(X_3) + E(X_4)$. To calculate $E(X_i)$, $1 \leq i \leq 4$, let A_{ij} be the event that the jth card in the ith pile is j ($1 \leq i \leq 4, 1 \leq j \leq 13$). Then by defining

$$X_{ij} = \begin{cases} 1 & \text{if } A_{ij} \text{ occurs} \\ 0 & \text{otherwise,} \end{cases}$$

we have that $X_i = \sum_{j=1}^{13} X_{ij}$. Now $P(A_{ij}) = 4/52 = 1/13$ implies that

$$E(X_{ij}) = 1 \cdot P(A_{ij}) + 0 \cdot P(A_{ij}^c) = P(A_{ij}) = \frac{1}{13}.$$

Hence

$$E(X_i) = E\left(\sum_{j=1}^{13} X_{ij} \right) = \sum_{j=1}^{13} E(X_{ij}) = \sum_{j=1}^{13} \frac{1}{13} = 1.$$

Thus on average there is one match in every pile. From this we get

$$E(X) = E(X_1) + E(X_2) + E(X_3) + E(X_4) = 1 + 1 + 1 + 1 = 4,$$

showing that on average there are a total of four matches.   ◆

Example 10.3   Exactly n married couples are living in a small town. What is the expected number of intact couples after m deaths occur among the couples? Assume that the deaths occur at random, there are no divorces, and there are no new marriages.

Solution: Let X be the number of intact couples after m deaths, and for $i = 1, 2, \ldots, n$ define

$$X_i = \begin{cases} 1 & \text{if the } i\text{th couple is left intact} \\ 0 & \text{otherwise.} \end{cases}$$

Then $X = X_1 + X_2 + \cdots + X_n$, and hence

$$E(X) = E(X_1) + E(X_2) + \cdots + E(X_n),$$

where

$$E(X_i) = 1 \cdot P(X_i = 1) + 0 \cdot P(X_i = 0) = P(X_i = 1).$$

The event $\{X_i = 1\}$ occurs if the ith couple is left intact, and thus all of the m deaths are among the remaining $n - 1$ couples. Knowing that the deaths occur at random among these individuals, we can write

$$P(X_i = 1) = \frac{\binom{2n - 2}{m}}{\binom{2n}{m}} = \frac{\dfrac{(2n - 2)!}{m! \, (2n - m - 2)!}}{\dfrac{(2n)!}{(2n - m)! \, m!}} = \frac{(2n - m)(2n - m - 1)}{2n(2n - 1)}.$$

Thus the desired quantity is equal to

$$E(X) = E(X_1) + E(X_2) + \cdots + E(X_n) = n \, \frac{(2n - m)(2n - m - 1)}{2n(2n - 1)}$$

$$= \frac{(2n - m)(2n - m - 1)}{2(2n - 1)}.$$

To have a numerical feeling for this interesting example, let $n = 1000$. Then the following table shows the effect of the number of deaths on the expected value of the number of intact couples.

m	100	300	600	900	1200	1500	1800
$E(X)$	902.48	722.44	489.89	302.38	159.88	62.41	9.95

This example was posed by Daniel Bernoulli (1700–1782).   ♦

Example 10.4   Dr. Windler's secretary accidentally threw a patient's file into the wastebasket. A few minutes later, the janitor cleaned the entire clinic, dumped the wastebasket containing the patient's file randomly into one of the seven garbage cans outside the clinic, and left. Determine the expected number of cans that Dr. Windler should empty to find the file.

Solution: Let X be the number of garbage cans that Dr. Windler should empty to find the patient's file. For $i = 1, 2, \ldots, 7$, let $X_i = 1$ if the patient's file is in the ith garbage can that Dr. Windler will empty, and $X_i = 0$, otherwise. Then

$$X = 1 \cdot X_1 + 2 \cdot X_2 + \cdots + 7 \cdot X_7,$$

and, therefore,

$$E(X) = 1 \cdot E(X_1) + 2 \cdot E(X_2) + \cdots + 7 \cdot E(X_7)$$

$$= 1 \cdot \frac{1}{7} + 2 \cdot \frac{1}{7} + \cdots + 7 \cdot \frac{1}{7} = 4. \quad ♦$$

Example 10.5   A box contains nine light bulbs, of which two are defective. What is the expected value of the number of light bulbs that one will have to test (at random and without replacement) to find both defective bulbs?

Solution: For $i = 1, 2, \cdots, 8$ and $j > i$, let $X_{ij} = j$ if the ith and jth light bulbs to be examined are defective, and $X_{ij} = 0$ otherwise. Then $\sum_{i=1}^{8} \sum_{j=i+1}^{9} X_{ij}$ is the number of light bulbs to be examined. Therefore,

$$E(X) = \sum_{i=1}^{8} \sum_{j=i+1}^{9} E(X_{ij}) = \sum_{i=1}^{8} \sum_{j=i+1}^{9} j \frac{1}{\binom{9}{2}}$$

$$= \frac{1}{36} \sum_{i=1}^{8} \sum_{j=i+1}^{9} j = \frac{1}{36} \sum_{i=1}^{8} \frac{90 - i^2 - i}{2} \approx 6.67,$$

where the next-to-last equality follows from

$$\sum_{j=i+1}^{9} j = \sum_{j=1}^{9} j - \sum_{j=1}^{i} j = \frac{9 \times 10}{2} - \frac{i(i+1)}{2} = \frac{90 - i^2 - i}{2}. \quad \blacklozenge$$

An elegant application of the corollary of Theorem 10.1 is that it can be used to calculate the expected values of random variables, such as binomial, negative binomial, and hypergeometric. The following examples demonstrate some applications.

Example 10.6   Let X be a binomial random variable with parameters (n, p). Recall that X is the number of successes in n independent Bernoulli trials. Thus, for $i = 1, 2, \ldots, n$, letting

$$X_i = \begin{cases} 1 & \text{if the } i\text{th trial is a success} \\ 0 & \text{otherwise,} \end{cases}$$

we get

$$X = X_1 + X_2 + \cdots + X_n, \tag{10.1}$$

where X_i is a Bernoulli random variable for $i = 1, 2, \ldots, n$. Now, since $\forall i, \ 1 \le i \le n$,

$$E(X_i) = 1 \cdot p + 0 \cdot (1 - p) = p,$$

(10.1) implies that

$$E(X) = E(X_1) + E(X_2) + \cdots + E(X_n) = np. \quad \blacklozenge$$

Example 10.7   Let X be a negative binomial random variable with parameters (r, p). Then in a sequence of independent Bernoulli trials each with success probability p, X is the number of trials until the rth success. Let X_1 be the number of trials until the first success, X_2 be the number of additional trials to get the second success, X_3 the number of additional ones to obtain the third success, and so on. Then clearly

$$X = X_1 + X_2 + \cdots + X_r,$$

where for $i = 1, 2, \ldots, n$, the random variable X_i is geometric with parameter p. This is because $P(X_i = n) = (1-p)^{n-1} p$ by the independence of the trials. Since $E(X_i) = 1/p$ $(i = 1, 2, \ldots, r)$,

$$E(X) = E(X_1) + E(X_2) + \cdots + E(X_r) = \frac{r}{p}.$$

This formula shows that, for example, in the experiment of throwing a fair die successively, on the average, it takes $5/(1/6) = 30$ trials to get five 6's. $\quad \blacklozenge$

Example 10.8 Let X be a hypergeometric random variable with probability mass function

$$p(x) = P(X = x) = \frac{\binom{D}{x}\binom{N-D}{n-x}}{\binom{N}{n}},$$

$$n \leq \min(D, N-D), \quad x = 0, 1, 2, \ldots, n.$$

Then X is the number of defective items among n items drawn at random and without replacement from an urn containing D defective and $N - D$ nondefective items. To calculate $E(X)$, let A_i be the event that the ith item drawn is defective. Also, for $i = 1, 2, \ldots, n$, let

$$X_i = \begin{cases} 1 & \text{if } A_i \text{ occurs} \\ 0 & \text{otherwise;} \end{cases}$$

then $X = X_1 + X_2 + \cdots + X_n$. Hence

$$E(X) = E(X_1) + E(X_2) + \cdots + E(X_n),$$

where for $i = 1, 2, \ldots, n$,

$$E(X_i) = 1 \cdot P(X_i = 1) + 0 \cdot P(X_i = 0) = P(X_i = 1) = P(A_i) = \frac{D}{N}.$$

This follows since the ith item can be any of the N items with equal probabilities. Therefore,

$$E(X) = \frac{nD}{N}.$$

This shows that, for example, the expected number of spades in a random bridge hand is $(13 \times 13)/52 = 13/4 \approx 3.25$. ◆

Remark 10.1 For $n = \infty$, Theorem 10.1 is not necessarily true. That is, $E\left(\sum_{i=1}^{\infty} X_i\right)$ might not be equal to $\sum_{i=1}^{\infty} E(X_i)$. To show this, we give a counterexample. For $i = 1, 2, 3, \ldots$, let

$$Y_i = \begin{cases} i & \text{with probability } 1/i \\ 0 & \text{otherwise,} \end{cases}$$

and $X_i = Y_{i+1} - Y_i$. Then since $E(Y_i) = 1$ for $i = 1, 2, \ldots$, $E(X_i) = E(Y_{i+1}) - E(Y_i) = 0$. Hence $\sum_{i=1}^{\infty} E(X_i) = 0$. However, since $\sum_{i=1}^{\infty} X_i = -Y_1$, we have that $E\left(\sum_{i=1}^{\infty} X_i\right) = E(-Y_1) = -1$. Therefore, $E\left(\sum_{i=1}^{\infty} X_i\right) \neq \sum_{i=1}^{\infty} E(X_i)$. ◆

It can be shown that, in general,

If $\sum_{i=1}^{\infty} E(|X_i|) < \infty$ *or* if, for all i, the random variables $X_1, X_1, \ldots$ are nonnegative [that is, $P(X_i \geq 0) = 1$ for $i \geq 1$], then

$$E\left(\sum_{i=1}^{\infty} X_i\right) = \sum_{i=1}^{\infty} E(X_i). \tag{10.2}$$

As an application of (10.2), we prove the following important theorem.

Theorem 10.2   *Let N be a discrete random variable with set of possible values* $\{1, 2, 3, \ldots\}$. *Then*

$$E(N) = \sum_{i=1}^{\infty} P(N \geq i).$$

Proof: For $i \geq 1$, let

$$X_i = \begin{cases} 1 & \text{if } N \geq i \\ 0 & \text{otherwise;} \end{cases}$$

then

$$\sum_{i=1}^{\infty} X_i = \sum_{i=1}^{N} X_i + \sum_{i=N+1}^{\infty} X_i = \sum_{i=1}^{N} 1 + \sum_{i=N+1}^{\infty} 0 = N.$$

Since $P(X_i \geq 0) = 1$, for $i \geq 1$,

$$E(N) = E\left(\sum_{i=1}^{\infty} X_i\right) = \sum_{i=1}^{\infty} E(X_i) = \sum_{i=1}^{\infty} P(N \geq i). \quad \blacklozenge$$

Note that Theorem 10.2 is the analog of the fact that if X is a continuous nonnegative random variable, then

$$E(X) = \int_0^{\infty} P(X > t) \, dt.$$

This is explained in Remark 6.4.

We now prove an important inequality called the *Cauchy-Schwarz inequality*.

Theorem 10.3   (Cauchy-Schwarz Inequality)   *For random variables X and Y,*

$$E(XY) \leq \sqrt{E(X^2)E(Y^2)}.$$

Proof: For all real numbers λ, $(X - \lambda Y)^2 \geq 0$. Hence, for all values of λ, $X^2 - 2XY\lambda + \lambda^2 Y^2 \geq 0$. Since nonnegative random variables have nonnegative expectations,

$$E(X^2 - 2XY\lambda + \lambda^2 Y^2) \geq 0,$$

which implies that

$$E(X^2) - 2E(XY)\lambda + \lambda^2 E(Y^2) \geq 0.$$

Rewriting this as a polynomial in λ of degree 2, we get

$$E(Y^2)\lambda^2 - 2E(XY)\lambda + E(X^2) \geq 0.$$

It is a well-known fact that if a polynomial of degree 2 is positive, its discriminant is negative. Therefore,

$$4\big[E(XY)\big]^2 - 4E(X^2)E(Y^2) \leq 0$$

or

$$\big[E(XY)\big]^2 \leq E(X^2)E(Y^2).$$

This gives

$$E(XY) \leq \sqrt{E(X^2)E(Y^2)}. \quad \blacklozenge$$

Corollary   *For a random variable X, $\big[E(X)\big]^2 \leq E(X^2)$.*

Proof: In Cauchy-Schwarz's inequality, let $Y = 1$; then

$$E(X) = E(XY) \leq \sqrt{E(X^2)E(1)} = \sqrt{E(X^2)};$$

thus $\big[E(X)\big]^2 \leq E(X^2)$. $\quad \blacklozenge$

★ Pattern Appearance[†]

Suppose that a coin is tossed independently and successively. We are interested in the expected number of tosses until a specific pattern is first obtained. For example, the expected number of tosses until the first appearance of HT, the first appearance of HHH, or the first appearance of, say, HHTHH. Similarly, suppose that in generating random numbers from the set $\{0, 1, 2, \ldots , 9\}$ independently and successively, we are interested

---

[†]This topic may be skipped without loss of continuity. If it is, then all exercises and examples in this and future chapters marked "(Pattern Appearance)" should be skipped as well.

in the expected number of digits to be generated until the first appearance of 453, the first appearance of 353, or the first appearance of, say, 88588.

In coin tossing, for example, we can readily calculate the expected number of tosses after the first appearance of a pattern until its second appearance. The same is true in generating random numbers and other similar experiments. As an example, shortly, we will show that it is not difficult to calculate the expected number of digits to be generated after the first appearance of a pattern until its second appearance. We use this to calculate the expected number of digits to be generated until the first appearance of the pattern. First, we introduce some notation and make some definitions:

Suppose that A is a pattern of certain characters. Let the number of random characters generated until A appears for the first time be denoted by $\rightarrow A$. Let the number of random characters generated after the appearance of a pattern A until the next appearance of a pattern B be denoted by $A \rightarrow B$.

Definition   *Suppose that A is a pattern of length n of certain characters. For $1 \le i < n$, let $A_{(i)}$ and $A^{(i)}$ be the first i and the last i characters of A, respectively. If $A_{(i)} = A^{(i)}$ for some i ($1 \le i < n$), we say that the pattern is **overlapping**. Otherwise, we say it is a pattern with **no self-overlap**. For an overlapping pattern A, if $A_{(i)} = A^{(i)}$, then i is said to be an **overlap number**.*

For example, in successive and independent tosses of a coin, let $A =$ HHTHH; we have

$$A_{(1)} = \text{H}, \quad A_{(2)} = \text{HH}, \quad A_{(3)} = \text{HHT}, \quad A_{(4)} = \text{HHTH};$$

$$A^{(1)} = \text{H}, \quad A^{(2)} = \text{HH}, \quad A^{(3)} = \text{THH}, \quad A^{(4)} = \text{HTHH}.$$

We see that A is overlapping with overlap numbers 1 and 2. As another example, in generating random digits from the set $\{0, 1, 2, \dots, 9\}$ independently and successively, consider the pattern $A = 45345$; we have

$$A_{(1)} = 4, \quad A_{(2)} = 45, \quad A_{(3)} = 453, \quad A_{(4)} = 4534;$$

$$A^{(1)} = 5, \quad A^{(2)} = 45, \quad A^{(3)} = 345, \quad A^{(4)} = 5345.$$

Since $A_{(2)} = A^{(2)}$, the pattern A is overlapping, and 2 is its only overlap number. As a third example, the pattern 333333 is overlapping. Its overlap numbers are 1, 2, 3, 4, and 5. The patterns 3334 and 44565 are patterns with no self-overlap.

Next, we make the following observations: For a pattern with no self-overlap such as 453, $E(\rightarrow 453) = E(453 \rightarrow 453)$. That is, the expected number of random numbers to be generated until the first 453 appears is the same as the expected number of random numbers to be generated after the first appearance of 453 until its second appearance. However, this does not apply to an overlapping pattern such as 353. To our surprise, $E(\to 353)$ is *larger* than $E(353 \to 353)$. This is because $E(353 \to 353) = E(3 \to 353)$ and

$$E(\to 353) = E(\to 3) + E(3 \to 353)$$
$$= E(\to 3) + E(353 \to 353).$$

That is, the expected number of digits to be generated until the pattern 353 appears is the expected number of digits to be generated until the first appearance of 3, plus the expected number of additional digits after that until 353 is obtained for the first time. In general, *for patterns of the same length, on average, overlapping ones occur later than those with no self-overlap.* For example, when random digits are generated one after another independently and successively, on average, 453 appears sooner than 353, and 353 appears sooner than 333. Note that

$$E(\to 333) = E(\to 3) + E(3 \to 33) + E(33 \to 333)$$
$$= E(\to 3) + E(33 \to 33) + E(333 \to 333).$$

In general, let A be an overlapping pattern. Let $i_1, i_2, i_3, \ldots, i_m$ be the overlap numbers of A arranged in (strictly) ascending order. Then

$$E(\to A) = E(\to A_{(i_1)}) + E(A_{(i_1)} \to A_{(i_2)}) + E(A_{(i_2)} \to A_{(i_3)})$$
$$+ \cdots + E(A_{(i_m)} \to A).$$

Clearly, the number of digits to be generated to obtain, say, the first 3, the first 5, or the first 8 are all geometric random variables with parameter $p = 1/10$. So

$$E(\to 3) = E(\to 5) = E(\to 8) = \frac{1}{p} = 10.$$

In general, for a pattern A, this and $E(A \to A)$ enable us to find $E(\to A)$. So all that remains is to find the expected number of digits to be generated after the first appearance of a pattern until its second appearance. To demonstrate one method of calculation, we will find $E(353 \to 353)$ using an *intuitive probabilistic* argument. To do so, in the process of generating random digits, let D_1 be the first random digit and D_2 be the first two random digits generated. For $i \geq 3$, let D_i be the $(i-2)$nd, $(i-1)$st, and ith random digits generated. For example, if the first 6 random digits generated are 507229, then $D_1 = 5$, $D_2 = 50$, $D_3 = 507$, $D_4 = 072$, $D_5 = 722$, and $D_6 = 229$. Let $X_1 = X_2 = 0$. For $i \geq 3$, let

$$X_i = \begin{cases} 1 & \text{if } D_i = 353 \\ 0 & \text{if } D_i \neq 353. \end{cases}$$

Then the number of appearances of 353 among the first n random digits generated is $\sum_{i=1}^{n} X_i$. Hence the proportion of the D_i's that are 353 is $(1/n) \sum_{i=1}^{n} X_i$ when n random digits are generated. Now $E(X_1) = E(X_2) = 0$, and for $i \geq 3$, $E(X_i) = E(X_3)$ since the X_i's are identically distributed. Thus the expected value of the fraction of D_i's that are 353 is

$$E\left(\frac{1}{n}\sum_{i=1}^{n}X_i\right) = \frac{1}{n}E\left(\sum_{i=1}^{n}X_i\right) = \frac{1}{n}\left[\sum_{i=1}^{n}E(X_i)\right] = \frac{1}{n}\cdot(n-2)E(X_3)$$

$$= \frac{n-2}{n}P(D_3 = 353) = \frac{n-2}{n}\left(\frac{1}{10}\right)^3 = \frac{n-2}{1000\,n}.$$

Thus the expected value of the fraction of D_i's in n random digits that are 353 is $(n-2)/(1000n)$. Now as $n \to \infty$, this expected value approaches $1/1000$, implying that in the long-run, on average, the fraction of D_i's that are 353 is $1/1000$. This and the fact that the expected number of random digits between any two consecutive 353's is the same, imply that the average number of random digits between two consecutive 353's is 1000. Hence $E(353 \to 353) = 1000$. A rigorous proof for this fact needs certain theorems of *renewal theory*, a branch of stochastic processes. See, for example, *Stochastic Modeling and the Theory of Queues*, by Ronald W. Wolff, Prentice Hall, 1989. We can use the argument above to show that, for example,

$$E(23 \to 23) = 100,$$
$$E(453 \to 453) = 1000,$$
$$E(333 \to 333) = 1000,$$
$$E(5732 \to 5732) = 10,000.$$

Therefore,

$$E(\to 453) = E(453 \to 453) = 1000,$$

$$\begin{aligned}E(\to 353) &= E(\to 3) + E(3 \to 353)\\ &= E(\to 3) + E(353 \to 353)\\ &= 10 + 1000 = 1010,\end{aligned}$$

$$\begin{aligned}E(\to 333) &= E(\to 3) + E(3 \to 33) + E(33 \to 333)\\ &= E(\to 3) + E(33 \to 33) + E(333 \to 333)\\ &= 10 + 100 + 1000 = 1110,\end{aligned}$$

and

$$\begin{aligned}E(\to 88588) &= E(\to 8) + E(8 \to 88) + E(88 \to 88588)\\ &= E(\to 8) + E(88 \to 88) + E(88588 \to 88588)\\ &= 10 + 100 + 100000 = 100110.\end{aligned}$$

An argument similar to the preceding will establish the following generalization.

Suppose that an experiment results in one of the outcomes a_1, a_2, ..., a_k, and $P(\{a_i\}) = p_i$, $i = 1, 2, \ldots, k$; $\sum_{i=1}^{k} p_i = 1$. Let a_{i_1}, a_{i_2}, ..., a_{i_ℓ}, be (not necessarily distinct) elements of $\{a_1, a_2, \ldots, a_k\}$. Then in successive and independent performances of this experiment, the expected number of trials after the first appearance of the pattern $a_{i_1} a_{i_2} \cdots a_{i_\ell}$ until its second appearances is $1/(p_{i_1} p_{i_2} \cdots p_{i_\ell})$.

By this generalization, for example, in successive independent flips of a fair coin,

$$E(\to \text{HH}) = E(\to \text{H}) + E(\text{H} \to \text{HH})$$
$$= E(\to \text{H}) + E(\text{HH} \to \text{HH})$$
$$= \frac{1}{1/2} + \frac{1}{(1/2)(1/2)} = 6,$$

whereas

$$E(\to \text{HT}) = E(\text{HT} \to \text{HT}) = \frac{1}{(1/2)(1/2)} = 4.$$

Similarly,

$$E(\to \text{HHHH}) = E(\to \text{H}) + E(\text{H} \to \text{HH}) + E(\text{HH} \to \text{HHH})$$
$$+ E(\text{HHH} \to \text{HHHH})$$
$$= E(\to \text{H}) + E(\text{HH} \to \text{HH}) + E(\text{HHH} \to \text{HHH})$$
$$+ E(\text{HHHH} \to \text{HHHH})$$
$$= 2 + 4 + 8 + 16 = 30,$$

whereas

$$E(\to \text{THHH}) = E(\text{THHH} \to \text{THHH}) = 16.$$

At first glance, it seems paradoxical that, on the average, it takes nearly twice as many flips of a fair coin to obtain the first HHHH as to encounter THHH for the first time. However, THHH is a pattern with no self-overlap, whereas HHHH is an overlapping pattern.

EXERCISES

A

1. Let the probability density function of a random variable X be given by

$$f(x) = \begin{cases} |x - 1| & \text{if } 0 \leq x \leq 2 \\ 0 & \text{otherwise.} \end{cases}$$

Find $E(X^2 + X)$.

2. A calculator is able to generate random numbers from the interval $(0, 1)$. We need five random numbers from $(0, 2/5)$. Using this calculator, how many independent random numbers should we generate, on average, to find the five numbers needed?

3. Let X, Y, and Z be three independent random variables such that $E(X) = E(Y) = E(Z) = 0$, and $\text{Var}(X) = \text{Var}(Y) = \text{Var}(Z) = 1$. Calculate $E[X^2(Y + 5Z)^2]$.

4. Let the joint probability density function of random variables X and Y be

$$f(x, y) = \begin{cases} 2e^{-(x+2y)} & \text{if } x \geq 0, \ y \geq 0 \\ 0 & \text{otherwise.} \end{cases}$$

Find $E(X)$, $E(Y)$, and $E(X^2 + Y^2)$.

5. A company puts five different types of prizes into their cereal boxes, one in each box and in equal proportions. If a customer decides to collect all five prizes, what is the expected number of the boxes of cereals that he or she should buy?

6. An absentminded professor wrote n letters and sealed them in envelopes without writing the addresses on the envelopes. Having forgotten which letter he had put in which envelope, he wrote the n addresses on the envelopes at random. What is the expected number of the letters addressed correctly?
Hint: For $i = 1, 2, \ldots, n$, let

$$X_i = \begin{cases} 1 & \text{if the } i\text{th letter is addressed correctly} \\ 0 & \text{otherwise.} \end{cases}$$

Calculate $E(X_1 + X_2 + \cdots + X_n)$.

7. A cultural society is arranging a party for its members. The cost of a band to play music, the amount that the caterer will charge, the rent of a hall to give the party, and other expenses (in dollars) are uniform random variables over the intervals $(1300, 1800)$, $(1800, 2000)$, $(800, 1200)$, and $(400, 700)$, respectively. If the number of party guests is a random integer from $(150, 200]$, what is the least amount that the society should charge each participant to have no loss, on average?

8. **(Pattern Appearance)**   Suppose that random digits are generated from the set $\{0, 1, \ldots, 9\}$ independently and successively. Find the expected number of digits to be generated until the pattern (a) 007 appears, (b) 156156 appears, (c) 575757 appears.

B

9. Solve the following problem posed by Michael Khoury, U.S. Mathematics Olympiad Member, in "The Problem Solving Competition," Oklahoma Publishing Company and the American Society for Communication of Mathematics, February 1999.

    Bob is teaching a class with n students. There are n desks in the classroom, numbered from 1 to n. Bob has prepared a seating chart, but the students have already seated themselves randomly. Bob calls off the name of the person who belongs in seat 1. This person vacates the seat he or she is currently occupying and takes his or her rightful seat. If this displaces a person already in the seat, that person stands at the front of the room until he or she is assigned a seat. Bob does this for each seat in turn. After k $(1 \leq k < n)$ names have been called, what is the expected number of students standing at the front of the room?

10. Let $\{X_1, X_2, \ldots, X_n\}$ be a sequence of independent random variables with $P(X_j = i) = p_i$ $(1 \leq j \leq n$ and $i \geq 1)$. Let $h_k = \sum_{i=k}^{\infty} p_i$. Using Theorem 10.2, prove that

$$E\big[\min(X_1, X_2, \ldots, X_n)\big] = \sum_{k=1}^{\infty} h_k^n.$$

11. A coin is tossed n times $(n > 4)$. What is the expected number of exactly three consecutive heads?
    Hint: Let E_1 be the event that the first three outcomes are heads and the fourth outcome is tails. For $2 \leq i \leq n - 3$, let E_i be the event that the outcome $(i - 1)$ is tails, the outcomes i, $(i + 1)$, and $(i + 2)$ are heads, and the outcome $(i + 3)$ is tails. Let E_{n-2} be the event that the outcome $(n - 3)$ is tails, and the last three outcomes are heads. Let

$$X_i = \begin{cases} 1 & \text{if } E_i \text{ occurs} \\ 0 & \text{otherwise.} \end{cases}$$

    Then calculate the expected value of an appropriate sum of X_i's.

12. Suppose that 80 balls are placed into 40 boxes at random and independently. What is the expected number of the empty boxes?

13. There are 25 students in a probability class. What is the expected number of birthdays that belong only to one student? Assume that the birthrates are constant throughout the year and that each year has 365 days.
Hint: Let $X_i = 1$ if the birthday of the ith student is not the birthday of any other student, and $X_i = 0$, otherwise. Find $E(X_1 + X_2 + \cdots + X_{25})$.

14. There are 25 students in a probability class. What is the expected number of the days of the year that are birthdays of at least two students? Assume that the birthrates are constant throughout the year and that each year has 365 days.

15. From an ordinary deck of 52 cards, cards are drawn at random, one by one, and *without* replacement until a heart is drawn. What is the expected value of the number of cards drawn?
Hint: See Exercise 9, Section 3.2.

16. **(Pattern Appearance)** In successive independent flips of a fair coin, what is the expected number of trials until the pattern THTHTTHTHT appears?

17. Let X and Y be nonnegative random variables with an arbitrary joint probability distribution function. Let

$$
I(x, y) = \begin{cases} 1 & \text{if } X > x, \quad Y > y, \\ 0 & \text{otherwise.} \end{cases}
$$

   (a) Show that

$$
\int_0^\infty \int_0^\infty I(x, y)\, dx\, dy = XY.
$$

   (b) By calculating expected values of both sides of part (a), prove that

$$
E(XY) = \int_0^\infty \int_0^\infty P(X > x, \ Y > y)\, dx\, dy.
$$

   Note that this is a generalization of the result explained in Remark 6.4.

18. Let $\{X_1, X_2, \ldots, X_n\}$ be a sequence of continuous, independent, and identically distributed random variables. Let

$$
N = \min\{n: X_1 \geq X_2 \geq X_3 \geq \cdots \geq X_{n-1}, X_{n-1} < X_n\}.
$$

   Find $E(N)$.

19. From an urn that contains a large number of red and blue chips, mixed in equal proportions, 10 chips are removed one by one and at random. The chips that are removed before the first red chip are returned to the urn. The first red chip, together with all those that follow, is placed in another urn that is initially empty. Calculate the expected number of the chips in the second urn.

20. Under what condition does Cauchy-Schwarz's inequality become equality?

10.2   COVARIANCE

In Sections 4.5 and 6.3 we studied the notion of the variance of a random variable X. We showed that $E\big[(X - E(X))^2\big]$, the variance of X, measures the average magnitude of the fluctuations of the random variable X from its expectation, $E(X)$. We mentioned that this quantity measures the dispersion, or spread, of the distribution of X about its expectation. Now suppose that X and Y are two jointly distributed random variables. Then Var(X) and Var(Y) determine the dispersions of X and Y independently rather than jointly. In fact, Var(X) measures the spread, or dispersion, along the x-direction, and Var(Y) measures the spread, or dispersion, along the y-direction in the plane. We now calculate Var($aX + bY$), the joint spread, or dispersion, of X and Y along the $(ax + by)$-direction for arbitrary real numbers a and b:

$$
\begin{aligned}
&\text{Var}(aX + bY) \\
&= E\big[(aX + bY) - E(aX + bY)\big]^2 \\
&= E\big[(aX + bY) - aE(X) - bE(Y)\big]^2 \\
&= E\big[a(X - E(X)) + b(Y - E(Y))\big]^2 \\
&= E\big[a^2(X - E(X))^2 + b^2(Y - E(Y))^2 + 2ab(X - E(X))(Y - E(Y))\big] \\
&= a^2\,\text{Var}(X) + b^2\,\text{Var}(Y) + 2ab E\big[(X - E(X))(Y - E(Y))\big]. \qquad (10.3)
\end{aligned}
$$

This formula shows that the joint spread, or dispersion, of X and Y can be measured in *any* direction $(ax + by)$ if the quantities Var(X), Var(Y), and $E\big[(X - E(X))(Y - E(Y))\big]$ are known. On the other hand, the joint spread, or dispersion, of X and Y depends on these three quantities. However, Var(X) and Var(Y) determine the dispersions of X and Y independently; therefore, $E\big[(X - E(X))(Y - E(Y))\big]$ is the quantity that gives information about the joint spread, or dispersion, of X and Y. It is called the **covariance** of X and Y, is denoted by Cov(X, Y), and determines how X and Y covary jointly. For example, by relation (10.3), if for random variables X, Y, and Z, Var(Y) = Var(Z) and $ab > 0$, then the joint dispersion of X and Y along the $(ax + by)$-direction is greater than the joint dispersion of X and Z along the $(ax + bz)$-direction if and only if Cov(X, Y) > Cov(X, Z).

Definition   *Let X and Y be jointly distributed random variables; then the **covariance** of X and Y is defined by*

$$
\text{Cov}(X, Y) = E\big[(X - E(X))(Y - E(Y))\big].
$$

Note that

$$
\mathbf{Cov}(X, X) = \mathbf{Var}(X).
$$

Also, by the Cauchy-Schwarz inequality (Theorem 10.3),

$$\text{Cov}(X, Y) = E\big[(X - E(X))(Y - E(Y))\big]$$
$$\leq \sqrt{E[X - E(X)]^2 E[Y - E(Y)]^2}$$
$$= \sqrt{\sigma_X^2 \sigma_Y^2} = \sigma_X \sigma_Y,$$

which shows that if $\sigma_X < \infty$ and $\sigma_Y < \infty$, then $\text{Cov}(X, Y) < \infty$.

Rewriting relation (10.3) in terms of $\text{Cov}(X, Y)$, we obtain the following important theorem:

Theorem 10.4   *Let a and b be real numbers; for random variables X and Y,*

$$\mathbf{Var}(aX + bY) = a^2\,\mathbf{Var}(X) + b^2\,\mathbf{Var}(Y) + 2ab\,\mathbf{Cov}(X, Y).$$

In particular, if $a = 1$ and $b = 1$, this gives

$$\mathbf{Var}(X + Y) = \mathbf{Var}(X) + \mathbf{Var}(Y) + 2\,\mathbf{Cov}(X, Y). \tag{10.4}$$

Similarly, if $a = 1$ and $b = -1$, it gives

$$\mathbf{Var}(X - Y) = \mathbf{Var}(X) + \mathbf{Var}(Y) - 2\,\mathbf{Cov}(X, Y). \tag{10.5}$$

Letting $\mu_X = E(X)$, and $\mu_Y = E(Y)$, an alternative formula for

$$\text{Cov}(X, Y) = E\big[(X - E(X))(Y - E(Y))\big]$$

is calculated by the expansion of $E\big[(X - E(X))(Y - E(Y))\big]$:

$$\text{Cov}(X, Y) = E\big[(X - \mu_X)(Y - \mu_Y)\big]$$
$$= E(XY - \mu_X Y - \mu_Y X + \mu_X \mu_Y)$$
$$= E(XY) - \mu_X E(Y) - \mu_Y E(X) + \mu_X \mu_Y$$
$$= E(XY) - \mu_X \mu_Y - \mu_Y \mu_X + \mu_X \mu_Y$$
$$= E(XY) - \mu_X \mu_Y = E(XY) - E(X)E(Y).$$

Therefore,

$$\mathbf{Cov}(X, Y) = E(XY) - E(X)E(Y). \tag{10.6}$$

Using this relation, we get

$$\begin{aligned}
\text{Cov}(aX + b, cY + d) &= E\big[(aX + b)(cY + d)\big] - E(aX + b)E(cY + d) \\
&= E(acXY + bcY + adX + bd) - \big[aE(X) + b\big]\big[cE(Y) + d\big] \\
&= ac\big[E(XY) - E(X)E(Y)\big] = ac\,\text{Cov}(X, Y).
\end{aligned}$$

Hence, for arbitrary real numbers a, b, c, d and random variables X and Y,

$$\textbf{Cov}(aX + b, cY + d) = ac\,\textbf{Cov}(X, Y), \tag{10.7}$$

which can be generalized as follows: Let a_i's and b_j's be constants. For random variables $X_1, X_2, \ldots, X_n$ and $Y_1, Y_2, \ldots, Y_m$,

$$\textbf{Cov}\left(\sum_{i=1}^{n} a_i X_i, \ \sum_{j=1}^{m} b_j Y_j\right) = \sum_{i=1}^{n}\sum_{j=1}^{m} a_i b_j \textbf{Cov}(X_i, Y_j). \tag{10.8}$$

(See Exercise 24.)

For random variables X and Y, $\text{Cov}(X, Y)$ might be positive, negative, or zero. It is positive if the expected value of $\big[X - E(X)\big]\big[Y - E(Y)\big]$ is positive, that is, if X and Y decrease together or increase together. It is negative if X increases while Y decreases, or vice versa. If $\text{Cov}(X, Y) > 0$, we say that X and Y are **positively correlated.** If $\text{Cov}(X, Y) < 0$, we say that they are **negatively correlated.** If $\text{Cov}(X, Y) = 0$, we say that X and Y are **uncorrelated.** For example, the blood cholesterol level of a person is positively correlated with the amount of saturated fat consumed by that person, whereas the amount of alcohol in the blood is negatively correlated with motor coordination. Generally, the more saturated fat a person ingests, the higher his or her blood cholesterol level will be. The more alcohol a person drinks, the poorer his or her level of motor coordination becomes. As another example, let X be the weight of a person before starting a health fitness program and Y be his or her weight afterward. Then X and Y are negatively correlated because the effect of fitness programs is that, usually, heavier persons lose weight, whereas lighter persons gain weight. The best examples for uncorrelated random variables are independent ones. If X and Y are independent, then

$$\text{Cov}(X, Y) = E(XY) - E(X)E(Y) = 0.$$

However, as the following example shows, the converse of this is not true; that is,

Two dependent random variables might be uncorrelated.

Example 10.9   Let X be uniformly distributed over $(-1, 1)$ and $Y = X^2$. Then

$$\text{Cov}(X, Y) = E(X^3) - E(X)E(X^2) = 0,$$

since $E(X) = 0$ and $E(X^3) = 0$. Thus the perfectly related random variables X and Y are uncorrelated.   ◆

Example 10.10   There are 300 cards in a box numbered 1 through 300. Therefore, the number on each card has one, two, or three digits. A card is drawn at random from the box. Suppose that the number on the card has X digits of which Y are 0. Determine whether X and Y are positively correlated, negatively correlated, or uncorrelated.

Solution: Note that, between 1 and 300, there are 9 one-digit numbers none of which is 0; there are 90 two-digit numbers of which 81 have no 0's, 9 have one 0, and none has two 0's; and there are 201 three-digit numbers of which 162 have no 0's, 36 have one 0, and 3 have two 0's. These facts show that as X increases so does Y. Therefore, X and Y are positively correlated. To show this mathematically, let $p(x, y)$ be the joint probability mass function of X and Y. Simple calculations will yield the following table for $p(x, y)$.

		y		
x	0	1	2	$p_X(x)$
1	9/300	0	0	9/300
2	81/300	9/300	0	90/300
3	162/300	36/300	3/300	201/300
$p_Y(y)$	252/300	45/300	3/300	

To see how we calculated the entries of the table, as an example, consider $p(3, 0)$. This quantity is 162/300 because there are 162 three-digit numbers with no 0's. Now from this table we have that

$$E(X) = 1 \cdot \frac{9}{300} + 2 \cdot \frac{90}{300} + 3 \cdot \frac{201}{300} = 2.91.$$

$$E(Y) = 0 \cdot \frac{252}{300} + 1 \cdot \frac{45}{300} + 2 \cdot \frac{3}{300} = 0.017.$$

$$E(XY) = \sum_{x=1}^{3} \sum_{y=0}^{2} xy p(x, y) = 2 \cdot \frac{9}{300} + 3 \cdot \frac{36}{300} + 6 \cdot \frac{3}{300} = 1.44.$$

Therefore,

$$\text{Cov}(X, Y) = E(XY) - E(X)E(Y) = 1.44 - (0.017)(2.91) = 1.39053 > 0,$$

which shows that X and Y are positively correlated.   ◆

Example 10.11   Ann cuts an ordinary deck of 52 cards and displays the exposed card. Andy cuts the remaining stack of cards and displays his exposed card. Counting jack, queen, and king as 11, 12, and 13, let X and Y be the numbers on the cards that Ann and Andy expose, respectively. Find $\text{Cov}(X, Y)$ and interpret the result.

Solution: Observe that the number of cards in Ann's stack after she cuts the deck, and the number of cards in Andy's stack after he cuts the remaining cards will not change the probabilities we are interested in. The problem is equivalent to choosing two cards at random and without replacement from an ordinary deck of 52 cards, and letting X be the number on one card and Y be the number on the other card. Let $p(x, y)$ be the joint probability mass function of X and Y. For $1 \leq x, y \leq 13$,

$$p(x, y) = P(X = x, Y = y) = P(Y = y \mid X = x)P(X = x)$$

$$= \begin{cases} \dfrac{1}{12} \cdot \dfrac{1}{13} = \dfrac{1}{156} & x \neq y \\ 0 & x = y. \end{cases}$$

Therefore,

$$p_X(x) = \sum_{\substack{y=1 \\ y \neq x}}^{13} p(x, y) = \frac{12}{156} = \frac{1}{13}, \quad x = 1, 2, \dots, 13;$$

$$p_Y(y) = \sum_{\substack{x=1 \\ x \neq y}}^{13} p(x, y) = \frac{12}{156} = \frac{1}{13}, \quad x = 1, 2, \dots, 13.$$

By these relations,

$$E(X) = \sum_{x=1}^{13} \frac{x}{13} = \frac{1}{13} \cdot \frac{13 \times 14}{2} = 7;$$

$$E(Y) = \sum_{y=1}^{13} \frac{y}{13} = \frac{1}{13} \cdot \frac{13 \times 14}{2} = 7.$$

By Theorem 8.1,

$$E(XY) = \sum_{x=1}^{13} \sum_{\substack{y=1 \\ y \neq x}}^{13} \frac{xy}{156} = \frac{1}{156} \sum_{x=1}^{13} \sum_{y=1}^{13} xy - \frac{1}{156} \sum_{x=1}^{13} x^2$$

$$= \frac{1}{156} \left(\sum_{x=1}^{13} x \right) \left(\sum_{y=1}^{13} y \right) - \frac{1}{156} \cdot 819$$

$$= \frac{1}{156} \cdot \frac{13 \times 14}{2} \cdot \frac{13 \times 14}{2} - \frac{819}{156} = \frac{287}{6}.$$

Therefore,

$$\text{Cov}(X, Y) = E(XY) - E(X)E(Y) = \frac{287}{6} - 49 = -\frac{7}{6}.$$

This shows that X and Y are negatively correlated. That is, if X increses, then Y decreases; if X decreases, then Y increases. These facts should make sense intuitively. ◆

Example 10.12   Let X be the lifetime of an electronic system and Y be the lifetime of one of its components. Suppose that the electronic system fails if the component does (but not necessarily vice versa). Furthermore, suppose that the joint probability density function of X and Y (in years) is given by

$$f(x, y) = \begin{cases} \dfrac{1}{49}e^{-y/7} & \text{if } 0 \le x \le y < \infty \\ 0 & \text{elsewhere.} \end{cases}$$

(a)   Determine the expected value of the remaining lifetime of the component when the system dies.

(b)   Find the covariance of X and Y.

Solution:

(a)   The remaining lifetime of the component when the system dies is $Y - X$. So the desired quantity is

$$E(Y - X) = \int_0^\infty \int_0^y (y - x)\frac{1}{49}e^{-y/7}\, dx\, dy$$

$$= \frac{1}{49}\int_0^\infty e^{-y/7}\left(y^2 - \frac{y^2}{2}\right) dy = \frac{1}{98}\int_0^\infty y^2 e^{-y/7}\, dy = 7,$$

where the last integral is calculated using integration by parts twice.

(b)   To find $\text{Cov}(X, Y) = E(XY) - E(X)E(Y)$, note that

$$E(XY) = \int_0^\infty \int_0^y (xy)\frac{1}{49}e^{-y/7}\, dx\, dy$$

$$= \frac{1}{49}\int_0^\infty ye^{-y/7}\left(\int_0^y x\, dx\right) dy$$

$$= \frac{1}{98}\int_0^\infty y^3 e^{-y/7}\, dy = \frac{14,406}{98} = 147,$$

where the last integral is calculated, using integration by parts three times. We also have

$$E(X) = \int_0^\infty \int_0^y x\frac{1}{49}e^{-y/7}\, dx\, dy = 7,$$

$$E(Y) = \int_0^\infty \int_0^y y\frac{1}{49}e^{-y/7}\, dx\, dy = 14.$$

Therefore, $\text{Cov}(X, Y) = 147 - 7(14) = 49$. Note that $\text{Cov}(X, Y) > 0$ is expected because X and Y are positively correlated. ◆

As Theorem 10.4 shows, one important application of the covariance of two random variables X and Y is that it enables us to find $\text{Var}(aX + bY)$ for constants a and b. By direct calculations similar to (10.3), that theorem is generalized as follows: Let $a_1, a_2, \ldots, a_n$ be real numbers; for random variables $X_1, X_2, \ldots, X_n$,

$$\text{Var}\left(\sum_{i=1}^{n} a_i X_i\right) = \sum_{i=1}^{n} a_i^2 \text{Var}(X_i) + 2 \sum \sum_{i<j} a_i a_j \text{Cov}(X_i, X_j). \qquad (10.9)$$

In particular, for $a_i = 1$, $1 \le i \le n$,

$$\text{Var}\left(\sum_{i=1}^{n} X_i\right) = \sum_{i=1}^{n} \text{Var}(X_i) + 2 \sum \sum_{i<j} \text{Cov}(X_i, X_j). \qquad (10.10)$$

By (10.9) and (10.10),

If $X_1, X_2, \ldots, X_n$ are *pairwise independent* or, more generally, *pairwise uncorrelated*, then

$$\text{Var}\left(\sum_{i=1}^{n} a_i X_i\right) = \sum_{i=1}^{n} a_i^2 \text{Var}(X_i) \qquad (10.11)$$

$$\text{Var}\left(\sum_{i=1}^{n} X_i\right) = \sum_{i=1}^{n} \text{Var}(X_i), \qquad (10.12)$$

the reason being that $\text{Cov}(X_i, X_j) = 0$, for all $i, j, i \ne j$.

Example 10.13 Let X be the number of 6's in n rolls of a fair die. Find $\text{Var}(X)$.

Solution: Let $X_i = 1$ if on the ith roll the die lands 6, and $X_i = 0$, otherwise. Then $X = X_1 + X_2 + \cdots + X_n$. Since $X_1, X_2, \ldots, X_n$ are independent,

$$\text{Var}(X) = \text{Var}(X_1) + \text{Var}(X_2) + \cdots + \text{Var}(X_n).$$

But for $i = 1, 2, \ldots, n$,

$$E(X_i) = 1 \cdot \frac{1}{6} + 0 \cdot \frac{5}{6} = \frac{1}{6},$$

$$E(X_i^2) = (1)^2 \cdot \frac{1}{6} + 0^2 \cdot \frac{5}{6} = \frac{1}{6},$$

and hence

$$\text{Var}(X_i) = \frac{1}{6} - \frac{1}{36} = \frac{5}{36}.$$

Therefore, $\text{Var}(X) = n(5/36) = (5n)/36.$ ◆

★ **Example 10.14   (Investment)**   Dr. Caprio has invested money in three uncorrelated assets; 25% in the first one, 43% in the second one, and 32% in the third one. The means of the annual rate of returns for these assets, respectively, are 10%, 15%, and 13%. Their standard deviations are 8%, 12%, and 10%, respectively. Find the mean and standard deviation of the annual rate of return for Dr. Caprio's total investment.

Solution: Let r be the annual rate of return for Dr. Caprio's total investment. Let r_1, r_2, and r_3 be the annual rate of returns for the first, second, and third assets, respectively. By Example 4.25,

$$r = 0.25r_1 + 0.43r_2 + 0.32r_3.$$

Thus

$$E(r) = 0.25E(r_1) + 0.43E(r_2) + 0.32E(r_3)$$
$$= (0.25)(0.10) + (0.43)(0.15) + (0.32)(0.13) = 0.1311.$$

Since the assets are uncorrelated, by (10.11),

$$\text{Var}(r) = (0.25)^2 \text{Var}(r_1) + (0.43)^2 \text{Var}(r_2) + (0.32)^2 \text{Var}(r_3)$$
$$= (0.25)^2(0.08)^2 + (0.43)^2(0.12)^2 + (0.32)^2(0.10)^2 = 0.004087.$$

Therefore, $\sigma_r = \sqrt{0.004087} = 0.064$. Hence Dr. Caprio should expect an annual rate of return of 13.11% with standard deviation 6.4%. Note that Dr. Caprio has reduced the standard deviation of his investments considerably by diversifying his investment; that is, by not putting all of his eggs in one basket.

In general,

If Dr. Caprio divides his money into n equal portions and invests them in n uncorrelated financial assets with identically distributed rates of returns $r_1, r_2, \ldots, r_n$, each with variance σ^2, then by (10.11),

$$\text{Var}(r) = \text{Var}\left(\frac{1}{n}r_1 + \frac{1}{n}r_2 + \cdots + \frac{1}{n}r_n\right)$$

$$= \frac{1}{n^2}\text{Var}(r_1) + \frac{1}{n^2}\text{Var}(r_2) + \cdots + \frac{1}{n^2}\text{Var}(r_n) = \frac{1}{n^2} \cdot n\sigma^2 = \frac{\sigma^2}{n}.$$

This shows that the more financial assets Dr. Caprio invests in, the less the variance of the return will be—an indication that diversification reduces the investment risks.   ♦

Example 10.15   Using relation (10.10), calculate the variance of a binomial random variable X with parameters (n, p).

Solution: Recall that X is the number of successes in n independent Bernoulli trials.

Thus for $i = 1, 2, \ldots, n$, letting

$$X_i = \begin{cases} 1 & \text{if the } i\text{th trial is a success} \\ 0 & \text{otherwise,} \end{cases}$$

we obtain

$$X = X_1 + X_2 + \cdots + X_n,$$

where X_i is a Bernoulli random variable for $i = 1, 2, \ldots, n$. Note that $E(X_i) = p$ and $\text{Var}(X_i) = p(1 - p)$. Since $\{X_1, X_2, \ldots, X_n\}$ is an independent set of random variables, by (10.12),

$$\text{Var}(X) = \text{Var}(X_1) + \text{Var}(X_2) + \cdots + \text{Var}(X_n) = np(1 - p). \quad \blacklozenge$$

Example 10.16   Using relation (10.10), calculate the variance of a negative binomial random variable X, with parameter (r, p).

Solution: Recall that in a sequence of independent Bernoulli trials, X is the number of trials until the rth success. Let X_1 be the number of trials until the first success, X_2 the number of additional trials to get the second success, X_3 the number of additional ones to obtain the third success, and so on. Then $X = X_1 + X_2 + \cdots + X_r$, where for $i = 1, 2, \ldots, r$, the random variable X_i is geometric with parameter p. Since $X_1, X_2, \ldots, X_r$ are independent,

$$\text{Var}(X) = \text{Var}(X_1) + \text{Var}(X_2) + \cdots + \text{Var}(X_r).$$

But $\text{Var}(X_i) = (1 - p)/p^2$, for $i = 1, 2, \ldots, r$ (see Section 5.3). Thus $\text{Var}(X) = r(1 - p)/p^2$. $\quad \blacklozenge$

Let F be the distribution function of a certain characteristic of the elements of some population. For example, let F be the distribution function of lifetimes of light bulbs manufactured by a company, the Scottish soldiers chest size, or the score on an achievement test of a random student. Let $X_1, X_2, \ldots, X_n$ be a random sample from the distribution F. For example, let $X_1, X_2, \ldots, X_n$ be the chest sizes of n Scottish soldiers selected at random and independently. Then the following lemma enables us to find the expected value and variance of the sample mean in terms of the expected value and variance of the distribution function F (i.e., the population mean and variance).

Lemma 10.1   Let $X_1, X_2, \ldots, X_n$ be a random sample of size n from a distribution F with mean μ and variance σ^2. Let $\bar{X}$ be the mean of the random sample. Then

$$E(\bar{X}) = \mu \quad \text{and} \quad \text{Var}(\bar{X}) = \frac{\sigma^2}{n}.$$

Proof: By Theorem 10.1,

$$E(\bar{X}) = E\left(\frac{X_1 + X_2 + \cdots + X_n}{n}\right) = \frac{1}{n}\big[E(X_1) + E(X_2) + \cdots + E(X_n)\big] = \frac{1}{n}\cdot n\mu = \mu.$$

Since $X_1, X_2, \ldots, X_n$ are independent random variables, by (10.11),

$$\text{Var}(\bar{X}) = \text{Var}\left(\frac{X_1 + X_2 + \cdots + X_n}{n}\right) = \frac{1}{n^2}\big[\text{Var}(X_1) + \text{Var}(X_2) + \cdots + \text{Var}(X_n)\big]$$

$$= \frac{1}{n^2}\cdot n\sigma^2 = \frac{\sigma^2}{n}. \quad \blacklozenge$$

EXERCISES

A

1. Ann cuts an ordinary deck of 52 cards and displays the exposed card. After Ann places her stack back on the deck, Andy cuts the same deck and displays the exposed card. Counting jack, queen, and king as 11, 12, and 13, let X and Y be the numbers on the cards that Ann and Andy expose, respectively. Find $\text{Cov}(X, Y)$.

2. Let the joint probability mass function of random variables X and Y be given by

$$p(x, y) = \begin{cases} \dfrac{1}{70}\, x(x + y) & \text{if } x = 1, 2, 3, \quad y = 3, 4 \\ 0 & \text{elsewhere.} \end{cases}$$

Find $\text{Cov}(X, Y)$.

3. Roll a balanced die and let the outcome be X. Then toss a fair coin X times and let Y denote the number of tails. Find $\text{Cov}(X, Y)$ and interpret the result.
   Hint: Let $p(x, y)$ be the joint probability mass function of X and Y. To save time, use the table for $p(x, y)$ constructed in Example 8.2.

4. Thieves stole four animals at random from a farm that had seven sheep, eight goats, and five burros. Calculate the covariance of the number of sheep and goats stolen.

5. In n independent Bernoulli trials, each with probability of success p, let X be the number of successes and Y the number of failures. Calculate $E(XY)$ and $\text{Cov}(X, Y)$.

6. For random variables X, Y, and Z, prove that

   (a) $\text{Cov}(X + Y, Z) = \text{Cov}(X, Z) + \text{Cov}(Y, Z)$.

   (b) $\text{Cov}(X, Y + Z) = \text{Cov}(X, Y) + \text{Cov}(X, Z)$.

7. Show that if X and Y are independent random variables, then for all random variables Z,
$$\text{Cov}(X, Y + Z) = \text{Cov}(X, Z).$$

8. For random variables X and Y, show that
$$\text{Cov}(X + Y, X - Y) = \text{Var}(X) - \text{Var}(Y).$$

9. Prove that
$$\text{Var}(X - Y) = \text{Var}(X) + \text{Var}(Y) - 2\,\text{Cov}(X, Y).$$

10. Let X and Y be two independent random variables.

    (a) Show that $X - Y$ and $X + Y$ are uncorrelated if and only if $\text{Var}(X) = \text{Var}(Y)$.

    (b) Show that $\text{Cov}(X, XY) = E(Y)\text{Var}(X)$.

11. Prove that if Θ is a random number from the interval $[0, 2\pi]$, then the dependent random variables $X = \sin\Theta$ and $Y = \cos\Theta$ are uncorrelated.

12. Let X and Y be the coordinates of a random point selected uniformly from the unit disk $\{(x, y): x^2 + y^2 \leq 1\}$. Are X and Y independent? Are they uncorrelated? Why or why not?

13. Mr. Jones has two jobs. Next year, he will get a salary raise of X thousand dollars from one employer and a salary raise of Y thousand dollars from his second. Suppose that X and Y are independent random variables with probability density functions f and g, respectively, where
$$f(x) = \begin{cases} 8x/15 & \text{if } 1/2 < x < 2 \\ 0 & \text{elsewhere,} \end{cases}$$
$$g(y) = \begin{cases} 6\sqrt{y}/13 & \text{if } 1/4 < y < 9/4 \\ 0 & \text{elsewhere.} \end{cases}$$

   What are the expected value and variance of the total raise that Mr. Jones will get next year?

14. Let X and Y be independent random variables with expected values μ_1 and μ_2, and variances σ_1^2 and σ_2^2, respectively. Show that
$$\text{Var}(XY) = \sigma_1^2\sigma_2^2 + \mu_1^2\sigma_2^2 + \mu_2^2\sigma_1^2.$$

15. A voltmeter is used to measure the voltage of voltage sources, such as batteries. Every time this device is used, a random error is made, independent of other measurements, with mean 0 and standard deviation σ. Suppose that we want to measure the voltages, V_1 and V_2, of two batteries. If a measurement with smaller error variance is preferable, determine which of the following methods should be used:

(a) To measure V_1 and V_2 separately.

(b) To measure $V = V_1 + V_2$ and $W = V_1 - V_2$, and then find V_1 and V_2 from $V_1 = (V + W)/2$ and $V_2 = (V - W)/2$. There are methods available for engineers to do these. For example, to measure $V_1 + V_2$, they attach batteries in series so that each battery pushes current in the same direction. To measure $V_1 - V_2$, they attach the batteries in series so that they push current in opposite directions.

Assume that the internal resistances of the batteries are negligible.

16. (Investment)   Mr. Ingham has invested money in three assets; 18% in the first asset, 40% in the second one, and 42% in the third one. Let r_1, r_2, and r_3 be the annual rate of returns for these three investments, respectively. For $1 \leq i, j \leq 3$, $\text{Cov}(r_i, r_j)$ is the ith element in the jth row of the following table. [Note that $\text{Var}(r_i) = \text{Cov}(r_i, r_i)$.]

	r_1	r_2	r_3
r_1	0.064	0.03	0.015
r_2	0.03	0.0144	0.021
r_3	0.015	0.021	0.01

Find the standard deviation of the annual rate of return for Mr. Ingham's total investment.

17. (Investment)   Mr. Kowalski has invested $50,000 in three uncorrelated financial assets: 25% in the first financial asset, 40% in the second one, and 35% in the third one. The annual rates of return for these assets, respectively, are normal random variables with means 12%, 15%, and 18%, and standard deviations 8%, 12%, and 15%. What is the probability that, after a year, Mr. Kowalski's net profit from this investment is at least $10,000? Ignore transaction costs and assume that there is no annual dividend.

18. Let X and Y have the following joint probability density function

$$f(x, y) = \begin{cases} 8xy & \text{if } 0 < x \leq y < 1 \\ 0 & \text{otherwise.} \end{cases}$$

(a) Calculate $\text{Var}(X + Y)$.

(b) Show that X and Y are not independent. Explain why this does not contradict Exercise 23 of Section 8.2.

19. Find the variance of a sum of n randomly and independently selected points from the interval $(0, 1)$.

20. Let X and Y be jointly distributed with joint probability density function

$$f(x, y) = \begin{cases} \dfrac{1}{2}x^3 e^{-xy-x} & \text{if } x > 0, \ y > 0 \\ \\ 0 & \text{otherwise.} \end{cases}$$

Determine if X and Y are positively correlated, negatively correlated, or uncorrelated.

Hint: Note that for all $a > 0$, $\int_0^\infty x^n e^{-ax}\, dx = n!/a^{n+1}$.

21. Let X be a random variable. Prove that $\text{Var}(X) = \min_t E\big[(X - t)^2\big]$.

Hint: Let $\mu = E(X)$ and look at the expansion of

$$E\big[(X - t)^2\big] = E\big[(X - \mu + \mu - t)^2\big].$$

B

22. Let S be the sample space of an experiment. Let A and B be two events of S. Let I_A and I_B be the indicator variables for A and B. That is,

$$I_A(\omega) = \begin{cases} 1 & \text{if } \omega \in A \\ 0 & \text{if } \omega \notin A, \end{cases}$$

$$I_B(\omega) = \begin{cases} 1 & \text{if } \omega \in B \\ 0 & \text{if } \omega \notin B. \end{cases}$$

Show that I_A and I_B are positively correlated if and only if $P(A \mid B) > P(A)$, and if and only if $P(B \mid A) > P(B)$.

23. Show that for random variables X, Y, Z, and W and constants a, b, c, and d,

$$\text{Cov}(aX + bY, cZ + dW)$$
$$= ac\,\text{Cov}(X, Z) + bc\,\text{Cov}(Y, Z) + ad\,\text{Cov}(X, W) + bd\,\text{Cov}(Y, W).$$

Hint: For a simpler proof, use the results of Exercise 6.

24. Prove the following generalization of Exercise 23:

$$\text{Cov}\left(\sum_{i=1}^n a_i X_i, \ \sum_{j=1}^m b_j Y_j \right) = \sum_{i=1}^n \sum_{j=1}^m a_i b_j \text{Cov}(X_i, Y_j).$$

25. A fair die is thrown n times. What is the covariance of the number of 1's and the number of 6's obtained?

Hint: Use the result of Exercise 24.

26. Show that if $X_1, X_2, \ldots, X_n$ are random variables and $a_1, a_2, \ldots, a_n$ are constants, then

$$\text{Var}\left(\sum_{i=1}^{n} a_i X_i \right) = \sum_{i=1}^{n} a_i^2 \text{Var}(X_i) + 2 \sum \sum_{i<j} a_i a_j \text{Cov}(X_i, X_j).$$

27. Let X be a hypergeometric random variable with probability mass function

$$p(x) = P(X = x) = \frac{\dbinom{D}{x} \dbinom{N-D}{n-x}}{\dbinom{N}{n}},$$

$$n \le \min(D, N-D), \quad x = 0, 1, 2, \ldots, n.$$

Recall that X is the number of defective items among n items drawn randomly and without replacement from a box containing D defective and $N - D$ nondefective items. Show that

$$\text{Var}(X) = \frac{nD(N-D)}{N^2}\left(1 - \frac{n-1}{N-1}\right).$$

Hint: Let A_i be the event that the ith item drawn is defective. Also for $i = 1, 2, \ldots, n$, let

$$X_i = \begin{cases} 1 & \text{if } A_i \text{ occurs} \\ 0 & \text{otherwise.} \end{cases}$$

Then $X = X_1 + X_2 + \cdots + X_n$.

28. Exactly n married couples are living in a small town. What is the variance of the surviving couples after m deaths occur among them? Assume that the deaths occur at random, there are no divorces, and there are no new marriages.

Note: This situation involves the Daniel Bernoulli problem discussed in Example 10.3.

10.3   CORRELATION

Although for random variables X and Y, $\text{Cov}(X, Y)$ provides information about how X and Y vary jointly, it has a major shortcoming: It is not independent of the units in which X and Y are measured. For example, suppose that for random variables X and Y, when measured in (say) centimeters, $\text{Cov}(X, Y) = 0.15$. For the same random variables, if we change the measurements to millimeters, then $X_1 = 10X$ and $Y_1 = 10Y$ will be the new observed values, and by relation (10.7) we get

$$\text{Cov}(X_1, Y_1) = \text{Cov}(10X, 10Y) = 100\,\text{Cov}(X, Y) = 15,$$

showing that $\text{Cov}(X, Y)$ is sensitive to the units of measurement. From Section 4.6 we know that for a random variable X, the standardized X, $X^* = \big[X - E(X)\big]/\sigma_X$, is independent of the units in which X is measured. Thus, to define a measure of association between X and Y, independent of the scales of measurements, it is appropriate to consider $\text{Cov}(X^*, Y^*)$ rather than $\text{Cov}(X, Y)$. Using relation (10.7), we obtain

$$\text{Cov}(X^*, Y^*) = \text{Cov}\Big(\frac{X - E(X)}{\sigma_X}, \frac{Y - E(Y)}{\sigma_Y}\Big)$$

$$= \text{Cov}\Big(\frac{1}{\sigma_X}X - \frac{1}{\sigma_X}E(X), \frac{1}{\sigma_Y}Y - \frac{1}{\sigma_Y}E(Y)\Big)$$

$$= \frac{1}{\sigma_X}\frac{1}{\sigma_Y}\text{Cov}(X, Y) = \frac{\text{Cov}(X, Y)}{\sigma_X\sigma_Y}.$$

Definition   *Let X and Y be two random variables with $0 < \sigma_X^2 < \infty$ and $0 < \sigma_Y^2 < \infty$. The covariance between the standardized X and the standardized Y is called the* **correlation coefficient** *between X and Y and is denoted by $\rho = \rho(X, Y)$. Therefore,*

$$\rho = \frac{\text{Cov}(X, Y)}{\sigma_X\sigma_Y}.$$

The quantity $\text{Cov}(X, Y)/(\sigma_X\sigma_Y)$ gives all the important information that $\text{Cov}(X, Y)$ provides about how X and Y covary and, at the same time, it is not sensitive to the scales of measurement. Clearly, $\rho(X, Y) > 0$ if and only if X and Y are positively correlated; $\rho(X, Y) < 0$ if and only if X and Y are negatively correlated; and $\rho(X, Y) = 0$ if and only if X and Y are uncorrelated. Moreover, $\rho(X, Y)$ roughly measures the amount and the sign of linear relationship between X and Y. It is -1 if $Y = aX + b$, $a < 0$, and $+1$ if $Y = aX + b$, $a > 0$. Thus $\rho(X, Y) = \pm 1$ in the case of perfect linear relationship and 0 in the case of independence of X and Y. The key to the proof of "$\rho(X, Y) = \pm 1$ if and only if $Y = aX + b$" is the following lemma.

Lemma 10.2   *For random variables X and Y with correlation coefficient $\rho(X, Y)$,*

$$\text{Var}\Big(\frac{X}{\sigma_X} + \frac{Y}{\sigma_Y}\Big) = 2 + 2\rho(X, Y);$$

$$\text{Var}\left(\frac{X}{\sigma_X} - \frac{Y}{\sigma_Y}\right) = 2 - 2\rho(X, Y).$$

Proof: We prove the first relation; the second can be shown similarly. By Theorem 10.4,

$$\text{Var}\left(\frac{X}{\sigma_X} + \frac{Y}{\sigma_Y}\right) = \frac{1}{\sigma_X^2}\text{Var}(X) + \frac{1}{\sigma_Y^2}\text{Var}(Y) + 2\frac{\text{Cov}(X, Y)}{\sigma_X \sigma_Y} = 2 + 2\rho(X, Y). \quad \blacklozenge$$

Theorem 10.5   *For random variables X and Y with correlation coefficient $\rho(X, Y)$,*

(a) $-1 \le \rho(X, Y) \le 1.$

(b)   *With probability 1, $\rho(X, Y) = 1$ if and only if $Y = aX + b$ for some constants a, b, $a > 0$.*

(c)   *With probability 1, $\rho(X, Y) = -1$ if and only if $Y = aX + b$ for some constants a, b, $a < 0$.*

Proof:

(a)   Since the variance of a random variable is nonnegative,

$$\text{Var}\left(\frac{X}{\sigma_X} + \frac{Y}{\sigma_Y}\right) \ge 0 \quad \text{and} \quad \text{Var}\left(\frac{X}{\sigma_X} - \frac{Y}{\sigma_Y}\right) \ge 0.$$

Therefore, by Lemma 10.2, $2 + 2\rho(X, Y) \ge 0$ and $2 - 2\rho(X, Y) \ge 0$. That is, $\rho(X, Y) \ge -1$ and $\rho(X, Y) \le 1$.

(b)   First, suppose that $\rho(X, Y) = 1$. In this case, by Lemma 10.2,

$$\text{Var}\left(\frac{X}{\sigma_X} - \frac{Y}{\sigma_Y}\right) = 2[1 - \rho(X, Y)] = 0.$$

Therefore, with probability 1,

$$\frac{X}{\sigma_X} - \frac{Y}{\sigma_Y} = c,$$

for some constant c. Hence, with probability 1,

$$Y = \frac{\sigma_Y}{\sigma_X}X - c\,\sigma_Y \equiv aX + b,$$

with $a = \sigma_Y/\sigma_X > 0$ and $b = -c\,\sigma_Y$. Next, assume that $Y = aX + b$, $a > 0$. We have that

$$\rho(X, Y) = \rho(X, aX + b) = \frac{\text{Cov}(X, aX + b)}{\sigma_X \sigma_{aX+b}}$$

$$= \frac{a\,\text{Cov}(X, X)}{\sigma_X\,(a\sigma_X)} = \frac{a\,\text{Var}(X)}{a\,\text{Var}(X)} = 1.$$

(c)   The proof of this statement is similar to that of part (b). $\blacklozenge$

Example 10.17   Show that if X and Y are continuous random variables with the joint probability density function

$$f(x, y) = \begin{cases} x + y & \text{if } 0 < x < 1, \ 0 < y < 1 \\ 0 & \text{otherwise,} \end{cases}$$

then X and Y are not linearly related.

Solution: Since X and Y are linearly related if and only if $\rho(X, Y) = \pm 1$ with probability 1, it suffices to prove that $\rho(X, Y) \neq \pm 1$. To do so, note that

$$E(X) = \int_0^1 \int_0^1 x(x + y)\, dx\, dy = \frac{7}{12},$$

$$E(XY) = \int_0^1 \int_0^1 xy(x + y)\, dx\, dy = \frac{1}{3}.$$

Also, by symmetry, $E(Y) = 7/12$; therefore,

$$\text{Cov}(X, Y) = E(XY) - E(X)E(Y) = \frac{1}{3} - \frac{7}{12}\frac{7}{12} = -\frac{1}{144}.$$

Similarly,

$$E(X^2) = \int_0^1 \int_0^1 x^2(x + y)\, dx\, dy = \frac{5}{12},$$

$$\sigma_X = \sqrt{E(X^2) - \left[E(X)\right]^2} = \sqrt{\frac{5}{12} - \left(\frac{7}{12}\right)^2} = \frac{\sqrt{11}}{12}.$$

Again, by symmetry, $\sigma_Y = \sqrt{11}/12$. Thus

$$\rho(X, Y) = \frac{\text{Cov}(X, Y)}{\sigma_X \sigma_Y} = \frac{-1/144}{\sqrt{11}/12 \cdot \sqrt{11}/12} = -\frac{1}{11} \neq \pm 1. \quad \blacklozenge$$

The following example shows that even if X and Y are dependent through a nonlinear relationship such as $Y = X^2$, still, statistically, there might be a strong linear association between X and Y. That is, $\rho(X, Y)$ might be very close to 1 or -1, indicating that the points (x, y) are tightly clustered around a line.

Example 10.18   Let X be a random number from the interval $(0, 1)$ and $Y = X^2$. The probability density function of X is

$$f(x) = \begin{cases} 1 & \text{if } 0 < x < 1 \\ 0 & \text{elsewhere,} \end{cases}$$

and for $n \geq 1$,

$$E(X^n) = \int_0^1 x^n \, dx = \frac{1}{n+1}.$$

Thus $E(X) = 1/2$, $E(Y) = E(X^2) = 1/3$,

$$\sigma_X^2 = \text{Var}(X) = E(X^2) - \left[E(X)\right]^2 = \frac{1}{3} - \frac{1}{4} = \frac{1}{12},$$

$$\sigma_Y^2 = \text{Var}(Y) = E(X^4) - \left[E(X^2)\right]^2 = \frac{1}{5} - \left(\frac{1}{3}\right)^2 = \frac{4}{45},$$

and, finally,

$$\text{Cov}(X, Y) = E(X^3) - E(X)E(X^2) = \frac{1}{4} - \frac{1}{2}\frac{1}{3} = \frac{1}{12}.$$

Therefore,

$$\rho(X, Y) = \frac{\text{Cov}(X, Y)}{\sigma_X \sigma_Y} = \frac{1/12}{1/2\sqrt{3} \cdot 2/3\sqrt{5}} = \frac{\sqrt{15}}{4} = 0.968. \quad \blacklozenge$$

★ **Example 10.19** **(Investment)**   Mr. Kowalski has invested money in two financial assets. Let r be the annual rate of return for his total investment. Let r_1 and r_2 be the annual rates of return for the first and second assets, respectively. Let $\sigma^2 = \text{Var}(r)$, $\sigma_1^2 = \text{Var}(r_1)$, and $\sigma_2^2 = \text{Var}(r_2)$. Prove that

$$\sigma^2 \leq \max\left(\sigma_1^2, \sigma_2^2\right).$$

In particular, by this inequality, if r_1 and r_2 are identically distributed, then $\sigma_1^2 = \sigma_2^2$ implies that

$$\sigma^2 \leq \sigma_1^2 = \sigma_2^2.$$

This is an extension of the result discussed in Example 10.14. It shows that, even if the financial assets are correlated, still diversification reduces the investment risk.

Proof: We will show that if $\sigma_2^2 \leq \sigma_1^2$, then $\sigma^2 \leq \sigma_1^2$. By symmetry, if $\sigma_1^2 \leq \sigma_2^2$, then $\sigma^2 \leq \sigma_2^2$. Therefore, σ^2 is either less than or equal to σ_1^2, or it is less than or equal to σ_2^2, implying that

$$\sigma^2 \leq \max\left(\sigma_1^2, \sigma_2^2\right).$$

To show that $\sigma_2^2 \leq \sigma_1^2$ implies that $\sigma^2 \leq \sigma_1^2$, let w_1 and w_2 be the fractions of Mr. Kowalski's investment in the first and second financial assets, respectively. Then $w_1 + w_2 = 1$ and

$$r = w_1 r_1 + w_2 r_2.$$

Let ρ be the correlation coefficient of r_1 and r_2. Then

$$\text{Cov}(r_1, r_2) = \rho\sigma_1\sigma_2.$$

We have

$$\text{Var}(r) = \text{Var}(w_1 r_1 + w_2 r_2) = \text{Var}(w_1 r_1) + \text{Var}(w_2 r_2) + 2\text{Cov}(w_1 r_1, w_2 r_2)$$
$$= w_1^2 \text{Var}(r_1) + w_2^2 \text{Var}(r_2) + 2w_1 w_2 \text{Cov}(r_1, r_2).$$

Noting that $w_1 + w_2 = 1$, $-1 \le \rho \le 1$, and $\sigma_2^2 \le \sigma_1^2$, this relation implies that

$$\sigma^2 = w_1^2 \sigma_1^2 + w_2^2 \sigma_2^2 + 2w_1 w_2 \rho\sigma_1\sigma_2$$
$$\le w_1^2 \sigma_1^2 + w_2^2 \sigma_1^2 + 2w_1 w_2 \sigma_1^2$$
$$= (w_1^2 + w_2^2 + 2w_1 w_2)\sigma_1^2$$
$$= (w_1 + w_2)^2 \sigma_1^2 = \sigma_1^2. \quad \blacklozenge$$

EXERCISES

A

1. Let X and Y be jointly distributed, with $\rho(X, Y) = 1/2$, $\sigma_X = 2$, $\sigma_Y = 3$. Find $\text{Var}(2X - 4Y + 3)$.

2. Let the joint probability density function of X and Y be given by

$$f(x, y) = \begin{cases} \sin x \sin y & \text{if } 0 \le x \le \pi/2, \quad 0 \le y \le \pi/2 \\ 0 & \text{otherwise.} \end{cases}$$

Calculate the correlation coefficient of X and Y.

3. A stick of length 1 is broken into two pieces at a random point. Find the correlation coefficient and the covariance of these pieces.

4. For real numbers α and β, let

$$\text{sgn}(\alpha\beta) = \begin{cases} 1 & \text{if } \alpha\beta > 0 \\ 0 & \text{if } \alpha\beta = 0 \\ -1 & \text{if } \alpha\beta < 0. \end{cases}$$

Prove that for random variables X and Y,

$$\rho(\alpha_1 X + \alpha_2, \beta_1 Y + \beta_2) = \rho(X, Y)\,\text{sgn}(\alpha_1\beta_1).$$

5. Is it possible that for some random variables X and Y, $\rho(X, Y) = 3$, $\sigma_X = 2$, and $\sigma_Y = 3$?

6. Prove that if $\text{Cov}(X, Y) = 0$, then

$$\rho(X + Y, X - Y) = \frac{\text{Var}(X) - \text{Var}(Y)}{\text{Var}(X) + \text{Var}(Y)}.$$

B

7. Show that if the joint probability density function of X and Y is

$$f(x, y) = \begin{cases} \dfrac{1}{2}\sin(x + y) & \text{if } 0 \le x \le \dfrac{\pi}{2}, \quad 0 \le y \le \dfrac{\pi}{2} \\ 0 & \text{elsewhere,} \end{cases}$$

then there exists no linear relation between X and Y. ∎

10.4   CONDITIONING ON RANDOM VARIABLES

An important application of conditional expectations is that ordinary expectations and probabilities can be calculated by conditioning on appropriate random variables. First, to explain this procedure, we state a definition.

Definition   *Let X and Y be two random variables. By $E(X|Y)$ we mean a function of Y that is defined to be $E(X \mid Y = y)$ when $Y = y$.*

Recall that a function of a random variable Y, say $h(Y)$, is defined to be $h(a)$ at all sample points at which $Y = a$. For example, if $Z = \log Y$, then at a sample point ω, where $Y(\omega) = a$, we have that $Z(\omega) = \log a$. In this definition $E(X \mid Y)$ is a function of Y, which at a sample point ω is defined to be $E(X \mid Y = y)$, where $y = Y(\omega)$. Since $E(X|Y)$ is defined only when $p_Y(y) > 0$, it is defined at all sample points ω, where $p_Y(y) > 0$, $y = Y(\omega)$. $E(X|Y)$, being a function of Y, is a random variable. Its expectation, whenever finite, is equal to the expectation of X. This extremely important property sometimes enables us to calculate expectations which are otherwise, if possible, very difficult to find. To see why the expected value of $E(X|Y)$ is $E(X)$, let X and Y be discrete random variables with sets of possible values A and B, respectively. Let $p(x, y)$ be the joint probability mass function of X and Y. On the one hand,

$$E(X) = \sum_{x \in A} x p_X(x) = \sum_{x \in A} x \sum_{y \in B} p(x, y)$$

$$= \sum_{y \in B} \sum_{x \in A} x p(x, y) = \sum_{y \in B} \sum_{x \in A} x p_{X|Y}(x|y) p_Y(y)$$

$$= \sum_{y \in B} \left(\sum_{x \in A} x p_{X|Y}(x|y) \right) p_Y(y)$$

$$= \sum_{y \in B} E(X \mid Y = y) P(Y = y), \tag{10.13}$$

showing that $E(X)$ is the weighted average of the conditional expectations of X (given $Y = y$) over all possible values of Y. On the other hand, we know that, for a real-valued function h,

$$E[h(Y)] = \sum_{y \in B} h(y) P(Y = y).$$

Applying this formula to $h(Y) = E(X|Y)$ yields

$$E[E(X|Y)] = \sum_{y \in B} E(X \mid Y = y) P(Y = y). \tag{10.14}$$

Comparing (10.13) and (10.14), we obtain

$$E(X) = E[E(X|Y)] = \sum_{y \in B} E(X \mid Y = y) P(Y = y). \tag{10.15}$$

Therefore, $E[E(X|Y)] = E(X)$ is a condensed way to say that $E(X)$ is the weighted average of the conditional expectations of X (given $Y = y$) over all possible values of Y. We have proved the following theorem for the discrete case.

Theorem 10.6   *Let X and Y be two random variables. Then*

$$E[E(X \mid Y)] = E(X).$$

Proof: We have already proven this for the discrete case. Now we will prove it for the case where X and Y are continuous random variables with joint probability density function, $f(x, y)$.

$$E[E(X|Y)] = \int_{-\infty}^{\infty} E(X|Y = y) f_Y(y) \, dy$$

$$= \int_{-\infty}^{\infty} \left(\int_{-\infty}^{\infty} x f_{X|Y}(x|y) \, dx \right) f_Y(y) \, dy$$

$$= \int_{-\infty}^{\infty} x \left(\int_{-\infty}^{\infty} f_{X|Y}(x|y) f_Y(y) \, dy \right) dx$$

$$= \int_{-\infty}^{\infty} x \left(\int_{-\infty}^{\infty} \frac{f(x,y)}{f_Y(y)} f_Y(y) \, dy \right) dx$$

$$= \int_{-\infty}^{\infty} x \left(\int_{-\infty}^{\infty} f(x,y) \, dy \right) dx$$

$$= \int_{-\infty}^{\infty} x f_X(x) \, dx = E(X). \quad \blacklozenge$$

Example 10.20   Suppose that $N(t)$, the number of people who pass by a museum at or prior to t, is a Poisson process having rate λ. If a person passing by enters the museum with probability p, what is the expected number of people who enter the museum at or prior to t?

Solution: Let $M(t)$ denote the number of people who enter the museum at or prior to t; then

$$E\big[M(t)\big] = E\big[E\big(M(t) \mid N(t)\big)\big] = \sum_{n=0}^{\infty} E\big[M(t) \mid N(t) = n\big] P\big(N(t) = n\big).$$

Given that $N(t) = n$, the number of people who enter the museum at or prior to t, is a binomial random variable with parameters n and p. Thus $E\big[M(t) \mid N(t) = n\big] = np$. Therefore,

$$E\big[M(t)\big] = \sum_{n=1}^{\infty} np \frac{e^{-\lambda t}(\lambda t)^n}{n!} = pe^{-\lambda t}\lambda t \sum_{n=1}^{\infty} \frac{(\lambda t)^{n-1}}{(n-1)!} = pe^{-\lambda t}\lambda t e^{\lambda t} = p\lambda t. \quad \blacklozenge$$

Example 10.21   Let X and Y be continuous random variables with joint probability density function

$$f(x,y) = \begin{cases} \dfrac{3}{2}(x^2 + y^2) & \text{if } 0 < x < 1, \quad 0 < y < 1 \\[2mm] 0 & \text{otherwise.} \end{cases}$$

Find $E(X|Y)$.

Solution: $E(X|Y)$ is a random variable that is defined to be $E(X \mid Y = y)$ when $Y = y$. First, we calculate $E(X \mid Y = y)$:

$$E(X \mid Y = y) = \int_0^1 x f_{X|Y}(x|y) \, dx = \int_0^1 x \frac{f(x,y)}{f_Y(y)} dx,$$

where

$$f_Y(y) = \int_0^1 \frac{3}{2}(x^2 + y^2) \, dx = \frac{3}{2}y^2 + \frac{1}{2}.$$

Hence

$$E(X \mid Y = y) = \int_0^1 x \frac{(3/2)(x^2 + y^2)}{(3/2)y^2 + 1/2} dx = \int_0^1 x \frac{3(x^2 + y^2)}{3y^2 + 1} dx$$

$$= \frac{3}{3y^2 + 1} \int_0^1 (x^3 + xy^2) \, dx = \frac{3(2y^2 + 1)}{4(3y^2 + 1)}.$$

Thus, if $Y = y$, then

$$E(X \mid Y = y) = \frac{3(2y^2 + 1)}{4(3y^2 + 1)}.$$

Now since the random variable $E(X|Y)$ coincides with $E(X \mid Y = y)$ if $Y = y$, we have

$$E(X|Y) = \frac{3(2Y^2 + 1)}{4(3Y^2 + 1)}. \quad \blacklozenge$$

Example 10.22   What is the expected number of random digits that should be generated to obtain three consecutive zeros?

Solution: Let X be the number of random digits to be generated until three consecutive zeros are obtained. Let Y be the number of random digits to be generated until the first nonzero digit is obtained. Then

$$E(X) = E\big[E(X|Y)\big] = \sum_{i=1}^{\infty} E(X \mid Y = i)P(Y = i)$$

$$= \sum_{i=1}^{3} E(X \mid Y = i)P(Y = i) + \sum_{i=4}^{\infty} E(X \mid Y = i)P(Y = i)$$

$$= \sum_{i=1}^{3} [i + E(X)]\Big(\frac{1}{10}\Big)^{i-1}\Big(\frac{9}{10}\Big) + \sum_{i=4}^{\infty} 3\Big(\frac{1}{10}\Big)^{i-1}\Big(\frac{9}{10}\Big),$$

which gives

$$E(X) = 1.107 + 0.999 \, E(X) + 0.003.$$

Solving this for $E(X)$, we find that $E(X) = 1110$. $\blacklozenge$

Example 10.23   Let X and Y be two random variables and f be a real-valued function from $\mathbf{R}$ to $\mathbf{R}$. Prove that

$$E\big[f(Y)X \mid Y\big] = f(Y)E(X|Y).$$

Proof: If $Y = y$, then $f(Y)E(X|Y)$ is $f(y)E(X \mid Y = y)$. We show that $E\big[f(Y)X|Y\big]$ is also equal to this quantity. Let $f_{X|Y}(x|y)$ be the conditional probability density function of X given that $Y = y$; then

$$E\big[f(Y)X \mid Y = y\big] = E\big[f(y)X|Y = y\big] = \int_{-\infty}^{\infty} f(y)xf_{X|Y}(x|y)\,dx$$

$$= f(y)\int_{-\infty}^{\infty} xf_{X|Y}(x|y)\,dx = f(y)E(X \mid Y = y). \quad \blacklozenge$$

Suppose that a certain airplane breaks down N times a year, where N is a random variable. If the repair time for the ith breakdown is X_i, then the total repair time for this airplane is the random variable $X_1 + X_2 + \cdots + X_N$. To find the expected length of time that, due to breakdowns, the plane cannot fly, we need to calculate $E\big(\sum_{i=1}^{N} X_i\big)$. What is different about this sum is that, not only is each of its terms a random variable, but the number of its terms is also a random variable. The expected values of such sums are calculated using the following theorem, discovered by Abraham Wald, a statistician who is best known for developing the theory of sequential statistical procedures during World War II.

Theorem 10.7   (Wald's Equation)   *Let $X_1, X_2, \ldots$ be independent and identically distributed random variables with the finite mean $E(X)$. Let $N > 0$ be an integer-valued random variable, independent of $\{X_1, X_2, \ldots\}$, with $E(N) < \infty$. Then*

$$E\Big(\sum_{i=1}^{N} X_i\Big) = E(N)E(X).$$

Proof: By (10.15),

$$E\Big(\sum_{i=1}^{N} X_i\Big) = E\Big[E\Big(\sum_{i=1}^{N} X_i \big| N\Big)\Big] = \sum_{n=1}^{\infty} E\Big(\sum_{i=1}^{N} X_i \big| N = n\Big) P(N = n), \quad (10.16)$$

where

$$E\Big(\sum_{i=1}^{N} X_i \big| N = n\Big) = E\Big(\sum_{i=1}^{n} X_i \big| N = n\Big) = E\Big(\sum_{i=1}^{n} X_i\Big) = \sum_{i=1}^{n} E(X_i) = nE(X),$$

since N is independent of $\{X_1, X_2, \ldots\}$. Hence, by (10.16),

$$E\Big(\sum_{i=1}^{N} X_i\Big) = \sum_{n=1}^{\infty} nE(X)P(N = n) = E(X)\sum_{n=1}^{\infty} nP(N = n) = E(X)E(N). \quad \blacklozenge$$

Example 10.24   Suppose that the average number of breakdowns for a certain airplane is 12.5 times a year. If the expected value of repair time is 7 days for each breakdown, and if the repair times are identically distributed, independent random variables, find the expected total repair time. Assume that repair times are independent of the number of breakdowns.

Solution: Let N be the number of breakdowns in a year and X_i be the repair time for the ith breakdown. Then, by Wald's equation, the expected total repair time is

$$E\left(\sum_{i=1}^{N} X_i\right) = E(N)E(X_i) = (12.5)(7) = 87.5. \quad \blacklozenge$$

The following theorem gives a formula, analogous to Wald's equation, for variance. We leave its proof as an exercise.

Theorem 10.8   *Let $\{X_1, X_2, \ldots\}$ be an independent and identically distributed sequence of random variables with finite mean $E(X)$ and finite variance $\text{Var}(X)$. Let $N > 0$ be an integer-valued random variable independent of $\{X_1, X_2, \ldots\}$ with $E(N) < \infty$ and $\text{Var}(N) < \infty$. Then*

$$\text{Var}\left(\sum_{i=1}^{N} X_i\right) = E(N)\text{Var}(X) + \left[E(X)\right]^2 \text{Var}(N).$$

We now explain a procedure for calculation of probabilities by conditioning on random variables. Let B be an event associated with an experiment and X be a discrete random variable with possible set of values A. Let

$$Y = \begin{cases} 1 & \text{if } B \text{ occurs} \\ 0 & \text{if } B \text{ does not occur.} \end{cases}$$

Then

$$E(Y) = E\big[E(Y|X)\big]. \tag{10.17}$$

But

$$E(Y) = 1 \cdot P(B) + 0 \cdot P(B^c) = P(B) \tag{10.18}$$

and

$$E\big[E(Y|X)\big] = \sum_{x \in A} E(Y \mid X = x)P(X = x)$$

$$= \sum_{x \in A} P(B \mid X = x)P(X = x), \tag{10.19}$$

where the last equality follows since

$$E(Y \mid X = x) = 1 \cdot P(Y = 1 \mid X = x) + 0 \cdot P(Y = 0 \mid X = x)$$

$$= \frac{P(Y = 1, X = x)}{P(X = x)} = \frac{P(B \text{ and } X = x)}{P(X = x)} = P(B \mid X = x).$$

Relations (10.17), (10.18), and (10.19) imply the following theorem.

Theorem 10.9   *Let B be an arbitrary event and X be a discrete random variable with possible set of values A; then*

$$P(B) = \sum_{x \in A} P(B \mid X = x) P(X = x). \tag{10.20}$$

If X is a continuous random variable, the relation analogous to (10.20) is

$$P(B) = \int_{-\infty}^{\infty} P(B \mid X = x) f(x) \, dx, \tag{10.21}$$

where f is the probability density function of X.

Theorem 10.9 shows that the probability of an event B is the weighted average of the conditional probabilities of B (given $X = x$) over all possible values of X. For the discrete case, this is a conclusion of Theorem 3.4, the most general version of the law of total probability.

Example 10.25   The time between consecutive earthquakes in San Francisco and the time between consecutive earthquakes in Los Angeles are independent and exponentially distributed with means $1/\lambda_1$ and $1/\lambda_2$, respectively. What is the probability that the next earthquake occurs in Los Angeles?

Solution: Let X and Y denote the times between now and the next earthquake in San Francisco and Los Angeles, respectively. Because of the memoryless property of exponential distribution, X and Y are exponentially distributed with means $1/\lambda_1$ and $1/\lambda_2$, respectively. To calculate $P(X > Y)$, the desired probability, we will condition on Y:

$$P(X > Y) = \int_0^\infty P(X > Y \mid Y = y) \lambda_2 e^{-\lambda_2 y} \, dy$$

$$= \int_0^\infty P(X > y) \lambda_2 e^{-\lambda_2 y} \, dy = \int_0^\infty e^{-\lambda_1 y} \lambda_2 e^{-\lambda_2 y} \, dy$$

$$= \lambda_2 \int_0^\infty e^{-(\lambda_1 + \lambda_2) y} \, dy = \frac{\lambda_2}{\lambda_1 + \lambda_2},$$

where $P(X > y)$ is calculated from

$$P(X > y) = 1 - P(X \leq y) = 1 - (1 - e^{-\lambda_1 y}) = e^{-\lambda_1 y}. \quad \blacklozenge$$

Example 10.26   Suppose that Z_1 and Z_2 are independent standard normal random variables. Show that the ratio $Z_1/|Z_2|$ is a Cauchy random variable. That is, $Z_1/|Z_2|$ is a random variable with the probability density function

$$f(t) = \frac{1}{\pi(1+t^2)}, \quad -\infty < t < \infty.$$

Solution: Let $g(x)$ be the probability density function of $|Z_2|$. To find $g(x)$, note that, for $x \geq 0$,

$$P(|Z_2| \leq x) = P(-x \leq Z_2 \leq x) = \int_{-x}^{x} \frac{1}{\sqrt{2\pi}} e^{-u^2/2}\, du = 2\int_{0}^{x} \frac{1}{\sqrt{2\pi}} e^{-u^2/2}\, du.$$

Hence

$$g(x) = \frac{d}{dx} P(|Z_2| \leq x) = \frac{2}{\sqrt{2\pi}} e^{-x^2/2}, \quad x \geq 0.$$

To find the probability density function of $Z_1/|Z_2|$, note that, by Theorem 10.9,

$$P\left(\frac{Z_1}{|Z_2|} \leq t\right) = 1 - P\left(\frac{Z_1}{|Z_2|} > t\right) = 1 - P(Z_1 > t|Z_2|)$$

$$= 1 - \int_{0}^{\infty} P(Z_1 > t|Z_2| \mid |Z_2| = x) \frac{2}{\sqrt{2\pi}} e^{-x^2/2}\, dx$$

$$= 1 - \int_{0}^{\infty} P(Z_1 > tx) \frac{2}{\sqrt{2\pi}} e^{-x^2/2}\, dx$$

$$= 1 - \int_{0}^{\infty} \left(\int_{tx}^{\infty} \frac{1}{\sqrt{2\pi}} e^{-u^2/2}\, du\right) \frac{2}{\sqrt{2\pi}} e^{-x^2/2}\, dx.$$

Now, by the fundamental theorem of calculus,

$$\frac{d}{dt} \int_{tx}^{\infty} \frac{1}{\sqrt{2\pi}} e^{-u^2/2}\, du = \frac{d}{dt}\left(1 - \int_{-\infty}^{tx} \frac{1}{\sqrt{2\pi}} e^{-u^2/2}\, du\right) = -x\frac{1}{\sqrt{2\pi}} e^{-t^2x^2/2}.$$

Therefore,

$$\frac{d}{dt} P\left(\frac{Z_1}{|Z_2|} \leq t\right) = \int_{0}^{\infty} \frac{x}{\sqrt{2\pi}} e^{-t^2x^2/2} \cdot \frac{2}{\sqrt{2\pi}} e^{-x^2/2}\, dx = \frac{2}{2\pi} \int_{0}^{\infty} x e^{-(t^2+1)x^2/2}\, dx.$$

Making the change of variable $y = (1+t^2)x^2/2$ yields

$$\frac{d}{dt} P\left(\frac{Z_1}{|Z_2|} \leq t\right) = \frac{1}{\pi(1+t^2)} \int_{0}^{\infty} e^{-y}\, dy = \frac{1}{\pi(1+t^2)}, \quad -\infty < t < \infty. \blacklozenge$$

Example 10.27   At the intersection of two remote roads, the vehicles arriving are either cars or trucks. Suppose that cars arrive at the intersection at a Poisson rate of λ per minute, and trucks arrive at a Poisson rate of μ per minute. Suppose that the arrivals are independent of each other. If we are given that the next vehicle arriving at this intersection is a car, find the expected value of the time until the next arrival.

Warning: Since cars and trucks arrive at this intersection independently, we might fallaciously think that, given the next arrival is a car, the expected value until the next arrival is $1/\lambda$.

Solution: Let T be the time until the next vehicle arrives at the intersection. Let X be the time until the next car arrives at the intersection, and Y be the time until the next truck arrives at the intersection. Let A be the event that the next vehicle arriving at this intersection is a car. Note that X and Y are independent exponential random variables with means $1/\lambda$ and $1/\mu$, respectively. We are interested in $E(T|A) = E(X \mid X < Y)$. To find this quantity, we will first calculate the probability distribution function of X given that $X < Y$:

$$P(X \leq t \mid X < Y) = \frac{P(X \leq t,\ X < Y)}{P(X < Y)},$$

where

$$P(X < Y) = \int_0^\infty P(X < Y \mid X = x) f_X(x)\, dx$$

$$= \int_0^\infty P(Y > x \mid X = x) f_X(x)\, dx = \int_0^\infty P(Y > x)\lambda e^{-\lambda x}\, dx$$

$$= \int_0^\infty e^{-\mu x}\lambda e^{-\lambda x}\, dx = \frac{\lambda}{\lambda + \mu},$$

and

$$P(X \leq t,\ X < Y) = P\big(X < \min\{t, Y\}\big)$$

$$= \int_0^\infty P\big(X < \min\{t, Y\} \mid X = x\big) f_X(x)\, dx$$

$$= \int_0^\infty P\big(\min\{t, Y\} > x \mid X = x\big) f_X(x)\, dx$$

$$= \int_0^t P(Y > x \mid X = x) f_X(x)\, dx$$

$$= \int_0^t P(Y > x) f_X(x)\, dx = \int_0^t e^{-\mu x}\lambda e^{-\lambda x}\, dx$$

$$= \frac{\lambda}{\lambda + \mu}[1 - e^{-(\lambda + \mu)t}].$$

Thus

$$P(X \leq t \mid X < Y) = \frac{\dfrac{\lambda}{\lambda + \mu}[1 - e^{-(\lambda+\mu)t}]}{\dfrac{\lambda}{\lambda + \mu}} = 1 - e^{-(\lambda+\mu)t}.$$

This result shows that, given that the next vehicle arriving is a car, the distribution of the time until the next arrival is exponential with parameter $\lambda + \mu$. Therefore,

$$E(T \mid A) = \frac{1}{\lambda + \mu}.$$

Similarly,

$$E(T \mid A^c) = \frac{1}{\lambda + \mu}. \quad \blacklozenge$$

Let X and Y be two given random variables. Define the new random variable **Var$(X|Y)$** by

$$\mathbf{Var}(X|Y) = E\big[\big(X - E(X|Y)\big)^2 \mid Y\big].$$

Then the formula analogous to $\text{Var}(X) = E(X^2) - \big[E(X)\big]^2$ is given by

$$\mathbf{Var}(X|Y) = E(X^2|Y) - E(X|Y)^2. \tag{10.22}$$

(See Exercise 20.) The following theorem shows that $\text{Var}(X)$ is the sum of the expected value of $\text{Var}(X|Y)$ and the variance of $E(X|Y)$.

Theorem 10.10 $\mathbf{Var}(X) = E\big[\mathbf{Var}(X|Y)\big] + \mathbf{Var}\big(E[X|Y]\big).$

Proof: By (10.22),

$$E\big[\text{Var}(X|Y)\big] = E\big[E(X^2|Y)\big] - E\big[E(X|Y)^2\big]$$
$$= E(X^2) - E\big[E(X|Y)^2\big].$$

By the definition of variance,

$$\text{Var}\big(E[X|Y]\big) = E\big[E(X|Y)^2\big] - \big(E\big[E(X|Y)\big]\big)^2$$
$$= E\big[E(X|Y)^2\big] - \big[E(X)\big]^2.$$

Adding these two equations, we have the theorem. $\blacklozenge$

Example 10.28   A fisherman catches fish in a large lake with lots of fish, at a Poisson rate of two per hour. If, on a given day, the fisherman spends randomly anywhere between 3 and 8 hours fishing, find the expected value and the variance of the number of fish he catches.

Solution: Let X be the number of hours the fisherman spends fishing. Then X is a uniform random variable over the interval $(3, 8)$. Label the time the fisherman begins fishing on the given day at $t = 0$. Let $N(t)$ denote the total number of fish caught at or prior to t. Then $\{N(t): t \geq 0\}$ is a Poisson process with parameter $\lambda = 2$. Assuming that X is independent of $\{N(t): t \geq 0\}$, we have

$$E[N(X) \mid X = t] = E[N(t)] = 2t.$$

This implies that

$$E[N(X) \mid X] = 2X.$$

Therefore,

$$E[N(X)] = E[E(N(X) \mid X)] = E(2X) = 2E(X) = 2 \cdot \frac{8+3}{2} = 11.$$

Similarly,

$$\text{Var}(N(X) \mid X = t) = \text{Var}(N(t)) = 2t.$$

Thus

$$\text{Var}(N(X) \mid X) = 2X.$$

By Theorem 10.10,

$$\text{Var}(N(X)) = E[\text{Var}(N(X) \mid X)] + \text{Var}(E[N(X) \mid X])$$
$$= E(2X) + \text{Var}(2X) = 2E(X) + 4\text{Var}(X)$$
$$= 2 \cdot \frac{8+3}{2} + 4 \cdot \frac{(8-3)^2}{12} = 19.33. \quad \blacklozenge$$

EXERCISES

A

1. A fair coin is tossed until two tails occur successively. Find the expected number of the tosses required.

    Hint: Let

    $$X = \begin{cases} 1 & \text{if the first toss results in tails} \\ 0 & \text{if the first toss results in heads,} \end{cases}$$

    and condition on X.

2. The orders received for grain by a farmer add up to X tons, where X is a continuous random variable uniformly distributed over the interval $(4, 7)$. Every ton of grain sold brings a profit of a, and every ton that is not sold is destroyed at a loss of $a/3$. How many tons of grain should the farmer produce to maximize his expected profit?

    Hint: Let $Y(t)$ be the profit if the farmer produces t tons of grain. Then

    $$E[Y(t)] = E\left[aX - \frac{a}{3}(t - X)\right]P(X < t) + E(at)P(X \geq t).$$

3. In a box, Lynn has b batteries of which d are dead. She tests them randomly and one by one. Every time that a good battery is drawn, she will return it to the box; every time that a dead battery is drawn, she will replace it by a good one.

    (a)   Determine the expected value of the number of good batteries in the box after n of them are checked.

    (b)   Determine the probability that on the nth draw Lynn draws a good battery.

    Hint: Let X_n be the number of good batteries in the box after n of them are checked. Show that

    $$E(X_n \mid X_{n-1}) = 1 + \left(1 - \frac{1}{b}\right)X_{n-1}.$$

    Then, by computing the expected value of this random variable, find a recursive relation between $E(X_n)$ and $E(X_{n-1})$. Use this relation and induction to prove that

    $$E(X_n) = b - d\left(1 - \frac{1}{b}\right)^n.$$

    Note that n should approach ∞ to get $E(X_n) = b$. For part (b), let E_n be the event that on the nth draw she gets a good battery. By conditioning on X_{n-1} prove that $P(E_n) = E(X_{n-1})/b$.

4. For given random variables Y and Z, let

    $$X = \begin{cases} Y & \text{with probability } p \\ Z & \text{with probability } 1 - p. \end{cases}$$

    Find $E(X)$ in terms of $E(Y)$ and $E(Z)$.

5. A typist, on average, makes three typing errors in every two pages. If pages with more than two errors must be retyped, on average how many pages must she type to prepare a report of 200 pages? Assume that the number of errors in a page is a Poisson random variable. Note that some of the retyped pages should be retyped, and so on.

Hint: Find p, the probability that a page should be retyped. Let X_n be the number of pages that should be typed at least n times. Show that $E(X_1) = 200p$, $E(X_2) = 200p^2, \ldots, E(X_n) = 200p^n$. The desired quantity is $E\left(\sum_{i=1}^{\infty} X_i\right)$, which can be calculated using relation (10.2).

6. In data communication, usually messages sent are combinations of characters, and each character consists of a number of bits. A bit is the smallest unit of information and is either 1 or 0. Suppose that the length of a character (in bits) is a geometric random variable with parameter p. Suppose that a message is combined of K characters, where K is a random variable with mean μ and variance σ^2. If the lengths of characters of a message are independent of each other and of K, and if it takes a sender 1000 bits per second to emit a message, find the expected value of T, the time it will take the sender to emit a message.

7. From an ordinary deck of 52 cards, cards are drawn at random, one by one and without replacement until a heart is drawn. What is the expected value of the number of cards drawn?

    Hint: Consider a deck of cards with 13 hearts and $39 - n$ nonheart cards. Let X_n be the number of cards to be drawn *before* the first heart is drawn. Let

$$Y = \begin{cases} 1 & \text{if the first card drawn is a heart} \\ 0 & \text{otherwise.} \end{cases}$$

By conditioning on Y, find a recursive relation between $E(X_n)$ and $E(X_{n+1})$. Use $E(X_{39}) = 0$ to show that $E(X_i) = (39 - i)/14$. The answer is $1 + E(X_0)$. (For a totally different solution see Exercise 15, Section 10.1.)

8. Suppose that X and Y are independent random variables with probability density functions f and g, respectively. Use conditioning technique to calculate $P(X < Y)$.

9. Prove that, for a Poisson random variable N, if the parameter λ is not fixed and is itself an exponential random variable with parameter 1, then

$$P(N = i) = \left(\frac{1}{2}\right)^{i+1}.$$

10. Suppose that X and Y represent the amount of money in the wallets of players A and B, respectively. Let X and Y be jointly uniformly distributed on the unit square $[0, 1] \times [0, 1]$. A and B each places his wallet on the table. Whoever has the smallest amount of money in his wallet, wins all the money in the other wallet. Let W_A be the amount of money that player A will win. Show that $E(W_A) = 0$. (For a history of this problem, see the *Wallet Paradox* in Exercise 22, Review Problems, Chapter 8.)

11. A fair coin is tossed successively. Let K_n be the number of tosses until n consecutive heads occur.

(a)   Argue that

$$E(K_n \mid K_{n-1} = i) = (i + 1)\frac{1}{2} + \left[i + 1 + E(K_n)\right]\frac{1}{2}.$$

(b)   Show that

$$E(K_n \mid K_{n-1}) = K_{n-1} + 1 + \frac{1}{2}E(K_n).$$

(c)   By finding the expected values of both sides of (b) find a recursive relation between $E(K_n)$ and $E(K_{n-1})$.

(d)   Note that $E(K_1) = 2$. Use this and (c) to find $E(K_n)$.

B

12.   In Rome, tourists arrive at a historical monument according to a Poisson process, on average, one every five minutes. There are guided tours that depart (a) whenever there is a group of 10 tourists waiting to take the tour, or (b) one hour has elapsed from the time the previous tour began. It is the policy of the Tourism Department that a tour will only run for less than 10 people if the last guided tour left one hour ago. If in any one hour period, there will always be tourists arriving to take the tour, find the expected value of the time between two consecutive tours.

13.   During an academic year, the admissions office of a small college receives student applications at a Poisson rate of 5 per day. It is a policy of this college to double its student recruitment efforts if no applications arrive for two consecutive business days. Find the expected number of business days until a time when the college needs to double its recruitment efforts. Do the admission officers need to worry about this policy at all?
Hint: Let X_1 be the time until the first application arrives. Let X_2 be the time between the first and second applications, and so forth. Let N be the first integer for which

$$X_1 \le 2, \ X_2 \le 2, \ \dots, \ X_N \le 2, \ X_{N+1} > 2.$$

The time that the admissions office has to wait before doubling its student recruitment efforts is $S_{N+1} = X_1 + X_2 + \cdots + X_{N+1}$. Find S_{N+1} by conditioning on N.

14.   Each time that Steven calls his friend Adam, the probability that Adam is available to talk with him is p independently of other calls. On average, after how many calls has Steven not missed Adam k consecutive times?

15.   Recently, Larry taught his daughter Emily how to play backgammon. To encourage Emily to practice this game, Larry decides to play with her until she wins two of the recent three games. If the probability that Emily wins a game is 0.35 independently of all preceding and future games, find the expected number of games to be played.

16. **(Genetics)**   Hemophilia is a sex-linked disease with normal allele H dominant to the mutant allele h. Kim and John are married, and John is phenotypically normal. Suppose that, in the entire population, the frequencies of H and h are 0.98 and 0.02, respectively. If Kim and John have four sons and three daughters, what is the expected number of their hemophilic children?

17.   A spice company distributes cinnamon in one-pound bags. Suppose that the Food and Drug Administration (FDA) considers more than 500 insect fragments in one bag excessive and hence unacceptable. To meet the standards of the FDA, the quality control division of the company begins inspecting the bags of the cinnamon at time $t = 0$ according to the following scheme. It inspects each bag with probability α until it encounters an unacceptable bag. At that point, the division inspects every single bag until it finds m consecutive acceptable bags. When this happens, the division has completed one inspection cycle. It then resumes its normal inspection process. Inspected bags that are found with excessive numbers of insect fragments are sent back for further cleaning. Let p be the probability that a bag is acceptable, independent of the number of insect fragments in other bags. Find the expected value of the number of bags inspected in one inspection cycle.

18.   Suppose that a device is powered by a battery. Since an uninterrupted supply of power is needed, the device has a spare battery. When the battery fails, the circuit is altered electronically to connect the spare battery and remove the failed battery from the circuit. The spare battery then becomes the working battery and emits a signal to alert the user to replace the failed battery. When that battery is replaced, it becomes the new spare battery. Suppose that the lifetimes of the batteries used are independent uniform random variables over the interval $(0, 1)$, where the unit of measurement is 1000 hours. For $0 < t \leq 1$, on average, how many batteries are changed by time t? How many are changed, on average, after 950 hours of operation?

19.   Let X and Y be continuous random variables. Prove that

$$E\big[(X - E(X|Y))^2\big] = E(X^2) - E\big[E(X|Y)^2\big].$$

Hint: Let $Z = E(X|Y)$. By conditioning on Y and using Example 10.23, first show that $E(XZ) = E(Z^2)$.

20.   Let X and Y be two given random variables. Prove that

$$\mathrm{Var}(X|Y) = E[X^2|Y] - E(X|Y)^2.$$

21.   Prove Theorem 10.8.

10.5   BIVARIATE NORMAL DISTRIBUTION

Let $f(x, y)$ be the joint probability density function of continuous random variables X and Y; f is called a **bivariate normal probability density function** if

(a)   The distribution function of X is normal. That is, the marginal probability density function of X is

$$f_X(x) = \frac{1}{\sigma_X \sqrt{2\pi}} \exp\left[-\frac{(x - \mu_X)^2}{2\sigma_X^2} \right], \qquad -\infty < x < \infty. \qquad (10.23)$$

(b)   The conditional distribution of Y, given that $X = x$, is normal for each $x \in (-\infty, \infty)$. That is, $f_{Y|X}(y|x)$ has a normal density for each $x \in \mathbf{R}$.

(c)   The conditional expectation of Y, given that $X = x$, $E(Y \mid X = x)$, is a linear function of x. That is, $E(Y \mid X = x) = a + bx$ for some $a, b \in \mathbf{R}$.

(d)   The conditional variance of Y, given that $X = x$, is constant. That is, $\sigma_{Y|X=x}^2$ is independent of the value of x.

As an example, let X be the height of a man and let Y be the height of his daughter. It is reasonable to assume, and is statistically verified, that the joint probability density function of X and Y satisfies (a) to (d) and hence is bivariate normal. As another example, let X and Y be grade point averages of a student in his or her freshman and senior years, respectively. Then the joint probability density function of X and Y is bivariate normal.

We now prove that if $f(x, y)$ satisfies (a) to (d), it must be of the following form:

$$f(x, y) = \frac{1}{2\pi \sigma_X \sigma_Y \sqrt{1 - \rho^2}} \exp\left[-\frac{1}{2(1 - \rho^2)} Q(x, y) \right], \qquad (10.24)$$

where ρ is the correlation coefficient of X and Y and

$$Q(x, y) = \left(\frac{x - \mu_X}{\sigma_X}\right)^2 - 2\rho \frac{x - \mu_X}{\sigma_X} \frac{y - \mu_Y}{\sigma_Y} + \left(\frac{y - \mu_Y}{\sigma_Y}\right)^2.$$

Figure 10.1 demonstrates the graph of $f(x, y)$ in the case where $\rho = 0$, $\mu_X = \mu_Y = 0$, and $\sigma_X = \sigma_Y = 1$. To prove (10.24), note that from Lemmas 10.3 and 10.4 (which follow) we have that for each $x \in \mathbf{R}$ the expected value and the variance of the normal density function $f_{Y|X}(y|x)$ are $\mu_Y + \rho(\sigma_Y/\sigma_X)(x - \mu_X)$ and $(1 - \rho^2)\sigma_Y^2$, respectively. Therefore, for every real x,

$$f_{Y|X}(y|x) = \frac{1}{\sigma_Y \sqrt{2\pi} \sqrt{1 - \rho^2}} \exp\left[-\frac{[y - \mu_Y - \rho(\sigma_Y/\sigma_X)(x - \mu_X)]^2}{2\sigma_Y^2(1 - \rho^2)} \right], \qquad (10.25)$$

$$-\infty < y < \infty.$$

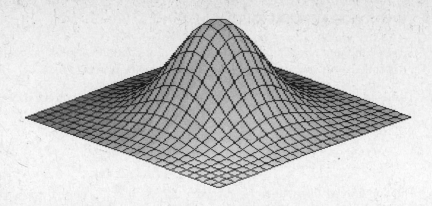

Figure 10.1   Bivariate normal probability density function.

Now $f(x, y) = f_{Y|X}(y|x) f_X(x)$, and $f_X(x)$ has the normal density given by (10.23). Multiplying (10.23) by (10.25), we obtain (10.24). So what remains to be shown is the proofs of the following general lemmas, which are valid even if the joint density function of X and Y is not bivariate normal.

Lemma 10.3   *Let X and Y be two random variables with probability density function $f(x, y)$. If $E(Y \mid X = x)$ is a linear function of x, that is, if $E(Y \mid X = x) = a + bx$ for some $a, b \in \mathbf{R}$, then*

$$E(Y \mid X = x) = \mu_Y + \rho \frac{\sigma_Y}{\sigma_X} (x - \mu_X).$$

Proof: By definition,

$$E(Y \mid X = x) = \int_{-\infty}^{\infty} y \frac{f(x, y)}{f_X(x)} \, dy = a + bx.$$

Therefore,

$$\int_{-\infty}^{\infty} y f(x, y) \, dy = a f_X(x) + b x f_X(x). \tag{10.26}$$

This gives

$$\int_{-\infty}^{\infty} \int_{-\infty}^{\infty} y f(x, y) \, dy \, dx = a \int_{-\infty}^{\infty} f_X(x) \, dx + b \int_{-\infty}^{\infty} x f_X(x) \, dx, \tag{10.27}$$

which is equivalent to

$$\mu_Y = a + b \mu_X. \tag{10.28}$$

Now, multiplying (10.26) by x and integrating both sides, we obtain

$$\int_{-\infty}^{\infty} \int_{-\infty}^{\infty} xyf(x, y) \, dy \, dx = a \int_{-\infty}^{\infty} xf_X(x) \, dx + b \int_{-\infty}^{\infty} x^2 f_X(x) \, dx,$$

which is equivalent to

$$E(XY) = a\mu_X + bE(X^2). \tag{10.29}$$

Solving (10.28) and (10.29) for a and b, we get

$$b = \frac{E(XY) - \mu_X \mu_Y}{E(X^2) - \mu_X^2} = \frac{\text{Cov}(X, Y)}{\sigma_X^2} = \frac{\rho \sigma_X \sigma_Y}{\sigma_X^2} = \frac{\rho \sigma_Y}{\sigma_X}$$

and

$$a = \mu_Y - \frac{\rho \sigma_Y}{\sigma_X} \mu_X.$$

Therefore,

$$E(Y \mid X = x) = a + bX = \mu_Y + \rho \frac{\sigma_Y}{\sigma_X}(x - \mu_X). \quad \blacklozenge$$

Lemma 10.4   *Let $f(x, y)$ be the joint probability density function of continuous random variables X and Y. If $E(Y \mid X = x)$ is a linear function of x and $\sigma_{Y|X=x}^2$ is constant, then*

$$\sigma_{Y|X=x}^2 = (1 - \rho^2)\sigma_Y^2.$$

Proof: We have that

$$\sigma_{Y|X=x}^2 = \int_{-\infty}^{\infty} \left[y - E(Y \mid X = x)\right]^2 f_{Y|X}(y|x) \, dy.$$

Multiplying both sides of this equation by $f_X(x)$ and integrating on x gives

$$\sigma_{Y|X=x}^2 \int_{-\infty}^{\infty} f_X(x) \, dx = \int_{-\infty}^{\infty} \int_{-\infty}^{\infty} \left[y - E(Y \mid X = x)\right]^2 f_{Y|X}(y|x) f_X(x) \, dy \, dx.$$

Now, since $f(x, y) = f_{Y|X}(y|x) f_X(x)$, and

$$E(Y \mid X = x) = \mu_Y + \rho \frac{\sigma_Y}{\sigma_X}(x - \mu_X),$$

we get

$$\sigma_{Y|X=x}^2 = \int_{-\infty}^{\infty} \int_{-\infty}^{\infty} \left[y - \mu_Y - \rho \frac{\sigma_Y}{\sigma_X}(x - \mu_X)\right]^2 f(x, y) \, dx \, dy,$$

or, equivalently,

$$\sigma_{Y|X=x}^2 = E\left[(Y - \mu_Y) - \rho\frac{\sigma_Y}{\sigma_X}(X - \mu_X)\right]^2.$$

Therefore,

$$\sigma_{Y|X=x}^2 = E\left[(Y - \mu_Y)^2\right] - 2\rho\frac{\sigma_Y}{\sigma_X}E\left[(Y - \mu_Y)(X - \mu_X)\right] + \rho^2\frac{\sigma_Y^2}{\sigma_X^2}E\left[(X - \mu_X)^2\right]$$

$$= \sigma_Y^2 - 2\rho\frac{\sigma_Y}{\sigma_X}\rho\sigma_X\sigma_Y + \rho^2\frac{\sigma_Y^2}{\sigma_X^2}\sigma_X^2$$

$$= (1 - \rho^2)\sigma_Y^2. \quad \blacklozenge$$

We know that if X and Y are independent random variables, their correlation coefficient is 0. We also know that, in general, the converse of this fact is not true. However, if X and Y have a bivariate normal distribution, the converse is also true. That is, $\rho = 0$ implies that X and Y are independent, which can be deduced from (10.24). For $\rho = 0$,

$$f(x, y) = \frac{1}{2\pi\sigma_X\sigma_Y}\exp\left[-\frac{(x - \mu_X)^2}{2\sigma_X^2} - \frac{(y - \mu_Y)^2}{2\sigma_Y^2}\right]$$

$$= \frac{1}{\sigma_X\sqrt{2\pi}}\exp\left[-\frac{(x - \mu_X)^2}{2\sigma_X^2}\right]\frac{1}{\sigma_Y\sqrt{2\pi}}\exp\left[-\frac{(y - \mu_Y)^2}{2\sigma_Y^2}\right]$$

$$= f_X(x)f_Y(y),$$

showing that X and Y are independent.

Example 10.29   At a certain university, the joint probability density function of X and Y, the grade point averages of a student in his or her freshman and senior years, respectively, is bivariate normal. From the grades of past years it is known that $\mu_X = 3$, $\mu_Y = 2.5$, $\sigma_X = 0.5$, $\sigma_Y = 0.4$, and $\rho = 0.4$. Find the probability that a student with grade point average 3.5 in his or her freshman year will earn a grade point average of at least 3.2 in his or her senior year.

Solution: The conditional probability density function of Y, given that $X = 3.5$, is normal with mean $2.5 + (0.4)(0.4/0.5)(3.5 - 3) = 2.66$ and standard deviation $(0.4)\sqrt{1 - (0.4)^2} = 0.37$. Therefore, the desired probability is calculated as follows:

$$P(Y \geq 3.2 \mid X = 3.5) = P\left(\frac{Y - 2.66}{0.37} \geq \frac{3.2 - 2.66}{0.37} \,\middle|\, X = 3.5\right)$$

$$= P(Z \geq 1.46 \mid X = 3.5) = 1 - \Phi(1.46) \approx 1 - 0.9279 = 0.0721,$$

where Z is a standard normal random variable. $\blacklozenge$

EXERCISES

1. Let X be the height of a man and Y the height of his daughter (both in inches). Suppose that the joint probability density function of X and Y is bivariate normal with the following parameters: $\mu_X = 71$, $\mu_Y = 60$, $\sigma_X = 3$, $\sigma_Y = 2.7$, and $\rho = 0.45$. Find the probability that the height of the daughter, of a man who is 70 inches tall, is at least 59 inches.

2. The joint probability density function of X and Y is bivariate normal with $\sigma_X = \sigma_Y = 9$, $\mu_X = \mu_Y = 0$, and $\rho = 0$. Find (a) $P(X \le 6, Y \le 12)$; (b) $P(X^2 + Y^2 \le 36)$.

3. Let the joint probability density function of X and Y be bivariate normal. For what values of α is the variance of $\alpha X + Y$ minimum?

4. Let $f(x, y)$ be a joint bivariate normal probability density function. Determine the point at which the maximum value of f is obtained.

5. Let the joint probability density function of two random variables X and Y be given by

    $$f(x, y) = \begin{cases} 2 & \text{if } 0 < y < x, 0 < x < 1 \\ 0 & \text{elsewhere.} \end{cases}$$

    Find $E(X \mid Y = y)$, $E(Y \mid X = x)$, and $\rho(X, Y)$.
    Hint: To find ρ, use Lemma 10.3.

6. Let Z and W be independent standard normal random variables. Let X and Y be defined by

    $$X = \sigma_1 Z + \mu_1,$$
    $$Y = \sigma_2[\rho Z + \sqrt{1 - \rho^2}\, W] + \mu_2,$$

    where $\sigma_1, \sigma_2 > 0$, $-\infty < \mu_1, \mu_2 < \infty$, and $-1 < \rho < 1$. Show that the joint probability density function of X and Y is bivariate normal and $\sigma_X = \sigma_1$, $\sigma_Y = \sigma_2$, $\mu_X = \mu_1$, $\mu_Y = \mu_2$, and $\rho(X, Y) = \rho$.
    Note: By this exercise, if the joint probability density function of X and Y is bivariate normal, X and Y can be written as sums of independent standard normal random variables.

7. Let the joint probability density function of random variables X and Y be bivariate normal. Show that if $\sigma_X = \sigma_Y$, then $X + Y$ and $X - Y$ are independent random variables.
    Hint: Show that the joint probability density function of $X + Y$ and $X - Y$ is bivariate normal with correlation coefficient 0.

REVIEW PROBLEMS

1. In a commencement ceremony, for the dean of a college to present the diplomas of the graduates, a clerk piles the diplomas in the order that the students will walk on the stage. However, the clerk mixes the last 10 diplomas in some random order accidentally. Find the expected number of the last 10 graduates walking on the stage who will receive their own diplomas from the dean of the college.

2. Let the probability density function of a random variable X be given by

    $$f(x) = \begin{cases} 2x - 2 & \text{if } 1 < x < 2 \\ 0 & \text{elsewhere.} \end{cases}$$

    Find $E(X^3 + 2X - 7)$.

3. Let the joint probability density function of random variables X and Y be

    $$f(x, y) = \begin{cases} \dfrac{3x^3 + xy}{3} & \text{if } 0 \le x \le 1, \quad 0 \le y \le 2 \\ 0 & \text{elsewhere.} \end{cases}$$

    Find $E(X^2 + 2XY)$.

4. In a town there are n taxis. A woman takes one of these taxis every day at random and with replacement. On average, how long does it take before she can claim that she has been in every taxi in the town?
    Hint: The final answer is in terms of $a_n = 1 + 1/2 + \cdots + 1/n$.

5. Determine the expected number of tosses of a die required to obtain four consecutive 6's.

6. Let the joint probability density function of X, Y, and Z be given by

    $$f(x, y, z) = \begin{cases} 8xyz & \text{if } 0 < x < 1, \quad 0 < y < 1, \quad 0 < z < 1 \\ 0 & \text{otherwise.} \end{cases}$$

    Find $\rho(X, Y)$, $\rho(X, Z)$, and $\rho(Y, Z)$.

7. Let X and Y be jointly distributed with $\rho(X, Y) = 2/3$, $\sigma_X = 1$, $\text{Var}(Y) = 9$. Find $\text{Var}(3X - 5Y + 7)$.

8. Let the joint probability mass function of discrete random variables X and Y be given by

    $$p(x, y) = \begin{cases} \dfrac{1}{25}(x^2 + y^2) & \text{if } (x, y) = (1, 1), (1, 3), (2, 3) \\ 0 & \text{otherwise.} \end{cases}$$

Find $\text{Cov}(X, Y)$.

9. Two dice are rolled. The sum of the outcomes is denoted by X and the absolute value of their difference by Y. Calculate the covariance of X and Y. Are X and Y uncorrelated? Are they independent?

10. Two green and two blue dice are rolled. If X and Y are the numbers of 6's on the green and on the blue dice, respectively, calculate the correlation coefficient of $|X - Y|$ and $X + Y$.

11. A random point (X, Y) is selected from the rectangle $[0, \pi/2] \times [0, 1]$. What is the probability that it lies below the curve $y = \sin x$?

12. Let the joint probability density function of X and Y be given by

$$f(x, y) = \begin{cases} e^{-x} & \text{if } 0 < y < x < \infty \\ 0 & \text{elsewhere.} \end{cases}$$

   (a) Find the marginal probability density functions of X and Y.

   (b) Determine the correlation coefficient of X and Y.

13. In terms of the means, variances, and the covariance of the random variables X and Y, find α and β for which $E(Y - \alpha - \beta X)^2$ is minimum. This is the **method of least squares**; it fits the "best" line $y = \alpha + \beta x$ to the distribution of Y.

14. Let the joint probability density function of X and Y be given by

$$f(x, y) = \begin{cases} ye^{-y(1+x)} & \text{if } x > 0, \quad y > 0 \\ 0 & \text{otherwise.} \end{cases}$$

   (a) Show that $E(X)$ does not exist.

   (b) Find $E(X|Y)$.

15. Bus A arrives at a station at a random time between 10:00 A.M. and 10:30 A.M. tomorrow. Bus B arrives at the same station at a random time between 10:00 A.M. and the arrival time of bus A. Find the expected value of the arrival time of bus B.

16. Let $\{X_1, X_2, X_3, \dots\}$ be a sequence of independent and identically distributed exponential random variables with parameter λ. Let N be a geometric random variable with parameter p independent of $\{X_1, X_2, X_3, \dots\}$. Find the distribution function of $\sum_{i=1}^{N} X_i$.

17. Slugger Bubble Gum Company markets its best-selling brand to young baseball fans by including pictures of current baseball stars in packages of its bubble gum. In the latest series, there are 20 players included, but there is no way of telling which player's picture is inside until the package of gum is purchased and opened.

If a young fan wishes to collect all 20 pictures, how many packages must he buy, on average? Assume that there is no trading of one player's picture for another and that the number of cards printed for each player is the same.

■

Chapter 11

Sums of Independent Random Variables and Limit Theorems

11.1 MOMENT-GENERATING FUNCTIONS

For a random variable X, we have demonstrated how important its first moment $E(X)$ and its second moment $E(X^2)$ are. For other values of n also, $E(X^n)$ is a valuable measure both in theory and practice. For example, letting $\mu = E(X)$, $\mu_X^{(r)} = E[(X - \mu)^r]$, we have that the quantity $\mu_X^{(3)}/\sigma_X^3$ is a measure of symmetry of the distribution of X. It is called the measure of **skewness** and is zero if the distribution of X is symmetric, negative if it is skewed to the left, and positive if it is skewed to the right (for examples of distributions that are, say, skewed to the right, see Figure 7.12). As another example, $\mu_X^{(4)}/\sigma_X^4$, the measure of **kurtosis**, indicates relative flatness of the distribution function of X. For a standard normal distribution function this quantity is 3. Therefore, if $\mu_X^{(4)}/\sigma_X^4 > 3$, the distribution function of X is more peaked than that of standard normal, and if $\mu_X^{(3)}/\sigma_X^3 < 3$, it is less peaked (flatter) than that of the standard normal. The moments of a random variable X give information of other sorts, too. For example, it can be proven that $E(|X|^k) < \infty$ implies that $\lim_{n \to \infty} n^k P(|X| > n) \to 0$, which shows that if $E(|X|^k) < \infty$, then $P(|X| > n)$ approaches 0 faster than $1/n^k$ as $n \to \infty$.

In this section we study moment-generating functions of random variables. These are real-valued functions with two major properties: They enable us to calculate the moments of random variables readily, and upon existence, they are unique. That is, no two different random variables have the same moment-generating function. Because of this, for proving that a random variable has a certain distribution function F, it is often shown that its moment-generating function coincides with that of F. This method enables us to prove many interesting facts, including the celebrated central limit theorem (see Section 11.5). Furthermore, Theorem 8.9, the convolution theorem, which is used to find

457

the distribution of the sum of *two* independent random variables, cannot be extended so easily to many random variables. This fact is a major motivation for the introduction of the moment-generating functions, which, by Theorem 11.3, can be used to find, almost readily, the distribution functions of sums of many independent random variables.

Definition   *For a random variable X, let*

$$M_X(t) = E(e^{tX}).$$

*If $M_X(t)$ is defined for all values of t in some interval $(-\delta, \delta)$, $\delta > 0$, then $M_X(t)$ is called the **moment-generating function** of X.*

Hence if X is a discrete random variable with set of possible values A and probability mass function $p(x)$, then

$$M_X(t) = \sum_{x \in A} e^{tx} p(x),$$

and if X is a continuous random variable with probability density function $f(x)$, then

$$M_X(t) = \int_{-\infty}^{\infty} e^{tx} f(x) \, dx.$$

Note that the condition that $M_X(t)$ be finite in some neighborhood of 0, $(-\delta, \delta)$, $\delta > 0$, is an important requirement. Without this condition some moments of X may not exist.

As we mentioned in the beginning and as can be guessed from its name, the moment-generating function of a random variable X can be used to calculate the moments of X. The following theorem shows that the moments of X can be found by differentiating $M_X(t)$ and evaluating the derivatives at $t = 0$.

Theorem 11.1   *Let X be a random variable with moment-generating function $M_X(t)$. Then*

$$E(X^n) = M_X^{(n)}(0),$$

where $M_X^{(n)}(t)$ is the nth derivative of $M_X(t)$.

Proof: We prove the theorem for continuous random variables. For discrete random variables, the proof is similar. If X is continuous with probability density function f, then

$$M_X'(t) = \frac{d}{dt} \left(\int_{-\infty}^{\infty} e^{tx} f(x) \, dx \right) = \int_{-\infty}^{\infty} x e^{tx} f(x) \, dx,$$

$$M_X''(t) = \frac{d}{dt} \left(\int_{-\infty}^{\infty} x e^{tx} f(x) \, dx \right) = \int_{-\infty}^{\infty} x^2 e^{tx} f(x) \, dx,$$

$$\vdots$$

$$M_X^{(n)}(t) = \int_{-\infty}^{\infty} x^n e^{tx} f(x) \, dx, \tag{11.1}$$

where we have assumed that the derivatives of these integrals are equal to the integrals of the derivatives of their integrands. This property is valid for sufficiently smooth densities. Letting $t = 0$ in (11.1), we get

$$M_X^{(n)}(0) = \int_{-\infty}^{\infty} x^n f(x)\, dx = E(X^n).$$

Note that since in some interval $(-\delta, \delta)$, $\delta > 0$, $M_X(t)$ is finite, we have that $M_X^{(n)}(0)$ exists for all $n \geq 1$. ♦

One immediate application of Theorem 11.1 is the following corollary.

Corollary   *The MacLaurin's series for $M_X(t)$ is given by*

$$M_X(t) = \sum_{n=0}^{\infty} \frac{M_X^{(n)}(0)}{n!} t^n = \sum_{n=0}^{\infty} \frac{E(X^n)}{n!} t^n. \tag{11.2}$$

Therefore, $E(X^n)$ is the coefficient of $t^n/n!$ in the MacLaurin's series representation of $M_X(t)$.

It is important to know that if M_X is to be finite, then the moments of all orders of X must be finite. But the converse need not be true. That is, all the moments may be finite and yet there is no neighborhood of 0, $(-\delta, \delta)$, $\delta > 0$, on which M_X is finite.

Example 11.1   Let X be a Bernoulli random variable with parameter p, that is,

$$P(X = x) = \begin{cases} 1 - p & \text{if } x = 0 \\ p & \text{if } x = 1 \\ 0 & \text{otherwise.} \end{cases}$$

Determine $M_X(t)$ and $E(X^n)$.

Solution: From the definition of a moment-generating function,

$$M_X(t) = E(e^{tX}) = (1 - p)e^{t \cdot 0} + pe^{t \cdot 1} = (1 - p) + pe^t.$$

Since $M_X^{(n)}(t) = pe^t$ for all $n > 0$, we have that $E(X^n) = M_X^{(n)}(0) = p$. ♦

Example 11.2   Let X be a binomial random variable with parameters (n, p). Find the moment-generating function of X, and use it to calculate $E(X)$ and $\text{Var}(X)$.

Solution: The probability mass function of X, $p(x)$, is given by

$$p(x) = \binom{n}{x} p^x q^{n-x}, \qquad x = 0, 1, 2, \dots, n, \quad q = 1 - p.$$

Hence

$$M_X(t) = E(e^{tX}) = \sum_{x=0}^{n} e^{tx} \binom{n}{x} p^x q^{n-x} = \sum_{x=0}^{n} \binom{n}{x} (pe^t)^x q^{n-x} = (pe^t + q)^n,$$

where the last equality follows from the binomial expansion (see Theorem 2.5). To find the mean and the variance of X, note that

$$M'_X(t) = npe^t(pe^t + q)^{n-1}$$
$$M''_X(t) = npe^t(pe^t + q)^{n-1} + n(n-1)(pe^t)^2(pe^t + q)^{n-2}.$$

Thus

$$E(X) = M'_X(0) = np$$
$$E(X^2) = M''_X(0) = np + n(n-1)p^2.$$

Therefore,

$$\text{Var}(X) = E(X^2) - \left[E(X)\right]^2 = np + n(n-1)p^2 - n^2p^2 = npq. \quad \blacklozenge$$

Example 11.3 Let X be an exponential random variable with parameter λ. Using moment-generating functions, calculate the mean and the variance of X.

Solution: The probability density function of X is given by

$$f(x) = \lambda e^{-\lambda x}, \qquad x \geq 0.$$

Thus

$$M_X(t) = E(e^{tX}) = \int_0^\infty e^{tx} \lambda e^{-\lambda x} \, dx = \lambda \int_0^\infty e^{(t-\lambda)x} \, dx.$$

Since the integral $\int_0^\infty e^{(t-\lambda)x} \, dx$ converges if $t < \lambda$, restricting the domain of $M_X(t)$ to $(-\infty, \lambda)$, we get $M_X(t) = \lambda/(\lambda - t)$. Thus $M'_X(t) = \lambda/(\lambda - t)^2$ and $M''_X(t) = (2\lambda)/(\lambda - t)^3$. We obtain $E(X) = M'_X(0) = 1/\lambda$ and $E(X^2) = M''_X(0) = 2/\lambda^2$. Therefore,

$$\text{Var}(X) = E(X^2) - \left[E(X)\right]^2 = \frac{2}{\lambda^2} - \frac{1}{\lambda^2} = \frac{1}{\lambda^2}. \quad \blacklozenge$$

Example 11.4 Let X be an exponential random variable with parameter λ. Using moment-generating functions, find $E(X^n)$, where n is a positive integer.

Solution: By Example 11.3,

$$M_X(t) = \frac{\lambda}{\lambda - t} = \frac{1}{1 - (t/\lambda)}, \qquad t < \lambda.$$

Now, on the one hand, by the geometric series theorem,

$$M_X(t) = \frac{1}{1 - (t/\lambda)} = \sum_{n=0}^{\infty} \left(\frac{t}{\lambda}\right)^n = \sum_{n=0}^{\infty} \left(\frac{1}{\lambda}\right)^n t^n,$$

and, on the other hand, by (11.2),

$$M_X(t) = \sum_{n=0}^{\infty} \frac{E(X^n)}{n!} t^n.$$

Comparing these two relations, we obtain

$$\frac{E(X^n)}{n!} = \left(\frac{1}{\lambda}\right)^n,$$

which gives $E(X^n) = n!/\lambda^n$.   ◆

Example 11.5   Let Z be a standard normal random variable.

(a)   Calculate the moment-generating function of Z.

(b)   Use part (a) to find the moment-generating function of X, where X is a normal random variable with mean μ and variance σ^2.

(c)   Use part (b) to calculate the mean and the variance of X.

Solution:

(a)   From the definition,

$$M_Z(t) = E(e^{tZ}) = \int_{-\infty}^{\infty} e^{tz} \frac{1}{\sqrt{2\pi}} e^{-z^2/2}\, dz = \frac{1}{\sqrt{2\pi}} \int_{-\infty}^{\infty} e^{tz - z^2/2}\, dz.$$

Since

$$tz - \frac{z^2}{2} = \frac{t^2}{2} - \frac{(z-t)^2}{2},$$

we obtain

$$M_Z(t) = \frac{1}{\sqrt{2\pi}} \int_{-\infty}^{\infty} \exp\left[\frac{t^2}{2} - \frac{(z-t)^2}{2}\right] dz$$

$$= \frac{1}{\sqrt{2\pi}} e^{t^2/2} \int_{-\infty}^{\infty} \exp\left[-\frac{(z-t)^2}{2}\right] dz.$$

Let $u = z - t$. Then $du = dz$ and

$$M_Z(t) = \frac{1}{\sqrt{2\pi}} e^{t^2/2} \int_{-\infty}^{\infty} e^{-u^2/2} \, du$$

$$= e^{t^2/2} \frac{1}{\sqrt{2\pi}} \int_{-\infty}^{\infty} e^{-u^2/2} \, du = e^{t^2/2},$$

where the last equality follows since the function $(1/\sqrt{2\pi})e^{-u^2/2}$ is the probability density function of a standard normal random variable, and hence its integral on $(-\infty, +\infty)$ is 1.

(b) Letting $Z = (X - \mu)/\sigma$, we have that Z is $N(0, 1)$ and $X = \sigma Z + \mu$. Thus

$$M_X(t) = E(e^{tX}) = E(e^{t\sigma Z + t\mu}) = E(e^{t\sigma Z} e^{t\mu})$$

$$= e^{t\mu} E(e^{t\sigma Z}) = e^{t\mu} M_Z(t\sigma) = e^{t\mu} \exp\left(\frac{1}{2} t^2 \sigma^2\right)$$

$$= \exp\left(t\mu + \frac{1}{2}\sigma^2 t^2\right).$$

(c) Differentiating $M_X(t)$, we obtain

$$M_X'(t) = (\mu + \sigma^2 t) \exp\left(t\mu + \frac{1}{2}\sigma^2 t^2\right),$$

which upon differentiation gives

$$M_X''(t) = (\mu + \sigma^2 t)^2 \exp\left(t\mu + \frac{1}{2}\sigma^2 t^2\right) + \sigma^2 \exp\left(t\mu + \frac{1}{2}\sigma^2 t^2\right).$$

Therefore,

$$E(X) = M_X'(0) = \mu \quad \text{and} \quad E(X^2) = M_X''(0) = \mu^2 + \sigma^2.$$

Thus

$$\text{Var}(X) = E(X^2) - \left[E(X)\right]^2 = \mu^2 + \sigma^2 - \mu^2 = \sigma^2. \quad \blacklozenge$$

Example 11.6   A positive random variable X is called **lognormal** with parameters μ and σ^2 if $\ln X \sim N(\mu, \sigma^2)$. Let X be a lognormal random variable with parameters μ and σ^2.

(a) For a positive integer r, calculate the rth moment of X.

(b) Use the rth moment of X to find $\text{Var}(X)$.

(c) In 1977, a British researcher demonstrated that if X is the loss from a large fire, then $\ln X$ is a normal random variable. That is, X is lognormal. Suppose that the expected loss due to fire in the buildings of a certain industry, in thousands of dollars, is 120 with standard deviation 36. What is the probability that the loss from a fire in such an industry is less than $100,000?

Solution:

(a)   Let $Y = \ln X$, then $X = e^Y$. Now $Y \sim N(\mu, \sigma^2)$ gives

$$E(X^r) = E(e^{rY}) = M_Y(r) = \exp\left(\mu r + \frac{1}{2}\sigma^2 r^2\right).$$

(b)   By part (a),

$$E(X) = e^{\mu + (1/2)\sigma^2},$$

$$E(X^2) = e^{2\mu + 2\sigma^2},$$

$$\text{Var}(X) = e^{2\mu + 2\sigma^2} - e^{2\mu + \sigma^2} = e^{2\mu + \sigma^2}\left(e^{\sigma^2} - 1\right).$$

(c)   Let X be the loss from the fire in the industry. Let the mean and standard deviation of $\ln X$ be μ and σ, respectively. By part (b),

$$\frac{\text{Var(X)}}{\left[E(X)\right]^2} = e^{\sigma^2} - 1.$$

Hence $e^{\sigma^2} - 1 = 36^2/120^2 = 0.09$, or, equivalently, $e^{\sigma^2} = 1.09$, which gives $\sigma^2 = \ln 1.09$, or $\sigma = \sqrt{\ln 1.09} = 0.294$. Therefore,

$$120 = e^\mu \cdot e^{(1/2)\sigma^2} = e^\mu \cdot \sqrt{1.09},$$

which gives $e^\mu = 114.939$, or $\mu = \ln(114.939) = 4.744$. Thus the probability we are interested in is

$$P(X < 100) = P(\ln X < \ln 100) = P(\ln X < 4.605)$$

$$= P\left(Z < \frac{4.605 - 4.744}{0.294}\right) = P(Z < -0.47) = 0.3192. \quad \blacklozenge$$

Another important property of a moment-generating function is its uniqueness, which we now state without proof.

Theorem 11.2   *Let X and Y be two random variables with moment-generating functions $M_X(t)$ and $M_Y(t)$. If for some $\delta > 0$, $M_X(t) = M_Y(t)$ for all values of t in $(-\delta, \delta)$, then X and Y have the same distribution.*

Note that the converse of Theorem 11.2 is trivially true.

Theorem 11.2 is both an important tool and a surprising result. It shows that a moment-generating function determines the distribution function uniquely. That is, if two random variables have the same moment-generating function, then they are identically distributed. As we mentioned before, because of the uniqueness property, in order that a random variable has a certain distribution function F, it suffices to show that its moment-generating function coincides with that of F.

Example 11.7   Let the moment-generating function of a random variable X be

$$M_X(t) = \frac{1}{7}e^t + \frac{3}{7}e^{3t} + \frac{2}{7}e^{5t} + \frac{1}{7}e^{7t}.$$

Since the moment-generating function of a discrete random variable with the probability mass function

i	1	3	5	7	Other values
$p(i)$	1/7	3/7	2/7	1/7	0

is $M_X(t)$ given previously, by Theorem 11.2, the probability mass function of X is $p(i)$. ♦

Example 11.8   Let X be a random variable with moment-generating function $M_X(t) = e^{2t^2}$. Find $P(0 < X < 1)$.

Solution: Comparing $M_X(t) = e^{2t^2}$ with $\exp\left[\mu t + (1/2)\sigma^2 t^2\right]$, the moment-generating function of $N(\mu, \sigma^2)$ (see Example 11.5), we have that $X \sim N(0, 4)$ by the uniqueness of the moment-generating function. Let $Z = (X - 0)/2$. Then $Z \sim N(0, 1)$, so

$$P(0 < X < 1) = P(0 < X/2 < 1/2) = P(0 < Z < 0.5)$$
$$= \Phi(0.5) - \Phi(0) \approx 0.6915 - 0.5 = 0.1915,$$

by Table 2 of the Appendix.   ♦

One of the most important properties of moment-generating functions is that they enable us to find distribution functions of sums of independent random variables. We discuss this important property in Section 11.2 and use it in Section 11.5 to prove the central limit theorem.

Often it happens that the moment-generating function of a random variable is known but its distribution function is not. In such cases Table 3 of the Appendix might help us identify the distribution.

EXERCISES

A

1. Let X be a discrete random variable with probability mass function $p(i) = 1/5$, $i = 1, 2, \ldots, 5$, zero elsewhere. Find $M_X(t)$.

2. Let X be a random variable with probability density function

$$f(x) = \begin{cases} 1/4 & \text{if } x \in (-1, 3) \\ 0 & \text{otherwise.} \end{cases}$$

   (a) Find $M_X(t)$, $E(X)$, and $\text{Var}(X)$.

   (b) Using $M_X(t)$, calculate $E(X)$.

   Hint: Note that by the definition of derivative,

   $$M_X'(0) = \lim_{h \to 0} \frac{M_X(h) - M_X(0)}{h}.$$

3. Let X be a discrete random variable with the probability mass function

   $$p(i) = 2\left(\frac{1}{3}\right)^i, \qquad i = 1, 2, 3, \ldots; \quad \text{zero elsewhere.}$$

   Find $M_X(t)$ and $E(X)$.

4. Let X be a continuous random variable with probability density function $f(x) = 2x$, if $0 \le x \le 1$, zero elsewhere. Find the moment-generating function of X.

5. Let X be a continuous random variable with the probability density function $f(x) = 6x(1 - x)$, if $0 \le x \le 1$, zero elsewhere.

   (a) Find $M_X(t)$.

   (b) Using $M_X(t)$, find $E(X)$.

6. Let X be a discrete random variable. Prove that $E(X^n) = M_X^{(n)}(0)$.

7. (a) Find $M_X(t)$, the moment-generating function of a Poisson random variable X with parameter λ.

   (b) Use $M_X(t)$ to find $E(X)$ and $\text{Var}(X)$.

8. Let X be a uniform random variable over the interval (a, b). Find the moment-generating function of X.

9. Let X be a geometric random variable with parameter p. Show that the moment-generating function of X is given by

$$M_X(t) = \frac{pe^t}{1 - qe^t}, \qquad q = 1 - p, \qquad t < -\ln q.$$

Use $M_X(t)$ to find $E(X)$ and $\text{Var}(X)$.

10. Let $M_X(t) = (1/21) \sum_{n=1}^{6} ne^{nt}$. Find the probability mass function of X.

11. Suppose that the moment-generating function of a random variable X is given by

$$M_X(t) = \frac{1}{3}e^t + \frac{4}{15}e^{3t} + \frac{2}{15}e^{4t} + \frac{4}{15}e^{5t}.$$

Find the probability mass function of X.

12. Let $M_X(t) = 1/(1 - t)$, $t < 1$ be the moment-generating function of a random variable X. Find the moment-generating function of the random variable $Y = 2X + 1$.

13. For a random variable X, $M_X(t) = \left[2/(2 - t)\right]^3$. Find $E(X)$ and $\text{Var}(X)$.

14. Suppose that the moment-generating function of X is given by

$$M_X(t) = \frac{e^t + e^{-t}}{6} + \frac{2}{3}, \qquad -\infty < t < \infty.$$

Find $E(X^r)$, $r \geq 1$.

15. Prove that the function $t/(1-t)$, $t < 1$, cannot be the moment-generating function of a random variable.

16. In each of the following cases $M_X(t)$, the moment-generating function of X, is given. Determine the distribution of X.

    (a) $M_X(t) = \left(\frac{1}{4}e^t + \frac{3}{4}\right)^7$.

    (b) $M_X(t) = e^t/(2 - e^t)$.

    (c) $M_X(t) = \left[2/(2 - t)\right]^r$.

    (d) $M_X(t) = \exp\left[3(e^t - 1)\right]$.

    Hint: Use Table 3 of the Appendix.

17. For a random variable X, $M_X(t) = (1/81)(e^t + 2)^4$. Find $P(X \leq 2)$.

18. Suppose that for a random variable X, $E(X^n) = 2^n$, $n = 1, 2, 3, \ldots$. Calculate the moment-generating function and the probability mass function of X.
    Hint: Use (11.2).

19. Let X be a uniform random variable over $(0, 1)$. Let a and b be two positive numbers. Using moment-generating functions, show that $Y = aX + b$ is uniformly distributed over $(b, a + b)$.

B

20. Let $Z \sim N(0, 1)$. Use $M_Z(t) = e^{t^2/2}$ to calculate $E(Z^n)$, where n is a positive integer.
Hint: Use (11.2).

21. Let X be a gamma random variable with parameters r and λ. Derive a formula for $M_X(t)$, and use it to calculate $E(X)$ and $\text{Var}(X)$.

22. Let X be a continuous random variable whose probability density function f is even; that is, $f(-x) = f(x)$, $\forall x$. Prove that (a) the random variables X and $-X$ have the same probability distribution function; (b) the function $M_X(t)$ is an even function.

23. Let X be a discrete random variable with probability mass function

$$p(i) = \frac{6}{\pi^2 i^2}, \qquad i = 1, 2, 3, \dots; \quad \text{zero elsewhere.}$$

Show that the moment-generating function of X does not exist.
Hint: Show that $M_X(t)$ is a divergent series on $(0, \infty)$. This implies that on no interval of the form $(-\delta, \delta)$, $\delta > 0$, $M_X(t)$ exists.

24. Suppose that $\forall n \geq 1$, the nth moment of a random variable X, is given by $E(X^n) = (n + 1)! \, 2^n$. Find the distribution of X.

25. Suppose that A dollars are invested in a bank that pays interest at a rate of X per year, where X is a random variable.

(a) Show that if a year is divided into k equal periods, and the bank pays interest at the end of each of these k periods, then after n such periods, with probability 1, the investment will grow to $A\left(1 + \dfrac{X}{k}\right)^n$.

(b) For an infinitesimal $\varepsilon > 0$, suppose that the interest is compounded at the end of each period of length ε. If $\varepsilon \to 0$, then the interest is said to be **compounded continuously.** Suppose that at time t, the investment has grown to $A(t)$. By demonstrating that

$$A(t + \varepsilon) = A(t) + A(t) \cdot \varepsilon X,$$

show that, with probability 1,

$$A'(t) = X A(t).$$

(c) Using part (b), prove that,

If the bank compounds interest continuously, then, on average, the money will grow by a factor of $M_X(t)$, the moment-generating function of the interest rate.

26. Let the joint probability mass function of $X_1, X_2, \ldots, X_r$ be multinomial with parameters n and $p_1, p_2, \ldots, p_r$ $(p_1 + p_2 + \cdots + p_r = 1)$. Find $\rho(X_i, X_j)$, $1 \leq i \neq j \leq r$.

Hint: Note that by Remark 9.3, X_i and X_j are binomial random variables and the joint marginal probability mass function of X_i and X_j is multinomial. To find $E(X_i X_j)$, calculate

$$M(t_1, t_2) = E(e^{t_1 X_i + t_2 X_j})$$

for all values of t_1 and t_2 and note that

$$E(X_i X_j) = \frac{\partial^2 M(0, 0)}{\partial t_1 \, \partial t_2}.$$

The function $M(t_1, t_2)$ is called the **moment-generating function of the joint distribution** of X_i *and* X_j.

∎

11.2   SUMS OF INDEPENDENT RANDOM VARIABLES

As mentioned before, the convolution theorem cannot be extended so easily to find the distribution function of sums of more than two independent random variables. For this reason, in this section, we study the distribution functions of such sums using moment-generating functions. To begin, we prove the following theorem, which together with the uniqueness property of moment-generating functions (see Theorem 11.2) are two of the main tools of this section.

Theorem 11.3   *Let $X_1, X_2, \ldots, X_n$ be independent random variables with moment-generating functions $M_{X_1}(t), M_{X_2}(t), \ldots, M_{X_n}(t)$. The moment-generating function of $X_1 + X_2 + \cdots + X_n$ is given by*

$$M_{X_1+X_2+\cdots+X_n}(t) = M_{X_1}(t) M_{X_2}(t) \cdots M_{X_n}(t).$$

Proof: Let $W = X_1 + X_2 + \cdots + X_n$; by definition,

$$
\begin{aligned}
M_W(t) = E(e^{tW}) &= E\big(e^{tX_1 + tX_2 + \cdots + tX_n}\big) \\
&= E\big(e^{tX_1} e^{tX_2} \cdots e^{tX_n}\big) \\
&= E(e^{tX_1}) E(e^{tX_2}) \cdots E(e^{tX_n}) \\
&= M_{X_1}(t) M_{X_2}(t) \cdots M_{X_n}(t),
\end{aligned}
$$

where the next-to-last equality follows from the independence of $X_1, X_2, X_3, \ldots, X_n$.

♦

We now prove that sums of independent binomial random variables are binomial. Let X and Y be two independent binomial random variables with parameters (n, p) and (m, p), respectively. Then X and Y are the numbers of successes in n and m independent Bernoulli trials with parameter p, respectively. Thus $X + Y$ is the number of successes in $n + m$ independent Bernoulli trials with parameter p. Therefore, $X + Y$ is a binomial random variable with parameters $(n+m, p)$. An alternative for this probabilistic argument is the following analytic proof.

Theorem 11.4   *Let $X_1, X_2, \ldots, X_r$ be independent binomial random variables with parameters (n_1, p), (n_2, p), $\ldots$, (n_r, p), respectively. Then $X_1 + X_2 + \cdots + X_r$ is a binomial random variable with parameters $n_1 + n_2 + \cdots + n_r$ and p.*

Proof: Let, as usual, $q = 1 - p$. We know that

$$M_{X_i}(t) = (pe^t + q)^{n_i}, \qquad i = 1, 2, 3, \ldots, n$$

(see Example 11.2). Let $W = X_1 + X_2 + \cdots + X_r$; then, by Theorem 11.3,

$$\begin{aligned}
M_W(t) &= M_{X_1}(t) M_{X_2}(t) \cdots M_{X_r}(t) \\
&= (pe^t + q)^{n_1} (pe^t + q)^{n_2} \cdots (pe^t + q)^{n_r} \\
&= (pe^t + q)^{n_1 + n_2 + \cdots + n_r}.
\end{aligned}$$

Since $(pe^t + q)^{n_1 + n_2 + \cdots + n_r}$ is the moment-generating function of a binomial random variable with parameters $(n_1 + n_2 + \cdots + n_r, p)$, the uniqueness property of moment-generating functions implies that $W = X_1 + X_2 + \cdots + X_r$ is binomial with parameters $(n_1 + n_2 + \cdots + n_r, p)$.   ◆

We just showed that if X and Y are independent binomial random variables with parameters (n, p) and (m, p), respectively, then $X + Y$ is a binomial random variable with parameters $(n+m, p)$. For large values of n and m, small p, and moderate $\lambda_1 = np$ and $\lambda_2 = mp$, Poisson probability mass functions with parameters λ_1, λ_2, and $\lambda_1 + \lambda_2$ approximate probability mass functions of X, Y, and $X + Y$, respectively. Therefore, it is reasonable to expect that for independent Poisson random variables X and Y with parameters λ_1 and λ_2, $X + Y$ is a Poisson random variable with parameter $\lambda_1 + \lambda_2$. The proof of this interesting fact follows for n random variables.

Theorem 11.5   *Let $X_1, X_2, \ldots, X_n$ be independent Poisson random variables with means $\lambda_1, \lambda_2, \ldots, \lambda_n$, respectively. Then $X_1 + X_2 + \cdots + X_n$ is a Poisson random variable with mean $\lambda_1 + \lambda_2 + \cdots + \lambda_n$.*

Proof: Let Y be a Poisson random variable with mean λ. Then

$$M_Y(t) = E(e^{tY}) = \sum_{y=0}^{\infty} e^{ty} \frac{e^{-\lambda} \lambda^y}{y!} = e^{-\lambda} \sum_{y=0}^{\infty} \frac{(e^t \lambda)^y}{y!}$$

$$= e^{-\lambda} \exp(\lambda e^t) = \exp\left[\lambda(e^t - 1)\right].$$

Let $W = X_1 + X_2 + \cdots + X_n$; then, by Theorem 11.3,

$$
\begin{aligned}
M_W(t) &= M_{X_1}(t) M_{X_2}(t) \cdots M_{X_n}(t) \\
&= \exp\left[\lambda_1(e^t - 1)\right]\exp\left[\lambda_2(e^t - 1)\right]\cdots\exp\left[\lambda_n(e^t - 1)\right] \\
&= \exp\left[(\lambda_1 + \lambda_2 + \cdots + \lambda_n)(e^t - 1)\right].
\end{aligned}
$$

Now, since $\exp\left[(\lambda_1 + \lambda_2 + \cdots + \lambda_n)(e^t - 1)\right]$ is the moment-generating function of a Poisson random variable with mean $\lambda_1 + \lambda_2 + \cdots + \lambda_n$, the uniqueness property of moment-generating functions implies that $X_1 + X_2 + \cdots + X_n$ is Poisson with mean $\lambda_1 + \lambda_2 + \cdots + \lambda_n$. ◆

Theorem 11.6   *Let $X_1 \sim N(\mu_1, \sigma_1^2)$, $X_2 \sim N(\mu_2, \sigma_2^2)$, ..., $X_n \sim N(\mu_n, \sigma_n^2)$ be independent random variables. Then*

$$
X_1 + X_2 + \cdots + X_n \sim N(\mu_1 + \mu_2 + \cdots + \mu_n, \sigma_1^2 + \sigma_2^2 + \cdots + \sigma_n^2).
$$

Proof: In Example 11.5 we showed that if X is normal with parameters μ and σ^2, then $M_X(t) = \exp\left[\mu t + (1/2)\sigma^2 t^2\right]$. Let $W = X_1 + X_2 + \cdots + X_n$; then

$$
\begin{aligned}
M_W(t) &= M_{X_1}(t) M_{X_2}(t) \cdots M_{X_n}(t) \\
&= \exp\left(\mu_1 t + \frac{1}{2}\sigma_1^2 t^2\right)\exp\left(\mu_2 t + \frac{1}{2}\sigma_2^2 t^2\right)\cdots\exp\left(\mu_n t + \frac{1}{2}\sigma_n^2 t^2\right) \\
&= \exp\left[(\mu_1 + \mu_2 + \cdots + \mu_n)t + \frac{1}{2}(\sigma_1^2 + \sigma_2^2 + \cdots + \sigma_n^2)t^2\right].
\end{aligned}
$$

This implies that

$$
X_1 + X_2 + \cdots + X_n \sim N(\mu_1 + \mu_2 + \cdots + \mu_n, \sigma_1^2 + \sigma_2^2 + \cdots + \sigma_n^2). ◆
$$

The technique used to show Theorems 11.4, 11.5, and 11.6 can also be used to prove the following and similar theorems.

- Sums of independent geometric random variables are negative binomial.

- Sums of independent negative binomial random variables are negative binomial.

- Sums of independent exponential random variables are gamma.

- Sums of independent gamma random variables are gamma.

These important theorems are discussed in the exercises.

If X is a normal random variable with parameters (μ, σ^2), then for $\alpha \in \mathbf{R}$, we can prove that $M_{\alpha X}(t) = \exp\left[\alpha\mu t + (1/2)\alpha^2\sigma^2 t^2\right]$, implying that $\alpha X \sim N(\alpha\mu, \alpha^2\sigma^2)$ (see Exercise 1). This and Theorem 11.6 imply the following important theorem.

Linear combinations of sets of independent normal random variables are normal:

Theorem 11.7   *Let* $\{X_1, X_2, \ldots, X_n\}$ *be a set of independent random variables and* $X_i \sim N(\mu_i, \sigma_i^2)$ *for* $i = 1, 2, \ldots, n$; *then for constants* $\alpha_1, \alpha_2, \ldots, \alpha_n$,

$$\sum_{i=1}^{n} \alpha_i X_i \sim N\left(\sum_{i=1}^{n} \alpha_i \mu_i, \ \sum_{i=1}^{n} \alpha_i^2 \sigma_i^2\right).$$

In particular, this theorem implies that if $X_1, X_2, \ldots, X_n$ are independent normal random variables all with the same mean μ, and the same variance σ^2, then $S_n = X_1 + X_2 + \cdots + X_n$ is $N(n\mu, n\sigma^2)$ and the sample mean, $\bar{X} = S_n/n$, is $N(\mu, \sigma^2/n)$. That is,

$$\bar{X} = \frac{1}{n} \sum_{i=1}^{n} X_i, \text{ the sample mean of } n \text{ independent } N(\mu, \sigma^2), \text{ is } N\left(\mu, \frac{\sigma^2}{n}\right).$$

Example 11.9   Suppose that the distribution of students' grades in a probability test is normal, with mean 72 and variance 25.

(a)   What is the probability that the average of grade of such a probability class with 25 students is 75 or more?

(b)   If a professor teaches two different sections of this course, each containing 25 students, what is the probability that the average of one class is at least three more than the average of the other class?

Solution:

(a)   Let $X_1, X_2, \ldots, X_{25}$ denote the grades of the 25 students. Then $X_1, X_2, \ldots, X_{25}$ are independent random variables all being normal, with $\mu = 72$ and $\sigma^2 = 25$. The average of the grades of the class, $\bar{X} = (1/25) \sum_{i=1}^{25} X_i$, is normal, with mean $\mu = 72$ and variance $\sigma^2/n = 25/25 = 1$. Hence

$$P(\bar{X} \geq 75) = P\left(\frac{\bar{X} - 72}{1} \geq \frac{75 - 72}{1}\right) = P(\bar{X} - 72 \geq 3) = 1 - \Phi(3) \approx 0.0013.$$

(b)   Let $\bar{X}$ and $\bar{Y}$ denote the means of the grades of the two classes. Then, as seen in part (a), $\bar{X}$ and $\bar{Y}$ are both $N(72, 1)$. Since a student does not take two sections of the same course, $\bar{X}$ and $\bar{Y}$ are independent random variables. By Theorem 11.7,

$\bar{X} - \bar{Y}$ is $N(0, 2)$. Hence, by symmetry,

$$
\begin{aligned}
P\big(|\bar{X} - \bar{Y}| > 3\big) &= P(\bar{X} - \bar{Y} > 3 \text{ or } \bar{Y} - \bar{X} > 3) \\
&= 2P(\bar{X} - \bar{Y} > 3) \\
&= 2P\left(\frac{\bar{X} - \bar{Y} - 0}{\sqrt{2}} > \frac{3 - 0}{\sqrt{2}}\right) \\
&= 2P\left(\frac{\bar{X} - \bar{Y}}{\sqrt{2}} > 2.12\right) = 2\big[1 - \Phi(2.12)\big] \approx 0.034. \quad \blacklozenge
\end{aligned}
$$

As we mentioned previously, by using moment-generating functions, it can be shown that a sum of n independent exponential random variables, each with parameter λ, is gamma with parameters n and λ (see Exercise 3). We now present an alternative proof for this important fact: Let X be a gamma random variable with parameters (n, λ), where n is a positive integer. Consider a Poisson process $\{N(t): t \geq 0\}$ with rate λ. $\{X_1, X_2, \ldots\}$, the set of interarrival times of this process $\big[X_i$ is the time between the $(i-1)$st event and the ith event, $i = 1, 2, \ldots\big]$, is an independent, identically distributed sequence of exponential random variables with $E(X_i) = 1/\lambda$ for $i = 1, 2, \ldots$. Since the distribution of X is the same as the time of the occurrence of the nth event,

$$
X = X_1 + X_2 + \cdots + X_n.
$$

Hence

A gamma random variable with parameters (n, λ) is the sum of n independent exponential random variables, each with mean $1/\lambda$, and vice versa.

Using moment-generating functions, we can also prove that if $X_1, X_2, \ldots, X_n$ are n independent gamma random variables with parameters (r_1, λ), (r_2, λ), $\ldots$, (r_n, λ), respectively, then $X_1 + X_2 + \cdots + X_n$ is gamma with parameters $(r_1 + r_2 + \cdots + r_n, \lambda)$ (see Exercise 5). The following example is an application of this fact. It gives an interesting class of gamma random variables with parameters $(r, 1/2)$, where r is not necessarily an integer.

Example 11.10   Office fire insurance policies by a certain company have a $1000 deductible. The company has received three claims, independent of each other, for damages caused by office fire. If reconstruction expenses for such claims are exponentially distributed, each with mean $45,000, what is the probability that the total payment for these claims is less than $120,000?

Solution: Let X be the total reconstruction expenses for the three claims in thousands of dollars; X is the sum of three independent exponential random variables, each with parameter \$45. Therefore, it is a gamma random variable with parameters 3 and $\lambda = 1/45$. Hence its probability density function is given by

$$f(x) = \begin{cases} \dfrac{1}{45} e^{-x/45} \dfrac{(x/45)^2}{2} = \dfrac{1}{182,250} x^2 e^{-x/45} & \text{if } x \geq 0 \\ 0 & \text{elsewhere.} \end{cases}$$

Considering the deductibles, the probability we are interested in is

$$P(X < 123) = \frac{1}{182,250} \int_0^{123} x^2 e^{-x/45}\, dx$$

$$= \frac{1}{182,250}(-45x^2 - 4050x - 182,250)e^{-x/45}\Big|_0^{123} = 0.5145. \quad \blacklozenge$$

Example 11.11   Let $X_1, X_2, \ldots, X_n$ be independent standard normal random variables. Then $X = X_1^2 + X_2^2 + \cdots + X_n^2$, referred to as **chi-squared random variable with n degrees of freedom**, is gamma with parameters $(n/2, 1/2)$. An example of such a gamma random variable is the error of hitting a target in n-dimensional Euclidean space when the error of each coordinate is individually normally distributed.

Proof: Since the sum of n independent gamma random variables, each with parameters $(1/2, 1/2)$, is gamma with parameters $(n/2, 1/2)$, it suffices to prove that for all i, $1 \leq i \leq n$, X_i^2 is gamma with parameters $(1/2, 1/2)$. To prove this assertion, note that

$$P(X_i^2 \leq t) = P(-\sqrt{t} \leq X_i \leq \sqrt{t}) = \Phi(\sqrt{t}) - \Phi(-\sqrt{t}).$$

Let the probability density function of X_i^2 be f. Differentiating this equation yields

$$f(t) = \left(\frac{1}{2\sqrt{t}} \frac{1}{\sqrt{2\pi}} e^{-t/2}\right) - \left(-\frac{1}{2\sqrt{t}} \frac{1}{\sqrt{2\pi}} e^{-t/2}\right).$$

Therefore,

$$f(t) = \frac{1}{\sqrt{2\pi t}} e^{-t/2} = \frac{(1/2)e^{-t/2}(t/2)^{1/2-1}}{\sqrt{\pi}}.$$

Since $\sqrt{\pi} = \Gamma(1/2)$ (see Exercise 7 of Section 7.4), this relation shows that X_i^2 is a gamma random variable with parameters $(1/2, 1/2)$. $\quad \blacklozenge$

Remark 11.1:   If $X_1, X_2, \ldots, X_n$ are independent random variables and, for $1 \leq i \leq n$, $X_i \sim N(\mu_i, \sigma_i^2)$, then, by Example 11.11, $\displaystyle\sum_{i=1}^{n} \left(\frac{X_i - \mu_i}{\sigma_i}\right)^2$ is gamma with parameters $(n/2, 1/2)$. This holds since $(X_i - \mu_i)/\sigma_i$ is standard normal for $1 \leq i \leq n$.

EXERCISES

A

1. Show that if X is a normal random variable with parameters (μ, σ^2), then for $\alpha \in \mathbf{R}$, we have that $M_{\alpha X}(t) = \exp\left[\alpha \mu t + (1/2)\alpha^2 \sigma^2 t^2\right]$.

2. Let $X_1, X_2, \dots, X_n$ be independent geometric random variables each with parameter p. Using moment-generating functions, prove that $X_1 + X_2 + \cdots + X_n$ is negative binomial with parameters (n, p).

3. Let $X_1, X_2, \dots, X_n$ be n independent exponential random variables with the identical mean $1/\lambda$. Use moment-generating functions to find the probability distribution function of $X_1 + X_2 + \cdots + X_n$.

4. Using moment-generating functions, show that the sum of n independent negative binomial random variables with parameters $(r_1, p), (r_2, p), \dots, (r_n, p)$ is negative binomial with parameters $(r, p), r = r_1 + r_2 + \cdots + r_n$.

5. Let $X_1, X_2, \dots, X_n$ be n independent gamma random variables with parameters $(r_1, \lambda), (r_2, \lambda), \dots, (r_n, \lambda)$, respectively. Use moment-generating functions to find the probability distribution function of $X_1 + X_2 + \cdots + X_n$.

6. The probability is 0.15 that a bottle of a certain soda is underfilled, independent of the amount of soda in other bottles. If machine one fills 100 bottles and machine two fills 80 bottles of this soda per hour, what is the probability that tomorrow, between 10:00 A.M. and 11:00 A.M., both of these machines will underfill exactly 27 bottles altogether?

7. Let X and Y be independent binomial random variables with parameters (n, p) and (m, p), respectively. Calculate $P(X = i \mid X + Y = j)$ and interpret the result.

8. Let X, Y, and Z be three independent Poisson random variables with parameters λ_1, λ_2, and λ_3, respectively. For $y = 0, 1, 2, \dots, t$, calculate $P(Y = y \mid X + Y + Z = t)$.

9. Mr. Watkins is at a train station, waiting to make a phone call. There is only one public telephone booth, and it is being used by someone. Another person ahead of Mr. Watkins is also waiting to call. If the duration of each telephone call is an exponential random variable with $\lambda = 1/8$, find the probability that Mr. Watkins should wait at least 12 minutes before being able to call.

10. Let $X \sim N(1, 2)$ and $Y \sim N(4, 7)$ be independent random variables. Find the probability of the following events: (a) $X + Y > 0$, (b) $X - Y < 2$, (c) $3X + 4Y > 20$.

11. The distribution of the IQ of a randomly selected student from a certain college is $N(110, 16)$. What is the probability that the average of the IQ's of 10 randomly selected students from this college is at least 112?

12. Vicki owns two department stores. Delinquent charge accounts at store 1 show a normal distribution, with mean $90 and standard deviation $30, whereas at store 2 they show a normal distribution with mean $100 and standard deviation $50. If 10 delinquent accounts are selected randomly at store 1 and 15 at store 2, what is the probability that the average of the accounts selected at store 1 exceeds the average of those selected at store 2?

13. Let the joint probability density function of X and Y be bivariate normal. Prove that any linear combination of X and Y, $\alpha X + \beta Y$, is a normal random variable.
Hint: Use Theorem 11.7 and the result of Exercise 6, Section 10.5.

14. Let X be the height of a man, and let Y be the height of his daughter (both in inches). Suppose that the joint probability density function of X and Y is bivariate normal with the following parameters: $\mu_X = 71$, $\mu_Y = 60$, $\sigma_X = 3$, $\sigma_Y = 2.7$, and $\rho = 0.45$. Find the probability that the man is at least 8 inches taller than his daughter.
Hint: Use the result of Exercise 13.

15. The capacity of an elevator is 2700 pounds. If the weight of a random athlete is normal with mean 225 pounds and standard deviation 25, what is the probability that the elevator can safely carry 12 random athletes?

16. The distributions of the grades of the students of probability and calculus at a certain university are $N(65, 418)$ and $N(72, 448)$, respectively. Dr. Olwell teaches a calculus section with 28 and a probability section with 22 students. What is the probability that the difference between the averages of the final grades of these two classes is at least 2?

17. Suppose that car mufflers last random times that are normally distributed with mean 3 years and standard deviation 1 year. If a certain family buys two new cars at the same time, what is the probability that (a) they should change the muffler of one car at least $1\frac{1}{2}$ years before the muffler of the other car; (b) one car does not need a new muffler for a period during which the other car needs two new mufflers?

B

18. An elevator can carry up to 3500 pounds. The manufacturer has included a safety margin of 500 pounds and lists the capacity as 3000 pounds. The building's management seeks to avoid accidents by limiting the number of passengers on the elevator. If the weight of the passengers using the elevator is $N(155, 625)$, what is the maximum number of passengers who can use the elevator if the odds against exceeding the rated capacity (3000 pounds) are to be greater than 10,000 to 3?

19. Let the joint probability mass function of $X_1, X_2, \ldots, X_r$ be multinomial, that is,

$$p(x_1, x_2, \ldots, x_r) = \frac{n!}{x_1! \, x_2! \cdots x_r!} \, p_1^{x_1} p_2^{x_2} \cdots p_r^{x_r},$$

where $x_1 + x_2 + \cdots + x_r = n$, and $p_1 + p_2 + \cdots + p_r = 1$. Show that for $k < r$, $X_1 + X_2 + \cdots + X_k$ has a binomial distribution.

20. Kim is at a train station, waiting to make a phone call. Two public telephone booths, next to each other, are occupied by two callers, and 11 persons are waiting in a single line ahead of Kim to call. If the duration of each telephone call is an exponential random variable with $\lambda = 1/3$, what are the distribution and the expectation of the time that Kim must wait until being able to call? ∎

11.3   MARKOV AND CHEBYSHEV INEQUALITIES

Thus far, we have seen that, to calculate probabilities, we need to know probability distribution functions, probability mass functions, or probability density functions. It frequently happens that, for some random variables, we cannot determine any of these three functions, but we can calculate their expected values and/or variances. In such cases, although we cannot calculate exact probabilities, using Markov and Chebyshev inequalities, we are able to derive bounds on probabilities. These inequalities have useful applications and significant theoretical values. Moreover, Chebyshev's inequality is a further indication of the importance of the concept of variance. Chebyshev's inequality was first discovered by the French mathematician Irénée Bienaymé (1796–1878). For this reason some authors call it the Chebyshev-Bienaymé inequality. In the middle of the nineteenth century, Chebyshev discovered the inequality independently, in connection with the laws of large numbers (see Section 11.4). He used it to give an elegant and short proof for the law of large numbers, discovered by James Bernoulli early in the eighteenth century. Since the usefulness and applicability of the inequality were demonstrated by Chebyshev, most authors call it Chebyshev's inequality.

Theorem 11.8   (Markov's Inequality)   *Let X be a nonnegative random variable; then for any $t > 0$,*

$$P(X \geq t) \leq \frac{E(X)}{t}.$$

Proof: We prove the theorem for a discrete random variable X with probability mass function $p(x)$. For continuous random variables the proof is similar. Let A be the set of possible values of X and $B = \{x \in A : x \geq t\}$. Then

$$E(X) = \sum_{x \in A} x p(x) \geq \sum_{x \in B} x p(x) \geq t \sum_{x \in B} p(x) = t P(X \geq t).$$

Thus

$$P(X \geq t) \leq \frac{E(X)}{t}. \quad \blacklozenge$$

Example 11.12   A post office, on average, handles 10,000 letters per day. What can be said about the probability that it will handle (a) at least 15,000 letters tomorrow; (b) less than 15,000 letters tomorrow?

Solution: Let X be the number of letters that this post office will handle tomorrow. Then $E(X) = 10,000$.

(a)   By Markov's inequality,

$$P(X \geq 15,000) \leq \frac{E(X)}{15,000} = \frac{10,000}{15,000} = \frac{2}{3}.$$

(b)   Using the inequality obtained in (a), we have

$$P(X < 15,000) = 1 - P(X \geq 15,000) \geq 1 - \frac{2}{3} = \frac{1}{3}. \quad \blacklozenge$$

Theorem 11.9   (Chebyshev's Inequality)   *If X is a random variable with expected value μ and variance σ^2, then for any $t > 0$,*

$$P\big(|X - \mu| \geq t\big) \leq \frac{\sigma^2}{t^2}.$$

Proof: Since $(X - \mu)^2 \geq 0$, by Markov's inequality

$$P\big((X - \mu)^2 \geq t^2\big) \leq \frac{E\big[(X - \mu)^2\big]}{t^2} = \frac{\sigma^2}{t^2}.$$

Chebyshev's inequality follows since $(X - \mu)^2 \geq t^2$ is equivalent to $|X - \mu| \geq t$. $\quad \blacklozenge$

Letting $t = k\sigma$ in Chebyshev's inequality, we get that

$$P\big(|X - \mu| \geq k\sigma\big) \leq 1/k^2.$$

That is,

The probability that X deviates from its expected value at least k standard deviations is less than $1/k^2$.

Thus, for example,

$$P(|X - \mu| \geq 2\sigma) \leq 1/4,$$

$$P(|X - \mu| \geq 4\sigma) \leq 1/16,$$

$$P(|X - \mu| \geq 10\sigma) \leq 1/100.$$

On the other hand,

$$P(|X - \mu| \geq k\sigma) \leq \frac{1}{k^2}$$

implies that

$$P(|X - \mu| < k\sigma) \geq 1 - 1/k^2.$$

Therefore,

The probability that X deviates from its mean less than k standard deviations is at least $1 - 1/k^2$.

In particular, this implies that, for any set of data, at least a fraction $1 - 1/k^2$ of the data are within k standard deviations on either side of the mean. Thus, for any data, at least $1 - 1/2^2 = 3/4$, or 75% of the data, lie within two standard deviations on either side of the mean. This implication of Chebyshev's inequality is true for any set of real numbers. That is, let $\{x_1, x_2, \ldots, x_n\}$ be a set of real numbers, and define

$$\bar{x} = \frac{1}{n} \sum_{i=1}^{n} x_i, \qquad s^2 = \frac{1}{n-1} \sum_{i=1}^{n} (x_i - \bar{x})^2;$$

then at least a fraction $1 - 1/k^2$ of the x_i's are between $\bar{x} - ks$ and $\bar{x} + ks$ (see Exercise 21).

Example 11.13   Suppose that, on average, a post office handles 10,000 letters a day with a variance of 2000. What can be said about the probability that this post office will handle between 8000 and 12,000 letters tomorrow?

Solution: Let X denote the number of letters that this post office will handle tomorrow. Then $\mu = E(X) = 10,000$, $\sigma^2 = \text{Var}(X) = 2000$. We want to calculate $P(8000 < X < 12,000)$. Since

$$P(8000 < X < 12,000) = P(-2000 < X - 10,000 < 2000)$$

$$= P(|X - 10,000| < 2000)$$

$$= 1 - P(|X - 10,000| \geq 2000),$$

by Chebyshev's inequality,

$$P(|X - 10,000| \geq 2000) \leq \frac{2000}{(2000)^2} = 0.0005.$$

Hence

$$P(8000 < X < 12,000) = P(|X - 10,000| < 2000) \geq 1 - 0.0005 = 0.9995.$$

Note that this answer is consistent with our intuitive understanding of the concepts of expectation and variance.   ♦

Example 11.14   A blind will fit Myra's bedroom's window if its width is between 41.5 and 42.5 inches. Myra buys a blind from a store that has 30 such blinds. What can be said about the probability that it fits her window if the average of the widths of the blinds is 42 inches with standard deviation 0.25?

Solution: Let X be the width of the blind that Myra purchased. We know that

$$P(|X - \mu| < k\sigma) \geq 1 - \frac{1}{k^2}.$$

Therefore,

$$P(41.5 < X < 42.5) = P(|X - 42| < 2(0.25)) \geq 1 - \frac{1}{4} = 0.75.   ♦$$

It should be mentioned that the bounds that are obtained on probabilities by Markov's and Chebyshev's inequalities are not usually very close to the actual probabilities. The following example demonstrates this.

Example 11.15   Roll a die and let X be the outcome. Clearly,

$$E(X) = \frac{1}{6}(1 + 2 + 3 + 4 + 5 + 6) = \frac{21}{6},$$

$$E(X^2) = \frac{1}{6}(1^2 + 2^2 + 3^2 + 4^2 + 5^2 + 6^2) = \frac{91}{6}.$$

Thus $\text{Var}(X) = 91/6 - 441/36 = 35/12$. By Markov's inequality,

$$P(X \geq 6) \leq \frac{21/6}{6} \approx 0.583.$$

By Chebyshev's inequality,

$$P\left(\left|X - \frac{21}{6}\right| \geq \frac{3}{2}\right) \leq \frac{35/12}{9/4} \approx 1.296,$$

a trivial bound because we already know that $P(|X - 21/6| \geq 3/2) \leq 1$. However, the exact values of these probabilities are much smaller than these bounds: $P(X \geq 6) = 1/6 \approx 0.167$ and

$$P\left(\left|X - \frac{21}{6}\right| \geq \frac{3}{2}\right) = P(X \leq 2 \text{ or } X \geq 5) = \frac{4}{6} \approx 0.667.   ♦$$

The following example is an elegant proof of the fact, shown previously, that if $\text{Var}(X) = 0$, then X is constant with probability 1. It is a good application of Chebyshev's inequality.

Example 11.16   Let X be a random variable with mean μ and variance 0. In this example, we will show that X is a constant. That is,

$$P(X = \mu) = 1.$$

To prove this, let $E_i = \{|X - \mu| < 1/i\}$; then

$$E_1 \supseteq E_2 \supseteq E_3 \supseteq \cdots \supseteq E_n \supseteq E_{n+1} \supseteq \cdots.$$

That is, $\{E_i, i = 1, 2, 3, \ldots\}$ is a decreasing sequence of events. Therefore,

$$\lim_{n \to \infty} E_n = \bigcap_{n=1}^{\infty} E_n = \bigcap_{n=1}^{\infty} \left\{ |X - \mu| < \frac{1}{n} \right\} = \{X = \mu\}.$$

Hence, by the continuity property of the probability function (Theorem 1.8),

$$\lim_{n \to \infty} P(E_n) = P(\lim_{n \to \infty} E_n) = P(X = \mu).$$

Now, by Chebyshev's inequality,

$$P\left(|X - \mu| \geq \frac{1}{n} \right) \leq \frac{\text{Var}(X)}{1/n^2} = 0,$$

implying that $\forall n \geq 1$,

$$P(E_n) = P\left(|X - \mu| < \frac{1}{n} \right) = 1.$$

Thus

$$P(X = \mu) = \lim_{n \to \infty} P(E_n) = 1. \quad \blacklozenge$$

Chebyshev's Inequality and Sample Mean

Let X_1, X_2, $\ldots$, X_n be a random sample of size n from a (continuous or discrete) distribution function F with mean μ and variance σ^2. Let $\bar{X}$ be the mean of the sample. Then, by Lemma 10.1,

$$E(\bar{X}) = E\left(\frac{X_1 + X_2 + \cdots + X_n}{n} \right) = \mu,$$

$$\text{Var}(\bar{X}) = \text{Var}\left(\frac{X_1 + X_2 + \cdots + X_n}{n} \right) = \frac{\sigma^2}{n}.$$

Applying Chebyshev's inequality to $\bar{X}$, we have that, for any $\varepsilon > 0$,

$$P(|\bar{X} - \mu| \geq \varepsilon) \leq \frac{\sigma^2}{\varepsilon^2 n}. \qquad (11.3)$$

Therefore, in other words, if $\{X_1, X_2, \dots, \}$ is a sequence of independent and identically distributed random variables with mean μ and variance σ^2, then, for all $n \geq 1$ and $\varepsilon > 0$, (11.3) holds. Inequality (11.3) has many important applications, some of which are discussed next.

Example 11.17   For the scores on an achievement test given to a certain population of students, the expected value is 500 and the standard deviation is 100. Let $\bar{X}$ be the mean of the scores of a random sample of 10 students. Find a lower bound for $P(460 < \bar{X} < 540)$.

Solution: Since

$$P(460 < \bar{X} < 540) = P(-40 < \bar{X} - 500 < 40)$$

$$= P(|\bar{X} - 500| < 40)$$

$$= 1 - P(|\bar{X} - 500| \geq 40),$$

by (11.3),

$$P(|\bar{X} - 500| \geq 40) \leq \frac{100^2}{(40)^2 10} = \frac{5}{8}.$$

Hence

$$P(460 < \bar{X} < 540) \geq 1 - \frac{5}{8} = \frac{3}{8}. \quad \blacklozenge$$

Example 11.18   A biologist wants to estimate ℓ, the life expectancy of a certain type of insect. To do so, he takes a sample of size n and measures the lifetime from birth to death of each insect. Then he finds the average of these numbers. If he believes that the lifetimes of these insects are independent random variables with variance 1.5 days, how large a sample should he choose to be at least 98% sure that his average is accurate within ± 0.2 (± 4.8 hours)?

Solution: For $i = 1, 2, \dots, n$, let X_i be the lifetime of the ith insect of the sample. We want to determine n, so that

$$P\left(-0.2 < \frac{X_1 + X_2 + \cdots + X_n}{n} - \ell < 0.2\right) \geq 0.98;$$

that is,

$$P(|\bar{X} - \ell| < 0.2) \geq 0.98.$$

This is equivalent to

$$P(|\bar{X} - \ell| \geq 0.2) \leq 0.02.$$

Since $E(X_i) = \ell$ and $\text{Var}(X_i) = 1.5$, by (11.3),

$$P(|\bar{X} - \ell| \geq 0.2) \leq \frac{1.5}{(0.2)^2 n} = \frac{37.5}{n}.$$

Therefore, all we need to do is to find n for which $37.5/n \leq 0.02$. This gives $n \geq 1875$. Hence the biologist should choose a sample of size 1875.   ♦

In statistics, an important application of Chebyshev's inequality is approximation of the parameter of a Bernoulli random variable. Suppose that, for a Bernoulli trial, p, the probability of success is unknown, and we are interested in estimating it. Suppose that, for small $\varepsilon > 0$ and small $\alpha > 0$, we want the probability to be at least $1 - \alpha$ that the error of estimation is less than ε. A simple way to do this is to repeat the Bernoulli trial independently and, for $i \geq 1$, let

$$X_i = \begin{cases} 1 & \text{if the } i\text{th trial is a success} \\ 0 & \text{otherwise.} \end{cases}$$

Then $(X_1 + X_2 + \cdots + X_n)/n$ is the average number of successes in the first n trials. The goal is to find the smallest n for which $\widehat{p} = \dfrac{X_1 + X_2 + \cdots + X_n}{n}$ satisfies

$$P(|\widehat{p} - p| < \varepsilon) \geq 1 - \alpha.$$

Note that, for $i \geq 1$, $E(X_i) = p$ and $\text{Var}(X_i) = p(1 - p)$. Applying (11.3) to the sequence $\{X_1, X_2, \ldots\}$ yields

$$P(|\widehat{p} - p| \geq \varepsilon) \leq \frac{p(1 - p)}{\varepsilon^2 n},$$

or, equivalently,

$$P(|\widehat{p} - p| < \varepsilon) \geq 1 - \frac{p(1 - p)}{\varepsilon^2 n}. \tag{11.4}$$

Thus we need to find the smallest n for which

$$1 - \frac{p(1 - p)}{\varepsilon^2 n} \geq 1 - \alpha,$$

or

$$n \geq \frac{p(1 - p)}{\varepsilon^2 \alpha}.$$

The difficulty with this inequality is that p is unknown. To overcome this, note that the function $f(p) = p(1 - p)$ is maximum for $p = 1/2$. Hence, for all $0 < p < 1$,

$$p(1 - p) \leq \frac{1}{4}.$$

Thus if we choose n to satisfy

$$n \geq \frac{1}{4\varepsilon^2\alpha},$$ (11.5)

then

$$n \geq \frac{1}{4\varepsilon^2\alpha} \geq \frac{p(1-p)}{\varepsilon^2\alpha},$$

and we have the desired inequality; namely, $P\big(|\widehat{p} - p| < \varepsilon\big) \geq 1 - \alpha$.

Example 11.19   To estimate the percentage of defective nails manufactured by a specific company, a quality control officer decides to take a random sample of nails produced by the company and calculate the fraction of those that are defective. If the officer wants to be sure, with a probability of at least 98%, that the error of estimation is less than 0.03, how large should the sample size be?

Solution: In (11.5), let $\varepsilon = 0.03$ and $\alpha = 0.02$; we have

$$n \geq \frac{1}{4(0.03)^2(0.02)} = 13,888.89.$$

Therefore, at least 13,889 nails should be tested. Note that if the quality control officer reduces the confidence level from 98% to 95%, then $\alpha = 0.05$, and we obtain $n \geq 5555.56$. That is, the number of nails to be tested, in that case, is $n = 5556$, a number considerably lower than the number required when the confidence level is to be 98%. ◆

Example 11.20   For a coin, p, the probability of heads is unknown. To estimate p, we flip the coin 3000 times and let $\widehat{p}$ be the fraction of times it lands heads up. Show that the probability is at least 0.90 that $\widehat{p}$ estimates p within ± 0.03.

Solution   By (11.4),

$$P\big(|\widehat{p} - p| < 0.03\big) \geq 1 - \frac{p(1-p)}{(0.03)^2 3000} \geq 1 - \frac{1}{4(0.03)^2 3000} = 0.907,$$

since $p(1-p) \leq 1/4$ implies that $-p(1-p) \geq -1/4$.   ◆

EXERCISES

A

1. According to the Bureau of Engraving and Printing,

    http://www.moneyfactory.com/document.cfm/18/106, December 10, 2003,

    the average life of a one-dollar Federal Reserve note is 22 months. Show that at most 37% of one-dollar bills last for at least five years.

2. Show that if for a nonnegative random variable X, $P(X < 2) = 3/5$, then $E(X) \geq 4/5$.

3. Let X be a nonnegative random variable with $E(X) = 5$ and $E(X^2) = 42$. Find an upper bound for $P(X \geq 11)$ using (a) Markov's inequality, (b) Chebyshev's inequality.

4. The average and standard deviation of lifetimes of light bulbs manufactured by a certain factory are, respectively, 800 hours and 50 hours. What can be said about the probability that a random light bulb lasts, at most, 700 hours?

5. Suppose that the average number of accidents at an intersection is two per day.

    (a) Use Markov's inequality to find a bound for the probability that at least five accidents will occur tomorrow.

    (b) Using Poisson random variables, calculate the probability that at least five accidents will occur tomorrow. Compare this value with the bound obtained in part (a).

    (c) Let the variance of the number of accidents be two per day. Use Chebyshev's inequality to find a bound on the probability that tomorrow at least five accidents will occur.

6. The average IQ score on a certain campus is 110. If the variance of these scores is 15, what can be said about the percentage of students with an IQ above 140?

7. The waiting period from the time a book is ordered until it is received is a random variable with mean seven days and standard deviation two days. If Helen wants to be 95% sure that she receives a book by certain date, how early should she order the book?

8. Show that for a nonnegative random variable X with mean μ, $P(X \geq 2\mu) \leq 1/2$.

9. Suppose that X is a random variable with $E(X) = \text{Var}(X) = \mu$. What does Chebyshev's inequality say about $P(X > 2\mu)$?

10. From a distribution with mean 42 and variance 60, a random sample of size 25 is taken. Let $\bar{X}$ be the mean of the sample. Show that the probability is at least 0.85 that $\bar{X} \in (38, 46)$.

11. The mean IQ of a randomly selected student from a specific university is μ; its variance is 150. A psychologist wants to estimate μ. To do so, for some n, she takes a sample of size n of the students at random and independently and measures their IQ's. Then she finds the average of these numbers. How large a sample should she choose to make at least 92% sure that the average is accurate within ± 3 points?

12. For a distribution, the mean of a random sample is taken as estimation of the expected value of the distribution. How large should the sample size be so that, with a probability of at least 0.98, the error of estimation is less than 2 standard deviations of the distribution?

13. To determine p, the proportion of time that an airline operator is busy answering customers, a supervisor observes the operator at times selected randomly and independently from other observed times. Let $X_i = 1$ if the ith time the operator is observed, he is busy; let $X_i = 0$ otherwise. For large values of n, if $(1/n) \sum_{i=1}^{n} X_i$'s are taken as estimates of p, for what values of n is the error of estimation at most 0.05 with probability 0.96?

14. For a coin, p, the probability of heads is unknown. To estimate p, for some n, we flip the coin n times independently. Let $\hat{p}$ be the proportion of heads obtained. Determine the value of n for which $\hat{p}$ estimates p within ± 0.05 with a probability of at least 0.94.

15. Let X be a random variable with mean μ. Show that if $E\left[(X - \mu)^{2n}\right] < \infty$, then for $\alpha > 0$,

$$P\left(|X - \mu| \geq \alpha\right) \leq \frac{1}{\alpha^{2n}} E\left[(X - \mu)^{2n}\right].$$

16. Let X be a random variable and k be a constant. Prove that

$$P(X > t) \leq \frac{E(e^{kX})}{e^{kt}}.$$

17. Prove that if the random variables X and Y satisfy $E\left[(X - Y)^2\right] = 0$, then with probability 1, $X = Y$.

18. Let X and Y be two randomly selected numbers from the set of positive integers $\{1, 2, \ldots, n\}$. Prove that $\rho(X, Y) = 1$ if and only if $X = Y$ with probability 1.
Hint: First prove that $E\left[(X - Y)^2\right] = 0$; then use Exercise 17.

19. Let X be a random variable; show that for $\alpha > 1$ and $t > 0$,

$$P\left(X \geq \frac{1}{t} \ln \alpha\right) \leq \frac{1}{\alpha} M_X(t).$$

B

20. Let the probability density function of a random variable X be

$$f(x) = \frac{x^n}{n!}e^{-x}, \qquad x \geq 0.$$

Show that

$$P(0 < X < 2n + 2) > \frac{n}{n+1}.$$

Hint: Note that $\int_0^\infty x^n e^{-x}\, dx = \Gamma(n+1) = n!$. Use this to calculate $E(X)$ and $\mathrm{Var}(X)$. Then apply Chebyshev's inequality.

21. Let $\{x_1, x_2, \dots, x_n\}$ be a set of real numbers and define

$$\bar{x} = \frac{1}{n}\sum_{i=1}^{n} x_i, \qquad s^2 = \frac{1}{n-1}\sum_{i=1}^{n}(x_i - \bar{x})^2.$$

Prove that at least a fraction $1 - 1/k^2$ of the x_i's are between $\bar{x} - ks$ and $\bar{x} + ks$.

Sketch of a Proof: Let N be the number of $x_1, x_2, \dots, x_n$ that fall in $A = [\bar{x} - ks, \bar{x} + ks]$. Then

$$s^2 = \frac{1}{n-1}\sum_{i=1}^{n}(x_i - \bar{x})^2 \geq \frac{1}{n-1}\sum_{x_i \notin A}(x_i - \bar{x})^2$$

$$\geq \frac{1}{n-1}\sum_{x_i \notin A}k^2 s^2 = \frac{n-N}{n-1}k^2 s^2.$$

This gives $(N-1)/(n-1) \geq 1 - (1/k^2)$. The result follows since $N/n \geq (N-1)/(n-1)$. ∎

11.4   LAWS OF LARGE NUMBERS

Let A be an event of some experiment that can be repeated. In Chapter 1 we mentioned that the mathematicians of the eighteenth and nineteenth centuries observed that, in a series of sequential or simultaneous repetitions of the experiment, the proportion of times that A occurs approaches a constant. As a result, they were motivated to define the probability of A to be the number $p = \lim_{n \to \infty} n(A)/n$, where $n(A)$ is the number of times that A occurs in the first n repetitions. We also mentioned that this relative frequency interpretation of probability, which to some extent satisfies one's intuition, is mathematically problematic and cannot be the basis of a rigorous probability theory. We now show that despite these facts, for repeatable experiments, the relative frequency interpretation is valid. It is the special case of one of the most celebrated theorems of probability and statistics: the **strong law of large numbers**. As a consequence of another important theorem called the **weak law of large numbers**, we have that, for sufficiently large n, it is very likely that $\left| \dfrac{n(A)}{n} - P(A) \right|$ is very small. Therefore, for repeatable experiments, the laws of large numbers confirm our intuitive interpretation of the concept of probability as relative frequency of the occurrence of an event. Moreover, they play a pivotal role in the study of probability and statistics. It should be noted that, even though the weak law of large numbers is mathematically weaker than the strong law of large numbers, strangely enough, it has more applications in certain areas of statistics.

Recall that a sequence $X_1, X_2, \ldots$ of random variables is called **identically distributed** if all of them have the same distribution function. If $X_1, X_2, \ldots$ are identically distributed, then since they have the same probability distribution function, their set of possible values is the same. Also, in the discrete case their probability mass functions, and in the continuous case their probability density functions, are identical. Moreover, in both cases X_i's have the same expectation, the same standard deviation, and the same variance.

Theorem 11.10   (Weak Law of Large Numbers)   *Let $X_1, X_2, X_3, \ldots$ be a sequence of independent and identically distributed random variables with $\mu = E(X_i)$ and $\sigma^2 = \mathrm{Var}(X_i) < \infty$, $i = 1, 2, \ldots$. Then $\forall \varepsilon > 0$,*

$$\lim_{n \to \infty} P\left(\left| \frac{X_1 + X_2 + \cdots + X_n}{n} - \mu \right| > \varepsilon \right) = 0.$$

Note that, in the standard statistical models, $(X_1 + X_2 + \cdots + X_n)/n$ is $\bar{X}$, the sample mean.

Let $\{X_1, X_2, \ldots, X_n\}$ be a random sample of a population. The weak law of large numbers implies that if the sample size is large enough ($n \to \infty$), then μ, the mean of the population, is very close to $\bar{X} = (X_1 + X_2 + \cdots + X_n)/n$. That is, for sufficiently large n, it is very likely that $|\bar{X} - \mu|$ is very small.

Note that the type of convergence in the weak law of large numbers is not the usual pointwise convergence discussed in calculus. This type of convergence is called **convergence in probability** and is defined as follows.

Definition   *Let $X_1, X_2, \ldots$ be a sequence of random variables defined on a sample space S. We say that X_n **converges to a random variable X in probability** if, for each $\varepsilon > 0$,*

$$\lim_{n \to \infty} P\left(|X_n - X| > \varepsilon \right) = 0.$$

So if X_n converges to X in probability, then for sufficiently large n it is very likely that $|X_n - X|$ is very small. By this definition, the weak law of large numbers states that if $X_1, X_2, \ldots$ is a sequence of independent and identically distributed random variables with $\mu = E(X_i)$ and $\sigma^2 = \mathrm{Var}(X_i) < \infty$, $i = 1, 2, \ldots$, then $(X_1 + X_2 + \cdots + X_n)/n$ converges to μ in probability. To prove Theorem 11.10, we use the fact that the variance of a sum of independent random variables is the sum of the variances of the random variables [see relation (10.12)].

Proof of Theorem 11.10:   Since

$$E\left(\frac{X_1 + X_2 + \cdots + X_n}{n}\right) = \frac{1}{n}\big[E(X_1) + E(X_2) + \cdots + E(X_n)\big] = \frac{1}{n}n\mu = \mu$$

and

$$\mathrm{Var}\left(\frac{X_1 + X_2 + \cdots + X_n}{n}\right) = \frac{1}{n^2}\big[\mathrm{Var}(X_1) + \mathrm{Var}(X_2) + \cdots + \mathrm{Var}(X_n)\big] = \frac{1}{n^2}n\sigma^2 = \frac{\sigma^2}{n},$$

by Chebyshev's inequality we get

$$P\left(\left|\frac{X_1 + X_2 + \cdots + X_n}{n} - \mu\right| > \varepsilon\right) \le \frac{\sigma^2/n}{\varepsilon^2} = \frac{\sigma^2}{\varepsilon^2 n}.$$

This shows that

$$\lim_{n\to\infty} P\left(\left|\frac{X_1 + X_2 + \cdots + X_n}{n} - \mu\right| > \varepsilon\right) = 0. \quad \blacklozenge$$

To show how the weak law of large numbers relates to the relative frequency interpretation of probability, in a repeatable experiment with sample space S, let A be an event of S. Suppose that the experiment is repeated independently, and let $X_i = 1$ if A occurs on the ith repetition, and $X_i = 0$ if A does not occur on the ith repetition. Then, for $i = 1, 2, \ldots$,

$$E(X_i) = 1 \cdot P(A) + 0 \cdot P(A^c) = P(A)$$

and

$$X_1 + X_2 + \cdots + X_n = n(A), \tag{11.6}$$

where $n(A)$, as before, is the number of times that A occurs in the first n repetitions of the experiment. Thus, according to the weak law of large numbers,

$$\lim_{n\to\infty} P\left(\left|\frac{n(A)}{n} - P(A)\right| > \varepsilon\right) = \lim_{n\to\infty} P\left(\left|\frac{X_1 + X_2 + \cdots + X_n}{n} - P(A)\right| > \varepsilon\right) = 0.$$

This shows that, for large n, the relative frequency of occurrence of A [that is, $n(A)/n$], is very likely to be close to $P(A)$.

It is important to know that, historically, the weak law of large numbers was shown in an ingenious way by Jacob Bernoulli (1654–1705) for Bernoulli random variables. Later, this law was generalized by Chebyshev for a certain class of sequences of identically distributed and independent random variables. Chebyshev discovered his inequality in this connection. Theorem 11.10 is Chebyshev's version of the weak law of large numbers. This is weaker than the theorem that the Russian mathematician Khintchine proved; that is, its conclusions are the same as Khintchine's result, but its assumptions are more restrictive. In Khintchine's proof, it is not necessary to assume that $\text{Var}(X_i) < \infty$.

Theorem 11.11   (Strong Law of Large Numbers)   *Let X_1, X_2, ... be an independent and identically distributed sequence of random variables with $\mu = E(X_i)$, $i = 1, 2, \ldots$. Then*

$$P\left(\lim_{n \to \infty} \frac{X_1 + X_2 + \cdots + X_n}{n} = \mu \right) = 1.$$

As before, in a repeatable experiment, let A be an event of the sample space S. Then, by (11.6),

$$\frac{X_1 + X_2 + \cdots + X_n}{n} = \frac{n(A)}{n},$$

which, according to the strong law of large numbers, converges to $P(A)$ with probability 1 as $n \to \infty$.

Note that, in Theorem 11.11, convergence of $(X_1 + X_2 + \cdots + X_n)/n$ to μ is neither the usual pointwise convergence studied in calculus, nor is it convergence in probability as discussed earlier. This type of convergence is called **almost sure convergence** and is defined as follows: Let X and the sequence $X_1, X_2, X_3, \ldots$ be random variables defined on a sample space S. Let V be the set of all points ω, in S, at which $X_n(\omega)$ converges to $X(\omega)$. That is,

$$V = \left\{ \omega \in S \colon \lim_{n \to \infty} X_n(\omega) = X(\omega) \right\}.$$

If $P(V) = 1$, then we say that X_n converges **almost surely** to X.

Therefore, almost sure convergence implies that, on a set whose probability is 1, at all sample points ω, $X_n(\omega) - X(\omega)$ gets smaller and smaller as n gets larger and larger. Whereas convergence in probability implies that the probability that $X_n - X$ remains nonnegligible gets smaller and smaller as n gets larger and larger. It is not hard to show that, if X_n converges to X almost surely, it converges to X in probability. However, the converse of this is not true (see Exercise 7).

To clarify the definition of almost sure convergence, we now give an example of a sequence that converges almost surely but is not pointwise convergent. Consider an experiment in which we choose a random point from the interval $[0, 1]$. The sample space of this experiment is $[0, 1]$; for all $\omega \in [0, 1]$, let $X_n(\omega) = \omega^n$ and $X(\omega) = 0$. Then, clearly,

$$V = \{\omega \in [0, 1]: \lim_{n \to \infty} X_n(\omega) = X(\omega)\} = [0, 1).$$

Since $P(V) = P([0, 1)) = 1$, X_n converges almost surely to X. However, this convergence is not pointwise because, at 1, X_n converges to 1 and not to $X(1) = 0$.

Example 11.21   At a large international airport, a currency exchange bank with only one teller is open 24 hours a day, 7 days a week. Suppose that at some time $t = 0$, the bank is free of customers and new customers arrive at random times $T_1, T_1 + T_2, T_1 + T_2 + T_3$, ..., where $T_1, T_2, T_3, \ldots$ are identically distributed and independent random variables with $E(T_i) = 1/\lambda$. When the teller is free, the service time of a customer entering the bank begins upon arrival. Otherwise, the customer joins the queue and waits to be served on a first-come, first-served basis. The customer leaves the bank after being served. The service time of the ith new customer is S_i, where $S_1, S_2, S_3, \ldots$ are identically distributed and independent random variables with $E(S_i) = 1/\mu$. Therefore, new customers arrive at the rate λ, and while the teller is busy, they are served at the rate μ. Show that if $\lambda < \mu$, that is, if new customers are served at a higher rate than they arrive, then with probability 1, eventually, for some period, the bank will be empty of customers again.

Solution: Suppose that the bank will never again be empty of customers. We will show a contradiction. Let $U_n = T_1 + T_2 + \cdots + T_n$. Then U_n is the time the nth new customer arrives. Let $Z_n = T_1 + S_1 + S_2 + \cdots + S_n$. Since the bank will never be empty of customers again, and customers are served on a first-come, first-served basis, Z_n is the departure time of the nth new customer. By the strong law of large numbers,

$$\lim_{n \to \infty} \frac{U_n}{n} = \frac{1}{\lambda}$$

and

$$\lim_{n \to \infty} \frac{Z_n}{n} = \lim_{n \to \infty} \left(\frac{T_1}{n} + \frac{S_1 + S_2 + \cdots + S_n}{n} \right)$$
$$= \lim_{n \to \infty} \frac{T_1}{n} + \lim_{n \to \infty} \frac{S_1 + S_2 + \cdots + S_n}{n} = 0 + \frac{1}{\mu} = \frac{1}{\mu}.$$

Clearly, the bank will never remain empty of customers again if and only if $\forall n$,

$$U_{n+1} < Z_n.$$

This implies that

$$\frac{U_{n+1}}{n} < \frac{Z_n}{n},$$

or, equivalently,

$$\frac{n+1}{n} \cdot \frac{U_{n+1}}{n+1} < \frac{Z_n}{n}.$$

Thus

$$\lim_{n\to\infty} \frac{n+1}{n} \cdot \frac{U_{n+1}}{n+1} \leq \lim_{n\to\infty} \frac{Z_n}{n}. \tag{11.7}$$

Since $\lim_{n\to\infty} \dfrac{n+1}{n} = 1$, and with probability 1, $\lim_{n\to\infty} \dfrac{U_{n+1}}{n+1} = \dfrac{1}{\lambda}$ and $\lim_{n\to\infty} \dfrac{Z_n}{n} = \dfrac{1}{\mu}$, (11.7) implies that $\dfrac{1}{\lambda} \leq \dfrac{1}{\mu}$ or $\lambda \geq \mu$. This contradicts the fact that $\lambda < \mu$. Hence, with probability 1, eventually, for some period, the bank will be empty of customers again. ◆

Example 11.22   Suppose that an "immortal monkey" is constantly typing on a word processor that is not breakable, lasts forever, and has infinite memory. Suppose that the keyboard of the wordprocessor has $m-1$ keys, a space bar for blank spaces, and separate keys for different symbols. If each time the monkey presses one of the m symbols (including the space bar) at random, and if at the end of each line and at the end of each page the word processor advances to a new line and a new page by itself, what is the probability that the monkey eventually will produce the complete works of Shakespeare in chronological order and with no errors?

Solution: The complete works of Shakespeare in chronological order and with no errors are the result of typing N specific symbols (including the space bar) for some large N. For $i = 1, 2, \ldots$, let A_i be the event that the symbols numbered $(i-1)N+1$ to number iN, typed by the monkey, form the complete works of Shakespeare, in chronological order with no errors. Also for $i = 1, 2, \ldots$, let

$$X_i = \begin{cases} 1 & \text{if } A_i \text{ occurs} \\ 0 & \text{otherwise;} \end{cases}$$

then $E(X_i) = P(A_i) = (1/m)^N$. Now, since $\{X_1, X_2, \ldots\}$ is a sequence of independent and identically distributed random variables, by the strong law of large numbers we have that

$$P\left(\lim_{n\to\infty} \frac{X_1 + X_2 + \cdots + X_n}{n} = \left(\frac{1}{m}\right)^N \right) = 1.$$

This relation shows that $\sum_{i=1}^{\infty} X_i = \infty$, the reason being that otherwise $\sum_{i=1}^{\infty} X_i < \infty$ implies that $\lim_{n\to\infty}(X_1+X_2+\cdots+X_n)/n$ is 0 and not $(1/m)^N > 0$. Now $\sum_{i=1}^{\infty} X_i = \infty$ implies that an infinite number of X_i's are equal to 1, meaning that infinitely many of A_i's will occur. Therefore, not only once, but an infinite number of times, the monkey will produce the complete works of Shakespeare in chronological order with no errors.

However, if the monkey types 200 symbols a minute, then for $m = 88$ and $N = 10^7$, it is easily shown that, on average, it will take him T years to produce the complete works of Shakespeare for the first time, where T is a number with tens of millions of digits.   ◆

Example 11.23   Suppose that $f : [0, 1] \to [0, 1]$ is a continuous function. We will now present a probabilistic method for evaluation of $\int_0^1 f(x)\, dx$, using the strong law of large numbers. Let $\{X_1, Y_1, X_2, Y_2, \ldots \}$ be a sequence of independent uniform random numbers from the interval $[0, 1]$. Then $\{(X_1, Y_1), (X_2, Y_2), \ldots \}$ is a sequence of independent random points from $[0, 1] \times [0, 1]$. For $i = 1, 2, 3, \ldots$, let

$$Z_i = \begin{cases} 1 & \text{if } Y_i < f(X_i) \\ 0 & \text{otherwise.} \end{cases}$$

Note that Z_i is 1 if (X_i, Y_i) falls under the curve $y = f(x)$, and is 0 if it falls above $y = f(x)$. Therefore, from the first n points selected, $Z_1 + Z_2 + \cdots + Z_n$ of them fall below the curve $y = f(x)$, and thus $\lim_{n \to \infty}(Z_1 + Z_2 + \cdots + Z_n)/n$ is the limit of the relative frequency of the points that fall below the curve $y = f(x)$. We will show that this limit converges to $\int_0^1 f(x)\, dx$. To do so, note that Z_i's are identically distributed and independent. Moreover, since g, the joint probability density function of (X_i, Y_i), $i \geq 0$, is given by

$$g(x, y) = \begin{cases} 1 & \text{if } 0 \leq x \leq 1, \quad 0 \leq y \leq 1 \\ 0 & \text{otherwise,} \end{cases}$$

we have that

$$E(Z_i) = P\big(Y_i < f(X_i)\big) = \iint\limits_{\{(x, y):\, y < f(x)\}} g(x, y)\, dx\, dy$$

$$= \int_0^1 \left(\int_0^{f(x)} dy \right) dx = \int_0^1 f(x)\, dx < \infty.$$

Hence, by the strong law of large numbers, with probability 1,

$$\lim_{n \to \infty} \frac{Z_1 + Z_2 + \cdots + Z_n}{n} = \int_0^1 f(x)\, dx.$$

This gives a new method for calculation of $\int_0^1 f(x)\, dx$. Namely, to find an approximate value for $\int_0^1 f(x)\, dx$ it suffices that, for a large n, we choose n random points from $[0, 1] \times [0, 1]$ and calculate m, the number of those falling below the curve $y = f(x)$. The quotient m/n is then approximately $\int_0^1 f(x)\, dx$. This is elaborated further in our discussion of Monte Carlo procedure and simulation in Section 13.5.   ◆

Example 11.24   Suppose that F, the probability distribution function of the elements of a class of random variables (say, a population), is unknown and we want to estimate it at a point x. For this purpose we take a random sample from the population; that is, we find independent random variables $X_1, X_2, \ldots, X_n$, each with distribution function F. Then we let $n(x) =$ the number of X_i's $\leq x$ and $\hat{F}_n(x) = n(x)/n$. Clearly, $\hat{F}_n(x)$ is the relative frequency of the number of data $\leq x$ [in statistics, $\hat{F}_n(x)$ is called the **empirical distribution function of the sample**]. We will show that $\lim_{n\to\infty} \hat{F}_n(x) = F(x)$. To do so, let

$$Y_i = \begin{cases} 1 & \text{if } X_i \leq x \\ 0 & \text{otherwise.} \end{cases}$$

Then Y_i's are identically distributed, independent, and

$$E(Y_i) = P(X_i \leq x) = F(x) \leq 1 < \infty.$$

Thus, by the strong law of large numbers,

$$\lim_{n\to\infty} \hat{F}_n(x) = \lim_{n\to\infty} \frac{Y_1 + Y_2 + \cdots + Y_n}{n} = E(Y_i) = F(x).$$

Therefore, for large n, $\hat{F}_n(x)$ is an approximation of $F(x)$. ◆

Example 11.25   A number between 0 and 1 is called **normal** if for every d, $0 \leq d \leq 9$, the "frequency" of the digit d in the decimal expansion of the number is 1/10. In the first version of the strong law of large numbers, in 1909, Émile Borel proved that, if X is a random number from the interval $(0, 1)$, then, with probability 1, X is a normal number. In other words, the probability that a number selected at random from $(0, 1)$ is not normal is 0—a fact known as **Borel's normal number theorem**. It is interesting to know that in spite of this fact, *constructing specific normal numbers is an extremely challenging problem.* To prove Borel's normal number theorem, let X be a random number from the interval $(0, 1)$. The probability that X has two distinct decimal expansions is 0, since the set of such numbers is countable (see Exercise 10, Section 1.7). Therefore, with probability 1, X has a unique decimal expansion $0.X_1 X_2 \cdots X_n \cdots$. For a fixed $d, 0 \leq d \leq 9$, let

$$Y_n = \begin{cases} 1 & \text{if } X_n = d \\ 0 & \text{if } X_n \neq d. \end{cases}$$

Then $\sum_{i=1}^{n} Y_i$ is the number of times that d appears among the first n digits of the decimal expansion of X. Therefore, $\dfrac{1}{n} \sum_{i=1}^{n} Y_i$ is the relative frequency of d. The fact that $\{Y_1, Y_2, \ldots\}$ is a sequence of identically distributed random variables with $\mu = E(Y_n) = P(X_n = d) = 1/10$ is trivial. If we show that this sequence is independent, then by the strong law of large numbers, with probability 1,

$$\lim_{n\to\infty} \frac{1}{n} \sum_{i=1}^{n} Y_n = \mu = \frac{1}{10},$$

showing that X is normal. Therefore, all that is left is to show that $\{Y_1, Y_2, \dots\}$ is a sequence of independent random variables. To do so, we need to prove that for any subsequence $\{Y_{i_1}, Y_{i_2}, \dots, Y_{i_m}\}$ of $\{Y_1, Y_2, \dots\}$, and any set of numbers $\alpha_i \in \{0, 1\}$, $1 \le i \le m$, we have

$$P(Y_{i_1} = \alpha_1, Y_{i_2} = \alpha_2, \dots, Y_{i_m} = \alpha_m) = P(Y_{i_1} = \alpha_1)P(Y_{i_2} = \alpha_2)\dots P(Y_{i_m} = \alpha_m).$$

For $1 \le i \le m$, let

$$\beta_i = \begin{cases} 9 & \text{if } \alpha_i = 0 \\ 1 & \text{if } \alpha_i = 1. \end{cases}$$

Then, by the generalized counting principle (Theorem 2.2),

$$P(Y_{i_1} = \alpha_1, Y_{i_2} = \alpha_2, \dots, Y_{i_m} = \alpha_m) = \frac{\beta_1 \beta_2 \cdots \beta_m}{10^m}$$

$$= \frac{\beta_1}{10} \cdot \frac{\beta_2}{10} \cdots \frac{\beta_m}{10} = P(Y_{i_1} = \alpha_1)P(Y_{i_2} = \alpha_2) \cdots P(Y_{i_m} = \alpha_m).$$

This establishes Borel's normal number theorem. ◆

Example 11.26　Suppose that there exist N families on the earth and that the maximum number of children a family has is c. Let α_j, $j = 0, 1, \dots, c$, $\sum_{j=0}^{c} \alpha_j = 1$, be the fraction of families with j children. Determine the fraction of the children in the world who are from a family with k ($k = 1, 2, \dots, c$) children.

Solution:　Let X_i denote the number of children of family i ($1 \le i \le N$) and

$$Y_i = \begin{cases} 1 & \text{if family } i \text{ has } k \text{ children} \\ 0 & \text{otherwise.} \end{cases}$$

Then the fraction of the children in the world who are from a family with k children is

$$\frac{k \sum_{i=1}^{N} Y_i}{\sum_{i=1}^{N} X_i} = \frac{k \dfrac{1}{N} \sum_{i=1}^{N} Y_i}{\dfrac{1}{N} \sum_{i=1}^{N} X_i}.$$

Now $(1/N) \sum_{i=1}^{N} Y_i$ is the fraction of the families with k children; therefore, it is equal to α_k. The quantity $(1/N) \sum_{i=1}^{N} X_i$ is the average number of children in a family; hence it is the same as $\sum_{j=0}^{c} j\alpha_j$. So the desired fraction is $k\alpha_k / \sum_{j=0}^{c} j\alpha_j$.

For large N, the strong law of large numbers can be used to show the same thing. Note that it is reasonable to assume that $\{X_1, X_2, \dots\}$ and hence $\{Y_1, Y_2, \dots\}$ are sequences of independent random variables. So we can write that

$$
\lim_{N \to \infty} \frac{k \dfrac{1}{N} \displaystyle\sum_{i=1}^{N} Y_i}{\dfrac{1}{N} \displaystyle\sum_{i=1}^{N} X_i} = \frac{k E(Y_i)}{E(X_i)} = \frac{k \alpha_k}{\displaystyle\sum_{j=0}^{c} j \alpha_j}. \quad \blacklozenge
$$

★ Proportion versus Difference in Coin Tossing

Suppose that a fair coin is flipped independently and successively. Let $n(H)$ be the number of times that heads occurs in the first n flips of the coin. Let $n(T)$ be the number of times that tails occurs in the first n flips. By the strong law of large numbers, with probability 1,

$$
\lim_{n \to \infty} \frac{n(H)}{n} = \lim_{n \to \infty} \frac{n(T)}{n} = \frac{1}{2}.
$$

That is, with probability 1, the *proportion* of heads and the *proportion* of tails, as $n \to \infty$, is 1/2. This result is subject to misinterpretation. Some might think that if a fair coin is tossed a very large number of times, then heads occurs as often as tails. For example, Bertrand Russell (1872–1970), in his *Religion and Science* (Oxford University Press, 1947, page 158), writes

> It is said (though I have never seen any good experimental evidence)
> that if you toss a penny a great many times, it will come heads about
> as often as tails. It is further said that this is not certain, but only
> extremely probable.

None of these statements is false. But they are subject to interpretations.

In Chapter 12, when discussing recurrent states of symmetric simple random walks, we will show that, in successive and independent flips of a fair coin, it happens that at times the number of heads obtained is equal to the number of tails obtained. In that chapter, we will show that, with probability 1, this happens infinitely often, but the expected number of tosses between two such consecutive times is ∞. Hence, for some large number of tosses, what Bertrand Russell quotes is accurate. However, in general, when we say heads occurs "about as often" as tails, it is important to note that the fuzzy term about as often is a relative term and should be interpreted only relative to the number of times the coin is flipped. Frequently, when a fair coin is flipped a large number of times, in the usual sense, one side will not come "about as often" as the other side. But relatively speaking, it does. For example, we will show that if a fair coin is flipped independently and successively 10^{20} times, then the probability exceeds 0.96 that the absolute value of the difference between the number of heads and the number of tails obtained is at least half a billion. This is a very large difference. Nevertheless when compared with the huge number of tosses, 10^{20}, it is insignificant. In fact, it is straightforward to show that in

10^{20} tosses of the coin, almost certainly, the difference between the number of heads and the number of tails is less than 40 billion. Considering the fact that 10^{20} is one hundred billion billion, relative to 10^{20}, 40 billion is really a small number. It is $\dfrac{1}{25 \times 10^9}$th of the number of times the coin is flipped. Therefore, relatively speaking, when the coin is flipped 10^{20} times, still "it will come heads about as often as tails."

We will end this section by showing that in 10^{20} flips of a fair coin, successively and independently, the probability exceeds 0.96 that the absolute value of the difference between the number of heads and the number of tails obtained is at least half a billion. Let X be the number of heads obtained. Clearly, X is a binomial random variable with parameters 10^{20} and $1/2$. Therefore,

$$E(X) = \frac{1}{2} \cdot 10^{20} \quad \text{and} \quad \sigma_X = \sqrt{10^{20}\left(\frac{1}{2}\right)\left(1 - \frac{1}{2}\right)} = \frac{1}{2} \cdot 10^{10}.$$

To show that

$$P\big(\big|X - (10^{20} - X)\big| \geq \frac{1}{2} \cdot 10^9\big) \geq 0.96,$$

we will use the DeMoivre-Laplace theorem, and we ignore the correction for continuity since its effect, when dealing with large numbers, is negligible. We have

$$P\big(\big|X - (10^{20} - X)\big| \geq \frac{1}{2} \cdot 10^9\big)$$

$$= P\big(\big|2X - 10^{20}\big| \geq \frac{1}{2} \cdot 10^9\big)$$

$$= P\big(2X - 10^{20} \geq \frac{1}{2} \cdot 10^9 \quad \text{or} \quad 2X - 10^{20} \leq -\frac{1}{2} \cdot 10^9\big)$$

$$= P\big(X \geq \frac{1}{2} \cdot 10^{20} + \frac{1}{4} \cdot 10^9\big) + P\big(X \leq \frac{1}{2} \cdot 10^{20} - \frac{1}{4} \cdot 10^9\big)$$

$$= P\Big(\frac{X - \frac{1}{2} \cdot 10^{20}}{\frac{1}{2} \cdot 10^{10}} \geq 0.05\Big) + P\Big(\frac{X - \frac{1}{2} \cdot 10^{20}}{\frac{1}{2} \cdot 10^{10}} \leq -0.05\Big)$$

$$\approx \int_{0.05}^{\infty} \frac{1}{\sqrt{2\pi}} e^{-x^2/2}\, dx + \int_{-\infty}^{-0.05} \frac{1}{\sqrt{2\pi}} e^{-x^2/2}\, dx$$

$$= 1 - \Phi(0.05) + \Phi(-0.05) = 2\big[1 - \Phi(0.05)\big]$$

$$\approx 2(1 - 0.5199) = 0.9602.$$

EXERCISES

A

1. Let $\{X_1, X_2, \ldots\}$ be a sequence of nonnegative independent random variables and, for all i, suppose that the probability density function of X_i is

$$f(x) = \begin{cases} 6x(1-x) & \text{if } 0 \leq x \leq 1 \\ 0 & \text{otherwise.} \end{cases}$$

Find

$$\lim_{n \to \infty} \frac{X_1 + X_2 + \cdots + X_n}{n}.$$

2. Let $\{X_1, X_2, \ldots\}$ be a sequence of independent, identically distributed random variables with positive expected value. Show that for all $M > 0$,

$$\lim_{n \to \infty} P(X_1 + X_2 + \cdots + X_n > M) = 1.$$

3. Let X be a nonnegative continuous random variable with probability density function $f(x)$. Define

$$Y_n = \begin{cases} 1 & \text{if } X > n \\ 0 & \text{otherwise.} \end{cases}$$

Prove that Y_n converges to 0 in probability.

4. Let $\{X_1, X_2, \ldots\}$ be a sequence of independent, identically distributed random variables. In other words, for all n, let $X_1, X_2, \ldots, X_n$ be a random sample from a distribution with mean $\mu < \infty$. Let $S_n = X_1 + X_2 + \cdots + X_n$, $\bar{X}_n = S_n/n$. Show that S_n grows at rate n. That is,

$$\lim_{n \to \infty} P\big(n(\mu - \varepsilon) \leq S_n \leq n(\mu + \varepsilon)\big) = 1.$$

B

5. Suppose that in Example 11.21 rather than customers being served in their arrival order, they are served on a last-come, first-served basis, or they are simply served in a random order. Show that if $\lambda < \mu$, then still with probability 1, eventually, for some period, the bank will be empty of customers again.

6. In Example 11.21, suppose that at $t = 0$ the bank is not free, there are $m > 0$ customers waiting in a queue to be served, and a customer is being served. Show that, with probability 1, eventually, for some period, the bank will be empty of customers.

7. For a positive integer n, let $\tau(n) = (2^k, i)$, where i is the remainder when we divide n by 2^k, the largest possible power of 2. For example, $\tau(10) = (2^3, 2)$, $\tau(12) = (2^3, 4)$, $\tau(19) = (2^4, 3)$, and $\tau(69) = (2^6, 5)$. In an experiment a point is selected at random from $[0, 1]$. For $n \geq 1$, $\tau(n) = (2^k, i)$, let

$$X_n = \begin{cases} 1 & \text{if the outcome is in } \left[\dfrac{i}{2^k}, \dfrac{i+1}{2^k} \right] \\ 0 & \text{otherwise.} \end{cases}$$

Show that X_n converges to 0 in probability while it does not converge at any point, let alone almost sure convergence. ∎

11.5   CENTRAL LIMIT THEOREM

Let X be a binomial random variable with parameters n and p. In Section 7.2 we discussed the De Moivre-Laplace theorem, which states that, for real numbers a and b, $a < b$,

$$\lim_{n \to \infty} P\left(a < \frac{X - np}{\sqrt{np(1-p)}} < b \right) = \frac{1}{\sqrt{2\pi}} \int_a^b e^{-t^2/2} \, dt.$$

We mentioned that, for values of n and p for which $np(1-p) \geq 10$, this theorem can be used to approximate binomial by standard normal. Since this approximation is very useful, it is highly desirable to obtain similar ones for distributions other than binomial. However, this form of the De Moivre-Laplace theorem does not readily suggest any generalizations. Therefore, we restate it in an alternative way that can be generalized. To do so, note that if X is a binomial random variable with parameters n and p, then X is the sum of n independent Bernoulli random variables, each with parameter p. That is, $X = X_1 + X_2 + \cdots + X_n$, where

$$X_i = \begin{cases} 1 & \text{if the } i\text{th trial is a success} \\ 0 & \text{if the } i\text{th trial is a failure.} \end{cases}$$

With this representation, the De Moivre-Laplace theorem is restated as follows:

Let $X_1, X_2, X_3, \ldots$ be a sequence of independent Bernoulli random variables each with parameter p. Then $\forall i$, $E(X_i) = p$, $\text{Var}(X_i) = p(1-p)$, and

$$\lim_{n \to \infty} P\left(\frac{X_1 + X_2 + \cdots + X_n - np}{\sqrt{np(1-p)}} \leq x \right) = \frac{1}{\sqrt{2\pi}} \int_{-\infty}^x e^{-y^2/2} \, dy.$$

It is this version of the De Moivre-Laplace theorem that motivates elegant general-izations. The first was given in 1887 by Chebyshev. But since its proof was not rigorous enough, his student Markov worked on the proof and made it both rigorous and simpler. In 1901, Lyapunov, another student of Chebyshev, weakened the conditions of Cheby-shev and proved the following generalization of the De Moivre-Laplace theorem, which is now called *the central limit theorem*, arguably the most celebrated theorem in both probability and statistics.

Theorem 11.12   (Central Limit Theorem)   *Let X_1, X_2, ... be a sequence of independent and identically distributed random variables, each with expectation μ and variance σ^2. Then the distribution of*

$$Z_n = \frac{X_1 + X_2 + \cdots + X_n - n\mu}{\sigma\sqrt{n}}$$

converges to the distribution of a standard normal random variable. That is,

$$\lim_{n\to\infty} P(Z_n \le x) = \lim_{n\to\infty} P\left(\frac{X_1 + X_2 + \cdots + X_n - n\mu}{\sigma\sqrt{n}} \le x\right)$$

$$= \frac{1}{\sqrt{2\pi}} \int_{-\infty}^{x} e^{-y^2/2}\, dy.$$

Remark 11.2   Note that Z_n is simply $X_1 + X_2 + \cdots + X_n$ standardized.   ◆

Remark 11.3   In Theorem 11.12, let $\bar{X}$ be the mean of the random variables X_1, X_2, ..., X_n. The central limit theorem is equivalent to

$$\lim_{n\to\infty} P\left(\frac{\bar{X} - \mu}{\sigma/\sqrt{n}} \le x\right) = \frac{1}{\sqrt{2\pi}} \int_{-\infty}^{x} e^{-y^2/2}\, dy.$$

It is this version of the central limit theorem that has enormous applications in statistics.
   ◆

The central limit theorem shows that many natural phenomena obey approximately a standard normal distribution. In natural processes, the changes that take place are often the result of a sum of many insignificant random factors. Although the effect of each of these factors, separately, may be ignored, the effect of their sum in general may not. Therefore, the study of distributions of sums of a large number of independent random variables is essential. In this connection, the central limit theorem shows that in an enormous number of random phenomena, the distribution functions of such sums are approximately normal. Some examples are the weight of a man; the height of a woman; the error made in a measurement; the position and velocity of a molecule of a gas; the quantity of diffused material; the growth distribution of plants, animals, or their organs;

and so on. The weight of a man, for example, is the result of many environmental and genetic factors, more or less unrelated but each contributing a small amount to his weight.

The proof to the central limit theorem given here is a standard one based on the following result of Lévy, which we state without proof.

Theorem 11.13　(Lévy Continuity Theorem)　*Let* X_1, X_2, ... *be a sequence of random variables with distribution functions* F_1, F_2, ... *and moment-generating functions* $M_{X_1}(t)$, $M_{X_2}(t)$, ..., *respectively. Let* X *be a random variable with distribution function* F *and moment-generating function* $M_X(t)$. *If for all values of* t, $M_{X_n}(t)$ *converges to* $M_X(t)$, *then at the points of continuity of* F, F_n *converges to* F.

Proof of Theorem 11.12　*(Central Limit Theorem):*　Let $Y_n = X_n - \mu$; then $E(Y_n) = 0$ and $\text{Var}(Y_n) = \sigma^2$. We want to prove that the sequence of distribution functions of the random variables $Z_n = (Y_1 + Y_2 + \cdots + Y_n)/(\sigma\sqrt{n})$, $n \geq 1$, converges to the distribution of Z. Since Y_1, Y_2, ... are identically distributed, they have the same moment-generating function M. By independence of Y_1, Y_2, ... , Y_n,

$$M_{Z_n}(t) = E\left[\exp\left(\frac{Y_1 + Y_2 + \cdots + Y_n}{\sigma\sqrt{n}}t\right)\right] = M_{Y_1 + Y_2 + \cdots + Y_n}\left(\frac{t}{\sigma\sqrt{n}}\right)$$

$$= M_{Y_1}\left(\frac{t}{\sigma\sqrt{n}}\right)M_{Y_2}\left(\frac{t}{\sigma\sqrt{n}}\right)\cdots M_{Y_n}\left(\frac{t}{\sigma\sqrt{n}}\right)$$

$$= \left[M\left(\frac{t}{\sigma\sqrt{n}}\right)\right]^n. \tag{11.8}$$

By the Lévy continuity theorem, it suffices to prove that $M_{Z_n}(t)$ converges to $\exp(t^2/2)$, the moment-generating function of Z. Equivalently, it is enough to show that

$$\lim_{n\to\infty} \ln M_{Z_n}(t) = \frac{t^2}{2}. \tag{11.9}$$

Let $h = t/(\sigma\sqrt{n})$; then $n = t^2/(\sigma^2 h^2)$. Thus (11.8) gives

$$\ln M_{Z_n}(t) = n\ln M(h) = \frac{t^2}{\sigma^2 h^2}\ln M(h) = \frac{t^2}{\sigma^2}\left[\frac{\ln M(h)}{h^2}\right].$$

Therefore,

$$\lim_{n\to\infty} \ln M_{Z_n}(t) = \frac{t^2}{\sigma^2}\lim_{h\to 0}\left[\frac{\ln M(h)}{h^2}\right]. \tag{11.10}$$

Since $M(0) = 1$, $\lim_{h\to 0}\left[\ln M(h)/h^2\right]$ is indeterminate. To determine its value, we apply l'Hôpital's rule twice. We get

$$\lim_{h\to 0}\left[\frac{\ln M(h)}{h^2}\right] = \lim_{h\to 0}\frac{M'(h)/M(h)}{2h} = \lim_{h\to 0}\frac{M'(h)}{2hM(h)}$$

$$= \lim_{h\to 0}\frac{M''(h)}{2M(h) + 2hM'(h)} = \frac{M''(0)}{2M(0)} = \frac{\sigma^2}{2},$$

where $M''(0) = \sigma^2$ since $E(Y_i) = 0$. Thus, by (11.10),

$$\lim_{n \to \infty} \ln M_{Z_n}(t) = \frac{t^2}{\sigma^2}\frac{\sigma^2}{2} = \frac{t^2}{2}.$$

This establishes (11.9) and proves the theorem.   ◆

We now show some of the many useful applications of the central limit theorem.

Example 11.27   The lifetime of a TV tube (in years) is an exponential random variable with mean 10. What is the probability that the average lifetime of a random sample of 36 TV tubes is at least 10.5?

Solution: The parameter of the exponential density function of the lifetime of a tube is $\lambda = 1/10$. For $1 \le i \le 36$, let X_i be the lifetime of the ith TV tube in the sample. Clearly, for $1 \le i \le 36$, $E(X_i) = 1/\lambda = 10$ and $\sigma_{X_i} = 1/\lambda = 10$. By the version of the central limit theorem expressed in Remark 11.3, the desired probability is approximately

$$P(\bar{X} \ge 10.5) = P\left(\frac{\bar{X} - 10}{10/\sqrt{36}} \ge \frac{10.5 - 10}{10/\sqrt{36}}\right) = P\left(\frac{\bar{X} - 10}{10/\sqrt{36}} \ge 0.30\right)$$

$$\approx 1 - \Phi(0.30) = 1 - 0.6179 = 0.3821.   ◆$$

Example 11.28   The time it takes for a student to finish an aptitude test (in hours) has the probability density function

$$f(x) = \begin{cases} 6(x - 1)(2 - x) & \text{if } 1 < x < 2 \\ 0 & \text{otherwise.} \end{cases}$$

Approximate the probability that the average length of time it takes for a random sample of 15 students to complete the test is less than 1 hour and 25 minutes.

Solution: For $1 \le i \le 15$, let X_i be the time that it will take for the ith student in the sample to complete the test. Let $\bar{X} = (1/15)\sum_{i=1}^{15} X_i$. Direct calculations show that, for $1 \le i \le 15$,

$$E(X_i) = \int_1^2 6x(x - 1)(2 - x)\,dx = 3/2 = 1.5,$$

$$E(X_i^2) = \int_1^2 6x^2(x - 1)(2 - x)\,dx = 23/10 = 2.3,$$

$$\sigma_{X_i} = \sqrt{(23/10) - (3/2)^2} = 1/\sqrt{20} = 0.2236.$$

Using the central limit theorem, the desired probability is approximately equal to

$$P(\bar{X} < 17/12) = P\left(\frac{\bar{X} - 1.5}{0.2236/\sqrt{15}} < \frac{(17/12) - 1.5}{0.2236/\sqrt{15}}\right)$$

$$= P\left(\frac{\bar{X} - 1.5}{0.2236/\sqrt{15}} < -1.44\right) \approx \Phi(-1.44) = 0.0749. \quad \blacklozenge$$

Example 11.29   If 20 random numbers are selected independently from the interval $(0, 1)$, what is the approximate probability that the sum of these numbers is at least eight?

Solution: Let X_i be the ith number selected, $i = 1, 2, \ldots, 20$. To calculate $P\left(\sum_{i=1}^{20} X_i \geq 8\right)$, we use the central limit theorem. Since $\forall i$, $E(X_i) = (0+1)/2 = 1/2$ and $\text{Var}(X) = (1-0)^2/12 = 1/12$,

$$P\left(\sum_{i=1}^{20} X_i \geq 8\right) = P\left(\frac{\sum_{i=1}^{20} X_i - 20(1/2)}{\sqrt{1/12}\,\sqrt{20}} \geq \frac{8 - 20(1/2)}{\sqrt{1/12}\,\sqrt{20}}\right)$$

$$= P\left(\frac{\sum_{i=1}^{20} X_i - 10}{\sqrt{5/3}} \geq -1.55\right) \approx 1 - \Phi(-1.55) \approx 0.9394. \quad \blacklozenge$$

Example 11.30   A biologist wants to estimate ℓ, the life expectancy of a certain type of insect. To do so, he takes a sample of size n and measures the lifetime from birth to death of each insect. Then he finds the average of these numbers. If he believes that the lifetimes of these insects are independent random variables with variance 1.5 days, how large a sample should he choose to be 98% sure that his average is accurate within ± 0.2 (± 4.8 hours)?

Solution: For $i = 1, 2, \ldots, n$, let X_i be the lifetime of the ith insect of the sample. We want to determine n, so that

$$P\left(-0.2 < \frac{X_1 + X_2 + \cdots + X_n}{n} - \ell < 0.2\right) \approx 0.98.$$

Since $E(X_i) = \ell$ and $\text{Var}(X_i) = 1.5$, by the central limit theorem,

$$P\left(-0.2 < \frac{\sum_{i=1}^{n} X_i}{n} - \ell < 0.2\right) = P\left((-0.2)n < \sum_{i=1}^{n} X_i - n\ell < (0.2)n\right)$$

$$= P\left(\frac{-(0.2)n}{\sqrt{1.5n}} < \frac{\sum_{i=1}^{n} X_i - n\ell}{\sqrt{1.5n}} < \frac{(0.2)n}{\sqrt{1.5n}}\right)$$

$$\approx \Phi\left[\frac{(0.2)n}{\sqrt{1.5n}}\right] - \Phi\left[\frac{(-0.2)n}{\sqrt{1.5n}}\right] = 2\Phi\left(\frac{0.2\sqrt{n}}{\sqrt{1.5}}\right) - 1.$$

Thus the quantity $2\Phi(0.2\sqrt{n}/\sqrt{1.5}) - 1$ should approximately equal 0.98; that is, $\Phi(0.2\sqrt{n}/\sqrt{1.5}) \approx 0.99$. From Table 1 of the Appendix, we find that $0.2\sqrt{n}/\sqrt{1.5} = 2.33$. This gives $n = 203.58$. Therefore, the biologist should choose a sample of size 204. ◆

Remark 11.4:   In the solution of Example 11.30, we have assumed implicitly that, for a sample of size 204, normal distribution is a good approximation. Although usually such assumptions are reasonable, the size of n depends on the distribution function of X_i's, which is not known in most cases. If the biologist wants to be absolutely sure about the results of an analysis, he or she should use Chebyshev's inequality instead. The results obtained by applying Chebyshev's inequality are safe but, in most cases, much larger than needed. For this example, the answer obtained by applying Chebyshev's inequality is 1875, a number much larger than the 204 obtained by using the central limit theorem. (See Example 11.18.)   ◆

Example 11.31   Suppose that, with equal probabilities, the value of a specific stock on any trading day increases 30% or decreases 25%, independent of the fluctuations of the stock value on the past and future trading days. Let

$$r_i = \begin{cases} 0.30 & \text{with probability } 1/2 \\ -0.25 & \text{with probability } 1/2. \end{cases}$$

Then r_i is the rate of return on the ith trading day. Schiller, a very conservative investor, invests A dollars in this stock thinking that

$$E(r_i) = 0.30 \cdot \frac{1}{2} + (-0.25) \cdot \frac{1}{2} = 0.025$$

implies that, on average, every day the value of his investment increases by 2.5% of its value on the previous day. The fact that, by the end of the first trading day, the value of his stock shares is $A + Ar_1 = A(1 + r_1)$, and

$$E\big[A(1 + r_1)\big] = A\big[1 + E(r_1)\big] = A(1.025) = 1.025A,$$

makes this investment attractive to Schiller. In this example, we show that Schiller is mistaken, and if he holds his shares of this stock long enough, eventually he will lose the entire value of his investment in this stock. To do so, note that at the end of the second trading day, the value of Schiller's stock shares is

$$A(1 + r_1) + A(1 + r_1)r_2 = A(1 + r_1)(1 + r_2);$$

at the end of the third trading day, its value is

$$A(1 + r_1)(1 + r_2) + A(1 + r_1)(1 + r_2)r_3 = A(1 + r_1)(1 + r_2)(1 + r_3);$$

and, in general, at the end of the nth trading day its value is

$$A(1 + r_1)(1 + r_2) \cdots (1 + r_n).$$

Let $Y_i = 1 + r_i$; then

$$Y_i = \begin{cases} 1.30 & \text{with probability } 1/2 \\ 0.75 & \text{with probability } 1/2. \end{cases}$$

The sequence $\{Y_1, Y_2, \dots\}$ is an independent sequence of identically distributed random variables. Using the central limit theorem, we will now find n, the number of trading days after which, with probability 0.99, the value of the stock decreases to 10% of its original value. That is, we will find n so that

$$0.99 \le P\big(A(1 + r_1)(1 + r_2) \cdots (1 + r_n) \le (0.10)A\big)$$
$$= P(Y_1 Y_2 \cdots Y_n \le 0.10) = P(\ln Y_1 + \ln Y_2 + \cdots + \ln Y_n \le \ln 0.10).$$

Note that $\ln 0.10 = -2.303$, and $\{\ln Y_1, \ln Y_2, \dots\}$ is a sequence of identically distributed and independent random variables with

$$E(\ln Y_i) = \ln(1.30) \cdot \frac{1}{2} + \ln(0.75) \cdot \frac{1}{2} = -0.127,$$

$$E\big[(\ln Y_i)^2\big] = \big[\ln(1.30)\big]^2 \cdot \frac{1}{2} + \big[\ln(0.75)\big]^2 \cdot \frac{1}{2} = 0.0758,$$

$$\text{Var}(\ln Y_i) = E\big[(\ln Y_i)^2\big] - \big[E(\ln Y_i)\big]^2 = (0.0758) - (-0.127)^2 = 0.0597$$

$$\sigma_{\ln Y_i} = \sqrt{0.0597} = 0.244.$$

Since by the central limit theorem,

$$P(\ln Y_1 + \ln Y_2 + \cdots + \ln Y_n \le -2.303)$$

$$= P\left(\frac{\ln Y_1 + \ln Y_2 + \cdots + \ln Y_n - n(-0.127)}{0.244\sqrt{n}} \le \frac{-2.303 - n(-0.127)}{0.244\sqrt{n}}\right)$$

$$\approx \Phi\left(\frac{-2.303 + 0.127n}{0.244\sqrt{n}}\right),$$

for this probability to be at least 0.99, by Table 2 of the Appendix, we must have

$$\frac{-2.303 + 0.127n}{0.244\sqrt{n}} = 2.33,$$

or, equivalently,

$$0.127n - 0.569\sqrt{n} - 2.303 = 0.$$

This is a quadratic equation in $\sqrt{n}$. Solving it for $\sqrt{n}$ yields $\sqrt{n} = 7.052$, or $n = 49.73$. This shows that, with probability 0.99, after 50 trading days, the value of the stock reduces to 10% of its original value in spite of the fact that the expected return, on each trading day, is positive. The assertion that, if Schiller holds his shares of the stock long enough, eventually he will lose the entire value of his investment in this stock follows from the preceding observation. ◆

Example 11.32   A (nonrigorous) proof of Stirling's formula (Theorem 2.7):

$$n! \sim \sqrt{2\pi n}\, n^n\, e^{-n}.$$

That is,

$$\lim_{n \to \infty} \frac{n!}{\sqrt{2\pi n}\, n^n\, e^{-n}} = 1.$$

To justify this theorem, let $\{X_1, X_2, \ldots\}$ be a sequence of independent and identically distributed Poisson random variables with $\lambda = 1$. Then $\mu = E(X_i) = 1$ and $\sigma^2 = \text{Var}(X_i) = 1$, $i \geq 1$. Let $S_n = \sum_{i=1}^{n} X_i$; then, by Theorem 11.5, S_n is Poisson with parameter n. Therefore,

$$P(S_n = n) = \frac{e^{-n} \cdot n^n}{n!}. \tag{11.11}$$

Since $\mu = \sigma = 1$, by the central limit theorem,

$$P(S_n = n) = P(n - 1 < S_n \leq n) = P\left(\frac{-1}{\sqrt{n}} < \frac{S_n - n}{\sqrt{n}} \leq 0\right)$$

$$= P\left(\frac{-1}{\sqrt{n}} < \frac{X_1 + X_2 + \cdots + X_n - n\mu}{\sigma\sqrt{n}} \leq 0\right) \approx \Phi(0) - \Phi\left(\frac{-1}{\sqrt{n}}\right).$$

By the mean value theorem of calculus, there exists $t_n \in (-1/\sqrt{n}, 0)$ such that

$$\Phi(0) - \Phi\left(\frac{-1}{\sqrt{n}}\right) = \Phi'(t_n)\left[0 - \left(-\frac{1}{\sqrt{n}}\right)\right] = \frac{1}{\sqrt{2\pi}} e^{-t_n^2/2} \cdot \frac{1}{\sqrt{n}} = \frac{1}{\sqrt{2\pi n}} e^{-t_n^2/2}.$$

Clearly, the sequence $\{t_n\}$ converges to 0 as $n \to \infty$. Therefore, for large n,

$$P(S_n = n) \approx \Phi(0) - \Phi\left(\frac{-1}{\sqrt{n}}\right) \approx \frac{1}{\sqrt{2\pi n}}. \tag{11.12}$$

Comparing (11.11) and (11.12), we have

$$\lim_{n \to \infty} \frac{1/\sqrt{2\pi n}}{e^{-n} \cdot n^n/n!} = 1,$$

or, equivalently,

$$\lim_{n \to \infty} \frac{n!}{\sqrt{2\pi n}\, n^n\, e^{-n}} = 1;$$

that is, $n! \sim \sqrt{2\pi n}\, n^n\, e^{-n}$. ◆

EXERCISES

A

1. What is the probability that the average of 150 random points from the interval (0, 1) is within 0.02 of the midpoint of the interval?

2. For the scores on an achievement test given to a certain population of students, the expected value is 500 and the standard deviation is 100. Let $\bar{X}$ be the mean of the scores of a random sample of 35 students from the population. Estimate $P(460 < \bar{X} < 540)$.

3. A random sample of size 24 is taken from a distribution with probability density function

$$f(x) = \begin{cases} \dfrac{1}{9}\left(x + \dfrac{5}{2}\right) & 1 < x < 3 \\ 0 & \text{otherwise.} \end{cases}$$

Let $\bar{X}$ be the sample mean. Approximate $P(2 < \bar{X} < 2.15)$.

4. A random sample of size n $(n \geq 1)$ is taken from a distribution with the following probability density function:

$$f(x) = \frac{1}{2}e^{-|x|}, \qquad -\infty < x < \infty.$$

What is the probability that the sample mean is positive?

5. Let $X_1, X_2, \ldots, X_n$ be independent and identically distributed random variables, and let $S_n = X_1 + X_2 + \cdots + X_n$. For large n, what is the approximate probability that S_n is between $E(S_n) - \sigma_{S_n}$ and $E(S_n) + \sigma_{S_n}$?

6. Each time that Jim charges an item to his credit card, he rounds the amount to the nearest dollar in his records. If he has used his credit card 300 times in the last 12 months, what is the probability that his record differs from the total expenditure by, at most, 10 dollars?

7. A physical quantity is measured 50 times, and the average of these measurements is taken as the result. If each measurement has a random error uniformly distributed over $(-1, 1)$, what is the probability that our result differs from the actual value by less than 0.25?

8. Suppose that, whenever invited to a party, the probability that a person attends with his or her guest is 1/3, attends alone is 1/3, and does not attend is 1/3. A company has invited all 300 of its employees and their guests to a Christmas party. What is the probability that at least 320 will attend?

9. Consider a distribution with mean μ and probability density function

$$f(x) = \begin{cases} \dfrac{1}{x \ln(3/2)} & 4 \le x \le 6 \\ 0 & \text{elsewhere.} \end{cases}$$

Determine the values of n for which the probability is at least 0.98 that the mean of a random sample of size n from the population is within ± 0.07 of μ?

B

10. An investor buys 1000 shares of the XYZ Corporation at $50.00 per share. Subsequently, the stock price varies by $0.125 (1/8) every day, but unfortunately it is just as likely to move down as up. What is the most likely value of his holdings after 60 days?

    Hint: First calculate the distribution of the *change* in the stock price after 60 days.

11. A fair coin is tossed successively. Using the central limit theorem, find an approximation for the probability of obtaining at least 25 heads before 50 tails.

12. Let $\{X_1, X_2, \ldots\}$ be a sequence of independent standard normal random variables. Let $S_n = X_1^2 + X_2^2 + \cdots + X_n^2$. Find

$$\lim_{n \to \infty} P(S_n \le n + \sqrt{2n}).$$

    Hint: See Example 11.11.

13. Let $\{X_1, X_2, \ldots\}$ be a sequence of independent Poisson random variables, each with parameter 1. By applying the central limit theorem to this sequence, prove that

$$\lim_{n \to \infty} \frac{1}{e^n} \sum_{k=0}^{n} \frac{n^k}{k!} = \frac{1}{2}.$$

■

REVIEW PROBLEMS

---

1. Yearly salaries paid to the salespeople employed by a certain company are normally distributed with mean $27,000 and standard deviation $4900. What is the probability that the average wage of a random sample of 10 employees of this company is at least $30,000?

2. The moment-generating function of a random variable X is given by

$$M_X(t) = \left(\frac{1}{3} + \frac{2}{3}e^t\right)^{10}.$$

Find $\text{Var}(X)$ and $P(X \geq 8)$.

3. The moment-generating function of a random variable X is given by

$$M_X(t) = \frac{1}{6}e^t + \frac{1}{3}e^{2t} + \frac{1}{2}e^{3t}.$$

Find the distribution function of X.

4. For a random variable X, suppose that $M_X(t) = \exp(2t^2 + t)$. Find $E(X)$ and $\text{Var}(X)$.

5. Let the moment-generating function of a random variable X be given by

$$M_X(t) = \begin{cases} \dfrac{1}{t}(e^{t/2} - e^{-t/2}) & \text{if } t \neq 0 \\ 1 & \text{if } t = 0. \end{cases}$$

Find the distribution function of X.

6. The moment-generating function of X is given by

$$M_X(t) = \exp\left(\frac{e^t - 1}{2}\right).$$

Find $P(X > 0)$.

7. The moment-generating function of a random variable X is given by

$$M_X(t) = \frac{1}{(1-t)^2}, \qquad t < 1.$$

Find the moments of X.

8. Suppose that in a community the distributions of heights of men and women (in centimeters) are $N(173, 40)$ and $N(160, 20)$, respectively. Calculate the probability that the average height of 10 randomly selected men is at least 5 centimeters larger than the average height of six randomly selected women.

9. Find the moment-generating function of a random variable X with **Laplace density function** defined by

$$f(x) = \frac{1}{2}e^{-|x|}, \qquad -\infty < x < \infty.$$

10. Let X and Y be independent Poisson random variables with parameters λ and μ, respectively.

   (a) Show that
   $$P(X + Y = n) = \sum_{i=0}^{n} P(X = i)P(Y = n - i).$$

   (b) Use part (a) to prove that $X + Y$ is a Poisson random variable with parameter $\lambda + \mu$.

11. Let $\bar{X}$ denote the mean of a random sample of size 28 from a distribution with $\mu = 1$ and $\sigma^2 = 4$. Approximate $P(0.95 < \bar{X} < 1.05)$.

12. In a clinical trial, the probability of success for a treatment is to be estimated. If the error of estimation is not allowed to exceed 0.01 with probability 0.94, how many patients should be chosen independently and at random for the treatment group?

13. For a coin, p, the probability of heads is unknown. To estimate p, we flip the coin 5000 times and let $\hat{p}$ be the fraction of times it lands heads up. Show that the probability is at least 0.98 that $\hat{p}$ estimates p within ± 0.05.

14. A psychologist wants to estimate μ, the mean IQ of the students of a university. To do so, she takes a sample of size n of the students and measures their IQ's. Then she finds the average of these numbers. If she believes that the IQ's of these students are independent random variables with variance 170, using the central limit theorem, how large a sample should she choose to be 98% sure that her average is accurate within ± 0.2?

15. Each time that Ed charges an expense to his credit card, he omits the cents and records only the dollar value. If this month he has charged his credit card 20 times, using Chebyshev's inequality, find an upper bound on the probability that his record shows at least $15 less than the actual amount charged.

16. In a multiple-choice test with false answers receiving negative scores, the mean of the grades of the students is 0 and its standard deviation is 15. Find an upper bound for the probability that a student's grade is at least 45.

17. A randomly selected book from Vernon's library is X centimeters thick, where $X \sim N(3, 1)$. Vernon has an empty shelf 87 centimeters long. What is the probability that he can fit 31 randomly selected books on it?

18. A fair die is rolled 20 times. What is the approximate probability that the sum of the outcomes is between 65 and 75?

19. Show that for a nonnegative random variable X with mean μ, we have that $\forall n$, $nP(X \geq n\mu) \leq 1$.

20. An ordinary deck of 52 cards is divided randomly into 26 pairs. Using Chebyshev's inequality, find an upper bound for the probability that, at most, 10 pairs consist of a black and a red card.

Hint: For $i = 1, 2, \ldots, 26$, let $X_i = 1$ if the ith red card is paired with a black card and $X_i = 0$ otherwise. Find an upper bound for

$$P\left(\sum_{i=1}^{26} X_i \leq 10\right).$$

■

Chapter 12

Stochastic Processes

12.1 INTRODUCTION

To study those real-world phenomena in which systems evolve randomly, we need probabilistic models rather than deterministic ones. Such systems are usually studied as a function of time, and their mathematical models are called *stochastic models*. The building blocks of stochastic models are **stochastic processes**, defined as sets of random variables $\{X_n : n \in I\}$ for a finite or countable index set I, or $\{X(t) : t \in T\}$ for an uncountable index set T. For example, let X_n be the number of customers served in a bank at the end of the nth working day. Then $\{X_n : n = 1, 2, \dots\}$ is a stochastic process. It is called a **discrete-time** stochastic process since its index set, $I = \{1, 2, \dots\}$, is countable. As another example, let $X(t)$ be the sum of the remaining service times of all customers being served in a bank at time t. Then $\{X(t) : t \geq 0\}$ is a stochastic process, and since its index set, $T = [0, \infty)$, is uncountable, it is called a **continuous-time** stochastic process. The set of all possible values of X_n's in the discrete-time case and $X(t)$'s in the continuous-time case is called the **state space** of the stochastic process, and it is usually denoted by S. The state space for the number of customers served in a bank at the end of the nth working day is $S = \{0, 1, 2, \dots\}$. The state space for the sum of the remaining service times of all customers being served in a bank at time t is $S = [0, \infty)$. Other examples of stochastic processes follow.

Example 12.1 For $n \geq 1$, let $X_n = 1$ if the nth fish caught in a lake by a fisherman is trout, and let $X_n = 0$ otherwise. Then $\{X_n : n = 1, 2, \dots\}$ is a discrete-time stochastic process with state space $\{0, 1\}$. ◆

Example 12.2 Suppose that there are three machines in a factory, each working for a random time that is exponentially distributed. When a machine fails, the repair time is also a random variable exponentially distributed. Let $X(t)$ be the number of functioning machines in the factory at time t. Then $\{X(t) : t \geq 0\}$ is a continuous-time stochastic process with state space $\{0, 1, 2, 3\}$. ◆

Example 12.3 The electrical grid supplies electric energy to m identical electric ovens whose heating elements cycle on and off as needed to maintain each oven's temperature at its desired level. Each oven's electrical needs are independent of the other ovens. Let $X(t)$ be the number of ovens with heating elements on at time t. Then $\{X(t): t \geq 0\}$ is a stochastic process with state space $S = \{0, 1, \dots, m\}$. ♦

Example 12.4   A cosmic particle entering the earth's atmosphere collides with air particles randomly and transfers kinetic energy to them. These in turn collide with other particles randomly transferring energy to them and so on. A shower of particles results. Let $X(t)$ be the number of particles in the shower t units of time after the cosmic particle enters the earth's atmosphere. Then $\{X(t): t \geq 0\}$ is a stochastic process with state space $S = \{1, 2, 3, \dots\}$. ♦

The theory of stochastic processes is a broad area, and it has applications in a wide spectrum of fields. In this chapter, we study some basic properties of four areas of stochastic processes: Poisson processes, discrete- and continuous-time Markov chains, and Brownian motions.

12.2   MORE ON POISSON PROCESSES

In Section 5.2, we showed how the Poisson distribution appears in connection with the study of sequences of random events occurring over time. Examples of such sequences of random events are accidents occurring at an intersection, β-particles emitted from a radioactive substance, customers entering a post office, and California earthquakes. Suppose that, starting from a time point labeled $t = 0$, we begin counting the number of events. Then for each value of t, we obtain a number denoted by $N(t)$, which is the number of events that have occurred during $[0, t]$. The set of random variables $\{N(t): t \geq 0\}$ is called a **counting process.** This is because $N(t)$ counts the number of events that have occurred at or prior to t. Theorem 5.2 indicates that if $N(0) = 0$, for all $t > 0$,

$$0 < P\big(N(t) = 0\big) < 1,$$

and $\{N(t): t \geq 0\}$ is stationary, possesses independent increments, and is orderly, then $N(t)$ is a Poisson random variable with parameter λt, where $\lambda = E\big[N(1)\big]$. Poisson processes can be generalized in numerous ways. For example, in a branch of counting processes called *renewal theory*, the time between two consecutive events, unlike Poisson processes, is not necessarily exponential. As another example, in *pure birth processes,* to be studied later, unlike Poisson processes, the probability of an event (a birth) at time t, depends on the size of $N(t)$, the population size at t. In this section, first, we will give a rigorous proof of Theorem 5.2 by studying counting processes further. Then we will present other important results for a better understanding of Poisson processes.

Definition: *A function* $f: \mathbf{R} \to \mathbf{R}$ *is called* $o(h)$ *if* $\displaystyle\lim_{h \to 0} \frac{f(h)}{h} = 0.$

If a function f is $o(h)$, we write $f(h) = o(h)$, and read "f is little o of h." If f is $o(h)$, then f approaches 0 "faster" than h. Clearly, the function $f(x) = x^r$ is $o(h)$ if $r > 1$. This is because

$$\lim_{h \to 0} \frac{f(h)}{h} = \lim_{h \to 0} \frac{h^r}{h} = \lim_{h \to 0} h^{r-1} = 0.$$

If $r < 1$, then $\lim_{h \to 0} f(h)/h$ does not exist. For $r = 1$, the limit is 1. Therefore, $f(x)$ is not $o(h)$ if $r \leq 1$.

Example 12.5 Show that if f and g are both $o(h)$ and c is a constant, then $f + g$ and cf are also $o(h)$.

Proof: Clearly,

$$\lim_{h \to 0} \frac{(f + g)(h)}{h} = \lim_{h \to 0} \frac{f(h)}{h} + \lim_{h \to 0} \frac{g(h)}{h} = 0 + 0 = 0.$$

Thus $f + g$ is $o(h)$. Also,

$$\lim_{h \to 0} \frac{(cf)(h)}{h} = \lim_{h \to 0} \frac{cf(h)}{h} = c \cdot 0 = 0,$$

showing that cf is $o(h)$ as well. ◆

Example 12.6 Show that if f is $o(h)$ and g is a bounded function, then fg is also $o(h)$.

Proof: We have that

$$\lim_{h \to 0} \frac{(fg)(h)}{h} = \lim_{h \to 0} \frac{f(h)g(h)}{h} = \lim_{h \to 0} \frac{f(h)}{h} g(h) = 0.$$

Thus $g(h) \cdot o(h) = o(h)$ if g is bounded. ◆

Example 12.7 Prove that $e^t = 1 + t + o(t)$.

Proof: We know that the Maclaurin series of e^t converges to e^t. Thus

$$e^t = 1 + t + \frac{t^2}{2!} + \frac{t^3}{3!} + \cdots + \frac{t^n}{n!} + \cdots .$$

Let

$$g(t) = \frac{t^2}{2!} + \frac{t^3}{3!} + \cdots + \frac{t^n}{n!} + \cdots ;$$

then $e^t = 1 + t + g(t)$. But $\lim_{t \to 0} g(t)/t = 0$. Hence $g(t) = o(t)$, and therefore, $e^t = 1 + t + o(t)$. ◆

As in Section 5.2, a counting process $\{N(t): t \geq 0\}$ is called **orderly** if

$$\lim_{h \to 0} \frac{P\big(N(h) > 1\big)}{h} = 0.$$

Hence $\{N(t): t \geq 0\}$ is orderly if $P\big(N(h) > 1\big) = o(h)$. This implies that as $h \to 0$, the probability of two or more events in the time interval $[0, h]$, $P\big(N(h) > 1\big)$, approaches 0 faster than h does. That is, if h is negligible, then $P\big(N(h) > 1\big)$ is even more negligible. We will now prove the following important fact:

> For a stationary counting process, $\{N(t): t \geq 0\}$, $P\big(N(h) > 1\big) = o(h)$ means that simultaneous occurrence of two or more events is impossible.

Proof: For an infinitesimal h, let A be the event that two or more events occur simultaneously in the time interval $[0, h]$. We will prove that $P(A) = 0$. Let n be an arbitrary positive integer and for $j = 0, 1, \ldots, n - 1$, let A_j be the event that the two or more events that occur in $[0, h]$ are in the subinterval $\left(\dfrac{jh}{n}, \dfrac{(j+1)h}{n}\right]$. Then $A \subseteq \bigcup_{j=0}^{n-1} A_j$. Hence

$$P(A) \leq P\left(\bigcup_{j=0}^{n-1} A_j\right) = \sum_{j=0}^{n-1} P(A_j).$$

Since the length of the time interval $\left(\dfrac{jh}{n}, \dfrac{(j+1)h}{n}\right]$ is h/n, the stationarity of the counting process implies that

$$P(A_j) = P\big(N(h/n) > 1\big), \qquad j = 0, 1, \ldots, n - 1.$$

Thus

$$P(A) \leq \sum_{j=0}^{n-1} P(A_j) = n P\big(N(h/n) > 1\big) = h \cdot \frac{P\big(N(h/n) > 1\big)}{h/n}. \tag{12.1}$$

Now (12.1) is valid for all values of n. Since $P\big(N(h/n) > 1\big) = o(h/n)$,

$$\lim_{n \to \infty} \frac{P\big(N(h/n) > 1\big)}{h/n} = 0.$$

Hence (12.1) implies that $P(A) \leq 0$. But $P(A) < 0$ is impossible. Therefore, $P(A) = 0$. ◆

For a Poisson process with parameter λ, let X be the time until the first event occurs. Then

$$P(X > t) = P\big(N(t) = 0\big) = \frac{e^{-\lambda t} \cdot (\lambda t)^0}{0!} = e^{-\lambda t}.$$

That is, X is an exponential random variable with parameter λ. The following lemma shows that this result is true even if $\{N(t): t \geq 0\}$ is not an orderly process.

Lemma 12.1   *Let $\{N(t): t \geq 0\}$ be a stationary counting process that possesses independent increments. Suppose that $N(0) = 0$ and, for all $t > 0$,*

$$0 < P\big(N(t) = 0\big) < 1.$$

Then, for any $t \geq 0$,

$$P\big(N(t) = 0\big) = e^{-\lambda t}$$

for some $\lambda > 0$.

Proof: Let $p = P\big(N(1) = 0\big)$. We are given that $0 < p < 1$. To begin, we show that for any nonnegative rational number x, $P\big(N(x) = 0\big) = p^x$. If $x = 0$, then $P\big(N(0) = 0\big) = p^0 = 1$ is consistent with what we are given. If $x \neq 0$, then $x = k/n$, where k and n are positive integers. Since $[0, 1]$ is the union of the nonoverlapping intervals $[0, 1/n]$, $(1/n, 2/n], \ldots, \big((n-1)/n, 1\big]$, by the independent-increments property and stationarity of $\{N(t): t \geq 0\}$:

$$p = P\big(N(1) = 0\big)$$
$$= P\big(N(1/n) = 0, \ N(2/n) - N(1/n) = 0, \ldots, N(1) - N((n-1)/n) = 0\big)$$
$$= \big[P\big(N(1/n) = 0\big)\big]^n.$$

Thus

$$P\big(N(1/n) = 0\big) = p^{1/n}.$$

Therefore,

$$P\big(N(k/n) = 0\big)$$
$$= P\big(N(1/n) = 0, \ N(2/n) - N(1/n) = 0, \ldots, N(k/n) - N((k-1)/n) = 0\big)$$
$$= \big[P\big(N(1/n) = 0\big)\big]^k = (p^{1/n})^k = p^{k/n}.$$

Hence, for a nonnegative rational number x,

$$P\big(N(x) = 0\big) = p^x. \tag{12.2}$$

Now we show that, for any positive irrational number t, also $P\big(N(t) = 0\big) = p^t$. From calculus, we know that any subinterval of real numbers contains infinitely many rational points. For $n = 1, 2, \ldots$, we choose a rational number x_n in $[t, t + (1/n)]$. The inequalities $t \leq x_n \leq t + \dfrac{1}{n}$ imply that

$$t \leq \lim_{n \to \infty} x_n \leq \lim_{n \to \infty} \left(t + \frac{1}{n}\right),$$

which yields $\lim_{n \to \infty} x_n = t$. Also, since $x_n - \dfrac{1}{n} \le t \le x_n$, the occurrence of the event $\{N(x_n) = 0\}$ implies the occurrence of $\{N(t) = 0\}$, and the occurrence of $\{N(t) = 0\}$ implies the occurrence of $\left\{N\left(x_n - \dfrac{1}{n}\right) = 0\right\}$. Hence

$$P\big(N(x_n) = 0\big) \le P\big(N(t) = 0\big) \le P\big(N(x_n - (1/n)) = 0\big).$$

Since x_n and $x_n - \dfrac{1}{n}$ are rational numbers, (12.2) and these inequalities imply that

$$p^{x_n} \le P\big(N(t) = 0\big) \le p^{x_n - \frac{1}{n}}.$$

Letting $n \to \infty$ and noting that $\lim_{n \to \infty} x_n = t$, this gives

$$p^t \le P\big(N(t) = 0\big) \le p^t.$$

Therefore,

$$P\big(N(t) = 0\big) = p^t. \tag{12.3}$$

Comparing this with (12.2), we see that (12.3) is valid for all rational and irrational t. Now because p satisfies $0 < p < 1$, if we let $\lambda = -\ln(p)$, then $\lambda > 0$, $p = e^{-\lambda}$, and (12.3) implies that

$$P\big(N(t) = 0\big) = (e^{-\lambda})^t = e^{-\lambda t}. \quad \blacklozenge$$

We are now ready to give a rigorous proof for Theorem 5.2, stated in this chapter as Theorem 12.1:

Theorem 12.1   *Let $\{N(t): t \ge 0\}$ be a Poisson process—that is, a counting process with $N(0) = 0$ for which, for all $t > 0$,*

$$0 < P\big(N(t) = 0\big) < 1$$

and is stationary, orderly, and possesses independent increments. Then there exists $\lambda > 0$ such that, for every $t > 0$, $N(t)$ is a Poisson random variable with parameter λt; that is,

$$P\big(N(t) = n\big) = \frac{e^{-\lambda t} \cdot (\lambda t)^n}{n!}.$$

Proof: By Lemma 12.1, there exists $\lambda > 0$ such that $P\big(N(t) = 0\big) = e^{-\lambda t}$. From the Maclaurin series of $e^{-\lambda t}$,

$$P\big(N(t) = 0\big) = e^{-\lambda t} = 1 - \lambda t + \frac{(\lambda t)^2}{2!} - \frac{(\lambda t)^3}{3!} + \cdots.$$

So

$$P(N(t) = 0) = 1 - \lambda t + o(t). \qquad (12.4)$$

(See Example 12.7.) Now since

$$P(N(t) = 0) + P(N(t) = 1) + P(N(t) > 1) = 1,$$

and $\{N(t): t \geq 1\}$ is an orderly process,

$$
\begin{aligned}
P(N(t) = 1) &= 1 - P(N(t) = 0) - P(N(t) > 1) \\
&= 1 - [1 - \lambda t + o(t)] - o(t) = \lambda t + o(t). \qquad (12.5)
\end{aligned}
$$

(See Example 12.5.) For $n = 0, 1, \ldots$, let $P_n(t) = P(N(t) = n)$. Then, for $n \geq 2$,

$$
\begin{aligned}
P_n(t + h) &= P(N(t + h) = n) \\
&= P(N(t) = n, \ N(t + h) - N(t) = 0) \\
&\quad + P(N(t) = n - 1, \ N(t + h) - N(t) = 1) \\
&\quad + \sum_{k=2}^{n} P(N(t) = n - k, N(t + h) - N(t) = k). \qquad (12.6)
\end{aligned}
$$

By the stationarity and independent-increments property of $\{N(t): t \geq 0\}$, it is obvious that

$$P(N(t) = n, \ N(t + h) - N(t) = 0) = P_n(t) P_0(h),$$

and

$$P(N(t) = n - 1, \ N(t + h) - N(t) = 1) = P_{n-1}(t) P_1(h).$$

We also have that

$$\sum_{k=2}^{n} P(N(t) = n - k, \ N(t + h) - N(t) = k) \leq$$

$$\sum_{k=2}^{n} P(N(t + h) - N(t) = k) = \sum_{k=2}^{n} P_k(h) = \sum_{k=2}^{n} o(h) = o(h),$$

by orderliness of $\{N(t): t \geq 0\}$. Substituting these into (12.6) yields

$$P_n(t + h) = P_n(t) P_0(h) + P_{n-1}(t) P_1(h) + o(h).$$

Therefore, by (12.4) and (12.5),

$$
\begin{aligned}
P_n(t + h) &= P_n(t)[1 - \lambda h + o(h)] + P_{n-1}(t)[\lambda h + o(h)] + o(h) \\
&= (1 - \lambda h) P_n(t) + \lambda h P_{n-1}(t) + o(h).
\end{aligned}
$$

(See Example 12.6.) This implies that

$$\frac{P_n(t+h) - P_n(t)}{h} = -\lambda P_n(t) + \lambda P_{n-1}(t) + \frac{o(h)}{h}.$$

Letting $h \to 0$, we obtain

$$\frac{d}{dt} P_n(t) = -\lambda P_n(t) + \lambda P_{n-1}(t),$$

which, by similar calculations, can be verified for $n = 1$ as well. Multiplying both sides of this equation by $e^{\lambda t}$, we have

$$e^{\lambda t} \left[\frac{d}{dt} P_n(t) + \lambda P_n(t) \right] = \lambda e^{\lambda t} P_{n-1}(t), \quad n \geq 1,$$

which is the same as

$$\frac{d}{dt} \left[e^{\lambda t} P_n(t) \right] = \lambda e^{\lambda t} P_{n-1}(t). \tag{12.7}$$

Since $P_0(t) = e^{-\lambda t}$, (12.7) gives

$$\frac{d}{dt} \left[e^{\lambda t} P_1(t) \right] = \lambda,$$

or, equivalently, for some constant c,

$$e^{\lambda t} P_1(t) = \lambda t + c,$$

where $P_1(0) = P\big(N(0) = 1\big) = 0$ implies that $c = 0$. Thus

$$P_1(t) = \lambda t e^{-\lambda t}.$$

To complete the proof, we now use induction on n. For $n = 1$, we just showed that $P_1(t) = \lambda t e^{-\lambda t}$. Let

$$P_{n-1}(t) = \frac{(\lambda t)^{n-1} e^{-\lambda t}}{(n-1)!}.$$

Then, by (12.7),

$$\frac{d}{dt} \left[e^{\lambda t} P_n(t) \right] = \lambda e^{\lambda t} \cdot \frac{(\lambda t)^{n-1} e^{-\lambda t}}{(n-1)!} = \frac{\lambda^n t^{n-1}}{(n-1)!},$$

which yields

$$e^{\lambda t} P_n(t) = \frac{(\lambda t)^n}{n!} + c,$$

for some constant c. But $P_n(0) = P\big(N(0) = n\big) = 0$ implies that $c = 0$. Thus

$$P_n(t) = \frac{(\lambda t)^n e^{-\lambda t}}{n!}. \quad \blacklozenge$$

In Section 5.2, we showed that the Poisson distribution was discovered to approximate the binomial probability mass function when the number of trials is large ($n \to \infty$), the probability of success is small ($p \to 0$), and the average number of successes remains a fixed quantity of moderate value ($np = \lambda$ for some constant λ). Then we showed that, on its own merit, the Poisson distribution appears in connection with the study of sequence of random events over time. The following interesting theorems show that the relation between Poisson and binomial is not restricted to the approximation mentioned previously.

Theorem 12.2   *Let $\{N(t)\colon t \geq 0\}$ be a Poisson process with parameter λ. Suppose that, for a fixed $t > 0$, $N(t) = n$. That is, we are given that n events have occurred by time t. Then, for u, $0 < u < t$, the number of events that have occurred at or prior to u is binomial with parameters n and u/t.*

Proof: We want to show that, for $0 \leq i \leq n$,

$$P\big(N(u) = i \mid N(t) = n\big) = \binom{n}{i}\left(\frac{u}{t}\right)^i\left(1 - \frac{u}{t}\right)^{n-i}.$$

By the independent-increments property of Poisson processes,

$$P\big(N(u) = i \mid N(t) = n\big) = \frac{P\big(N(u) = i,\, N(t) = n\big)}{P\big(N(t) = n\big)}$$

$$= \frac{P\big(N(u) = i,\, N(t) - N(u) = n - i\big)}{P\big(N(t) = n\big)}$$

$$= \frac{P\big(N(u) = i\big)P\big(N(t) - N(u) = n - i\big)}{P\big(N(t) = n\big)}$$

$$= \frac{P\big(N(u) = i\big)P\big(N(t - u) = n - i\big)}{P\big(N(t) = n\big)}$$

(since stationarity implies that the number of events between u and t has the same distribution as the number of events between 0 and $t - u$)

$$= \frac{\dfrac{e^{-\lambda u} \cdot (\lambda u)^i}{i!} \cdot \dfrac{e^{-\lambda(t-u)} \cdot \big[\lambda(t - u)\big]^{n-i}}{(n - i)!}}{\dfrac{e^{-\lambda t}(\lambda t)^n}{n!}}$$

$$= \frac{n!}{i!\,(n - i)!} \cdot \frac{u^i\,(t - u)^{n-i}}{t^n} = \binom{n}{i}\frac{u^i\,(t - u)^{n-i}}{t^i\,t^{n-i}}$$

$$= \binom{n}{i}\left(\frac{u}{t}\right)^i\left(1 - \frac{u}{t}\right)^{n-i}. \quad \blacklozenge$$

Suppose that, at a certain district of a large city, cars and buses are the only vehicles allowed to enter. Suppose that cars cross a specific intersection, located at that district, at a Poisson rate of λ and, independently, buses cross the same intersection at a Poisson rate of μ. If we are given that n vehicles have crossed the intersection by time t, the following theorem shows that the distribution of the number of cars among the n vehicles is binomial with parameters n and $\lambda/(\lambda + \mu)$.

Theorem 12.3   *Let $N_1(t)$ and $N_2(t)$ be independent Poisson processes with parameters λ and μ, respectively. The conditional distribution of $N_1(t)$ given that $N_1(t) + N_2(t) = n$ is binomial with parameters n and $\lambda/(\lambda + \mu)$.*

Proof: For each t, the independent random variables $N_1(t)$ and $N_2(t)$ are Poisson with parameters λt and μt, respectively. By Theorem 11.5, $N_1(t) + N_2(t)$ is Poisson with parameter $\lambda t + \mu t = (\lambda + \mu)t$. For $1 \leq i \leq n$, this and the independence of $N_1(t)$ and $N_2(t)$ yield

$$P\big(N_1(t) = i \mid N_1(t) + N_2(t) = n\big)$$

$$= \frac{P\big(N_1(t) = i, \ N_1(t) + N_2(t) = n\big)}{P\big(N_1(t) + N_2(t) = n\big)}$$

$$= \frac{P\big(N_1(t) = i, \ N_2(t) = n - i\big)}{P\big(N_1(t) + N_2(t) = n\big)}$$

$$= \frac{P\big(N_1(t) = i\big) P\big(N_2(t) = n - i\big)}{P\big(N_1(t) + N_2(t) = n\big)}$$

$$= \frac{\dfrac{e^{-\lambda t} \cdot (\lambda t)^i}{i!} \cdot \dfrac{e^{-\mu t} \cdot (\mu t)^{n-i}}{(n-i)!}}{\dfrac{e^{-(\lambda + \mu)t} \cdot \big[(\lambda + \mu)t\big]^n}{n!}}$$

$$= \frac{n!}{i!\,(n-i)!} \cdot \frac{\lambda^i \cdot \mu^{n-i}}{(\lambda + \mu)^n} = \binom{n}{i} \frac{\lambda^i \cdot \mu^{n-i}}{(\lambda + \mu)^i \cdot (\lambda + \mu)^{n-i}}$$

$$= \binom{n}{i}\Big(\frac{\lambda}{\lambda + \mu}\Big)^i \Big(\frac{\mu}{\lambda + \mu}\Big)^{n-i} = \binom{n}{i}\Big(\frac{\lambda}{\lambda + \mu}\Big)^i \Big(1 - \frac{\lambda}{\lambda + \mu}\Big)^{n-i}. \ \blacklozenge$$

We will now discuss a useful property of the Poisson process that connects it to the uniform distribution. That property gives further insight into the distribution of the times at which Poisson events occur. Suppose that calls arrive on Ann's cellular phone at a Poisson rate of λ. Every time that Ann attends a lecture, she turns off her cellular phone. Suppose that, after a lecture, Ann turns on her cellular phone and receives a message that, during the last t minutes that her phone was off, n calls arrived on her cellular phone.

However, no information is given concerning the times that the calls arrived. To find the distributions of the arrival times of the calls, choose n independent points, $X_1, X_2, \ldots,$ X_n, from the interval $[0, t]$ at random. Let $X_{(1)}$ be the smallest value in $\{X_1, \ldots, X_n\}$, $X_{(2)}$ be the second smallest value, and, in general, $X_{(k)}$ $(1 \leq k \leq n)$ be the kth smallest value in $\{X_1, \ldots, X_n\}$. The following theorem shows that $X_{(1)}$ has the same distribution as the arrival time of the first call, $X_{(2)}$ has the same distribution as the arrival time of the second call, and so on. In other words, the order statistics of $X_1, X_2, \ldots, X_n$ can be thought of as the arrival times of the calls.

Theorem 12.4   *Let $\{N(t): t \geq 0\}$ be a Poisson process with parameter λ, and suppose that, for a fixed t, we are given that $N(t) = n$. For $1 \leq i \leq n$, let S_i be the time of the ith event. Then the joint probability density function of $S_1, S_2, \ldots, S_n$ given that $N(t) = n$ is given by*

$$f_{S_1,\ldots,S_n|N(t)}(t_1, \ldots, t_n \mid n) = \frac{n!}{t^n}, \quad 0 < t_1 < t_2 < \cdots < t_n < t.$$

Let $X_1, X_2, \ldots, X_n$ be n independent random points from the interval $[0, t]$. For $1 \leq i \leq n$, the probability density function of X_i is

$$f(x) = \begin{cases} 1/t & \text{if } 0 \leq x \leq t \\ 0 & \text{elsewhere.} \end{cases}$$

Hence, by Theorem 9.7, $f_{12 \cdots n}$, the joint probability density function of $X_{(1)}, X_{(2)}, \ldots,$ $X_{(n)}$, the order statistics of $X_1, X_2, \ldots, X_n$, is given by

$$f_{12 \cdots n}(t_1, \ldots, t_n) = \begin{cases} n!/t^n & \text{if } 0 < t_1 < \cdots < t_n < t \\ 0 & \text{otherwise.} \end{cases}$$

Comparing this with Theorem 12.4, we have that, given $N(t) = n$, the order statistics of $X_1, X_2, \ldots, X_n$ can be thought of as $S_1, S_2, \ldots, S_n$.

Proof of Theorem 12.4:   Let $F_{S_1,\ldots,S_n|N(t)}$ be the joint probability distribution function of $S_1, S_2, \ldots, S_n$ given that $N(t) = n$. Then, for $0 < t_1 < t_2 \cdots < t_n < t$,

$$F_{S_1,\ldots,S_n|N(t)}(t_1, t_2, \ldots, t_n \mid n) = P\big(S_1 \leq t_1, S_2 \leq t_2, \ldots, S_n \leq t_n \mid N(t) = n\big)$$

$$= \frac{P(S_1 \leq t_1, S_2 \leq t_2, \ldots, S_n \leq t_n, N(t) = n)}{P\big(N(t) = n\big)}.$$

Note that the event $\{S_1 \leq t_1, S_2 \leq t_2, \ldots, S_n \leq t_n, N(t) = n\}$ occurs if and only if exactly one event occurs in each of the intervals $[0, t_1], (t_1, t_2], \ldots, (t_{n-1}, t_n]$, and no events occur in $(t_n, t]$. Thus, by the independent-increments property and stationarity of Poisson processes,

$$F_{S_1,\ldots,S_n|N(t)}(t_1, t_2, \ldots, t_n \mid n)$$

$$= \frac{\lambda t_1 e^{-\lambda t_1} \cdot \lambda(t_2 - t_1)e^{-\lambda(t_2 - t_1)} \cdots \lambda(t_n - t_{n-1})e^{-\lambda(t_n - t_{n-1})} \cdot e^{-\lambda(t - t_n)}}{\dfrac{e^{-\lambda t} \cdot (\lambda t)^n}{n!}}$$

$$= \frac{n!}{t^n} t_1(t_2 - t_1) \cdots (t_n - t_{n-1}).$$

The theorem follows from

$$f_{S_1,\ldots,S_n|N(t)}(t_1, \ldots, t_n \mid n) = \frac{\partial^n}{\partial t_1 \partial t_2 \cdots \partial t_n} F_{S_1,\ldots,S_n|N(t)}(t_1, \ldots, t_n \mid n),$$

and the following fact, which we will verify by induction.

$$\frac{\partial^n}{\partial t_1 \partial t_2 \cdots \partial t_n} t_1(t_2 - t_1) \cdots (t_n - t_{n-1}) = 1.$$

Observe that, for $n = 2$,

$$\frac{\partial^2}{\partial t_1 \partial t_2} t_1(t_2 - t_1) = 1.$$

Suppose that

$$\frac{\partial^{n-1}}{\partial t_1 \partial t_2 \cdots \partial t_{n-1}} t_1(t_2 - t_1) \cdots (t_{n-1} - t_{n-2}) = 1.$$

Then

$$\frac{\partial^n}{\partial t_1 \partial t_2 \cdots \partial t_n} t_1(t_2 - t_1) \cdots (t_n - t_{n-1})$$

$$= \frac{\partial^{n-1}}{\partial t_1 \partial t_2 \cdots \partial t_{n-1}} \left[\frac{\partial}{\partial t_n} t_1(t_2 - t_1) \cdots (t_{n-1} - t_{n-2})(t_n - t_{n-1}) \right]$$

$$= \frac{\partial^{n-1}}{\partial t_1 \partial t_2 \cdots \partial t_{n-1}} t_1(t_2 - t_1) \cdots (t_{n-1} - t_{n-2}) = 1. \quad \blacklozenge$$

Note that, by Theorem 12.4, the conditional distribution of the times at which the events occur, given that $N(t) = n$, does not depend on λ. In particular, *if $N(t) = 1$, by Theorem 12.4, the time at which the event occurred is a uniform random variable over the interval $(0, t)$.*

For $1 \le i \le n$, we will now find the conditional probability density function of S_i given that $N(t) = n$. Let X be a uniform random variable over the interval $[0, t]$. The probability density function of X is given by

$$f(x) = \begin{cases} 1/t & \text{if } 0 \le x \le t \\ 0 & \text{elsewhere.} \end{cases}$$

Let F be the probability distribution function of X. Then, for $0 \le x \le t$,

$$F(x) = P(X \le x) = \frac{x}{t}.$$

Thus, given that $N(t) = n$, by Theorems 12.4 and 9.5, for $1 \le i \le n$, *the probability density function of S_i, the time that the ith event occurred is given by*

$$f_{S_i|N(t)}(x|n) = \frac{n!}{(i-1)!\,(n-i)!}\, \frac{1}{t}\left(\frac{x}{t}\right)^{i-1}\left(1 - \frac{x}{t}\right)^{n-i}, \quad 0 \le x \le t. \qquad (12.8)$$

Example 12.8   Suppose that jobs arrive at a file server at a Poisson rate of 3 per minute. If two jobs arrived within one minute, between 10:00 and 10:01, what is the probability that the first job arrived before 20 seconds past 10:00 and the second job arrived before 40 seconds past 10:00?

Solution: Label the time point 10:00 as $t = 0$. Then $t = 60$ corresponds to 10:01. Let S_1 and S_2 be the arrival times of the first and second jobs, respectively. Given that $N(60) = 2$, by Theorem 12.4, the joint probability density function of S_1 and S_2 is

$$f_{S_1, S_2|N(60)}(t_1, t_2 \mid 2) = \begin{cases} 2/3600 & \text{if } 0 < t_1 < t_2 < 60 \\ 0 & \text{elsewhere.} \end{cases}$$

Therefore,

$$P\big(S_1 < 20,\ S_2 < 40 \mid N(60) = 2\big) = \int_0^{20} \left(\int_{t_1}^{40} \frac{2}{3600}\, dt_2 \right) dt_1$$

$$= \int_0^{20} \frac{1}{1800}(40 - t_1)\, dt_1 = \frac{1}{3}. \quad \blacklozenge$$

An important area for applications of stochastic processes is *queueing theory*. Here, we will explain what a queueing system is and introduce notations that we will use throughout the rest of this chapter.

What Is a Queueing System?

A **queueing system** is a facility in which "customers" arrive and take their place in a waiting line for service if all of the servers are busy. If there is no queue and a server is free, then customers will be served immediately upon arrival. Otherwise, the customer waits in a line until his or her turn for service. First-come, first-served is a common service order. However, depending on the system, customers might be served on a last-come, first-served basis, in a random order, or based on some preassigned priority rule. A queueing system might have one waiting line, as in banks, or several waiting lines, as in grocery stores. In general, customers depart the system when they are served.

In examples of queueing systems such as banks, post offices, and hospital emergency rooms, "customers" are people. However, queueing systems are mathematical models for a wide range of phenomena, and depending on the application, "customer" may be anything from a broken-down machine waiting to be repaired at a maintenance station, to an airplane in a runway of an airport waiting to take off, to a computer program that is waiting in a storage buffer to be served by the central processing unit. A queueing system might have finite or infinite capacity for customers. If it has a finite capacity, then customers may get lost forever, or they may return with certain probabilities. Service times might have identical distributions, or they may depend on the number of customers in the system. Sometimes the system serves at a faster rate when it reaches a certain capacity. Customers may be served one at a time, as in a post office, or in batches, as in an elevator. Similarly, the arrivals may occur one at a time as in a Poisson process, or in bulks, say, when buses of tourists arrive at a market. A queueing system may have one, several, or infinitely many servers. A customer may need to be served only at one station, or there might be a number of service stages for each or some customers. These are a few models of queueing systems. You can use your imagination to think of many other systems, realistic or theoretical.

For a queueing system, we number customers $C_1, C_2, \ldots$ in the order of their arrival. Let the interarrival time between C_n and C_{n+1} be T_n, and the service time of C_n be S_n, $n = 1, 2, \ldots$. By the **$GI/G/c$ queueing system**, we mean that $\{T_1, T_2, \ldots\} \cup \{S_1, S_2, \ldots\}$ is an independent set of random variables, the random variables within each of the sequences $\{T_1, T_2, \ldots\}$ and $\{S_1, S_2, \ldots\}$ are identically distributed, there is one waiting line, and customers are served in the order of their arrival by c servers operating in parallel. Therefore, the service times are independent and identically distributed, the interarrival times are independent and identically distributed, and the service times are independent of the interarrival times, and hence independent of the arrival stream, in general. The interarrival and service probability distribution functions will be denoted by A and G, respectively. Let $1/\lambda = E(T_n)$ and $1/\mu = E(S_n)$ and $\rho = \lambda/c\mu$. We assume throughout that $\lambda/c\mu < 1$. Under this assumption the queue is stable. Note that the customers are arriving at a rate of λ and are being served at a rate of $c\mu$. If $\lambda > c\mu$, then the queue will increase without bound as time passes by. If $\lambda = c\mu$, then in the long run, the system will not be stable. In 1972, Whitt showed that if $P(T_n - S_n > 0) > 0$, then, with probability 1, customers find the system empty infinitely often.[†] At epochs where an arrival finds the system empty, the process probabilistically restarts itself, and the system is regenerative.

In the $GI/G/c$ queueing system just explained, both the interarrival and service times have general distributions. An **$M/G/c$ queueing system** is a $GI/G/c$ system in which the arrival stream is a Poisson process and the service times have general distributions. Therefore, the interarrival times are exponential random variables with mean $1/\lambda$. Similarly, the queueing system **$GI/M/c$** is a special type of $GI/G/c$ in which the interarrival times have general distributions, but the service times are exponentially

---

[†]Whitt, W. [1972] Embedded Renewal Processes in the $GI/G/s$ Queue. *J. Appl. Prob.*, **9**, 650–658.

distributed. It must be clear that, for example, by an $M/M/2$ queueing system we mean a $GI/G/c$ system in which the interarrival times are exponential with mean $1/\lambda$, the service times are exponential with mean $1/\mu$, and there are two servers operating in parallel. Therefore, the arrival stream is a Poisson process. In the notation "$GI/G/c$," we use GI for "general distribution and independent interarrival times" and G for "general distribution." For exponential interarrival times or service times, we use M, some believe because of the "memoryless" property of exponential random variables. Others believe that M, the initial of "Markov," is used in honor of the Russian mathematician Markov.

A $GI/G/c$ queueing system is denoted by $D/G/c$ if the interarrival times are a constant d and the service times have a general distribution. Similarly, a $GI/D/c$ queueing system is a $GI/G/c$ system in which the interarrival times have a general distribution, but the service times are a constant d. Hence in a $D/D/1$ system both the interarrival and service times are constants. Obviously, $D/D/1$ is a deterministic system. For constant service times or interarrival times, the notation used is D for "deterministic."

PASTA: Poisson Arrivals See Time Averages

For a queueing system, let $U(t)$ be the number of customers present at time t. Let $A(t)$ be the number of customers found in the system by an arrival at time t. In general, the probability distributions of $U(t)$ and $A(t)$ are not identical. For example, consider a queueing system in which customers arrive at times $0, 5, 10, 15, \ldots$. Suppose that the service time of each customer is 2.5 minutes. Then at a random time t,

$$P\big(A(t) = 0\big) = 1, \qquad P\big(A(t) = 1\big) = 0,$$

whereas

$$P\big(U(t) = 0\big) = P\big(U(t) = 1\big) = \frac{1}{2}.$$

One of the most interesting properties of the Poisson processes is that, if for a system, customers arrive according to a Poisson process, then, for $t > 0$, $A(t)$ and $U(t)$ are identically distributed. To show this, let $N(t)$ be the total number of customers arrived at or prior to t [note the difference between $U(t)$ and $N(t)$]. Then, for $n \geq 0$,

$$P\big(A(t) = n\big) = \lim_{\varepsilon \to 0} P\big(U(t) = n \mid N(t + \varepsilon) - N(t) = 1\big)$$

$$= \lim_{\varepsilon \to 0} \frac{P\big(U(t) = n, \, N(t + \varepsilon) - N(t) = 1\big)}{P\big(N(t + \varepsilon) - N(t) = 1\big)}$$

$$= \lim_{\varepsilon \to 0} \frac{P\big(N(t + \varepsilon) - N(t) = 1 \mid U(t) = n\big) P\big(U(t) = n\big)}{P\big(N(t + \varepsilon) - N(t) = 1\big)}.$$

Now

$$P\big(N(t + \varepsilon) - N(t) = 1 \mid U(t) = n\big) = P\big(N(t + \varepsilon) - N(t) = 1\big),$$

since, by the independent-increments property of the Poisson processes, the number of arrivals in $(t, t + \varepsilon]$ is independent of how many customers have arrived earlier or how they have arrived. Hence, for $n \geq 0$,

$$P\big(A(t) = n\big) = \lim_{\varepsilon \to 0} P\big(U(t) = n\big) = P\big(U(t) = n\big).$$

Intuitively, the fact that the arrival in $(t, t + \varepsilon]$ is independent of $N(t)$ and hence $U(t)$ makes the arrival time t no different from any other time. That is why $A(t)$ and $U(t)$ are identically distributed.

The property we just explained was proved, in general, in 1982, by Berkeley professor Ronald W. Wolff, stated as "PASTA: Poisson Arrivals See Time Averages."[†] Wolff showed that

> For any system in which the arrival stream is a Poisson process, the proportion of the arrivals who **find** the system in a specific state is equal to the proportion of time that the system **is** in that state.

EXERCISES

1. For a Poisson process with parameter λ, show that, for all $\varepsilon > 0$,

    $$P\left(\left| \frac{N(t)}{t} - \lambda \right| \geq \varepsilon \right) \to 0,$$

    as $t \to \infty$. This shows that, for a large t, $N(t)/t$ is a good estimate for λ.

2. The number of accidents at an intersection is a Poisson process $\big\{N(t) : t \geq 0\big\}$ with rate 2.3 per week. Let X_i be the number of injuries in accident i. Suppose that $\{X_i\}$ is a sequence of independent and identically distributed random variables with mean 1.2 and standard deviation 0.7. Furthermore, suppose that the number of injuries in each accident is independent of the number of accidents that occur at the intersection. Let $Y(t) = \sum_{i=1}^{N(t)} X_i$; then $Y(t)$, the total number of injuries from the accidents at that intersection, at or prior to t, is said to be a **compound Poisson process.** Find the expected value and the standard deviation of $Y(52)$, the total number of injuries in a year.

3. When Linda walks from home to work, she has to cross the street at a certain point. Linda needs a gap of 15 seconds in the traffic to cross the street at that point. Suppose that the traffic flow is a Poisson process, and the mean time between two consecutive cars passing by Linda is 7 seconds. Find the expected value of the time Linda has to wait before she can cross the street.

[†]Wolff, R. W. [1982]. Poisson Arrivals See Time Averages. *Oper. Res.,* **30,** 223–231.

Hint: Let X_1 be the time between Linda's arrival at the point and the first car passing by her. Let X_2 be the time between the first and second cars passing Linda, and so forth. Let N be the first integer for which

$$X_1 \leq 15, \ X_2 \leq 15, \ \ldots, \ X_N \leq 15, \ X_{N+1} > 15.$$

The time Linda has to wait before being able to cross the street is 0 if $N = 0$ (i.e., $X_1 > 15$) and is $S_N = X_1 + X_2 + \cdots + X_N$ otherwise. Calculate S_N by conditioning on N.

4. Suppose that a fisherman catches fish at a Poisson rate of 2 per hour. We know that yesterday he began fishing at 9:00 A.M., and by 1:00 P.M. he caught 6 fish. What is the probability that he caught the first fish before 10:00 A.M.?

5. A wire manufacturing company has inspectors to examine the wire for fractures as it comes out of a machine. The number of fractures is distributed in accordance with a Poisson process, having one fracture on the average for every 60 meters of wire. One day an inspector had to take an emergency phone call, so she left her post for a few minutes. Later on, she was informed that, while she was away from her post, 200 meters of wire was manufactured with three fractures. What is the probability that the consecutive fractures, among this portion of wire manufactured, were at least 60 meters apart?
Hint: Let $N(t)$ be the number of fractures in the first t meters of wire manufactured *after* the inspector left her post; $\{N(t): t \geq 0\}$ is a Poisson process with rate 60.

6. Let $\{N(t): t \geq 0\}$ be a Poisson process. For $k \geq 1$, let S_k be the time that the kth event occurs. Show that

$$E[S_k \mid N(t) = n] = \frac{kt}{n+1}.$$

7. Recall that an $M/M/1$ queueing system is a $GI/G/1$ system in which there is one server, customers arrive according to a Poisson process with rate λ, and service times are exponential with mean $1/\mu$. For an $M/M/1$ system, each time a customer arrives at or a customer departs from the system, we say that a *transition* occurs. For a fixed $t > 0$, let A be the event that there will be no transitions during the next t minutes. Let B be the event that the next transition is an arrival. Show that A and B are independent. Note that this phenomenon might be counterintuitive.

8. Recall that an $M/M/1$ queueing system is a $GI/G/1$ system in which there is one server, customers arrive according to a Poisson process with rate λ, and service times are exponential with mean $1/\mu$. For an $M/M/1$ queueing system,

   (a) show that the number of arrivals during a period in which a customer is being served is geometric with parameter $\mu/(\lambda + \mu)$;

(b) suppose that there are n customers waiting in the queue, and a customer is being served. Find the probability mass function of the number of new customers arriving by the time that all of these $n + 1$ customers are served.

9. Customers arrive at a bank at a Poisson rate of λ. Let $M(t)$ be the number of customers who enter the bank by time t only to make deposits to their accounts. Suppose that, independent of other customers, the probability is p that a customer enters the bank only to make a deposit. Show that $\{M(t): t \geq 0\}$ is a Poisson process with parameter λp.

10. There are k types of shocks identified that occur, independently, to a system. For $1 \leq i \leq k$, suppose that shocks of type i occur to the system at a Poisson rate of λ_i. Find the probability that the nth shock occurring to the system is of type i, $1 \leq i \leq n$.

    Hint: For $1 \leq i \leq k$, let $N_i(t)$ be the number of type i shocks occurring to the system at or prior to t. It can be readily seen that $N_1(t) + N_1(t) + \cdots + N_k(t)$ is a Poisson process at a rate of $\lambda = \lambda_1 + \lambda_2 + \cdots + \lambda_k$. Merging these Poisson processes to create a new Poisson process is called **superposition of Poisson processes**. For $1 \leq i \leq k$, let V_i be the time of the first shock of type i occurring to the system. Argue that the desired probability is $P\big(V_i = \min(V_1, V_2, \ldots, V_k)\big)$. Then calculate this probability by conditioning on V_i. ∎

12.3 MARKOV CHAINS

Consider the gambler's ruin problem (Example 3.14) in which two gamblers play the game of "heads or tails." Each time a fair coin lands heads up, player A wins \$1 from player B, and each time it lands tails up, player B wins \$1 from player A. Suppose that, initially, player A has a dollars and player B has b dollars. Let X_i be the amount of money that player A has after i games. Clearly, $X_0 = a$, and $X_1, X_2, \ldots$ are discrete random variables. One of the main properties of the sequence $\{X_i\}$ is that, if at a certain step n we know the value of X_n, then, from that step on, the amount of money player A will have depends only on X_n and not on the values of $X_0, X_1, \ldots, X_{n-1}$. That is, given the present amount of money player A has, the future of his fortune is independent of the amounts of money he had, at each step of the game, in the past. The subject of Markov chains is the study of such processes; namely, the processes in which *given the present state, the future is independent of the past* of the process. There are abundant examples in natural, mathematical, and social sciences that possess this property. Such processes are called **Markov chains**, and they are studied in **discrete steps**. In the gambler's ruin problem, each step is one game. The **state** of the Markov chain in step n is the value of X_n. Another example of a Markov chain is the number of customers present at a bank at time n. Since we study Markov chains in discrete steps, in such a case, by "time n," we really mean "step n." In this context, each step corresponds to a transition in the bank; that is, an arrival or a departure. Suppose that, at time n, there are 5 customers in the bank. If the next transition is an arrival, then at time $n + 1$ there will be 6 customers in the bank. However, if the next transition is a departure, then at time $n + 1$ there will be 4 customers in the bank.

Markov chains were introduced by the Russian Mathematician, Andrei Andreyevich Markov (1856–1922), a student of Chebyshev and a graduate of Saint Petersburg University (1878). During his fruitful life, Markov did extensive research in several branches of mathematics including probability theory. He was also interested in poetry and did some work in poetic styles. The topic of Markov chains is one of his ever-lasting contributions to the world of knowledge. He invented that branch of stochastic processes in an attempt to generalize the strong law of large numbers to cases in which the random variables are not independent. Markov was a political activist and is known for protesting Czar's intervention to stop the writer Maxim Gorky (1868–1936) from assuming an elective position at Saint Petersburg Academy.

Definition   *A stochastic process $\{X_n : n = 0, 1, \dots\}$ with a finite or countably infinite state space S is said to be a **Markov chain**, if for all i, j, i_0, $\dots$, $i_{n-1} \in S$, and $n = 0, 1, 2, \dots$,*

$$P(X_{n+1} = j \mid X_n = i, X_{n-1} = i_{n-1}, \dots, X_0 = i_0) = P(X_{n+1} = j \mid X_n = i).$$
(12.9)

The elements of the state space S are not necessarily nonnegative integers (or numbers). However, for simplicity, it is a common practice to label the elements of S by nonnegative integers. If S is finite, the Markov chain is called a **finite Markov chain** or a **finite-state Markov chain**. If S is infinite, it is called an **infinite Markov chain** or an **infinite-state Markov chain**.

The main property of a Markov chain, expressed by (12.9), is called the **Markovian property** of the Markov chain. Thus, by the Markovian property,

> Given the state of the Markov chain at present (X_n), its future state (X_{n+1}) is independent of the past states ($X_{n-1}, \dots, X_1, X_0$).

A Markov chain with state space S is said to have **stationary transition probabilities,** if, for all $i, j \in S$, the probability of a transition from state i to state j, in one step, does not depend on the time that the transition will occur—that is, if $P(X_{n+1} = j \mid X_n = i)$ is independent of n. In this book, unless otherwise stated, we assume that Markov chains have stationary transition probabilities. For $i, j \in S$, let

$$p_{ij} = P(X_{n+1} = j \mid X_n = i);$$

then the matrix

$$P = \begin{pmatrix} p_{00} & p_{01} & p_{02} & \cdots \\ p_{10} & p_{11} & p_{12} & \cdots \\ p_{20} & p_{21} & p_{22} & \cdots \\ \vdots & & & \end{pmatrix},$$

sometimes simply denoted by (p_{ij}), is called the **transition probability matrix** of the Markov chain $\{X_n : n = 0, 1, \ldots\}$. Note that the jth entry in the ith row is the probability of a transition from state $i - 1$ to state $j - 1$. For $i \in S$, note that the probability of a transition *from* state i is $\sum_{j=0}^{\infty} p_{ij}$. Hence we must have $\sum_{j=0}^{\infty} p_{ij} = 1$; that is, the sum of the elements of each row of the transition probability matrix is 1. However, the sum of the elements of a column is not necessarily 1. In matrix theory, if all of the entries of a matrix are nonnegative and the sum of the entries of each row is 1, then it is called a **Markov matrix**. This is because for such a matrix P, one can construct a sample space, associate probabilities to all events of the sample space, and then define a Markov chain over the sample space in such a way that its transition probability matrix is P.

Example 12.9   At an intersection, a working traffic light will be out of order the next day with probability 0.07, and an out-of-order traffic light will be working the next day with probability 0.88. Let $X_n = 1$ if on day n the traffic light will work; $X_n = 0$ if on day n it will not work. Then $\{X_n : n = 0, 1, \ldots\}$ is a Markov chain with state space $\{0, 1\}$ and transition probability matrix

$$P = \begin{pmatrix} 0.12 & 0.88 \\ 0.07 & 0.93 \end{pmatrix}. \quad \blacklozenge$$

Example 12.10   Suppose that a mouse is moving inside the maze shown in Figure 12.1, from one cell to another, in search of food. When at a cell, the mouse will move to one of the adjoining cells randomly. For $n \geq 0$, let X_n be the cell number the mouse will visit after having changed cells n times. Then $\{X_n : n = 0, 1, \ldots\}$ is a Markov chain with state space $\{1, 2, \ldots, 9\}$ and transition probability matrix

$$P = \begin{pmatrix} 0 & 1/2 & 0 & 1/2 & 0 & 0 & 0 & 0 & 0 \\ 1/3 & 0 & 1/3 & 0 & 1/3 & 0 & 0 & 0 & 0 \\ 0 & 1/2 & 0 & 0 & 0 & 1/2 & 0 & 0 & 0 \\ 1/3 & 0 & 0 & 0 & 1/3 & 0 & 1/3 & 0 & 0 \\ 0 & 1/4 & 0 & 1/4 & 0 & 1/4 & 0 & 1/4 & 0 \\ 0 & 0 & 1/3 & 0 & 1/3 & 0 & 0 & 0 & 1/3 \\ 0 & 0 & 0 & 1/2 & 0 & 0 & 0 & 1/2 & 0 \\ 0 & 0 & 0 & 0 & 1/3 & 0 & 1/3 & 0 & 1/3 \\ 0 & 0 & 0 & 0 & 0 & 1/2 & 0 & 1/2 & 0 \end{pmatrix}. \quad \blacklozenge$$

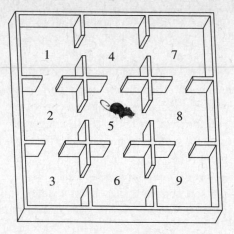

Figure 12.1   The moving mouse of Example 12.10.

Example 12.11   Experience shows that, whether or not a bidder is successful on a bid, depends on the successes and failures of his previous two bids. If his last two bids were successful, his next bid will be successful with probability 0.5. If only one of his last two bids was successful, the probability is 0.6 that the next bid will be successful. Finally, if none of the last two bids were successful, the probability is 0.7 that the next one will be successful.

Suppose that each step corresponds to a new bid. If, at step n, the state of the process is defined based on whether bid n is a success or a failure, then, for this problem, the model obtained is not a Markov chain. To construct a mathematical model using Markov chains, let a successful bid be denoted by s and an unsuccessful bid be denoted by f. Define X_n to be sf, if at step n the last bid was a failure and the bid before that was a success. Define ss, fs, and ff similarly. Let state $0 = sf$, state $1 = ss$, state $2 = fs$, and state $3 = ff$. The following transition probability matrix shows that $\{X_n : n \geq 2\}$ is a Markov chain. That is, given the present state, we can determine the transition probabilities to the next step no matter what the course of transitions in the past was.

$$P = \begin{pmatrix} 0 & 0 & 0.6 & 0.4 \\ 0.5 & 0.5 & 0 & 0 \\ 0.4 & 0.6 & 0 & 0 \\ 0 & 0 & 0.7 & 0.3 \end{pmatrix}.$$

To see that, for example, $P(X_{n+1} = 2 \mid X_n = 0) = 0.6$ (that is, the probability of a one-step transition from sf to fs is 0.6), note that the probability is 0.6 that the last bid was a failure and the next bid will be successful, given that the last bid was a failure and the one before that was successful. As another example, $P(X_{n+1} = 3 \mid X_n = 2) = 0$ since it is impossible to move from fs to ff in one step. If it is given that the last bid was a success and the bid before that was a failure, then the probability is 0 that the last bid was a failure and the next bid will be a failure as well.   ♦

Example 12.12   (One-Dimensional Random Walks)   In a large town, Kennedy Avenue is a long north-south avenue with many intersections. A drunken man is wandering along the avenue and does not really know which way he is going. He is currently at an intersection O somewhere in the middle of the avenue. Suppose that, at the end of each block, he either goes north with probability p, or he goes south with probability $1 - p$. A mathematical model for the movements of the drunken man is known as a **one-dimensional random walk**. Let α be an integer or $-\infty$, β be an integer or $+\infty$, and $\alpha < \beta$. Consider a stochastic process $\{X_n : n = 0, 1, \dots\}$ with state space $\{\alpha, \alpha + 1, \dots, \beta\}$, a subset of the integers, finite or infinite. Suppose that if, for some n, $X_n = i$, then $X_{n+1} = i + 1$ with probability p, and $X_{n+1} = i - 1$ with probability $1 - p$. That is, one-step transitions are possible only from a state i to its adjacent states $i - 1$ and $i + 1$. Clearly $\{X_n : n = 0, 1, \dots\}$ is a Markov chain with transition probabilities

$$
p_{ij} = \begin{cases} p & \text{if } j = i + 1 \\ 1 - p & \text{if } j = i - 1 \\ 0 & \text{otherwise.} \end{cases}
$$

Such Markov chains are called *one-dimensional random walks*. As an example of a random walk, consider the gambler's ruin problem, Example 3.14, which was also discussed at the beginning of this section. Let X_n be player A's fortune after n games. Then $\{X_n : n = 0, 1, \dots\}$ is a Markov chain. Suppose that, at some time n, $X_n = i$. On each play, gambler A wins \$1 from B with probability p, $0 < p < 1$, and loses \$1 to B with probability $1 - p$. Therefore, gambler A's fortune either increases by \$1, moving to the adjacent state $i + 1$, or decreases by \$1, moving to the adjacent state $i - 1$. This is a finite-state Markov chain with state space $\{0, 1, \dots, a + b\}$. Note that player A's fortune cannot exceed $a + b$, in which case B is ruined, or go below 0, in which case A is ruined. The transition probability matrix of this Markov chain is

$$
P = \begin{pmatrix} 1 & 0 & 0 & 0 & \cdots & 0 & 0 & 0 \\ 1 - p & 0 & p & 0 & \cdots & 0 & 0 & 0 \\ 0 & 1 - p & 0 & p & \cdots & 0 & 0 & 0 \\ \vdots & & & & & & & \\ 0 & 0 & 0 & 0 & \cdots & 1 - p & 0 & p \\ 0 & 0 & 0 & 0 & \cdots & 0 & 0 & 1 \end{pmatrix}.
$$

The theory of random walks is a well-developed field with diverse applications. Historically, the study of random walks began in 1912 by George Pólya. Here is the story of how he was motivated to study random walks in words of Gerald Alexanderson:[†]

[†]Gerald Alexanderson,*The Random Walks of George pólya,* The Mathematical Association of America (MAA Spectrum), 2000, page 51.

There was a particular wooded area near the hotel where he [Pólya] enjoyed taking extended walks while thinking about mathematics. He would carry pencil and paper so he could jot down ideas as they came to him. Some students also lived at the Kurhaus [hotel] and Pólya got to know some of them. One day while out on his walk he encountered one of these students strolling with his fiancée. Somewhat later their paths crossed again and even later he encountered them once again. He was embarrassed and worried that the couple would conclude that he was somehow arranging these encounters. This caused him to wonder how likely it was that walking randomly through paths in the woods, one would encounter others similarly engaged. This lead to one of his most famous discoveries, his 1921 paper on random walk, a phrase used for the first time by Pólya. The concept is critical in the development of the investigation of Brownian Motions.   ◆

Example 12.13   The computers in the Writing Center of a college are inspected at the end of each semester. If a computer needs minor repairs, it will be fixed. If the computer has crashed, it will be replaced with a new one. For $k \geq 0$, let $p_k > 0$ be the probability that a new computer needs to be replaced after k semesters. For a computer in use at the end of the nth semester, let X_n be the number of additional semesters it will remain functional. Let Y be the lifetime, in semesters, of a new computer installed in the lab. Then

$$X_{n+1} = \begin{cases} X_n - 1 & \text{if } X_n \geq 1 \\ Y - 1 & \text{if } X_n = 0 \end{cases}$$

shows that $\{X_n : n = 0, 1, \dots\}$ is a Markov chain with transition probabilities

$$p_{0j} = P(X_{n+1} = j \mid X_n = 0) = P(Y = j + 1) \doteq p_{j+1}, \quad j \geq 0,$$

and for $i \geq 1$,

$$p_{ij} = P(X_{n+1} = j \mid X_n = i) = \begin{cases} 1 & \text{if } j = i - 1 \\ 0 & \text{if } j \neq i - 1. \end{cases}$$

Therefore, the transition probability matrix for the Markov chain $\{X_n : n = 0, 1, \dots\}$ is

$$P = \begin{pmatrix} p_1 & p_2 & p_3 & \cdots \\ 1 & 0 & 0 & \cdots \\ 0 & 1 & 0 & \cdots \\ 0 & 0 & 1 & \cdots \\ \vdots & & & \end{pmatrix}. \quad ◆$$

Example 12.14   Recall that an $M/D/1$ queue is a $GI/G/1$ queueing system in which there is one server, the arrival process is Poisson with rate λ, and the service times of the customers are a constant d. For an $M/D/1$ queue, let X_n be the number of customers the nth departure leaves behind. Let Z_n be the number of arrivals during the time that the nth customer is being served. By the stationarity property of Poisson processes, $\{Z_n\}$ is a sequence of identically distributed random variables with probability mass function

$$p(k) = P(Z_n = k) = \frac{e^{-\lambda d}(\lambda d)^k}{k!}, \quad k = 0, 1, \ldots.$$

Clearly,

$$X_{n+1} = \max(X_n - 1, 0) + Z_{n+1}.$$

This relation shows that $\{X_n: n = 0, 1, \ldots\}$ is a Markov chain with state space $\{0, 1, 2, \ldots\}$ and transition probability matrix

$$P = \begin{pmatrix} p(0) & p(1) & p(2) & p(3) & p(4) & \cdots \\ p(0) & p(1) & p(2) & p(3) & p(4) & \cdots \\ 0 & p(0) & p(1) & p(2) & p(3) & \cdots \\ 0 & 0 & p(0) & p(1) & p(2) & \cdots \\ 0 & 0 & 0 & p(0) & p(1) & \cdots \\ & & \vdots & & & \end{pmatrix}. \quad \blacklozenge$$

Example 12.15   (Ehrenfest Chain)   Suppose that there are N balls numbered $1, 2, \ldots, N$ distributed among two urns randomly. At time n, a number is selected at random from the set $\{1, 2, \ldots, N\}$. Then the ball with that number is found in one of the two urns and is moved to the other urn. Let X_n denote the number of balls in urn I after n transfers. It should be clear that $\{X_n: n = 0, 1, \ldots\}$ is a Markov chain with state space $\{0, 1, 2, \ldots, N\}$ and transition probability matrix

$$P = \begin{pmatrix} 0 & 1 & 0 & 0 & 0 & \cdots & 0 & 0 & 0 \\ 1/N & 0 & (N-1)/N & 0 & 0 & \cdots & 0 & 0 & 0 \\ 0 & 2/N & 0 & (N-2)/N & 0 & \cdots & 0 & 0 & 0 \\ 0 & 0 & 3/N & 0 & (N-3)/N & \cdots & 0 & 0 & 0 \\ \vdots & & & & & & & & \\ 0 & 0 & 0 & 0 & 0 & \cdots & (N-1)/N & 0 & 1/N \\ 0 & 0 & 0 & 0 & 0 & \cdots & 0 & 1 & 0 \end{pmatrix}.$$

This chain was introduced by the physicists Paul and T. Ehrenfest in 1907 to explain some paradoxes in connection with the study of thermodynamics on the basis of kinetic theory. It is noted that Einstein had said of Paul Ehrenfest (1880–1933) that Ehrenfest was the best physics teacher he had ever known. $\blacklozenge$

The entries of the transition probability matrix give the probabilities of moving from one state to another in one step. Let

$$p_{ij}^n = P(X_{n+m} = j \mid X_m = i), \quad n, m \geq 0.$$

Then p_{ij}^n is the probability of moving from state i to state j in n steps. Since the Markov chains we study in this book have stationary transition probabilities, p_{ij}^n's do not depend on m. The matrix

$$\boldsymbol{P}^{(n)} = \begin{pmatrix} p_{00}^n & p_{01}^n & p_{02}^n & \cdots \\ p_{10}^n & p_{11}^n & p_{12}^n & \cdots \\ p_{20}^n & p_{21}^n & p_{22}^n & \cdots \\ \vdots & & & \end{pmatrix}$$

is called the **n-step transition probability matrix**. Clearly, $\boldsymbol{P}^{(0)}$ is the identity matrix. That is, $p_{ij}^0 = 1$ if $i = j$, and $p_{ij}^0 = 0$ if $i \neq j$. Also, $\boldsymbol{P}^{(1)} = \boldsymbol{P}$, the transition probability matrix of the Markov chain.

Warning: We should be very careful not to confuse the n-step transition probability p_{ij}^n with $(p_{ij})^n$, which is the quantity p_{ij} raised to the power n. In general, $p_{ij}^n \neq (p_{ij})^n$.

For a transition from state i to state j in $n + m$ steps, the Markov chain will have to enter some state k, along the way, after n transitions, and then move from k to j in the remaining m steps. This observation leads us to the following celebrated equations called the **Chapman-Kolmogorov equations**:

$$p_{ij}^{n+m} = \sum_{k=0}^{\infty} p_{ik}^n p_{kj}^m. \tag{12.10}$$

The equations (12.10) can be proved rigorously by applying the law of total probability, Theorem 3.4, to the sequence of mutually exclusive events $\{X_n = k\}$, $k \geq 0$:

$$p_{ij}^{n+m} = P(X_{n+m} = j \mid X_0 = i)$$

$$= \sum_{k=0}^{\infty} P(X_{n+m} = j \mid X_n = k, X_0 = i) P(X_n = k \mid X_0 = i)$$

$$= \sum_{k=0}^{\infty} P(X_{n+m} = j \mid X_n = k) P(X_n = k \mid X_0 = i)$$

$$= \sum_{k=0}^{\infty} p_{kj}^m p_{ik}^n = \sum_{k=0}^{\infty} p_{ik}^n p_{kj}^m.$$

Note that in (12.10), p_{ij}^{n+m} is the ijth entry of the matrix $\boldsymbol{P}^{(n+m)}$, p_{ik}^n is the ikth entry of the matrix $\boldsymbol{P}^{(n)}$, and p_{kj}^m is the kjth entry of the matrix $\boldsymbol{P}^{(m)}$. As we know, from the definition of the product of two matrices, the defining relation for the ijth entry of the product of matrices $\boldsymbol{P}^{(n)}$ and $\boldsymbol{P}^{(m)}$ is identical to (12.10). Hence the Chapman-Kolmogorov equations, in matrix form, are

$$\boldsymbol{P}^{(n+m)} = \boldsymbol{P}^{(n)} \cdot \boldsymbol{P}^{(m)},$$

which implies that

$$P^{(2)} = P^{(1)} \cdot P^{(1)} = P \cdot P = P^2,$$

$$P^{(3)} = P^{(2)} \cdot P^{(1)} = P^2 \cdot P = P^3,$$

and, in general, by induction,

$$P^{(n)} = P^{(n-1)} \cdot P^{(1)} = P^{n-1} \cdot P = P^n.$$

We have shown the following:

The n-step transition probability matrix is equal to the one-step transition probability matrix raised to the power of n.

Example 12.16   For the Markov chain of Example 12.9, the two-step transition probability matrix is given by

$$P^{(2)} = P^2 = \begin{pmatrix} 0.12 & 0.88 \\ 0.07 & 0.93 \end{pmatrix} \begin{pmatrix} 0.12 & 0.88 \\ 0.07 & 0.93 \end{pmatrix} = \begin{pmatrix} 0.076 & 0.924 \\ 0.0735 & 0.9265 \end{pmatrix}.$$

This shows that, for example, an out-of-order traffic light will be working the day after tomorrow with probability 0.924. Similarly, a working traffic light will be out of order the day after tomorrow with probability 0.0735. Also, to emphasize that, in general, $p_{ij}^n \neq (p_{ij})^n$, note that $(p_{10})^2 = 0.0049$ *is not* equal to $p_{10}^2 = 0.0735$. The matrix

$$P^{(6)} = p^6 = \begin{pmatrix} 0.0736842 & 0.926316 \\ 0.0736842 & 0.926316 \end{pmatrix}$$

shows that, whether or not the traffic light is working today, in six days, the probability that it will be working is 0.926316, and the probability that it will be out of order is 0.0736842. Later on, we will show that, for certain Markov chains, after a large number of transitions, the probability of visiting a certain state becomes independent of the initial state of the Markov chain.   ◆

Example 12.17   Direct calculations show that the 5-step transition probability matrix for the Markov chain of Example 12.10, in which a mouse is in a maze searching for food, is given by

$$P^{(5)} = P^5 = \begin{pmatrix} 0 & 5/18 & 0 & 5/18 & 0 & 2/9 & 0 & 2/9 & 0 \\ 5/27 & 0 & 5/27 & 0 & 1/3 & 0 & 4/27 & 0 & 4/27 \\ 0 & 5/18 & 0 & 2/9 & 0 & 5/18 & 0 & 2/9 & 0 \\ 5/27 & 0 & 4/27 & 0 & 1/3 & 0 & 5/27 & 0 & 4/27 \\ 0 & 1/4 & 0 & 1/4 & 0 & 1/4 & 0 & 1/4 & 0 \\ 4/27 & 0 & 5/27 & 0 & 1/3 & 0 & 4/27 & 0 & 5/27 \\ 0 & 2/9 & 0 & 5/18 & 0 & 2/9 & 0 & 5/18 & 0 \\ 4/27 & 0 & 4/27 & 0 & 1/3 & 0 & 5/27 & 0 & 5/27 \\ 0 & 2/9 & 0 & 2/9 & 0 & 5/18 & 0 & 5/18 & 0 \end{pmatrix}.$$

This matrix shows that, for example, if the mouse is in cell 4 at a certain time, then after changing cells five times, the mouse will be in cell 5 with probability 1/3, in cell 7 with probability 5/27, and in cell 9 with probability 4/27. Calculating P^{10}, we can see that after changing cells 10 times, the probability is 0 that the mouse will end up in any of the cells 5, 7, or 9.   ◆

Example 12.18   Consider the random walk of Example 12.12. In the model for the Gambler's ruin problem, suppose that player A's initial fortune is $3 and player B's initial fortune is $1. Furthermore, suppose that player A wins $1 from B with probability 0.6 and loses $1 to B with probability 0.4. Let X_n be player A's fortune after n games. Then the transition probability matrix of the Markov chain $\{X_n : n = 1, 2, \dots\}$ is

$$P = \begin{pmatrix} 1 & 0 & 0 & 0 & 0 \\ 0.4 & 0 & 0.6 & 0 & 0 \\ 0 & 0.4 & 0 & 0.6 & 0 \\ 0 & 0 & 0.4 & 0 & 0.6 \\ 0 & 0 & 0 & 0 & 1 \end{pmatrix}.$$

Direct calculations show that

$$P^{(10)} = P^{10} = \begin{pmatrix} 1 & 0 & 0 & 0 & 0 \\ 0.575 & 0.013 & 0 & 0.019 & 0.393 \\ 0.3 & 0 & 0.025 & 0 & 0.675 \\ 0.117 & 0.0085 & 0 & 0.0127 & 0.862 \\ 0 & 0 & 0 & 0 & 1 \end{pmatrix}.$$

Therefore, given that gambler A's initial fortune is $3, after 10 games, the probability that A has, say, $2 is 0; the probability that his fortune is $3 is 0.0127; the probability that he wins the game (his fortune is $4) is 0.862; and the probability that he loses the game (his fortune is $0) is 0.117.   ◆

Let $\{X_n : n = 0, 1, \dots\}$ be a Markov chain with its transition probability matrix given. The following theorem shows that if the probability mass function of X_0 is known, then, for all $n \geq 1$, we can find the probability mass function of X_n.

Theorem 12.5   *Let $\{X_n : n = 0, 1, \dots\}$ be a Markov chain with transition probability matrix $P = (p_{ij})$. For $i \geq 0$, let $p(i) = P(X_0 = i)$ be the probability mass function of X_0. Then the probability mass function of X_n is given by*

$$P(X_n = j) = \sum_{i=0}^{\infty} p(i) p_{ij}^n, \quad j = 0, 1, \dots.$$

Proof: Applying the law of total probability, Theorem 3.4, to the sequence of mutually exclusive events $\{X_0 = i\}$, $i \geq 0$, we have

$$P(X_n = j) = \sum_{i=0}^{\infty} P(X_n = j \mid X_0 = i) P(X_0 = i) = \sum_{i=0}^{\infty} p_{ij}^n p(i) = \sum_{i=0}^{\infty} p(i) p_{ij}^n. \quad ◆$$

Example 12.19   Suppose that, in Example 12.10, where a mouse is in a maze searching for food, initially, it is equally likely that the mouse is in any of the 9 cells. That is,

$$p(i) = P(X_0 = i) = \frac{1}{9}, \qquad 1 \le i \le 9.$$

Then, using the matrix P^5, calculated in Example 12.17, we can readily find the probability that the mouse is in cell j, $1 \le j \le 9$, after 5 transitions. For example,

$$P(X_5 = 4) = \sum_{i=1}^{9} p(i) p_{i4}^5 = \frac{1}{9} \sum_{i=1}^{9} p_{i4}^5$$

$$= \frac{1}{9}\left(\frac{5}{18} + 0 + \frac{2}{9} + 0 + \frac{1}{4} + 0 + \frac{5}{18} + 0 + \frac{2}{9}\right) = 0.139. \quad \blacklozenge$$

Classifications of States of a Markov Chain

Let $\{X_n : n = 0, 1, \dots\}$ be a Markov chain with state space S and transition probability matrix P. A state j is said to be **accessible** from state i if there is a positive probability that, starting from i, the Markov chain will visit state j after a finite number of transitions. If j is accessible from i, we write $i \to j$. Therefore, $i \to j$ if for some $n \ge 0$, $p_{ij}^n > 0$. If i and j are accessible from each other, then we say that i and j **communicate** and write $i \leftrightarrow j$. Clearly, communication is a *relation* on the state space of the Markov chain. We will now show that this relation is an *equivalence relation*. That is, it is reflexive, symmetric, and transitive.

Reflexivity: For all $i \in S$, $i \leftrightarrow i$ since $p_{ii}^0 = 1 > 0$.

Symmetry: If $i \leftrightarrow j$, then $j \leftrightarrow i$. This follows from the definition of i and j being accessible from each other.

Transitivity: If $i \leftrightarrow j$ and $j \leftrightarrow k$, then $i \leftrightarrow k$. To show this, we will establish that $i \to k$. The proof that $k \to i$ is similar. Now $i \to j$ implies that there exists $n \ge 0$ such that $p_{ij}^n > 0$; $j \to k$ implies that there exists $m \ge 0$ such that $p_{jk}^m > 0$. By the Chapman-Kolmogorov equations,

$$p_{ik}^{n+m} = \sum_{\ell=0}^{\infty} p_{i\ell}^n p_{\ell k}^m \ge p_{ij}^n p_{jk}^m > 0,$$

showing that $i \to k$.

As we know, an equivalence relation on a set divides that set into a collection of disjoint subsets, called *equivalence classes*, or simply *classes*. For a Markov chain, the equivalence relation defined by *communication* divides the state space into a collection of disjoint classes, where each class contains all of those elements of the state space that communicate with each other. Therefore, the states that communicate with each other belong to the same class. If all of the states of a Markov chain communicate with each other, then there is only one class. In such a case, the Markov chain is called **irreducible**.

To specify the classes of a Markov chain and to determine which states communicate and which do not, it is often helpful to draw a graph with the states as vertices. A directed line between i and j is drawn if $p_{ij} > 0$. Such a geometric representation is called a **transition graph**.

Example 12.20   Consider a Markov chain with transition probability matrix

$$P = \begin{pmatrix} 0 & 2/7 & 0 & 5/7 & 0 \\ 5/6 & 0 & 1/6 & 0 & 0 \\ 0 & 0 & 0 & 2/5 & 3/5 \\ 0 & 0 & 1/2 & 0 & 1/2 \\ 0 & 0 & 0 & 0 & 1 \end{pmatrix}.$$

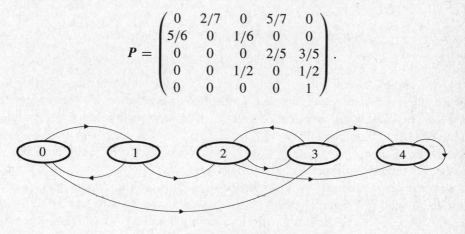

Figure 12.2   Transition graph of Example 12.20.

Figure 12.2 is a transition graph for this Markov chain. It shows that states 0 and 1 communicate so they belong to the same class. State 2 is accessible from 1, but not vice versa. So 2 does not belong to the class of 0 and 1. States 3 and 2 communicate. Therefore, 3 does not belong to the class of 0 and 1 either. States 2 and 3 belong to the same class. State 4 is accessible from states 0, 1, 2, and 3, but no state is accessible from 4. So 4 belongs to a class by itself. Thus this Markov chain consists of three classes: $\{0, 1\}$, $\{2, 3\}$, and $\{4\}$.   ♦

In Example 12.20, note that, for state 4, $p_{44} = 1$. That is, once the process enters 4, it will stay there forever. Such states are called **absorbing**. In general, state i of a Markov chain is absorbing if $p_{ii} = 1$.

Example 12.21   A Markov chain with transition probability matrix

$$p = \begin{pmatrix} 0 & 1 & 0 & 0 \\ 0 & 0 & 1 & 0 \\ 0 & 0 & 0 & 1 \\ 1/8 & 0 & 7/8 & 0 \end{pmatrix}$$

is irreducible. As Figure 12.3 shows, all four states 0, 1, 2, and 3 communicate.   ◆

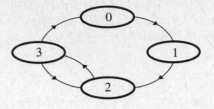

Figure 12.3   Transition graph of Example 12.21.

For a Markov chain $\{X_n : n = 0, 1, \ldots\}$, let f_{ii}^n be the probability that, starting from state i, the process will return to state i, for the first time, after *exactly* n transitions. Let f_i be the probability that, starting from state i, the process will return to state i after a finite number of transitions. Clearly, $f_i = \sum_{n=1}^{\infty} f_{ii}^n$. If $f_i = 1$, then the state i is called **recurrent**. Suppose that, starting from i, the process returns to i with probability 1. Since each time at i the process probabilistically restarts itself, the first return to i implies a second return to i, and so on. Therefore, if state i is recurrent, then the process enters i, with probability 1, infinitely many times. State i is called **transient** if $f_i < 1$—that is, if, starting from i, there is a positive probability that the process does not return to i. Starting from i, for a transient state i, the probability that the process returns to i exactly n times is $(f_i)^n (1 - f_i)$, $n \geq 0$. Thus the number of returns to i is a geometric random variable with parameter $1 - f_i$, and hence the average number of returns to i is $1/(1 - f_i) < \infty$. Therefore, if i is transient, then the average number of returns to i is finite. The following theorem gives a tool to determine whether a state is transient or it is recurrent.

Theorem 12.6   *For a Markov chain* $\{X_n : n = 0, 1, \ldots\}$ *with transition probability matrix* $P = (p_{ij})$,

(a) $\sum_{n=1}^{\infty} p_{ii}^n = \infty$ *if and only if state* i *is recurrent.*

(b) $\sum_{n=1}^{\infty} p_{ii}^n < \infty$ *if and only if state* i *is transient.*

Proof: Clearly, (a) and (b) are equivalent. We will show that $\sum_{n=1}^{\infty} p_{ii}^n$ is the average number of returns to state i, given that $X_0 = i$. This will establish the theorem for the following reasons. If i is recurrent, then the number of returns to state i is infinite; hence $\sum_{n=1}^{\infty} p_{ii}^n = \infty$. The contrapositive of this statement will then be true as well. That is, if $\sum_{n=1}^{\infty} p_{ii}^n < \infty$, then i is transient. Now suppose that i is transient; then the average number of returns to i is finite; hence $\sum_{n=1}^{\infty} p_{ii}^n < \infty$. This statement being true implies that its contrapositive is also true. That is, if $\sum_{n=1}^{\infty} p_{ii}^n = \infty$, then i is recurrent.

Therefore, what is left to show is that, given $X_0 = i$, $\sum_{n=1}^{\infty} p_{ii}^n$ is the average number of returns to state i. To prove this, for $n \geq 1$, let

$$Z_n = \begin{cases} 1 & \text{if } X_n = i \\ 0 & \text{if } X_n \neq i. \end{cases}$$

The number of returns to state i is $\sum_{n=1}^{\infty} Z_n$, and we have

$$E\left(\sum_{n=1}^{\infty} Z_n \mid X_0 = i\right) = \sum_{n=1}^{\infty} E(Z_n \mid X_0 = i) = \sum_{n=1}^{\infty} P(X_n = i \mid X_0 = i) = \sum_{n=1}^{\infty} p_{ii}^n. \quad \blacklozenge$$

We will now discuss some of the most important properties of recurrent and transient states.

(a) *If state i is transient, then the process will return to i only a finite number of times.*

This follows from the facts that, the number of returns to i is a nonnegative random variable, and the expected number of returns to a transient state is finite. To prove this directly, note that, starting from i, for a transient state i, the probability that the process will return to state i exactly n times is $f_i^n(1 - f_i)$, $n \geq 0$. Therefore, the probability that at some point the process will never return to i is

$$\sum_{n=0}^{\infty} f_i^n(1 - f_i) = (1 - f_i) \sum_{n=0}^{\infty} f_i^n = (1 - f_i) \cdot \frac{1}{1 - f_i} = 1.$$

(b) *A finite-state Markov chain has at least one recurrent state.*

Suppose not; then all states are transient, and hence each state will be entered a finite number of times. Suppose that there are n states, and they are entered $\alpha_1, \alpha_2,$ $\ldots,$ and α_n times, respectively. Then, after $\alpha_1 + \alpha_2 + \cdots + \alpha_n$ transitions, there will be no state to enter anymore; a contradiction. Thus at least one recurrent state must exist.

(c) *Recurrence is a class property. That is, if state i is recurrent and state j communicates with state i, then state j is also recurrent.*

Since i and j communicate, there exist n and m so that $p_{ij}^n > 0$ and $p_{ji}^m > 0$. Since i is recurrent, $\sum_{k=1}^{\infty} p_{ii}^k = \infty$. For $k \geq 1$, applying Chapman-Kolmogorov equations repeatedly yields

$$p_{jj}^{n+m+k} = \sum_{\ell=0}^{\infty} p_{j\ell}^m p_{\ell j}^{n+k} \geq p_{ji}^m p_{ij}^{n+k} = p_{ji}^m \sum_{\ell=0}^{\infty} p_{i\ell}^k p_{\ell j}^n \geq p_{ji}^m p_{ii}^k p_{ij}^n.$$

Hence

$$\sum_{k=1}^{\infty} p_{jj}^{n+m+k} \geq \sum_{k=1}^{\infty} p_{ji}^m p_{ii}^k p_{ij}^n = p_{ji}^m p_{ij}^n \sum_{k=1}^{\infty} p_{ii}^k = \infty,$$

since $p_{ji}^m > 0$, $p_{ij}^n > 0$, and $\sum_{k=1}^{\infty} p_{ii}^k = \infty$. This implies that $\sum_{k=1}^{\infty} p_{jj}^{n+m+k} = \infty$, which gives

$$\sum_{k=1}^{\infty} p_{jj}^k \geq \sum_{k=1}^{\infty} p_{jj}^{n+m+k} = \infty,$$

or $\sum_{k=1}^{\infty} p_{jj}^k = \infty$. Hence j is recurrent as well.

(d) *Transience is a class property. That is, if state i is transient, and state j communicates with state i, then state j is also transient.*

Suppose that j is not transient; then it is recurrent. Since j communicates with i, by property (c), i must also be recurrent; a contradiction. Therefore, state j is transient as well.

(e) *In an irreducible Markov chain, either all states are transient, or all states are recurrent. In a reducible Markov chain, the elements of each class are either all transient, or they are all recurrent. In the former case the class is called a **transient class**; in the latter case it is called a **recurrent class**.*

These are all immediate results of the facts that transience and recurrence are class properties.

(f) *In a finite irreducible Markov chain, all states are recurrent.*

By property (b), we know that a finite Markov chain has at least one recurrent state. Since each of the other states communicates with that state, the states are all recurrent.

(g) *Once a Markov chain enters a recurrent class R, it will remain in R forever.*

Suppose that the process leaves a state $i \in R$ and enters a state $j \notin R$. If this happens, then we must have $p_{ij} > 0$. However, since i is recurrent, eventually the process will have to return to i. This makes i accessible from j as well. Therefore, we must also have $p_{ji} > 0$. But then $p_{ij} > 0$ and $p_{ji} > 0$ imply that i communicates with j, implying that $j \in R$, a contradiction. Hence once in a recurrent class, the Markov chain cannot leave that class. This discussion also leads us to the observation that

For a recurrent class R, if $i \in R$ and $j \notin R$, then $p_{ij} = 0$.

Definition   *Let i be a recurrent state of a Markov chain. The state i is called* ***positive recurrent*** *if the expected number of transitions between two consecutive returns to i is finite. If a recurrent state i is not positive recurrent, then it is called* ***null recurrent***.

It can be shown that positive recurrence and null recurrence are both class properties. That is, if state i is positive recurrent, and $i \leftrightarrow j$, then state j is also positive recurrent. Similarly, if i is null recurrent, and $i \leftrightarrow j$, then j is also null recurrent. It can also be shown that

> In a finite-state Markov chain, if a state is recurrent, then it is positive recurrent.

For proofs of these theorems, a good reference is *Markov Chains: Gibbs Fields, Monte Carlo Simulation, and Queues*, by Pierre Brémaud (Springer-Verlag, New York Inc., 1999).

Example 12.22   (Random Walks Revisited)   For random walks, introduced in Example 12.12, to find out whether or not the states are recurrent, we need to recall the following concepts and theorems from calculus. Let $a_n \in R$, and consider the series $\sum_{n=1}^{\infty} a_n$. If $\lim_{n \to \infty} \sqrt[n]{|a_n|} < 1$, then $\sum_{n=1}^{\infty} a_n$ is convergent. If $\lim_{n \to \infty} \sqrt[n]{|a_n|} > 1$, then $\sum_{n=1}^{\infty} a_n$ is divergent. This is called the *root test*. If $\lim_{n \to \infty} \sqrt[n]{|a_n|} = 1$, then the root test is inconclusive. Let $\{a_n\}$ and $\{b_n\}$ be sequences of real numbers. We say that $\{a_n\}$ and $\{b_n\}$ are *equivalent* and write $a_n \sim b_n$ if $\lim_{n \to \infty}(a_n/b_n) = 1$. Therefore, if $a_n \sim b_n$, then, by the definition of the limit of a sequence, $\forall \varepsilon > 0$, there exists a positive integer N, such that $\forall n > N$,

$$\left| \frac{a_n}{b_n} - 1 \right| < \varepsilon.$$

For positive sequences $\{a_n\}$ and $\{b_n\}$ this is equivalent to $b_n(1 - \varepsilon) < a_n < b_n(1 + \varepsilon)$, which implies that

$$(1 - \varepsilon) \sum_{n=N+1}^{\infty} b_n < \sum_{n=N+1}^{\infty} a_n < (1 + \varepsilon) \sum_{n=N+1}^{\infty} b_n.$$

Hence, if $a_n \sim b_n$, $a_n > 0$, and $b_n > 0$, then $\sum_{n=1}^{\infty} a_n$ converges if and only if $\sum_{n=1}^{\infty} b_n$ converges. Now consider the random walk of Example 12.12 in which $\{X_n : n = 0, 1, \ldots\}$ is a Markov chain with state space $\{0, \pm 1, \pm 2, \ldots\}$. If for some n, $X_n = i$, then $X_{n+1} = i + 1$ with probability p, and $X_{n+1} = i - 1$ with probability $1 - p$. That is, one-step transitions are possible only from a state i to its adjacent states $i - 1$ and $i + 1$. For convenience, let us say that the random walk moves to the *right* every time the process makes a transition from some state i to its adjacent state $i + 1$, and the random walk moves to the *left* every time it makes a transition from some state i to its adjacent state $i - 1$. A random walk with state space $\{0, \pm 1, \pm 2, \ldots\}$ in which, at each step, the process either moves to the right with probability p or moves to the left with probability $1 - p$ is said to be a **simple random walk**. It should be clear that all of the states of a simple random walk are accessible from each other. Therefore, a simple random walk is irreducible, and hence its states are all transient, or all recurrent. Consider state 0; if we show that 0 is recurrent, then all states are recurrent. Likewise, if we show that 0 is transient, then all states are transient. To investigate whether or not 0 is recurrent, we will examine the convergence or divergence of the sequence $\sum_{n=1}^{\infty} p_{00}^n$. By Theorem 12.6, state 0 is recurrent if and only if $\sum_{n=1}^{\infty} p_{00}^n$ is ∞. To calculate p_{00}^n, note that for the Markov chain to return back to 0, it is necessary that for every transition to the right, there is a transition to the left. Thus it is impossible to move from 0 to 0 in an odd number of transitions. Hence, for $n \geq 1$, $p_{00}^{2n+1} = 0$. However, if starting from 0, in $2n$ transitions the Markov chain makes exactly n transitions to the right and n transitions to the left, then it will return to 0. Since in $2n$ transitions, the number of transitions to the right is a binomial random variable with parameters $2n$ and p, we have

$$p_{00}^{2n} = \binom{2n}{n} p^n (1 - p)^n, \quad n \geq 1.$$

Now 0 is recurrent if and only if

$$\sum_{n=1}^{\infty} p_{00}^{2n} = \sum_{n=1}^{\infty} \frac{(2n)!}{n! \, n!} \, p^n (1 - p)^n$$

is ∞. By Theorem 2.7 (Stirling's formula), $n! \sim \sqrt{2\pi n} \cdot n^n \cdot e^{-n}$. This implies that

$$\frac{(2n)!}{n! \, n!} \sim \frac{\sqrt{4\pi n} \cdot (2n)^{2n} \cdot e^{-2n}}{(\sqrt{2\pi n} \cdot n^n \cdot e^{-n})^2} = \frac{4^n}{\sqrt{\pi n}}.$$

Hence

$$\frac{(2n)!}{n! \, n!} \, p^n (1 - p)^n \sim \frac{4^n}{\sqrt{\pi n}} \, p^n (1 - p)^n,$$

and $\sum_{n=1}^{\infty} p_{00}^{2n}$ is convergent if and only if $\displaystyle\sum_{n=1}^{\infty} \frac{4^n}{\sqrt{\pi n}} \, p^n (1 - p)^n$ is convergent. Applying the root test to this series, and noting that $(\sqrt{\pi n})^{1/n} \to 1$, we find that

$$\lim_{n \to \infty} \sqrt[n]{\frac{4^n}{\sqrt{\pi n}} \, p^n (1 - p)^n} = 4p(1 - p).$$

By calculating the root of the derivative of $f(p) = 4p(1 - p)$, we obtain that the maximum of this function occurs at $p = 1/2$. Thus, for $p < 1/2$ and $p > 1/2$, $4p(1 - p) < f(1/2) = 1$; hence the series $\displaystyle\sum_{n=1}^{\infty} \frac{4^n}{\sqrt{\pi n}} \, p^n (1 - p)^n$ converges. This implies that $\sum_{n=1}^{\infty} p_{00}^{2n}$ is also convergent, implying that 0 is transient. For $p = 1/2$, the root test is inconclusive. However, in that case, the series reduces to $\sum_{n=1}^{\infty} \dfrac{1}{\sqrt{\pi n}}$, which we know from calculus to be divergent. This shows that $\sum_{n=1}^{\infty} p_{00}^n$ is also divergent; hence 0 is recurrent. It can be shown that 0 is null recurrent. That is, starting from 0, even though, with probability 1, the process will return to 0, the expected number of transitions for returning to 0 is ∞. If $p = 1/2$, the simple random walk is called **symmetric**. To summarize, in this example, we have discussed the following important facts.

> For a nonsymmetric simple random walk, all states are transient. For a symmetric simple random walk, all states are null recurrent.

Suppose that a fair coin is flipped independently and successively. Let $n(H)$ be the number of times that heads occurs in the first n flips of the coin. Let $n(T)$ be the number of times that tails occurs in the first n flips. Let $X_0 = 0$ and, for $n \geq 1$, let $X_n = n(H) - n(T)$. Then $\{X_n \colon n \geq 0\}$ is a symmetric simple random walk with state space $\{0, \pm 1, \pm 2, \ldots\}$ in which, every time a heads occurs, the process moves to the "right" with probability 1/2, and every time a tails occurs, the process moves to the "left" with probability 1/2. Clearly, $X_n = i$ if, in step n, the number of heads minus the number of tails is i. Thus, starting from 0, the process will return to 0 every time that the number of heads is equal to the number of tails. Applying the results discussed previously to the random walk $X_n = n(H) - n(T)$, $n \geq 0$, we have the following celebrated result:

> In successive and independent flips of a fair coin, it happens, with probability 1, that at times the number of heads obtained is equal to the number of tails obtained. This happens infinitely often, but the expected number of tosses between two such consecutive times is ∞.

Studying higher-dimensional random walks is also of major interest. For example, a **two-dimensional symmetric random walk** is a Markov chain with state space

$$\{(i, j) \colon i = 0, \pm 1, \pm 2, \ldots, j = 0, \pm 1, \pm 2, \ldots\}$$

in which if for some n, $X_n = (i, j)$, then X_{n+1} will be one of the states $(i + 1, j)$, $(i - 1, j)$, $(i, j + 1)$, or $(i, j - 1)$ with equal probabilities. If the process moves from (i, j) to $(i + 1, j)$, we say that it has moved to the *right*. Similarly, if it moves from (i, j), to $(i - 1, j)$, $(i, j + 1)$, or $(i, j - 1)$, we say that it has moved to the *left*, *up*, or *down*, respectively. A **three-dimensional symmetric random walk** is a Markov chain with state space

$$\{(i, j, k) \colon i = 0, \pm 1, \pm 2, \ldots, j = 0, \pm 1, \pm 2, \ldots, k = 0, \pm 1, \pm 2, \ldots\}$$

in which if for some n, $X_n = (i, j, k)$, then X_{n+1} will be one of the following six states with equal probabilities: $(i + 1, j, k)$, $(i - 1, j, k)$, $(i, j + 1, k)$, $(i, j - 1, k)$, $(i, j, k + 1)$, $(i, j, k - 1)$. It can be shown that

All of the states of a symmetric two-dimensional random walk are null recurrent, and all of the states of a three-dimensional symmetric random walk are transient.   ♦

Example 12.23   The purpose of this example is to state, without proof, a few facts about $M/D/1$ queues that might help enhance our intuitive understanding of recurrence and transience in Markov chains. Consider the $M/D/1$ queue of Example 12.14 in which the arrival process is Poisson with rate λ, and the service times of the customers are a constant d. Let X_n be the number of customers the nth departure leaves behind. We showed that, if $p(k)$ is the probability of k arrivals during a service time, then $\{X_n : n = 0, 1, \dots\}$ is a Markov chain with the transition probability matrix given in Example 12.14. Let $L = \sum_{k=0}^{\infty} k p(k)$; L is the expected number of customers arriving during a service period. If $L > 1$, the size of the queue will grow without bound. If $L = 1$, the system will be unstable, and if $L < 1$, the system is stable. Clearly, the Markov chain $\{X_n : n = 0, 1, \dots\}$ is irreducible since all of its states are accessible from each other. It can be shown that this Markov chain is positive recurrent if $L < 1$, null recurrent if $L = 1$, and transient if $L > 1$.   ♦

Example 12.24   (Branching Processes)   Suppose that before death an organism produces j offspring with probability α_j, $(j \geq 0)$ independently of other organisms. Let X_0 be the size of the initial population of such organisms. The number of all offspring of the initial population, denoted by X_1, is the population size at the first generation. All offspring of the first generation form the second generation, and the population size at the second generation is denoted by X_2, and so on. The stochastic process $\{X_n : n = 0, 1, \dots\}$ is a Markov chain with state space $\{0, 1, 2, \dots\}$. It is called a **branching process** and was introduced by Galton in 1889 when studying the extinction of family names. Therefore, in Galton's study, each "organism" is a family name and an "offspring" is a male child. Let $P = (p_{ij})$ be the transition probability matrix of the Markov chain $\{X_n : n = 0, 1, \dots\}$. Clearly, $p_{00} = 1$. Hence 0 is recurrent. Since the number of offspring of an organism is independent of the number of offspring of other organisms, we have that $p_{i0} = \alpha_0^i$. If $\alpha_0 > 0$, then $p_{i0} > 0$ implying that all states other than 0 are transient. Since a transient state is entered only a finite number of times, for every positive integer N, the set $\{1, 2, \dots, N\}$ is entered a finite number of times. That is, the population sizes of the future generations either get larger than N, for any positive integer N, and hence will increase with no bounds, or else extinction will eventually occur. Let μ be the expected number of offspring of an organism. Then $\mu = \sum_{i=0}^{\infty} i \alpha_i$. Let K_i be the number of offspring of the ith organism of the $(n-1)$st generation. Then $X_n = \sum_{i=1}^{X_{n-1}} K_i$. Since $\{K_1, K_2, \dots\}$ is an independent sequence of random variables, and is independent of X_{n-1}, by Wald's equation (Theorem 10.7),

$$E(X_n) = E(K_1)E(X_{n-1}) = \mu E(X_{n-1}).$$

Let $X_0 = 1$; this relation implies that

$$E(X_1) = \mu,$$
$$E(X_2) = \mu E(X_1) = \mu^2,$$
$$E(X_3) = \mu E(X_2) = \mu^3,$$
$$\vdots$$
$$E(X_n) = \mu^n.$$

We will prove that if $\mu < 1$ (that is, if the expected number of the offspring of an organism is less than 1), then, with probability 1, eventually extinction will occur. To do this, note that

$$P(X_n \geq 1) = \sum_{i=1}^{\infty} P(X_n = i) \leq \sum_{i=1}^{\infty} i P(X_n = i) = E(X_n) = \mu^n.$$

Hence

$$\lim_{n \to \infty} P(X_n \geq 1) \leq \lim_{n \to \infty} \mu^n = 0,$$

which gives that $\lim_{n \to \infty} P(X_n = 0) = 1$. This means that, with probability 1, extinction will eventually occur. This result can be proved for $\mu = 1$ as well. For $\mu > 1$, let A be the event that extinction will occur, given that $X_0 = 1$. Let $p = P(A)$; then

$$p = P(A) = \sum_{i=0}^{\infty} P(A \mid X_1 = i) P(X_1 = i) = \sum_{i=0}^{\infty} p^i \alpha_i.$$

It can be shown that p is the smallest positive root of the equation $x = \sum_{i=0}^{\infty} \alpha_i x^i$ (see Exercise 31). For example, if $\alpha_0 = 1/8$, $\alpha_1 = 3/8$, $\alpha_2 = 1/2$, and $\alpha_i = 0$ for $i > 2$, then $\mu = 11/8 > 1$. Therefore, for such a branching process, p, the probability that extinction will occur, satisfies $x = (1/8) + (3/8)x + (1/2)x^2$. The smallest positive root of this quadratic equation is 1/4. Thus, for organisms that produce no offspring with probability 1/8, one offspring with probability 3/8, and two offspring with probability 1/2, the probability is 1/4 that, starting with one organism, the population of future generations of that organism eventually dies out. ◆

★ **Example 12.25** **(Genetics)**   Recall that in organisms having two sets of chromosomes, called diploid organisms, which we all are, each hereditary character of each individual is carried by a pair of genes. A gene has alternate alleles which usually are dominant A or recessive a, so that the possible pairs of genes are AA, Aa (same as aA), and aa. Suppose that the zeroth generation of a diploid organism consists of *two* individuals of opposite sex of the entire population who are randomly mated. Let the first generation be *two* opposite sex offspring of the zeroth generation who are randomly mated, the second generation be *two* opposite sex offspring of the first generation who are randomly mated, and so on. For $n \geq 0$, define X_n to be $aa \times aa$ if the two individuals of the nth generation both are aa. Define X_n to be $Aa \times aa$ if one individual of the nth generation is AA and the other one is Aa. Define X_n to be $Aa \times aa$, $Aa \times Aa$, $AA \times Aa$, and $AA \times AA$ similarly. Let

State $0 = aa \times aa$ State $1 = AA \times aa$ State $2 = Aa \times aa$

State $3 = Aa \times Aa$ State $4 = AA \times Aa$ State $5 = AA \times AA$.

The following transition probability matrix shows that $\{X_n : n \geq 0\}$ is a Markov chain. That is, given the present state, we can determine the transition probabilities to the next step no matter what the course of transitions in the past was.

$$
P = \begin{pmatrix}
1 & 0 & 0 & 0 & 0 & 0 \\
0 & 0 & 0 & 1 & 0 & 0 \\
1/4 & 0 & 1/2 & 1/4 & 0 & 0 \\
1/16 & 1/8 & 1/4 & 1/4 & 1/4 & 1/16 \\
0 & 0 & 0 & 1/4 & 1/2 & 1/4 \\
0 & 0 & 0 & 0 & 0 & 1
\end{pmatrix}.
$$

To see that, for example, the probabilities of one-step transitions from step 3 to various steps are given by the numbers in the fourth row, suppose that X_n is $Aa \times Aa$. Then an offspring of the nth generation is AA with probability 1/4, Aa with probability 1/2, and aa with probability 1/4. Therefore, X_{n+1} is $aa \times aa$ with probability $(1/4)(1/4) = 1/16$. It is $AA \times aa$ if the father is AA and the mother is aa or vice versa. So X_{n+1} is $AA \times aa$ with probability $2(1/4)(1/4) = 1/8$. Similarly, X_{n+1} is $Aa \times aa$ with probability $2(1/2)(1/4) = 1/4$, it is $Aa \times Aa$ with probability $(1/2)(1/2) = 1/4$, $AA \times Aa$ with probability $2(1/4)(1/2) = 1/4$, and $AA \times AA$ with probability $(1/4)(1/4) = 1/16$.

It should be clear that the finite-state Markov chain above is reducible, and its communication classes are $\{0\}$, $\{1, 2, 3, 4\}$ and $\{5\}$. Clearly, states 0 and 5 are absorbing. Therefore, the classes $\{0\}$ and $\{5\}$ are positive recurrent. The class $\{1, 2, 3, 4\}$ is transient.

For the Markov chain $\{X_n : n \geq 0\}$, starting from state $i \in \{1, 2, 3, 4\}$, absorption to state 0 ($aa \times aa$) will occur at the nth generation if the state at the $(n-1)$st generation is either 2 ($Aa \times aa$) or 3 ($Aa \times Aa$), and the next transition is into state 0. Therefore, the probability of absorption to $aa \times aa$ at the nth generation is

$$
p_{i2}^{n-1} p_{20} + p_{i3}^{n-1} p_{30} = \frac{1}{4} p_{i2}^{n-1} + \frac{1}{16} p_{i3}^{n-1},
$$

where p_{i2}^{n-1} and p_{i3}^{n-1} are the $i2$-entry and $i3$-entry of the matrix P^{n-1}. Similarly, starting from $i \in \{1, 2, 3, 4\}$, the probability of absorption to state $AA \times AA$ at the nth generation is

$$
p_{i3}^{n-1} p_{35} + p_{i4}^{n-1} p_{45} = \frac{1}{16} p_{i3}^{n-1} + \frac{1}{4} p_{i4}^{n-1}. \quad \blacklozenge
$$

★ Absorption Probability[†]

For a Markov chain $\{X_n : n = 0, 1, \ldots\}$, let j be an absorbing state; that is, a state for which $p_{jj} = 1$. To find x_j, the probability that the Markov chain will eventually be absorbed into state j, one useful technique is the **first-step analysis**, in which relations between x_i's are found, by conditioning on the possible first-step transitions, using the law of total probability. It is often possible to solve the relations obtained for x_i's. Examples follow.

Example 12.26   (Gambler's Ruin Problem Revisited)   Consider the gambler's ruin problem (Example 3.14) in which two gamblers play the game of "heads or tails." For $0 < p < 1$, consider a coin that lands heads up with probability p and lands tails up with probability $q = 1 - p$. Each time the coin lands heads up, player A wins \$1 from player B, and each time it lands tails up, player B wins \$1 from A. Suppose that, initially, player A has a dollars and player B has b dollars. Let X_i be the amount of money that player A will have after i games. Clearly, $X_0 = a$ and, as discussed in the beginning of this section, $\{X_n : n = 0, 1, \ldots\}$ is a Markov chain. Its state space is $\{0, 1, \ldots, a, a + 1, \ldots, a + b\}$. Clearly, states 0 and $a + b$ are absorbing. Player A will be ruined if the Markov chain is absorbed into state 0. Player B will be ruined if the Markov chain is absorbed into $a + b$. Obviously, A wins if B is ruined and vice versa. To find the probability that the Markov chain will be absorbed into, say, state 0, for $i \in \{0, 1, \ldots, a + b\}$, suppose that, instead of a dollars, initially player A has i dollars. Let x_i be the probability that A will be ruined given that he begins with i dollars. To apply first-step analysis, note that A will win the first game with probability p and will lose the first game with probability $q = 1 - p$. By the law of total probability,

$$x_i = x_{i+1} \cdot p + x_{i-1} \cdot q, \qquad i = 1, 2, \ldots, a + b - 1.$$

For $p \neq q$, solving these equations with the boundary conditions $x_0 = 1$ and $x_{a+b} = 0$ yields

$$x_i = \frac{1 - (p/q)^{a+b-i}}{1 - (p/q)^{a+b}}, \qquad i = 0, 1, \ldots, a + b.$$

(See Example 3.14 for details.) For $p = q = 1/2$, solving the preceding equations directly gives

$$x_i = \frac{a + b - i}{a + b}, \qquad i = 0, 1, \ldots, a + b.$$

Therefore, the probability that the Markov chain is absorbed into state 0—that is, player A is ruined—is

$$x_a = \begin{cases} \dfrac{1 - (p/q)^b}{1 - (p/q)^{a+b}} & \text{if } p \neq q \\[2mm] \dfrac{b}{a + b} & \text{if } p = q = 1/2. \end{cases}$$

[†]This subsection can be skipped without loss of continuity.

Similar calculations show that the probability that the Markov chain is absorbed into state $a + b$ and, hence, player B is ruined, is

$$\frac{1 - (q/p)^a}{1 - (q/p)^{a+b}},$$

for $p \neq q$, and $a/(a + b)$ for $p = q = 1/2$.   ◆

Example 12.27   In Example 12.10, where a mouse is in a maze searching for food, suppose that a cat is hiding in cell 9, and there is a piece of cheese in cell 1 (see Figure 12.1). Suppose that if the mouse enters cell 9, the cat will eat him. We want to find the probability that the mouse finds the cheese before being eaten by the cat. For $n \geq 0$, let X_n be the cell number the mouse will visit after having changed cells n times. Then $\{X_n : n = 0, 1, \ldots\}$ is a Markov chain with absorbing state 9. To find the desired probability using first-step analysis, for $i = 1, 2, \ldots 9$, let x_i be the probability that, starting from cell i, the mouse enters cell 1 before entering cell 9. Applying the law of total probability repeatedly, we obtain

$$\begin{cases} x_2 = (1/3)x_1 + (1/3)x_3 + (1/3)x_5 \\ x_3 = (1/2)x_2 + (1/2)x_6 \\ x_4 = (1/3)x_1 + (1/3)x_5 + 1/3)x_7 \\ x_5 = (1/4)x_2 + (1/4)x_4 + (1/4)x_6 + (1/4)x_8 \\ x_6 = (1/3)x_3 + (1/3)x_5 + (1/3)x_9 \\ x_7 = (1/2)x_4 + (1/2)x_8 \\ x_8 = (1/3)x_5 + (1/3)x_7 + (1/3)x_9. \end{cases}$$

Solving this system of equations with boundary conditions $x_1 = 1$ and $x_9 = 0$, we obtain

$$\begin{cases} x_2 = x_4 = 2/3 \\ x_3 = x_5 = x_7 = 1/2 \\ x_6 = x_8 = 1/3. \end{cases}$$

Therefore, for example, if the mouse is in cell 3, the probability is 1/2 that he finds the cheese before being eaten by the cat.

Now forget about the cheese altogether, and let x_i be the probability that the mouse will eventually be eaten by the cat if, initially, he is in cell i. Then x_i's satisfy all of the preceding seven equations and the following equation as well:

$$x_1 = (1/2)x_2 + (1/2)x_4.$$

Solving these eight equations in eight unknowns with the boundary condition $x_9 = 1$, we obtain

$$x_1 = x_2 = \cdots = x_8 = 1,$$

showing that no matter which cell initially the mouse is in, the cat will eventually eat the mouse. That is, absorption to state 9 is inevitable.   ◆

★ **Example 12.28**   **(Genetics; The Wright-Fisher Model)**   Suppose that, in a population of N diploid organisms with alternate dominant allele A and recessive allele a, the population size remains constant for all generations. Under random mating, for $n = 0, 1, \ldots$, let X_n be the number of A alleles in the nth generation. Clearly, $\{X_n : n = 0, 1, \ldots\}$ is a Markov chain with state space $\{0, 1, \ldots, 2N\}$ and, given $X_n = i$, X_{n+1} is a binomial random variable with parameters $2N$ and $i/2N$. (Note that there are $2N$ alleles altogether.) Hence, for $0 \le i, j \le 2N$, the transition probabilities are given by

$$p_{ij} = \binom{2N}{j} \left(\frac{i}{2N}\right)^j \left(1 - \frac{i}{2N}\right)^{2N-j}.$$

Furthermore,

$$E(X_{n+1} \mid X_n = i) = 2N \cdot \frac{i}{2N} = i.$$

Hence

$$E(X_{n+1} \mid X_n) = X_n.$$

This implies that

$$E(X_{n+1}) = E\big[E(X_{n+1} \mid X_n)\big] = E(X_n),$$

showing that the expected value of the A alleles is the same for all generations.

The Markov chain $\{X_n : n = 0, 1, \ldots\}$ has two absorbing states, 0 and $2N$. Absorbing to 0 means permanent loss of allele A from the population. Obviously, if for some generation, all individuals are aa, then, from that generation on, the individuals of all future generations will also be aa. Similarly, absorbing to $2N$ means permanent loss of allele a from the population and, hence, all of its future generations. Let x_i be the probability that the Markov chain will be absorbed into 0 or $2N$ given that, initially, the number of A alleles is i. By the law of total probability,

$$\begin{cases} x_i = \displaystyle\sum_{j=0}^{2N} \binom{2N}{j} \left(\frac{i}{2N}\right)^j \left(1 - \frac{i}{2N}\right)^{2N-j} x_j, & 1 \le i \le 2N - 1, \\ x_0 = x_{2N} = 1. \end{cases}$$

It is readily seen that the solution to this system of $2N+1$ equations in $2N+1$ unknowns is

$$x_0 = x_1 = \cdots = x_{2N} = 1.$$

This is true since by the binomial expansion,

$$\sum_{j=0}^{2N} \binom{2N}{j} \left(\frac{i}{2N}\right)^{2N} \left(1 - \frac{i}{2N}\right)^{2N-j} = \left(\frac{i}{2N} + 1 - \frac{i}{2N}\right)^{2N} = 1.$$

Having $x_0 = x_1 = \cdots = x_{2N} = 1$ shows that no matter what the number of A alleles in the initial population is, the Markov chain will eventually be absorbed into state 0 or state $2N$ with probability 1. We will now find y_i, the probability that the Markov chain will be absorbed into state $2N$ given that $X_0 = i$, $0 \le i \le 2N$. Note that $(y_0, y_1, \ldots, y_{2N})$ is the solution to the following system of linear equations obtained by the law of total probability:

$$
\begin{cases}
y_0 = 0 \\
y_i = \displaystyle\sum_{j=0}^{2N} \binom{2N}{j}\left(\frac{i}{2N}\right)^j \left(1 - \frac{i}{2N}\right)^{2N-j} y_j, \quad 1 \le i \le 2N - 1 \\
y_{2N} = 1.
\end{cases}
$$

If we show that, for $0 \le i \le N$, $y_i = i/2N$ satisfy the equations of this system, we are done. The fact that, for $i = 0$ and $2N$, $y_i = i/2N$ is trivial. For $1 \le i \le 2N - 1$,

$$
\sum_{j=0}^{2N} \binom{2N}{j}\left(\frac{i}{2N}\right)^j \left(1 - \frac{i}{2N}\right)^{2N-j} y_j = \sum_{j=0}^{2N} \binom{2N}{j}\left(\frac{i}{2N}\right)^j \left(1 - \frac{i}{2N}\right)^{2N-j} \cdot \frac{j}{2N}
$$

$$
= \frac{1}{2N} \sum_{j=0}^{2N} j \binom{2N}{j}\left(\frac{i}{2N}\right)^j \left(1 - \frac{i}{2N}\right)^{2N-j}
$$

$$
= \frac{1}{2N} \cdot 2N \cdot \frac{i}{2N} = \frac{i}{2N} = y_i,
$$

where

$$
\sum_{j=0}^{2N} j \binom{2N}{j}\left(\frac{i}{2N}\right)^j \left(1 - \frac{i}{2N}\right)^{2N-j} = 2N \cdot \frac{i}{2N}
$$

follows since it is the expected value of a binomial random variable with parameters $2N$ and $i/2N$. Hence, if initially the number of A alleles is i, the probability that eventually all of the individuals in the population are AA is $i/2N$. The probability that eventually they are all aa is $1 - (i/2N)$. ◆

Period

Let $\{X_n : n = 0, 1, \ldots\}$ be a Markov chain with state space S and transition probability matrix $P = (p_{ij})$. For $i \in S$, suppose that, starting from i, there is *only* a positive probability to return to i after $n_1, n_2, \ldots$ transitions. Then $p_{ii}^{n_1} > 0$, $p_{ii}^{n_2} > 0, \ldots$. However, $p_{ii}^n = 0$ if $n \notin \{n_1, n_2, \ldots\}$. Let d be the greatest common divisor of $n_1, n_2, \ldots$. Then d is said to be the **period** of i. Clearly, $p_{ii}^n = 0$ *if n is not a multiple of d*. However, not for all multiples of d, a return to the state i has a positive probability. It can be shown that, for a state i with period d, there is a positive integer N, depending on i, such that $p_{ii}^{nd} > 0$ for all $n \ge N$. That is, for sufficiently large multiples of d, a return to state i always has a positive probability. A state with period 1 is said to be **aperiodic**. Clearly, *if $p_{ii} > 0$, then i is aperiodic.* If $p_{ii}^n = 0$ for all $n \geq 1$, then the period of i is defined to be 0. It can be shown that period is a class property. That is,

If $i \leftrightarrow j$, then i and j have the same period.

(See Exercise 28.)

Example 12.29   As discussed in Example 12.22, in a simple random walk, the set of all positive integers $n \geq 1$ for which $p_{00}^n > 0$ is $\{2, 4, 6, \dots\}$. Since the greatest common divisor of this set is 2, the period of state 0, and hence any other state, is 2.   ♦

Example 12.30   Consider a Markov chain $\{X_n : n = 0, 1, \dots\}$ with state space $\{0, 1, \dots, m - 2, m - 1\}$ and transition graph given by Figure 12.4. Clearly, the set of all integers $n \geq 1$ for which $p_{00}^n > 0$ is $\{m, 2m, 3m, \dots\}$. Since the greatest common divisor of this set is m, the period of 0 is m. Since this Markov chain is irreducible, the period of any other state is also m.   ♦

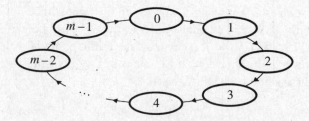

Figure 12.4   Transition graph of Example 12.30.

Example 12.31   Consider a Markov chain with state space $\{0, 1, 2, 3\}$, and transition graph given by Figure 12.5. The set of all integers $n \geq 1$ for which $p_{00}^n > 0$ is $\{4, 6, 8, \dots\}$. For example, $p_{00}^6 > 0$, since it is possible to return to 0 in 6 transitions: $0 \to 1 \to 2 \to 3 \to 2 \to 3 \to 0$. The greatest common divisor of $\{4, 6, 8, \dots\}$ is 2. So the period of 0 is 2 while $p_{00}^2 = 0$.   ♦

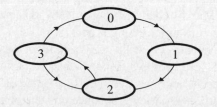

Figure 12.5   Transition graph of Example 12.31.

Example 12.32   Consider a Markov chain with state space $\{0, 1, 2\}$ and transition graph given by Figure 12.6. The set of all integers $n \geq 1$ for which $p_{00}^n > 0$ is $\{2, 3, 4, \dots\}$. Since the greatest common divisor of this set is 1, the period of 0 is 1. This Markov chain is irreducible, so the periods of states 1 and 2 are also 1. Therefore, this is an irreducible aperiodic Markov chain.   ◆

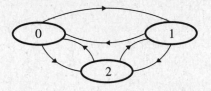

Figure 12.6   Transition graph of Example 12.32.

Steady-State Probabilities

Let us revisit Example 12.9 in which a working traffic light will be out of order the next day with probability 0.07, and an out-of-order traffic light will be working the next day with probability 0.88. Let $X_n = 1$, if on day n the traffic light will work; $X_n = 0$, if on day n it will not work. We showed that $\{X_n : n = 0, 1 \dots\}$ is a Markov chain with state space $\{0, 1\}$ and transition probability matrix

$$P = \begin{pmatrix} 0.12 & 0.88 \\ 0.07 & 0.93 \end{pmatrix}.$$

Direct calculations yield

$$P^{(6)} = P^6 = \begin{pmatrix} 0.0736842 & 0.926316 \\ 0.0736842 & 0.926316 \end{pmatrix}.$$

This shows that, whether or not the traffic light is working today, in six days, the probability that it will be working is 0.926316, and the probability that it will be out of order is 0.0736842. This phenomenon is not accidental. For certain Markov chains, after a large number of transitions, the probability of entering a specific state becomes independent of the initial state of the Markov chain. Mathematically, this means that for such Markov chains $\lim_{n \to \infty} p_{ij}^n$ converges to a limiting probability that is independent of the initial state i. For some Markov chains, these limits either cannot exist or they do not converge to limiting probabilities. For example, suppose that a Markov chain is reducible and has two recurrent classes R_1 and R_2. Let $j \in R_2$; then $\lim_{n \to \infty} p_{ij}^n$ might converge to a limiting probability for $i \in R_2$. However, since no state of R_1 is accessible from any state of R_2 and vice versa, for $i \in R_1$, $p_{ij}^n = 0$, for all $n \geq 1$. That is, it is possible for the sequence $\{p_{ij}^n\}_{n=1}^{\infty}$ to converge to a probability for $i \in R_2$. Since it converges to

0 for $i \in R_1$, it is impossible for $\{p_{ij}^n\}_{n=1}^{\infty}$ to converge to a limiting probability that is independent of state i. Thus we have established that, for p_{ij}^n to converge, we need to have an irreducible recurrent Markov chain. It turns out that two more conditions will be necessary. The recurrence must be positive recurrence, and the irreducible Markov chain needs to be aperiodic. To see that this latter condition is necessary, consider the Ehrenfest chain of Example 12.15. There are N balls numbered $1, 2, \ldots, N$ distributed among two urns randomly. At time n, a number is selected at random from the set $\{1, 2 \ldots, N\}$. Then the ball with that number is found in one of the two urns and is moved to the other urn. Let X_n denote the number of balls in urn I after n transfers. $\{X_n : n = 0, 1, \ldots\}$ is an irreducible Markov chain with state space $\{0, 1, \ldots, N\}$. Suppose that $X_0 = 3$; then the possible values for X_i's are given in the following table.

X_0	X_1	X_2	X_3	X_4	X_5	$\ldots$
3	2, 4	1, 3, 5	0, 2, 4, 6	1, 3, 5, 7	0, 2, 4, 6, 8	$\ldots$

Now, for p_{ij}^n to converge to a limiting probability that is independent of i, we need that p_{jj}^n also converge to the same quantity. However, the preceding table should clarify that, for $0 \le j \le N$, $p_{jj}^n = 0$ for any odd n. This makes it impossible for p_{jj}^n, and hence p_{ij}^n, to converge to a limiting probability that is independent of i. The reason for this phenomenon is that the irreducible Ehrenfest Markov chain has period 2 and is not aperiodic.

In general, it can be shown that for an irreducible, positive recurrent, aperiodic Markov chain $\{X_n : n = 0, 1, \ldots\}$ with state space $\{0, 1, 2, \ldots\}$ and transition probability matrix $\boldsymbol{P} = (p_{ij})$, $\lim_{n \to \infty} p_{ij}^n$ exists and is independent of i. The limit is denoted by π_j, and $\sum_{j=0}^{n} \pi_j = 1$. In such a case, we can assume that

$$\lim_{n \to \infty} P(X_n = j) = \pi_j.$$

Since, by conditioning on X_n,

$$P(X_{n+1} = j) = \sum_{i=0}^{\infty} P(X_{n+1} = j \mid X_n = i) P(X_n = i) = \sum_{i=0}^{\infty} p_{ij} P(X_n = i),$$

as $n \to \infty$, we must have

$$\pi_j = \sum_{i=0}^{\infty} p_{ij} \pi_i, \quad j \ge 0. \tag{12.11}$$

This system of equations along with $\sum_{j=0}^{\infty} \pi_j = 1$ enable us to find the limiting probabilities π_j. Let $\boldsymbol{\Pi} = \begin{pmatrix} \pi_0 \\ \pi_1 \\ \vdots \end{pmatrix}$, and let $\boldsymbol{P}^T$ be the transpose of the transition probability matrix $\boldsymbol{P}$; then equations (12.11) in matrix form are

$$\boldsymbol{\Pi} = \boldsymbol{P}^T \boldsymbol{\Pi}.$$

In particular, for a finite-state Markov chain with space $\{0, 1, \ldots, n\}$, this equation is

$$
\begin{pmatrix} \pi_0 \\ \pi_1 \\ \vdots \\ \pi_n \end{pmatrix} = \begin{pmatrix} p_{00} & p_{10} & \cdots & p_{n0} \\ p_{01} & p_{11} & \cdots & p_{n1} \\ \vdots & & & \\ p_{0n} & p_{1n} & \cdots & p_{nn} \end{pmatrix} \begin{pmatrix} \pi_0 \\ \pi_1 \\ \vdots \\ \pi_n \end{pmatrix}.
$$

If for a Markov chain, for each $j \geq 0$, $\lim_{n \to \infty} p_{ij}^n$ exists and is independent of i, we say that the Markov chain is in **equilibrium** or **steady state**. The limits $\pi_j = \lim_{n \to \infty} p_{ij}^n$, $j \geq 0$, are called the **stationary probabilities** of the Markov chain. We will now show that π_j, the stationary probability that the Markov chain is in state j, is the long-run proportion of the number of transitions to state j. To see this, let

$$
Z_k = \begin{cases} 1 & \text{if } X_k = j \\ 0 & \text{if } X_k \neq j. \end{cases}
$$

Then $\dfrac{1}{n} \sum_{k=0}^{n-1} Z_k$ is the average number of visits to state k between times 0 and $n - 1$. We have

$$
E\left(\frac{1}{n} \sum_{k=0}^{n-1} Z_k \;\Big|\; X_0 = i \right) = \frac{1}{n} \sum_{k=0}^{n-1} E(Z_k \mid X_0 = i)
$$

$$
= \frac{1}{n} \sum_{k=0}^{n-1} P(X_k = j \mid X_0 = i) = \frac{1}{n} \sum_{k=0}^{n-1} p_{ij}^k.
$$

From calculus, we know that, if a sequence of real numbers $\{a_n\}_{n=0}^{\infty}$ converges to ℓ, then $\dfrac{1}{n} \sum_{k=0}^{n-1} a_k$ converges to ℓ as well. Since $\{p_{ij}^n\}$ converges to π_j, we have that $\dfrac{1}{n} \sum_{k=0}^{n-1} p_{ij}^k$ also converges to π_j, showing that the proportion of times the Markov chain enters j, in the long-run, is π_j. Now suppose that for some j, π_j is, say, 1/5. Then, in the long-run, the fraction of visits to state j is 1/5. Therefore, on average, the number of transitions between two consecutive visits to state j must be 5. This simple observation can be proved in general: $1/\pi_j$ is the expected number of transitions between two consecutive visits to state j. We summarize the preceding discussion in the following theorem.

Theorem 12.7   *Let $\{X_n \colon n = 0, 1, \ldots\}$ be an irreducible, positive recurrent, aperiodic Markov chain with state space $\{0, 1, \ldots\}$ and transition probability matrix $\mathbf{P} = (p_{ij})$. Then, for each $j \geq 0$, $\lim_{n \to \infty} p_{ij}^n$ exists and is independent of i. Let $\pi_j = \lim_{n \to \infty} p_{ij}^n$, $j \geq 0$ and $\boldsymbol{\pi} = \begin{pmatrix} \pi_0 \\ \pi_1 \\ \vdots \end{pmatrix}$. We have*

(a) $\Pi = P^T \Pi$, and $\sum_{j=0}^{\infty} \pi_j = 1$. *Furthermore, these equations determine the stationary probabilities, $\pi_0, \pi_1, \ldots$, uniquely.*

(b) π_j *is the long-run proportion of the number of transitions to state j, $j \geq 0$.*

(c) *The expected number of transitions between two consecutive visits to state j is $1/\pi_j$, $j \geq 0$.*

Remark 12.1   If $\{X_n : n = 0, 1, \ldots\}$ is an irreducible, positive recurrent Markov chain, but it is *periodic*, then the system of equations, $\Pi = P^T \Pi$, $\sum_{j=0}^{\infty} \pi_j = 1$, still has a *unique* solution. However, for $j \geq 0$, π_j is no longer the limiting probability that the Markov chain is in state j. It is the long-run *proportion* of the number of visits to state j. ◆

Remark 12.2   The property that the limiting probability $\lim_{n \to \infty} p_{ij}^n$ exists and is independent of the initial state i is called **ergodicity**. Any Markov chain with this property is called **ergodic**. ◆

Remark 12.3   The equations obtained from $\Pi = P^T \Pi$, in some sense, are balancing equations when the Markov chain is in steady state. For each i, they basically equate the "probability flux" out of i to the "probability flux" into i. For a Markov chain with n states, it should be clear that, if the probability flux is balanced for $n - 1$ states, then it has to be balanced for the remaining state as well. That is why $\Pi = P^T \Pi$ gives one redundant equation, and that is why we need the additional equation, $\sum_{j=0}^{\infty} \pi_j = 1$, to calculate the unique π_j's. ◆

Example 12.33   On a given day, a retired English professor, Dr. Charles Fish, amuses himself with *only* one of the following activities: reading (activity 1), gardening (activity 2), or working on his book about a river valley (activity 3). For $1 \leq i \leq 3$, let $X_n = i$ if Dr. Fish devotes day n to activity i. Suppose that $\{X_n : n = 1, 2, \ldots\}$ is a Markov chain, and depending on which of these activities he chooses on a certain day, the probability of engagement in any one of the activities on the next day is given by the transition probability matrix

$$P = \begin{pmatrix} 0.30 & 0.25 & 0.45 \\ 0.40 & 0.10 & 0.50 \\ 0.25 & 0.40 & 0.35 \end{pmatrix}.$$

Find the proportion of days Dr. Fish devotes to each activity.

Solution: Clearly, the Markov chain is irreducible, aperiodic, and recurrent. Since it is finite state, it is also positive recurrent. Let π_1, π_2, and π_3 be the proportion of days Dr. Fish devotes to reading, gardening, and writing, respectively. Then, by Theorem 12.7,

π_1, π_2, and π_3 are obtained from solving the system of equations

$$\begin{pmatrix} \pi_1 \\ \pi_2 \\ \pi_3 \end{pmatrix} = \begin{pmatrix} 0.30 & 0.40 & 0.25 \\ 0.25 & 0.10 & 0.40 \\ 0.45 & 0.50 & 0.35 \end{pmatrix} \begin{pmatrix} \pi_1 \\ \pi_2 \\ \pi_3 \end{pmatrix}$$

along with $\pi_1 + \pi_2 + \pi_3 = 1$. The preceding matrix equation gives the following system of equations:

$$\begin{cases} \pi_1 = 0.30\pi_1 + 0.40\pi_2 + 0.25\pi_3 \\ \pi_2 = 0.25\pi_1 + 0.10\pi_2 + 0.40\pi_3 \\ \pi_3 = 0.45\pi_1 + 0.50\pi_2 + 0.35\pi_3. \end{cases}$$

By choosing any two of these equations along with the relation $\pi_1 + \pi_2 + \pi_3 = 1$, we obtain a system of 3 equations in 3 unknowns. Solving that system yields $\pi_1 = 0.306163$, $\pi_2 = 0.272366$, and $\pi_3 = 0.421471$. Therefore, Dr. Charles Fish devotes approximately 31% of the days to reading, 27% of the days to gardening, and 42% to writing. ◆

Example 12.34   An engineer analyzing a series of digital signals generated by a testing system observes that only 1 out of 15 highly distorted signals follows a highly distorted signal, with no recognizable signal between, whereas 20 out of 23 recognizable signals follow recognizable signals, with no highly distorted signal between. Given that only highly distorted signals are not recognizable, find the fraction of signals that are highly distorted.

Solution: For $n \geq 1$, let $X_n = 1$, if the nth signal generated is highly distorted; $X_n = 0$, if the nth signal generated is recognizable. Then $\{X_n : n = 0, 1, \ldots\}$ is a Markov chain with state space $\{0, 1\}$ and transition probability matrix

$$P = \begin{pmatrix} 20/23 & 3/23 \\ 14/15 & 1/15 \end{pmatrix}.$$

Furthermore, the Markov chain is irreducible, positive recurrent, and aperiodic. Let π_0 be the fraction of signals that are recognizable, and let π_1 be the fraction of signals that are highly distorted. Then, by Theorem 12.7, π_0 and π_1 satisfy

$$\begin{pmatrix} \pi_0 \\ \pi_1 \end{pmatrix} = \begin{pmatrix} 20/23 & 14/15 \\ 3/23 & 1/15 \end{pmatrix} \begin{pmatrix} \pi_0 \\ \pi_1 \end{pmatrix},$$

which gives the following system of equations:

$$\begin{cases} \pi_0 = \dfrac{20}{23}\pi_0 + \dfrac{14}{15}\pi_1 \\ \pi_1 = \dfrac{3}{23}\pi_0 + \dfrac{1}{15}\pi_1. \end{cases}$$

By choosing one of these equations along with $\pi_0 + \pi_1 = 1$, we obtain a system of two equations in two unknowns. Solving that system yields $\pi_0 = 322/367 \approx 0.877$, and

$\pi_1 = 45/367 \approx 0.123$. Therefore, approximately 12.3% of the signals generated by the testing system are highly distorted. ◆

Example 12.35 A Markov chain $\{X_n : n = 0, 1, \dots\}$ with state space $\{0, 1, \dots, k-1\}$ and transition probability matrix $P = (p_{ij})$ is called **doubly stochastic** if, in addition to the sum of the entries of each row of P being 1, the sum of the entries of each column of P is 1 as well—that is, if $\sum_{i=0}^{k-1} p_{ij} = 1$, for all j. Show that, for an irreducible, aperiodic, doubly stochastic, positive recurrent Markov chain, the limiting probabilities are $\pi_i = 1/k$, for $0 \le i \le k - 1$.

Solution: The system of equations $\Pi = P^T \Pi$, $\sum_{j=0}^{k-1} \pi_j = 1$ has exactly one solution.

If we show that $\Pi = \begin{pmatrix} 1/k \\ 1/k \\ \vdots \\ 1/k \end{pmatrix}$ satisfies this system, we have the theorem. Obviously,

$\pi_i = 1/k, 0 \le i \le k-1$, satisfy $\sum_{j=0}^{k-1} \pi_j = 1$. We can verify readily that $\sum_{i=0}^{k-1} p_{ij} = 1$, $0 \le j \le k-1$, imply that

$$\begin{pmatrix} 1/k \\ 1/k \\ \vdots \\ 1/k \end{pmatrix} = \begin{pmatrix} p_{00} & p_{10} & \cdots & p_{(k-1)0} \\ p_{01} & p_{11} & \cdots & p_{(k-1)1} \\ \vdots & & & \\ p_{0(k-1)} & p_{1(k-1)} & \cdots & p_{(k-1)(k-1)} \end{pmatrix} \begin{pmatrix} 1/k \\ 1/k \\ \vdots \\ 1/k \end{pmatrix}.$$

Thus the conditions are satisfied, and the theorem is proved. ◆

Example 12.36   For an English course, there are four popular textbooks dominating the market. The English department of an institution allows its faculty to teach only from these four textbooks. Each year, Professor Rosemary O'Donoghue adopts the same book she was using the previous year with probability 0.64. The probabilities of her changing to any of the other three books are equal. Find the proportion of years Professor O'Donoghue uses each book.

Solution: For $1 \le i \le 4$, let $X_n = i$ if on year n Professor O'Donoghue teaches from book i. Then $\{X_n : n = 0, 1, 2, \dots\}$ is a Markov chain with state space $\{1, 2, 3, 4\}$ and transition probability matrix

$$P = \begin{pmatrix} 0.64 & 0.12 & 0.12 & 0.12 \\ 0.12 & 0.64 & 0.12 & 0.12 \\ 0.12 & 0.12 & 0.64 & 0.12 \\ 0.12 & 0.12 & 0.12 & 0.64 \end{pmatrix}.$$

Clearly, $\{X_n : n = 0, 1, 2, \dots\}$ is irreducible, aperiodic, and positive recurrent. Moreover, it is doubly stochastic. That is, in addition to the sum of the entries of each row of P being 1, the sum of the entries of each column of P is 1 as well. Therefore, by Example 12.35, for $1 \le i \le 4$, the proportion of years Professor O'Donoghue uses book i is 1/4. ◆

EXERCISES

A

1. In a community, there are N male and M female residents, N, $M > 1000$. Suppose that in a study, people are chosen at random and are asked questions concerning their opinion with regard to a specific issue. Let $X_n = 1$ if the nth person chosen is female, and $X_n = 0$ otherwise. Is $\{X_n : n = 1, 2, \ldots\}$ a Markov chain? Why or why not?

2. For a Markov chain $\{X_n : n = 0, 1, \ldots\}$ with state space $\{0, 1, 2, \ldots\}$ and transition probability matrix $P = (p_{ij})$, let p be the probability mass function of X_0; that is,

    $$p(i) = P(X_0 = i), \qquad i = 0, 1, 2, \ldots .$$

    Find the probability mass function of X_n.

3. Consider a circular random walk in which six points $1, 2, \ldots, 6$ are placed, in a clockwise order, on a circle. Suppose that one-step transitions are possible only from a point to its adjacent points with equal probabilities. Starting from 1, (a) find the probability that in 4 transitions the Markov chain returns to 1; (b) find the probability that in 5 transitions the Markov chain enters to an adjacent point of 1, namely, to 2 or 6.

4. Let $\{X_n : n = 0, 1, \ldots\}$ be a Markov chain with state space $\{0, 1, 2\}$ and transition probability matrix

    $$P = \begin{pmatrix} 1/2 & 1/4 & 1/4 \\ 2/3 & 1/3 & 0 \\ 0 & 0 & 1 \end{pmatrix}.$$

    Starting from 0, what is the probability that the process never enters 1?

5. On a given day, Emmett drives to work (state 1), takes the train (state 2), or hails a taxi (state 3). Let $X_n = 1$ if he drives to work on day n, $X_n = 2$ if he takes the train on day n, and $X_n = 3$ if he hails a taxi on that day. Suppose that $\{X_n : n = 1, 2, \ldots\}$ is a Markov chain, and depending on how Emmett went to work the previous day, the probability of choosing any one of the means of transportation is given by the following transition probability matrix:

    $$P = \begin{pmatrix} 1/6 & 2/3 & 1/6 \\ 1/2 & 1/3 & 1/6 \\ 2/5 & 1/2 & 1/10 \end{pmatrix}.$$

(a)   Given that Emmett took the train today and every day in the last five days, what is the probability that he will not take the train to work tomorrow?

(b)   If Emmett took the train to work today, what is the probability that he will not take the train to work tomorrow and the day after tomorrow?

6.   Consider an Ehrenfest chain with 5 balls (see Example 12.15). If the probability mass function of X_0, the initial number of balls in urn I, is given by

$$P(X_0 = i) = \frac{i}{15}, \quad 0 \le i \le 5,$$

find the probability that, after 6 transitions, urn I has 4 balls.

7.   The following is the transition probability matrix of a Markov chain with state space $\{0, 1, 2, 3, 4\}$. Specify the classes, and determine which classes are transient and which are recurrent.

$$P = \begin{pmatrix} 2/5 & 0 & 0 & 3/5 & 0 \\ 1/3 & 1/3 & 0 & 1/3 & 0 \\ 0 & 0 & 1/2 & 0 & 1/2 \\ 1/4 & 0 & 0 & 3/4 & 0 \\ 0 & 0 & 1/3 & 0 & 2/3 \end{pmatrix}.$$

8.   A fair die is tossed repeatedly. The maximum of the first n outcomes is denoted by X_n. Is $\{X_n : n = 1, 2, \ldots\}$ a Markov chain? Why or why not? If it is a Markov chain, calculate its transition probability matrix, specify the classes, and determine which classes are recurrent and which are transient.

9.   Construct a transition probability matrix of a Markov chain with state space $\{1, 2, \ldots, 8\}$ in which $\{1, 2, 3\}$ is a transient class having period 3, $\{4\}$ is an aperiodic transient class, and $\{5, 6, 7, 8\}$ is a recurrent class having period 2.

10.   The following is the transition probability matrix of a Markov chain with state space $\{1, 2, \ldots, 7\}$. Starting from state 6, find the probability that the Markov chain will eventually be absorbed into state 4.

$$P = \begin{pmatrix} 0.3 & 0.7 & 0 & 0 & 0 & 0 & 0 \\ 0.3 & 0.2 & 0.5 & 0 & 0 & 0 & 0 \\ 0 & 0 & 0 & 0.6 & 0.4 & 0 & 0 \\ 0 & 0 & 0 & 1 & 0 & 0 & 0 \\ 0 & 0 & 1 & 0 & 0 & 0 & 0 \\ 0.1 & 0.3 & 0.1 & 0 & 0.2 & 0.2 & 0.1 \\ 0 & 0 & 0 & 0 & 0 & 0 & 1 \end{pmatrix}.$$

11.   On a given vacation day, a sportsman goes horseback riding (activity 1), sailing (activity 2), or scuba diving (activity 3). Let $X_n = 1$ if he goes horseback riding

on day n, $X_n = 2$ if he goes sailing on day n, and $X_n = 3$ if he goes scuba diving on that day. Suppose that $\{X_n : n = 1, 2, \ldots\}$ is a Markov chain, and depending on which one of these activities the sportsman chooses on a vacation day, his probability of engagement in any one of the activities on the next vacation day is given by the following transition probability matrix:

$$P = \begin{pmatrix} 0.20 & 0.30 & 0.50 \\ 0.32 & 0.15 & 0.53 \\ 0.60 & 0.13 & 0.27 \end{pmatrix}.$$

Find the long-run probability that, on a randomly selected vacation day, the sportsman sails.

12. An observer at a lake notices that when fish are caught, only 1 out of 9 trout is caught after another trout, with no other fish between, whereas 10 out of 11 nontrout are caught following nontrout, with no trout between. Assuming that all fish are equally likely to be caught, what fraction of fish in the lake is trout?

13. Three players play a game in which they take turns and draw cards from an ordinary deck of 52 cards, successively, at random and with replacement. Player I draws cards until an ace is drawn. Then player II draws cards until a diamond is drawn. Next, player III draws cards until a face card is drawn. At that point, the deck will be given to player I and the game continues. Determine the long-run proportion of cards drawn by each player.

14. For Example 12.10, where a mouse is moving inside the given maze, find the probability that the mouse is in cell i, $1 \le i \le 9$, at a random time in the future.

15. Consider an Ehrenfest chain with 5 balls (see Example 12.15). Find the expected number of balls transferred between two consecutive times that an urn becomes empty.

16. Seven identical balls are randomly distributed among two urns. Step 1 of a game begins by flipping a fair coin. If it lands heads up, urn I is selected; otherwise, urn II is selected. In step 2 of the game, a ball is removed randomly from the urn selected in step 1. Then the coin is flipped again. If it lands heads up, the ball will be placed in urn I. Otherwise, it will be placed in urn II. If this game is played successively, what are the long-run probability mass functions of the number of balls in urns I and II?

17. In Example 12.13, at the Writing Center of a college, $p_k > 0$ is the probability that a new computer needs to be replaced after k semesters. For a computer in use at the end of the nth semester, let X_n be the number of additional semesters it remains functional. Then $\{X_n : n = 0, 1, \ldots\}$ is a Markov chain with transition

probability matrix

$$P = \begin{pmatrix} p_1 & p_2 & p_3 & \cdots \\ 1 & 0 & 0 & \cdots \\ 0 & 1 & 0 & \cdots \\ 0 & 0 & 1 & \cdots \\ \vdots & & & \end{pmatrix}.$$

Show that $\{X_n : n = 0, 1, \ldots\}$ is irreducible, recurrent, and aperiodic. Find the long-run probability that a computer selected randomly at the end of a semester will last at least k additional semesters.

18. Mr. Gorfin is a movie buff who watches movies regularly. His son has observed that whether Mr. Gorfin watches a drama or not depends on the previous two movies he has watched with the following probabilities: 7/8 if the last two movies he watched were both dramas, 1/2 if exactly one of them was a drama, and 1/8 if none of them was a drama.

(a) Mr. Gorfin watched four movies last weekend. If the first two were dramas, what is the probability that the fourth one was a drama as well?

(b) What is the long-run probability that Mr. Gorfin will watch two dramas in a row?

19. A fair die is tossed repeatedly. We begin studying the outcomes after the first 6 occurs. Let the first 6 be called the zeroth outcome, let the first outcome after the first six, whatever it is, be called the first outcome, and so forth. For $n \geq 1$, define $X_n = i$ if the last 6 before the nth outcome occurred i tosses ago. Thus, for example, if the first 7 outcomes after the first 6 are 1, 4, 6, 5, 3, 1, and 6, then $X_1 = 1$, $X_2 = 2$, $X_3 = 0$, $X_4 = 1$, $X_5 = 2$, $X_6 = 3$, and $X_7 = 0$. Show that $\{X_n : n = 1, 2, \ldots\}$ is a Markov chain, and find its transition probability matrix. Furthermore, show that $\{X_n : n = 1, 2, \ldots\}$ is an irreducible, positive recurrent, aperiodic Markov chain. For $i \geq 1$, find π_i, the long-run probability that the last 6 occurred i tosses ago.

20. Alberto and Angela play backgammon regularly. The probability that Alberto wins a game depends on whether he won or lost the previous game. It is p for Alberto to win a game if he lost the previous game, and p to lose a game if he won the previous one.

(a) For $n > 1$, show that if Alberto wins the first game, the probability is $\dfrac{1}{2} + \dfrac{1}{2}(1 - 2p)^n$ that he will win the nth game.

(b) Find the expected value of the number of games Alberto will play between two consecutive wins.

21. Let $\{X_n : n = 0, 1, \ldots\}$ be a random walk with state space $\{0, 1, 2, \ldots\}$ and

transition probability matrix

$$P = \begin{pmatrix} 1-p & p & 0 & 0 & 0 & 0 & \cdots \\ 1-p & 0 & p & 0 & 0 & 0 & \cdots \\ 0 & 1-p & 0 & p & 0 & 0 & \cdots \\ 0 & 0 & 1-p & 0 & p & 0 & \cdots \\ 0 & 0 & 0 & 1-p & 0 & p & \cdots \\ \vdots & & & & & & \end{pmatrix},$$

where $0 < p < 1$. (See Example 12.12.) Determine those values of p for which, for each $j \geq 0$, $\lim_{n \to \infty} p_{ij}^n$ exists and is independent of i. Then find the limiting probabilities.

B

22. Carl and Stan play the game of "heads or tails," in which each time a coin lands heads up, Carl wins \$1 from Stan, and each time it lands tails up, Stan wins \$1 from Carl. Suppose that, initially, Carl and Stan have the same amount of money and, as necessary, will be funded equally so that they can continue playing indefinitely. If the coin is not fair and lands heads up with probability 0.46, what is the probability that Carl will ever get ahead of Stan?

23. Let $\{X_n : n = 0, 1, \dots\}$ be a Markov chain with state space $\mathcal{S}$. For $i_0, i_1, \dots, i_n$, $j \in \mathcal{S}, n \geq 0$, and $m > 0$, show that

$$P(X_{n+m} = j \mid X_0 = i_0, X_1 = i_1, \dots, X_n = i_n) = P(X_{n+m} = j \mid X_n = i_n).$$

24. For a two-dimensional symmetric random walk, defined in Example 12.22, show that $(0, 0)$ is recurrent and conclude that all states are recurrent.

25. Recall that an $M/M/1$ queueing system is a $GI/G/c$ system in which customers arrive according to a Poisson process with rate λ, and service times are exponential with mean $1/\mu$. For an $M/M/1$ queueing system, each time that a customer arrives to the system or a customer departs from the system, we say that a *transition* occurs. Let X_n be the number of customers in the system immediately after the nth transition. Show that $\{X_n : n = 1, 2, \dots\}$ is a Markov chain, and find its probability transition matrix. Find the period of each state of the Markov chain.

26. Show that if P and Q are two transition probability matrices with the same number of rows, and hence columns, then PQ is also a transition probability matrix. Note that this implies that if P is a transition probability matrix, then so is P^n for any positive integer n.

27. Consider a Markov chain with state space $\mathcal{S}$. Let $i, j \in \mathcal{S}$. We say that state j is accessible from state i in n steps if there is a path

$$i = i_1, \ i_2, \ i_3, \ \dots, \ i_n = j$$

with $i_1, i_2, \ldots, i_n \in S$ and $p_{i_m i_{m+1}} > 0, 1 \leq m \leq n - 1$. Show that if S is finite having K states, and j is accessible from i, then j is accessible from i in K or fewer steps.

Hint: Use the *Pigeonhole Principle*: If $n > K$ pigeons are placed into K pigeonholes, then at least one pigeonhole is occupied by two or more pigeons.

28. Let $\{X_n : n = 0, 1, \ldots\}$ be a Markov chain with state space S and probability transition matrix $P = (p_{ij})$. Show that periodicity is a class property. That is, for $i, j \in S$, if i and j communicate with each other, then they have the same period.

29. Every Sunday, Bob calls Liz to see if she will play tennis with him on that day. If Liz has not played tennis with Bob since i Sundays ago, the probability that she will say yes to him is $i/k, k \geq 2, i = 1, 2, \ldots, k$. Therefore, if, for example, Liz does not play tennis with Bob for $k - 1$ consecutive Sundays, then she will play with him the next Sunday with probability 1. On the nth Sunday, after Bob calls Liz, the number of times Liz has said no to Bob since they last played tennis is denoted by X_n. For $0 \leq i \leq k - 1$, find the long-run probability that Liz says no to Bob for i consecutive Sundays. (Note that the answer is not $1/k$.)

30. Consider the gambler's ruin problem (Example 3.14) in which two gamblers play the game of "heads or tails." Each time a fair coin lands heads up, player A wins \$1 from player B, and each time it lands tails up, player B wins \$1 from A. Suppose that, initially, player A has a dollars and player B has b dollars. We know that eventually either player A will be ruined in which case B wins the game, or player B will be ruined in which case A wins the game. Let T be the duration of the game. That is, the number of times A and B play until one of them is ruined. Find $E(T)$.

31. Consider the branching process of Example 12.24. In that process, before death an organism produces j ($j \geq 0$) offspring with probability α_j. Let $X_0 = 1$, and let μ, the expected number of offspring of an organism, be greater than 1. Let p be the probability that extinction will occur. Show that p is the smallest positive root of the equation $x = \sum_{i=0}^{\infty} \alpha_i x^i$.

32. In this exercise, we will outline a third technique for solving Example 3.31: We draw cards, one at a time, at random and successively from an ordinary deck of 52 cards with replacement. What is the probability that an ace appears before a face card?

Hint: Consider a Markov chain $\{X_n : n = 1, 2, \ldots\}$ with state space $\{1, 2, 3\}$ and transition probability matrix

$$P = \begin{pmatrix} 9/13 & 1/13 & 3/13 \\ 0 & 1 & 0 \\ 0 & 0 & 1 \end{pmatrix}.$$

The relation between the problem we want to solve and the Markov chain $\{X_n : n = 1, 2, \ldots\}$ is as follows: As long as a non-ace, non-face card is drawn, the Markov chain remains in state 1. If an ace is drawn before a face card, it enters the absorbing state 2 and will remain there indefinitely. Similarly, if a face card is drawn before an ace, the process enters the absorbing state 3 and will remain there forever. Let A_n be the event that the Markov chain moves from state 1 to state 2 in n steps. Show that

$$A_1 \subseteq A_2 \subseteq \cdots \subseteq A_n \subseteq A_{n+1} \subseteq \cdots ,$$

and calculate the desired probability $P\left(\bigcup_{n=1}^{\infty} A_n\right)$ by applying Theorem 1.8:

$$P\left(\bigcup_{n=1}^{\infty} A_n\right) = \lim_{n \to \infty} P(A_n) = \lim_{n \to \infty} p_{12}^n.$$

33. For a simple random walk $\{X_n : n = 0, \pm 1, \pm 2, \ldots \}$, discussed in Examples 12.12 and 12.22, show that

$$P(X_n = j \mid X_0 = i) = \left(\frac{n}{\dfrac{n + j - i}{2}} \right) p^{(n+j-i)/2}(1 - p)^{(n-j+i)/2}$$

if $n + j - i$ is an even nonnegative integer for which $\dfrac{n + j - i}{2} \le n$, and it is 0 otherwise.

■

The subject of the next section is continuous-time Markov chains. In this book, as previously mentioned, by a Markov chain we always mean a discrete-time Markov chain. Whenever we refer to continuous-time Markov chains, we shall always specify them by their complete title.

12.4   CONTINUOUS-TIME MARKOV CHAINS

Suppose that a certain machine is either operative or out of order. Let $X(t) = 0$ if at time t the machine is out of order, and let $X(t) = 1$ if it is operative at time t. Then $\{X(t) : t \ge 0\}$ is a stochastic process with state space $\mathcal{S} = \{0, 1\}$, but it is not a (discrete-time) Markov chain. For a stochastic process to be a Markov chain, after entering a state i, it must move to another state j, in one unit of time, with probability p_{ij}. This is not the case for $X(t)$ defined above. After $X(t)$ enters a state, it remains there for a period of time that is a random variable and then moves to the other state. The main property of a Markov chain, expressed by (12.9), is that given the state of the Markov chain at present, its future state is independent of the past states. In this section, we will study those stochastic processes $\{X(t) : t \ge 0\}$ that possess this property but when entering a state will remain in that state for a random period before moving to another state. Such processes are called **continuous-time Markov chains** and have a wide range of important applications in communication systems, biological systems, computer networks, various industrial engineering and operations research models, and other areas of engineering and sciences.

If a stochastic process $\{X(t): t \geq 0\}$ with a finite or countably infinite state space $\mathcal{S}$ (often labeled by nonnegative integers) possesses the **Markovian property**; that is, if it is a continuous-time Markov chain, then for all $s \geq 0$, given the state of the Markov chain at present $(X(s))$, its future $(X(u), \ u > s)$ is independent of its past $(X(u), \ 0 \leq u < s)$. Let $i, j \in \mathcal{S}$ and, for $u \geq 0$, let $x_u \in \mathcal{S}$. Then, by the Markovian property, for $s, t > 0$,

$$P\big(X(s+t) = j \mid X(s) = i, \ X(u) = x_u \text{ for } 0 \leq u < s\big) = P\big(X(s+t) = j \mid X(s) = i\big).$$

Similarly,

$$P\big(X(u) = x_u \text{ for } u > s \mid X(s) = i, X(u) = x_u \text{ for } 0 \leq u < s\big)$$
$$= P\big(X(u) = x_u \text{ for } u > s \mid X(s) = i\big).$$

Just as in Markov chains, throughout the book, unless otherwise explicitly specified, we assume that continuous-time Markov chains are **time homogeneous**. That is, they have **stationary transition probabilities** $\big[$i.e., $P\big(X(s+t) = j \mid X(s) = i\big)$ does not depend on $s\big]$. In other words, for all $s > 0$,

$$P\big(X(s+t) = j \mid X(s) = i\big) = P\big(X(t) = j \mid X(0) = i\big).$$

For a Markov chain, a transition from a state i to itself is possible. However, for a continuous-time Markov chain, a transition will occur *only* when the process moves from one state to another. At a time point labeled $t = 0$, suppose that a continuous-time Markov chain enters a state i, and let Y be the length of time it will remain in that state before moving to another state. Then, for $s, t \geq 0$,

$$P\big(Y > s + t \mid Y > s\big)$$
$$= P\big(X(u) = i \text{ for } s < u \leq s + t \mid X(s) = i, \ X(u) = i \text{ for } 0 \leq u < s\big)$$
$$= P\big(X(u) = i \text{ for } s < u \leq s + t \mid X(s) = i\big) = P(Y > t).$$

The relation $P(Y > s + t \mid Y > s) = P(Y > t)$ shows that Y is memoryless. Hence it is an exponential random variable. A rigorous mathematical proof of this fact is beyond the scope of this book. However, the justification just presented should intuitively be satisfactory. The expected value of Y, the length of time the process will remain in state i, is denoted by $1/v_i$. To sum up, a continuous-time Markov chain is a stochastic process $\{X(t): t \geq 0\}$ with a finite or countably infinite state space $\mathcal{S}$ and the following property: Upon entering a state i, the process will remain there for a period of time, which is exponentially distributed with mean $1/v_i$. Then it will move to *another* state j with probability p_{ij}. Therefore, for all $i \in \mathcal{S}$, $p_{ii} = 0$.

For a discrete-time Markov chain, we defined p_{ij}^n to be the probability of moving from state i to state j in n steps. The quantity analogous to p_{ij}^n in the continuous case is

$p_{ij}(t)$, defined to be the probability of moving from state i to state j in t units of time. That is,

$$p_{ij}(t) = P(X(s+t) = j \mid X(s) = i), \quad i, j \in \mathcal{S}; \; s, t \geq 0.$$

It should be clear that $p_{ij}(0) = 1$ if $i = j$; $p_{ij}(0) = 0$ if $i \neq j$. Furthermore, $p_{ij}(t) \geq 0$ and $\sum_{j=0}^{\infty} p_{ij}(t) = 1$.

Clearly, for a transition from state i to state j in $s+t$ units of time, the continuous-time Markov chain will have to enter some state k, along the way, after s units of time, and then move from k to j in the remaining t units of time. This observation leads us to the following celebrated equations, the continuous analog of (12.10), called the **Chapman-Kolmogorov Equations** for continuous-time Markov chains:

$$p_{ij}(s+t) = \sum_{k=0}^{\infty} p_{ik}(s) p_{kj}(t). \tag{12.12}$$

The equations (12.12) can be proved rigorously by applying the law of total probability, Theorem 3.4, to the sequence of mutually exclusive events $\{X(s) = k\}$, $k \geq 0$:

$$p_{ij}(s+t) = P(X(s+t) = j \mid X(0) = i)$$

$$= \sum_{k=0}^{\infty} P(X(s+t) = j \mid X(0) = i, \, X(s) = k) P(X(s) = k \mid X(0) = i)$$

$$= \sum_{k=0}^{\infty} P(X(s+t) = j \mid X(s) = k) P(X(s) = k \mid X(0) = i)$$

$$= \sum_{k=0}^{\infty} p_{kj}(t) p_{ik}(s) = \sum_{k=0}^{\infty} p_{ik}(s) p_{kj}(t).$$

Remark 12.4   Note that, for a continuous-time Markov chain, while p_{ii} is 0, $p_{ii}(t)$ is not necessarily 0. This is because, in t units of time, the process might leave i, enter other states and then return to i.

Example 12.37   Suppose that a certain machine operates for a period which is exponentially distributed with parameter λ. Then it breaks down, and it will be in a repair shop for a period, which is exponentially distributed with parameter μ. Let $X(t) = 1$ if the machine is operative at time t; $X(t) = 0$ if it is out of order at that time. Then $\{X(t): t \geq 0\}$ is a continuous-time Markov chain with $\nu_0 = \mu$, $\nu_1 = \lambda$, $p_{00} = p_{11} = 0$, $p_{01} = p_{10} = 1$. ♦

Example 12.38   Let $\{N(t): t \geq 0\}$ be a Poisson process with rate λ. Then $\{N(t): t \geq 0\}$ is a stochastic process with state space $\mathcal{S} = \{0, 1, 2, \dots\}$, with the property that upon entering a state i, it will remain there for an exponential amount of time with mean $1/\lambda$ and then will move to state $i + 1$ with probability 1. Hence $\{N(t): t \geq 0\}$ is a continuous-time Markov chain with $\nu_i = \lambda$ for all $i \in S$, $p_{i(i+1)} = 1$; $p_{ij} = 0$, if $j \neq i + 1$. Furthermore, for all $t, s > 0$,

$$p_{ij}(t) = P\big(N(s+t) = j \mid N(s) = i\big) = \begin{cases} 0 & \text{if } j < i \\ \dfrac{e^{-\lambda t} \cdot (\lambda t)^{j-i}}{(j-i)!} & \text{if } j \geq i. \end{cases} \quad \blacklozenge$$

For a discrete-time Markov chain, to find p_{ij}^n, the probability of moving from state i to state j in n steps, we calculated the n-step transition probability matrix, which is equal to the one-step transition probability matrix raised to the power n. For a continuous-time Markov chain, to find $p_{ij}(t)$'s, the quantities analogous to p_{ij}^n's in the discrete case, we show that they satisfy two sets of systems of differential equations called **Kolmogorov forward equations** and **Kolmogorov backward equations**. Sometimes we will be able to find $p_{ij}(t)$ by solving those systems of differential equations. To derive the Kolmogorov forward and backward equations, we first need some preliminaries. Note that a continuous-time Markov chain $\{X(t): t \geq 0\}$ remains in state i for a period that is exponentially distributed with mean $1/\nu_i$. Therefore, it leaves the state i at the rate of ν_i. Since p_{ij} is the probability that the process leaves i and enters j, $\nu_i p_{ij}$ is the rate at which the process leaves i and enters j. This quantity is denoted by q_{ij} and is called the **instantaneous transition rate**. Since $q_{ij} = \nu_i p_{ij}$ is the rate at which the process leaves state i and enters j, $\sum_{i=0}^{\infty} q_{ij} = \sum_{i=0}^{\infty} \nu_i p_{ij}$ is the rate at which the process enters j.

Let $\{N_i(t): t \geq 0\}$ be a Poisson process with rate ν_i. It should be clear that, for an infinitesimal h,

$$p_{ii}(h) = P\big(N_i(h) = 0\big).$$

By (12.4),

$$P\big(N_i(h) = 0\big) = 1 - \nu_i h + o(h).$$

Thus

$$p_{ii}(h) = 1 - \nu_i h + o(h),$$

or, equivalently,

$$\frac{1 - p_{ii}(h)}{h} = \nu_i - \frac{o(h)}{h}.$$

Therefore,

$$\lim_{h \to 0} \frac{1 - p_{ii}(h)}{h} = \nu_i. \tag{12.13}$$

Similarly, let $\{N_{ij}: t \geq 0\}$ be a Poisson process with rate q_{ij}. Then it should be clear that, for an infinitesimal h,

$$p_{ij}(h) = P\big(N_{ij}(h) = 1\big).$$

By (12.5),

$$P\big(N_{ij}(h) = 1\big) = q_{ij}h + o(h).$$

Therefore,

$$p_{ij}(h) = q_{ij}h + o(h),$$

or, equivalently,

$$\frac{p_{ij}(h)}{h} = q_{ij} + \frac{o(h)}{h}.$$

This gives

$$\lim_{h \to 0} \frac{p_{ij}(h)}{h} = q_{ij}. \tag{12.14}$$

We are now ready to present the Kolmogorov forward and backward equations. In the following theorem and discussions, we assume that for all $i, j \in S$, the functions $p_{ij}(t)$ and their derivatives satisfy appropriate regularity conditions so that we can interchange the order of two limits as well as the order of a limit and a sum.

Theorem 12.8   *Let $\big\{X(t): t \geq 0\big\}$ be a continuous-time Markov chain with state space S. Then, for all states $i, j \in S$ and $t \geq 0$, we have the following equations:*

(a) **Kolmogorov's Forward Equations:**

$$p'_{ij}(t) = \sum_{k \neq j} q_{kj} p_{ik}(t) - v_j p_{ij}(t).$$

(b) **Kolmogorov's Backward Equations:**

$$p'_{ij}(t) = \sum_{k \neq i} q_{ik} p_{kj}(t) - v_i p_{ij}(t).$$

Proof: We will show part (a) and leave the proof of part (b), which is similar to the proof of part (a), as an exercise. Note that, by the Chapman-Kolmogorov equations,

$$p_{ij}(t + h) - p_{ij}(t) = \sum_{k=0}^{\infty} p_{ik}(t) p_{kj}(h) - p_{ij}(t)$$

$$= \sum_{k \neq j} p_{ik}(t) p_{kj}(h) + p_{ij}(t) p_{jj}(h) - p_{ij}(t)$$

$$= \sum_{k \neq j} p_{ik}(t) p_{kj}(h) + p_{ij}(t)\big[p_{jj}(h) - 1\big].$$

Thus

$$\frac{p_{ij}(t + h) - p_{ij}(t)}{h} = \sum_{k \neq j} \frac{p_{kj}(h)}{h} p_{ik}(t) + p_{ij}(t) \frac{p_{jj}(h) - 1}{h}.$$

Letting $h \to 0$, by (12.13) and (12.14), we have

$$p'_{ij}(t) = \sum_{k \neq j} q_{kj} p_{ik}(t) - v_j p_{ij}(t). \quad \blacklozenge$$

Example 12.39   Passengers arrive at a train station according to a Poisson process with rate λ and wait for a train to arrive. Independently, trains arrive at the same station according to a Poisson process with rate μ. Suppose that each time a train arrives, all the passengers waiting at the station will board the train. The train then immediately leaves the station. If there are no passengers waiting at the station, the train will not wait until passengers arrive. Suppose that at time 0, there is no passenger waiting for a train at the station. For $t > 0$, find the probability that at time t also there is no passenger waiting for a train.

Solution: Clearly, the interarrival times between consecutive passengers arriving at the station are independent exponential random variables with mean $1/\lambda$, and the interarrival times between consecutive trains arriving at the station are independent exponential random variables with mean $1/\mu$. Let $X(t) = 1$, if there is at least one passenger waiting in the train station for a train, and let $X(t) = 0$, otherwise. Due to the memoryless property of exponential random variables, when a train leaves the station, it will take an amount of time, exponentially distributed with mean $1/\lambda$, until a passenger arrives at the station. Then the period from that passenger's arrival time until the next train arrives is exponential with mean $1/\mu$. Therefore, $\{X(t): t \geq 0\}$ is a stochastic process with state space $\{0, 1\}$ that will remain in state 0 for an exponential amount of time with mean $1/\lambda$. Then it will move to state 1 and will remain in that state for an exponential length of time with mean $1/\mu$. At that point another change of state to state 0 will occur, and the process continues. Hence $\{X(t): t \geq 0\}$ is a continuous-time Markov chain for which $v_0 = \lambda$, $v_1 = \mu$, and

$$p_{01} = p_{10} = 1,$$

$$p_{00} = p_{11} = 0,$$

$$q_{10} = v_1 p_{10} = v_1 = \mu,$$

$$q_{01} = v_0 p_{01} = v_0 = \lambda.$$

To calculate the desired probability, $p_{00}(t)$, note that by Kolmogorov's forward equations,

$$p'_{00}(t) = q_{10} p_{01}(t) - v_0 p_{00}(t)$$

$$p'_{01}(t) = q_{01} p_{00}(t) - v_1 p_{01}(t),$$

or, equivalently,

$$p'_{00}(t) = \mu p_{01}(t) - \lambda p_{00}(t) \quad \quad (12.15)$$

$$p'_{01}(t) = \lambda p_{00}(t) - \mu p_{01}(t).$$

Adding these two equations yields

$$p_{00}'(t) + p_{01}'(t) = 0.$$

Hence

$$p_{00}(t) + p_{01}(t) = c.$$

Since $p_{00}(0) = 1$ and $p_{01}(0) = 0$, we have $c = 1$, which gives $p_{01}(t) = 1 - p_{00}(t)$. Substituting this into equation (12.15), we obtain

$$p_{00}'(t) = \mu\big[1 - p_{00}(t)\big] - \lambda p_{00}(t),$$

or, equivalently,

$$p_{00}'(t) + (\lambda + \mu)p_{00}(t) = \mu.$$

The usual method to solve this simple differential equation is to multiply both sides of the equation by $e^{(\lambda+\mu)t}$. That makes the left side of the equation the derivative of a product of two functions, hence possible to integrate. We have

$$e^{(\lambda+\mu)t} p_{00}'(t) + (\lambda + \mu)e^{(\lambda+\mu)t} p_{00}(t) = \mu e^{(\lambda+\mu)t},$$

which is equivalent to

$$\frac{d}{dt}\big[e^{(\lambda+\mu)t} p_{00}(t)\big] = \mu e^{(\lambda+\mu)t}.$$

Integrating both sides of this equation gives

$$e^{(\lambda+\mu)t} p_{00}(t) = \frac{\mu}{\lambda + \mu}e^{(\lambda+\mu)t} + c,$$

where c is a constant. Using $p_{00}(0) = 1$ implies that $c = \lambda/(\lambda + \mu)$. So

$$e^{(\lambda+\mu)t} p_{00}(t) = \frac{\mu}{\lambda + \mu}e^{(\lambda+\mu)t} + \frac{\lambda}{\lambda + \mu},$$

and hence, $p_{00}(t)$, the probability we are interested in, is given by

$$p_{00}(t) = \frac{\mu}{\lambda + \mu} + \frac{\lambda}{\lambda + \mu}e^{-(\lambda+\mu)t}. \quad \blacklozenge$$

Steady-State Probabilities

In Section 12.3, for a discrete-time Markov chain, we showed that $\lim_{n\to\infty} p_{ij}^n$ exists and is independent of i if the Markov chain is irreducible, positive recurrent, and aperiodic. Furthermore, we showed that if, for $j \geq 0$, $\pi_j = \lim_{n\to\infty} p_{ij}^n$, then $\pi_j = \sum_{i=0}^{\infty} p_{ij}\pi_i$, and $\sum_{j=0}^{\infty} \pi_j = 1$. For each state j, the limiting probability, π_j, which is the long-run probability that the process is in state j is also the long-run proportion of the number of transitions to state j. Similar results are also true for continuous-time Markov chains. Suppose that $\{X(t): t \geq 0\}$ is a continuous-time Markov chain with state space $\mathcal{S}$. Suppose that, for each $i, j \in \mathcal{S}$, there is a positive probability that, starting from i, the process eventually enters j. Furthermore, suppose that, starting from i, the process will return to i with probability 1, and the expected number of transitions for a first return to i is finite. Then, under these conditions, $\lim_{t \to \infty} p_{ij}(t)$ exists and is independent of i. Let $\pi_j = \lim_{t \to \infty} p_{ij}(t)$. Then π_j is the long-run probability that the process is in state j. It is also the proportion of time the process is in state j. Note that if $\lim_{t \to \infty} p_{ij}(t)$ exists, then

$$\lim_{t \to \infty} p'_{ij}(t) = \lim_{t \to \infty} \lim_{h \to 0} \frac{p_{ij}(t+h) - p_{ij}(t)}{h}$$

$$= \lim_{h \to 0} \lim_{t \to \infty} \frac{p_{ij}(t+h) - p_{ij}(t)}{h}$$

$$= \lim_{h \to 0} \frac{\pi_j - \pi_j}{h} = 0.$$

Now consider Kolmogorov's forward equations:

$$p'_{ij}(t) = \sum_{k \neq j} q_{kj} p_{ik}(t) - v_j p_{ij}(t).$$

Letting $t \to \infty$ gives

$$0 = \sum_{k \neq j} q_{kj} \pi_k - v_j \pi_j,$$

or, equivalently,

$$\sum_{k \neq j} q_{kj} \pi_k = v_j \pi_j. \tag{12.16}$$

Observe that $q_{kj} \pi_k$ is the rate at which the process departs state k and enters state j. Hence $\sum_{k \neq j} q_{kj} \pi_k$ is the rate at which the process enters state j. $v_j \pi_j$ is, clearly, the rate at which the process departs state j. Thus we have shown that, for all $j \in \mathcal{S}$, if the continuous-time Markov chain is in steady state, then

total rate of transitions to state j = total rate of transitions from state j.

Since equations (12.16) equate the total rate of transitions *to* state j with the total rate of transitions *from* state j, they are called **balance equations**. Just as in discrete case, for a continuous-time Markov chain with n states, if the input rates are equal to the output rates for $n - 1$ states, the balance equation must be valid for the remaining state as well. Hence equations (12.16) give one redundant equation, and therefore, to calculate π_j's, we need the additional equation $\sum_{j=0}^{\infty} \pi_j = 1$.

Example 12.40   Consider Example 12.39; let π_0 be the long-run probability that there is no one in the train station waiting for a train to arrive. Let π_1 be the long-run probability that there is at least one person in the train station waiting for a train. For the continuous-time Markov chain $\{X(t): t \geq 0\}$ of that example, the balance equations are

State	Input rate to	=	Output rate from
0	$\mu\pi_1$	=	$\lambda\pi_0$
1	$\lambda\pi_0$	=	$\mu\pi_1$

As expected, these two equations are identical. Solving the system of two equations in two unknowns

$$\begin{cases} \lambda\pi_0 = \mu\pi_1 \\ \pi_0 + \pi_1 = 1, \end{cases}$$

we obtain $\pi_0 = \mu/(\lambda + \mu)$ and $\pi_1 = \lambda/(\lambda + \mu)$. Recall that in Example 12.39, we showed that

$$p_{00}(t) = \frac{\mu}{\lambda + \mu} + \frac{\lambda}{\lambda + \mu} e^{-(\lambda+\mu)t}.$$

The results obtained can also be found by noting that

$$\pi_0 = \lim_{t \to \infty} p_{00}(t) = \frac{\mu}{\lambda + \mu}, \quad \pi_1 = 1 - \pi_0 = \frac{\lambda}{\lambda + \mu}. \quad \blacklozenge$$

Example 12.41   In Ponza, Italy, a man is stationed at a specific port and can be hired to give sightseeing tours with his boat. If the man is free, it takes an interested tourist a time period, exponentially distributed with mean $1/\mu_1$, to negotiate the price and the type of tour. Suppose that the probability is α that a tourist does not reach an agreement and leaves. For those who decide to take a tour, the duration of the tour is exponentially distributed with mean $1/\mu_2$. Suppose that tourists arrive at this businessman's station according to a Poisson process with parameter λ and request service only if he is free. They leave the station otherwise. If the negotiation times, the duration of the tours, and the arrival times of the tourists at the station are independent random variables, find the proportion of time the businessman is free.

Solution: Let $X(t) = 0$ if the businessman is free, let $X(t) = 1$ if he is negotiating, and let $X(t) = 2$ if he is giving a sightseeing tour. Clearly, $\{X(t): t \geq 0\}$ is a continuous-time Markov chain with state space $\{0, 1, 2\}$. Let π_0, π_1, and π_2 be the long-run proportion of time the businessman is free, negotiating, and giving a tour, respectively. The balance equations for $\{X(t): t \geq 0\}$ are as follows:

State	Input rate to	=	Output rate from
0	$\alpha\mu_1\pi_1 + \mu_2\pi_2$	=	$\lambda\pi_0$
1	$\lambda\pi_0$	=	$\mu_1\pi_1$
2	$(1-\alpha)\mu_1\pi_1$	=	$\mu_2\pi_2$

By these equations, $\pi_1 = (\lambda/\mu_1)\pi_0$ and $\pi_2 = (1-\alpha)(\lambda/\mu_2)\pi_0$. Substituting π_1 and π_2 in $\pi_0 + \pi_1 + \pi_2 = 1$, we obtain

$$\pi_0 = \frac{\mu_1\mu_2}{\mu_1\mu_2 + \lambda\mu_2 + \lambda\mu_1(1-\alpha)}. \quad \blacklozenge$$

Example 12.42   Johnson Medical Associates has two physicians on call, Drs. Dawson and Baick. Dr. Dawson is available to answer patients' calls for time periods that are exponentially distributed with mean 2 hours. Between those periods, he takes breaks, each of which being an exponential amount of time with mean 30 minutes. Dr. Baick works independently from Dr. Dawson, but with similar work patterns. The time periods she is available to take patients' calls and the times she is on break are exponential random variables with means 90 and 40 minutes, respectively. In the long run, what is the proportion of time in which neither of the two doctors is available to take patients' calls?

Solution: Let $X(t) = 0$ if neither Dr. Dawson nor Dr. Baick is available to answer patients' calls. Let $X(t) = 2$ if both of them are available to take the calls; $X(t) = d$ if Dr. Dawson is available to take the calls and Dr. Baick is not; $X(t) = b$ if Dr. Baick is available to take the calls but Dr. Dawson is not. Clearly, $\{X(t): t \geq 0\}$ is a continuous-time Markov chain with state space $\{0, 2, d, b\}$. Let π_0, π_2, π_d, and π_b be the long-run proportions of time the process is in the stats 0, 2, d, and b, respectively. The balance equations for $\{X(t): t \geq 0\}$ are

State	Input rate to	=	Output rate from
0	$(1/2)\pi_d + (2/3)\pi_b$	=	$2\pi_0 + (3/2)\pi_0$
2	$(3/2)\pi_d + 2\pi_b$	=	$(1/2)\pi_2 + (2/3)\pi_2$
d	$(2/3)\pi_2 + 2\pi_0$	=	$(1/2)\pi_d + (3/2)\pi_d$
b	$(1/2)\pi_2 + (3/2)\pi_0$	=	$(2/3)\pi_b + 2\pi_b$

Solving any three of these equations along with $\pi_0 + \pi_2 + \pi_d + \pi_b = 1$, we obtain $\pi_0 = 4/65$, $\pi_2 = 36/65$, $\pi_d = 16/65$, and $\pi_b = 9/65$. Therefore, the proportion of time none of the two doctors is available to take patients' calls is $\pi_0 = 4/65 \approx 0.06$. $\quad\blacklozenge$

Birth and Death Processes

Let $X(t)$ be the number of individuals in a population of living organisms at time t. Suppose that members of the population may give birth to new individuals, and they may die. Furthermore, suppose that, (i) if $X(t) = n$, $n \geq 0$, then the time until the next birth is exponential with parameter λ_n; (ii) if $X(t) = n$, $n > 0$, then the time until the next death is exponential with parameter μ_n; and (iii) births occur independently of deaths. For $n > 0$, given $X(t) = n$, let T_n be the time until the next birth and S_n be the time until the next death. Then T_n and S_n are independent exponential random variables with means $1/\lambda_n$ and $1/\mu_n$, respectively. Under the conditions above, each time in state n, the process $\{X(t) : t \geq 0\}$ will remain in that state for a period of length $\min(T_n, S_n)$. Since

$$P\big(\min(T_n, S_n) > x\big) = P(T_n > x, \, S_n > x)$$

$$= P(T_n > x)P(S_n > x) = e^{-\lambda_n x}e^{-\mu_n x} = e^{-(\lambda_n + \mu_n)x},$$

we have that $\min(S_n, T_n)$ is an exponential random variable with mean $1/(\lambda_n + \mu_n)$. For $n = 0$, the time until the next birth is exponential with mean $1/\lambda_0$. Thus, each time in state 0, the process will remain in 0 for a length of time that is exponentially distributed with mean $1/\lambda_0$. Letting $\mu_0 = 0$, we have shown that, for $n \geq 0$, if $X(t) = n$, then the process remains in state n for a time period that is exponentially distributed with mean $1/(\lambda_n + \mu_n)$. Then, for $n > 0$, the process leaves state n and either enters state $n + 1$ (if a birth occurs), or enters state $n - 1$ (if a death occurs). For $n = 0$, after an amount of time exponentially distributed with mean $1/\lambda_0$, the process will enter state 1 with probability 1. These facts show that $\{X(t) : t \geq 0\}$ is a continuous-time Markov chain with state space $\{0, 1, 2, \ldots\}$ and $\nu_n = \lambda_n + \mu_n$, $n \geq 0$. It is called a **birth and death process**. For $n \geq 0$, the parameters λ_n and μ_n are called the **birth and death rates**, respectively. If for some i, $\lambda_i = 0$, then, while the process is in state i, no birth will occur. Similarly, if $\mu_i = 0$, then, while the process is in state i, no death will occur. We have

$$p_{n(n+1)} = P(S_n > T_n) = \int_0^\infty P(S_n > T_n \mid T_n = x)\lambda_n e^{-\lambda_n x}\, dx$$

$$= \int_0^\infty P(S_n > x)\lambda_n e^{-\lambda_n x}\, dx = \int_0^\infty e^{-\mu_n x} \cdot \lambda_n e^{-\lambda_n x}\, dx$$

$$= \lambda_n \int_0^\infty e^{-(\lambda_n + \mu_n)x}\, dx = \frac{\lambda_n}{\lambda_n + \mu_n},$$

$$p_{n(n-1)} = 1 - \frac{\lambda_n}{\lambda_n + \mu_n} = \frac{\mu_n}{\lambda_n + \mu_n}.$$

The terms *birth* and *death* are broad abstract terms and apply to appropriate events in various models. For example, if $X(t)$ is the number of customers in a bank at time t, then every time a new customer arrives, a birth occurs. Similarly, every time a customer leaves the bank, a death occurs. As another example, suppose that, in a factory, there are a number of operating machines and a number of out-of-order machines being repaired. For such a case, a birth occurs every time a machine is repaired and begins to operate. Similarly, a death occurs every time that a machine breaks down.

If for a birth and death process, $\mu_n = 0$ for all $n \geq 0$, then it is called a **pure birth process**. A Poisson process with rate λ is a pure birth process with birth rates $\lambda_n = \lambda$, $n \geq 0$. Equivalently, a pure birth process is a generalization of a Poisson process in which the occurrence of an event at time t depends on the total number of events that have occurred by time t. A **pure death process** is a birth and death process in which $\lambda_n = 0$ for $n \geq 0$. Note that a pure death process is eventually absorbed in state 0.

For a birth and death process with birth rates $\{\lambda_n\}_{n=0}^{\infty}$ and death rates $\{\mu_n\}_{n=1}^{\infty}$, let π_n be the limiting probability that the population size is n, $n \geq 0$. The balance equations [equations (12.16)] for this family of continuous-time Markov chains are as follows:

State	Input rate to	=	Output rate from
0	$\mu_1 \pi_1$	=	$\lambda_0 \pi_0$
1	$\mu_2 \pi_2 + \lambda_0 \pi_0$	=	$\lambda_1 \pi_1 + \mu_1 \pi_1$
2	$\mu_3 \pi_3 + \lambda_1 \pi_1$	=	$\lambda_2 \pi_2 + \mu_2 \pi_2$
$\vdots$	$\vdots$		$\vdots$
n	$\mu_{n+1} \pi_{n+1} + \lambda_{n-1} \pi_{n-1}$	=	$\lambda_n \pi_n + \mu_n \pi_n$
$\vdots$	$\vdots$		$\vdots$

By the balance equation for state 0,

$$\pi_1 = \frac{\lambda_0}{\mu_1} \pi_0.$$

Considering the fact that $\mu_1 \pi_1 = \lambda_0 \pi_0$, the balance equation for state 1 gives $\mu_2 \pi_2 = \lambda_1 \pi_1$, or, equivalently,

$$\pi_2 = \frac{\lambda_1}{\mu_2} \pi_1 = \frac{\lambda_0 \lambda_1}{\mu_1 \mu_2} \pi_0.$$

Now, considering the fact that $\mu_2 \pi_2 = \lambda_1 \pi_1$, the balance equation for state 2 implies that $\mu_3 \pi_3 = \lambda_2 \pi_2$, or, equivalently,

$$\pi_3 = \frac{\lambda_2}{\mu_3} \pi_2 = \frac{\lambda_0 \lambda_1 \lambda_2}{\mu_1 \mu_2 \mu_3} \pi_0.$$

Continuing this argument, for $n \geq 1$, we obtain

$$\pi_n = \frac{\lambda_{n-1}}{\mu_n} \pi_{n-1} = \frac{\lambda_0 \lambda_1 \cdots \lambda_{n-1}}{\mu_1 \mu_2 \cdots \mu_n} \pi_0. \tag{12.17}$$

Using $\sum_{n=0}^{\infty} \pi_n = 1$, we have

$$\pi_0 + \sum_{n=1}^{\infty} \frac{\lambda_0 \lambda_1 \cdots \lambda_{n-1}}{\mu_1 \mu_2 \cdots \mu_n} \pi_0 = 1.$$

Solving this equation yields

$$\pi_0 = \frac{1}{1 + \sum_{n=1}^{\infty} \frac{\lambda_0 \lambda_1 \cdots \lambda_{n-1}}{\mu_1 \mu_2 \cdots \mu_n}}. \qquad (12.18)$$

Hence

$$\pi_n = \frac{\lambda_0 \lambda_1 \cdots \lambda_{n-1}}{\mu_1 \mu_2 \cdots \mu_n \left(1 + \sum_{n=1}^{\infty} \frac{\lambda_0 \lambda_1 \cdots \lambda_{n-1}}{\mu_1 \mu_2 \cdots \mu_n}\right)}, \quad n \geq 1. \qquad (12.19)$$

Clearly, for π_n, $n \geq 0$, to exist, we need to have

$$\sum_{n=1}^{\infty} \frac{\lambda_0 \lambda_1 \cdots \lambda_{n-1}}{\mu_1 \mu_2 \cdots \mu_n} < \infty. \qquad (12.20)$$

It can be shown that if this series is convergent, then the limiting probabilities exist.

For $m \geq 1$, note that for a **finite-state** birth and death process with state space $\{0, 1, \ldots, m\}$, the balance equations are as follows:

State	Input rate to	=	Output rate from
0	$\mu_1 \pi_1$	=	$\lambda_0 \pi_0$
1	$\mu_2 \pi_2 + \lambda_0 \pi_0$	=	$\lambda_1 \pi_1 + \mu_1 \pi_1$
2	$\mu_3 \pi_3 + \lambda_1 \pi_1$	=	$\lambda_2 \pi_2 + \mu_2 \pi_2$
$\vdots$	$\vdots$		
m	$\lambda_{m-1} \pi_{m-1}$	=	$\mu_m \pi_m$

Solving any $m - 1$ of these equations along with $\pi_0 + \pi_1 + \cdots + \pi_m = 1$, we obtain

$$\pi_n = \frac{\lambda_0 \lambda_1 \cdots \lambda_{n-1}}{\mu_1 \mu_2 \cdots \mu_n} \pi_0, \quad 1 \leq n \leq m; \qquad (12.21)$$

$$\pi_0 = \frac{1}{1 + \sum_{n=1}^{m} \frac{\lambda_0 \lambda_1 \cdots \lambda_{n-1}}{\mu_1 \mu_2 \cdots \mu_n}}; \qquad (12.22)$$

$$\pi_n = \frac{\lambda_0 \lambda_1 \cdots \lambda_{n-1}}{\mu_1 \mu_2 \cdots \mu_n \left(1 + \sum_{n=1}^{m} \frac{\lambda_0 \lambda_1 \cdots \lambda_{n-1}}{\mu_1 \mu_2 \cdots \mu_n}\right)}, \quad 1 \leq n \leq m. \tag{12.23}$$

Example 12.43   Recall that an $M/M/1$ queueing system is a $GI/G/1$ system in which there is one server, customers arrive according to a Poisson process with rate λ, and service times are exponential with mean $1/\mu$. For an $M/M/1$ queueing system, let $X(t)$ be the number of customers in the system at t. Let customers arriving to the system be births, and customers departing from the system be deaths. Then $\{X(t): t \geq 0\}$ is a birth and death process with state space $\{0, 1, 2, \ldots\}$, birth rates $\lambda_n = \lambda$, $n \geq 0$, and death rates $\mu_n = \mu$, $n \geq 1$. For $n \geq 0$, let π_n be the proportion of time that there are n customers in the queueing system. Letting $\rho = \lambda/\mu$, by (12.20), the system is stable and the limiting probabilities, π_n's, exist if and only if

$$\sum_{n=1}^{\infty} \frac{\lambda^n}{\mu^n} = \sum_{n=1}^{\infty} \rho^n < \infty.$$

We know that the geometric series $\sum_{n=1}^{\infty} \rho^n$ converges if and only if $\rho < 1$. Therefore, the queue is stable and the limiting probabilities exist if and only if $\lambda < \mu$. That is, if and only if the arrival rate to the system is less than the service rate. Under this condition, by (12.18),

$$\pi_0 = \frac{1}{1 + \sum_{n=1}^{\infty} \rho^n} = \frac{1}{1 + \frac{\rho}{1 - \rho}} = 1 - \rho,$$

and by (12.17),

$$\pi_n = \frac{\lambda^n}{\mu^n} \pi_0 = \rho^n (1 - \rho), \quad n \geq 0.$$

Let L be the number of customers in the system at a random future time. Let L_Q be the number of customers in the system waiting in line to be served at a random future time. We have

$$E(L) = \sum_{n=0}^{\infty} n\rho^n (1 - \rho) = (1 - \rho) \sum_{n=1}^{\infty} n\rho^n = (1 - \rho) \cdot \frac{\rho}{(1 - \rho)^2} = \frac{\rho}{1 - \rho},$$

$$E(L_Q) = \sum_{n=0}^{\infty} n\rho^{n+1} (1 - \rho) = \rho(1 - \rho) \sum_{n=1}^{\infty} n\rho^n = \rho(1 - \rho) \cdot \frac{\rho}{(1 - \rho)^2} = \frac{\rho^2}{1 - \rho}.$$

Note that $E(L)$ is not $E(L_Q) + 1$. This is because when the system is empty L and L_Q are both 0. That is, with probability $1 - \rho$, we have that $L = L_Q = 0$. ◆

Example 12.44   On a campus building, there are m offices of similar sizes with identical air conditioners. The electrical grid supplies electric energy to the air conditioners whose thermostats turn on and off in each individual office as needed to maintain each office's temperature at the desired level of 76° Fahrenheit, independent of the other offices. Suppose that a thermostat remains on or off for exponential amounts of times with means $1/\mu$ and $1/\lambda$, respectively. Find the long-run probability that there will be i, $0 \le i \le m$, thermostats on at the same time.

Solution: Let $X(t)$ be the number of thermostats that are on at time t. We say that a birth occurs each time that a thermostat turns on, and a death occurs each time that a thermostat turns off. It should be clear that $\{X(t) : t \ge 0\}$ is a birth and death process with state space $\{0, 1, \dots, m\}$, and birth and death rates, respectively, given by $\lambda_i = (m-i)\lambda$ and $\mu_i = i\mu$ for $i = 0, 1, \dots, m$. To find π_0, first we will calculate the following sum:

$$\sum_{i=1}^{m} \frac{\lambda_0 \lambda_1 \cdots \lambda_{i-1}}{\mu_1 \mu_2 \cdots \mu_i} = \sum_{i=1}^{m} \frac{(m\lambda)\big[(m-1)\lambda\big]\big[(m-2)\lambda\big]\cdots\big[(m-i+1)\lambda\big]}{\mu(2\mu)(3\mu)\cdots(i\mu)}$$

$$= \sum_{i=1}^{m} \frac{{}_mP_i\, \lambda^i}{i!\, \mu^i} = \sum_{i=1}^{m} \binom{m}{i}\Big(\frac{\lambda}{\mu}\Big)^i$$

$$= -1 + \sum_{i=0}^{m} \binom{m}{i}\Big(\frac{\lambda}{\mu}\Big)^i 1^{m-i} = -1 + \Big(1 + \frac{\lambda}{\mu}\Big)^m,$$

where ${}_mP_i$ is the number of i-element permutations of a set containing m objects. Hence, by (12.22),

$$\pi_0 = \Big(1 + \frac{\lambda}{\mu}\Big)^{-m} = \Big(\frac{\lambda + \mu}{\mu}\Big)^{-m} = \Big(\frac{\mu}{\lambda + \mu}\Big)^m.$$

By (12.21),

$$\pi_i = \frac{\lambda_0 \lambda_1 \cdots \lambda_{i-1}}{\mu_1 \mu_2 \cdots \mu_i} \pi_0 = \frac{{}_mP_i\, \lambda^i}{i!\, \mu^i} \pi_0$$

$$= \binom{m}{i}\Big(\frac{\lambda}{\mu}\Big)^i \Big(\frac{\mu}{\lambda + \mu}\Big)^m = \binom{m}{i}\Big(\frac{\lambda}{\mu}\Big)^i \Big(\frac{\mu}{\lambda + \mu}\Big)^i \Big(\frac{\mu}{\lambda + \mu}\Big)^{m-i}$$

$$= \binom{m}{i}\Big(\frac{\lambda}{\lambda + \mu}\Big)^i \Big(1 - \frac{\lambda}{\lambda + \mu}\Big)^{m-i}, \quad 0 \le i \le m.$$

Therefore, in steady-state, the number of thermostats that are on is binomial with parameters m and $\lambda/(\lambda + \mu)$.   ◆

Let $\{X(t) : t \ge 0\}$ be a birth and death process with birth rates $\{\lambda_i\}_{i=0}^{\infty}$ and death rates $\{\mu_i\}_{i=1}^{\infty}$. For $i \ge 0$, let H_i be the time, starting from i, until the process enters state $i + 1$ for the first time. If we can calculate $E(H_i)$ for $i \ge 0$, then, for $j > i$, $\sum_{n=i}^{j-1} E(H_n)$ is the expected length of time, starting from i, it will take the process to enter state j

for the first time. The following lemma gives a useful recursive relation for computing $E(H_i)$, $i \geq 0$.

Lemma 12.2   *Let* $\{X(t): t \geq 0\}$ *be a birth and death process with birth rates* $\{\lambda_n\}_{n=0}^{\infty}$ *and death rates* $\{\mu_n\}_{n=1}^{\infty}$; $\mu_0 = 0$. *For* $i \geq 0$, *let* H_i *be the time, starting from* i, *until the process enters* $i + 1$ *for the first time. Then*

$$E(H_i) = \frac{1}{\lambda_i} + \frac{\mu_i}{\lambda_i} E(H_{i-1}), \quad i \geq 1.$$

Proof: Clearly, starting from 0, the time until the process enters 1 is exponential with parameter λ_0. Hence $E(H_0) = 1/\lambda_0$. For $i \geq 1$, starting from i, let $Z_i = 1$ if the next event is a birth, and let $Z_i = 0$ if the next event is a death. By conditioning on Z_i, we have

$$E(H_i) = E(H_i \mid Z_i = 1)P(Z_i = 1) + E(H_i \mid Z_i = 0)P(Z_i = 0)$$

$$= \frac{1}{\lambda_i} \cdot \frac{\lambda_i}{\lambda_i + \mu_i} + \left[E(H_{i-1}) + E(H_i) \right] \frac{\mu_i}{\lambda_i + \mu_i}, \quad (12.24)$$

where $E(H_i \mid Z_i = 0) = E(H_{i-1}) + E(H_i)$ follows, since the next transition being a death will move the process to $i - 1$. Therefore, on average, it will take the process $E(H_{i-1})$ units of time to enter state i from $i - 1$, and $E(H_i)$ units of time to enter $i + 1$ from i. Solving (12.24) for $E(H_i)$, we have the lemma.   ◆

Example 12.45   Consider an $M/M/1$ queueing system in which customers arrive according to a Poisson process with rate λ and service times are exponential with mean $1/\mu$. Let $X(t)$ be the number of customers in the system at t. In Example 12.43, we showed that $\{X(t): t \geq 0\}$ is a birth and death process with birth rates $\lambda_n = \lambda$, $n \geq 0$, and death rates $\mu_n = \mu$, for $n \geq 1$. Suppose that $\lambda < \mu$. Under this condition, the queueing system is stable and the limiting probabilities exist. For $i \geq 0$, let H_i be the time, starting from i, until the process enters state $i + 1$ for the first time. Clearly, $E(H_0) = 1/\lambda$, and by Lemma 12.2,

$$E(H_i) = \frac{1}{\lambda} + \frac{\mu}{\lambda} E(H_{i-1}), \quad i \geq 1.$$

Hence

$$E(H_1) = \frac{1}{\lambda} + \frac{\mu}{\lambda^2} = \frac{1}{\lambda}\left(1 + \frac{\mu}{\lambda}\right),$$

$$E(H_2) = \frac{1}{\lambda} + \frac{\mu}{\lambda}\left[\frac{1}{\lambda}\left(1 + \frac{\mu}{\lambda}\right)\right] = \frac{1}{\lambda}\left[1 + \left(\frac{\mu}{\lambda}\right) + \left(\frac{\mu}{\lambda}\right)^2\right].$$

Continuing this process, we obtain

$$E(H_i) = \frac{1}{\lambda}\left[1 + \left(\frac{\mu}{\lambda}\right) + \left(\frac{\mu}{\lambda}\right)^2 + \cdots + \left(\frac{\mu}{\lambda}\right)^i\right]$$

$$= \frac{1}{\lambda} \cdot \frac{(\mu/\lambda)^{i+1} - 1}{(\mu/\lambda) - 1} = \frac{(\mu/\lambda)^{i+1} - 1}{\mu - \lambda}, \quad i \geq 0.$$

Now suppose that there are i customers in the queueing system. For $j > i$, the expected length of time until the system has j customers is

$$\sum_{n=i}^{j-1} E(H_n) = \sum_{n=i}^{j-1} \frac{(\mu/\lambda)^{n+1} - 1}{\mu - \lambda} = \frac{1}{\mu - \lambda}\left[\sum_{n=i}^{j-1}\left(\frac{\mu}{\lambda}\right)^{n+1}\right] - \frac{j - i}{\mu - \lambda}$$

$$= \frac{1}{\mu - \lambda}\left(\frac{\mu}{\lambda}\right)^{i+1}\left[1 + \left(\frac{\mu}{\lambda}\right) + \cdots + \left(\frac{\mu}{\lambda}\right)^{j-i-1}\right] - \frac{j - i}{\mu - \lambda}$$

$$= \frac{1}{\mu - \lambda}\left(\frac{\mu}{\lambda}\right)^{i+1}\left[\frac{(\mu/\lambda)^{j-i} - 1}{(\mu/\lambda) - 1}\right] - \frac{j - i}{\mu - \lambda}$$

$$= \frac{\mu}{(\mu - \lambda)^2}\left(\frac{\mu}{\lambda}\right)^i\left[(\mu/\lambda)^{j-i} - 1\right] - \frac{j - i}{\mu - \lambda}. \quad \blacklozenge$$

EXERCISES

A

1. Let $\{X(t): t \geq 0\}$ be a continuous-time Markov chain with state space S. Show that for $i, j \in S$ and $t \geq 0$,

$$p'_{ij}(t) = \sum_{k \neq i} q_{ik} p_{kj}(t) - v_i p_{ij}(t).$$

In other words, prove Kolmogorov's backward equations.

2. The director of the study abroad program at a college advises one, two, or three students at a time depending on how many students are waiting outside his office. The time for each advisement session, regardless of the number of participants, is exponential with mean $1/\mu$, independent of other advisement sessions and the arrival process. Students arrive at a Poisson rate of λ and wait to be advised only if two or less other students are waiting to be advised. Otherwise, they leave. Upon the completion of an advisement session, the director will begin a new session if there are students waiting outside his office to be advised. Otherwise, he begins

a new session when the next student arrives. Let $X(t) = f$ if the director of the study abroad program is free and, for $i = 0, 1, 2, 3$, let $X(t) = i$ if an advisement session is in process and there are i students waiting outside to be advised. Show that $\{X(t): t \geq 0\}$ is a continuous-time Markov chain and find π_f, π_0, π_1, π_2, and π_3, the steady-state probabilities of this process.

3. Taxis arrive at the pick up area of a hotel at a Poisson rate of μ. Independently, passengers arrive at the same location at a Poisson rate of λ. If there are no passengers waiting to be put in service, the taxis wait in a queue until needed. Similarly, if there are no taxis available, passengers wait in a queue until their turn for a taxi service. For $n \geq 0$, let $X(t) = (n, 0)$ if there are n passengers waiting in the queue for a taxi. For $m \geq 0$, let $X(t) = (0, m)$ if there are m taxis waiting in the queue for passengers. Show that $\{X(t): t \geq 0\}$ is a continuous-time Markov chain, and write down the balance equations for the states of this Markov chain. You do not need to find the limiting probabilities.

4. An $M/M/\infty$ **queueing system** is similar to an $M/M/1$ system except that it has infinitely many servers. Therefore, all customers will be served upon arrival, and there will not be a queue. Examples of infinite-server systems are service facilities that provide self-service such as libraries. Consider an $M/M/\infty$ queuing system in which customers arrive according to a Poisson process with rate λ, and service times are exponentially distributed with mean $1/\lambda$. Find the long-run probability mass function and the expected value of the number of customers in the system.

5. **(Erlang's Loss System)**   Each operator at the customer service department of an airline can serve only one call. There are c operators, and the incoming calls form a Poisson process with rate λ. The time it takes to serve a customer is exponential with mean $1/\mu$, independent of other customers and the arrival process. If all operators are busy serving other customers, the additional incoming calls are rejected. They do not return and are called *lost calls*.

   (a) In the long-run, what proportion of calls are lost?

   (b) Suppose that $\lambda = \mu$. How many operators should the airline hire so that the probability that a call is lost is at most 0.004?

6. In Example 12.41, is the continuous-time Markov chain $\{X(t): t \geq 0\}$ a birth and death process?

7. Consider an $M/M/1$ queuing system in which customers arrive according to a Poisson process with rate λ, and service times are exponential with mean $1/\lambda$. We know that, in the long run, such a system will not be stable. For $i \geq 0$, suppose that at a certain time there are i customers in the system. For $j > i$, find the expected length of time until the system has j customers.

8. There are m machines in a factory operating independently. The factory has k $(k < m)$ repairpersons, and each repairperson repairs one machine at a time.

Suppose that (i) each machine works for a time period that is exponentially distributed with mean $1/\mu$, then it breaks down; (ii) the time that it takes to repair an out-of-order machine is exponential with mean $1/\lambda$, independent of repair times for other machines; and (iii) at times when all repair persons are busy repairing machines, the newly broken down machines will wait for repair. Let $X(t)$ be the number of machines operating at time t. Show that $\{X(t): t \geq 0\}$ is a birth and death process and find the birth and death rates.

9. In Springfield, Massachusetts, people drive their cars to a state inspection center for annual safety and emission certification at a Poisson rate of λ. For $n \geq 1$, if there are n cars at the center either being inspected or waiting to be inspected, the probability is $1 - \alpha_n$ that an additional driver will not join the queue and will leave. A driver who joins the queue has a *patience time* that is exponentially distributed with mean $1/\gamma$. That is, if the car's inspection turn does not occur within the patience time, the driver will leave. Suppose that cars are inspected one at a time, inspection times are independent and identically distributed exponential random variables with mean $1/\mu$, and they are independent of the arrival process and patience times. Let $X(t)$ be the number of cars being or waiting to be inspected at t. Find the birth and death rates of the birth and death process $\{X(t): t \geq 0\}$.

10. **(Birth and Death with Immigration)**   Consider a population of a certain colonizing species. Suppose that each individual produces offspring at a Poisson rate λ as long as it lives. Moreover, suppose that new individuals immigrate into the population at a Poisson rate of γ. If the lifetime of an individual in the population is exponential with mean $1/\mu$, starting with no individuals, find the expected length of time until the population size is 3.

11. Consider a pure death process with $\mu_n = \mu, n > 0$. For $i, j \geq 0$, find $p_{ij}(t)$.

12. Johnson Medical Associates has two physicians on call practicing independently. Each physician is available to answer patients' calls for independent time periods that are exponentially distributed with mean $1/\lambda$. Between those periods, the physician takes breaks for independent exponential amounts of time each with mean $1/\mu$. Suppose that the periods a physician is available to answer the calls are independent of the periods the physician is on breaks. Let $X(t)$ be the number of physicians on break at time t. Show that $\{X(t): t \geq 0\}$ can be modeled as a birth and death process. Then write down the list of all the Kolmogorov backward equations.

13. Recall that an $M/M/c$ queueing system is a $GI/G/c$ system in which there are c servers, customers arrive according to a Poisson process with rate λ, and service times are exponential with mean $1/\mu$. Suppose that $\rho = \lambda/(c\mu) < 1$; hence the queueing system is stable. Find the long-run probability that there are no customers in the system.

B

14. Let $\{N(t): t \geq 0\}$ be a Poisson process with rate λ. By Example 12.38, the process $\{N(t): t \geq 0\}$ is a continuous-time Markov chain. Hence it satisfies equations (12.12), the Chapman-Kolmogorov equations. Verify this fact by direct calculations.

15. **(The Yule Process)**   A cosmic particle entering the earth's atmosphere collides with air particles and transfers kinetic energy to them. These in turn collide with other particles transferring energy to them and so on. A shower of particles results. Suppose that the time that it takes for each particle to collide with another particle is exponential with parameter λ. Find the probability that t units of time after the cosmic particle enters the earth's atmosphere, there are n particles in the shower it causes.

16. **(Tandem or Sequential Queueing System)**   In a computer store, customers arrive at a cashier desk at a Poisson rate of λ to pay for the goods they want to purchase. If the cashier is busy, then they wait in line until their turn on a first-come, first-served basis. The time it takes for the cashier to serve a customer is exponential with mean $1/\mu_1$, independent of the arrival process. After being served by the cashier, customers join a second queue and wait until their turn to receive the goods they have purchased. When a customer's turn begins, it take an exponential period with mean $1/\mu_2$ to be served, independent of the service times of other customers and service times and arrival times at the cashier's desk. Clearly, the first station, cashier's desk, is an $M/M/1$ queueing system. The departure process of the first station forms an arrival process for the second station, delivery desk. Show that the second station is also an $M/M/1$ queueing system and is independent of the queueing system at the first station.
Hint: For $i \geq 0$, $j \geq 0$, by the state (i, j) we mean that there are i customers in the first queueing system and j customers in the second one. Let $\pi_{(i,j)}$ be the long-run probability that there are i customers at the cashier's desk and j customers at the delivery desk. Write down the set of balance equations for the entire process and show that

$$\pi_{(i,j)} = \left(\frac{\lambda}{\mu_1}\right)^i \left(1 - \frac{\lambda}{\mu_1}\right) \left(\frac{\lambda}{\mu_2}\right)^j \left(1 - \frac{\lambda}{\mu_2}\right), \quad i, j \geq 0,$$

satisfy the balance equations. If this is shown, by Example 12.43, we have shown that $\pi_{(i,j)}$ is the product of the long-run probabilities that the first queueing system is $M/M/1$ and has i customers, and the second queueing system is $M/M/1$ and has j customers.

17. **(Birth and Death with Disaster)**   Consider a population of a certain colonizing species. Suppose that each individual produces offspring at a Poisson rate of λ as long as it lives. Furthermore, suppose that the natural lifetime of an individual in the population is exponential with mean $1/\mu$ and, regardless of the population size, individuals die at a Poisson rate of γ because of disasters occurring independently of natural deaths and births. Let $X(t)$ be the population size at time t. If $X(0) = n$ $(n > 0)$, find $E\big[X(t)\big]$.

Hint: Using (12.5), calculate $E\big[X(t + h) \mid X(t) = m\big]$ for an infinitesimal h. Then find

$$E\big[X(t + h)\big] = E\Big[E\big[X(t + h) \mid X(t)\big]\Big].$$

Letting $h \to 0$ in $\dfrac{E\big[X(t + h)\big] - E\big[X(t)\big]}{h}$, show that $E\big[X(t)\big]$ satisfies a first-order linear differential equation. Solve that equation.

18.   Let $\big\{X(t) : t \geq 0\big\}$ be a birth and death process with birth rates $\big\{\lambda_n\big\}_{n=0}^{\infty}$ and death rates $\big\{\mu_n\big\}_{n=1}^{\infty}$. Show that if $\displaystyle\sum_{k=1}^{\infty} \dfrac{\mu_1 \mu_2 \cdots \mu_k}{\lambda_1 \lambda_2 \cdots \lambda_k} = \infty$, then, with probability 1, eventually extinction will occur.

■

12.5   BROWNIAN MOTION

To show that all material is made up from molecules, in 1827, the English Botanist Robert Brown (1773–1851) studied the motion of pollen particles in a container of water. He observed that, even though liquid in a container may appear to be motionless, particles suspended in the liquid, under constant and incessant random collisions with nearby particles, undergo unceasing motions in a totally erratic way. In 1905, unaware of Robert Brown's work, Albert Einstein (1879–1955) presented the first mathematical description of this phenomenon using laws of physics. After Einstein, much work was done by many scientists to advance the physical theory of the motion of particles in liquid. However, it was in 1923 that Norbert Wiener (1894–1964) was able to formulate Robert Brown's observations with mathematical rigor.

Suppose that liquid in a cubic container is placed in a coordinate system, and at time 0, a particle is at $(0, 0, 0)$, the origin. Unless otherwise specified, for mathematical simplicity, we assume that the container is unbounded from all sides and is full of liquid. Let $\big(X(t), Y(t), Z(t)\big)$ be the position of the particle after t units of time. We will find the distribution functions of $X(t)$, $Y(t)$, and $Z(t)$. To do so, we will find the distribution function of $X(t)$. It should be clear that the distribution functions of $Y(t)$ and $Z(t)$ are similarly calculated and are identical to that of the distribution function of $X(t)$. Observe that, in infinitesimal lengths of times during the time period $(0, t)$, $X(t)$ is a sum of infinitely many small movements in the direction of the x-coordinate. One way to find it is to divide the time interval t into $n = [t/h]$ subintervals, where the length of each subinterval is h, for some infinitesimal h, and $[t/h]$ is the greatest integer less than or equal to t/h. Suppose that, for infinitesimal $\delta > 0$, in each of these time subintervals, the x-coordinate of the particle moves to the right δ units with probability 1/3, moves to the left δ units with probability 1/3, and does not move with probability 1/3. For $i \geq 1$, let

$$X_i = \begin{cases} \delta & \text{with probability 1/3} \\ -\delta & \text{with probability 1/3} \\ 0 & \text{with probability 1/3.} \end{cases}$$

Then $\{X_i : i \geq 1\}$ is a random walk with $E(X_i) = 0$ and

$$\text{Var}(X_i) = E(X_i^2) = \frac{1}{3} \cdot \delta^2 + \frac{1}{3} \cdot (-\delta)^2 + \frac{1}{3} \cdot 0^2 = \frac{2\delta^2}{3}.$$

For large n, it should be clear that $X(t)$ and $\sum_{i=1}^{n} X_i$ have approximately the same distribution. Since the motions of the particle are totally erratic, $X_1, X_2, \ldots$ are independent random variables. Hence, by the central limit theorem, for large n, the distribution of $\sum_{i=1}^{n} X_i$ is approximately $N\left(0, \frac{2\delta^2}{3}n\right)$ or $N\left(0, \frac{2\delta^2}{3h}t\right)$. For the limiting behavior of the random walk $\{X_i : i \geq 1\}$ to closely approximate the x-coordinate, $X(t)$, of the position of the particle after t units of time, we need to let $h \to 0$ and $\delta \to 0$ in a way that $(2\delta^2)/(3h)$ remains a constant σ^2. Doing this, we obtain that, for all $t > 0$, $X(t)$ is a normal random variable with mean 0 and variance $\sigma^2 t$. The same argument can be used to show that $Y(t)$, the y-coordinate, and $Z(t)$, the z-coordinate of the position of the particle after t units of time, are also normal with mean 0 and variance $\sigma^2 t$. It can be shown that

$X(t)$, $Y(t)$, and $Z(t)$ are independent random variables.

Moreover, physical observations show that they possess stationary and independent increments. For the process $\{X(t): t \geq 0\}$, by **stationary increments** we mean that, for $s < t$ and $h \in (-\infty, \infty)$, the random variables $X(t) - X(s)$ and $X(t + h) - X(s + h)$ are identically distributed. Similarly, as previously defined, we say that $\{X(t): t \geq 0\}$ possesses independent increments if for $s \leq t \leq u \leq v$, we have that $X(t) - X(s)$ and $X(v) - X(u)$ are independent random variables. The properties of the processes $\{X(t): t \geq 0\}$, $\{Y(t): t \geq 0\}$ and $\{Z(t): t \geq 0\}$ obtained by studying the x, y, and z coordinates of the positions of a particles's motion in a liquid motivate the following definition.

Definition   *A stochastic process $\{X(t): t \geq 0\}$ with state space $\mathcal{S} = (-\infty, \infty)$ that possesses stationary and independent increments is said to be a **Brownian motion** if $X(0) = 0$ and, for $t > 0$, $X(t)$ is a normal random variable with mean 0 and variance $\sigma^2 t$, for some $\sigma > 0$.*

Brownian motions are also called **Wiener processes**. σ^2 is called the **variance parameter** of the Brownian motion. For $t > 0$, the probability density function of $X(t)$

is denoted by $\phi_t(x)$. Therefore,

$$\phi_t(x) = \frac{1}{\sigma\sqrt{2\pi t}} \exp\left[-\frac{x^2}{2\sigma^2 t}\right]. \tag{12.25}$$

For a Brownian motion $\{X(t): t \geq 0\}$, note that by stationarity of the increments:

For $s, t \geq 0$, $X(t+s) - X(s)$ is $N(0, \sigma^2 t)$.

Brownian motion is an important area in stochastic processes. It has diverse applications in various fields. Even though its origin is in the studies of motions of particles in liquid or gas, it is used in quantum mechanics, stock market fluctuations, and testing of goodness of fit in statistics.

Let $\{X(t): t \geq 0\}$ be a Brownian motion with variance parameter σ^2. Let $W(t) = X(t)/\sigma$. It is straightforward to show that $\{W(t): t \geq 0\}$ is a Brownian motion with variance parameter 1. A Brownian motion with variance parameter 1 is called a **standard Brownian motion**. For simplicity, we sometimes prove theorems about a Brownian motion with variance parameter 1 and then use the transformation $W(t) = X(t)/\sigma$ to find the more general theorem.

Intuitively, it is not difficult to see that the graph of the x-coordinates of the positions of a particle's motion in a liquid, as a function of t, is continuous. This is true for the y- and z-coordinates as well. It is interesting to know that these functions are *everywhere continuous but nowhere differentiable*. Therefore, being at no point smooth, they are extremely kinky. The study of nowhere differentiability of such graphs is difficult and beyond the scope of this book. However, from a mechanical point of view, nowhere differentiability is a result of the assumption that the mass of the particle in motion is 0.

We will now discuss, step by step, some of the most important properties of Brownian motions.

The Conditional Probability Density Function of $X(t)$ Given that $X(0) = x_0$

In Section 12.3, we showed that for a discrete-time Markov chain, p_{ij}^n, the probability of moving from state i to state j in n steps satisfies the Chapman-Kolmogorov equations. In Section 12.4, we proved that $p_{ij}(t)$, the probability that a continuous-time Markov chain moves from state i to state j in t units of time satisfies the Kolmogorov forward and backward equations. To point out the parallel between these and analogous results in Brownian motions, let $f_{t|0}(x|x_0)$ be the conditional probability density function of $X(t)$ given that $X(0) = x_0$.[†] This is the function analogous to $p_{ij}(t)$ for continuous-time Markov chain, and p_{ij}^n for the discrete-time Markov chain. The functions $f_{t|0}(x|x_0)$ are called **transition probability density functions** of the Brownian motion. They should not be confused with $f_{X|Y}(x|y)$, the conditional probability density function of a random variable X given that $Y = y$.

[†]Even though, in general, we adopt the convention of $X(0) = 0$, this choice is not necessary. It is not contradictory to assume that $X(0) = x_0$, for some point $x_0 \neq 0$.

Note that, by definition of a probability density function, for $u \in (-\infty, \infty)$,

$$P\big(X(t) \le u \mid X(0) = x_0\big) = \int_{-\infty}^{u} f_{t|0}(x|x_0)\, dx.$$

Also note that, since Brownian motions possess stationary increments, the probability density function of $X(t + t_0)$ given that $X(t_0) = x_0$ is $f_{t|0}(x|x_0)$ as well. By selecting appropriate scales, Albert Einstein showed that $f_{t|0}(x|x_0)$ satisfies the partial differential equation

$$\frac{\partial f}{\partial t} = \frac{1}{2}\sigma^2 \frac{\partial^2 f}{\partial x^2}, \tag{12.26}$$

called the **backward diffusion equation**. Considering the facts that $f_{t|0}(x|x_0) \ge 0$, $\int_{-\infty}^{\infty} f_{t|0}(x|x_0)\, dx = 1$, and $\lim_{t \to 0} f_{t|0}(x|x_0) = 0$ if $x \ne x_0$, it is straightforward to see that the unique solution to the backward diffusion equation under these conditions is

$$f_{t|0}(x|x_0) = \frac{1}{\sigma\sqrt{2\pi t}} \exp\left[-\frac{(x - x_0)^2}{2\sigma^2 t}\right]. \tag{12.27}$$

The Joint Probability Density Function of $X(t_1), X(t_2), \ldots, X(t_n)$

Next, for $t_1 < t_2$, let $f(x_1, x_2)$ be the joint probability density function of $X(t_1)$ and $X(t_2)$. To find this important function, let $U = X(t_1)$ and $V = X(t_2) - X(t_1)$. Applying Theorem 8.8, we will first find the joint probability density function of U and V. Observe that the system of equations

$$\begin{cases} x_1 = u \\ x_2 - x_1 = v \end{cases}$$

has the unique solution $x_1 = u$, $x_2 = u + v$. Since

$$J = \begin{vmatrix} \dfrac{\partial x_1}{\partial u} & \dfrac{\partial x_1}{\partial v} \\[2mm] \dfrac{\partial x_2}{\partial u} & \dfrac{\partial x_2}{\partial v} \end{vmatrix} = \begin{vmatrix} 1 & 0 \\ 1 & 1 \end{vmatrix} = 1 \ne 0,$$

we have that, $g(u, v)$, the joint probability density function of U and V is given by

$$g(u, v) = f(u, u + v)|J| = f(u, u + v).$$

By this relation, $f(x_1, x_2) = g(x_1, x_2 - x_1)$. Since U and V are independent random variables, $g(x_1, x_2 - x_1)$ is the product of the marginal probability density functions of

$X(t_1)$ and $X(t_2) - X(t_1)$, which by the stationarity and independence of the increments of Brownian motion processes, are $N(0, \sigma^2 t_1)$ and $N(0, \sigma^2(t_2 - t_1))$, respectively. Hence

$$f(x_1, x_2) = \frac{1}{\sigma\sqrt{2\pi t_1}} \exp\left(-\frac{x_1^2}{2\sigma^2 t_1}\right) \cdot \frac{1}{\sigma\sqrt{2\pi(t_2 - t_1)}} \exp\left[-\frac{(x_2 - x_1)^2}{2\sigma^2(t_2 - t_1)}\right],$$

or, equivalently,

$$f(x_1, x_2) =$$
$$\frac{1}{2\sigma^2\pi\sqrt{t_1(t_2 - t_1)}} \exp\left[-\frac{1}{2\sigma^2}\left(\frac{x_1^2}{t_1} + \frac{(x_2 - x_1)^2}{t_2 - t_1}\right)\right], \quad -\infty < x_1, x_2 < \infty.$$

$$(12.28)$$

For $t_1 < t_2 < \cdots < t_n$, let $f(x_1, x_2, \ldots, x_n)$ be the joint probability density function of $X(t_1), X(t_2), \ldots, X(t_n)$. Let $g(x_1, x_2, \ldots, x_n)$ be the joint probability density function of $X(t_1), X(t_2) - X(t_1), \ldots, X(t_n) - X(t_{n-1})$. An argument similar to the one presented for the case $n = 2$, shows that

$$f(x_1, x_2, \ldots, x_n) = g(x_1, x_2 - x_1, \ldots, x_n - x_{n-1}).$$

Since the random variables $X(t_1), X(t_2) - X(t_1), \ldots, X(t_n) - X(t_{n-1})$ are independent, $X(t_1) \sim N(0, \sigma^2 t)$ and, for $2 \le i \le n$, $X(t_i) - X(t_{i-1}) \sim N(0, \sigma^2(t_i - t_{i-1}))$. Thus

$$f(x_1, x_2, \ldots, x_n) =$$
$$\frac{\exp\left[-\frac{1}{2\sigma^2}\left(\frac{x_1^2}{t_1} + \sum_{i=1}^{n-1}\frac{(x_{i+1} - x_i)^2}{t_{i+1} - t_i}\right)\right]}{\sigma^n\sqrt{(2\pi)^n t_1(t_2 - t_1)\cdots(t_n - t_{n-1})}}, \quad -\infty < x_1, x_2, \ldots, x_n < \infty.$$

$$(12.29)$$

For $t_1 < t < t_2$, the Conditional Probability Density Function of $X(t)$ Given That $X(t_1) = x_1$ and $X(t_2) = x_2$

Let us begin with the special case in which $t_1 = 0$, $x_1 = 0$, $t_2 = u_1$, and $x_2 = 0$. In such a case, the x-coordinate of the particle under consideration in the liquid, which at time 0 was at 0, is again at 0 at time u_1. For $0 < u < u_1$, we want to find the conditional probability density function of $X(u)$ given that $X(u_1) = 0$. [The condition $X(0) = 0$ is automatically assumed for all Brownian motions unless otherwise stated.] Let $f(x, y)$ be the joint probability density function of $X(u)$ and $X(u_1)$. Then, by (12.28),

$$f(x, y) =$$
$$\frac{1}{2\sigma^2\pi\sqrt{u(u_1 - u)}} \exp\left[-\frac{1}{2\sigma^2}\left(\frac{x^2}{u} + \frac{(y - x)^2}{u_1 - u}\right)\right], \quad -\infty < x, y < \infty. \quad (12.30)$$

Let $f_{X(u)|X(u_1)}(x|0)$ be the conditional probability density function of $X(u)$ given that $X(u_1) = 0$. Then, by (8.18) and (12.25),

$$f_{X(u)|X(u_1)}(x|0) = \frac{f(x, 0)}{f_{X(u_1)}(0)} = \frac{f(x, 0)}{\phi_{u_1}(0)} = \sigma\sqrt{2\pi u_1}f(x, 0).$$

Thus, by (12.30),

$$f_{X(u)|X(u_1)}(x|0) = \sigma\sqrt{2\pi u_1} \cdot \frac{1}{2\sigma^2\pi\sqrt{u(u_1 - u)}} \exp\left[-\frac{1}{2\sigma^2}\left(\frac{x^2}{u} + \frac{x^2}{u_1 - u}\right)\right]$$

$$= \frac{1}{\sigma\sqrt{2\pi}}\sqrt{\frac{u_1}{u(u_1 - u)}} \cdot \exp\left[-\frac{1}{2\sigma^2} \cdot \frac{u_1}{u(u_1 - u)}x^2\right], \quad -\infty < x < \infty,$$

which is the probability density function of a normal random variable with mean 0 and variance $\dfrac{\sigma^2 u(u_1 - u)}{u_1}$. Therefore, for $0 < u < u_1$,

$$E[X(u) \mid X(0) = 0 \text{ and } X(u_1) = 0] = 0 \qquad (12.31)$$

and

$$\mathbf{Var}[X(u) \mid X(0) = 0 \text{ and } X(u_1) = 0] = \frac{\sigma^2 u(u_1 - u)}{u_1}. \qquad (12.32)$$

Now, for $t_1 < t < t_2$, to find the conditional probability density function of $X(t)$ given that $X(t_1) = x_1$ and $X(t_2) = x_2$, define

$$\widetilde{X}(t) = X(t + t_1) - x_1 - \frac{t}{t_2 - t_1}(x_2 - x_1).$$

This is equivalent to

$$\widetilde{X}(t - t_1) = X(t) - x_1 - \frac{t - t_1}{t_2 - t_1}(x_2 - x_1). \qquad (12.33)$$

Clearly, for $t = t_1$, we obtain

$$\widetilde{X}(0) = X(t_1) - x_1 = 0,$$

and

$$\widetilde{X}(t_2 - t_1) = X(t_2) - x_1 - (x_2 - x_1) = 0.$$

Since $t_1 < t < t_2$ implies that $0 < t - t_1 < t_2 - t_1$, in (12.31) and (12.32) letting $u_1 = t_2 - t_1$ and $u = t - t_1$, for the process $\widetilde{X}(t)$ we have

$$E[\widetilde{X}(t - t_1) \mid \widetilde{X}(0) = 0 \text{ and } \widetilde{X}(t_2 - t_1) = 0] = 0$$

and

$$\text{Var}\big[\widetilde{X}(t - t_1) \mid \widetilde{X}(0) = 0 \text{ and } \widetilde{X}(t_2 - t_1) = 0\big]$$

$$= \frac{\sigma^2(t - t_1)\big[(t_2 - t_1) - (t - t_1)\big]}{t_2 - t_1} = \frac{\sigma^2(t_2 - t)(t - t_1)}{t_2 - t_1}.$$

Now, by (12.33),

$$0 = E\big[\widetilde{X}(t - t_1) \mid \widetilde{X}(0) = 0 \text{ and } \widetilde{X}(t_2 - t_1) = 0\big]$$

$$= E\big[X(t) \mid X(t_1) = x_1 \text{ and } X(t_2) = x_2\big] - x_1 - \frac{t - t_1}{t_2 - t_1}(x_2 - x_1).$$

Hence

$$E\big[X(t) \mid X(t_1) = x_1 \text{ and } X(t_2) = x_2\big] = x_1 + \frac{x_2 - x_1}{t_2 - t_1}(t - t_1),$$

which is the equation of a line passing through (t_1, x_1) and (t_2, x_2). Similarly, by (12.33),

$$\text{Var}\big[X(t) \mid X(t_1) = x_1 \text{ and } X(t_2) = x_2\big]$$

$$= \text{Var}\big[\widetilde{X}(t - t_1) \mid \widetilde{X}(0) = 0 \text{ and } \widetilde{X}(t_2 - t_1) = 0\big] = \sigma^2\frac{(t_2 - t)(t - t_1)}{t_2 - t_1}.$$

We have shown the following important theorem:

Theorem 12.9   *For $t_1 < t < t_2$, the conditional probability density function of $X(t)$ given that $X(t_1) = x_1$ and $X(t_2) = x_2$ is normal with mean $x_1 + \dfrac{x_2 - x_1}{t_2 - t_1}(t - t_1)$ and variance $\sigma^2\dfrac{(t_2 - t)(t - t_1)}{t_2 - t_1}$.*

Example 12.46   Suppose that liquid in a container is placed in a coordinate system, and at time 0, a pollen particle suspended in the liquid is at $(0, 0, 0)$, the origin. Let $X(t)$ be the x-coordinate of the position of the pollen after t minutes. Suppose that $\{X(t) : t \geq 0\}$ is a Brownian motion with variance parameter 4 and, after one minute, the x-coordinate of the pollen's position is 2.

(a)   What is the probability that after 2 minutes it is between 0 and 1?

(b)   What is the expected value and variance of the x-coordinate of the position of pollen after 30 seconds?

Solution:

(a)   The desired probability is

$$P\big(0 < X(2) < 1 \mid X(1) = 2\big) = P\big(-2 < X(2) - X(1) < -1 \mid X(1) = 2\big)$$

$$= P\big(-2 < X(2) - X(1) < -1\big),$$

by the independent-increments property of Brownian motion. Since $X(2) - X(1)$ is normal with mean 0 and variance $(2 - 1)\sigma^2 = 4$, letting $Z \sim N(0, 1)$, we have

$$P\left(-2 < X(2) - X(1) < -1\right) = P\left(\frac{-2 - 0}{2} < Z < \frac{-1 - 0}{2}\right)$$

$$= P(-1 < Z < -0.5)$$

$$= \Phi(-0.5) - \Phi(-1) = 0.1523.$$

(b)   In Theorem 12.9, let $t_1 = 0$, $t_2 = 1$, $x_1 = 0$, $x_2 = 2$. We have

$$E\left[X(1/2) \mid X(0) = 0 \text{ and } X(1) = 2\right] = 0 + \frac{2 - 0}{1 - 0}\left(\frac{1}{2} - 0\right) = 1$$

$$\text{Var}\left[X(1/2) \mid X(0) = 0 \text{ and } X(1) = 2\right] = 4 \cdot \frac{\left(1 - \frac{1}{2}\right)\left(\frac{1}{2} - 0\right)}{1 - 0} = 1. \quad \blacklozenge$$

First Passage Time Distribution

Let $\{X(t): t \geq 0\}$ be a Brownian motion with variance parameter σ^2. The **first passage time** to a point α, also called the **time of hitting α first**, denoted by T_α, is the minimum of the set $\{t: X(t) = \alpha\}$. We will first find $P(T_\alpha \leq t)$, $-\infty < t < \infty$, for $\alpha > 0$. Then we will calculate it for $\alpha < 0$. For $\alpha > 0$, note that, by Theorem 3.4,

$$P\left(X(t) \geq \alpha\right) = P\left(X(t) \geq \alpha \mid T_\alpha < t\right)P(T_\alpha < t) + P\left(X(t) \geq \alpha \mid T_\alpha = t\right)P(T_\alpha = t)$$

$$+ P\left(X(t) \geq \alpha \mid T_\alpha > t\right)P(T_\alpha > t). \tag{12.34}$$

Since $\alpha > 0$, and the graph of $X(t)$, as a function of t, is continuous, we have that $P\left(X(t) \geq \alpha \mid T_\alpha > t\right) = 0$. Since T_α is a continuous random variable, $P(T_\alpha = t) = 0$. Observe that after the process hits α, for $t > T_\alpha$, by symmetry, it is equally likely that $X(t) \geq \alpha$ and $X(t) \leq \alpha$. Hence

$$P\left(X(t) \geq \alpha \mid T_\alpha < t\right) = \frac{1}{2}.$$

Incorporating these facts into (12.34), we obtain

$$P\left(X(t) \geq \alpha\right) = \frac{1}{2}P(T_\alpha < t).$$

Since $X(t) \sim N(0, \sigma^2 t)$, we have

$$P(T_\alpha \leq t) = P(T_\alpha < t) = 2P\left(X(t) \geq \alpha\right) = \frac{2}{\sigma\sqrt{2\pi t}} \int_\alpha^\infty e^{-x^2/(2t\sigma^2)} \, dx.$$

Letting $\dfrac{x}{\sigma\sqrt{t}} = y$, this reduces to

$$P(T_\alpha \le t) = \frac{2}{\sqrt{2\pi}} \int_{\alpha/(\sigma\sqrt{t})}^{\infty} e^{-y^2/2}\,dy = 2\Big[1 - \Phi\Big(\frac{\alpha}{\sigma\sqrt{t}}\Big)\Big]. \qquad (12.35)$$

To find $P(T_\alpha \le t)$ for $\alpha < 0$, note that T_α and $T_{-\alpha}$ are identically distributed. Hence

$$P(T_\alpha \le t) = P(T_{-\alpha} \le t) = 2\Big[1 - \Phi\Big(-\frac{\alpha}{\sigma\sqrt{t}}\Big)\Big]. \qquad (12.36)$$

Putting (12.35) and (12.36) together, we have that for all $\alpha \in (-\infty, \infty)$, the distribution function of the first passage time to α is given by

$$P(T_\alpha \le t) = 2\Big[1 - \Phi\Big(\frac{|\alpha|}{\sigma\sqrt{t}}\Big)\Big]. \qquad (12.37)$$

As in the case of symmetric random walks, it can be shown that, even though, with probability 1, $\{X(t): t \ge 0\}$ eventually will hit α, none of the moments of the first passage time to α is finite.

The Maximum of a Brownian Motion

As before, let T_α be the first passage time to α. By the continuity of the paths of Brownian motions, for $\alpha > 0$, the event $\max\limits_{0 \le s \le t} X(s) \ge \alpha$ occurs if and only if $T_\alpha \le t$. Therefore,

$$P\Big(\max_{0 \le s \le t} X(s) \ge \alpha\Big) = P(T_\alpha \le t) = 2\Big[1 - \Phi\Big(\frac{\alpha}{\sigma\sqrt{t}}\Big)\Big],$$

or, equivalently,

$$P\Big(\max_{0 \le s \le t} X(s) \le \alpha\Big) = 1 - 2\Big[1 - \Phi\Big(\frac{\alpha}{\sigma\sqrt{t}}\Big)\Big] = 2\Phi\Big(\frac{\alpha}{\sigma\sqrt{t}}\Big) - 1.$$

Thus the distribution function of $\max\limits_{0 \le s \le t} X(s)$ is given by

$$P\Big(\max_{0 \le s \le t} X(s) \le x\Big) = \begin{cases} 2\Phi\Big(\dfrac{x}{\sigma\sqrt{t}}\Big) - 1 & \text{if } x \ge 0 \\[2mm] 0 & \text{if } x < 0. \end{cases} \qquad (12.38)$$

The Zeros of Brownian Motion

Let $\{X(t): t \ge 0\}$ be a Brownian motion with variance parameter σ^2. In this subsection, we will find the probability that $X(t) = 0$ at least once in the time interval (t_1, t_2), $0 < t_1 < t_2$. The event that $X(t) = 0$ for at least one t in this interval is $\bigcup\limits_{t_1 < t < t_2} \{X(t) = 0\}$.

To begin with, note that if $X(t_1) = a > 0$, then, by the continuity of the Brownian motion's paths, $\bigcup_{t_1 < t < t_2} \{X(t) = 0\}$ occurs if and only if $\min_{t_1 < t < t_2} X(t) \leq 0$. Hence, for $a > 0$,

$$P\left(\bigcup_{t_1 < t < t_2} \{X(t) = 0\} \mid X(t_1) = a \right) = P\left(\min_{t_1 < t < t_2} X(t) \leq 0 \mid X(t_1) = a \right). \quad (12.39)$$

For $a < 0$, we consider the Brownian motion $\{-X(t): t \geq 0\}$. Since, for all $t \geq 0$, $X(t)$ and $-X(t)$ are identically distributed, we have

$$P\left(\bigcup_{t_1 < t < t_2} \{X(t) = 0\} \mid X(t_1) = a \right) = P\left(\bigcup_{t_1 < t < t_2} \{-X(t) = 0\} \mid -X(t_1) = a \right)$$

$$= P\left(\bigcup_{t_1 < t < t_2} \{X(t) = 0\} \mid X(t_1) = -a \right),$$

which, by (12.39), is equal to $P\left(\min_{t_1 < t < t_2} X(t) \leq 0 \mid X(t_1) = -a \right)$ since $-a > 0$. This and (12.39) imply that, for all $a \in (-\infty, \infty)$,

$$P\left(\bigcup_{t_1 < t < t_2} \{X(t) = 0\} \mid X(t_1) = a \right) = P\left(\min_{t_1 < t < t_2} X(t) \leq 0 \mid X(t_1) = |a| \right). \quad (12.40)$$

Now, again, using the fact that $X(t)$ and $-X(t)$ are identically distributed, and

$$-\min_{t_1 < t < t_2} -X(t) = \max_{t_1 < t < t_2} X(t),$$

we have

$$P\left(\min_{t_1 < t < t_2} X(t) \leq 0 \mid X(t_1) = |a| \right) = 1 - P\left(\min_{t_1 < t < t_2} X(t) > 0 \mid X(t_1) = |a| \right)$$

$$= 1 - P\left(\min_{t_1 < t < t_2} -X(t) > 0 \mid -X(t_1) = |a| \right)$$

$$= 1 - P\left(-\min_{t_1 < t < t_2} -X(t) < 0 \mid -X(t_1) = |a| \right)$$

$$= 1 - P\left(\max_{t_1 < t < t_2} X(t) < 0 \mid X(t_1) = -|a| \right)$$

$$= P\left(\max_{t_1 < t < t_2} X(t) \geq 0 \mid X(t_1) = -|a| \right)$$

$$= P\left(\max_{t_1 < t < t_2} X(t) \geq |a| \mid X(t_1) = 0 \right)$$

$$= P\left(\max_{0 < t < t_2 - t_1} X(t) \geq |a| \mid X(0) = 0 \right)$$

$$= 2\left[1 - \Phi\left(\frac{|a|}{\sigma \sqrt{t_2 - t_1}} \right) \right], \quad (12.41)$$

where the last equality follows from (12.38). Now, on the one hand, by conditioning on $X(t_1)$ and by relations (12.40) and (12.41), we obtain

$$P\left(\bigcup_{t_1 < t < t_2} \{X(t) = 0\}\right) = \int_{-\infty}^{\infty} P\left(\bigcup_{t_1 < t < t_2} \{X(t) = 0\} \mid X(t_1) = a\right)\phi_{t_1}(a)\, da$$

$$= \int_{-\infty}^{\infty} P\left(\min_{t_1 < t < t_2} X(t) \le 0 \mid X(t_1) = |a|\right)\phi_{t_1}(a)\, da$$

$$= \int_{-\infty}^{\infty} 2\left[1 - \Phi\left(\frac{|a|}{\sigma\sqrt{t_2 - t_1}}\right)\right]\phi_{t_1}(a)\, da. \qquad (12.42)$$

On the other hand, letting $Z \sim N(0, 1)$, we get

$$P\big(X(t_2) - X(t_1) > |X(t_1)|\big) = \int_{-\infty}^{\infty} P\big(X(t_2) - X(t_1) > |X(t_1)| \mid X(t_1) = a\big)\phi_{t_1}(a)\, da$$

$$= \int_{-\infty}^{\infty} P\big(X(t_2) - X(t_1) > |a|\big)\phi_{t_1}(a)\, da$$

$$= \int_{-\infty}^{\infty} P\left(Z > \frac{|a|}{\sigma\sqrt{t_2 - t_1}}\right)\phi_{t_1}(a)\, da$$

$$= \int_{-\infty}^{\infty} \left[1 - \Phi\left(\frac{|a|}{\sigma\sqrt{t_2 - t_1}}\right)\right]\phi_{t_1}(a)\, da. \qquad (12.43)$$

Comparing (12.42) and (12.43) yields

$$P\left(\bigcup_{t_1 < t < t_2} \{X(t) = 0\}\right) = 2P\big(X(t_2) - X(t_1) > |X(t_1)|\big).$$

Let Z_1 and Z_2 be two independent standard normal random variables. Then $X(t_2) - X(t_1)$ is identically distributed with $\sigma\sqrt{t_2 - t_1}\,Z_1$ and $X(t_1)$ is identically distributed with $\sigma\sqrt{t_1}\,Z_2$. Hence

$$P\left(\bigcup_{t_1 < t < t_2} \{X(t) = 0\}\right) = 2P\big(\sigma\sqrt{t_2 - t_1}\,Z_1 > |\sigma\sqrt{t_1}\,Z_2|\big) = 2P\left(\frac{Z_1}{|Z_2|} > \sqrt{\frac{t_1}{t_2 - t_1}}\right).$$

By Example 10.26, $Z_1/|Z_2|$ is a Cauchy random variable. Therefore,

$$P\left(\bigcup_{t_1 < t < t_2} \{X(t) = 0\}\right) = 2 \int_{\sqrt{t_1/(t_2 - t_1)}}^{\infty} \frac{1}{\pi(1 + x^2)}\, dx$$

$$= \frac{2}{\pi}\Big[\arctan x\Big]_{\sqrt{t_1/(t_2 - t_1)}}^{\infty} = \frac{2}{\pi}\left[\frac{\pi}{2} - \arctan\sqrt{\frac{t_1}{t_2 - t_1}}\right]$$

$$= 1 - \frac{2}{\pi}\arctan\sqrt{\frac{t_1}{t_2 - t_1}}. \qquad (12.44)$$

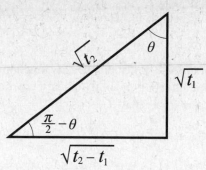

Figure 12.7   Triangle of Theorem 12.10.

Now draw a right triangle, such as the one in Figure 12.7, with hypotenuse of length $\sqrt{t_2}$ and sides of lengths $\sqrt{t_1}$ and $\sqrt{t_2 - t_1}$. Let the angle adjacent to the side of length $\sqrt{t_1}$ be θ radians. Then

$$\tan\left(\frac{\pi}{2} - \theta\right) = \frac{\sqrt{t_1}}{\sqrt{t_2 - t_1}},$$

which implies that

$$\arctan\sqrt{\frac{t_1}{t_2 - t_1}} = \frac{\pi}{2} - \theta,$$

or

$$1 - \frac{2}{\pi}\arctan\sqrt{\frac{t_1}{t_2 - t_1}} = \frac{2}{\pi}\theta = \frac{2}{\pi}\arccos\sqrt{\frac{t_1}{t_2}}.$$

Considering this identity and (12.44), we have shown the following theorem:

Theorem 12.10   *For a Brownian motion $\{X(t)\colon t \geq 0\}$ with variance parameter σ^2, the probability that $X(t) = 0$ at least once in the time interval (t_1, t_2), $0 < t_1 < t_2$, is*

$$\frac{2}{\pi}\arccos\sqrt{\frac{t_1}{t_2}}.$$

Brownian Motion with Drift

Let $\{W(t)\colon t \geq 0\}$ be a standard Brownian motion. That is, a Brownian motion with variance parameter 1. For $\mu \in (-\infty, \infty)$ and $\sigma > 0$, the process

$$X(t) = \mu t + \sigma W(t), \quad t \geq 0,$$

is called the **Brownian motion with drift parameter μ and variance parameter σ^2**. Note that for such a process, it is still the case that $X(0) = 0$ and $\{X(t)\colon t \geq 0\}$ possesses stationary and independent increments. However, for $t > 0$, $X(t)$ is normal with mean μt and variance $\sigma^2 t$.

For a Brownian motion $\{X(t) : t \geq 0\}$ with drift parameter μ and variance parameter σ^2, let $f_{t|0}(x|x_0)$ be the conditional probability density function of $X(t)$ given that $X(0) = x_0$, or, equivalently, the conditional probability density function of $X(t + t_0)$ given that $X(t_0) = x_0$. To find $f_{t|0}(x|x_0)$, note that

$$P\big(X(t) \leq x \mid X(0) = x_0\big) = P\big(\mu t + \sigma W(t) \leq x \mid \sigma W(0) = x_0\big)$$

$$= P\Big(W(t) \leq \frac{x - \mu t}{\sigma} \mid W(0) = \frac{x_0}{\sigma}\Big)$$

$$= P\Big(W(t) - W(0) \leq \frac{x - x_0 - \mu t}{\sigma} \mid W(0) = \frac{x_0}{\sigma}\Big).$$

Let $Z \sim N(0, 1)$; since $W(t) - W(0)$ and $W(0)$ are independent random variables, and $W(t) - W(0)$ is $N(0, t)$, this equation gives

$$P\big(X(t) \leq x \mid X(0) = x_0\big) = P\Big(W(t) - W(0) \leq \frac{x - x_0 - \mu t}{\sigma}\Big)$$

$$= P\Bigg(Z \leq \frac{\dfrac{x - x_0 - \mu t}{\sigma} - 0}{\sqrt{t}}\Bigg) = \Phi\Big(\frac{x - x_0 - \mu t}{\sigma \sqrt{t}}\Big).$$

Differentiating this equation with respect to x yields

$$f_{t|0}(x|x_0) = \frac{1}{\sigma \sqrt{t}} \cdot \frac{1}{\sqrt{2\pi}} \exp\Big[-\Big(\frac{x - x_0 - \mu t}{\sigma \sqrt{t}}\Big)^2 \Big/ 2\Big]$$

$$= \frac{1}{\sigma \sqrt{2\pi t}} \exp\Big[\frac{-(x - x_0 - \mu t)^2}{2\sigma^2 t}\Big],$$

which is the probability density function of a normal random variable with mean $x_0 + \mu t$ and variance $\sigma^2 t$.

Geometric Brownian Motion

As mentioned before, particles suspended in a liquid are constantly and incessantly under random collisions with nearby particles. As a result, they undergo unceasing motions in a totally erratic way, and the changes in each coordinate of the positions of a particle, in infinitesimal time periods, form a Brownian motion. In finance, random fluctuations of prices of shares of stocks, or other assets, are a result of an enormity of political, social, economical, and financial factors independently assessed by a large number of traders. For that reason financial scholars and consultants use Brownian motions as models for prices of assets in general and shares of stocks in particular. However, since such prices are nonnegative and in the long run follow exponential growth or decay curves, they use a type of a stochastic process that is called **geometric Brownian motion**.

Let $\{W(t): t \geq 0\}$ be a standard Brownian motion. Let $X(t) = \mu t + \sigma W(t)$ so that $\{X(t): t \geq 0\}$ is a Brownian motion with drift parameter μ and variance parameter σ^2. The process $\{V(t): t \geq 0\}$ defined by

$$V(t) = v_0 \exp\left[X(t)\right] = v_0 \exp\left[\mu t + \sigma W(t)\right]$$

is said to be a **geometric Brownian motion** with drift parameter μ and variance parameter σ^2. Clearly, $W(0) = 0$ implies that $V(0) = v_0$.

Stock prices are often assumed to be Markov processes, since, usually, the price of a stock in the future depends on its present value only. The past history of the value of a stock and how it evolved to form the current price does not play a major role on the price of the stock in the future. In the preceding geometric model, if $V(t)$ is the price of some stock at t, then, for $t_0 < t_1 < \cdots < t_n$, the relations

$$\frac{V(t_i)}{V(t_{i-1})} = \exp\left[\mu(t_i - t_{i-1})\right] \cdot \exp\left[\sigma[W(t_i) - W(t_{i-1})]\right]$$

show that

$$\frac{V(t_1)}{V(t_0)}, \ \frac{V(t_2)}{V(t_1)}, \ \ldots, \ \frac{V(t_n)}{V(t_{n-1})},$$

the ratio changes of the price of the stock over nonoverlapping time intervals, are independent random variables.

We will now calculate the expected value and variance of $V(t)$ given that $V(0) = v_0$. To do so, let $Z \sim N(0, 1)$. We will first show that, for $\alpha \in (-\infty, \infty)$,

$$E(e^{\alpha Z}) = e^{\alpha^2/2}. \tag{12.45}$$

This follows from the following calculations:

$$E(e^{\alpha Z}) = \int_{-\infty}^{\infty} e^{\alpha x} \cdot \frac{1}{\sqrt{2\pi}} e^{-x^2/2} \, dx$$

$$= e^{\alpha^2/2} \int_{-\infty}^{\infty} \frac{1}{\sqrt{2\pi}} e^{-\frac{1}{2}\alpha^2 + \alpha x - \frac{1}{2}x^2} \, dx$$

$$= e^{\alpha^2/2} \int_{-\infty}^{\infty} \frac{1}{\sqrt{2\pi}} e^{-\frac{1}{2}(x-\alpha)^2} \, dx = e^{\alpha^2/2},$$

where $\int_{-\infty}^{\infty} \frac{1}{\sqrt{2\pi}} e^{-\frac{1}{2}(x-\alpha)^2} \, dx = 1$, since $\frac{1}{\sqrt{2\pi}} e^{-\frac{1}{2}(x-\alpha)^2}$ is the probability density function of a normal random variable with mean α and variance 1.

Using (12.45) and the fact that $Z = W(t)/\sqrt{t} \sim N(0, 1)$, calculation of $E\left[V(t)\right]$ and $\text{Var}\left[V(t)\right]$ is straightforward. We have

$$E\left[V(t)\right] = E\left[v_0 e^{\mu t + \sigma W(t)}\right] = v_0 e^{\mu t} E\left(e^{\sigma \sqrt{t} Z}\right) = v_0 e^{\mu t + \frac{1}{2}\sigma^2 t}.$$

Also,

$$E[V(t)^2] = E[v_0^2 e^{2\mu t + 2\sigma W(t)}] = v_0^2 e^{2\mu t} E(e^{2\sigma \sqrt{t} Z}) = v_0^2 e^{2\mu t + 2\sigma^2 t}.$$

Therefore,

$$\text{Var}[V(t)] = E[V(t)^2] - E[V(t)]^2 = v_0^2 e^{2\mu t + \sigma^2 t}(e^{\sigma^2 t} - 1).$$

Example 12.47   Let $V(t)$ be the price of a stock, per share, at time t. Suppose that the stock's current value, per share, is \$40 with drift parameter \$2 per year and variance parameter 4.41. If $\{V(t): t \geq 0\}$ is a geometric Brownian motion, what is the probability that after 18 months the stock price, per share, exceeds \$65?

Solution: Noting that $\sqrt{4.41} = 2.1$, we have

$$V(t) = 40e^{2t + 2.1W(t)},$$

where $\{W(t): t \geq 0\}$ is a standard Brownian motion. Hence $W(t) \sim N(0, t)$. The desired probability is

$$\begin{aligned}
P(V(1.5) > 65) &= P(40e^{2(1.5) + 2.1W(1.5)} > 65) \\
&= P(e^{2.1W(1.5)} > 0.0809) = P(W(1.5) > -1.197) \\
&= P\left(\frac{W(1.5) - 0}{\sqrt{1.5}} > -\frac{1.197}{\sqrt{1.5}}\right) = P(Z > -0.98) \\
&= 1 - \Phi(-0.98) = \Phi(0.98) = 0.8365. \quad \blacklozenge
\end{aligned}$$

EXERCISES

A

1. Suppose that liquid in a container is placed in a coordinate system, and at time 0, a pollen particle suspended in the liquid is at $(0, 0, 0)$, the origin. Let $Z(t)$ be the z-coordinate of the position of the pollen after t minutes. Suppose that $\{Z(t): t \geq 0\}$ is a Brownian motion with variance parameter 9. Suppose that after 5 minutes the z-coordinate of the pollen's position is 0 again.

    (a)   What is the probability that after 10 minutes it is between $-1/2$ and $1/2$?

(b)   If after seven minutes the z-coordinate of the pollen's position is -1, find the expected value and variance of the z-coordinate of the position of pollen after six minutes.

2.   Let $\{X(t): t \geq 0\}$ be a Brownian motion with variance parameter σ^2. Show that, for all $t > 0$, $|X(t)|$ and $\max\limits_{0 \leq s \leq t} X(s)$ are identically distributed.

3.   Let $\{X(t): t \geq 0\}$ be a Brownian motion with variance parameter σ^2. For $\varepsilon > 0$, show that $\lim\limits_{t \to 0} P\left(\dfrac{|X(t)|}{t} > \varepsilon\right) = 1$, whereas $\lim\limits_{t \to \infty} P\left(\dfrac{|X(t)|}{t} > \varepsilon\right) = 0$.

4.   Let $\{X(t): t \geq 0\}$ be a Brownian motion with variance parameter σ^2. Let T_α be the time of hitting α first. Let $Y \sim N(0, \sigma^2/\alpha^2)$. Show that, for $\alpha > 0$, T_α and $1/Y^2$ are identically distributed.

5.   Let $\{X(t): t \geq 0\}$ be a Brownian motion with variance parameter σ^2. For a fixed $t > 0$, let T be the smallest zero greater than t. Find the probability distribution function of T.

6.   Let $\{X(t): t \geq 0\}$ be a Brownian motion with variance parameter σ^2. As we know, for t_1 and t_2, $t_1 < t_2$, the random variables $X(t_1)$ and $X(t_2)$ are not independent. Find the distribution of $X(t_1) + X(t_2)$.

7.   Let $\{X(t): t \geq 0\}$ be a Brownian motion with variance parameter σ^2. For $u > 0$, show that
$$E\big[X(t+u) \mid X(t)\big] = X(t).$$
Therefore, for $s > t$, $E\big[X(s) \mid X(t)\big] = X(t)$.

8.   Let $\{X(t): t \geq 0\}$ be a Brownian motion with variance parameter σ^2. For $u > 0$, $t \geq 0$, find $E\big[X(t)X(t+u)\big]$.

9.   **(Reflected Brownian Motion)**   Suppose that liquid in a cubic container is placed in a coordinate system in such a way that the bottom of the container is placed on the xy-plane. Therefore, whenever a particle reaches the xy-plane, it cannot cross the bottom of the container. So it reverberates back to the nonnegative side of the z-axis. Suppose that at time 0, a particle is at $(0, 0, 0)$, the origin. Let $V(t)$ be the z-coordinate of the particle after t units of time. Find $E\big[V(t)\big]$, $\text{Var}\big[V(t)\big]$, and $P\big(V(t) \leq z \mid V(0) = z_0\big)$.

Hint: Let $\{Z(t): t \geq 0\}$ be a Brownian motion with variance parameter σ^2. Note that
$$V(t) = \begin{cases} Z(t) & \text{if } Z(t) \geq 0 \\ -Z(t) & \text{if } Z(t) < 0. \end{cases}$$
The process $\{V(t): t \geq 0\}$ is called **reflected Brownian motion.**

10.   Suppose that liquid in a cubic container is placed in a coordinate system. Suppose that at time 0, a particle is at $(0, 0, 0)$, the origin. Let $(X(t), Y(t), Z(t))$ be the

coordinates of the particle after t units of time, and assume that $X(t)$, $Y(t)$, and $Z(t)$ are independent Brownian motions, each with variance parameter σ^2. Let $D(t)$ be the distance of the particle from the origin after t units of time. Find $E[D(t)]$.

11. Let $V(t)$ be the price of a stock, per share, at time t. Suppose that the stock's current value, per share, is \$95.00 with drift parameter $-\$2$ per year and variance parameter 5.29. If $\{V(t): t \geq 0\}$ is a geometric Brownian motion, what is the probability that after 9 months the stock price, per share, is below \$80?

■

REVIEW PROBLEMS

---

1. Jobs arrive at a file server at a Poisson rate of 3 per minute. If 10 jobs arrived within 3 minutes, between 10:00 and 10:03, what is the probability that the last job arrived after 40 seconds past 10:02?

2. A Markov chain with transition probability matrix $P = (p_{ij})$ is called *regular*, if for some positive integer n, $p_{ij}^n > 0$ for all i and j. Let $\{X_n: n = 0, 1, \ldots\}$ be a Markov chain with state space $\{0, 1\}$ and transition probability matrix

$$P = \begin{pmatrix} 0 & 1 \\ 1 & 0 \end{pmatrix}$$

Is $\{X_n: n = 0, 1, \ldots\}$ regular? Why or why not?

3. Show that the following matrices are the transition probability matrix of the same Markov chain with elements of the state space labeled differently.

$$P_1 = \begin{pmatrix} 2/5 & 0 & 0 & 3/5 & 0 \\ 1/3 & 1/3 & 0 & 1/3 & 0 \\ 0 & 0 & 1/2 & 0 & 1/2 \\ 1/4 & 0 & 0 & 3/4 & 0 \\ 0 & 0 & 1/3 & 0 & 2/3 \end{pmatrix}, \quad P_2 = \begin{pmatrix} 2/5 & 3/5 & 0 & 0 & 0 \\ 1/4 & 3/4 & 0 & 0 & 0 \\ 0 & 0 & 1/2 & 1/2 & 0 \\ 0 & 0 & 1/3 & 2/3 & 0 \\ 1/3 & 1/3 & 0 & 0 & 1/3 \end{pmatrix}.$$

4. Let $\{X_n: n = 0, 1, \ldots\}$ be a Markov chain with state space $\{0, 1\}$ and transition probability matrix

$$P = \begin{pmatrix} 2/5 & 3/5 \\ 1/3 & 2/3 \end{pmatrix}$$

Starting from 0, find the expected number of transitions until the first visit to 1.

5. The following is the transition probability matrix of a Markov chain with state space $\{1, 2, 3, 4, 5\}$. Specify the classes and determine which classes are transient and which are recurrent.

$$P = \begin{pmatrix} 0 & 0 & 0 & 0 & 1 \\ 0 & 1/3 & 0 & 2/3 & 0 \\ 0 & 0 & 1/2 & 0 & 1/2 \\ 0 & 0 & 0 & 1 & 0 \\ 0 & 0 & 2/5 & 0 & 3/5 \end{pmatrix}.$$

6. A fair die is tossed repeatedly. Let X_n be the number of 6's obtained in the first n tosses. Show that $\{X_n : n = 1, 2, \dots \}$ is a Markov chain. Then find its transition probability matrix, specify the classes and determine which are recurrent and which are transient.

7. On a given vacation day, a sportsman either goes horseback riding (activity 1), or sailing (activity 2), or scuba diving (activity 3). For $1 \le i \le 3$, let $X_n = i$, if the sportsman devotes vacation day n to activity i. Suppose that $\{X_n : n = 1, 2, \dots \}$ is a Markov chain, and depending on which of these activities the sportsman chooses on a certain vacation day, the probability of engagement in any one of the activities on the next vacation day is given by the transition probability matrix

$$P = \begin{pmatrix} 0.20 & 0.30 & 0.50 \\ 0.32 & 0.15 & 0.53 \\ 0.60 & 0.13 & 0.27 \end{pmatrix}.$$

We know that the sportsman did not go scuba diving on the first day of his vacation. What is the probability that he did not go scuba diving on the second and third vacation days either?

8. Construct a transition probability matrix of a Markov chain with state space $\{1, 2, \dots , 8\}$ in which $\{1, 2, 3, 4\}$ is transient having period 4, $\{5\}$ is aperiodic transient, and $\{6, 7, 8\}$ is recurrent having period 3.

9. In a golfball production line, a golfball produced with no logo is called defective; all other golfballs are called good. A quality assurance engineer performing statistical process control observes that only 1 out of 12 defective golfballs is produced after another defective ball, with no good balls between, whereas 15 out of 18 good balls are produced following good balls, with no defective balls between. Find the fraction of golfballs produced with no logo.

10. An urn contains 7 red, 11 blue, and 13 yellow balls. Carmela, Daniela, and Lucrezia play a game in which they take turns and draw balls from the urn, successively, at random and with replacement. Suppose that the colors red, blue, and yellow are assigned to Carmela, Daniela, and Lucrezia, respectively. The only rule governing the game is that, as long as a player draws her own color, she will keep drawing

balls from the urn. When a ball of another player's color is drawn, that player will begin drawing balls, and the game will continue. Determine the long-run proportion of balls drawn by each player.

11. On a given vacation day, Francesco either plays golf (activity 1) or tennis (activity 2). For $i = 1, 2$, let $X_n = i$, if Francesco devotes vacation day n to activity i. Suppose that $\{X_n : n = 1, 2, \ldots\}$ is a Markov chain, and depending on which of the two activities he chooses on a certain vacation day, the probability of engagement in any one of the activities on the next vacation day is given by the transition probability matrix

$$P = \begin{pmatrix} 0.30 & 0.70 \\ 0.58 & 0.42 \end{pmatrix}$$

Find the long-run probability that, on a randomly selected vacation day, Francesco plays tennis.

12. Passengers arrive at a train station according to a Poisson process with rate λ and, independently, trains arrive at the same station according to another Poisson process, but with the same rate λ. Suppose that each time a train arrives, all of the passengers waiting at the station will board the train. Let $X(t) = 1$, if there is at least one passenger waiting at the train station for a train; let $X(t) = 0$, otherwise. Let $N(t)$ be the number of times the continuous-time Markov chain $\{X(t) : t \geq 0\}$ changes states in $[0, t]$. Find the probability mass function of $N(t)$.

13. Consider a **parallel system** consisting of two components denoted by 1 and 2. Such a system functions if and only if at least one of its components functions. Suppose that each component functions for a time period which is exponentially distributed with mean $1/\lambda$ independent of the other component. There is only one repairperson available to repair failed components, and repair times are independent exponential random variables with mean $1/\mu$. Find the long-run probability that the system works.

14. In a factory, there are m operating machines and s machines used as spares and ready to operate. The factory has k repairpersons, and each repairperson repairs one machine at a time. Suppose that, (i) each machine works, independent of other machines, for a time period which is exponentially distributed with mean $1/\mu$, then it breaks down; (ii) the time that it takes to repair an out-of-order machine is exponential with mean $1/\lambda$, independent of repair times for other machines; (iii) at times when all repairpersons are busy repairing machines, the newly broken down machines will wait for repair; and (iv) when a machine breaks down, one of the spares will be used unless there is no spare machine available. Let $X(t)$ be the number of machines operating or ready to operate at time t. Show that $\{X(t) : t \geq 0\}$ is a birth and death process and find the birth and death rates.

15. There are m machines in a factory operating independently. Each machine works for a time period that is exponentially distributed with mean $1/\mu$. Then it breaks

down. The time that it takes to repair an out-of-order machine is exponential with mean $1/\lambda$, independent of repair times for and operating times of other machines. For $0 \leq i \leq m$, find the long-run proportion of time that there are exactly i machines operating.

16. In Springfield, Massachusetts, people drive their cars to a state inspection center for annual safety and emission certification at a Poisson rate of λ. For $n \geq 0$, if there are n cars at the center either being inspected or waiting to be inspected, the probability is $1/(n + 1)$ that an additional driver will join the queue. Hence the probability is $n/(n + 1)$ that he or she will not join the queue and will leave. Suppose that inspection times are independent and identically distributed exponential random variables with mean $1/\mu$ and they are independent of the arrival process. For $n \geq 0$, find the probability that in the long run, at a random time, there are n cars at the center either being inspected or waiting to be inspected.

17. Consider a population of a certain colonizing species. Suppose that each individual produces offspring at a Poisson rate of λ as long as it lives, and the time until the first individual arrives is exponential with mean $1/\lambda$. If the lifetime of an individual in the population is exponential with mean $1/\mu$, starting with a single individual, find the expected length of time until the population size is 5.

18. **(Death Process with Immigration)**   Consider a population of size n, $n \geq 0$, of a certain species in which individuals do not reproduce. However, new individuals immigrate into the population at a Poisson rate of γ. If the lifetime of an individual in the population is exponential with mean $1/\mu$, find π_i, $i \geq 0$, the long-run probability that the population size is i.

19. Suppose that liquid in a cubic container is placed in a coordinate system, and at time 0, a pollen particle suspended in the liquid is at $(0, 0, 0)$, the origin. Let $Y(t)$ be the y-coordinate of the position of the pollen after t minutes. Suppose that the process $\{Y(t): t \geq 0\}$ is a Brownian motion with variance parameter σ^2. For $0 < s < t$, find $E[Y(s) \mid Y(t) = y]$ and $\text{Var}[Y(s) \mid Y(t) = y]$.

20. Let $\{X(t): t \geq 0\}$ be a Brownian motion with variance parameter σ^2. For $t, s > 0$, find $E[X(s)X(t)]$.

21. Let $\{X(t): t \geq 0\}$ be a Brownian motion with variance parameter σ^2. For $t > 0$, let T be the smallest zero greater than t, and let U be the largest zero smaller than t. For $x < t < y$, find $P(U < x \text{ and } T > y)$.

22. Let $V(t)$ be the price of a stock, per share, at time t. Suppose that $\{V(t): t \geq 0\}$ is a geometric Brownian motion with drift parameter \$3 per year and variance parameter 27.04. What is the probability that the price of this stock is at least twice its current price after two years?

Chapter 13

Simulation

13.1 INTRODUCTION

Solving a scientific or an industrial problem usually involves mathematical analysis and/or simulation. To perform a simulation, we repeat an experiment a large number of times to assess the probability of an event or condition occurring. For example, to estimate the probability of at least one 6 occurring within four rolls of a die, we may do a large number of experiments rolling four dice and calculate the number of times that at least one 6 is obtained. Similarly, to estimate the fraction of time that, in a certain bank all the tellers are busy, we may measure the lengths of such time intervals over a long period X, add them, and then divide by X. Clearly, in simulations, the key to reliable answers is *to perform the experiment a large number of times or over a long period of time, whichever is applicable.* Since manually this is almost impossible, simulations are carried out by computers. Only computers can handle millions of operations in short periods of time.

To simulate a problem that involves random phenomena, generating random numbers from the interval $(0, 1)$ is essential. In almost every simulation of a probabilistic model, we will need to select random points from the interval $(0, 1)$. For example, to simulate the experiment of tossing a fair coin, we draw a random number from $(0, 1)$. If it is in $(0, 1/2)$, we say that the outcome is heads, and if it is in $[1/2, 1)$, we say that it is tails. Similarly, in the simulation of die tossing, the outcomes 1, 2, 3, 4, 5, and 6, respectively, correspond to the events that the random point from $(0, 1)$ is in $(0, 1/6)$, $[1/6, 1/3)$, $[1/3, 1/2)$, $[1/2, 2/3)$, $[2/3, 5/6)$, and $[5/6, 1)$.

As discussed in Section 1.7, choosing a random number from a given interval is, in practice, impossible. In real-world problems, to perform simulation we use **pseudorandom numbers** instead. To generate n pseudorandom numbers from a uniform distribution on an interval (a, b), we take an initial value $x_0 \in (a, b)$, called the **seed,** and construct a function ψ so that the sequence $\{x_1, x_2, \ldots, x_n\} \subset (a, b)$ obtained recursively from

$$x_{i+1} = \psi(x_i), \qquad 0 \le i \le n - 1, \tag{13.1}$$

satisfies certain statistical tests for randomness. (Choosing the tests and constructing the function ψ are complicated matters beyond the scope of this book.) The function ψ takes a seed and generates a sequence of pseudorandom numbers in the interval (a, b). Clearly, in any pseudorandom number generating process, the numbers generated are rounded to a certain number of decimal places. Therefore, ψ can only generate a finite number of pseudorandom numbers, which implies that, eventually, some x_j will be generated a second time. From that point on, by (13.1), a pitfall is that the same sequence of numbers that appeared after x_j's first appearance will reappear. Beyond that point, numbers are not effectively random. One important aspect of the construction of ψ is that the second appearance of any of the x_j's is postponed as long as possible.

It should be noted that passing certain statistical tests for randomness does not mean that the sequence $\{x_1, x_2, \ldots, x_n\}$ is a randomly selected sequence from (a, b) in the true mathematical sense discussed in Section 1.7. It is quite surprising that there are deterministic real-valued functions ψ that, for each i, generate an x_{i+1} that is completely determined by x_i, and yet the sequence $\{x_1, x_2, \ldots, x_n\}$ passes certain statistical tests for randomness.

For convenience, throughout this chapter, in practical problems, by random number we simply mean pseudorandom number. In general, for good choices of ψ, the generated numbers are usually sufficiently random for practical purposes.

Most of the computer languages and some scientific computer software are equipped with subroutines that generate random numbers from intervals and from sets of integers. In *Mathematica*, from Wolfram Research, Inc. (http://www.wolfram.com), the command **Random[   ]** selects a random number from $(0, 1)$, **Random[Real, {a, b}]** chooses a random number from the interval (a, b), and **Random [Integer, {m, m + n}]** picks up an integer from $\{m, m + 1, m + 2, \ldots, m + n\}$ randomly. However, there are computer languages that are not equipped with a subroutine that generates random numbers, and there are a few that are equipped with poor algorithms. As mentioned, it is difficult to construct good random number generators. However, an excellent reference for such algorithms is *The Art of Computer Programming*, Volume 2, *Seminumerical Algorithms*, 3rd edition, by Donald E. Knuth (Addison-Wesley, 1998).

At this point, let us emphasize that the main goal of scientists and engineers is always to solve a problem mathematically. It is a mathematical solution that is accurate, exact, and completely reliable. Simulations cannot take the place of a rigorous mathematical solution. They are widely used (a) to find good estimations for solutions of problems that either cannot be modeled mathematically or whose mathematical models are too difficult to solve; (b) to get a better understanding of the behavior of a complicated phenomenon; and/or (c) to obtain a mathematical solution by acquiring insight into the nature of the problem, its functions, and the magnitude and characteristics of its solution. Intuitively, it is clear why the results that are obtained by simulations are good. Theoretically, most of them can be justified by the strong law of large numbers, as discussed in Section 11.4.

For a simulation in each of the following examples we have presented an algorithm, in English, that can be translated into any programming language. Therefore, readers may use these algorithms to write and execute their own programs in their favorite computer languages.

Example 13.1   Two numbers are selected at random and without replacement from the set $\{1, 2, 3, \ldots, \ell\}$. Write an algorithm for a computer simulation of approximating the probability that the difference of these numbers is at least k.

Solution: For a large number of times, say n, each time choose two distinct random numbers, a and b, from $\{1, 2, 3, \ldots, \ell\}$ and check to see if $|a - b| \geq k$. In all of these n experiments, let m be the number of those in which $|a - b| \geq k$. Then m/n is the desired approximation. An algorithm follows.

STEP

1:      Set $i = 1$;
2:      Set $m = 0$;
3:      While $i \leq n$, do steps 4 to 7.
4:      Generate a random number a from $\{1, 2, \ldots, \ell\}$.
5:      Generate a random number b from $\{1, 2, \ldots, \ell\}$.
6:      If $b = a$, Goto step 5.
7:      If $|a - b| \geq k$, then

                Set $m = m + 1$;
                Set $i = i + 1$;
                Goto step 3.

      else

                Set $i = i + 1$;
                Goto step 3.

8:      Set $p = m/n$;
9:      Output (p).

STOP

Since the exact answer to this problem, obtained by analytical methods, is the quantity $\dfrac{(\ell - k)(\ell - k + 1)}{2\ell(\ell - 1)}$, any simulation result should be close to this number.   ♦

Example 13.2   An urn contains 25 white and 35 red balls. Balls are drawn from the urn successively and without replacement. Write an algorithm to approximate by simulation the probability that at some instant the number of red and white balls drawn are equal (a tie occurs).

Solution: For a large number of times n, we will repeat the following experiment and count m, the number of times that at some instant a tie occurs. The desired quantity is m/n. To begin, let $m = 0$, $w = 25$, and $r = 35$. Generate a random number from $(0, 1)$. If it belongs to $\left(0, \dfrac{w}{w + r}\right]$, a white ball is selected. Thus change w to $w - 1$. Otherwise, a red ball is drawn; thus change r to $r - 1$. Repeat this procedure and check each time to see if $25 - w = 35 - r$. If at some instance these two quantities are equal, change m

to $m + 1$ and start a new experiment. Otherwise, continue until either $r = 0$ or $w = 0$. An algorithm follows.

STEP

1:    Set $m = 0$;

2:    Set $i = 1$;

3:    While $i \leq n$, do steps 4 to 10.

4:    Set $w = 25$;

5:    Set $r = 35$;

6:    While $r \geq 0$ and $w \geq 0$ do steps 7 to 10.

7:    Generate a random number x from $(0, 1)$.

8:    If $x \leq w/(w + r)$, then
            Set $w = w - 1$;
        else
            Set $r = r - 1$;

9:    If $25 - w = 35 - r$, then
            Set $m = m + 1$;
            Set $i = i + 1$;
            Goto step 3.

10:    If $r = 0$ or $w = 0$, then
            Set $i = i + 1$;
            Goto step 3.

11:    Set $p = m/n$;

12:    Output (p).

STOP   ◆

EXERCISES

---

1.   A die is rolled successively until, for the first time, a number appears three consecutive times. By simulation, determine the approximate probability that it takes at least 50 rolls before we accomplish this event.

2.   In an election, the Democratic candidate obtained 3586 votes and the Republican candidate obtained 2958. Use simulations to find the approximate probability that the Democratic candidate was ahead during the entire process of counting the votes.
     Answer: Approximately 0.096.

3.   The probability that a bank refuses to finance an applicant is 0.35. Using simulation, find the approximate probability that of 300 applicants more than 100 are refused.

4. There are five urns, each containing 10 white and 15 red balls. A ball is drawn at random from the first urn and put into the second one. Then a ball is drawn at random from the second urn and put into the third one and the process is continued. By simulation, determine the approximate probability that the last ball is red. *Answer:* 0.6.

5. In a small town of 1000 inhabitants, someone gossips to a random person, who in turn tells the story to another random person, and so on. Using simulation, calculate the approximate probability that before the story is told 150 times, it is returned to the first person.

6. A city has n taxis numbered 1 through n. A statistician takes taxis numbered 31, 50, and 112 on three random occasions. Based on this information, determine, using simulation, which of the numbers 112 through 120 is a better estimate for n. *Hint:* For each n ($112 \leq n \leq 120$), repeat the following experiment a large number of times: Choose three random numbers from $\{1, 2, \ldots, n\}$ and check to see if they are 31, 50, and 112.

7. Suppose that an airplane passenger whose itinerary requires a change, of airplanes in Ankara, Turkey, has a 4% chance, independently, of losing each piece of his or her luggage. Suppose that the probability of losing each piece of luggage in this way is 5% at Da Vinci airport in Rome, 5% at Kennedy airport in New York, and 4% at O'Hare airport in Chicago. Dr. May travels from Bombay to San Francisco with three suitcases. He changes airplanes in Ankara, Rome, New York, and Chicago. Using simulation, find the approximate probability that one of Dr. May's suitcases does not reach his destination with him.

■

13.2   SIMULATION OF COMBINATORIAL PROBLEMS

Suppose that $X(1)$, $X(2)$, $\ldots$, $X(n)$ are n objects numbered from 1 to n in a convenient way. For example, suppose that $X(1)$, $X(2)$, $\ldots$, $X(52)$ are the cards of an ordinary deck of 52 cards and they are assigned numbers 1 to 52 in some convenient order. One of the most common problems in combinatorial simulation is to find an efficient procedure for choosing m ($m \leq n$) distinct objects randomly from $X(1)$, $X(2)$, $\ldots$, $X(n)$. To solve this problem, we choose a random integer from $\{1, 2, \ldots, n\}$ and call it n_1. Clearly, $X(n_1)$ is a random object from $\{X(1), X(2), \ldots, X(n)\}$. To choose a second random object distinct from $X(n_1)$, we exchange $X(n)$ and $X(n_1)$ and then choose a random object from $\{X(1), X(2), \ldots, X(n-1)\}$, as before. That is, we choose a random integer, n_2, from $\{1, 2, \ldots, n-1\}$ and then exchange $X(n_2)$ and $X(n-1)$. At this point, $X(n)$ and $X(n-1)$ are two distinct random objects from the original set of objects. Continuing this procedure, after m selections the set

$$\{X(n), X(n-1), X(n-2), \dots, X(n-m+1)\}$$

consists of m distinct random objects from $\{X(1), X(2), \dots, X(n)\}$. In particular, if $m = n$, the *ordered set* $\{X(n), X(n-1), X(n-2), \dots, X(1)\}$ is a permutation of $\{X(1), X(2), \dots, X(n)\}$. An algorithm follows.

Algorithm 13.1:

STEP

1:	Set $i = 1$;
2:	While $i \leq m$, do steps 3 to 5.
3:	Choose a random integer k from $\{1, 2, \dots, n-i+1\}$.
4:	Set $Z = X(k)$;
	Set $X(k) = X(n-i+1)$;
	Set $X(n-i+1) = Z$;
5:	Set $i = i + 1$;
6:	Output $\{X(n), X(n-1), X(n-2), \dots, X(n-m+1)\}$.

STOP   ◆

Example 13.3   In a commencement ceremony, for the dean of a college to present the diplomas of the graduates, a clerk piles the diplomas in the order that the students will walk on the stage. However, the clerk mixes the last 10 diplomas in some random order accidentally. Write an algorithm to approximate, using simulation, the probability that at least one of the last 10 graduates who will walk on the stage receives his or her own diploma from the dean.

Solution: Number the last 10 graduates who will walk on the stage 1 through 10. Let the diploma of graduate i be numbered i, $1 \leq i \leq 10$. If $X(i)$ denotes the diploma that the dean of the college gives to the graduate i, then $X(i)$ is a random integer between 1 and 10. Thus $X(1), X(2), \dots, X(10)$, are numbered $1, 2, \dots, 10$ in some random order, and graduate i receives his or her diploma if $X(i) = i$. One way to simulate this problem is that for a large n, we generate n random orders for 1 through 10, find m, the number of those in which there is at least one i with $X(i) = i$, and then divide m by n. To do so, we present the following algorithm, which puts the numbers 1 through 10 in random order n times and calculates m and $p = m/n$.

STEP

1:	Set $m = 0$;
2:	Set $i = 1$;
3:	While $i \leq n$, do steps 4 to 10.
4:	Generate a random permutation $\{X(10), X(9), \dots, X(1)\}$ of $\{1, 2, 3, \dots, 10\}$ (see Algorithm 13.1).

5:        Set $k = 0$;
6:        Set $j = 1$;
7:        While $j \leq 10$, do step 8.
8:        If $X(j) = j$, then
                    Set $j = 11$;
                    Set $k = 1$;
          else
                    Set $j = j + 1$.
9:        Set $m = m + k$;
10:       Set $i = i + 1$;
11:       Set $p = m/n$;
12:       Output (p).
STOP

A simulation program based on this algorithm was executed for several values of n. The corresponding values of p were as follows:

n	p
10	0.7000
1,000	0.6090
10,000	0.6340
100,000	0.6326

For comparison, it is worthwhile to mention that the mathematical solution of this problem, up to four decimal points, gives 0.6321 (see Example 2.24).   ◆

EXERCISES

Warning: In some combinatorial probability problems, the desired quantity is very small. For this reason, in simulations, the number of experiments should be very large to get good approximations.

1. Suppose that 18 customers stand in a line at a boxoffice, nine with $5 bills and nine with $10 bills. Each ticket costs $5, and the box office has no money initially. Write a simulation program to calculate an approximate value for the probability that none of the customers has to wait for change.

2. A cereal company puts exactly one of its 20 prizes into every box of its cereals at random.

    (a)   Julie bought 50 boxes of cereals from this company. Write a simulation program to calculate an approximate value of the probability that she gets all 20 prizes.

(b)   Suppose that Jim wants to have at least a 50% chance of getting all 20 prizes. Write a simulation program to calculate the approximate value of the minimum number of cereal boxes that he should buy.

3.   Nine students lined up in some order and had their picture taken. One year later, the same students lined up again in random order and had their picture taken. Using simulation, find an approximate probability that, the second time, no student was standing next to the one next to whom he or she was standing the previously.

4.   Use simulation to find an approximate value for the probability that, in a class of 50, exactly four students have the same birthday and the other 46 students all have different birthdays. Assume that the birth rates are constant throughout the year and that each year has 365 days.

5.   Using simulation, find the approximate probability that at least four students of a class of 87 have the same birthday. Assume that the birth rates are constant throughout the year and that each year has 365 days.
Answer: Approximately 0.4998.

6.   Suppose that two pairs of the vertices of a regular 10-gon are selected at random and connected. Using simulations, find the approximate probability that they do not intersect.
Hint: The exact value of the desired probability is

$$\frac{\frac{1}{11}\binom{20}{10}}{\binom{20}{2}\binom{18}{2}} \approx 0.58.$$

7.   Every day during the last 10 days, an executive wrote seven letters and his secretary prepared seven envelopes for the letters. The secretary kept envelopes of each day in chronological order in a packet, but she forgot to insert the letters in the envelopes. Today, the secretary finds out that all 70 letters and all 10 packets of envelopes are mixed up. If she selects packets of envelopes at random and inserts randomly chosen letters into the envelopes of each packet, what is the approximate probability, obtained by using simulation, that none of the letters is addressed correctly? Note that the envelopes within a packet are left in chronological order but the packets themselves are mixed up.
Answer: Approximately 0.53416. For an analytic discussion of this problem, see the article by Steve Fisk in the April 1988 issue of *Mathematics Magazine*.

13.3   SIMULATION OF CONDITIONAL PROBABILITIES

Let S be the sample space of an experiment and A and B be two events of S with $P(B) > 0$. To find approximate values for $P(A \mid B)$, using computer simulation, we either reduce the sample space to B and then calculate $P(AB)$ in the reduced sample space, or we use the formula $P(A \mid B) = P(AB)/P(B)$. In the latter case, we perform the experiment a large number of times, find n and m, the number of times in which B and AB occur, respectively, and then divide m by n. In the former case, $P(AB)$ is estimated simply by performing the experiment n times in the reduced sample space, for a large n, finding m, the number of those in which AB occurs, and dividing m by n. We illustrate this method by Example 13.4, and the other method by Example 13.5.

Example 13.4   From an ordinary deck of 52 cards, 10 cards are drawn at random. If exactly four are hearts, write an algorithm to approximate by simulation the probability of at least two spades.

Solution: The problem in the reduced sample space is as follows: A deck of cards consists of 13 spades, 13 clubs, and 13 diamonds, but only 9 hearts. If six cards are drawn at random, what is the probability that none of them is a heart and that two or more are spades? To simulate this problem, suppose that hearts are numbered 1 to 9, spades 10 to 22, clubs 23 to 35, and diamonds 36 to 48. For a large number of times, say n, each time choose six distinct random integers between 1 and 48 and check to see if there are no numbers between 1 and 9 and at least two numbers between 10 and 22. Let m be the number of those draws satisfying both conditions; then m/n is the desired approximation. For sufficiently large n, the final answer of any accurate simulation is close to 0.42. An algorithm follows:

STEP

1:	Set $k = 1$;
2:	Set $m = 0$;
3:	While $k \le n$, do steps 4 to 13.
4:	Choose six distinct numbers $X(1), X(2), \ldots, X(6)$ randomly from $\{1, 2, \ldots, 48\}$ (see Algorithm 13.1, Section 13.2).
5:	For $j = 1$ to 6, do steps 6 and 7.
6:	For $l = 1$ to 9, do step 7.
7:	If $X(j) = l$, then
	Set $k = k + 1$;
	Goto step 3.
8:	Set $s = 0$;
9:	For $j = 1$ to 6, do steps 10 to 12.
10:	For $l = 10$ to 22, do steps 11 and 12.
11:	If $X(j) = l$, then
	Set $s = s + 1$;

12:        If $s = 2$, then
              Set $m = m + 1$;
              Set $k = k + 1$;
              Goto step 3.
13:        Set $k = k + 1$; Goto step 3.
14:        Set $p = m/n$.
15:        Output (p).
STOP   ◆

Example 13.5 (Laplace's Law of Succession) Suppose that u urns are numbered
0 through $u - 1$, and that the ith urn contains i red and $u - 1 - i$ white balls, $0 \leq i \leq u - 1$.
An urn is selected at random and then its balls are removed one by one, at random and
with replacement. If the first m balls are all red, write an algorithm to find by simulation
an approximate value for the probability that the $(m + 1)$st ball removed is also red.

Solution: Let A be the event that the first m balls drawn are all red, and let B be the event
that the $(m + 1)$st ball drawn is red. We are interested in $P(B \mid A) = P(BA)/P(A)$.
To find an approximate value for $P(B \mid A)$ by simulation, for a large n, perform this
experiment n times. Let a be the number of those experiments in which the first m draws
are all red, and let b be the number of those in which the first $m + 1$ draws are all red.
Then, to obtain the desired approximation, divide b by a.

  To write an algorithm, introduce i to count the number of experiments, and a and b
to count, respectively, the numbers of those in which the first m and the first $m + 1$ draws
are all red. Initially, set $i = 1$, $a = 0$, and $b = 0$. Each experiment begins with choosing
t, a random number from $(0, 1)$. If $t \in (0, 1/u]$, set $k = 0$, meaning that urn number
0 is selected; if $t \in (1/u, 2/u]$, set $k = 1$, meaning that urn number 1 is selected; and
so on. If $k = 0$, urn 0 is selected. Since urn 0 has no red balls, and hence neither A
nor AB occurs, do not change the values of a and b; simply change i to $i + 1$ and start
a new experiment. If $k = u - 1$, urn number $u - 1$, which contains only red balls, is
selected. In such a case both A and AB will occur; therefore, set $i = i + 1$, $a = a + 1$,
and $b = b + 1$ and start a new experiment. For $k = 1, 2, 3, \ldots, u - 2$, the urn that
is selected has $u - 1$ balls, of which exactly k are red. Start choosing random numbers
from $(0, 1)$ to represent drawing balls from this urn. If a random number is less than
$k/(u - 1)$, a red ball is drawn. Otherwise, a white ball is drawn. If all the first m draws
are red, set $a = a + 1$. In this case, draw another ball if it is red, then set $b = b + 1$ as
well. If among the first m draws, a white ball is removed at any draw, then only change
i to $i + 1$ and start a new experiment. If the first m draws are red but the $(m + 1)$st one is
not, change i to $i + 1$ and a to $a + 1$, but keep b unchanged and start a new experiment.
When $i = n$, all the experiments are complete. The approximate value of the desired
probability is then equal to $p = b/a$. An algorithm follows:

STEP

1:      Set $i = 1$;
2:      Set $a = 0$;
3:      Set $b = 0$;
4:      While $i \leq n$, do steps 5 to 20.
5:      Generate a random number t from $(0, 1)$.
6:      Set $r = 1$;
7:      While $r \leq u$, do step 8.
8:      If $t \leq r/u$, then
                Set $k = r - 1$;
                Set $r = u + 1$;
        else
                Set $r = r + 1$;
9:      If $k = 0$, then
                Set $i = i + 1$;
                Goto step 4.
10:     If $k = u - 1$, then
                Set $i = i + 1$;
                Set $a = a + 1$;
                Set $b = b + 1$;
                Goto step 4.
11:     Set $j = 0$;
12:     While $j < m$, do steps 13 and 14.
13:     Generate a random number r from $(0, 1)$.
14:     If $r < k/(u - 1)$, then
                Set $j = j + 1$;
        else
                Set $j = m + 1$;
15:     If $j \neq m$, then
                Set $i = i + 1$;
                Goto step 4.
16:     If $j = m$, then do steps 17 to 19.
17:     Set $a = a + 1$;
18:     Generate a random number r from $(0, 1)$.
19:     If $r < k/(u - 1)$, then
                Set $b = b + 1$;
20:     Set $i = i + 1$;
21:     Set $p = b/a$.
22:     Output (p).

STOP

A sample run of a simulation program based on this algorithm gives us the following results for $n = 1000$ and $m = 5$:

u	2	10	50	100	200	500	1000	2000	5000
p	1	0.893	0.886	0.852	0.852	0.848	0.859	0.854	0.853

It can be shown that, if the number of urns is large, the answer is approximately $(m + 1)/(m + 2) = 6/7 \approx 0.857$ (see Exercise 44 of Section 3.5). Hence the results of these simulations are quite good.   ◆

EXERCISES

1. An ordinary deck of 52 cards is dealt among A, B, C, and D, 13 each. If A and B have a total of six hearts and five spades, using computer simulation find an approximate value for the probability that C has two of the remaining seven hearts and three of the remaining eight spades.

2. From families with five children, *a family* is selected at random and found to have a boy. Using computer simulation, find the approximate value of the probability that the family has three boys and two girls and that the middle child is a boy. Assume that, in a five-child family, all gender distributions have equal probabilities.

3. In Example 13.5, Laplace's law of succession, suppose that, among the first m balls drawn, r are red and $m - r$ are white. Using computer simulation, find the approximate probability that the $(m + 1)$st ball is red.

4. Targets A and B are placed on a wall. It is known that for every shot the probabilities of hitting A and hitting B with a missile are, respectively, 0.3 and 0.4. If target A was not hit in an experiment, use simulation to calculate an approximate value for the probability that target B was hit. (The exact answer is $4/7$. Hence the result of simulation should be close to $4/7 \approx 0.571$.)   ◼

13.4   SIMULATION OF RANDOM VARIABLES

Let X be a random variable over the sample space S. To approximate $E(X)$ by simulation, we use the strong law of large numbers. That is, for a large n, we repeat the experiment over which X is defined independently n times. Each time we will find the value of X. We then add the n values obtained and divide the result by n. For example, in the experiment of rolling a die, let X be the following random variable:

$$X = \begin{cases} 1 & \text{if the outcome is 6} \\ 0 & \text{if the outcome is not 6.} \end{cases}$$

To find $E(X)$ by simulation, for a large number n, we generate n random numbers from $(0, 1)$ and find m, the number of those that are in $(0, 1/6]$. The sum of all the values of X that are obtained in these n experiments is m. Hence m/n is the desired approximation. A more complicated example is the following:

Example 13.6   A fair coin is flipped successively until, for the first time, four consecutive heads are obtained. Write an algorithm for a computer simulation to determine the approximate value of the average number of trials that are required.

Solution: Let H stand for the outcome heads. We present an algorithm that repeats the experiment of flipping the coin until the first HHHH, n times. In the algorithm, j is the variable that counts the number of experiments. It is initially 1; each time that an experiment is finished, the value of j is updated to $j+1$. When $j = n$, the last experiment is performed. For every experiment we begin choosing random numbers from $(0, 1)$. If an outcome is in $(0, 1/2)$, heads is obtained; otherwise, tails is obtained. The variable i counts the number of successive heads. Initially, it is 0; every time that heads is obtained, one unit is added to i, and every time that the outcome is tails, i becomes 0. Therefore, an experiment is over when, for the first time, i becomes 4. The variable k counts the number of trials until the first HHHH, m is the sum of all the k's for the n experiments, and a is the average number of trials until the first HHHH. Therefore, $a = m/n$. An algorithm follows.

STEP

1:	Set $m = 0$;
2:	Set $j = 0$;
3:	While $j < n$, do steps 4 to 11.
4:	Set $k = 0$;
5:	Set $i = 0$;
6:	While $i < 4$, do steps 7 to 9.
7:	Set $k = k + 1$;
8:	Generate a random number r from $(0, 1)$.
9:	If $r < 0.5$, then
	$\quad$ Set $i = i + 1$;
	else
	$\quad$ Set $i = 0$;
10:	Set $m = m + k$;
11:	Set $j = j + 1$;
12:	Set $a = m/n$;
13:	Output (a).

STOP

We execute a simulation program based on this algorithm for several values of n. The corresponding values obtained for a were as follows:

n	a	n	a
5	23.80	5,000	29.90
50	27.22	50,000	30.00
500	29.28	500,000	30.03

It can be shown that the exact answer to this problem is 30 (see the subsection "Pattern Appearance" in Section 10.1). Therefore, these simulation results for $n \geq 5000$ are quite good. ♦

To simulate a Bernoulli random variable X with parameter p, generate a random number from $(0, 1)$. If it is in $(0, p]$, let $X = 1$; otherwise, let $X = 0$. To simulate a binomial random variable Y with parameters (n, p), choose n independent random numbers from $(0, 1)$, and let Y be the number of those that lie in $(0, p]$. To simulate a geometric random variable T with parameter p, we may keep choosing random points from $(0, 1)$ until for the first time a number selected lies in $(0, p]$. The number of random points selected is a simulation of T. However, this method of simulating T is very inefficient. By the following theorem, T can be simulated most efficiently, namely, by just choosing one random point from $(0, 1)$.

Theorem 13.1 *Let r be a random number from the interval $(0, 1)$ and let*

$$T = 1 + \left[\frac{\ln r}{\ln(1 - p)} \right], \qquad 0 < p < 1,$$

where by $[x]$ we mean the largest integer less than or equal to x. Then T is a geometric random variable with parameter p.

Proof: For all $n \geq 1$,

$$P(T = n) = P\left(1 + \left[\frac{\ln r}{\ln(1 - p)} \right] = n \right)$$

$$= P\left(\left[\frac{\ln r}{\ln(1 - p)} \right] = n - 1 \right) = P\left(n - 1 \leq \frac{\ln r}{\ln(1 - p)} < n \right)$$

$$= P\big((n - 1)\ln(1 - p) \geq \ln r > n\ln(1 - p) \big)$$

$$= P\big(\ln(1 - p)^{n-1} \geq \ln r > \ln(1 - p)^n \big)$$

$$= P\big((1 - p)^{n-1} \geq r > (1 - p)^n \big) = (1 - p)^{n-1} - (1 - p)^n$$

$$= (1 - p)^{n-1}\big[1 - (1 - p) \big] = (1 - p)^{n-1}p,$$

which shows that T is geometric with parameter p.   ♦

Therefore,

> To simulate a geometric random variable T with parameter p, $0 < p < 1$, all we must do is to choose a point r at random from $(0, 1)$ and let $T = 1 + \left[\ln r / \ln(1 - p)\right]$.

Let X be a negative binomial random variable with parameters (n, p), $0 < p < 1$. By Example 10.7, X is a sum of n independent geometric random variables. Thus

> To simulate X, a negative binomial random variable with parameters (n, p), $0 < p < 1$, it suffices to choose n independent random numbers $r_1, r_2, \ldots, r_n$ from $(0, 1)$ and let

$$X = \sum_{i=1}^{n} \left(1 + \left[\frac{\ln r_i}{\ln(1 - p)}\right]\right).$$

We now explain simulation of continuous random variables. The following theorem is a key to simulation of many special continuous random variables.

Theorem 13.2   *Let X be a continuous random variable with probability distribution function F. Then $F(X)$ is a uniform random variable over $(0, 1)$.*

Proof: Let S be the sample space over which X is defined. The functions $X : S \to \mathbf{R}$ and $F : \mathbf{R} \to [0, 1]$ can be composed to obtain the random variable $F(X) : S \to [0, 1]$. Clearly,

$$P\big(F(X) \leq t\big) = \begin{cases} 1 & \text{if } t \geq 1 \\ 0 & \text{if } t \leq 0. \end{cases}$$

Let $t \in (0, 1)$; it remains to prove that $P\big(F(X) \leq t\big) = t$. To do so, note that since F is continuous, $F(-\infty) = 0$, and $F(\infty) = 1$, the inverse image of t, $F^{-1}(\{t\})$, is nonempty. We know that F is nondecreasing; since F is not necessarily strictly increasing, $F^{-1}(\{t\})$ might have more than one element. For example, if F is the constant t on some interval $(a, b) \subseteq (0, 1)$, then $F(x) = t$ for all $x \in (a, b)$, implying that (a, b) is contained in $F^{-1}(\{t\})$. Let $x_0 = \inf\big\{x : F(x) > t\big\}$. Then $F(x_0) = t$ and $F(x) \leq t$ if and only if $x \leq x_0$. Therefore,

$$P\big(F(X) \leq t\big) = P(X \leq x_0) = F(x_0) = t.$$

We have shown that

$$P\big(F(X) \leq t\big) = \begin{cases} 0 & \text{if } t \leq 0 \\ t & \text{if } 0 \leq t \leq 1 \\ 1 & \text{if } t \geq 1, \end{cases}$$

meaning that $F(X)$ is uniform over $(0, 1)$. ◆

Based on this theorem, to simulate a continuous random variable X with *strictly increasing* probability distribution function F, it suffices to generate a random number u from $(0, 1)$ and then solve the equation $F(t) = u$ for t. The solution t to this equation, $F^{-1}(u)$, is unique because F is strictly increasing. It is a simulation of X. For example, let X be an exponential random variable with parameter λ. Since $F(t) = 1 - e^{-\lambda t}$ and the solution to $1 - e^{-\lambda t} = u$ is $t = -1/\lambda \ln(1-u)$, we have that, for any random number u from $(0, 1)$, the quantity $-1/\lambda \ln(1 - u)$ is a simulation of X. But, if u is a random number from $(0, 1)$, then $(1 - u)$ is also a random number from $(0, 1)$. Therefore,

$$-\frac{1}{\lambda} \ln\left[1 - (1 - u)\right] = -\frac{1}{\lambda} \ln u$$

is also a simulation of X. So

> To simulate an exponential random variable X with parameter λ, it suffices to generate a random number u from $(0, 1)$ and then calculate $-(1/\lambda)\ln u$.

The method described is called the **inverse transformation method** and is good whenever the equation $F(t) = u$ can be solved without complications.

As we know, there are close relations between Poisson processes and exponential random variables and also between exponential and gamma random variables. These relations enable us to use the results obtained for simulation of the exponential case [e.g., to simulate Poisson processes and gamma random variables with parameters (n, λ), n being a positive integer]. Let X be a gamma random variable with such parameters. Then $X = X_1 + X_2 + \cdots + X_n$, where X_i's $(i = 1, 2, \ldots, n)$ are independent exponential random variables each with parameter λ. Hence

> If $u_1, u_2, \ldots, u_n$ are n independent random numbers generated from $(0, 1)$, then
>
> $$\sum_{i=1}^{n} -\frac{1}{\lambda} \ln u_i = -\frac{1}{\lambda} \ln(u_1 u_2 \cdots u_n)$$
>
> is a simulation of a gamma random variable with parameters (n, λ).

To simulate a Poisson process $\{N(t) : t \geq 0\}$ with parameter λ, note that $N(t)$ is the number of "events" that have occurred at or prior to t. Let X_1 be the time of the first event, X_2 be the elapsed time between the first and second events, X_3 be the elapsed time between the second and third events, and so on. Then the sequence $\{X_1, X_2, \ldots\}$ is an independent sequence of exponential random variables, each with parameter λ. Clearly,

$$N(t) = \max\{n : X_1 + X_2 + \cdots + X_n \leq t\}.$$

Let $U_1, U_2, \ldots$ be independent uniform random variables from $(0, 1)$; then

$$N(t) = \max\left\{n: \; -\frac{1}{\lambda}\ln U_1 - \frac{1}{\lambda}\ln U_2 - \cdots - \frac{1}{\lambda}\ln U_n \le t\right\}$$

$$= \max\left\{n: \; -\ln(U_1 U_2 \cdots U_n) \le \lambda t\right\}$$

$$= \max\left\{n: \; \ln(U_1 U_2 \cdots U_n) \ge -\lambda t\right\}$$

$$= \max\{n: \; U_1 U_2 \cdots U_n \ge e^{-\lambda t}\}.$$

This implies that

$$N(t) + 1 = \min\{n: \; U_1 U_2 \cdots U_n < e^{-\lambda t}\}.$$

We have the following:

Let $\{N(t): t \ge 0\}$ be a Poisson process with rate λ. To simulate $N(t)$, we keep generating random numbers $u_1, u_2, \ldots$ from $(0, 1)$ until, for the first time, $u_1 u_2 \cdots u_n$ is less than $e^{-\lambda t}$. The number of random numbers generated minus 1 is a simulation of $N(t)$.

Let N be a Poisson random variable with parameter λ. Since N has the same distribution as $N(1)$, to simulate N all we have to do is to simulate $N(1)$ as explained previously.

The inverse transformation method is not appropriate for simulation of normal random variables. This is because there is no closed form for F, the distribution function of a normal random variable, and hence, in general, $F(t) = u$ cannot be solved for t. However, other methods can be used to simulate normal random variables. Among them the following, introduced by Box and Muller, is perhaps the simplest. It is based on the next theorem, proved in Example 8.27.

Theorem 13.3 *Let V and W be two independent uniform random variables over $(0, 1)$. Then the random variables*

$$Z_1 = \cos(2\pi V)\sqrt{-2\ln W}$$

and

$$Z_2 = \sin(2\pi V)\sqrt{-2\ln W}$$

are independent standard normal random variables.

Based on this theorem,

To simulate the standard normal random variable Z, we may generate two independent random numbers v and w from $(0, 1)$ and let $z = \cos(2\pi v)\sqrt{-2\ln w}$. The number z is then a simulation of Z.

The advantage of Box and Muller's method is that it can generate two independent standard normal random variables at the same time. Its disadvantage is that it is not efficient.

As for an arbitrary normal random variable, suppose that $X \sim N(\mu, \sigma^2)$. Then, since $Z = (X - \mu)/\sigma$ is standard normal, we have that $X = \sigma Z + \mu$. Thus

> If X is a normal random variable with parameters μ and σ^2, to simulate X, we may generate two independent random numbers v and w from $(0, 1)$ and let $t = \sigma \cos(2\pi v)\sqrt{-2\ln w} + \mu$. The quantity t is then a simulation of X.

Example 13.7   Passengers arrive at a train station according to a Poisson process, with parameter λ. If the arrival time of the next train is uniformly distributed over the interval $(0, T)$, write an algorithm to approximate by simulation the probability that by the time the next train arrives there are at least N passengers at the station. *Solution:*   For a

large positive integer n, we generate n independent random numbers $t_1, t_2, \ldots, t_n$ from $(0, T)$. Then for each t_i we simulate a Poisson random variable X_i with parameter λt_i. Finally, we find m, the number of X_i's that are greater than or equal to N. The desired approximation is $p = m/n$. An algorithm follows.

STEP

1:	Set $m = 0$.
2:	Do steps 3 to 5 n times.
3:	Generate a random number t from $(0, T)$.
4:	Generate a Poisson random variable X with parameter λt.
5:	If $X \geq N$, then set $m = m + 1$.
6:	Set $p = m/n$.
7:	Output (p).

STOP   ◆

Example 13.8   Let us assume that the time between any two earthquakes in town A and the time between any two earthquakes in town B are exponentially distributed with means $1/\lambda_1$ and $1/\lambda_2$, respectively. Write an algorithm to approximate by simulation the probability that the next earthquake in town B occurs prior to the next earthquake in town A.

Solution: Let X and Y denote the times between now and the next earthquakes in town A and town B, respectively. For a large n, let n be the total number of simulations of the times of the next earthquakes in town A and town B. If m is the number of those simulations in which $X > Y$, then $p = P(X > Y)$ is approximately m/n. An algorithm follows.

STEP

1:      $m = 0$.
2:      Do steps 3 to 5, n times.
3:      Generate an exponential random variable X with parameter λ_1.
4:      Generate an exponential random variable Y with parameter λ_2.
5:      If $X > Y$, then set $m = m + 1$.
6:      Set $p = m/n$.
7:      Output (p).

STOP   ♦

Example 13.9   The grade distributions of students in calculus, statistics, music, computer programming, and physical education are, respectively, $N(70, 400)$, $N(75, 225)$, $N(80, 400)$, $N(75, 400)$, and $N(85, 100)$. Write an algorithm to approximate by simulation the probability that the median of the grades of a randomly selected student taking these five courses is at least 75.

Solution: Let $X(1)$, $X(2)$, $X(3)$, $X(4)$, and $X(5)$ be the grades of the randomly selected student in calculus, statistics, music, computer programming, and physical education, respectively. Let X be the median of these grades. We will simulate $X(1)$ through $X(5)$, n times. Each time we calculate X by sorting $X(1)$, $X(2)$, $X(3)$, $X(4)$, and $X(5)$ and letting $X = X(3)$. Then we check to see if X is at least 75. In all these n simulations, let m be the total number of X's that are at least 75; m/n is the desired approximation. An algorithm follows.

STEP

1:      Set $m = 0$.
2:      Do steps 3 to 12 n times.
3:      Generate $X(1) \sim N(70, 400)$.
4:      Generate $X(2) \sim N(75, 225)$.
5:      Generate $X(3) \sim N(80, 400)$.
6:      Generate $X(4) \sim N(75, 400)$.
7:      Generate $X(5) \sim N(85, 100)$.
8:      Do steps 9 and 10 for $L = 1$ to 3.
9:      Do step 10 for j = 1 to 5-L.
10:     If $X(j) > X(j + 1)$, then
               Set $K = X(j + 1)$;
               Set $X(j + 1) = X(j)$;
               Set $X(j) = K$.
11:     Set $X = X(3)$.
12:     If $X \geq 75$, set $m = m + 1$.
13:     Set $p = m/n$.
14:     Output (p).

STOP   ♦

EXERCISES

1. Let X be a random variable with probability distribution function

$$F(x) = \begin{cases} \dfrac{x-3}{x-2} & \text{if } x \geq 3 \\ \\ 0 & \text{elsewhere.} \end{cases}$$

Develop a method to simulate X.

2. Explain how a random variable X with the following probability density function can be simulated:

$$f(x) = e^{-2|x|}, \qquad -\infty < x < +\infty.$$

3. Explain a procedure for simulation of lognormal random variables. A random variable X is called **lognormal** with parameters μ and σ^2 if $\ln X \sim N(\mu, \sigma^2)$

4. Use the result of Example 11.11 to explain how a gamma random variable with parameters $(n/2, 1/2)$, n being a positive integer, can be simulated.

5. It can be shown that the median of $(2n + 1)$ random numbers from the interval $(0, 1)$ is a beta random variable with parameters $(n+1, n+1)$. Use this property to simulate a beta random variable with parameters $(n + 1, n + 1)$, n being a positive integer.

6. Suppose that in a community the distributions of the heights of men and women, in centimeters, are $N(173, 40)$ and $N(160, 20)$, respectively. Write an algorithm to calculate by simulation the approximate value of the probability that (a) a wife is taller than her husband; (b) a husband is at least 10 centimeters taller than his wife.

7. Mr. Jones is at a train station, waiting to make a phone call. There are two public telephone booths next to each other and occupied by two persons, say A and B. If the duration of each telephone call is an exponential random variable with $\lambda = 1/8$, using simulation, approximate the probability that among Mr. Jones, A, and B, Mr. Jones is not the last person to finish his call.

8. The distributions of students' grades for probability and calculus at a certain university are, respectively, $N(65, 400)$ and $N(72, 450)$. Dr. Olwell teaches a calculus class with 28 and a probability class with 22 students. Write an algorithm to simulate the probability that the difference between the averages of the final grades of

the classes of Dr. Olwell is at least 2.

Hint: Note that if $X_1, X_2, \ldots, X_n$ are all $N(\mu, \sigma^2)$, then

$$\bar{X} = \frac{X_1 + X_2 + \cdots + X_n}{n} \sim N\left(\mu, \frac{\sigma^2}{n}\right).$$

■

13.5   MONTE CARLO METHOD

As explained in Section 8.1, if S is a subset of the plane with area $A(S)$ and R is a subset of S with area $A(R)$, the probability that a random point from S falls in R is equal to $A(R)/A(S)$. This important fact gives an excellent algorithm, called the **Monte Carlo method**, for finding the area under a bounded curve $y = f(x)$ by simulation. Suppose that we want to estimate I, the area under $y = f(x)$, from $x = a$ to $x = b$ of Figure 13.1. To do so, we first construct a rectangle $[a, b] \times [0, d]$ that includes the region under $y = f(x)$ from a to b as a subset. Then for a large integer n, we choose n random points from $[a, b] \times [0, d]$ and count m, the number of those that lie below the curve $y = f(x)$. Now the probability that a random point from $[a, b] \times [0, d]$ is under the curve $y = f(x)$ is approximately m/n. Thus

$$\frac{m}{n} \approx \frac{\text{area under } f(x) \text{ from } a \text{ to } b}{\text{area of the rectangle } [a, b] \times [0, d]} = \frac{I}{(b-a)(d-0)} = \frac{I}{(b-a)d},$$

and hence

$$I \approx \frac{md(b-a)}{n}.$$

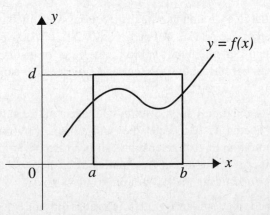

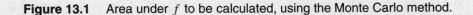

Figure 13.1   Area under f to be calculated, using the Monte Carlo method.

This method of simulation was introduced by the Hungarian-born American mathematician John von Neumann (1903–1957), and Stanislaw Ulam (1909–1984). They used it during World War II to study the extent that neutrons can travel through various materials. Since their studies were classified, von Neumann gave it the code name **Monte Carlo method**. An algorithm for this procedure is as follows.

STEP

1:      Set $m = 0$;
2:      Set $i = 1$;
3:      While $i \leq n$, do steps 4 to 6.
4:      Generate a random number x from $[a, b]$.
5:      Generate a random number y from $[0, d]$.
6:      If $y < f(x)$, then
        Set $i = i + 1$;
        Set $m = m + 1$;
    else
        Set $i = i + 1$.
7:      Set $I = \big[md(b - a)\big]/n$.
8:      Output (I).

STOP   ♦

Example 13.10   (Buffon's Needle Problem Revisited)   In Example 8.14, we explained one of the most interesting problems of geometric probability, **Buffon's needle problem:** A plane is ruled with parallel lines a distance d apart. If a needle of length ℓ, $\ell < d$, is tossed at random onto the plane, then the probability that the needle intersects one of the parallel lines is $2\ell/\pi d$. This formula has been used throughout the history of probability to determine approximate values for π. For fixed and predetermined values of d and ℓ, $\ell < d$, rule a plane with parallel lines a distant d apart, and toss a needle of length ℓ, n times; then count m, the number of times that the needle intersects a line. Then m/n determines an approximate value for the probability that the needle intersects a line, provided that n is a large number. Thus $2\ell/\pi d \approx m/n$ and hence $\pi \approx 2n\ell/md$. In 1850, Rudolf Wolf, a Swiss scientist, conducted such an experiment. With 5000 tosses of a needle he got 3.1596 for π. In 1855, another mathematician, Smith, did the same experiment with 3204 tosses and got 3.1553. In 1860, De Morgan repeated the experiment; with 600 tosses he got 3.137 for π. In 1894 and 1901, Fox and Lazzarini conducted similar experiments; with 1120 and 3408 tosses, respectively, they got the values 3.1419 and 3.1415929. It should be noted that the latter results are subject to skepticism, because it can be shown that the probability of getting such close approximations to π for such low numbers of tosses of a needle is extremely small. For details, see N. T. Gridgeman's paper "Geometric Probability and Number π," *Scripta Mathematika*, 1960, Volume 25, Number 3.

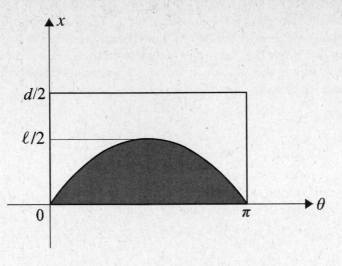

Figure 13.2 The curve $x = (\ell/2)\sin\theta$.

To find more accurate estimates for π, using the result of Buffon's needle problem, we may use the Monte Carlo procedure. As we explained in Example 8.14, every time that the needle is tossed, a point (Θ, X) is selected from the square $[0, \pi] \times [0, d/2]$, and the needle intersects the line if and only if $X/\sin\Theta < \ell/2$. Therefore, if for a large number n, we select n random points from the square $[0, \pi] \times [0, d/2]$ and find m, the number of those that fall below the curve $x = (\ell/2)\sin\theta$ (see Figure 13.2), then m/n is approximately the probability that the needle intersects a line; that is, $m/n \approx 2\ell/\pi d$, which gives $\pi \approx 2n\ell/md$. ♦

EXERCISES

1. Three concentric circles of radii 2, 3, and 4 are the boundaries of regions that form a circular target. If a person fires a shot at random at the target, use simulation to approximate the probability that the bullet lands in the middle region.
   Hint: To simulate a random point located inside a circle of radius r centered at the origin, choose random points (X, Y) inside the square $[-r, r] \times [-r, r]$ until you obtain one that satisfies $X^2 + Y^2 < r^2$.

2. Using Monte Carlo procedure, write a program to estimate

$$\Phi(3) - \Phi(-1) = \int_{-1}^{3} \frac{1}{\sqrt{2\pi}} e^{-x^2/2} \, dx.$$

   Then run your program for $n = 10,000$ and compare the answer with that obtained from Tables 1 and 2 of the Appendix.

3. In Buffon's needle problem, let $\ell = 1$ and $d = 2$, and use Monte Carlo procedure to estimate π with $n = 10,000$. Then do the same for $\ell = 1$ and $d = 3/2$, then for $\ell = 1$ and $d = 5/4$.

∎

Appendix Tables

1. Area under the Standard Normal Distribution to the Left of Z_0: Negative z_0

2. Area under the Standard Normal Distribution to the Left of Z_0: Positive z_0

3. Expectations, Variances, and Moment-Generating Functions of Important Distributions

Table 1   Area under the Standard Normal Distribution to the Left of z_0: Negative z_0

$$\Phi(z_0) = P(Z \leq z_0) = \frac{1}{\sqrt{2\pi}} \int_{-\infty}^{z_0} e^{-x^2/2}\, dx$$

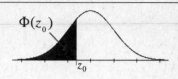

Note that for $z_0 \leq -3.90$, $\Phi(z_0) = P(Z \leq z_0) \approx 0$.

z_0	0	1	2	3	4	5	6	7	8	9
−3.8	.0001	0001	0001	0001	0001	0001	0001	0001	0001	0001
−3.7	.0001	0001	0001	0001	0001	0001	0001	0001	0001	0001
−3.6	.0002	0002	0001	0001	0001	0001	0001	0001	0001	0001
−3.5	0002	0002	0002	0002	0002	0002	0002	0002	0002	0002
−3.4	.0003	.0003	.0003	.0003	.0003	.0003	.0003	.0003	.0003	.0002
−3.3	.0005	.0005	.0005	.0004	.0004	.0004	.0004	.0004	.0004	.0003
−3.2	.0007	.0007	.0006	.0006	.0006	.0006	.0006	.0005	.0005	.0005
−3.1	.0010	.0009	.0009	.0009	.0008	.0008	.0008	.0008	.0007	.0007
−3.0	.0013	.0013	.0013	.0012	.0012	.0011	.0011	.0011	.0010	.0010
−2.9	.0019	.0018	.0018	.0017	.0016	.0016	.0015	.0015	.0014	.0014
−2.8	.0026	.0025	.0024	.0023	.0023	.0022	.0021	.0021	.0020	.0019
−2.7	.0035	.0034	.0033	.0032	.0031	.0030	.0029	.0028	.0027	.0026
−2.6	.0047	.0045	.0044	.0043	.0041	.0040	.0039	.0038	.0037	.0036
−2.5	.0062	.0060	.0059	.0057	.0055	.0054	.0052	.0051	.0049	.0048
−2.4	.0082	.0080	.0078	.0075	.0073	.0071	.0069	.0068	.0066	.0064
−2.3	.0107	.0104	.0102	.0099	.0096	.0094	.0091	.0089	.0087	.0084
−2.2	.0139	.0136	.0132	.0129	.0125	.0122	.0119	.0116	.0113	.0110
−2.1	.0179	.0174	.0170	.0166	.0162	.0158	.0154	.0150	.0146	.0143
−2.0	.0228	.0222	.0217	.0212	.0207	.0202	.0197	.0192	.0188	.0183
−1.9	.0287	.0281	.0274	.0268	.0262	.0256	.0250	.0244	.0239	.0233
−1.8	.0359	.0351	.0344	.0336	.0329	.0322	.0314	.0307	.0301	.0294
−1.7	.0446	.0436	.0427	.0418	.0409	.0401	.0392	.0384	.0375	.0367
−1.6	.0548	.0537	.0526	.0516	.0505	.0495	.0485	.0475	.0465	.0455
−1.5	.0668	.0655	.0643	.0630	.0618	.0606	.0594	.0582	.0571	.0559
−1.4	.0808	.0793	.0778	.0764	.0749	.0735	.0721	.0708	.0694	.0681
−1.3	.0968	.0951	.0934	.0918	.0901	.0885	.0869	.0853	.0838	.0823
−1.2	.1151	.1131	.1112	.1093	.1075	.1056	.1038	.1020	.1003	.0985
−1.1	.1357	.1335	.1314	.1292	.1271	.1251	.1230	.1210	.1190	.1170
−1.0	.1587	.1562	.1539	.1515	.1492	.1469	.1446	.1423	.1401	.1379
−0.9	.1841	.1814	.1788	.1762	.1736	.1711	.1685	.1660	.1635	.1611
−0.8	.2119	.2090	.2061	.2033	.2005	.1977	.1949	.1922	.1894	.1867
−0.7	.2420	.2389	.2358	.2327	.2296	.2266	.2236	.2206	.2177	.2148
−0.6	.2743	.2709	.2676	.2643	.2611	.2578	.2546	.2514	.2483	.2451
−0.5	.3085	.3050	.3015	.2981	.2946	.2912	.2877	.2843	.2810	.2776
−0.4	.3446	.3409	.3372	.3336	.3300	.3264	.3228	.3192	.3156	.3121
−0.3	.3821	.3783	.3745	.3707	.3669	.3632	.3594	.3557	.3520	.3483
−0.2	.4207	.4168	.4129	.4090	.4052	.4013	.3974	.3936	.3897	.3859
−0.1	.4602	.4562	.4522	.4483	.4443	.4404	.4364	.4325	.4286	.4247
−0.0	.5000	.4960	.4920	.4880	.4840	.4801	.4761	.4721	.4681	.4641

Table 2   Area under the Standard Normal Distribution to the Left of z_0: Positive z_0

$$\Phi(z_0) = P(Z \le z_0) = \frac{1}{\sqrt{2\pi}} \int_{-\infty}^{z_0} e^{-x^2/2}\, dx$$

z_0	0	1	2	3	4	5	6	7	8	9
.0	.5000	.5040	.5080	.5120	.5160	.5199	.5239	.5279	.5319	.5359
.1	.5398	.5438	.5478	.5517	.5557	.5596	.5636	.5675	.5714	.5753
.2	.5793	.5832	.5871	.5910	.5948	.5987	.6026	.6064	.6103	.6141
.3	.6179	.6217	.6255	.6293	.6331	.6368	.6406	.6443	.6480	.6517
.4	.6554	.6591	.6628	.6664	.6700	.6736	.6772	.6808	.6844	.6879
.5	.6915	.6950	.6985	.7019	.7054	.7088	.7123	.7157	.7190	.7224
.6	.7257	.7291	.7324	.7357	.7389	.7422	.7454	.7486	.7517	.7549
.7	.7580	.7611	.7642	.7673	.7703	.7734	.7764	.7794	.7823	.7852
.8	.7881	.7910	.7939	.7967	.7995	.8023	.8051	.8078	.8106	.8133
.9	.8159	.8186	.8212	.8238	.8264	.8289	.8315	.8340	.8365	.8389
1.0	.8413	.8438	.8461	.8485	.8508	.8531	.8554	.8577	.8599	.8621
1.1	.8643	.8665	.8686	.8708	.8729	.8749	.8770	.8790	.8810	.8830
1.2	.8849	.8869	.8888	.8907	.8925	.8944	.8962	.8980	.8997	.9015
1.3	.9032	.9049	.9066	.9082	.9099	.9115	.9131	.9147	.9162	.9177
1.4	.9192	.9207	.9222	.9236	.9251	.9265	.9279	.9292	.9306	.9319
1.5	.9332	.9345	.9357	.9370	.9382	.9394	.9406	.9418	.9429	.9441
1.6	.9452	.9463	.9474	.9484	.9495	.9505	.9515	.9525	.9535	.9545
1.7	.9554	.9564	.9573	.9582	.9591	.9599	.9608	.9616	.9625	.9633
1.8	.9641	.9649	.9656	.9664	.9671	.9678	.9686	.9693	.9699	.9706
1.9	.9713	.9719	.9726	.9732	.9738	.9744	.9750	.9756	.9761	.9767
2.0	.9772	.9778	.9783	.9788	.9793	.9798	.9803	.9808	.9812	.9817
2.1	.9821	.9826	.9830	.9834	.9838	.9842	.9846	.9850	.9854	.9857
2.2	.9861	.9864	.9868	.9871	.9875	.9878	.9881	.9884	.9887	.9890
2.3	.9893	.9896	.9898	.9901	.9904	.9906	.9909	.9911	.9913	.9916
2.4	.9918	.9920	.9922	.9925	.9927	.9929	.9931	.9932	.9934	.9936
2.5	.9938	.9940	.9941	.9943	.9945	.9946	.9948	.9949	.9951	.9952
2.6	.9953	.9955	.9956	.9957	.9959	.9960	.9961	.9962	.9963	.9964
2.7	.9965	.9966	.9967	.9968	.9969	.9970	.9971	.9972	.9973	.9974
2.8	.9974	.9975	.9976	.9977	.9977	.9978	.9979	.9979	.9980	.9981
2.9	.9981	.9982	.9982	.9983	.9984	.9984	.9985	.9985	.9986	.9986
3.0	.9987	.9987	.9987	.9988	.9988	.9889	.9889	.9889	.9990	.9990
3.1	.9990	.9991	.9991	.9991	.9992	.9992	.9992	.9992	.9993	.9993
3.2	.9993	.9993	.9994	.9994	.9994	.9994	.9994	.9995	.9995	.9995
3.3	.9995	.9995	.9995	.9996	.9996	.9996	.9996	.9996	.9996	.9997
3.4	.9997	.9997	.9997	.9997	.9997	.9997	.9997	.9997	.9997	.9998
3.5	.9998	.9998	.9998	.9998	.9998	.9998	.9998	.9998	.9998	.9998
3.6	.9998	.9998	.9999	.9999	.9999	.9999	.9999	.9999	.9999	.9999
3.7	.9999	.9999	.9999	.9999	.9999	.9999	.9999	.9999	.9999	.9999
3.8	.9999	.9999	.9999	.9999	.9999	.9999	.9999	.9999	.9999	.9999

Note that for $z_0 > 3.89$, $\Phi(z_0) = P(Z \le z_0) \approx 1$.

Table 3 Expectations, Variances, and Moment-Generating Functions of Important Distributions

Distribution of X	Probability Function of X	Range of X	$E(X)$	$\text{Var}(X)$	$M_X(t)$
Binomial with parameters n and p	$\binom{n}{x} p^x (1-p)^{n-x}$	$x = 0, 1, \ldots, n$	np	$np(1-p)$	$(pe^t + 1 - p)^n$
Poisson with parameter $\lambda > 0$	$\dfrac{e^{-\lambda}\lambda^x}{x!}$	$x = 0, 1, 2, \ldots$	λ	λ	$\exp\left[\lambda(e^t - 1)\right]$
Geometric with parameter p	$p(1-p)^{x-1}$	$x = 1, 2, \ldots$	$\dfrac{1}{p}$	$\dfrac{1-p}{p^2}$	$\dfrac{pe^t}{1-(1-p)e^t}$
Negative binomial with parameters r and p	$\binom{x-1}{r-1} p^r (1-p)^{x-r}$	$x = r, r+1, r+2, \ldots$	$\dfrac{r}{p}$	$\dfrac{r(1-p)}{p^2}$	$\left[\dfrac{pe^t}{1-(1-p)e^t}\right]^r$
Uniform over (a, b)	$\dfrac{1}{b-a}$	$a < x < b$	$\dfrac{a+b}{2}$	$\dfrac{(b-a)^2}{12}$	$\dfrac{e^{tb} - e^{ta}}{t(b-a)}$
Normal with parameters μ and σ^2	$\dfrac{1}{\sqrt{2\pi}}\exp\left[-\dfrac{(x-\mu)^2}{2\sigma^2}\right]$	$-\infty < x < \infty$	μ	σ^2	$\exp\left(\mu t + \dfrac{\sigma^2 t^2}{2}\right)$
Exponential with parameter λ	$\lambda e^{-\lambda x}$	$0 \le x < \infty$	$\dfrac{1}{\lambda}$	$\dfrac{1}{\lambda^2}$	$\dfrac{\lambda}{\lambda - t}$
Gamma with parameters r and λ	$\dfrac{\lambda e^{-\lambda x}(\lambda x)^{r-1}}{\Gamma(r)}$	$x \ge 0$	$\dfrac{r}{\lambda}$	$\dfrac{r}{\lambda^2}$	$\left(\dfrac{\lambda}{\lambda - t}\right)^r$

Answers to Odd-Numbered Exercises

Section 1.2

1. (a) Yes; (b) No. **3.** $\{x: 0 < x < 20\}$; $\{1, 2, 3, \ldots, 19\}$. **5.** One 1 and one even; One 1 and one odd; Both even or both belong to $\{3, 5\}$. **7.** $S = \left\{x: 7 \leq x \leq 9\frac{1}{6}\right\}$; $\left\{x: 7 \leq x \leq 7\frac{1}{4}\right\} \cup \left\{x: 7\frac{3}{4} \leq x \leq 8\frac{1}{4}\right\} \cup \left\{x: 8\frac{3}{4} \leq x \leq 9\frac{1}{6}\right\}$. **9.** For $1 \leq i \leq 3$, $1 \leq j \leq 3$, by $a_i b_j$ we mean passenger a gets off at hotel i and passenger b gets off at hotel j. The answers are $\{a_i b_j: 1 \leq i \leq 3, \ 1 \leq j \leq 3\}$ and $\{a_1 b_1, a_2 b_2, a_3 b_3\}$, respectively. **11.** (a) $AB^c C^c$; (b) $A \cup B \cup C$; (c) $A^c B^c C^c$; (d) $ABC^c \cup AB^c C \cup A^c BC$; (e) $AB^c C^c \cup A^c B^c C \cup A^c BC^c$; (f) $(A - B) \cup (B - A) = (A \cup B) - AB$. **13.** (a) True; (b) false; (c) true; (d) true. **19.** $\bigcap_{m=1}^{\infty} \bigcup_{n=m}^{\infty} A_n$.

Section 1.4

1. No! **3.** 0.9673. **5.** 35%. **7.** He was not. **9.** 15/29. **11.** (a) 1/13; (b) 8/13; (c) 9/13. **13.** 65% and 10%, respectively. **17.** $\sum_{i=1}^{n} p_{ij}$. **19.** (a) 1/8; (b) 5/24; (c) 5/24. **21.** 4/7. **25.** 0.172. **27.** 99/200. **31.** Neither Q nor R is a probability on S.

Section 1.7

1. 2/3. **3.** (a) False; (b) false. **5.** 0. **7.** No, it is not! **13.** Let $E_t = (0, 1) - \{t\}$, $0 < t < 1$.

Review Problems for Chapter 1

1. 0.54. **3.** (a) False; (b) true. **5.** $\{x_1 x_2 \cdots x_n: n \geq 1, x_i \in \{H, T\}; \ x_i \neq x_{i+1}, 1 \leq i \leq n - 2; \ x_{n-1} = x_n\}$. **9.** 20%. **11.** $1 - P(A) - P(B) + P(AB)$. **13.** Denote a box of books from publisher i by a_i, $i = 1, 2, 3$. Then the sample space is $S = \{x_1 x_2 x_3 x_4 x_5 x_6: \text{two } x_i\text{'s are } a_1, \text{two are } a_2, \text{and the remaining two are } a_3\}$. The desired event is $\{x_1 x_2 x_3 x_4 x_5 x_6 \in S: x_5 = x_6\}$. **15.** 0.571.

17. (a) $U_i^c D_i^c$; (b) $U_1 U_2 \cdots U_n$; (c) $(U_1^c D_1^c) \cup \cdots \cup (U_n^c D_n^c)$; (d) $(U_1 D_2 U_3^c D_3^c) \cup (U_1 U_2^c D_2^c D_3) \cup (D_1 U_2 U_3^c D_3^c) \cup (D_1 U_2^c D_2^c U_3)$; (e) $D_1^c D_2^c \cdots D_n^c$.   **19.** 173/216.

Section 2.2

1. 900,000; 427,608.   **3.** Yes, it is.   **5.** 0.00000024.   **7.** 1/6.
9. 0.00000000093.   **11.** Yes, there are.   **13.** 36.   **15.** 625; 505.
17. 0.274.   **19.** $1 - \left[(N-1)^n / N^n \right]$.   **21.** 30.43%.   **23.** 0.469.
25. 0.868.   **27.** 0.031.   **29.** 0.067.

Section 2.3

1. 0.0417.   **3.** 56.   **5.** (a) 531,441; (b) 924; (c) 27,720.   **7.** 3,491,888,400.
9. 20,160.   **11.** 0.985.   **13.** 0.000333.   **15.** $m!/(n+m)!$ **17.** 0.000054.
19. 0.962.   **21.** $n!/n^n$.   **23.** 326,998,056.   **25.** 0.003.   **27.** 3.23×10^{-16}; 2.25×10^{-6}.

Section 2.4

1. 38,760.   **3.** 6,864,396,000.   **5.** n/N.   **7.** 560.   **9.** 0.318.   **11.** 71,680.
13. (a) 0.246; (b) 0.623.   **15.** (a) 0.228; (b) 0.00084.   **17.** (a) 3^n; (b) $(x+1)^n$.
19. 0.00151.   **21.** 0.3157.   **23.** 0.023333.   **25.** $-7,560$.   **27.** 4.47×10^{-28}.
29. 36.   **31.** (a) $n_k = \binom{n_{k-1}}{2} + n_{k-1}(n + n_1 + \cdots + n_{k-1})$; (b) approximately 710×10^{63}.   **33.** 0.000346.   **35.** 0.218.   **37.** (a) $(1/n^m) \sum_{i=0}^{n} (-1)^i \binom{n}{i} (n-i)^m$; (b) $(1/n^m) \binom{n}{r} \sum_{i=0}^{n-r} (-1)^i \binom{n-r}{i} (n-r-i)^m$.   **43.** 0.264.   **45.** $(2n^2 + 4n + 3)/[3(n+1)^3]$.
47. 16,435,440.   **51.** $\left[1/(10 \times 2^{2n}) \right] \left[(5 + 3\sqrt{5})(1 + \sqrt{5})^n + (5 - 3\sqrt{5})(1 - \sqrt{5})^n \right]$.

Section 2.5

1. (a) $1/\sqrt{\pi n}$; (b) $\sqrt{2}/4^n$.

Review Problems for Chapter 2

1. 127.   **3.** 0.278.   **5.** 7560.   **7.** 82,944.   **9.** In one way.   **11.** 34.62%.
13. 43,046,721.   **15.** 0.467.   **17.** 0.252.   **19.** 531,441.   **21.** 0.379.
23. $(N - n + 1)/\binom{N}{n}$.   **25.** 369,600.   **27.** 0.014.

Section 3.1

1. 60%.   **3.** 0.625.   **5.** 2/3.   **7.** 2/5.   **9.** 0.239.   **11.** (a) 1/14; (b) 13/42; (c) 3/7.   **17.** 0.0263.   **19.** 0.615.   **21.** 0.883.   **23.** The crucial difference is reflected in the implicit assumption that both girls cannot be Mary.

Section 3.2

1. 0.1625.   **3.** 0.72.   **5.** (a) 0.00216; (b) 0.00216.   **7.** It makes no difference if you draw first, last, or anywhere in the middle.   **9.** 1/14.   **11.** 0.

Section 3.3

1. 0.02625.   **3.** 0.633.   **5.** 1/4.   **7.** 0.7055.   **9.** 0.034.   **11.** Yes, they are.
13. 0.4.   **17.** $1/(n-1)$.   **19.** 0.0383   **21.** 1/3.   **23.** To maximize the chances of freedom, one urn must contain 1 green and 0 red balls, and the other one must contain $N - 1$ green and N red balls.

Section 3.4

1. 3/5.   **3.** 0.87.   **5.** 0.1463.   **7.** 0.084.   **9.** 0.056.   **11.** 0.21.   **13.** 0.69.
15. 0.61.

Section 3.5

1. You should not agree!   **3.** Neither Mia is right nor Jim.   **5.** 0.00503.
7. (a) 0.526; (b) 0.076.   **11.** 0.9936.   **17.** 0.994.   **19.** 0.226.
23. (a) $1 - [(n-1)/n]^n$; (b) this approaches $1 - (1/e)$.   **25.** No, it is not.   **27.** 0.2.
29. $2p^4 - p^6$.   **33.** 0.5614.   **35.** 13/14.   **37.** 2/5.
39. $1 - \prod_{i=1}^{n}(1 - p_i) - \sum_{i=1}^{n}\left[p_i \prod_{\substack{j=1 \\ j \neq i}}^{n}(1 - p_j)\right]$.   **41.** $2p^2 + 2p^3 - 5p^4 + p^5$.
43. (a) 1/8; (b) 1/8; (c) 13/16.

Section 3.6

1. 1.   **3.** The parents are rr and Rr.   **5.** The observed data shows that the genes are linked.   **7.** 40.71%   **9.** 0.17.   **11.** $p/6$.

Review Problems for Chapter 3

1. 0.347.   **3.** 65%.   **5.** (a) 0.783; (b) 0.999775; (c) 0.217; (d) 0.000225.
7. 0.0796.   **9.** 1/6.   **11.** 0.35.   **13.** 0.03.   **15.** 0.36.   **17.** 0.748.
19. 0.308; R, B, and S are *not* independent. They are conditionally independent given that Kevin is prepared, and they are conditionally independent given that Kevin is unprepared.

Section 4.2

1. Possible values are 0, 1, 2, 3, 4, and 5. Probabilities associated with these values are 6/36, 10/36, 8/36, 6/36, 4/36, and 2/36, respectively.   **3.** $(1 - p)^{i-1}p$ if

$1 \leq i \leq N$; $(1-p)^N$ if $i = 0$.   **5.** The answers are 1/2, 1/6, 1/4, 1/2, 0, and 1/3, respectively.   **7.** (a) 1/33; (b) 25/33; (c) 9/33; (d) 5/6.   **11.** F is a distribution function.
13. $F(t) = 0$ if $t < 0$; $t/45$ if $0 \leq t < 45$; 1 if $t \geq 45$.   **15.** 0.277.   **17.** $F(t) = 0$
if $t < 0$; $t/(1-t)$ if $0 \leq t < 1/2$; 1 if $t \geq 1/2$.   **19.** It is $F(t)$ if $t < 5$; 1 if $t \geq 5$.

Section 4.3

1. $F(x) = 0$ if $x < 1$, 1/15 if $1 \leq x < 2$, 3/15 if $2 \leq x < 3$, 6/15 if $3 \leq$
$x < 4$, 10/15 if $4 \leq x < 5$, and 1 if $x \geq 5$.   **3.** Possible values are 2, 3, ...,
12. Probabilities associated with these values are 1/36, 2/36, 3/36, 4/36, 5/36, 6/36,
5/36, 4/36, 3/36, 2/36, 1/36, respectively.   **5.** (a) $(9/10)^{i-1}(1/10)$, $i = 1, 2, \ldots$;
(b) $(9/10)^{(j-3)/2}(1/10)$, $j = 3, 5, 7, \ldots$.   **7.** (a) 1/15; (b) 1/15; (c) 8; (d) $2/[n(n+1)]$;
(e) $6/[n(n+1)(2n+1)]$.   **9.** $F(x) = 0$ if $x < 0$; $1 - (1/4)^{n+1}$ if $n \leq x < n+1$,
$n = 0, 1, 2, \ldots$.   **11.** $(1/2)^{n-1}$, $n \geq 2$.   **13.** Possible values are 0, 1, 2, 3, 4, 5.
Probabilities associated with these values are 42/1001, 252/1001, 420/1001, 240/1001,
45/1001, 2/1001, respectively.   **15.** $(5/6)^{n-1} - 5(4/6)^{n-1} + 10(3/6)^{n-1} - 10(2/6)^{n-1} +$
$5(1/6)^{n-1}$, $n \geq 6$.

Section 4.4

1. Yes, of course.   **3.** 0.86.   **5.** No, it is not.   **7.** Either 4 or 5 magazines should
be ordered.   **9.** (b) 0, 44/27, 349/27, respectively.   **11.** 5/8, 31/8, 23/8, respectively.   **13.** 3.98.   **15.** $\sum_{k=1}^{c} \sum_{j=k}^{c}(k\alpha_j)/j$.   **17.** $[N^{n+1} - \sum_{i=1}^{N}(i-1)^n]/N^n$.
19. $[(n+1)^2]/(2n-1)$.

Section 4.5

1. The first business.   **3.** 3.   **5.** They are $(N+1)/2$, $(N^2-1)/12$, and $\sqrt{(N^2-1)/12}$,
respectively.   **7.** 36.   **9.** X is more concentrated about 0 than Y is.

Section 4.6

1. Mr. Norton should hire the salesperson who worked in store 2.

Review Problems for Chapter 4

1. It is 1/45 for $i = 1, 2, 16, 17$; 2/45 for $i = 3, 4, 14, 15$; 3/45 for $i = 5, 6, 12, 13$;
4/45 for $i = 7, 8, 10, 11$; and 5/45 for $i = 9$.   **3.** 35.   **5.** 1.067.   **7.** They are 16,
16, 0.013, and 0.008, respectively.   **9.** (a) e^{-2t}; (b) $e^{-2t}[1 + 2t + 2t^2 + (4t^3/3)]$ and
$1 - e^{-2t} - 2te^{-2t}$, respectively.   **11.** (a) $\binom{j-1}{3}/\binom{52}{4}$. (b) 1/13. (c) 42.4.

Section 5.1

1. 0.087.   **3.** 0.054.   **5.** 0.33.   **7.** $p(x) = \binom{4}{x}(0.60)^x(0.4)^{4-x}$, $x = 0, 1,$
2, 3, 4; $q(y) = \binom{4}{(y-1)/2}(0.60)^{(y-1)/2}(0.4)^{(9-y)/2}$, $y = 1, 3, 5, 7, 9$.   **9.** 0.108.

11. They are $np + p$ and $np + p - 1$.   **13.** 0.219.   **15.** $k = 4$ and the maximum probability is 0.238.   **17.** 0.995.   **19.** It is preferable to send in a single parcel.
21. (a) $Np(1 - p)^{N-1}$; (c) $1/e$.   **23.** $(1/2)^n \sum_{i=0}^{\left[\frac{n+(b-a)}{2}\right]} \binom{n}{i}$, where $\left[\frac{n+(b-a)}{2}\right]$ is the greatest integer less than or equal to $\frac{n+(b-a)}{2}$.   **27.** 0.000028.   **29.** 90,072.
31. 0.054.   **33.** $1/\left[\sum_{i=0}^{n} \binom{n}{i}(\frac{n-i}{n})^k\right]$.

Section 5.2

1. 0.9502.   **3.** 0.594.   **5.** 0.823.   **7.** 0.063.   **9.** 83%.   **11.** 0.21.
13. 0.87.   **15.** 0.13.   **17.** 0.325.   **19.** 0.873716.   **21.** The least integer greater than or equal to $-N \ln(1 - \alpha)/M$.   **23.** (a) $(1/2)(1 + e^{-2\lambda\alpha})$. (b) $(1/2)(1 - e^{-2\lambda\alpha})$
25. 0.035.

Section 5.3

1. $\{NNN, DNN, NDN, NND, NDD, DND, DDN\}$.   **3.** (a) 12; (b) 0.07.
5. 0.055.   **7.** 0.42.   **9.** x/n.   **11.** 49.   **13.** $(1/2)^{N+M-m+1}\left[\binom{N+M-m}{N} + \binom{N+M-m}{M}\right]$.   **15.** 0.74.   **17.** $p^{[t/2]+1}$, where $[t/2]$ is the greatest integer less than or equal to $t/2$.   **19.** $1/(1 - p)^{N-1}$.   **21.** 254.80.   **23.** Negative binomial with parameters k and p.   **25.** $\binom{2N-M-1}{N-1}(1/2)^{2N-M-1}$.   **27.** 1/7.   **29.** (a) D/N;
(b) $(D - 1)/(N - 1)$.

Review Problems for Chapter 5

1. 0.0009.   **3.** 1.067.   **5.** 0.179.   **7.** 0.244.   **9.** 0.285.   **11.** 2 and 2, respectively.   **13.** $p/(1 - p)$.   **15.** 0.772.   **17.** 0.91.   **19.** 0.10111.
21. (a) 0.000368; (b) 0.000224.   **23.** (a) $(\frac{w}{w+b})^{n-1}(\frac{b}{w+b})$; (b) $(\frac{w}{w+b})^{n-1}$.   **25.** k/n.

Section 6.1

1. (a) 3; (b) 0.78.   **3.** (a) 6; (b) 0 if $x < 1$, $-2x^3 + 9x^2 - 12x + 5$ if $1 \leq x < 2$, and 1 if $x \geq 2$; (c) 5/32 and 1/2, respectively.   **5.** (a) $1/\pi$; (b) 0 if $x < -1$, $\frac{1}{\pi} \arcsin x + \frac{1}{2}$ if $-1 \leq x < 1$, 1 if $x \geq 1$.   **7.** (b) They are symmetric about 3 and 1, respectively.
9. 0.3327.   **11.** $\alpha = 1/2$; $\beta = 1/\pi$; $f(x) = 2/\left[\pi(4 + x^2)\right]$, $-\infty < x < \infty$.

Section 6.2

1. The probability density function of Y is $(1/12)y^{-2/3}$ if $-8 \leq y < 8$; 0, otherwise. The probability density function of Z is $(1/8)z^{-3/4}$ if $0 < z \leq 16$; 0, otherwise.
3. $f_Y(y) = \left[2/(3\sqrt[3]{y})\right] \exp\left(-y^{2/3}\right)$ if $y \in (0, \infty)$; 0, elsewhere. $f_Z(z) = 1, z \in (0, 1)$; 0, elsewhere.   **5.** The probability density function of Y is $(3\lambda/2)\sqrt{y}e^{-\lambda y \sqrt{y}}$ if $y \geq 0$; 0, otherwise.   **7.** The probability density function of Z is $1/\pi$ if $-\pi/2 < z < \pi/2$; 0, elsewhere.

Section 6.3

1. (a) 8.   **3.** The muffler of company B.   **5.** 0.   **7.** $(5/9)h\big((t - 32)/1.8\big)$; $1.8 \int_{-\infty}^{\infty} xh(x)\,dx + 32$.   **9.** 8.4.   **13.** (a) c_1 is arbitrary, $c_n = n^{-1/(n-1)}$; (b) ∞ if $n = 1$, and $n^{(n-2)/(n-1)}/(n - 1)$ if $n > 1$; (c) $c_n e^{-nt}$, $t \geq \ln c_n$; (d) it exists if and only if $m - n < -1$.   **19.** $[(2\alpha - \alpha^2)/\lambda^2] + [(2\beta - \beta^2)/\mu^2] - [2\alpha\beta/\lambda\mu]$.

Review Problems for Chapter 6

1. $1/y^2$ if $y \in (1, \infty)$; 0, elsewhere.   **3.** $11/(5\sqrt{5})$.   **5.** No!
7. $[15(1 - \sqrt[4]{y})^2]/(2\sqrt[4]{y})$ if $y \in (0, 1)$; 0, elsewhere.   **9.** Valid.   **11.** 0.3.

Section 7.1

1. 3/7.   **3.** $a = -6$; $b = 6$.   **5.** 1/2.   **7.** 1/3.   **9.** a/b.   **11.** $[nX]$ is a random number from the set $\{0, 1, 2, \dots, n - 1\}$.   **13.** $g(X) = 3X^2 - 4X + 1$.

Section 7.2

1. 0.5948.   **5.** They are all equal to 0.   **7.** 0.2%, 4.96%, 30.04%, 45.86%, and 18.94%, respectively.   **9.** 0.4796.   **11.** 0.388.   **13.** The median is μ.   **15.** No, it is not correct.   **17.** 671.   **19.** $2\Phi(t/\sigma) - 1$ if $t \geq 0$; 0, otherwise. The expected value is $\sigma\sqrt{2/\pi}$.   **21.** 49.9.   **23.** $e^{\alpha^2/2}$ **25.** $\frac{1}{\sigma t\sqrt{2\pi}} \exp[-(\ln t - \mu)^2/(2\sigma^2)]$, $t \geq 0$; 0, otherwise.   **27.** 10 copies.   **29.** Yes, absolutely!   **33.** 15.93 and 0.29, respectively.   **35.** $1/(2\sqrt{\lambda})$.

Section 7.3

1. 0.0001234.   **3.** $\exp(-y - e^{-y})$, $-\infty < y < \infty$.   **5.** (a) 0.1535; (b) 0.1535.
7. (a) $e^{-\lambda t}$; (b) $e^{-\lambda t} - e^{-\lambda s}$.   **9.** $323.33.   **11.** It is a Poisson random variable with parameter λt.   **13.** (a) 1/2.

Section 7.4

3. 0.3208.   **5.** Two hours and 35 minutes.   **7.** $(2n)!\sqrt{\pi}/(4^n \cdot n!)$.   **9.** Howard's waiting time in the queue is gamma with parameters 2 and $n\lambda$.

Section 7.5

1. Yes, it is.   **3.** $c = 1260$; $E(X) = 5/11$; $\text{Var}\,X = 5/242$.   **5.** 0.17.   **7.** 0.214.
9. Only beta density functions with $\alpha = \beta$ are symmetric, and they are symmetric about 1/2.

Section 7.6

1. (a) 0.392; (b) 0.0615.

Review Problems for Chapter 7

1. 5/12.   **3.** 0.248.   **5.** 0.999999927.   **7.** $(1/\sqrt{2\pi}) \int_{-\infty}^{x/2.5} \exp\left[-(t-4)^2/2\right] dt$.
9. 0.99966.   **11.** 0.51%, 12.2%, 48.7%, 34.23%, and 4.36%, respectively.   **13.** It is uniform over the interval $(-5/10^{k+1}, 5/10^{k+1})$.   **15.** 0.89.   **17.** $g(X) = 3X^2 - 1$.

Section 8.1

1.   (a) 2/9; (b) $p_X(x) = x/3$, $x = 1, 2$; $p_Y(y) = 2/(3y)$, $y = 1, 2$; (c) 2/3;
(d) $E(X) = 5/3$, $E(Y) = 4/3$.   **3.** (a) $k = 1/25$; (b) $p_X(x)$ is 12/25 if $x = 1$, 13/25 if $x = 2$; $p_Y(y)$ is 2/25 if $y = 1$, 23/25 if $y = 3$; (c) $E(X) = 38/25$, $E(Y) = 71/25$.
5. $[\binom{7}{x}\binom{8}{y}\binom{5}{4-x-y}]/\binom{20}{4}$, $0 \le x \le 4, 0 \le y \le 4, 0 \le x + y \le 4$.   **7.** $p(1, 1) = 0$,
$p(1, 0) = 0.30$, $p(0, 1) = 0.50$, $p(0, 0) = 0.20$.   **9.** (a) $f_X(x) = 2x$, $0 \le x \le 1$;
$f_Y(y) = 2(1 - y)$, $0 \le y \le 1$; (b) $E(X) = 2/3$, $E(Y) = 1/3$; (c) they are 1/4, 1/2,
and 0, respectively.   **11.** $f_X(x) = e^{-x}$, $x > 0$; $f_Y(y) = (1/2)y$, $0 < y < 2$.
13. (a) $f(x, y) = 6$ if $(x, y) \in R$; 0, elsewhere. (b) $f_X(x) = 6x(1 - x), 0 < x < 1$;
$f_Y(y) = 6(\sqrt{y} - y)$, $0 < y < 1$; (c) $E(X) = 1/2$, $E(Y) = 2/5$.   **15.** $\ell/3$.
17. $\pi/8$.   **21.** 3/4.   **23.** $a/2b$.   **25.** $f(x) = 1 - |x|$, $-1 \le x \le 1$.
27. $n(1/2)^{n-1}$.

Section 8.2

1. No, they are not.   **3.** 4/81 and 4/27, respectively.   **5.** 0.0179.   **7.** $(1/2)^{2n}\binom{2n}{n}$.
9. No, they are not.   **11.** No, they are not.   **13.** 1.   **15.** 3/2.   **17.** $f(t) = -\ln t$, $0 < t < 1$; 0, elsewhere.   **19.** 0.22639.

Section 8.3

1. $p_{X|Y}(x|y) = (x^2 + y^2)/(2y^2 + 5)$, $x = 1, 2$, $y = 0, 1, 2$. The desired conditional probability and conditional expectation are 5/7 and 12/7, respectively.   **3.** It is
$\binom{x-6}{2}(1/2)^{x-5}$, $x = 8, 9, 10, \ldots$.   **7.** $1/(y+1)$.   **9.** 10/33.   **11.** The joint probability density function of X and Y is $1/(20y)$ if $20 < x < 20 + (2y)/3$ and $0 < y < 30$.
It is 0 otherwise.   **13.** Given that $X = x$, $Y - x$ is binomial with parameters p and
$n - m$.   **15.** Given that $N(s) = k$, $N(t) - k$ is Poisson with parameter $\lambda(t - s)$.
19. (a) $f(x, y) = 2$ if $x \ge 0$, $y \ge 0$, $x + y \le 1$; 0, otherwise; (b) $f_{X|Y}(x|y) = 1/(1 - y)$ if $0 \le x \le 1 - y$, $0 \le y < 1$. (c) $E(X|Y = y) = (1 - y)/2$, $0 < y < 1$.
21. $[1/F(s)]\int_0^s tf(t) dt$.

Section 8.4

1. $g(u, v) = (1/4)e^{-(u+v)/2}$, $v > 0, u > 0$.   **3.** $(1/2\pi)re^{-r^2/2}$, $0 < \theta < 2\pi$, $r > 0$.   **7.** $g(u, v) = (1/v)e^{-u}$, $u > 0, 1 < v < e^u$.   **9.** (a) $g(u, v) = [\lambda^{r_1+r_2}u^{r_1+r_2-1}e^{-\lambda u}v^{r_1-1}(1 - v)^{r_2-1}]/[\Gamma(r_1)\Gamma(r_2)]$, $u > 0, 0 < v < 1$; (c) U is gamma with parameters $r_1 + r_2$ and λ; V is beta with parameters r_1 and r_2.

Review Problems for Chapter 8

1. (a) 0.69; (b) 2.05 and 3.38.   **3.** $\left[\binom{13}{x}\binom{26}{9-x}\right]/\binom{39}{9}$, $0 \le x \le 9$.   **5.** $\left[\binom{13}{x}\binom{13}{6-x}\right]/\binom{26}{6}$, $0 \le x \le 6$.   **7.** Yes, $E(XY) = E(X)E(Y)$.   **9.** $(r_2^2 - r_3^2)/r_1^2$.   **11.** 2/5.   **13.** $1/(n + 1)$, $n \ge 1$.   **15.** (a) $c = 12$; (b) they are not independent.   **17.** 0 if $t < 0$; 1 if $0 \le t < 1/2$; $1/(4t^2)$ if $t \ge 1/2$.   **19.** 0.2775.   **21.** $E(Y|X = x) = 0$, $E(X|Y = y) = (1 - y)/2$ if $-1 < y < 0$, and $(1 + y)/2$ if $0 < y < 1$.

Section 9.1

1. $\left[\binom{13}{h}\binom{13}{d}\binom{13}{c}\binom{13}{s}\right]/\binom{52}{13}$, $h + d + c + s = 13, 0 \le h, d, c, s \le 13$.   **3.** (a) $p_{X,Y}(x, y) = xy/54$, $x = 4, 5$, $y = 1, 2, 3$. $p_{Y,Z}(y, z) = yz/18$, $y = 1, 2, 3$, $z = 1, 2$. $p_{X,Z}(x, z) = xz/27$, $x = 4, 5$, $z = 1, 2$. (b) 35/9.   **5.** They are not independent.   **7.** (a) They are independent. (b) $\lambda_1\lambda_2\lambda_3 \exp(-\lambda_1 x - \lambda_2 y - \lambda_3 z)$. (c) $\lambda_1\lambda_2/[(\lambda_2 + \lambda_3)(\lambda_1 + \lambda_2 + \lambda_3)]$.   **9.** $[1 - (r^2/R^2)]^n$, $r < R$; 0, otherwise.   **11.** Yes, it is.   **13.** X is exponential with parameter $\lambda_1 + \lambda_2 + \cdots + \lambda_n$.   **15.** $1/(n\lambda)$.   **17.** $p_1p_7(p_2p_4 + p_3p_4 - p_2p_3p_4 + p_2p_5p_6 + p_3p_5p_6 - p_2p_3p_5p_6 - p_2p_4p_5p_6 - p_3p_4p_5p_6 + p_2p_3p_4p_5p_6)$.   **23.** They are not independent.   **25.** 1/2880.

Section 9.2

1. 0.26172.   **3.** $1 - \left(1 - e^{-3\lambda^2}\right)^4$.   **5.** Let $p_1 = \sum_{l=0}^{k-1} \binom{n}{l}p^l(1 - p)^{n-l}$, $p_2 = \sum_{l=k+1}^{n} \binom{n}{l}p^l(1 - p)^{n-l}$. The answer is $P(X_{(i)} = k) = 1 - \sum_{j=i}^{m} \binom{m}{j}p_1^j(1 - p_1)^{m-j} - \sum_{j=m-i+1}^{m} \binom{m}{j}p_2^j(1 - p_2)^{m-j}$.   **9.** (a) $\int_{-\infty}^{\infty} n(n - 1)f(v - r)f(v)[F(v) - F(v - r)]^{n-2} dv$, $r > 0$. (b) $n(n - 1)r^{n-2}(1 - r)$, $0 < r < 1$.

Section 9.3

1. 0.028.   **3.** 0.171.   **5.** (a) 0.8125; (b) 0.135, (c) 0.3046.   **7.** 0.117.   **9.** 0.009033.

Review Problems for Chapter 9

1. $\binom{20}{b}\binom{30}{r}\binom{50}{g}/\binom{100}{20}$, $b + r + g = 20$, $0 \le b, r, g \le 20$.   **3.** $\left[1 - (\pi/6)(r/a)^3\right]^n$.   **5.** 0.00135.   **7.** $\bar{F}_1(t)\bar{F}_2(t) \cdots \bar{F}_n(t)$.   **9.** 0.2775.

Section 10.1

1. 13/6.   **3.** 26.   **5.** 11.42.   **7.** $28.49.   **9.** $(n - k)k/n$.   **11.** $n/32$.
13. 23.41   **15.** 3.786.   **19.** 9.001.

Section 10.2

1. 0.   **3.** 35/24.   **5.** $n(n - 1)p(1 - p)$ and $-np(1 - p)$, respectively.
13. $2796 and $468, respectively.   **15.** The second method is preferable.
17. 0.2611.   **19.** $n/12$.   **25.** $-n/36$.

Section 10.3

1. 112.   **3.** They are -1 and $-1/12$, respectively.   **5.** No! it is not.   **7.** Show
that $\rho(X, Y) = -0.248 = \pm 1$.

Section 10.4

1. 6.   **5.** 47.28.   **7.** 3.786.   **11.** $2(2^n - 1)$.   **13.** It will take 4,407.286 business
days, on average, until there is a lapse of two days between two consecutive applications.
Since an academic year is 9 months long the admission officers should not be concerned
about this rule at all.   **15.** 7.805.   **17.** $[(1 - \alpha)p^m + \alpha]/[\alpha p^m (1 - p)]$.

Section 10.5

1. 0.5987.   **3.** $-\rho(X, Y)(\sigma_Y/\sigma_X)$.   **5.** They are $(1 + y)/2, 0 < y < 1$; $x/2$,
$0 < x < 1$; and 1/2, respectively.

Review Problems for Chapter 10

1. 1.   **3.** 511/270.   **5.** 1554.   **7.** 174.   **9.** $\text{Cov}(X, Y) = 0$; X and Y
are uncorrelated but not independent.   **11.** $2/\pi$.   **13.** $\alpha = \mu_Y - \rho(\sigma_Y/\sigma_X)\mu_X$,
$\beta = \rho(\sigma_Y/\sigma_X)$.   **15.** 7.5 minutes past 10:00 A.M.   **17.** 71.9548.

Section 11.1

1. $(1/5)(e^t + e^{2t} + e^{3t} + e^{4t} + e^{5t})$.   **3.** $2e^t/(3 - e^t), t < \ln 3$. $E(X) = 3/2$.
5. (a) $M_X(t) = (12/t^3)(1 - e^t) + (6/t^2)(1 + e^t), t = 0$; $M_X(t) = 1$ if $t = 0$.
(b) $E(X) = 1/2$.   **7.** (a) $M_X(t) = \exp[\lambda(e^t - 1)]$. (b) $E(X) = \text{Var}(X) = \lambda$.
9. $E(X) = 1/p$, $\text{Var}(X) = (1 - p)/p^2$.   **11.** $p(1) = 5/15$, $p(3) = 4/15$,
$p(4) = 2/15$, $p(5) = 4/15$; $p(i) = 0$, $i \in \{1, 3, 4, 5\}$.   **13.** $E(X) = 3/2$,
$\text{Var}(X) = 3/4$.   **17.** 8/9.   **21.** $M_X(t) = [\lambda/(\lambda - t)]^r$, $E(X) = r/\lambda$, $\text{Var}(X) = r/\lambda^2$.

Section 11.2

3. The sum is gamma with parameters n and λ.   **5.** The sum is gamma with parameters $r_1 + r_2 + \cdots + r_n$ and λ.   **7.** The answer is $\binom{n}{i}\binom{m}{j-i}/\binom{n+m}{j}$.   **9.** 0.558.   **11.** 0.0571   **15.** 1/2.   **17.** (a) 0.289; (b) 0.0836.

Section 11.3

3. (a) 0.4545; (b) 0.472.   **5.** (a) 0.4; (b) 0.053; (c) 0.222.   **7.** 16 days earlier.   **9.** It is $\leq 1/\mu$.   **11.** A sample of size 209.   **13.** 2500 or higher.

Section 11.4

1. The limit is 1/2 with probability 1.

Section 11.5

1. 0.6046.   **3.** 0.4806.   **5.** 0.6826.   **7.** 0.9778.   **9.** 378 or larger   **11.** 0.9938.

Review Problems for Chapter 11

1.   0.0262   **3.**   The distribution function of X is 0 for $t < 1$, 1/6 for $1 \leq t < 2$, 1/2 for $2 \leq t < 3$, and 1 for $t \geq 3$.   **5.** X is uniform over $(-1/2, 1/2)$.   **7.** $E(X^n) = (-1)^{n+1}(n+1)!$.   **9.** $1/(1 - t^2)$, $-1 < t < 1$.   **11.** 0.1034.   **15.** 0.0667.   **17.** 0.1401.

Section 12.2

3. 38 seconds.   **5.** 0.064.

Section 12.3

1. No, it is not.   **3.** (a) 3/8; (b) 11/16.   **5.** (a) 2/3; (b) 1/4.   **7.** There are two recurrent classes, $\{0, 3\}$ and $\{2, 4\}$, and one transient class, $\{1\}$.

9.
$$\begin{pmatrix} 0 & 0 & 1/2 & 0 & 1/2 & 0 & 0 & 0 \\ 1 & 0 & 0 & 0 & 0 & 0 & 0 & 0 \\ 0 & 1 & 0 & 0 & 0 & 0 & 0 & 0 \\ 0 & 0 & 1/3 & 2/3 & 0 & 0 & 0 & 0 \\ 0 & 0 & 0 & 0 & 0 & 2/5 & 0 & 3/5 \\ 0 & 0 & 0 & 0 & 1/2 & 0 & 1/2 & 0 \\ 0 & 0 & 0 & 0 & 0 & 3/5 & 0 & 2/5 \\ 0 & 0 & 0 & 0 & 1/3 & 0 & 2/3 & 0 \end{pmatrix}$$
11. 0.20.   **13.** $\pi_1 = 0.61, \pi_2 = 0.19$,

and $\pi_3 = 0.20$.   **15.** 15.5.   **17.** $\pi_0 = 1/[\,1 + \sum_{i=1}^{\infty}(1 - p_1 - p_2 - \cdots - p_i)]$, $\pi_k = (1 - p_1 - p_2 - \cdots - p_k)/[\,1 + \sum_{i=1}^{\infty}(1 - p_1 - p_2 - \cdots - p_i)]$, $k \geq 1$.   **19.** $\pi_i = (5/6)^i(1/6)$, $i = 0, 1, 2, \ldots$.   **21.** For $p < 1/2$ the limiting probabilities exist and are $\pi_i = [\,p/(1 - p)]^i[1 - [\,p/(1 - p)]]$, $i = 0, 1, 2, \ldots$.

25. $P = \begin{pmatrix} 0 & 1 & 0 & 0 & 0 & \cdots \\ \frac{\mu}{\lambda+\mu} & 0 & \frac{\lambda}{\lambda+\mu} & 0 & 0 & \cdots \\ 0 & \frac{\mu}{\lambda+\mu} & 0 & \frac{\lambda}{\lambda+\mu} & 0 & \cdots \\ 0 & 0 & \frac{\mu}{\lambda+\mu} & 0 & \frac{\lambda}{\lambda+\mu} & \cdots \\ & \vdots & & & & \end{pmatrix}$. The period of each state is 2.

29. $\pi_0 = 2/(k+1)$; $\pi_i = [2(k-i)]/[k(k+1)]$, $i = 0, 1, \ldots, k-1$.

Section 12.4

3. $\mu\pi_{(1,0)} + \lambda\pi_{(0,1)} = \lambda\pi_{(0,0)} + \mu\pi_{(0,0)}$; $\mu\pi_{(n+1,0)} + \lambda\pi_{(n-1,0)} = \lambda\pi_{(n,0)} + \mu\pi_{(n,0)}$, $n \geq 1$; $\lambda\pi_{(0,m+1)} + \mu\pi_{(0,m-1)} = \lambda\pi_{(0,m)} + \mu\pi_{(0,m)}$, $m \geq 1$.
5. (a) $[(1/c!)(\lambda/\mu)^c]/\left[\sum_{n=0}^{c}(1/n!)(\lambda/\mu)^n\right]$;   (b) At least five operators.
7. $[j(j+1)-i(i+1)]/(2\lambda)$.   **9.** $\lambda_0 = \lambda$; $\lambda_n = \alpha_n\lambda$, $n \geq 1$; $\mu_0 = 0$; $\mu_n = \mu+(n-1)\gamma$, $n \geq 1$.   **11.** $p_{00}(t) = 1$; for $i > 0$, $p_{i0}(t) = \sum_{k=i}^{\infty}\left[[e^{-\mu t}(\mu t)^k]/k!\right]$; for $0 < j \leq i$, $p_{ij}(t) = [e^{-\mu t}(\mu t)^{i-j}]/(i-j)!$.
13. $\pi_0 = [c!\,(1-\rho)]/\left[c!\,(1-\rho)\sum_{n=0}^{c}[(1/n!)(\lambda/\mu)^n] + c^c\rho^{c+1}\right]$.   **15.** $p_{1n}(t) = e^{-\lambda t}(1-e^{-\lambda t})^{n-1}$, $n \geq 1$.   **17.** $ne^{(\lambda-\mu)t} + [\gamma/(\lambda-\mu)][1-e^{(\lambda-\mu)t}]$.

Section 12.5

1. (a) 0.056; (b) -0.5 and 4.5, respectively.   **5.** 0, if $x \leq t$; $(2/\pi)\arccos\sqrt{t/x}$, if $x \geq t$.   **9.** $\sigma\sqrt{2t/\pi}$, $\sigma^2 t[1-(2/\pi)]$, and $\Phi[(z+z_0)/\sigma\sqrt{t}] + \Phi[(z-z_0)/\sigma\sqrt{t}] - 1$, respectively.   **11.** 0.7486.

Review Problems for Chapter 12

1. 0.692.   **3.** If states 0, 1, 2, 3, and 4 are renamed 0, 4, 2, 1, and 3, respectively, then the transition probability matrix P_1 will change to P_2.   **5.** It consists of two recurrent classes $\{3, 5\}$ and $\{4\}$,   and   two   transient   classes $\{1\}$ and $\{2\}$.
7.   0.4715.   **9.**   Approximately 15% have no logos.   **11.**   0.55.
13. $[\mu(2\lambda+\mu)]/(2\lambda^2+2\lambda\mu+\mu^2)$.   **15.** In steady state, the number of machines that are   operating   is   binomial   with   parameters m and $\lambda/(\lambda+\mu)$.
17. $(25\lambda^4 + 34\lambda^3\mu + 30\lambda^2\mu^2 + 24\lambda\mu^3 + 12\mu^4)/(12\lambda^5)$.   **19.** The desired quantities are $(s/t)y$ and $\sigma^2(t-s)(s/t)$, respectively.   **21.** $1 - (2/\pi)\arccos\sqrt{x/y}$.

Index

The following *abbreviations* are used in this index: r.v. for random variable, p.d.f. for probability density function, p.m.f. for probability mass function, and m.g.f. for moment-generating function.